SCIENCEPLUS®

Technology and Society

Annotated Teacher's Edition

Project Directors

International: **Charles McFadden**
Professor of Science Education
The University of New Brunswick
Fredericton, New Brunswick

National: **Robert E. Yager**
Professor of Science Education
The University of Iowa
Iowa City, Iowa

This new United States edition has been adapted from prior work by the Atlantic Science Curriculum Project, an international project linking teaching, curriculum development, and research in science education.

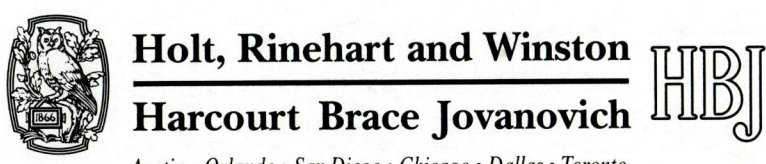

Holt, Rinehart and Winston
Harcourt Brace Jovanovich   HBJ
Austin • Orlando • San Diego • Chicago • Dallas • Toronto

ACKNOWLEDGEMENTS

Project Advisors

Herbert Brunkhorst
Director, Institute for Science
 Education
California State University, San
 Bernardino
San Bernardino, California

David L. Cross
Science Consultant
Lansing School District
Lansing, Michigan

Jerry Hayes
Associate Director, Scientific Outreach
Teacher's Academy, Mathematics
 and Science
Chicago, Illinois

William C. Kyle, Jr.
Director, School Mathematics and
 Science Center
Purdue University
West Lafayette, Indiana

Mozell Lang
Science Education Specialist
Michigan Department of Education
Lansing, Michigan

Project Authors

Earl S. Morrison (Author in Chief)
Nan Armour
Allan Hammond
John Haysom

Alan Moore (Associate Author in Chief)
Elinor Nicoll
Muriel Smyth

Project Associates

We wish to thank the hundreds of science educators, teachers, and administrators from the scores of universities, high schools, and middle schools who have contributed to the success of *SciencePlus*.

Copyright © 1993 by Holt, Rinehart and Winston, Inc.

All rights reserved. No part of this publication may be reproduced or transmitted in any form or by any means, electronic or mechanical, including photocopy, recording, or any information storage and retrieval system, without permission in writing from the publisher.

Requests for permission to make copies of any part of the work should be mailed to: Permissions Department, Holt, Rinehart and Winston, Inc., 8th Floor, Orlando, Florida 32887

Acknowledgements: See page S158, which is an extension of the copyright page.

SciencePlus® is a registered trademark of Harcourt Brace Jovanovich, Inc., licensed to Holt, Rinehart and Winston, Inc.

Printed in the United States of America

ISBN 0-03-074962-X

7 8 9 125 96 95 94

ATE Writers

Nancy Straus (Writer in Chief)
David Stienecker
Sharon Kahkonen
Iris Kane

ATE Consultants

Shirley Gholston Key (Multicultural Advisor)
Allan Cobb (Content Reviewer)

Field-Test Teachers and Sites

Charles Bissell
Gompers Secondary School
San Diego, California

Leslie Blanscet
Olive Peirce Middle School
Ramona, California

Steven E. Byrd
Florida State University School
Tallahassee, Florida

Suzette Carroll
Edison Junior High School
Los Angeles, California

Daisy Century
Sulzberger Junior High School
Philadelphia, Pennsylvania

Sharon Cox
Stidwell Junior High School
Sandpoint, Idaho

Ken Crease
White Mountain Junior High School
Rock Springs, Wyoming

Matt Keller
Rancho Cotate High School
Rohnert Park, California

Frank Lucio
Olive Peirce Middle School
Ramona, California

Naomi Lyall
Sunnyvale Junior High
Sunnyvale, California

Kim McConathy
Deep Water Junior High School
Pasadena, Texas

Bruce Metz
White Mountain Junior High School
Rock Springs, Wyoming

Nancy Miller
Bell Junior High School
San Diego, California

David Mooney
Upland High School
Upland, California

Mary Mund
Priest River Junior High School
Priest River, Idaho

Cindy Murray
Upland High School
Upland, California

Sally Pisani
Olive Peirce Middle School
Ramona, California

David Reynolds
Olive Peirce Middle School
Ramona, California

June Turnquist
Fremont High School
Sunnyvale, California

We also wish to acknowledge the contributions of the many Canadian field-test teachers who have helped to shape the *SciencePlus* program.

SourceBook Consultants

Robert H. Allers
Earth Science Teacher
Vernon-Verona-Sherrill
  High School
Verona, New York

Linda Butler
Lecturer, Division of
  Biological Sciences
The University of Texas at Austin
Austin, Texas

Christopher J. Chiaverina
Physics Instructor
New Trier Township High School
Winnetka, Illinois

Juan Cotera
Architect
Austin, Texas

Edmund J. Escudero
Chemistry Teacher
Summit Country Day School
Cincinnati, Ohio

Roger H. Kolar
Architect
Austin, Texas

Maureen Lemke
Biology Laboratory Instructor
Goodnight Jr. High School
San Marcos, Texas

Contents

SciencePlus Owner's Manual

The SciencePlus Philosophy **T14**

The SciencePlus Method **T17**
Building a Better Understanding of Science T17
Themes in Science T18

Components of SciencePlus **T21**
The *Pupil's Edition* T21
The *SourceBook* T23
The *Annotated Teacher's Edition* T24
The *Teacher's Resource Binder* T27
Other Teaching Aids T27

Science Discovery Videodisc Program T28

Using SciencePlus **T29**
Communicating Science T29
Keeping a Journal T30
Process Skills T31
Critical Thinking T32
Environmental Awareness T33
Cooperative Learning T34
Concept Mapping T36
Multicultural Instruction T38

Crossing Disciplines T38
Meeting Individual Needs T39
Science, Technology, and Society T42
Materials and Equipment T43
SciencePlus Teacher's Network T43

Assessing Student Performance **T44**
A Comprehensive Approach to Assessment T44
Assessing Science Projects T48
Developing Assessment Items T49

Unit Interleaf Information

Unit 1 Interactions 1A
Unit 2 Diversity of Living Things 63A
Unit 3 Solutions 129A
Unit 4 Force and Motion 191A
Unit 5 Structures and Design 261A
Unit 6 The Restless Earth 315A
Unit 7 Toward the Stars 375A
Unit 8 Growing Plants 445A

To the Student x
About Safety xii

Unit 1 Interactions 2
THE PLAYERS AND THEIR ROLES 5
   1   You and Your Environment 5
   2   The Living Players 10
   3   The Movement of Energy in a Community 16
   4   Who Eats What? 20
   5   The Nonliving Players 26

PUTTING IT ALL TOGETHER 36
   6   Natural Changes 36
   7   Humans: Boss or Co-Worker? 50

Making Connections 58

Science in Action:
   Spotlight on Veterinary Science 60

Science and Technology:
   Oceanography 62

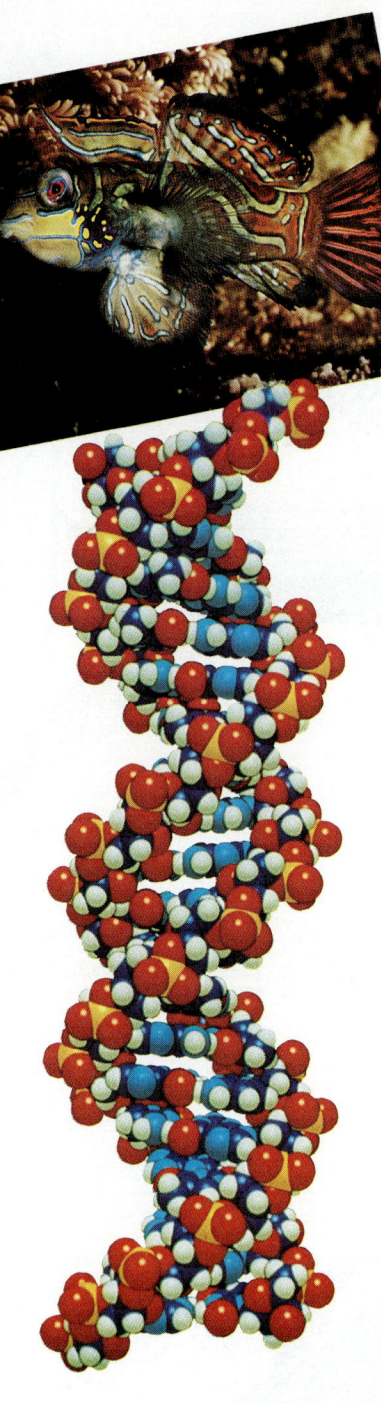

Unit 2 Diversity of Living Things 64

SUCH VARIETY! 66
1 Living Things in All Shapes and Sizes 66
2 Why Is There Diversity? 74

BEGINNINGS AND ENDINGS 86
3 The Origins of Diversity 86
4 The Loss of Diversity 90

KEEPING TRACK 94
5 Making Sense of Diversity 94
6 Ordering Diversity 112
7 Life Cycles: Diverse Stages 116

Making Connections 124

Science in Action:
Spotlight on Medicine 126

Science and Technology:
Extinction in the Modern World 128

Unit 3 Solutions 130

A SOLUTION OR NOT? 132
1 Solutions and Non-Solutions 132
2 Parts of a Solution—Solutes and Solvents 140
3 Soluble, Not Very Soluble, or Insoluble? 148

SEPARATING SOLUTE AND SOLVENT 152
4 The Salt Industry 152
5 Desalination and Distillation 156
6 Another Solution 160
7 Sugar Solutions 164

CONCENTRATION OF SOLUTIONS 168
    8 Getting Colder 168
    9 How Much Solute? 170
   10 Saturation 176
   11 Solutions in the Environment 179
   12 Growing Crystals 183

Making Connections 186

Science in Action:
Spotlight on Chemistry 188

Science and the Arts:
The Perfume Maker's Art 190

Unit 4 Force and Motion 192

UNDERSTANDING FORCES 194
    1 Forces at Work 194
    2 Different Kinds of Forces 200
    3 What Makes Things Fall? 204

MEASURING FORCES ... 210
    4 The Tools to Use ... 210
    5 Estimating the Size of a Force ... 216
    6 How Strong Are You? ... 220

FRICTIONAL FORCE ... 225
    7 Take a Look at Friction ... 225
    8 Causes and Consequences of Friction ... 231

GOING THROUGH THE MOTIONS ... 236
    9 Thought Experiments ... 236
    10 Puffball: A Game of Force and Motion ... 242
    11 What Is Inertia? ... 246
    12 Forces in Pairs ... 251

Making Connections ... 256

Science in Action:
    Spotlight on Electrical Engineering ... 258

Science and Technology:
    Rocket Science ... 260

Unit 5 Structures and Design ... 262

THE SCIENCE OF STRUCTURE ... 264
    1 Planning the Structure ... 264
    2 Response of Structures to Force ... 272
    3 Engineering Structures ... 282

THE ART OF DESIGN ... 293
    4 Beauty in Structures ... 293
    5 A Question of the Environment ... 304

Making Connections ... 310

Science in Action:
    Spotlight on Architecture ... 312

Science and the Arts:
    An Architectural Marvel **314**

Unit 6 The Restless Earth **316**

SHAKE, RATTLE, AND FLOW **318**
    1 Making Mountains **318**
    2 The Earth Breaks Apart **325**
    3 Volcanoes—Holes in the Earth **332**

THE ROLE OF ROCKS **341**
    4 Rocks in the Past **341**
    5 Hot Rocks **346**
    6 Rocks From Sediments **351**
    7 Changed Rocks **355**

FOSSILS—RECORDS OF THE PAST **360**
    8 Rocks Reveal a Story **360**
    9 Telling Time With Rocks **365**

Making Connections **370**

Science in Action:
    Spotlight on Geology **372**

Science and the Arts:
    Jewelry **374**

T9

Unit 7 Toward the Stars 376

THE OBSERVERS 378
1 What Is Astronomy? 378
2 An Ancient Science 382

THE EARTH MOVES 392
3 A Scientific Revolution 392
4 Motions and Their Effects 398

EXPLORING THE SOLAR SYSTEM 406
5 Visitors from Space 406
6 The Space Probes 411
7 Colonizing the Solar System 416

OUR UNIVERSE 422
8 Messages from the Stars 422
9 Earth's Place in the Universe 429
10 A Likely Beginning 436

Making Connections 440

Science in Action:
Spotlight on Astronomy 442

Science and Technology:
The Most Powerful Objects in the Universe 444

Unit 8 Growing Plants 446

GARDEN INGREDIENTS 448
1 A Garden in Space 448
2 The Beginning of the Garden 450
3 Breaking Ground 454

INNER ACTIONS OF PLANTS 463
4 Water, Water, Everywhere! 463
5 Giving Plants a Hand 469

6	Flowers and Pollination	**473**
7	Plants from Plant Parts	**479**

PLANTS IN YOUR ENVIRONMENT **484**
8	Medicine or Poison?	**484**
9	Make a Green World!	**486**

Making Connections **494**

Science in Action:
Spotlight on Agricultural Engineering **496**

Science and Technology:
Genetic Engineering **498**

CONCEPT MAPPING **500**
GLOSSARY **S145**
INDEX **S150**

SourceBook **501**

Unit 1 .... **S1**
1.1	The Biosphere	**S2**
1.2	Succession and the Biomes	**S8**
1.3	Humans and the Environment	**S17**

Unit 2 .... **S23**
2.1	The Diversity of Modern Life	**S24**
2.2	The Evolution of Diversity	**S35**

Unit 3 .... **S45**
3.1	Solutions, Suspensions, and Colloids	**S46**
3.2	Acids, Bases, and Salts	**S51**

Unit 4 .... **S61**
4.1	Motion and Force	**S62**
4.2	Laws of Motion	**S68**
4.3	Gravitation	**S74**

Unit 5 .... **S77**
5.1	Architecture in Other Cultures	**S78**

Unit 6 .... **S89**
6.1	Rocks and the Rock Cycle	**S90**
6.2	Rock-Forming Minerals	**S99**
6.3	Stories in Rocks	**S105**

Unit 7 .... **S111**
7.1	Upward and Outward	**S112**
7.2	Stars: A Universe of Suns	**S118**
7.3	One Small Step	**S126**

Unit 8 .... **S133**
8.1	Plant Structure	**S134**
8.2	Uses of Plants	**S141**

SCIENCEPLUS

Owner's Manual

The SciencePlus Philosophy

Welcome to SciencePlus, an innovative approach to science education. SciencePlus is unlike any science program you have used before. It is designed from the ground up to teach science in precisely the way that students learn best—by thinking, talking, and writing about what they do and discover. SciencePlus is activity- and inquiry-based. In other words, it is both hands-on and minds-on. SciencePlus is lively, engaging, and relevant to the student's world. SciencePlus is loaded with thought-provoking activities designed to challenge students' thinking skills while introducing them to realistic methods of science.

SciencePlus works for students and teacher alike. Students enjoy and benefit from its active, varied approach, while teachers find it to be teachable under real conditions. Laboratory-type activities require about 30–40 percent of class time. The remainder of class time is taken up by a rich variety of learning activities.

SciencePlus is tailored to meet the needs of middle school-aged students. It accommodates every type of student you are likely to encounter: the basic student, who requires substantial guidance; the average student, who is not yet fully-equipped with abstract-thinking skills; and the advanced student, who needs only to be pointed in the right direction to succeed.

At every stage, the SciencePlus program emphasizes concept and skill development over memorization of facts. By doing science activities and then thinking about the results, students learn the whys and hows, not just the whats and whens, of science. Ultimately, students come to see science as a system for making sense of the world.

SciencePlus is lively, engaging and relevant to the student's world.

Origins

The SciencePlus program was originally developed by the Atlantic Science Curriculum Project (ASCP) of Canada to replace the traditional recall-based curriculum, which had proven to be ineffective. SciencePlus represents a ground-breaking effort, the culmination of many years of labor by dozens of talented, dedicated science educators.

The SciencePlus development team was guided every step of the way by the latest insights into how children actually learn. The SciencePlus program has been thoroughly tested on real students in realistic settings, refined, and then retested. The result is a program that works! Teachers using SciencePlus have reported dramatic gains in student comprehension and retention of scientific concepts. Above all, students enjoy using SciencePlus and develop a heightened interest in science.

A Continuing Tradition

The American Edition of SciencePlus continues the tradition of excellence begun in Canada. Many of the recommendations of Project 2061 and NSTA Scope, Sequence, and Coordination have been employed in order to make the program even better. In addition, SciencePlus has been made easier to follow and more culturally and regionally relevant. Exciting new features have also been added to highlight the linkages among science, technology, and society.

An Interactive, Effective Program

Science*Plus* employs proven teaching strategies: guided and open-ended investigations, small-group discussions, exploratory writing and reflective reading tasks, games, picture and word puzzles, and independent long-range projects. This variety helps motivate and maintain the interest of students and teachers alike.

Science*Plus* develops scientific process skills as an essential goal. As the curriculum progresses, the students will gradually master increasingly complex tasks. For example, students will move from directed to open-ended inquiry, and from reading and completing tables and graphs to constructing them from experimental data they have collected on their own.

In general, each of the units in Science*Plus* is self-contained and may be taught as a separate instructional module. Science*Plus* contains a balance of physical, biological, earth/space, and environmental science topics.

Guiding Principles

The guiding principles of Science*Plus* are simple and few:

❖ Anyone can learn science

The popular image of science as the private domain of the super-intelligent is wrong and damaging. Science is for everyone. Children exposed to science for the first time take to it naturally. It is only later, after the science kits and fun activities have been abandoned in favor of fill-in-the-blank worksheets and recall drills, that love for science is replaced by fear and dread. Science*Plus* can rekindle the sense of wonder and fascination that lies dormant within your students.

❖ Science is a natural endeavor

Whether we realize it or not, each of us applies science nearly every day. Hardly a day passes that we don't ask ourselves "How does this work?" or "Why does that happen?" or "What happens if...?" Scientists differ from other people only in that it is their profession, rather than their avocation, to figure out "how," "when," and "why."

Unfortunately, stereotypes about science and scientists abound. Many students feel that only "nerds" or "geeks" enjoy science. This falsehood may do as much to turn people away from science as any curricular shortcoming. Science*Plus* actively refutes these stereotypes. It portrays science as a rewarding, quintessentially human undertaking. Scientists are portrayed as normal people, not aloof geniuses who talk in equations.

❖ Science is its own reward

There is no feeling quite like the thrill of discovery or the sense of accomplishment that comes from rising to a difficult challenge. Science can be thought of as a voyage into the unknown. This voyage can be exciting and rewarding for all.

Aims

Science*Plus* is designed to help you further develop each of the following in your students:
- Understanding of the interrelationships among science, technology, and society.
- Understanding of important science concepts, processes, and ideas
- Use of higher-order thinking skills
- Ability to solve problems and apply scientific principles
- Commitment to environmental protection
- Interest in independent study of scientific topics
- Social skills
- Communication skills

To accomplish these goals, a wide variety of teaching strategies are employed. The common denominator among these is their emphasis on *doing*. At all times, students are to be active and involved.

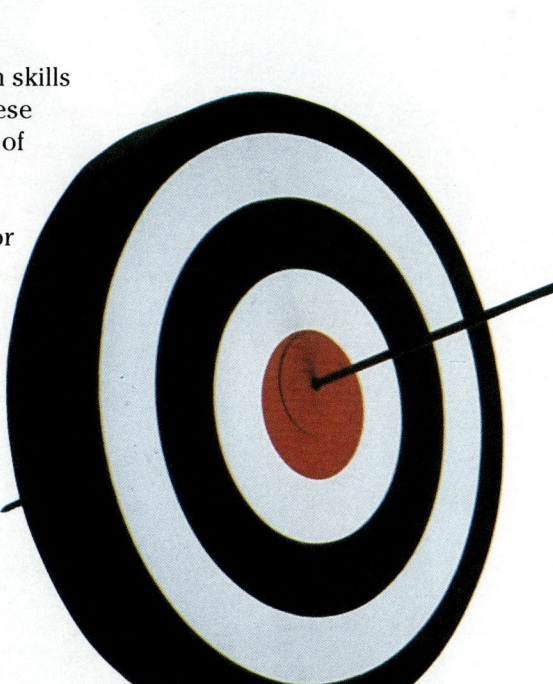

*Science*Plus* can rekindle the sense of wonder and fascination that lies dormant within your students.*

To prepare students for the challenges of the future, they must become science literate.

Science, Technology, and Society

Science and technology are flip sides of the same coin; each supports the other. Neither should be studied in isolation. Science*Plus* explores the relationship between science and technology, and the effect of both on our society as a whole. Even people who never again set foot in a laboratory after leaving school can benefit from an understanding of science and technology and how both relate to each other and to society at large.

Our society, complex as it is, will become even more so in the years to come. Science and technology will play roles in every aspect of life. In the future, "high-tech" will be more than a catch phrase, it will permeate every aspect of life. To prepare students for the challenges of the future, they must become science literate. They must be given the tools to become responsible and productive individuals in a highly technological world.

Science for All

Science*Plus* is designed to put the "process" back into science education and, in so doing, provide students with the intellectual skills they need to truly understand and apply science. Now, as never before, a thorough grounding in science is absolutely essential; without it, students—the citizens of tomorrow—cannot expect to be fully conversant in and responsive to the complex issues of the twenty-first century.

No program can teach itself. You, with your energy, enthusiasm, and ability, are the key to a successful outcome. Science*Plus* will help you help the students develop all the skills they need to learn independently. Science*Plus* fosters a spirit of joint exploration—students with teacher and with one another. **Let the journey begin.**

The SciencePlus Method

Building a Better Understanding of Science

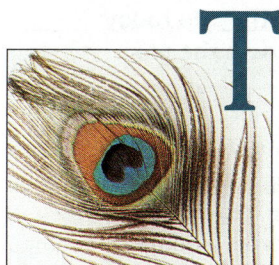

The SciencePlus program is based on the Constructivist Learning Model (CLM). Constructivism is based as much on common sense as on the results of research. With the CLM, students "construct" an understanding of concepts step by step. Students begin by identifying what they already know about a topic. Any misconceptions they may have about a topic are exposed at this point. Identifying these misconceptions is a critical part of the process. Next, students do hands-on activities to experience the subject matter directly. Their experiences cause them to amend, add to, or scrap altogether the mental model they already have of the subject in question.

Constructivism is based on a few key steps:

1. Invitation

The Invitation stimulates students' curiosity and engages their interest. At this stage, students note the unexpected, pose questions, or define a problem.

2. Exploration

Explorations engage students in the search for solutions or explanations. Students look for alternative sources of information, collect and evaluate data, and clarify their findings through discussion and debate.

3. Proposing explanations and solutions

At the conclusion of the Explorations, students propose their response to the problem or question posed in the Invitation. The class is exposed to a variety of possible responses, and students have the opportunity to consider each.

4. Taking action

Students make decisions about a course of action based on the various proposals offered. If the class reaches consensus, then this stage may bring about closure of the lesson. It may happen, though, that this stage identifies new questions to explore.

▶ *For an in-depth discussion of Constructivism, see "The Constructivist Learning Model," by Bob Yager,* **The Science Teacher,** *September 1991*

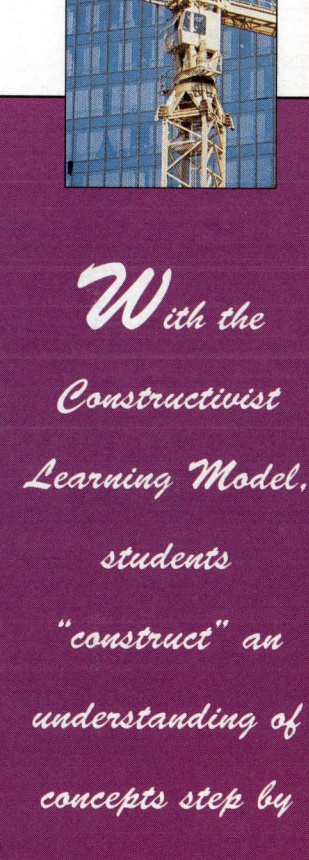

With the Constructivist Learning Model, students "construct" an understanding of concepts step by step.

T17

Themes in Science

The traditional division of science into three branches—life, earth, and physical sciences—often leads students to believe that nature is similarly arranged. Too often, students view science as a system of separate, nonrelated abstractions, or as a compilation of facts and difficult-sounding terms. But science is simply the study of nature, and there are certain underlying principles, or themes, that unite the study of all areas of science. The themes provided here are not meant to replace the traditional teaching of scientific disciplines, but rather to create a framework for the unification of these disciplines.

The themes are intended to integrate facts and ideas, and to provide a context for discussing the textual matter in a meaningful way. You can employ these themes as an organizational tool to reinforce understanding of the subject matter, rather than the rote memorization of facts. The following six themes are emphasized in Science*Plus*.

> *One subject can be addressed from the viewpoint of many different themes.*

Energy

Energy is the ability to put matter into motion. Energy is what makes the universe and everything in it dynamic and ever-changing. The study of dynamic systems in any field of science requires an understanding of energy: its origins, how it flows through systems, how it is converted from one form to another, and how it is conserved. Energy provides the basis for all interactions, whether chemical, biological, or physical. Thematically, energy connects all disciplines.

Evolution

Evolution can be broadly described as change over time. In the evolutionary sense, change is not a simple alteration, but rather a progression of such alterations over the continuum of time. The entire universe and all of its inhabitants can thus be studied from the viewpoint of historical change. This theme can be used to understand why things exist as they do today by understanding how they were in the past.

Patterns of Change

Change through time can be analyzed in connection to its rates and patterns. Changes can be predictable or unpredictable, regular or irregular, cyclical or unidirectional. Change can affect both simple and complex systems. Knowing different patterns of change provides students with an understanding of natural systems and their underlying mechanisms, and can help them predict what will happen in the future based on this understanding.

Scale and Structure

Scale and Structure provides a means for understanding the nature of matter and systems within the natural world. Structure defines how matter, a system, or a process fits together; scale determines the level at which this structure is examined. Scale and structure provide a means for understanding the underlying framework that gives order to matter.

Stability

Stability is constancy within a system. It can refer to the way in which systems as a whole do not change, or it can refer to a state of static equilibrium in which the forces causing change within a system are balanced. Stability also relates to the predictability of change within nature.

Systems and Interactions

Systems and Interactions provides a basis for the study of all matter from the simplest level to the most complex forms. A system can be defined as an interacting group of forces, bodies, or substances. How these systems interact gives vital information about the individual components of the system being studied.

The following table summarizes the application of themes throughout the Science*Plus* program.

UNIT	Energy	Evolution	Patterns of Change	Scale and Structure	Stability	Systems and Interactions
Level Green						
Science and Technology			◆		◆	
Patterns of Living Things		◆	◆	◆	◆	
It's a Small World			◆	◆		◆
Investigating Matter	◆			◆	◆	
Chemical Changes			◆	◆		◆
Energy and You	◆		◆			◆
Temperature and Heat	◆			◆		◆
Our Changing Earth		◆		◆	◆	
Level Red						
Interactions	◆		◆			◆
Diversity of Living Things		◆	◆	◆		
Solutions			◆	◆		◆
Force and Motion	◆				◆	◆
Structures and Design				◆	◆	◆
The Restless Earth		◆	◆	◆		
Toward the Stars		◆	◆	◆		
Growing Plants			◆	◆		◆
Level Blue						
Life Processes	◆			◆	◆	
Machines, Work, and Energy	◆				◆	◆
Oceans and Climates		◆	◆			◆
Electromagnetic Systems	◆			◆		◆
Sound	◆	◆				◆
Light	◆		◆		◆	
Particles	◆		◆	◆		
Continuity of Life		◆	◆	◆		

Using the Themes

A major strength of the thematic approach is that seemingly different processes, structures, or systems can be shown to have underlying similarities. Although many thematic organizations are possible, each Unit Interleaf in this *Annotated Teacher's Edition* suggests at least two major themes that can be discussed in relation to the unit material. Focus questions are also provided that can be used to promote discussion and an understanding of how the themes relate to the text material.

Although major themes have been identified in each Unit Interleaf, it is up to you to decide which themes you feel are most appropriate. The direction that your class discussion takes will most likely guide you in your choices.

In your class discussions, use the themes to provide a framework of understanding for your students. For example, whether you are studying photosynthesis or the way in which the forces of nature have shaped the physical appearance of the earth, the theme of energy can be discussed. Similarly, one subject can be addressed from the viewpoint of many different themes. For example, in discussing an organism such as a zebra, energy can be applied in a discussion of how the zebra takes in food from the environment; evolution can be discussed in relation to how the zebra's structures are adaptations to its environment; and patterns of change can be introduced by a discussion of how the zebra's migration habits are based on the seasons.

At the end of a unit, the themes can provide a helpful source of review and a way to consolidate the material presented. In addition, questions provided within the text as well as in the *Teacher's Edition* are thematically based.

In your class discussions, use the themes to provide a framework of understanding for your students.

Components of SciencePlus

The Pupil's Edition

SciencePlus is no ordinary textbook. SciencePlus is a student-friendly text; lively, abundantly illustrated with clever, colorful illustrations, and loaded with engaging activities. Every effort has been taken to make this text the sort of book that students will actually want to use.

such as communicating ideas and sharing responsibility. In addition, the cooperative groups not only make science more interactive and more fun, but also provide valuable opportunities to develop socialization skills.

▶ **For more information on cooperative learning, see pages T34 and T35 of this *Annotated Teacher's Edition*.**

Units, Sections, Lessons

SciencePlus contains eight units, which are further divided into sections and lessons on closely related subject matter. Each lesson includes a wide variety of activities and explorations designed to develop the lesson content.

Explorations

Scattered throughout each unit are a series of Explorations—hands-on, inquiry-based activities. These Explorations allow students to see scientific principles in action. The Explorations are essential for inducing real learning in students. As students do the Explorations, they have the opportunity to compare their mental models of scientific principles to the real things. As weaknesses are exposed, students adjust their thinking to accommodate what they have learned.

Most of the Explorations are designed to be done cooperatively in small groups. In this way, the Explorations model real scientific experiences, in which scientists work together to solve problems. By working in cooperative groups, students also develop important skills

Many Explorations can be completed within a single class period. Others are more involved and may require several class periods to complete. Most of the supplies needed for the Explorations consist of very common equipment and materials. In most cases, they can be easily gathered from the home. For help in gathering supplies, the second page of each parent letter in the TRB provides a listing of the materials you will need to teach the unit. This page makes it easy for you to ask parents to donate materials. This will help you keep your budget low and at the same time get parents involved in the SciencePlus experience.

Assessment

SciencePlus contains a variety of methods for checking student learning.

◆ Challenge Your Thinking

To make sure your students comprehend the new information, each section concludes with *Challenge Your Thinking* questions. These questions challenge students to apply newly learned material in a variety of ways. Many of the questions are like brain-teasers in that they are unusual and highly creative. Because the questions are not simply recall, students actually find them fun to figure out.

◆ Making Connections

Each unit concludes with *Making Connections*, the SciencePlus equivalent of a chapter review. The *Making Connections* pages consist of four parts:

- *The Big Ideas*
- *Checking Your Understanding*
- *Reading* Plus
- *Updating Your Journal*

Making Connections differs in a number of key respects from a traditional chapter review. To begin with, **The Big Ideas** asks students to formulate their own summary of the unit, using a list of questions as a guide. Following this, **Checking Your Understanding** poses a selection of comprehensive questions designed to gauge students' understanding of the unit's subject matter.

The third part of the review pages consists of a short description of the corresponding unit in the *SourceBook*. This part, called **Reading Plus**, refers students to the appropriate pages in the *SourceBook* and whets their appetite for further exploration.

The review pages conclude with an invitation to students to rewrite their answers to the **For Your Journal** questions located at the beginning of the unit. This part, called *Updating Your Journal*, gives students the opportunity to confront and discard any misconceptions they may have had at the outset of the unit.

All answers to the *Challenge Your Thinking* and *Making Connections* questions are located in the extended margins of this Teacher's Edition.

Special Features

A common complaint among students is that the material they learn is not relevant. The end-of-unit features in Science*Plus* help show students how science is an integral part of their lives.

In keeping with the spirit of Science*Plus*, each of the features are interactive, with thought-provoking questions and research ideas that encourage students to explore the topic further.

◆ ### Science in Action

The *Science in Action* features consist of interviews with scientists and other science-related professionals. As students read the interviews, they learn a lot more than just the requirements of each career. They learn that scientists are likable and interesting people, very much like themselves. They also learn about how and why the people got involved in their careers in the first place. Students will learn about real experiences and what the scientists like most—and least—about their work. They may even learn about the scientists' personal career aspirations as well as their greatest disappointments.

◆ ### Science and Technology

New developments in science-related technologies are showcased in the *Science and Technology* features. Written in language that students can easily understand, *Science and Technology* is designed to captivate student interest at the same time that it informs. The impact of these new technologies on our everyday lives show students how relevant scientific research can be, both from a scientific standpoint as well as a social standpoint.

◆ ### Science and the Arts

People tend to think of art and science as polar opposites, when in fact they have much in common. The connection between science and society is reinforced by the *Science and the Arts* features. These features show students how artists have been inspired by nature and how scientific methods or principles have enhanced or empowered their work. *Science and the Arts* also demonstrates how disciplined, orderly thinking can be a useful asset to creative individuals in many different fields.

The SourceBook™

As an added resource, each level of the Science*Plus* program includes a *SourceBook,* which provides additional information related to the units of Science*Plus*. The *SourceBook* is available in two formats. It is either bound into the back of the *Pupil's Edition,* or it is bound as a separate booklet.

Regardless of which format you choose, the intended use of the *SourceBook* is the same. It is designed as a companion to the Science*Plus* textbook. Each *SourceBook* unit should be used only after the unit in the textbook has been successfully completed. This sequence will ensure that the students will have mastered the fundamental concepts before they start adding details.

The Science*Plus SourceBook* consists of eight units of resource information that corresponds to the eight units of the text itself. Each *SourceBook* unit extends beyond the material presented in the text, providing an excellent resource with which students can add depth to their understanding of the topics presented in Science*Plus*.

Students are directed to use the *SourceBook* in each *Making Connections* review. A brief summary of the corresponding *SourceBook* unit is included under the heading Reading *Plus*. Page numbers are also provided for ease of use.

Once students refer to the *SourceBook* unit, they will find a brief introduction that includes an activity such as writing a newspaper article, constructing a model, or making a collage. Several questions are then provided to help direct the students' thinking as they read the unit.

The Annotated Teacher's Edition

The SciencePlus Annotated Teacher's Edition will help you achieve the full potential of SciencePlus. The Teacher's Edition consists of two major parts: the Unit Interleaf and the Extended Margins. Each Unit Interleaf consists of a four-page insert preceding each unit. The Extended Margins provide on-page annotations and teaching suggestions.

Using the Unit Interleaf

For ease of use, each Unit Interleaf is divided into three parts: *Teaching the Unit*, *Meeting Individual Needs*, and *Resources*. The information provided under these headings will help you make the best use of your time in planning and organizing your instruction.

Teaching the Unit

This portion of the Unit Interleaf is devoted to giving you the necessary background and planning information about the unit. Extensive information is also provided about incorporating themes into the unit instruction.

❖ **Unit Overview**
To quickly bring you up to speed, the *Unit Overview* provides a section-by-section synopsis of what the students will be asked to do in completing the unit.

❖ **Using the Themes**
Suggestions are provided here for relating the unit information to the appropriate science themes. For each theme emphasized in the unit, a focus question is provided to help you stimulate your students to think along thematic lines.

❖ **Using the *Science Discovery* Videodiscs**
Each SciencePlus unit is correlated directly to *Science Discovery*, a videodisc program that has been designed specifically for SciencePlus. The *Using the Science Discovery Videodiscs* section of the Unit Interleaf provides information on the videodisc resources that apply to the unit. The *Science Discovery* videodisc program is discussed in greater detail beginning on page T28.

❖ **Using the SciencePlus SourceBook**
Here you will find a brief overview of what is contained in the corresponding unit of the *SourceBook*. For more information about the *SourceBook*, see page T23.

❖ **Planning Chart**
The extensive *Planning Chart* provides a graphic overview of the unit, making it easy to prepare lesson plans, set realistic schedules, and preview and select materials from the ancillary package. The *Planning Chart* provides the following information for each section and lesson of the unit.

- Page numbers for easy reference
- Estimated completion time in class periods
- Process skills emphasized
- Exploration and assessment items
- Ancillary materials and features, including *Science Discovery* videodisc resources

Meeting Individual Needs

Today's classrooms are places of diversity—populated by students with different ability levels and from a variety of ethnic backgrounds and cultures. The information and suggestions provided here will help you meet the challenge of individualizing your instruction to meet the needs of all your students. The information and suggestions are organized under the following headings.

❖ **Gifted Students**

Here you will find suggestions on how to keep your advanced and gifted students actively engaged in learning. Suggestions include long-term projects, open-ended activities, research projects, and challenges to students' creativity.

❖ **LEP Students**

Here you will find suggestions on how to teach students who may be just beginning to learn English.

❖ **At-Risk Students**

Suggestions are provided for helping you tailor your lessons to provide a successful experience for students who, for any of a number of reasons, are not motivated to learn about science. Recommendations are provided on how to keep such students motivated in their learning of science.

❖ **Cross-Disciplinary Focus**

Under this heading you will find activities that cross the boundaries between the various specialties of science, as well as between science and other disciplines such as art, history, and geography. By doing these activities, students will experience the inter-relatedness of science and other areas of study. These activities will help students see the significance of science to many areas of study.

❖ **Cross-Cultural Focus**

Your classes probably have students from many different cultural backgrounds. This, of course, has a major impact on how students learn and what they are able to relate to. Under this heading you will find activities and suggestions that serve to highlight cultural diversity and show its positive influence on science as well as other disciplines. These activities will help you add depth to your instruction by showing students how culture and science are integrated to the benefit of us all.

Resources

A list of teacher and student resources are provided for the unit, including appropriate software and media.

Today's classrooms are places of diversity—populated by students with different ability levels and from a variety of ethnic backgrounds and cultures.

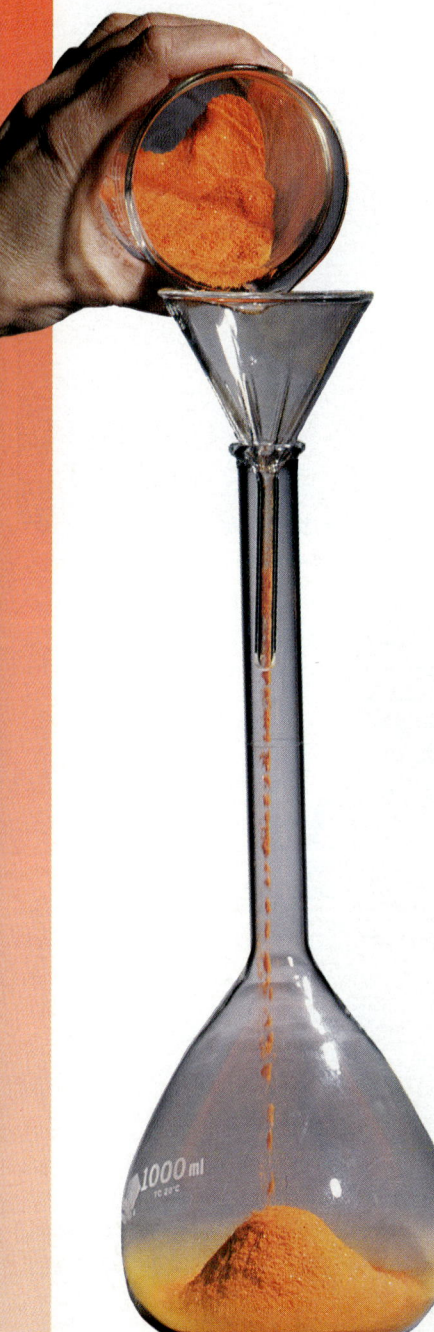

Using the Extended Margins

The entire *Pupil's Edition* of Science*Plus* has been reduced in size and placed on the pages of this *Annotated Teacher's Edition*. The margins around the reduced pages have been filled with many teaching suggestions and extensive commentary to help you teach each lesson with maximum effectiveness. The following headings will help you locate desired information quickly.

Unit Opener

Each unit opener has the following teacher's information to help you quickly engage your students in the subject of the unit.

❖ **Unit Focus** provides suggestions for introducing students to the unit in an active way. The *Unit Focus* poses thought-provoking questions that will help you crystallize your students' thinking.

❖ **About the Photograph** provides background information about the unit photograph: why it was chosen, how it represents the subject matter of the unit, and items of interest that it contains.

❖ **Using the Photograph** provides suggestions for challenging students to use their powers of observation and reasoning to derive knowledge from the unit opener photograph. Suggestions help you start your students thinking and asking questions about the unit topic.

❖ **About Your Journal** identifies the intent of the *For Your Journal* activity.

Lesson

Each Lesson of the unit has extensive commentary and teacher's notes to help you in planning and teaching the unit. The lesson information is divided into three parts: *Getting Started, Teaching Strategies,* and *Follow Up*. In addition a *Lesson Organizer* is provided at the beginning of each lesson.

❖ **Getting Started**
The *Getting Started* information provides a short summary of the lesson and a listing of the main ideas, or main concepts, that are covered in the unit.

❖ **Lesson Organizer**
A *Lesson Organizer* is also included at the beginning of each lesson. The *Lesson Organizer* gives a quick reference of the lesson objectives, process skills, new terms, materials needed, and estimated time requirement for completing the lesson. Resource information in the *Teacher's Resource Binder* and the *Science Discovery Videodiscs* is also listed here for easy reference.

❖ **Teaching Strategies**
Teaching Strategies provide helpful hints, highlight points of interest, suggest additional activities, and pose interesting questions for students to answer. *Teaching Strategies* also alert you to any special materials and preparation that might be required to carry out the lesson and to anticipate possible outcomes and questions.

❖ **Answers to Questions**
All questions that have been posed in the course of the lessons, either within the running text or in the review materials, are answered in the expanded margins for your convenience. In only a few instances, where the answer requires a graph or chart, will you have to turn pages to refer to the answer.

❖ **Follow Up**
The *Follow Up* provides closure to the lesson and consists of two parts: *Assessment* and *Extension*. The *Assessment* poses one or two questions or problems suitable for demonstrating mastery of the material in the lesson. The *Extension* provides problems or questions that complement and extend the subject matter of the lesson.

Review Pages

All questions appearing in each *Challenge Your Thinking* and *Making Connections* are fully answered in the Extended Margins.

Features

The end-of-unit features *Science in Action, Science and Technology,* and *Science and the Arts* are all accompanied by commentary designed to enhance the value of each feature.

The Teacher's Resource Binder

The *Teacher's Resource Binder* (*TRB*) contains a wide variety of useful resource materials that are designed to supplement and extend the subject matter of *SciencePlus*.

Unit Worksheets are loose-leaf blackline masters conveniently organized by unit.

◆ *Home Connection* is a two-page parent letter that introduces the unit of study to the parents and includes home activities to encourage parents' participation. The last page lists the supplies needed to do the Explorations. This page makes it easy for you to invite donations in order to keep your budget low.

◆ *Worksheets* contain activities such as charts, graphs, puzzles, and games that clarify and reinforce concepts. *Activity Worksheets* provide additional activities to add depth to your instruction. *Resource Worksheets* provide blank charts, graphs, and puzzles directly from the textbook.

◆ *Sample Assessment Items* provide sample test items for each unit. Because they consist of a wide variety of question types, you will find items suitable for students of all ability levels. Sample answers are provided.

Videodisc Resources describes how *Science Discovery* videodiscs can be used in the *SciencePlus* program. Written materials include lesson plans, barcodes, and student worksheets. For more information about the *Science Discovery* Videodisc program, see page T-28 of this *Annotated Teacher's Edition*.

Science Sites are maps that show scientific points of interest in each of the 50 states, Canada, Mexico, and other continents. Five questions on the back of each map help integrate the life, earth, and physical sciences.

ESL Spanish includes blackline masters that consist of translated parent letters, unit summaries, and a full glossary.

Posters provide two colorful reference sources that highlight safety and equipment in the laboratory.

Materials Guide provides a ready reference of all the supplies and equipment that are required to teach *SciencePlus*. The *Materials Guide* contains unit by unit materials lists as well as a master materials list.

Other Teaching Aids

The *SciencePlus* program includes the following support materials to make your teaching both effective and efficient.

Teaching Transparencies—which include approximately 20 four-color *Resource Transparencies* and *Teacher's Notes* and approximately 50 two-color *Graphic Organizer Transparencies*. The Graphic Organizer Transparencies consists of charts and graphs from the *Pupil's Edition*.

English/Spanish Audiocassettes—which provide important preview information for each unit in both English and Spanish.

Unit Tests—which include blackline-master tests for each unit of the *Pupil's Edition*. Four tests are included for each unit—three tests of increasing difficulty and one activity-based test for assessing student performance.

Test Generators (IBM®, Macintosh®)—which include test items, including graphics, for the *Pupil's Edition* and *SourceBook*. The *Test Generators* also allow you to add your own questions.

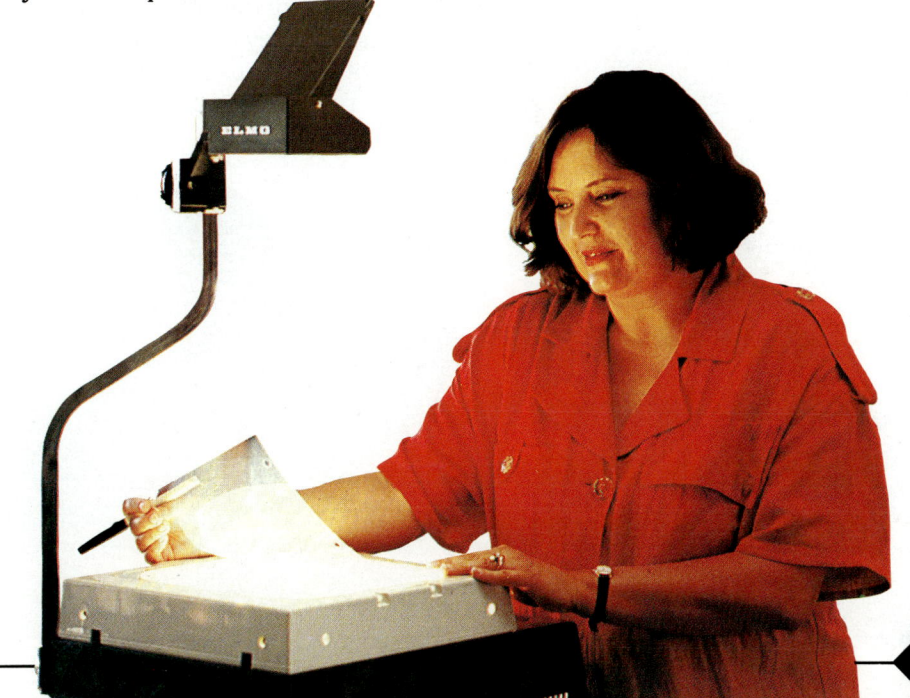

Science Discovery Videodisc Program

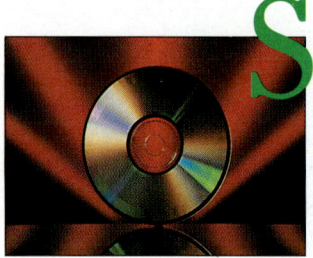

Science Discovery is a comprehensive videodisc program that blends imagination and state-of-the-art technology into an exciting program that will add a dynamic facet to your science teaching. The program includes:

- *Science Sleuths* videodisc
- *Science Sleuths Teacher's Guide*
- *Science Sleuths Resource Directory*
- *Image and Activity Bank* videodisc
- *Image and Activity Bank Directory*
- *Image and Activity Bank Quick Reference Card*

Science Sleuths

The *Science Sleuths* videodisc consists of 24 open-ended science mysteries, one for each unit of the Science*Plus* textbook series. Each mystery begins with a dramatization in which a problem is presented and the students are asked to serve as consultants (sleuths) to figure it out. Problems include such mysteries as *Exploding Lawn Mowers, Dead Fish on Union Lake,* and *The Misplaced Fossil.* The videodisc then becomes a "videophone" that connects students to the laboratory, where they can request information such as the testing of samples, experiments, interviews, tables, newspaper clippings, news broadcasts, and many other bits of information from an extensive menu.

The students work as a class or in small groups to explore the mystery, develop hypotheses, and support their position using the data from the disc. They try to solve the mystery using as few of the video segments as possible. In this way they either challenge themselves or other groups to see who can solve the mystery most efficiently.

In working as Science Sleuths, students improve their problem-solving and reasoning skills while applying their science knowledge from the textbook. Students work in a realistic scientific mode in which information comes from a variety of sources. Students must also judge the accuracy of each source and separate raw data from interpretation and inference.

Science Sleuths Teacher's Guide

The *Science Sleuths Teacher's Guide* contains the teaching plans for using the videodisc.

Science Sleuths Resource Directory

The *Science Sleuths Resource Directory* contains the barcodes and frame numbers for using each mystery. There are five Directory sets so that you can have five working cooperative groups at a time.

Image and Activity Bank

The *Image and Activity Bank* videodisc consists of a still and motion image database designed to reinforce and extend concepts presented in Science*Plus*. The still images include hundreds of photographs and computer graphics related to life, earth, and physical sciences. The motion images include demonstrations, experiments, and selected motion footage.

Image and Activity Bank Directory

The *Image and Activity Directory* provides descriptions, frame numbers, and barcodes for the *Image and Activity Bank* videodisc. A separate Reference Card provides barcodes and frame numbers for selected topics.

> ⟹ *For complete directions on how to use Science Discovery with SciencePlus, open the Teacher's Resource Binder to the Videodisc Resource section.*

Using SciencePlus

Communicating Science

One of the most important skills that students can acquire is the ability to communicate what they have learned, both orally and in writing. Science*Plus* challenges students to develop and communicate their mastery of new ideas in novel ways—for example, by writing for one another or for some audience other than the teacher. Students' comprehension is enhanced when they are called upon to reformulate in their own words what they have learned.

Throughout Science*Plus*, students are called on to communicate what they have learned in many different ways. Students may be called on to interpret a passage, illustrate a paragraph, write a headline, label a diagram, or write a caption for a photo or drawing. These strategies complement the inquiry approach followed throughout Science*Plus*.

Reading and Writing in the Classroom

The time spent in helping students prepare to read is critical in fostering comprehension. Two strategies are particularly important in the pre-reading phase of instruction: building on prior knowledge and establishing a purpose for reading.

The Power of Prior Knowledge

The amount of prior knowledge that students have about a topic directly influences their comprehension of that topic. The more students know about something, the easier it is for them to grasp new information about it. Helping students identify the information they already have about a topic before reading assists them in relating the new information to existing knowledge.

Research strongly suggests that students tend to retain misconceptions they may have about a topic. If the text information seems to conflict with their preconceptions, students may ignore or reject new information. It is therefore extremely important to identify these misconceptions so that they may be dispelled.

Reading to Understand

One way to establish a purpose for reading is to make a study guide with questions that students can answer as they read. You can also help students learn to make their own study guides. First, teach students to preview a unit by looking at all the unit headings, illustrative material, and terms and phrases in boldface or italics. This technique helps students gain a feel for the unit and helps them build a basic structure for the new information. Then show them how to use the captions, headings, and highlighted words to devise study-guide questions to answer after reading the unit. Following this method, students will read with a purpose in mind, a purpose of their own devising.

Writing to Understand

Studies show that writing is an effective tool for improving reading. As students write, they are creating a text for others to read. The most important advice to give students about scientific writing is to strive for clarity and accuracy. These characteristics can often be achieved with simple vocabulary and short sentences. One useful approach might be to have them imagine that they are writing for a younger audience.

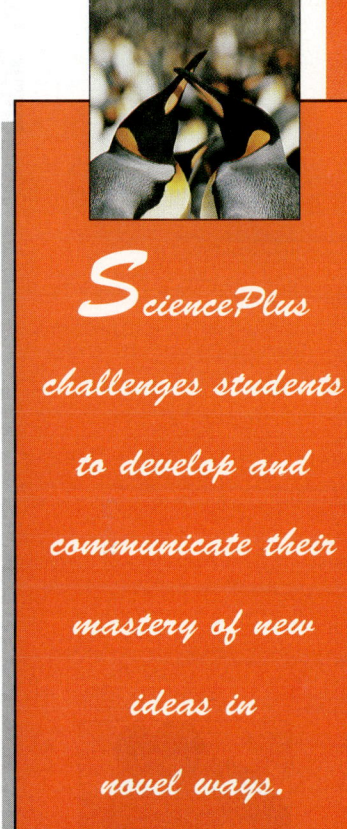

*Science*Plus *challenges students to develop and communicate their mastery of new ideas in novel ways.*

Keeping a Journal

One highly successful tool for improving students' performance in science is the Journal. The SciencePlus approach to learning depends heavily on the Journal. The Journal has many functions. First and foremost, it is an ongoing record of students' learning. Students begin the study of a new topic by recording prior knowledge of that topic. Any misconceptions that students may have are thus exposed. As the lesson progresses, students record any and all new findings. In many cases, students find that what they learn through their activities contradicts their preconceptions.

Much of the work that students do is recorded in their Journals. The Journal is a constant reminder to students that learning is occurring. Students can look back and compare their early work with later work to see and take pride in the progress they have made. To supplement their other work, you may also want to ask students to briefly summarize what they have learned each week. This makes a very handy capsule history of their work.

What makes a good Journal? Insofar as is possible, the Journal should be neat and easy to follow. Students may organize their Journals in any of a number of ways, chronologically, by unit, or by lesson, to name a few. Some kind of heading should set off each major entry.

A spiral-bound notebook or hard-bound lab-type notebook makes a good Journal. Or you may make copies of the sample Journal pages in the *Teacher's Resource Binder* and distribute them to your students to use if you wish. These Journal pages are located at the beginning of the Unit Worksheets for Unit 1.

The most important advice to give students about scientific writing is to strive for clarity and accuracy.

Process Skills

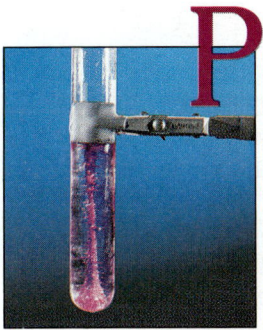

Process skills are a means for learning and are essential to the conduct of science. For this reason, Science*Plus* is strongly process-oriented.

Perhaps the best way to teach process skills is to let students carry out scientific investigations and then point out to them the process skills they use in the course of the investigations. The teacher's *Planning Chart* at the beginning of each unit and the *Lesson Organizer* at the beginning of each lesson identify the process skills that are highlighted in the corresponding text.

Science*Plus* makes regular use of many different process skills as highlighted below. These and other process skills are called upon in Science*Plus* in virtually every lesson.

Observing

An observation is simply a record of a sensory experience. Observations are made using all five senses. Scientists use observation skills in collecting data.

Communicating

Communicating is the process of sharing information with others. Communication can take many different forms: oral, written, nonverbal, or symbolic. Communication is essential to the conduct of science, given its collaborative nature.

Measuring

Measuring is the process of making observations that can be stated in numerical terms. In Science*Plus*, all measurements are carried out in SI units.

Comparing

Comparing involves assessing different objects, events, or outcomes for similarity. This skill allows students to recognize any commonality that exists between seemingly different situations. A companion skill to comparing is contrasting, in which objects, events, or outcomes are evaluated according to their differences.

Contrasting

Contrasting involves evaluating the ways in which objects, events, or outcomes are different. Contrasting is a way of finding subtle differences between otherwise similar objects, events, or outcomes.

Organizing

Organizing is the process of arranging data into a logical order so it is easier to analyze and understand. The organizing process includes sequencing and grouping, and classifying data by making tables and charts, plotting graphs, and labeling diagrams

Classifying

Classifying involves grouping items into like categories. Items can be classified at many different levels, from the very general to the very specific.

Analyzing

The ability to analyze is critical in science. Students analyze to determine relationships between events, to identify the separate components of a system, to diagnose causes, and to determine the reliability of data.

Inferring

Inferring is the process by which conclusions are drawn based on reasoning or past experience.

Hypothesizing

Hypothesizing is the process of developing explanations for events that can then be tested. Testing either supports a hypothesis or refutes it.

Predicting

Predicting is the process of stating in advance the result that will be obtained from testing a hypothesis. A prediction that is accurate tends to support the hypothesis.

Critical Thinking

Critical-thinking skills are essential for making sense of large amounts of information. Too often, science lessons leave students with a set of facts and little ability to integrate those facts into a comprehensible whole. Requiring students to think critically as they learn improves their comprehension and increases their motivation.

Loosely defined, critical thinking is the ability to make sense of new information based on a set of criteria. Critical-thinking skills draw upon higher-order thinking processes, especially synthesis and evaluation skills. Critical thinking takes a number of different forms.

Validating Facts

This type of critical thinking involves judging the validity of information presented as fact. Too often, people will accept as valid almost any statement, no matter how outrageous, as long as it comes from a supposedly authoritative source. It is important for scientists to treat all untested data with suspicion, no matter how reasonable it may seem.

Students may validate facts in a number of ways: by observing, by testing, by rigorously examining the logic of the so-called fact. Science*Plus* presents students with many opportunities to critically evaluate facts and hypotheses.

Making Generalizations

A scientist must often be able to identify similarities among disparate events. Generalizations are drawn based on a limited set of observations but can be applied across an entire class of phenomena. One does not have to test every substance known to make the generalization that solid substances melt when heated. Generalizations allow scientists to make predictions. Once the rule is known, future outcomes can be forecasted with a high degree of confidence.

It is important that students base their generalizations on an adequate amount of information. A generalization that is formed too quickly may be wrong or incomplete, or could lead the student down a dead-end path.

Making Decisions

Many students would not regard science as a field requiring decision-making skills. But in fact, scientists must make decisions routinely in the course of their work. Any time a scientist works through a problem or develops a model, a whole series of decisions must be made. A single faulty decision can throw the entire process into disarray. Making informed decisions requires knowledge, experience, and good judgment.

Interpreting Information

Having all the information in the world is useless unless one also has the tools to interpret that information. Scientists must know how to separate the meaningful information from the "noise." Information can come in any form detectable by the five senses. It is important that scientists and students alike interpret information to determine its meaning, validity, and usefulness.

Requiring students to think critically as they learn improves their comprehension and increases their motivation.

Environmental Awareness

No species affects its surroundings as dramatically as does the human species. Thanks to recent highly publicized events—Chernobyl, destruction of the rain forests, the depletion of the ozone layer, and the greenhouse effect among them—people have come to realize the global impact that human actions can have. It is incumbent upon the educational system to promote environmental awareness among students. Science*Plus* addresses environmental issues in a way that students can easily grasp.

Environmental issues run the gamut, from global to local. While large-scale problems get headlines, they can be hard to grasp for many students, who may have never directly observed their impact. In most cases it is best to introduce your students to local issues to start building their awareness. Local issues are not only more relevant to their lives, but are more likely to lead to direct involvement.

Environmental awareness serves two purposes: it promotes understanding of the living world and the place of humans within it, and it produces a positive change in students' behavior toward the environment. Science*Plus* pursues both goals. You may involve students directly in environmental issues by using the suggested activities in the text and in this *Annotated Teacher's Edition*.

*Science*Plus *addresses environmental issues in a way that students can easily grasp.*

Cooperative Learning

Cooperative learning is a learning technique in which students work in heterogeneous groups and earn recognition, rewards, and sometimes grades based on the academic performance of their groups. Cooperative learning is the cornerstone of the Science*Plus* approach. Discussions, explorations, research projects, games and puzzles—all are designed to foster group participation.

Benefits of Cooperative Learning

One of the benefits of cooperative learning is that it models the real scientific experience in which scientists work together, not in isolation, to solve difficult problems. Students working in groups learn about the joys as well as the frustrations involved in scientific inquiry. With cooperative learning, the classroom becomes a hothouse of ideas and novel solutions.

Another benefit of cooperative learning is that students sharpen social skills and develop a sense of confidence in their own abilities. Students also channel their youthful energies into constructive tasks, while satisfying their fundamental need for social interaction.

Establishing cooperative learning in the classroom requires you to relinquish some control, for the students themselves become responsible for building their knowledge. Working in groups to probe and investigate ideas, answer questions, and draw conclusions about observations allows students to discover and discuss concepts in their own language. When students learn through cooperation, the knowledge derived becomes their *own*, not just a loan of your ideas or those from the textbook.

Cooperative learning is also a good icebreaker for students of different ethnic or socioeconomic backgrounds. When students join forces to achieve a common goal, they almost invariably feel more positively about each other. Students come to recognize the commonality that cuts across all boundaries of class, race, and gender.

Cooperative learning also provides an excellent vehicle for students of differing ability levels to work together in a positive way. Basic students can interact successfully with average and advanced students, and in so doing, learn that they, too, have something to offer.

Cooperative learning empowers and involves students. It raises thier self-esteem because they are learning something on their own through cooperation rather than being handed pre-packaged knowledge. It helps students become self-sufficient, self-directed, lifelong learners. In a cooperative learning environment, students are less dependent on you for knowledge.

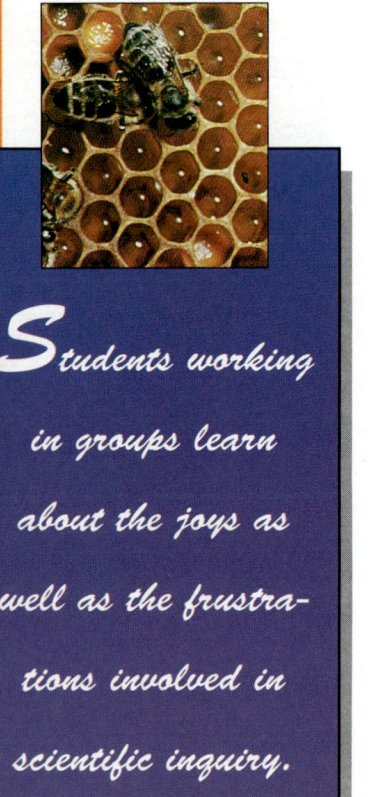

Students working in groups learn about the joys as well as the frustrations involved in scientific inquiry.

Using Cooperative Learning

In preparing for a cooperative learning activity in your class, consider the following guidelines.

Group size Cooperative learning groups may vary in size. Use small groups (2–4 members) for students unaccustomed to this learning style. Use larger groups for experienced students.

Group goals Students need to understand what is expected of them. Identify the group goal, whether it be to master specific objectives or to create a product such as a chart, report, or illustration. Identify and explain the specific cooperative skills required for each activity.

Individual accountability Each group member should have some specific responsibility that contributes to the learning of all group members. At the same time each group member should reach a certain minimum level of mastery.

Positive interdependence A learning activity becomes cooperative only when everyone realizes that no group member can be successful unless all group members are successful. This encourages positive interdependence: students working with one another to achieve group goals. Assign each student some meaningful role, or allow students to do this themselves. Or have each student in a group become an "expert" on some topic. Each member could then prepare and teach a lesson to the other group members or to the entire class.

Concept Mapping

Too often, students are able to master the individual elements of a topic without truly grasping the "big picture." If students fail to understand how the elements fit together or relate to one another, they cannot truly comprehend the topic. Concept mapping is a very effective method of helping students see how individual ideas or elements connect to form a larger whole. Concept maps are a highly effective tool for helping students make those logical connections.

The most effective concept maps are those that students construct on their own. Used in this way, concept maps are both a self-teaching system and a diagnostic tool. To construct a proper concept map, the student must first examine closely his or her mental model of the topic at hand. Any flaws or shortcomings in that model will be reflected in the concept map.

Concept maps are flexible; they can be simple or highly detailed, linear or branched, hierarchical or cross-linked, or can contain all of these major elements. Students can construct their own maps from scratch or can finish incomplete maps. Concept maps can take almost any form as long as they are logically arranged.

Concept maps are a highly effective tool for helping students make logical connections.

Making Concept Maps

▶ *The steps involved in making a concept map are outlined below. To provide guidance to your students in making concept maps, direct them to page 500 in the Pupil's Edition.*

1. Make a list of the concepts to be mapped. Concepts are signified by a noun or short phrase equivalent to a noun.
2. Choose the most general—the main—idea. Write it down and circle it.
3. Select from the list the concept most directly related to the main idea. Place it underneath the main idea and circle it. If two or more concepts bear the same relationship to the main idea, they should be placed at the same level.
4. Draw a line between the related concepts, leaving a space for a short action phrase that shows how the concepts are related. These are linkages.
5. Continue in this way until every concept in the list is accounted for.

The simple concept map below shows the relationship among the following terms: *plants, photosynthesis, carbon dioxide, water,* and *sun's energy*. More detailed maps are shown on the facing page.

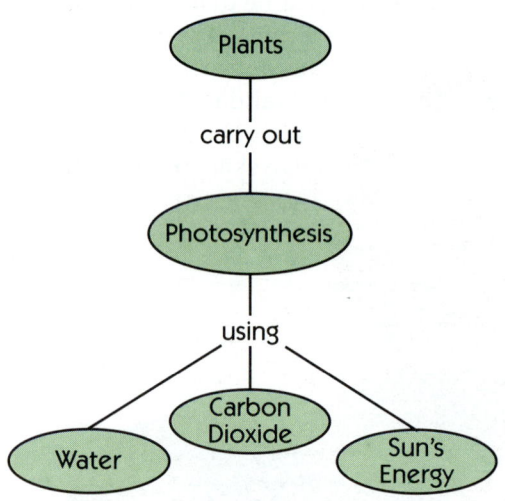

For any given topic, there is no single "correct" concept map. Not all maps are equally valid, however. Good concept maps have most or all of the following characteristics:

- start with a single, general concept—a big idea—and work down to more specific ideas
- represent each concept with a noun or short phrase—each of which appears only once
- link concepts with linkage words or short phrases
- show cross-linkages where appropriate
- consist of more than a single path
- include examples where appropriate

Using Concept Maps

Concept maps can be applied in many ways.
- to gauge prior knowledge of a topic
- end-of-lesson/section/unit evaluation
- pre-test review
- to help summarize special presentations such as films, videos, or guest speakers
- as an aid to note-taking
- for reteaching

You may also want to use partially completed concept maps as pop quizzes or as devices for summarizing particularly difficult class sessions. Also, be sure to use the concept map in each *Making Connections* review.

Evaluating Concept Maps

Again, there is no single correct concept map. However, you should consider the following criteria as you evaluate your students' concept maps.
- how comprehensive the map is (are all relationships shown?)
- how clearly concepts are linked (proper relationship between concepts, use of linkage terms between all concepts)
- overall clarity of presentation (could the map be simpler? is it redundant? is it logically arranged? are linkage terms used properly?)

Used properly, concept maps can increase comprehension, improve retention, and sharpen study skills in your students. They are a valuable addition to any student's arsenal of learning strategies.

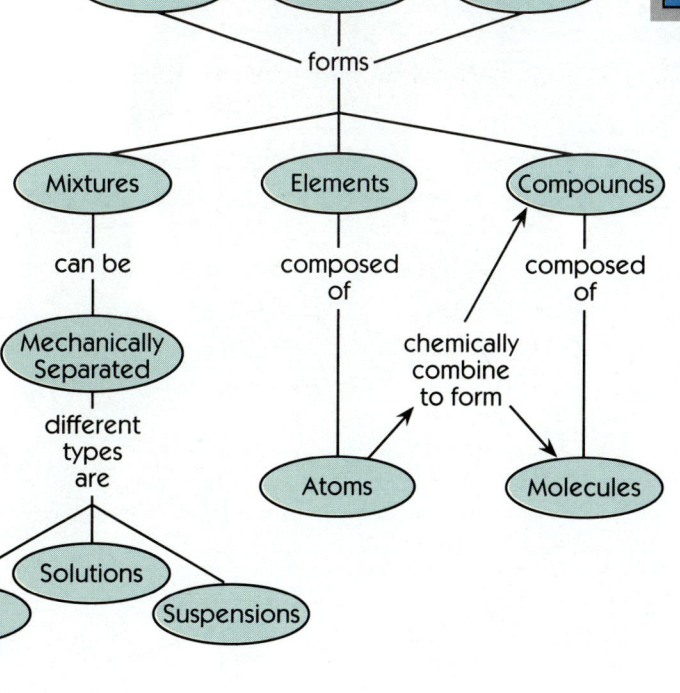

The most effective concept maps are those that students construct on their own.

Multicultural Instruction

Science is for everyone, and SciencePlus is designed to serve the multiethnic and multicultural classrooms of today. Students, regardless of their ethnic backgrounds, will not have to look hard to find positive role models on the pages of SciencePlus. In addition, content is provided that shows events, concepts, and issues from diverse ethnic and cultural perspectives. As students work through SciencePlus, they will come to understand that science is a human endeavor that has been advanced by the contributions of many cultures and ethnic groups.

To add depth to your multicultural instruction, each Unit Interleaf contains a section called Cross-Cultural Focus. This information provides activities to help you focus on cultural diversity, and in so doing help you highlight the individuality and contributions of different ethnic groups.

Crossing Disciplines

Throughout SciencePlus, students are asked to look at things from many different viewpoints. For example, students are asked to write a headline for a science story, to write advertising copy highlighting geological features for a travel agency, and even to make dioramas of animal habitats and ways of life. In so doing these activities, students see first-hand the connections between science and a variety of other disciplines including history, geography, social studies, and others. And the students even have fun making these connections.

Connections are also made naturally between the various science disciplines. Since SciencePlus is not organized according to life, earth, and physical sciences, these disciplines are integrated throughout the program. Each discipline comes into play as needed in covering the main concepts in SciencePlus. In this way, science disciplines are blended together so the emphasis is on student comprehension of the "big picture," rather than isolated components of different areas of science.

Additional cross-disciplinary suggestions are provided in each Unit Interleaf of this *Annotated Teacher's Edition*. The suggestions include a variety of activities that span science as well as non-science disciplines.

Meeting Individual Needs

Obviously, to teach effectively you must be able to reach every individual in your class. This is seldom easy, given the diverse nature of most of today's classrooms. In addition, certain students present special challenges. Dealing adequately with these students requires special preparation and strategies. In many cases a minimal amount of preparation is sufficient to make the classroom a place where all can learn. Some of the more common situations you are likely to encounter are discussed below.

At-Risk Students

At-risk students are those who, for any of a number of reasons, are liable to perform poorly and who have a high probability of dropping out. Science*Plus* is engaging and interesting throughout, appealing to all students. Throughout Science*Plus*, clear, easy-to-read prose and straightforward, attractive graphics reduce the potential for students to grow bored. The style of Science*Plus* is intentionally friendly and unintimidating. Field-testing has shown that the performance of at-risk students in science increases substantially when working with Science*Plus*.

Additional activities and teaching suggestions for at-risk students are provided in each Unit Interleaf of this *Teacher's Edition*.

LEP Students

Because Science*Plus* places so much emphasis on *doing* science, rather than reading about it, the program is ideal for students who are not wholly proficient in English. Science is a universal language—the language of curiosity and logical reasoning. Many Science*Plus* activities are easy to follow and require a minimum of reading. Lengthy explanations are seldom called for. You need only to get students started in the right direction; thereafter their intuition and common sense take over. The cooperative approach emphasized in Science*Plus* helps to give LEP students the extra support they need.

Additional activities and teaching suggestions for LEP students are found in each Unit Interleaf of this *Teacher's Edition*. The *Teacher's Resource Binder* also contains ESL information, including blackline masters of parent letters translated into Spanish, as well as unit summaries and unit glossaries in Spanish.

Also available are *English/Spanish Audiocassettes*, which provide important preview information to assist ESL students and students who are auditory learners.

Gifted Students

The difficulty of teaching gifted students lies in keeping them interested, motivated, and challenged. Gifted students who are inadequately challenged may become bored, withdrawn, or even openly disruptive. Science*Plus* includes many activities suitable for even the highest-performing student. Open-ended activities, in particular, are especially suited for gifted students.

The Science*Plus* approach emphasizes creative problem solving. In many cases there is no single right answer to a problem or question, so students' answers can reflect their individual abilities. This approach is ideal for gifted students, as they may extend the activities to fit their interests and talents.

Additional activities and teaching suggestions for gifted students are included in each Unit Interleaf of this *Teacher's Edition*.

In many cases a minimal amount of preparation is sufficient to make the classroom a place where all can learn.

T39

Physically Impaired Students

Make your classroom as easy to move about in as possible. Remove or bypass any obvious barriers. Encourage your students to physically assist impaired students. If the student uses a wheelchair, make the aisles wide enough to accommodate the chair. Make sure that the student can reach any equipment he or she needs. You may wish to enlist the aid of other students in the class to physically assist the disabled student as necessary.

As much as possible, adapt the classroom to make it possible for physically impaired students to engage in the same activities as other students. Use a mobile demonstration table so that it can be moved to different areas of the room for maximum visibility.

Visually Impaired Students

Seat students with marginal vision near the front of the room to maximize their view of both you and the chalkboard, or assign a student to make copies of what you write. You could also assign a student to explain all visual materials in detail as they are presented.

Students who are completely blind should be allowed to become familiar with the classroom layout before the first class begins. Promptly inform these students of any changes to your classroom layout. Whenever possible, provide blind students Braille or taped versions of all printed materials. Blind students may also use hand-held devices for converting written text into speech.

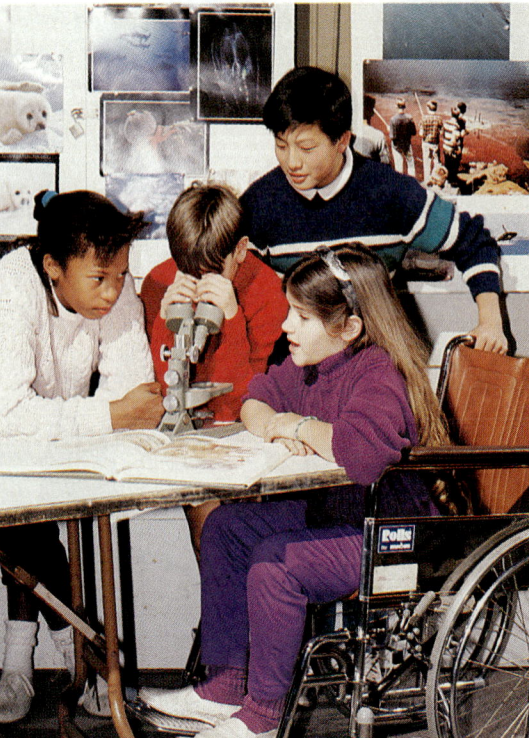

Hearing-Impaired Students

If you have partially hearing-impaired students in your class, remember to always face the class while speaking. Minimize classroom noise and arrange seating in a circle or semicircle so that hearing-impaired students can see others. This arrangement facilitates speechreading. Speak in simple, direct language and avoid digressions or sudden changes in topic. During class discussions, periodically summarize what students are saying and repeat students' questions before answering them. Use visual media such as filmstrips, overhead projectors, and close-captioned films when appropriate. You might arrange a buddy system in which another student provides copies of notes about activities and assignments.

A student who is completely deaf may require a sign language interpreter. If so, let the student and the interpreter determine the most convenient seating arrangement. When asking the student a question, be sure to look at the student, not at the interpreter. If the student also has a speech impairment, group assignments for oral reports may be advisable.

Speech-Impaired Students

Mainstreaming speech-impaired students is generally not very difficult. Patience is essential when dealing with speech-impaired students, however. For example, resist the temptation to finish sentences for a student who stutters. At the same time, do not show impatience. Also pay attention to nonverbal cues, such as facial expression and body language. Be supportive and encouraging. You need not leave the speech-impaired student out of the normal classroom give-and-take. For example, you may call on a speech-impaired student to answer a question and then allow the student to write out his or her response on the chalkboard or overhead projector. Use multisensory materials whenever possible to create a more comfortable learning environment for the speech-impaired student.

Learning-Disabled Students

Learning disabilities are any disorders that obstruct a person's listening, reasoning, communication, or mathematical abilities, and range from mild to severe. An estimated two percent of all adolescents have some type of learning disability. Learning disabilities are the most common type of disability. Provide a supportive and structured environment in which rules and assignments are clearly stated. Use familiar words and short, simple sentences. Repeat or rephrase your instructions as needed.

Students may require extra time to complete exams or assignments, the amount of extra time being dependent on the severity of their disability. Some students may need to tape-record lectures and answers to exam questions. For those who have difficulty organizing materials, you might provide chapter or lecture outlines for them to fill in. Peer tutors, who work with learning-disabled students on specific assignments and review materials, can be effective.

Computer-assisted instruction is an extremely useful tool for some learning-disabled students. This mode of instruction can even help these students develop good learning skills. For learning-disabled students, computers serve as a tireless instructor with unlimited patience. In addition, students receive simplified directions, proceed in small, manageable steps, and receive immediate reinforcement and feedback with computerized instruction.

Students with Behavioral Disorders

Behavioral disorders are emotional or behavioral disturbances that hinder a student's overall functioning. The behaviorally impaired may exhibit any of a variety of behaviors, ranging from extreme aggression to complete passivity.

Obviously, no single teaching strategy can accommodate all behavioral disorders. In addition, behavioral psychologists disagree on the best way to deal with students who have behavior disorders. As a general rule, try to be fair and consistent, yet flexible in your dealings with behaviorally disabled students. Make sure to clearly state rules and expectations. Reinforce desirable behavior or even approximations of such behavior, and ignore or mildly admonish undesirable behavior.

Because learning disabilities often accompany behavior disorders, you might also wish to refer to the guidelines for learning disabilities.

Computer-assisted instruction is an extremely useful tool for some learning-disabled students.

Science, Technology, and Society

STS is an approach to teaching science in which the impact of scientific, technological, and social matters on each other is explored. STS teaches science from the context of the human experience, and in so doing leads students to think of science as a social endeavor. STS emphasizes personal involvement in science. Students become active participants in the scientific experience. For this reason Science*Plus* incorporates an STS approach.

STS contrasts with the traditional "basic science" approach to science education, in which students follow a carefully sequenced program of the basic models, concepts, laws, and theories of science. The fundamental flaw of the basic science approach is that the student is more or less a passive participant in the learning process.

There are three parts to STS:

S T S science concept and skill development, and knowledge of the nature of science

This component of STS introduces science as a system for learning about the natural world and gives students the foundation they need to actually practice science in and out of the classroom. The ultimate goal of STS, and of Science*Plus,* is to turn students into scientists, at least for the duration of their science education. To accomplish this, students first learn the methods of science and the skills that scientists draw on. Whenever possible, the major ideas of science are not simply presented, but are introduced from the standpoint of those who developed them. In this way, students come to see the reasoning that went into the development of these ideas.

S T S knowledge of the relationship of science and technology and engagement in science-based problem solving design

Students who understand the real-world applications of science are better able to appreciate and enjoy it. This component of STS reinforces the practical value of science. Students see that science is a system for solving practical problems. Students themselves become practical scientists—first identifying problems and then developing solutions to them. Students learn to analyze, to plan, to organize, to design, and to refine models and designs.

S T S engagement in science-related social issues and attention to science as a social institution

This component of STS deals with the ways in which science serves human needs. The benefits may be tangible or intangible, but either way, they are real. To emphasize the social responsibility of science, scientists are shown to be concerned about the impact of their work on society as a whole. It sometimes happens that advances in science and technology lead to thorny ethical issues. Such issues are often used as a focus for discussion and investigation. From these investigations, students draw conclusions and form reasoned opinions.

Studies have shown that students begin the study of science full of curiosity and enthusiasm. But after a few years of a traditional curriculum, the curiosity is squelched and the enthusiasm has all but vanished. By contrast, under the STS approach student interest builds through the years, and is maintained long after formal study ends. The key to this success is the active involvement in science that STS imparts. The end products of the STS approach are students who appreciate and understand science and who are equipped to deal sensibly with the complex issues that will become commonplace in the decades ahead.

> *For a fuller discussion of STS, See the NSTA position paper, July 1990*

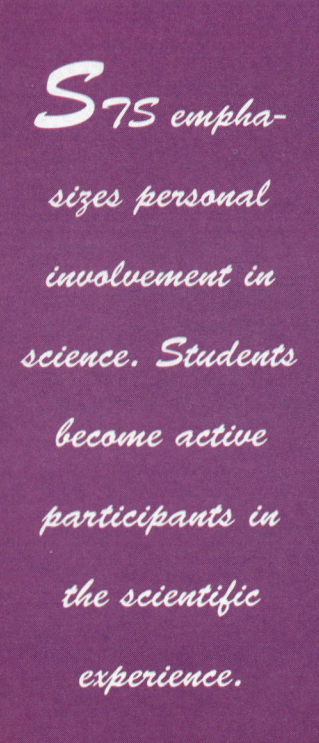

STS emphasizes personal involvement in science. Students become active participants in the scientific experience.

Materials and Equipment

Science*Plus* is designed to be teachable even by those with a limited budget for materials. Most activities use common household items, which can be brought to class by students, parents, or otherwise easily obtained. The *Teacher's Resource Binder* contains a *Materials Guide*, which is a comprehensive guide to the materials and equipment you will need in carrying out the activities of Science*Plus*. The *Materials Guide* features both a master list of materials and a unit-by-unit list. In addition, the second page of the parent letters contained in the *Teacher's Resource Binder* lists the supplies needed for each unit. This page makes it easy for you to invite donations from parents to keep your budget low.

Science*Plus* Teacher's Network

The *SciencePlus Teacher's Network (SPTN)* is part of the Atlantic Science Curriculum Project (ASCP), linking teaching, curriculum development, and research in science education. The ASCP, which produced the Canadian version of *SciencePlus: Technology and Society*, has existed for over fifteen years, beginning as a collaborative, grass-roots activity to improve science teaching in the middle/junior high schools. The goal of the SPTN is to continue this collaborative effort, linking teachers with teachers through newsletters, conferences, and small group and regional meetings.

The **Science*Plus* Communicator** is the official U.S. publication of the SPTN. The **Science*Plus* Communicator** offers opportunities for teachers to comment on science teaching and current issues, research, assessment, and science, technology, and society applications.

Assessing Student Performance

A Comprehensive Approach to Assessment

Developing a strategy for assessing student progress is an important step in realizing the goals of Science*Plus*. Students pay the most attention to those aspects of a lesson on which they know they will be graded. Teachers who want their students to be successful should therefore teach with continual assessment in mind.

In Science*Plus*, there is no distinct boundary between teaching and assessing. Every suggested assessment activity, including testing, is designed to teach. This emphasis can help correct the preoccupation with measuring and sorting students. The suggestions here are intended to aid you in your primary task: teaching.

You may use the assessment aids available with Science*Plus* to measure students' mastery of the concepts and processes of each unit. It is not recommended that you rely on these items alone, however. Assessment should be ongoing and should measure performance in every area. The quality of student class work, homework, lab work, and Journal entries should all be factors in assigning grades, in addition to performance on exams.

The authors strongly discourage relying on recall-based assessment strategies. Teachers who rely heavily on such assessment strategies may find it difficult at first to adopt new methods of assessment. However, once the transition is made, the reward—in the form of improved student performance and motivation—will have more than offset the inconvenience. As you work with Science*Plus*, you will find that it provides ample strategies and opportunities for assessing students in a variety of ways.

Assessing Understanding of Concepts

Throughout Science*Plus* are many opportunities for students to demonstrate their understanding of specific concepts. *For Your Journal*, at the beginning of each unit, encourages students to express through writing what they already know about specific concepts that will be covered in the unit. Then, after students work through the unit, they are given the opportunity to refine their journal entries in *Updating Your Journal*, at the end of the unit. By viewing their updated journal entries, you can get a good idea of the students' understanding of the main concepts of the unit.

Science*Plus* is generously supplied with questions and other activities that can serve to check students' understanding of the concepts developed by each lesson. *Challenge Your Thinking* is suitable for assessing student understanding of concepts at the section level, while *Making Connections* assesses students' understanding of concepts at the unit level.

Assessing Scientific, Psychomotor, and Communication Skills

In Science*Plus*, knowledge and understanding are tightly linked to the development of important process skills such as observing, measuring, graphing, writing, predicting, inferring, analyzing, and hypothesizing. All learning tasks are designed to help develop these skills. The teacher can assess such skill development by inspecting student work and by observing student performance.

The sample tables below are suitable as models for evaluating student performance.

Assessing Environmental Awareness

Science*Plus* is written with a commitment to environmental awareness. Many activities are included that promote such awareness. The teacher is provided with suggestions on extending this theme through creative projects, clean-up or recycling projects, and so on. Tasks such as these promote environmental consciousness. The care students take in carrying out these activities is a measure of their awareness of environmental issues.

Assessing Scientific Behavior

BEHAVIOR	Poor	Satisfactory	Good	Very Good	Excellent
Cooperates with others in small-group work					
Observes and records observations					
etc.					

Assessing Technical Skills

TASK	Yes	No	Uncertain
Is able to read thermometer correctly			
Is able to use spring scale to measure force			
etc.			

Assessing Scientific Attitudes

It can be useful to survey your students on the types of science-related hobbies and interests they pursue outside of class. In a direct way, this provides feedback on the success of your school's science program. A successful science program is reflected in a student body with outside interests in science. Ask your students to keep a tally of any science-related activities they undertake outside of class. These could include reading or writing about science and technology, science-related projects, visits to museums, attendance at lectures on science and technology topics, and viewing science programs on television.

For developing and assessing individual student interest and attitude in science, the assignment of elective reading and independent projects is essential. In addition to the numerous project ideas and extension activities included in the units of Science*Plus*, the *Science in Action* feature provides a range of project ideas from which students may choose. Students may also want to read further about a topic in the *SourceBook* or other science-related book or magazine.

Student work on elective projects should count as a significant part of overall assessment. It provides the surest indication of a student's interest and proficiency in science, especially the student's ability to study, plan, and research independently.

A successful science program is reflected in a student body with outside interests in science.

A Balanced Assessment

The authors of Science*Plus* recommend achieving a balance between the different forms of assessment. As a general rule, the following proportions are suggested:

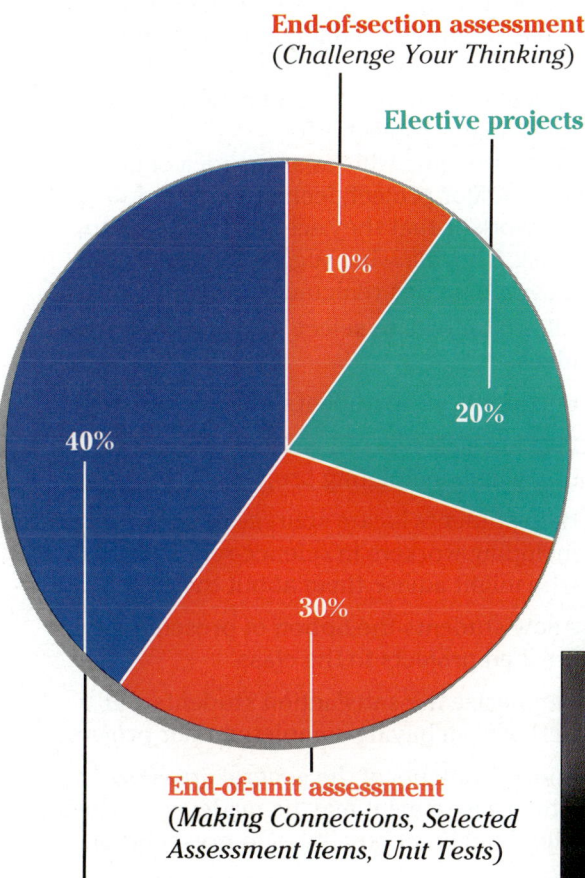

End-of-section assessment
(*Challenge Your Thinking*) — 10%

Elective projects — 20%

End-of-unit assessment
(*Making Connections, Selected Assessment Items, Unit Tests*) — 30%

Continuous assessment of assigned work and in-class performance — 40%

The Goals of Assessment

Obviously, tests should reinforce your instructional goals. Tests dominated by multiple-choice, matching, fill-in-the-blank, and short answer items have been very common in the past because of an emphasis on objectivity and ease in grading. However, such tests alone will not support the goals of Science*Plus*. By using various means to assess student progress, the teacher can test less frequently and more meaningfully.

Testing should be more than a means for assessing students' progress; it should also be an opportunity for students to learn. To reduce students' anxiety, give them ample time to prepare for and take the test. Careful, deliberate thought, rather than a superficial slapdash approach, should be encouraged in science. Quicker students and those who need additional challenge can be occupied with bonus questions while the others complete the required test items. Reviewing the test after it is completed provides yet another opportunity for students to learn.

Assessing Science Projects

SciencePlus offers students abundant opportunities for independent investigation. Many open-ended, curiosity-stimulating questions are posed for students. Some of these questions are natural starting points for science projects. Students using ScoiencePlus have often developed successful science fair projects based on questions in the text.

The *Science in Action* feature is another source of possible project ideas. Many techniques are suggested for getting started, reporting results, and distributing the work of the projects in an efficient way. The *Teacher's Edition* is a source for project ideas as well.

> *Do not allow preconceived notions about how the project should be done detract from the students' interest, enjoyment, and satisfaction in doing original work.*

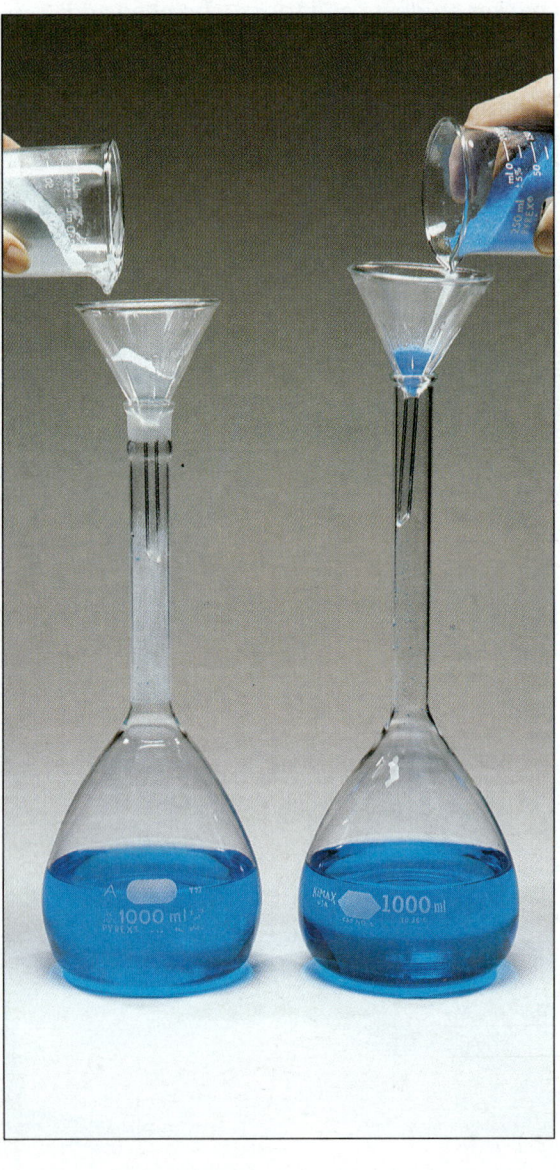

Undertaking a major project provides students with a host of positive experiences. Students learn to organize, plan, and piece together many separate ideas and information into a coherent whole. Undertaking a major project also allows students to experience the sense of accomplishment that comes from tackling and completing a difficult task. It might even be argued that no science education is complete without having undertaken and completed a major project.

Many students will resist the idea of undertaking a major project, feeling that it is too much work or that they are simply not up to the task. The following suggestions may help you overcome this reluctance:

- Allow students to select their own project ideas.
- Encourage your students to be creative.
- Provide a clear set of guidelines for developing and completing projects.
- Help students locate sources of information, including workers in science-related fields who might advise them about their projects.
- Allow students the option of presenting their finished projects to the class.
- Emphasize the satisfaction students will derive from having completed their projects.
- Inform students of the general areas on which assessment may be made: e. g., scientific thought, originality, and presentation.
- Do not emphasize the details of assessment. "Scoring points" should not be a major incentive.

Do not allow preconceived notions about how the project should be done detract from the students' interest, enjoyment, and satisfaction in doing original work. Rather than forcing all students to fit their project work into a mold suited to scientific research, establish three sets of criteria described in the tables on pages T50 to T52.

Developing Assessment Items

When developing tests, you should bear in mind the sort of skill required to answer each assessment item. The manner of testing determines what is learned: tests that require tick-mark responses teach tick-marking. Tests that require verbal, graphic, illustrative, and numeric responses develop writing, speaking, graphing, drawing, and mathematical skills. A superior test draws upon as many skills as possible.

The assessment/development model below is designed to help you construct the kind of comprehensive tests that will further the educational goals of Science*Plus*. The model features four categories (verbal, graphic, illustrative, and numeric) and twelve kinds of items. A set of test items, developed in accordance with the model, is provided in each unit of the *Teacher's Resource Binder*.

> *The manner of testing determines what is learned: tests that require tick-mark responses teach tick-marking.*

✓ Assessment Item Development Model

✓ Verbal

Word Usage: Words given are to be used in a prescribed situation.

Correction/Completion: Incomplete or incorrect sentences and paragraphs are given for completion or correction.

Short Essay: Information is given and/or a question is posed for short essay response.

Short Responses: Answers to these questions require a tick mark, a line, a single word, phrase, or sentence.

✓ Graphic

Graphs for Interpretation: A graph of a relationship between two variables is given for interpretation.

Graphs for Correction or Completion: An incomplete or incorrect graph is given for completion or correction.

Graphing Data: Data are given to be graphed.

✓ Illustrative

Illustration for Interpretation: Illustrations (drawings or photographs) are presented for interpretation.

Illustrations for Correction or Completion: An incorrect or incomplete illustration is given for correction.

Answering by Illustrations: A question is asked for which a drawing is the expected answer.

✓ Numeric

Data for Interpretation: A data table is given for interpretation.

Numerical Problems: A problem is given for a numerical solution.

The tables below provide an objective approach to evaluating students' independent projects.

Table 1

Criteria for Evaluating Student Reports and Presentations

Scientific Thought (possible 40 points)

Complete understanding of topic; topic extensively researched; variety of primary and secondary sources used and cited; proper and effective use of scientific vocabulary and terminology.	Good understanding of topic; topic well researched; a variety of sources used and cited; good use of scientific vocabulary and terminology.	Acceptable understanding of topic; adequate research evident; sources cited; adequate use of scientific terms.	Poor understanding of topic; inadequate research; little use of scientific terms.	Lacks an understanding of topic; very little research, if any; incorrect use of scientific terms.
40 39 38 37 36	35 34 33 32 31	30 29 28 27 26	25 24 23 22 21	20 15 10

Oral Presentation (possible 30 points)

Clear, concise, engaging presentation, well supported by use of multisensory aids; scientific content effectively communicated to peer group.	Well-organized, interesting, confident presentation supported by multisensory aids; scientific content communicated to peer group.	Presentation acceptable; only modestly effective in communicating science content to peer group.	Presentation lacks clarity and organization; ineffective in communicating science content to peer group.	Poor presentation; does not communicate science content to peer group.
30 29 28 27	26 25 24 23	22 21 20 19	18 17 16	15 10 5

Exhibit or Display (possible 30 points)

Exhibit layout self-explanatory and successfully incorporates a multisensory approach; creative use of materials.	Layout logical, concise, and can be followed easily; materials used in exhibit appropriate and effective.	Acceptable layout of exhibit; materials used appropriately.	Organization of layout could be improved; materials could have been chosen better.	Exhibit layout lacks organization and is difficult to understand; poor and ineffective use of materials.
30 29 28 27	26 25 24 23	22 21 20 19	18 17 16	15 10 5

Criteria for Evaluating Student Technology Projects

Scientific Technical Thought (possible 40 points)

Featuring an attempted design solution to technical problem; the problem is significant and stated clearly; the solution reveals creative thought and imagination; underlying technical and scientific principles are very well understood.	Featuring an attempted design solution to a technical problem; the solution may be a standard one for similar problems; underlying technical and scientific principles are recognized and understood.	Featuring a working model; underlying technical and scientific principles are well understood; model is built from a standard blueprint or design.	A model built from a standard blueprint or design or from a kit; underlying technical and scientific principles are recognized but not necessarily understood.	A model built from a kit; underlying technical and scientific principles are not recognized or understood.
40 39 38 37 36	35 34 33 32 31	30 29 28 27 26	25 24 23 22 21	20 15 10

Presentation (possible 30 points)

Clear, concise, confident presentation; proper and effective use of vocabulary and terminology; complete understanding of topic; able to extrapolate.	Well-organized, clear presentation; good use of scientific vocabulary and terminology; good understanding of topic.	Presentation acceptable; adequate use of scientific terms; acceptable understanding of topic.	Presentation lacks clarity and organization; little use of scientific terms and vocabulary; poor understanding of topic.	Poor presentation; cannot explain topic; scientific terminology lacking or confused; lacks understanding of topic.
30 29 28 27	26 25 24 23	22 21 20 19	18 17 16	15 10 5

Exhibit (possible 30 points)

Exhibit layout self-explanatory and successfully incorporates a good sensory approach; creative and very effective use of material.	Layout logical, concise, and easy to follow; materials used in exhibit appropriate and effective.	Acceptable layout of exhibit; materials used appropriately.	Organization of layout could be improved; materials could have been chosen better.	Layout lacks organization and is difficult to understand; poor and ineffective use of materials.
30 29 28 27	26 25 24 23	22 21 20 19	18 17 16	15 10 5

Criteria for Evaluating Student Experimental Research Projects

Scientific Thought (possible 40 points)

An attempt to design and conduct an experiment or project with all important variables controlled.	An attempt to design and conduct an experiment or project, but with inadequate control of significant variables.

40   38   38   37   36   35   30   25   15   10   5

Originality (possible 16 points)

Original, resourceful, novel approach; creative design and use of equipment.	Imaginative, extension of standard approach and use of equipment.	Standard approach and good treatment of current topic.	Incomplete and unimaginative use of resources.	Lack of creativity in both topic and resources.

16   15   14   13   12   11   10   9   8   7   6   5   4   2

Presentation (possible 24 points)

Clear, concise, confident presentation; proper and effective use of vocabulary and terminology; complete understanding of topic; able to extrapolate.	Well-organized, clear presentation; good use of scientific vocabulary and terminology; good understanding of topic.	Presentation acceptable; adequate use of scientific terms; acceptable understanding of topic.	Presentation lacks clarity and organization; little use of scientific vocabulary; poor understanding of topic.	Poor presentation; cannot explain topic; scientific terminology and vocabulary lacking or confused; lacks understanding of topic.

24   23   22   21   20   19   18   17   16   15   14   13   12   11   10   9   8   4

Exhibit (possible 20 points)

Exhibit layout self-explanatory and successfully incorporates a multi-sensory approach; creative and very effective use of materials.	Layout logical, concise, and can be followed easily; materials used appropriate and effective.	Acceptable layout; materials used appropriately.	Organization of layout could be improved; materials could have been chosen better.	Exhibit layout lacks organization and is difficult to understand; poor and ineffective use of materials.

20   19   18   17   16   15   14   13   12   11   10   8   6

SCIENCEPLUS

Technology and Society

To The Student

*T*his book was written with you in mind! There are many things to try, to create, and to investigate — both in and out of class. There are stories to be read, articles to think about, puzzles to be solved, and even games to play.

GET INVOLVED!

The best way to learn is by doing. In the words of an old Chinese proverb:

> Tell me — I will forget.
>
> Show me — I may remember.
>
> Involve me — I will understand.

The authors of this book want **you** to **get involved** in science.

The activities in this book will allow you to make some basic and important scientific discoveries on your own. You will be acting much like the early investigators in science who, without expensive or complicated equipment, contributed so much to our knowledge.

What these early investigators had, and had in abundance, was **curiosity** and **imagination**. If you have these qualities, you are in good company! And if you develop sharp scientific skills, who knows, you might make your own contributions to science someday.

THE LEADING EDGE

Scientists are usually interested in understanding things that happen in **nature**. However, the discoveries that scientists make are often used by inventors and engineers. The end result is our most sophisticated **technology**, including such things as computers, laser discs, nuclear reactors, and instant global communication.

There is an interaction between science and technology. Science makes technology possible. On the other hand, the products of technology are used to make further scientific discoveries. In fact, much of the scientific work that is done today has become so technically complicated and expensive that no one person can do it entirely alone. But make no mistake, the creative ideas for even the most highly technical and expensive scientific work still come from **individuals**.

GO FOR IT!

Science is a process of discovery: a trek into the unknown. The skills you develop as you do the activities in this book — like observing, experimenting, and explaining observations and ideas — are the skills you will need to be a part of science in the future. There is a universe of scientific exploration and discovery awaiting those who take the challenge.

Keep a Journal

A Journal is an important tool in creative work. In this book, you will be asked to keep a Journal of your thoughts, observations, experiments, and conclusions. As you develop your Journal, you will see your own ideas taking shape over time. This is often the way scientists arrive at new discoveries. You too may log some discoveries as you develop your own Journal.

About Safety

Science investigations and experiments should be both enjoyable and safe. If you follow the safety guidelines listed here, as well as any others mentioned in the explorations, you should have no problems. You should **always** follow these guidelines, even when you think that there is little or no danger.

The major causes of laboratory accidents are carelessness, lack of attention, and inappropriate behavior. These all spring from a person's **attitude**. With a proper attitude and **consistent** safety habits, you should be able to feel quite comfortable and at home in a science laboratory.

Safety Guidelines

● Eye Safety

Wear goggles when handling acids or bases, using an open flame, or performing any other activity that could harm the eyes. If a substance gets into your eyes, wash them with plenty of water and notify your teacher at once. Never place a chemical substance near your unprotected eyes. Never use direct sunlight to illuminate a microscope.

● Safety Equipment

Know the location of all safety equipment, such as fire extinguishers and first aid kits.

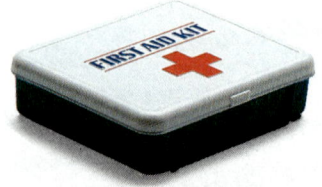

● Neatness

Keep work areas free of all unnecessary books and papers. Tie back long, loose hair and button or roll up loose sleeves when working with chemicals or near a flame.

● Chemicals

Chemicals and other dangerous substances can be dangerous if they are handled carelessly. When handling certain chemicals, such as acids or bases, you should protect your eyes and clothes with safety glasses and an apron.

Heat

Whenever possible, use an electric hot plate instead of an open flame. If you must use an open flame to heat a glass container, shield the flame with a wire screen that has a ceramic center. When heating chemicals in a test tube, do not point the test tube toward anyone.

Electricity

Be cautious around electrical wiring. When using a microscope with a lamp, do not place its cord where it can cause someone to trip. Do not let cords hang over a table edge in a way that will cause equipment to fall if a cord is pulled. Do not use equipment with frayed cords.

Glassware

Examine all glassware before using. Glass containers for heating should be made of heat-resistant material. Never use cracked or chipped glassware. Use caution when inserting glass tubing into a rubber stopper. Hold the tubing and stopper with a towel to prevent puncture wounds.

Never taste chemicals, unless you are specifically instructed to do so.

Never pour water into a strong acid or base. The mixture produces heat. Sometimes the heat causes splattering. The correct procedure is to pour the acid or base slowly *into the water*. This way the mixture will stay cool.

If any solution is spilled, wash it off with plenty of water. If a strong acid or base is spilled, neutralize it first with an agent such as baking soda (for acids) or boric acid (for bases).

If you are instructed to note the odor of a substance, wave the fumes toward your nose with your hand rather than putting your nose close to the source.

Sharp/Pointed Objects

Use knives, razor blades, and other sharp instruments with care. Do not use double-edged razor blades in the laboratory.

Cleanup

Wash your hands immediately after handling hazardous materials. Before leaving the laboratory, clean up all work areas. Put away all equipment and supplies. Make sure water, gas, burners, and electric hot plates are turned off.

The instructions for the explorations in this book will include warning statements where necessary. In addition, you will find one or more of the following safety symbols when a procedure requires specific caution.

Wear Safety Goggles

 Wear a Laboratory Apron

 Flame/ Heat

 Sharp/ Pointed Object

 Dangerous Chemical/ Poison

 Corrosive Substance

UNIT 1: Interactions

✷ Teaching the Unit

☀ Unit Overview

In this unit, students explore the relationships among plants, animals, and the environment. In the section *The Players and Their Roles,* students identify some of the special relationships that exist among living things, such as commensalism, mutualism, and parasitism. They observe how producers, consumers, and decomposers facilitate the flow of energy through a community. The final section, *Putting It All Together,* explores some of the changes that occur within a biological community. Students look at some of the causes of population changes that occur within biological communities. The section concludes by providing students with an opportunity to investigate the interactions between people and the environment.

☀ Using the Themes

The unifying themes emphasized in this unit are **Systems and Interactions, Patterns of Change,** and **Energy.** The following information will help you weave these themes into your teaching plan. A focus question is provided with each theme as a discussion tool to help you tie the information in the unit together.

Systems and Interactions is the dominant theme throughout the unit because of the emphasis placed on the interactions and interrelationships of biological systems.

Focus question: *What kinds of interactions occur between plants and animals in the environment?*

Plants and animals must interact in order to survive. Plants provide food and shelter for animals, while animals aid in pollination and providing fertilizer for plants. Students should recognize that certain kinds of plants and animals are found together in a particular place because each organism is adapted to that environment and performs a particular role within the community of organisms.

Patterns of Change is an integral part of Lessons 3, 4, and 5 as each of these lessons examines how changes occur within communities. The study of succession in Lesson 6 gives students a visual perspective on change over time in the environment.

Focus Question: *How do living organisms change when their environment changes?*

One example is the pattern of change that occurs in response to an increase in a food source or the loss of a predator within a community. Another example is succession within a community, which shows a predictable, orderly pattern of change in which one group of organisms is gradually replaced by another.

Energy is developed in Lessons 3 and 4. Observations and discussions focus on how energy flows through biological communities in food chains and food webs.

Focus Question: *Why is energy important to an organism?*

Students should relate acquiring energy from food with having the energy to find shelter, reproduce, raise young, and perform other necessary functions. The way plants use energy to convert sunlight into food, to transport nutrients, and to grow is an excellent example of the flow of energy.

☀ Using the *Science Discovery* Videodiscs

Disc 1 *Science Sleuths, Dead Fish on Union Lake*
The 4Cs, a civic organization, has coordinated the cleanup of a polluted urban lake and its surrounding parks. Suddenly, the fish in the lake are dying in large numbers. The 4Cs think someone is again polluting the lake. The Science Sleuths must analyze the evidence and determine the reason why the fish are dying.

Disc 2 *Image and Activity Bank*
A variety of still images, short videos, and activities are available for you to use as you teach this unit. See the *Videodisc Resources* section of the **Teacher's Resource Binder** for detailed instructions.

☀ Using the *SciencePlus SourceBook*

Unit 1 expands upon the concepts of populations and communities, succession, and the role of humans in the environment. Interactions among organisms are explored in a variety of ways. Students study characteristics of populations, abiotic resources, biomes, pollution, and conservation.

PLANNING CHART

SECTION AND LESSON	PG.	TIME*	PROCESS SKILLS	EXPLORATION AND ASSESSMENT	PG.	RESOURCES AND FEATURES
Unit Opener	2		observing, discussing	For Your Journal	3	Science Sleuths: *Dead Fish on Union Lake* Videodisc Activity Sheets TRB: Home Connection
THE PLAYERS AND THEIR ROLES	5			Challenge Your Thinking	35	
1 You and Your Environment	5	2	observing, predicting, classifying, summarizing	Exploration 1	6	
2 The Living Players	10	2 to 3	observing, analyzing, inferring, classifying			Image and Activity Bank 1–2 TRB: Resource Worksheet 1–1
3 The Movement of Energy in a Community	16	2	observing, analyzing, inferring, classifying			Image and Activity Bank 1–3 Resource Transparency 1–1
4 Who Eats What?	20	2 to 3	observing, classifying, inferring, analyzing diagrams	Exploration 2	23	Image and Activity Bank 1–4 TRB: Activity Worksheets 1–2 and 1–4 Resource Worksheet 1–3 Resource Transparency 1–2 Graphic Organizer Transparencies 1–3 and 1–4
5 The Nonliving Players	26	4	observing, designing experiments, drawing conclustions, reading a graph	Exploration 3 Exploration 4	27 30	Image and Activity Bank 1–5 TRB: Resource Worksheets 1–5, 1–8, and 1–9 Activity Worksheets 1–6 and 1–7 Graphic Organizer Transparencies 1–5, 1–6, and 1–7
PUTTING IT ALL TOGETHER	36			Challenge Your Thinking	57	
6 Natural Changes	36	5 to 6	observing, interpreting graphs, collecting and organizing data, drawing conclusions	Exploration 5 Exploration 6	38 44	Image and Activity Bank 1–6
7 Humans: Boss or Co-Worker?	50	4	observing, analyzing, inferring, drawing conclusions			Image and Activity Bank 1–7 Graphic Organizer Transparency 1–8
End of Unit	58		applying, analyzing, evaluating, summarizing	Making Connections TRB: Sample Assessment Items	58	TRB: Resource Worksheet 1–10 Graphic Organizer Transparency 1–9 Science in Action, p. 60 Science and Technology, p. 62 *SourceBook*, pp. S1–S22

*Time given in number of class periods.

✵ Meeting Individual Needs

☀ Gifted Students

1. Explain that all the living things in a biological community, or ecosystem, make up its *biomass.* Scientists know that an ecosystem can support only a certain amount of biomass. Challenge students to do some research on biomass and design a biomass pyramid that shows how total biomass changes from the producer level to higher levels of consumers.
2. Have students imagine that a forest fire has destroyed an entire forest community. In another forest, toxic waste from a chemical plant has killed all of the organisms. Ask: In which community will the process of succession take longer? Explain. Do you think that the destruction of an area is ever beneficial? Do some research to find out.

☀ LEP Students

1. Point out to students that in this unit they learned many terms that relate to the interrelationships and roles of organisms in biological communities. Suggest that they make a list of the terms and use them to create a bilingual dictionary. Suggest that students add illustrations to clarify their definitions. Invite them to make a cover for their bilingual dictionary and to think of a title. When they are finished, review their work for science content, with minimal emphasis on language proficiency.
2. Remind students that in this unit they learned how to construct a food web to show the interrelationships between plants and animals in a community. Suggest that they make a diagram of a food web to show the interrelationships in a biological community of their native country. Invite volunteers to share their finished diagrams with the class and read their information in English and their native language.
3. At the beginning of each unit, give Spanish-speaking students a copy of the *Spanish Glossary* from the *Teacher's Resource Binder.* Also, let Spanish-speaking students listen to the *English/Spanish Audiocassettes.*

☀ At-Risk Students

Students have several opportunities for hands-on involvement with the environment. In *Exploration 1,* the string game gives the student firsthand experience with interactions in the environment. To reinforce the lessons on food chains, suggest that students construct a food web by using pictures of plants and animals cut from magazines. Another experience that motivates students is the investigation of the rotting-log community in *Exploration 5* in which students carefully observe and analyze the environment around them. The activities on tolerance offer an opportunity to get students involved in a project with visible, rewarding results. For further reinforcement on the concepts of tolerance, *Exploration 3* and the photographs in Lesson 5 are useful.

☀ Cross-Disciplinary Focus

Social Studies
Explain to students that there are several different kinds of biological communities in the world called *biomes.* Biomes have particular climates and populations of plants and animals. Suggest that students make a map of the world and show where the different biomes are located.

Language Arts
Some students may enjoy setting up a reading center by gathering together materials that focus on the interrelationships of plants and animals in biological communities.
- Have students create their own anthology of poetry. They should pick a few topics from the unit and write poems about them. The poems could be organized into a booklet.
- Include newspaper and magazine articles on the ways plants and animals live and on different kinds of biological communities.
- Include riddles that emphasize how animals and plants interact.
- Include wildlife stories, both fiction and nonfiction, about animal adventures.
- Include picture books on biological communities.

Health
Remind students that pollution affects all living things. Suggest they do some research to discover how different kinds of pollution may affect people's health. Some of the areas they may wish to investigate include: air pollution, toxic waste, nuclear waste, water pollution, and pesticide contamination.

Art
Students that have access to cameras might enjoy taking pictures for use in a photo essay about how animals and plants interrelate in biological communities. Allow students some flexibility in how they interpret the concepts. Arrange the finished essays around the classroom in a manner that suggests an art gallery.

Cross-Cultural Focus

World Cuisine

Remind students that animals, and people, get their energy from the food they eat. Assign each student a different country, and have them do research to find out what kinds of food people eat in different parts of the world. Encourage them to share unusual recipes with the class. If there is sufficient interest on the subject, suggest that students compile an international cookbook.

What's the Environment Like?

Point out to students that people come from many different kinds of environments. Suggest that they find out what the environment is like in a different part of the world. Students should include information on how different cultures respond to their environments by investigating such topics as clothing, shelter, and type of crops grown. Have them use this information to write travel brochures that describe different environments for tourists. Students should illustrate their travel brochures and use them in a bulletin-board display.

If you have students in class from different countries, invite them to describe the environment of their native area.

 # Resources

 ## Bibliography for Teachers

Bates, Marston. *The Forest and the Sea.* New York, NY: Lyons & Burford Publishers, Inc., 1988.
Benarde, Melvin. *Our Precarious Habitat.* New York, NY: John Wiley & Sons, Inc., 1989.
Croall, Stephen. *Ecology for Beginners.* New York, NY: Pantheon Books, 1982.
Odum, Eugene P. *Basic Ecology.* Troy, MO: Saunders College Publishing, 1983.
Sabin, Francene. *Ecosystems & Food Chains.* Mahwah, NJ: Troll Associates, 1985.
Sandler, Shiphrah. *Discovering Ecology.* Buffalo, NY: DOK Publishers, 1982.
Steger, Will, and Jon Bowermaster. *Saving the Earth: A Citizen's Guide to Environmental Action.* New York, NY: Alfred A. Knopf, 1990.

 ## Bibliography for Students

Booth, Basil. *Temperate Forests.* Englewood Cliffs, NJ: Silver Burdett Press, 1983.
Cochrane, Jennifer. *Urban Ecology.* New York, NY: The Bookwright Press, 1988.
Elkington, John, Julia Hailes, Douglas Hill, and Joel Makower. *Going Green, A Kid's Handbook to Saving the Planet.* New York, NY: Penguin Books USA, Inc., 1990.
Forsyth, Adrian. *Journey Through a Tropical Jungle.* New York, NY: Simon & Schuster, 1988.
Gay, Kathlyn. *Acid Rain.* New York, NY: Franklin Watts Inc., 1983.

 ## Films, Videotapes, Software, and Other Media

Acid Rain.
  Videotape.
  Focus Media, Inc.
  839 Stewart Avenue
  Garden City, NY 11530
Animal Populations: Nature's Checks and Balances.
  Film and Videotape.
  Encyclopedia Britannica Educational Corporation
  425 North Michigan Ave.
  Chicago, IL 60611
The Greenhouse Effect.
  Videotape.
  Focus Media, Inc.
  839 Stewart Avenue
  Garden City, NY 11530
Hot Enough for You?
  Videotape.
  Coronet Film & Video
  108 Wilmot Road
  Deerfield, IL 60015
Take a Look at the Ecological Cycles.
  Software.
  For IBM PC
  IBM
  1133 Westchester Ave.
  White Plains, NY 10601
Weeds to Trees.
  Software.
  For Apple II Family.
  MECC
  3490 Lexington Avenue North
  St. Paul, MN 55126

UNIT 1

Unit Focus

Ask students to describe their environment. Encourage them to identify as many factors, or components, as they can. Then ask students what roles they play in their environment, and record their responses on the chalkboard. *(student, athlete, cousin, brother, sister, friend, consumer, and so on)* When students have exhausted their ideas, point out that everyone plays many roles in their environment. Explain that in this unit, they will learn about the roles played by the plants and animals that make up different biological communities.

About the Photograph

This photograph of a honeybee on a zinnia represents several kinds of interactions that occur in nature. The bee is gathering food (nectar and pollen) that it will take back to its hive, where the food will be eaten by other bees. Honey produced by the hive may be eaten by other animals as well. As bees move from flower to flower, they carry pollen with them. This allows plants to reproduce. Since both species benefit from the interactions, the relationship between the bee and the zinnia plant is known as mutualism.

Unit 1 Interactions

THE PLAYERS AND THEIR ROLES
1. You and Your Environment, p. 5
2. The Living Players, p. 10
3. The Movement of Energy in a Community, p. 16
4. Who Eats What? p. 20
5. The Nonliving Players, p. 26

PUTTING IT ALL TOGETHER
6. Natural Changes, p. 36
7. Humans: Boss or Co-Worker? p. 50

For Your Journal

Write a paragraph summarizing your thoughts about each of the following:
1. How do environments change over time, and what causes them to change?
2. In what ways do people influence the environment?
3. How are all of the things in this picture related?
4. Is every relationship in this picture visible? Explain.

✺ Using the Photograph

Have students describe what is happening in the photograph. Ask students to infer if the bee is harming the zinnia, helping it, or neither.

Point out that honeybees are social insects that form colonies in which different types of bees perform specialized tasks. Have students compare how bees live with how humans live. Then ask: What would happen to the honeybees if there were no more zinnias? no more flowers? *(Without zinnias, the honeybees would have one less food resource. If there were no flowers at all, the honeybees would suffer a serious food shortage resulting in their extinction.)* Ask: What would happen to zinnias if there were no more honeybees? *(Without honeybees, the zinnias would have a much more difficult time reproducing. The zinnia population would shrink.)*

✺ About Your Journal

Students should answer the Journal questions to the best of their abilities.

These questions are designed to serve two functions: to help students recognize that they do indeed have prior knowledge about the topic, and to help them identify any misconceptions they may have. In the course of studying the unit, these misconceptions should be dispelled.

Interactions 3

LESSON 1

✷ Getting Started

In this introductory lesson, students explore the concept that there are many interactions that exist within an environment. The lesson begins with a game that reinforces that all things in the environment interact with each other. As the lesson continues, students are introduced to biotic and abiotic factors in the environment, and are encouraged to think of a definition for each of these terms. The lesson concludes by having students take a look at their schoolyard environment and classify each of the factors they identify as either biotic or abiotic.

Main Ideas

1. All things in the environment interact with one another.
2. Those things in the environment that are alive, or were once alive, are the biotic factors of the environment.
3. Those things in the environment that are not alive, or never were, are the abiotic factors of the environment.

LESSON 1 ORGANIZER

Objectives

By the end of the lesson, students should be able to:
1. Identify interactions taking place in the environment.
2. Appreciate the degree to which living and nonliving things depend upon one another.
3. Classify parts of the environment as either biotic or abiotic.

Process Skills

observing, predicting, classifying, summarizing

New Terms

Environment—the physical surroundings of an organism, including all the conditions and circumstances that affect its development.
Interaction—a relationship between parts of the environment.

THE PLAYERS AND THEIR ROLES

1 You and Your Environment

Your **environment** is everything that surrounds you—trees, grass, soil, air, water, animals, insects, and other people too. How important do you think the environment is for your life? Could you live without it? Are there some parts you could live without? Are different parts of the environment related to other parts? Give these questions some thought.

To help your thinking, there is a game on page 7 that will help you find some answers.

✸ Teaching Strategies

Have students read the lesson introduction silently or call on a volunteer to read it aloud. Then invite students to identify factors in their classroom environment. Keep track of their responses on the chalkboard. Help students recognize that everything in the classroom is part of their environment. Involve them in a discussion of the questions in the lesson introduction. Accept all reasonable responses. Point out to students that they will gain important insights into these questions by completing *Exploration 1*.

Biotic factors—the parts of the environment that are living or were once alive.

Abiotic factors—the parts of the environment that are nonliving.

Materials

Exploration 1: balls of string or yarn, small index cards, markers, tape

Time Required

two class periods

The Players and Their Roles 5

Exploration 1

Explain to students that they will gain new insights by doing this "real-life simulation." Direct their attention to the illustration on page 6, and encourage them to predict what is happening. Then have them read page 7 to see if their predictions were correct, and to discover how to play the game.

If possible, perform the activity outdoors where students have room to spread out. Before the game begins, have each student select one part of the environment that they will represent. To ensure that a variety of environmental factors are selected, list some factors on the chalkboard for students to select. Ask students to identify the part of the environment they represent by writing it on a small index card. The index cards should be taped to their clothes as shown in the illustration.

Divide the class into groups of 10 to 15 students. Be sure to have several skeins or balls of inexpensive string or yarn ready for the activity.

You may wish to take a few minutes to use one of the groups to demonstrate how the game should be played. Then have students begin. Offer any helpful hints to keep the game moving.

After each group has had time to form its "web," have one student let go of his or her string. Encourage students to describe what happens when this part of the environment is disconnected. *(Students should recognize that when one part of the environment is disconnected, it affects or even destroys the rest of the environment.)*

After students have had time to play the game, have them return to the classroom to discuss the questions that refer to the illustration on page 7.

EXPLORATION 1

Relationships in Nature

6 Interactions

You Will Need

- one index card per student
- a ball of string or yarn

What to Do

Choose something from the environment that you'd like to be. You might be the wind, the sun, an ant, a daisy, the soil, or any one of many, many other things. Each person should be a different thing. Use a small card to label what you are.

The first person will hold on to the end of the string and pass the ball to another student in the circle with whom she or he can be "related." The first student will then explain to the whole group what this relationship is.

For example, the "daisy" student holds the end of the string. He or she passes the ball to the "water" student and says, "I need water to grow." The "water" student holds on to the ball of string, passes it to the "fish" student, and says, "I am your home." This continues, one move at a time, showing relationships in the circle.

You may end up holding portions of the string coming from and going to many different things. Now, what happens if one part of the environment is removed? Test your prediction. Have the person representing this part let go of the string.

What part(s) of the environment would you want not to let go of? Which parts seem to be the most important for maintaining the relationships in the circle? Look at the connections at air and water. Why are there so many?

The drawing above shows the connections a group of students made in playing the game. What does the circle look like? Does any living thing in the circle exist alone? Would removing mosquitoes from the circle make any difference? How?

Why does the string connect the earthworm and the soil? The answer is quite simple: the earthworm lives in the soil and gets its food there. The earthworm and the soil *interact*. In the game, wherever one part of the environment is connected by the string to another part, there is an **interaction**. What interaction takes place between the earthworm and the bird? between the air and the soil?

Questions

1. What are some of the different ways in which the connected pairs interact? In other words, describe the various ways in which one member of a pair can need the other member.
2. Are *all* of the possible interactions shown in the diagram?

In this unit, you're going to learn about many different kinds of interactions.

Answers to

In-Text Questions

- Most students will probably agree that air and water are the most important parts of the environment. Most living things require air and water, therefore many strings show these interactions.
- The circle looks like a spider's web.
- No living thing in the circle exists alone. Everything is connected directly or indirectly to everything else.
- If mosquitoes were removed from the circle, some birds and fish would lose a source of food. There would be one less organism dependent upon water and air.
- The bird interacts with the earthworm because it uses the earthworm for food.
- Air is an important part of soil. It provides oxygen and other gases to plants and to animals that live in the soil.

Answers to

Questions

1. Challenge students to chose a pair of factors and describe their interaction. Interactions can include use as food source, place where an organism lives, provider of nutrients, and so on.
2. Not all possible interactions are shown in the diagram. Invite students to think of some interactions that are not shown. For example, the earthworm could be connected to the daisy because it helps to supply the roots of the daisy with food and air.

The Players and Their Roles

Biotic and Abiotic

This activity provides an opportunity for students to summarize the concept of environmental interaction. Have students read the page silently, and give them time to think about the questions. Then involve them in a discussion of their responses. *(Most students will probably agree that the list of biotic factors consists of living things and the list of abiotic factors consists of nonliving things.)*

Have students read the first paragraph on page 9. Allow them to develop their own definitions for the terms *biotic* and *abiotic*. Their responses should indicate an understanding that biotic factors are the parts of the environment that are living or were once alive. Abiotic factors are the parts of the environment that are not alive and never have been.

Divide the class into small groups, and ask them to make a list of the biotic and abiotic factors in their schoolyard. Then, in their Journals, ask them to draw a diagram showing the various factors arranged in a circle. Students should draw lines between the factors that are connected. When this activity is completed, have the groups share and discuss their ideas as a class. This will provide students with the additional experience they need to add further connections to the diagram produced by Ms. Wilkie's class.

Biotic and Abiotic

Ms. Wilkie's class played the interaction string game. Look at the list of the parts they chose. These parts are divided into two groups: **biotic** and **abiotic**. What do all of the biotic parts have in common? The abiotic parts?

Biotic	Abiotic
perch	air
spider	temperature
bacteria	water
insect	sand
deer	rocks
crow	light
maple tree	
girl	
dandelion	
Venus' flytrap	

Some connections made by Ms. Wilkie's class:

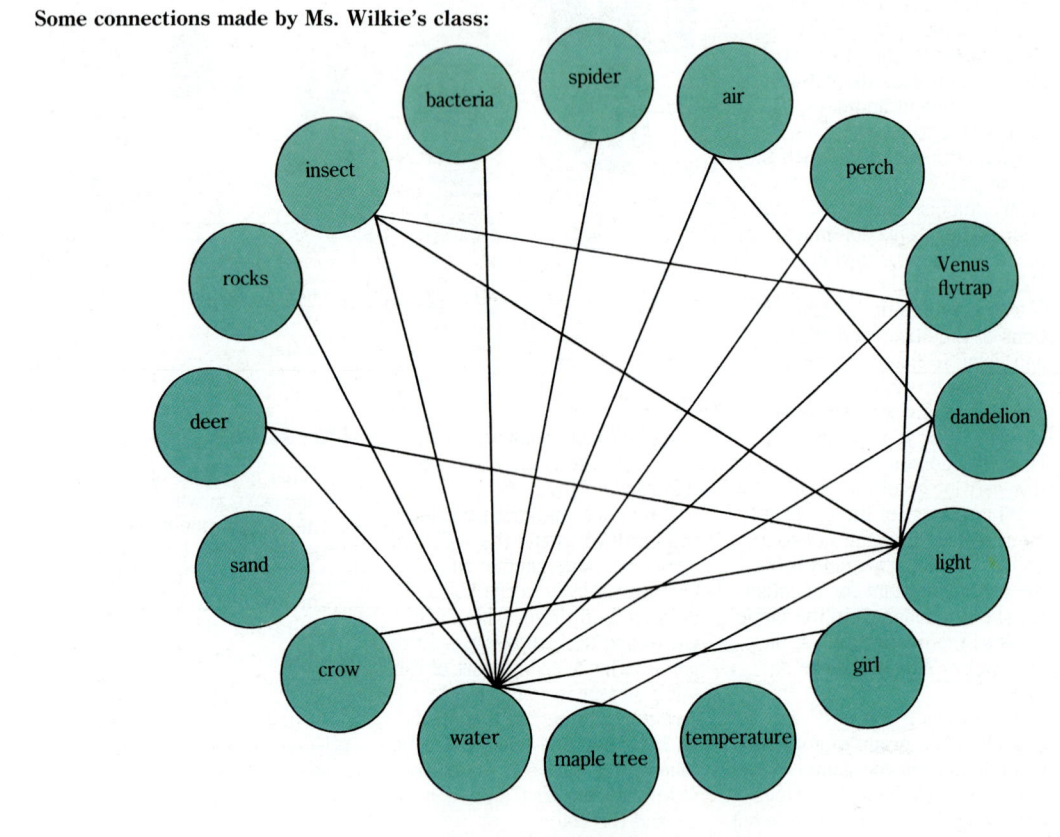

8   Interactions

If you say that the biotic parts of the environment are living things, you are right. But here are a few more examples of biotic parts of the environment:

- A rotting log
- A dead animal
- A bone

Together with a classmate, devise a new statement that best describes *biotic* and *abiotic*.

Look around your schoolyard. Which parts of the schoolyard environment are biotic? Which parts are abiotic? Arrange the various parts in a circle and connect the ones that you think interact. Draw lines only for those interactions that you are sure of.

Now you know something more about this unit: it provides a look at various biotic and abiotic parts of your environment and at some of the interactions between them.

By the way, is the interaction string game of Ms. Wilkie's class complete? What parts of that environment (both biotic and abiotic) are not connected, but might be? Give a reason for each new connection you propose.

Some Detective Work

Take a close look at the large drawing on pages 4 and 5. How many examples of interaction can you find? Classify each interaction under one of the following headings:

- Animal–Animal
- Animal–Plant
- Plant–Plant
- Living–Nonliving
- Living–Once Living

What abiotic parts of the environment did you include in your lists?

Which photographs show an abiotic part of the environment? Which show a biotic part?

Follow Up

Assessment

1. Display pictures of different environments and have students write brief descriptions of the interactions taking place in each one. Ask students to classify the interactions as animal–animal, animal–plant, plant–plant, living–nonliving, or living–once living.

2. Have students cut out pictures that illustrate environmental interaction from magazines and newspapers. Have them arrange their pictures on construction paper and write a caption describing each one. Display the posters around the classroom.

Extension

1. Suggest that students do some research to discover what an *ecosystem* is. Have them make a drawing or diagram of one, and identify some of the interactions that take place there. Possibilities include a desert, pond, or rain forest ecosystem.

2. Some students may enjoy doing field research. Suggest that they select one area to observe for an hour or two. The area might be their backyard, a park, or the schoolyard. They should keep track of all the interactions that they observe or can infer that take place. Have them share their observations with the class by presenting an oral summary.

Answers to
Some Detective Work

Allow students to include mushrooms and other fungi as "plant-like" living things, and microscopic organisms as "plant-like" or "animal-like." Some possibilities include:

- *Animal—Animal*
  fish/dragonfly
  snake/frog
  mouse/insect
- *Animal—Plant*
  deer/grass
  butterfly/flower
  mountain lion/tree
  bees/tree
- *Plant—Plant*
  bracket fungi/tree
  small plants/large trees
- *Living—Nonliving*
  fish/water
  deer/water
  deer/land
  algae/sunlight
- *Living—Once Living*
  plants/decaying plants
  bacteria/dead leaves

Abiotic parts include: water, rocks, air, light, and temperature.

Answer to caption: Students should recognize that the rocks and stream represent abiotic factors and the tree, the bird, and the person represent biotic factors.

The Players and Their Roles

LESSON 2

✹ Getting Started

In this lesson, students explore different types of biotic relationships within a community. The lesson begins with a discussion of the difference between a niche and a habitat. Then students are presented with examples of commensalism, mutualism, and parasitism. The lesson concludes with students classifying biotic relationships according to the type of interaction they exemplify.

Main Ideas

1. An organism's habitat is the place where it lives.
2. An organism's niche is its way of life in its habitat, including its biotic and abiotic relationships.
3. Commensalism, mutualism, and parasitism are examples of three kinds of biotic relationships.

✹ Teaching Strategies

Communities of Living Things

Ask students to think about the kinds of relationships they have with different people. Invite them to suggest words that describe these relationships. For example, friendly, supportive, unpleasant, helpful, loving, and so on.

Explain that just as people have different kinds of relationships with each other, organisms in the natural world interact with each other in many different ways, too. Some of these interactions help an organism, while others may harm an organism.

Have students read page 10 silently. Involve them in a discussion of the difference between *niche* and *habitat*.

---

2 The Living Players

Communities of Living Things

You would probably think it strange to find orchids growing in the Arctic, or polar bears roaming the beaches of Florida. It would also seem a bit odd to find a mouse chasing a cat, or carrots attacking a rabbit!

Your experiences have shown you that every organism has its own way of living. It exists in the environment that best suits its needs. Also, every living thing relates to other organisms around it in specific ways. These relationships between an organism, its physical surroundings, and its neighbors make up an organism's **niche**. Its niche is more than just the place where an organism lives. The niche is the organism's entire way of life: it indicates the organism's *role* in the community where it lives. The *place* where an organism lives is called its **habitat**. For instance, your habitat might be San Antonio, Sacramento, Portland—any city or town. Your niche (right now) is student, son or daughter, nephew or niece, member of the school choir or basketball team, and so on. Being a student in school, for example, reveals a great deal about your way of life, your environment, and those with whom you interact.

Do you understand the difference between *habitat* and *niche*? In your Journal, complete this dictionary definition:

A ___?___ is the way of life that an organism adopts to survive in a particular ___?___.

LESSON 2 ORGANIZER

Objectives

By the end of the lesson, students should be able to:
1. Describe the difference between a habitat and a niche.
2. Define and compare commensalism, mutualism, and parasitism.
3. Identify examples of commensalism, mutualism, and parasitism.

Process Skills

observing, analyzing, inferring, classifying

New Terms

Niche—an organism's way of life, including its relationship with other organisms and with its physical surroundings.

Copy this table into your Journal. Complete it by describing the niches of the organisms listed as thoroughly as you can.

Organism	Habitat	Description of Niche	
		How It Uses Other Things (Living and Nonliving)	How It Is Used by Other Things in Its Environment
robin	Open areas, lawns, bushes, trees, air, or ground	Uses bush or tree for a nesting site or for protection. Uses twigs and grass for nest. Eats worms and insects for food.	Eaten by larger birds, snakes, or cats. Eggs eaten by other birds. Parasites—mites and lice—live in its feathers.
turtle			
spruce tree			
beetle			

Answers to
In-Text Questions

The sentence at the end of page 10 should read, "A *niche* is the way of life that an organism adopts to survive in a particular *habitat*."

Direct students' attention to the table on page 11. Review the sample entry. After students complete the table, involve them in a discussion of their ideas. Answers may vary, possible responses include:

Turtle
- lives in and around water areas, such as streams, ponds, lakes, and rivers
- eats insects, minnows, and tadpoles; suns itself on rocks and floating logs
- may be food for such animals as raccoons, bears, and otters; its eggs are eaten by birds

Spruce trees
- grow in northern temperate climates with cool summers and cold winters
- get nutrients from the soil; use the sun's energy to make food
- birds build nests in spruce trees and may use the seeds for food; people use spruce trees for lumber

Beetles
- live on plants and under rocks
- use plants and other insects for food; bury their eggs in the soil
- some birds eat beetles; some plants use beetles to spread their pollen

---

Habitat—the place where an organism lives.
Commensalism—an association between two organisms in which one benefits and the other neither benefits nor is harmed by the relationship.
Mutualism—an association between two organisms in which both benefit.
Parasitism—an association between two organisms in which one (the parasite) benefits and one (the host) is harmed.

Materials
Communities of Living Things: Journal

Teacher's Resource Binder
Resource Worksheet 1–1

Science Discovery Videodisc
Disc 2, Image and Activity Bank, 1–2

Time Required
two to three class periods

The Players and Their Roles 11

Types of Interactions in Communities

"One-Way" Benefits

Direct students' attention to the photograph and have them identify what it shows. Encourage them to discuss the relationship between the robins and the tree. Accept all reasonable comments. Have students read the section silently or call on volunteers to read it aloud.

Involve students in a discussion of the example of the barnacles and the horseshoe crab. Help them recognize that by riding on the back of crabs, barnacles avoid overcrowding and increase their food resources. Students should recognize that this is an example of commensalism because one organism, the barnacle, benefits from the relationship while the other organism, the horseshoe crab, neither benefits nor is harmed. Other examples of commensalism: remoras riding on sharks, Spanish moss growing on trees, and squirrels living in trees.

"Two-Way" Benefits

Have students speculate about the relationship between the bee and the flower. Accept all reasonable responses. Then call on a volunteer to read the first two paragraphs aloud. Students should recognize that the bee is helped by the flower because the flower provides it with food (nectar). The flower is helped by the bee when the bee distributes the flower's pollen. This relationship is different from commensalism because both organisms benefit.

Encourage students to articulate the difference between commensalism and mutualism. Have them read the example about the lichens and discuss how it demonstrates mutualism. You may wish to review the terms fungus, algae, and photosynthesis after students have read about them in their texts. Other examples of mutualism: ants eating the wax from peony buds while protecting the plant from other insects, and butterflies drinking the nectar from flowers while helping to disperse pollen.

12 Interactions

Types of Interactions in Communities

"One-Way" Benefits

The scene in the photo at the right is probably familiar to you. Robins like these might even live in your back yard, where trees provide a convenient place for building their nests. Let's analyze the relationship between a robin and a tree.

In this situation, two different kinds of organisms live closely together. One organism—the robin—benefits or gains from the relationship. The robin benefits because the tree provides a place for its nest. The other organism—the tree—is neither helped nor harmed by the nest. This kind of relationship is called **commensalism.**

Now let's look at another example of commensalism. Adult barnacles cannot move from place to place on their own. But they often attach themselves to the shell of a horseshoe crab and travel with the crab. The crab is unaffected by the presence of the barnacles getting the free ride. How do the barnacles benefit from traveling with the crab? Why is this an example of commensalism?

Think of still other examples of commensalism.

"Two-Way" Benefits

The bee and the flowering plant provide an example of another kind of relationship. Can you think how this relationship might differ from commensalism? Here are the facts:

- The bee gets nectar (a liquid food that it uses to make honey) from the flower.
- While the bee is feeding from the flower, some of the flower's pollen sticks to the bee's body.
- When the bee flies to another flower, it deposits some of the pollen on that flower. *Pollination* is part of the process by which a flowering plant produces seeds.

How is the bee helped in this relationship? How is the plant helped? A relationship between two organisms in which both benefit is called **mutualism.**

Next, consider the *lichens,* which are plant-like organisms that can be found almost everywhere. They may look like grayish-green scales on rocks and on the bark of trees. One kind is called "old man's beard" because that's what it looks like. It hangs from tree branches. Another kind of lichen can be found on the ground in the shape of little red-tipped, vertical sticks or stems. This type is called "matchstick lichen" or "British soldiers."

But a lichen is not a single organism. It is really two kinds of

Various lichens

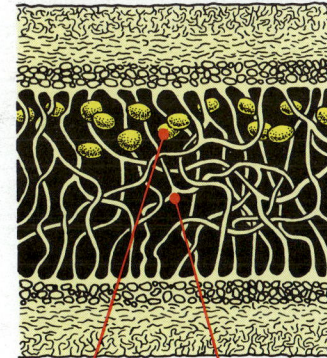

organisms living together for mutual benefit. You need a microscope to see both organisms.

One organism in lichen is a *fungus* (plural, fungi) with many thread-like branches. These branches provide both protection and moisture for green, single-celled organisms called *algae* (singular, alga). The algae make food by a process called *photosynthesis*. The fungi in lichen do not make their own food. Instead, they use the food made by the algae.

Name as many other examples of mutualism as you can.

Benefit and Harm

Compare commensalism and mutualism with a relationship called **parasitism**. In parasitism, one organism—the *parasite*—lives on and obtains its food from another organism—the *host*. However, in this relationship, the host is harmed instead of helped. Sometimes the host is even killed by the parasite.

Do research on the following pairs of living things that illustrate parasitism:

- People and tapeworms
- Trichina worms and pigs
- Rust and wheat
- Spruce trees and spruce budworms

Which organisms are the hosts? Which are the parasites?

A diagram showing fungi and algae that make up lichen, as seen with a microscope.

alga   fungus

Problems to Consider

Provide students with time to complete their diagrams. Then call on several volunteers to share and describe their diagrams to the class. Encourage students to research specific examples to illustrate their diagrams. To extend the exercise, have students use their diagrams to make bulletin board displays, or have groups of students use their ideas to create murals.

Allow students time to review all of the examples and record their ideas in their Journals. As a class, have them share and discuss their responses.

Answers to
In-Text Questions

Students' responses should be similar to the following:
- **(a)** Mutualism—the cow's digestion is aided by the bacteria. The bacteria benefit from the cow by having a place to live and food to eat.
- **(b)** Commensalism—the suckerfish benefits from the shark. The shark neither benefits nor is harmed by the suckerfish.
- **(c)** Commensalism—the orchids grow on other plants for support. The other plants neither benefit nor are harmed by the orchids.
- **(d)** Mutualism—the protozoa help the termites digest wood. The termites provide the protozoa with food and a place to live.
- **(e)** Parasitism—the dodder plant benefits from the wheat. The wheat is destroyed by the dodder plant.
- **(f)** Parasitism—the ichneumon wasp larvae benefit from other insects' eggs or larvae. The insect eggs or larvae are destroyed.

(Answers continue on next page)

Problems to Consider

1. Make three diagrams that will help you remember these three relationships: commensalism, mutualism, and parasitism.
2. Examine each of the pictures on this page and the next, and read the accompanying facts. Then identify the type of relationship being exhibited in each case.

a

The stomach of a cow is home to a type of bacteria that digests the grass eaten by the cows.

b

A suckerfish attaches itself to a shark by a sucker on the top of its head. It gets a free ride and shares some of the scraps of food left over from what the shark has eaten.

c

Orchids grow on other plants for mechanical support only.

d

Small organisms called protozoa live inside the body of a termite and digest wood for the termite and for themselves.

e

Dodder, a plant with no leaves, winds around wheat and other plants for both support and food.

f
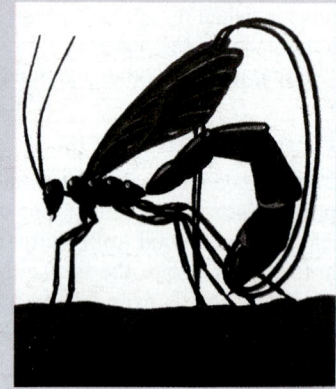
Ichneumon wasps pierce the eggs or larvae of other insects and lay their own eggs inside. Once hatched, the larvae of the ichneumon wasps then feed on the other insects' eggs or larvae.

14   Interactions

g

Lungworms thrive at the expense of the bighorn sheep that live in the Rocky Mountains.

h

Mistletoe grows on trees such as oak, obtaining both support and food while weakening the tree.

i

People carry burdock seeds (burs) on their clothing when walking through the brush. The seeds are eventually deposited somewhere else.

j

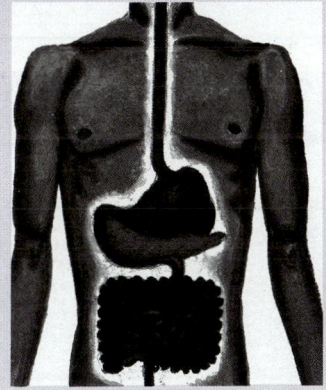

Certain bacteria find a comfortable shelter in the human digestive tract. The bacteria help remove water from indigestible material. They also help produce vitamin K, which is important in blood clotting.

k

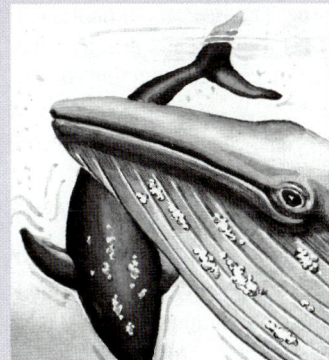

Barnacles often take trips on whales and end up in a new part of the ocean.

l

Certain ants actually "herd" aphids. The ants move the aphids from place to place and protect them from predators. The ants collect the sweet liquid (honeydew) produced by the aphids.

Follow Up

Assessment

1. Students may enjoy drawing cartoons to illustrate commensalism, mutualism, and parasitism. Ask them to use the animal and plant examples from the lessons in their cartoons. Use the finished cartoons to create a bulletin-board display.
2. Suggest that students make dioramas to show the habitat, including biotic and abiotic relationships, for a particular organism. Display the dioramas in the classroom for others to enjoy.

Extension

1. Point out to students that the terms *community, population,* and *habitat* each have a scientific meaning in regard to groups of organisms living together. Suggest that they research to discover what these terms mean in relationship to the environment. Have them present their information to the class in an oral report.
2. Have students make a list of several organisms that live in the schoolyard or in a neighborhood park. Then have them make a table similar to the one on page 11 using the organisms from their list. When students finish, have them share the information with their classmates.

Answers continued

(g) Parasitism—the bighorn sheep give the lungworms a place to live. The sheep are destroyed by the lungworms.

(h) Parasitism—the tree provides the mistletoe with support and food. The mistletoe weakens the tree.

(i) Commensalism—the burdock plant benefits from people. People neither benefit nor are harmed by the burdock plant.

(j) Mutualism—the people give the bacteria a place to live and food to eat. The bacteria aid humans' digestion and produce vitamin K.

(k) Commensalism—the barnacles use the whales to move from place to place. The whales neither benefit nor are harmed by the barnacles.

(l) Mutualism—the aphids receive food and a place to live. The ants benefit by getting food from the aphids.

The Players and Their Roles

Lesson 3

Getting Started

This lesson explores the flow of energy through a community of living things. The lesson begins by drawing on students' prior knowledge to explain the many roles organisms play in a community. Roles such as producer, consumer, decomposer, predator, prey, and scavenger are presented. The lesson concludes by having students reinforce their understanding of interactions by completing a relationship puzzle that consolidates the concepts covered so far in the unit.

Main Ideas

1. Organisms exist and interact in communities.
2. All living things depend on sunlight as their initial source of energy.
3. Green plants are called producers because they use the sun's energy to make food.
4. Consumers are organisms that eat other organisms.
5. Decomposers are organisms that break down dead plant and animal organisms into substances that enrich the soil.

3 The Movement of Energy in a Community

You live in a community. Your community is made up of your family, neighbors, friends, pets, and all the other living things in your neighborhood. Similarly, a forest contains many kinds of organisms. In a forest, some organisms cooperate, some compete, and some never even interact with each other. We can call this a forest, or woodland, community.

Every community of living things, whether in a forest or in a city, needs energy for its members to live. You know that an automobile gets its energy from burning a fuel, usually gasoline. Well, a community of living things also gets its energy from "burning" fuel, but it gets energy in a different way and in a different form: food.

Think about how energy "moves" through a forest community. Some animals get their energy (or food) by eating other animals. Some animals eat plants. But what do plants eat? What is their source of energy?

Energy → Plants → Animals → Animals

A few plants, such as the pitcher plant and the Venus' flytrap, eat animals (insects). But this is not the usual niche that plants occupy. So how do plants get their energy? You read earlier about **photosynthesis**—the process by which green plants change carbon dioxide (a gas they get from the air) and water (from the soil) into food. This food supplies energy not only for the plants, but also for the animals that eat the plants. The process of photosynthesis also *requires* some energy. Green plants use energy from sunlight to manufacture their food, as you can see from this energy diagram:

Sun's Energy → Plants → Animals → Animals

Types of Roles in a Community

Green plants play a particular and valuable role in a community. They make food energy, which they use themselves. But this food energy is also distributed throughout the community when plants are eaten by other organisms.

Because green plants produce food for themselves and for others, they are called **producers.** This is their niche.

The organisms that eat green plants and other organisms are called **consumers.** Why?

Which of the organisms on this page and the next are producers? Which are consumers?

LESSON 3 ORGANIZER

Objectives

By the end of the lesson, students should be able to:

1. Describe the differences between producers, consumers, and decomposers.
2. Identify the differences between herbivores, carnivores, and omnivores.
3. Identify how predator/prey relationships facilitate the flow of energy through a community.
4. Explain the importance of scavengers in a community.
5. Explain how energy flows through a community of living things.

Process Skills

observing, analyzing, inferring, classifying

New Terms

Photosynthesis—the process of using sunlight, water, and carbon dioxide to make food.

Producers—green plants that produce food for themselves and for others.

16 Interactions

✦ Teaching Strategies

Before students begin reading, involve them in a discussion about their community. Have them name some of the things that make up their community, such as friends, neighbors, relatives, strangers, stores, landmarks, and so on. Then have students silently read the lesson introduction.

Encourage students to identify some other kinds of communities that exist in the natural world, such as a pond community, a desert community, a mountain community. Help students recognize that energy flows through every community in a similar way.

Types of Roles in a Community

To help students understand that all energy in a community originally comes from the sun, have them trace something they ate back to its original energy source. Organisms that eat plants or other organisms are called consumers because they must eat (consume) food in order to get energy. Have students identify which of the organisms in the photographs are producers and which are consumers.

Answer to caption: *Producers:* grasses, trees, shrubs, and the plants that produce fruits and vegetables. *Consumers:* leaf hopper, hawk, rabbit, deer, mountain lion, snake, lizard, mouse, fox, shrew, and human.
(Continues on next page)

Consumers—organisms that depend upon other organisms for food.
Herbivores—consumers of only plants.
Carnivores—consumers of only meat.
Omnivores—consumers that eat plants and animals.
Predator—a consumer that hunts or captures other organisms (prey) for food.
Prey—an organism that is hunted or captured and eaten by another consumer.

Scavenger—a consumer of dead or decaying plants and animals.
Decomposers—organisms that break down dead plant and animal organisms into substances that enrich the soil.

Materials
none

Teacher's Resource Binder
Resource Transparency 1–1

Science Discovery Videodisc
Disc 2, Image and Activity Bank, 1–3

Time Required
two class periods

The Players and Their Roles 17

(Types of Roles in a Community continued)

Help students recognize that the term *omnivore* means an organism that eats both plants and animals. Then have them look back at the pictures on pages 16 and 17 and identify the animals that are herbivores, carnivores, and omnivores.

Herbivores: leaf hopper, rabbit, and deer

Carnivores: hawk, mountain lion, snake, lizard, fox, and shrew

Omnivores: the person and the mouse

The energy diagram is:
Sunlight → Producer → Herbivore → Carnivore

Consumer Relations

Call on a volunteer to read the first paragraph aloud. Encourage students to identify some predator/prey relationships. Examples are given here.

- The cat is the predator; a mouse is its prey.
- The lion is the predator; a zebra is its prey.
- The owl is the predator; a shrew is its prey.

Have students continue reading until the end of the third paragraph. Ask them to discuss the question. Help them recognize that lines could be drawn from the earthworm to all of the organisms in the diagram because all organisms die and provide organic material for the worm to eat.

Provide students with time to examine the pictures on the page. Call on volunteers to identify each of the animals, and to suggest where each might get its food.

Answers to
In-Text Questions

- The eagle catches its food, usually small mammals, on land. It also eats some dead animals.
- The shark gets its food in the ocean. It eats aquatic animals and scavenges on dead animals and refuse.
- The hyena eats carcasses of dead animals that it finds on the African plain.
- The seagull eats fish near ocean shores.
- The crayfish is a fresh-water dweller that eats snails, small fish, tadpoles, and insect larvae.

Both the deer and the fox are consumers. But because their diets are different, they are different types of consumers. The deer is called a **herbivore** (plant eater) because its main food is plants. The fox is called a **carnivore** (meat eater) because its main food is other animals. Which name fits humans? If you eat a turkey sandwich with lettuce and tomato, you are eating parts of both plants and animals. That would make you an **omnivore.** What do you think the term means? Make an energy diagram like the ones on page 16 to contain the following items: producer, sunlight, carnivore, herbivore.

Consumer Relations

Some consumers, such as the fox, kill other animals for food. The fox is a **predator,** and its food is its **prey.** What other examples of the predator-prey relationship can you think of?

Unlike the fox, some consumers do not simply eat and run. The tapeworm (a parasite) can live comfortably in a human (a host) for long periods of time, consuming food that the person eats. Certain parasites spend their entire lives in a single host. Other parasites spend part of their lives in one host and the rest of their lives in a second host. Still other organisms are parasites for part of their lives but are free-living the rest of the time.

Some animals wait until their food source is already dead instead of killing it. A vulture eating a dead bird is called a **scavenger.** An earthworm eating dead organic material in the soil is another example of a scavenger. Look back at the interaction diagram on page 6. What lines could you draw to show the earthworm's role as a scavenger? (Remember that all of the organisms in the diagram will eventually be dead.)

At right are some pictures of other consumers. Can you name them? Where might they get their food?

Not all living things are killed and eaten by consumers. Nor do scavengers eat every last bit of the dead organisms they feed on. What happens to the remains? Something must happen; otherwise the earth would be covered with the bodies of dead plants and animals. Can you imagine what this situation would be like?

Fortunately, some consumers clean up the earth by breaking down plant and animal bodies into substances that enrich the soil, which then supports the growth of new plants. Such consumers are called **decomposers.** Mushrooms growing on tree stumps and the black fuzzy mold on tomatoes are examples of decomposers that you can see easily.

The most common decomposers, however, are ones you can't see without a microscope. They are found in almost all environments on land and water. These are **bacteria.**

Have students read the rest of the material, continuing onto page 19.

Involve students in a discussion about decomposers. Help them realize how important decomposers are to the environment. Encourage students to offer examples of decomposers at work. *(Possible examples: a rotting log, spoiled food, sour milk, a decaying animal, a dead tree, the compost pile, and so on.)*

Now let's look at an example of where many decomposers are at work. Have you ever seen a compost pile—a mixture of leaves, grass, potato peelings, soil, and other biotic materials? If so, you may have noticed a matted layer of fine white threads on top of the pile. These threads are parts of fungi. Molds and mushrooms are also fungi. These organisms obtain their food from dead plants and animals. They also give off substances that break down the dead material into soluble substances that, in turn, can be absorbed by new plants.

Deeper down in a compost pile is a whole world of bacteria also feeding on the dead material and making it usable by plants.

A Relationship Puzzle

Below is a list of interactions. Can you match each numbered event on the left with the type of interaction it represents?

1. A bird eats a caterpillar.
2. A mosquito sucks blood.
3. Mold grows on oranges.
4. An aphid lives on the leaves of trees.
5. A toad eats a fly.
6. A millipede eats dead plants.
7. A dog has fleas.
8. A human kills and eats a deer.
9. Bacteria live on a dead sparrow.
10. A sowbug eats rotting wood and leaves.

(a) predator-prey
(b) parasite-host
(c) scavenger and its food
(d) decomposer and its food

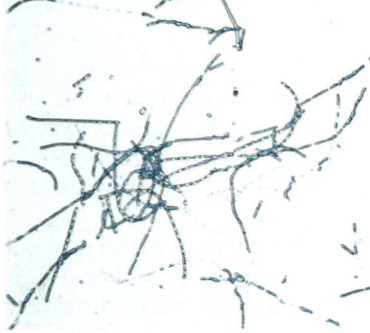

Bacteria *(above)* and fungi *(below)* both help break down, or decompose, dead plants and animals. Both are called decomposers.

Tell Me About It!

So far in this unit, you have learned many new terms that are used to describe an organism's niche. Take another look at the picture on pages 4 and 5. Suppose you were asked to describe this scene to a friend who is blind. Use your new scientific vocabulary to describe the picture to your friend. Tell your friend what is happening or is about to happen, using such words as these:

- producer
- consumer
- herbivore
- carnivore
- parasite
- host
- predator
- prey
- scavenger
- decomposer
- mutualism
- commensalism
- energy

Answers to

Tell Me About It!

Students' descriptions may vary, but should include some or all of the following relationships:

- Producer: plants of all sizes, including the algae.
- Consumer: animals, the magnified amoeba, and other protozoa.
- Herbivore: butterfly, rabbit, bees, and deer.
- Carnivore: eagle, snake, fish, robin, frog, and mountain lion.
- Parasite: bracket fungi on the tree.
- Host: tree on which the bracket fungi are living.
- Predator—prey: snake/frog, robin/worm, fish/dragonfly, eagle/mouse, and mouse/insect
- Scavenger: ants and possibly the eagle.
- Decomposer: mushrooms and magnified bacteria.
- Mutualism: butterfly pollinating a flower and getting nectar at the same time.
- Commensalism: the bird in the tree.
- Energy: flow of food from one organism to another.

Follow Up

Assessment

1. Have students make poster diagrams showing the flow of energy through a community of living things.
2. Have students write *Who Am I?* riddles for a herbivore, a carnivore, an omnivore, a predator, and a prey. Encourage them to exchange riddles with each other.

Extension

1. Take students on a field trip to identify producers, consumers, decomposers, and scavengers. Ask them to keep a list of their observations in their Journals.
2. Students may enjoy working in groups to make murals of a forest, pond, or desert community. Their murals should depict the relationships and roles of the various organisms living in that community.

Answers to

A Relationship Puzzle

1. (a) predator—prey
2. (b) parasite—host
3. (d) decomposer
4. (b) parasite—host
5. (a) predator—prey
6. (c) scavenger
7. (b) parasite—host
8. (a) predator—prey
9. (d) decomposer
10. (c) scavenger

The Players and Their Roles 19

LESSON 4

✷ Getting Started

In this lesson, students explore the movement of energy through a community of living things. Food chains are introduced as a simple way of showing how energy moves from producers to higher and higher levels of consumers. As the lesson continues, students realize that a complex system of food chains, called a food web, exists in the natural world.

Main Ideas

1. A food chain shows the flow of energy, in the form of food, from one organism to another.
2. A food web is a way of showing a complex system of food chains.
3. When one organism in a food web is removed, many other organisms are affected.

✷ Teaching Strategies

Ask students where they get their energy. *(from the food they eat)* Then have them identify some of their favorite foods. Keep a list of their suggestions on the chalkboard. Point to several of the items and ask students to identify the organisms that they come from and where those organisms got their energy. Help students recognize that the energy they receive from food has already passed through several other organisms before it reaches them. Then have students read the lesson introduction.

4 Who Eats What?

If you stand or sit still in a field or in the woods and watch closely, you can usually spot a number of animals. What are you most likely to find them doing? Did you say that they would be gathering or eating food? It's true—most animals spend a great deal of time looking for and consuming food, which is their source of energy.

Each type of animal prefers certain kinds of foods. And in most cases, each animal becomes food for other animals. Patterns of relationships among living things are easily found if you look for them.

Food Chains

Look at the pictures below and think about which organisms would complete this statement:

Plants are eaten by ___?___, which are eaten by ___?___, which are eaten by ___?___, which are eaten by ___?___.

LESSON 4 ORGANIZER

Objectives

By the end of the lesson, students should be able to:
1. Explain the difference between a food chain and a food web.
2. Diagram food chains and food webs.
3. Describe how a change in one part of a food web affects other parts of the web.
4. Identify the organisms in a food chain or a food web as either producers or consumers, and describe the role that they play in their community.

Process Skills

observing, classifying, inferring, analyzing diagrams

These relationships can be shown in a linked diagram or written out as a "chain." Such a diagram or statement describes a **food chain.** It shows how certain living things depend on one another for food energy. Each organism can be called a "link" in the chain. The food chain for the organisms shown on page 20 would have five links and would look like this:

Arrows are used to show the direction in which food energy moves. What would happen to the other organisms if the green plants were removed?

Green plants begin almost all food chains. Is it therefore true to say, "All members of food chains that begin with plants depend on the sun"?

Once again, study the food chain shown above and see if you can answer the following questions.

1. Which links in the food chain are producers? Which are consumers?
2. The grasshopper is called a *primary consumer*. Why?
3. Animals that eat animals are called *secondary consumers*. Why? Which organisms in the food chain above are the secondary consumers?
4. Which consumers are herbivores? Which consumers are carnivores?

At the end of the chain is the hawk. It apparently has no natural enemies in this situation. In this case, the hawk is called the *top carnivore*.

Construct a food chain showing what may happen or is happening in the diagram below. Identify the producer(s) and any primary or secondary consumers. Which organism is the top carnivore?

Food Chains

Direct students' attention to the photographs and ask them to identify each one. The sentence at the bottom of page 20 should read: *"Plants* are eaten by *grasshoppers,* which are eaten by *frogs,* which are eaten by *snakes,* which are eaten by *hawks."*

Have students read the first paragraph on page 21 and examine the diagram of the food chain. Students should realize that if the green plants were removed, all of the other organisms would die because the energy producer for the food chain would no longer exist. All members of a food chain depend on the sun's energy because it is the source of energy for green plants.

Answers to
In-Text Questions

1. The green plants are producers. The other organisms in the food chain are consumers.
2. The grasshopper is called a primary consumer because it is the first consumer to eat the green plants (the producers).
3. The frog, snake, and hawk are secondary consumers because they are not the first consumers to eat the green plants.
4. The grasshopper is a herbivore. The frog, snake, and hawk are carnivores.
    The food chain in the drawing can be depicted as: tree leaf → caterpillar → bird → cat. The producer is the tree. The primary consumer is the caterpillar. The secondary consumers are the bird and the cat. The cat is the top carnivore.

New Terms

Food chain—shows how certain living things depend on one another for food energy.
Food web—shows as many food relationships as possible between living things in an area.

Materials

Food Chains: Journal
Food Webs: Journal
Exploration 2: Journal

Teacher's Resource Binder

Activity Worksheets 1–2 and 1–4
Resource Worksheet 1–3
Resource Transparency 1–2
Graphic Organizer Transparencies 1–3 and 1–4

Science Discovery Videodisc

Disc 2, Image and Activity Bank, 1–4

Time Required

two to three class periods

The Players and Their Roles 21

Complicating the Food Chain: Food Webs

Have students read the paragraphs at the top of the page. Then direct their attention to the list of organisms in the right-hand margin. Ask students to identify which of the organisms in the list are:
- *producers*—grass, leaves
- *primary consumers*—earthworm, rabbit, grasshopper, and other insects
- *secondary consumers*—robin, snake, hawk, mouse
- *top carnivore*—hawk

Provide students with time to draw and complete the diagram in their Journals. Remind them that the arrows show the direction that the energy flows.

Answers to
In-Text Questions

Arrows may be drawn from:
- the grasshopper to the hawk, robin, and snake.
- the earthworms to the robin and mouse.
- the leaves to the rabbit, grasshopper, and other insects.
- the mouse to the snake and the hawk.
- the snake to the hawk.
- other insects to the snake, robin, and hawk.
- the robin to the hawk.
- the rabbit to the hawk and to the snake.
- the grass to the rabbit, grasshopper, and other insects.

Ask students why there are no arrows leading away from the hawk. *(The flow of energy stops with the hawk because it is the top carnivore.)* Ask students why there are no arrows leading to the leaves or the grass. *(They are producers, and their energy comes from the sun.)*

Have students identify several food chains in the food web. Answers may vary, possible responses include:
- grass, rabbit, hawk;
- leaves, grasshopper, snake, hawk;
- leaves, earthworms, robin, hawk;
- grass, grasshopper, snake, hawk.

Help students conclude that a food web is a number of interconnected food chains within a single community of living things. Students should recognize that all of the organisms in this food web could be found in a forest community.

Complicating the Food Chain: Food Webs

Real situations are not as simple or straightforward as the food chains you have studied so far. Robins get energy from organisms other than just grasshoppers; snakes eat other animals besides mice. If you were to include all of the food for every single animal in an area, your food chain would be very complicated.

Let's examine what it might look like. At the right is more information about the animals you have been discussing. (Keep in mind that even this expanded food list is far from complete.)

Circles containing the names of the various organisms are randomly arranged on the page. Sketch this arrangement in your Journal. Draw arrows showing the movement of food energy from one organism to another. A few arrows have already been drawn. (Arrows can cross one another.)

Animal	Food
earthworm	leaves (dead)
grasshopper	grass, leaves
robin	grasshoppers, other insects, earthworms
snake	mice, insects, young rabbits
hawk	snakes, insects, mice, rabbits
rabbit	grass, leaves

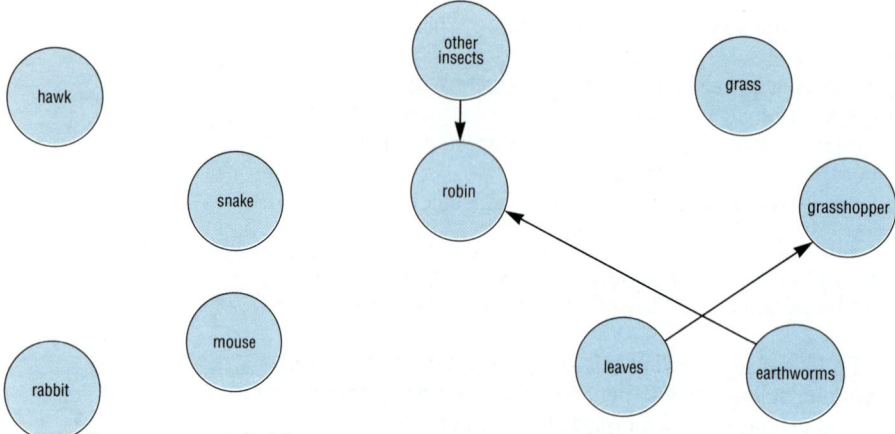

This kind of diagram is called a **food web.** A food web shows as many food relationships as possible between living things in an area. In what kind of area (environment) might you find the animals named above? Since there are many more animals than these in such an environment, you can see that a food web including all of them would be very large. It would show many links and a complex pattern of connections. To be complete, a food web must include all scavengers, parasites, and decomposers as well. In the following Exploration, you'll be analyzing a complex food web.

EXPLORATION 2

Analysis of a Terrestrial Food Web

Organism	Food
hawk	squirrels, grasshoppers, mice
coyote	squirrels, grasshoppers, mice, deer
bobcat	squirrels, mice, deer
squirrel	seeds, tree buds
grasshopper	grass
mouse	seeds, grass
deer	tree buds and twigs, grass
fungi and bacteria	hawks, coyotes, bobcats, squirrels grasshoppers, mice, deer, seeds, buds, twigs, grass

In your Journal, draw a food web for the 13 organisms (or parts of organisms) listed below.

hawk   seeds
bobcat   twigs
coyote   grass
squirrel   buds
grasshopper   fungi and
mouse   bacteria
deer

List the producers, the primary consumers (herbivores), and the secondary consumers (carnivores).

Did you notice all the arrows going to the fungi and bacteria? You may have wondered how all of these animals and plants could be food for fungi and bacteria. What role do fungi and bacteria play in this community?

Suppose that due to disease, the mice were removed from the community. What effect would this have? The chart below will help you in considering this question. Copy the chart into your Journal and complete it.

When mice are removed . . .	
These animals have lost a source of food:	These are the remaining food sources for each of the animals in the first square:
These organisms (or parts of organisms) are less likely to be eaten:	These primary consumers are left:

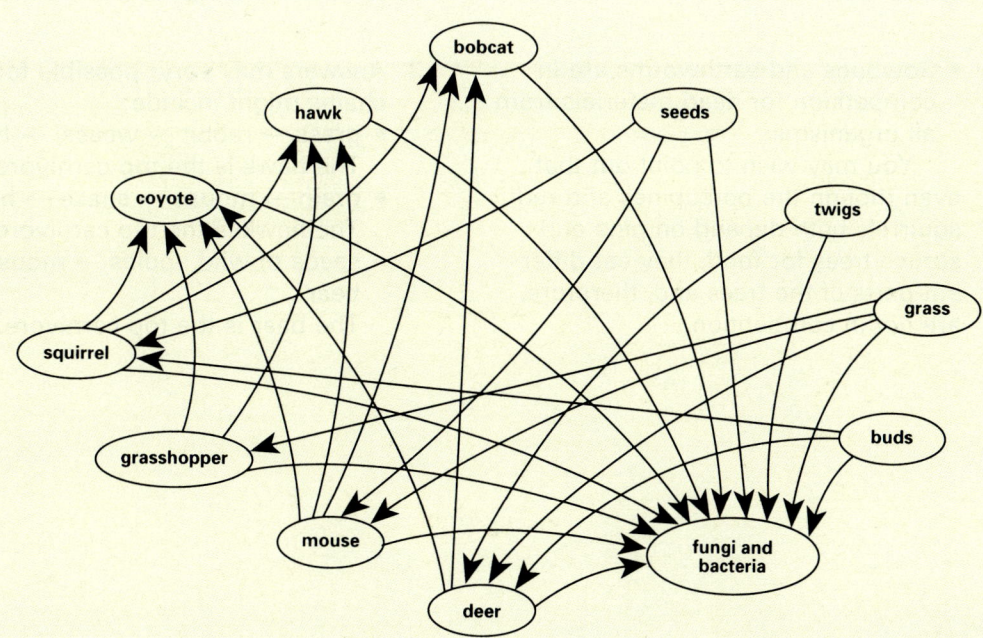

Exploration 2

Provide students with time to draw their food webs. Call on one or two volunteers to draw their webs on the chalkboard. Involve students in a discussion of the webs, and help them to resolve any differences of opinion on how they should be drawn.

Remind students that mice can be both primary and secondary consumers in a food web. In this Exploration, the mouse is a primary consumer. Discuss with the class what might happen to the food web if the mouse were a secondary consumer as well. For example, the mouse could eat grasshoppers along with seeds and grass. Students should realize that if mice as secondary consumers are removed, then grasshoppers are less likely to be eaten. The secondary consumers left would be the hawk, coyote, and bobcat.

Answers to
In-Text Questions

- A sample food web is shown below.
- *Producers* are grass, trees, buds, seeds, and twigs. They receive their energy from the sun.
- *Primary consumers* (herbivores) are the squirrel, deer, mouse, and grasshopper. They get their energy by eating producers.
- *Secondary consumers* (carnivores) are the hawk, coyote, and bobcat. They get energy from eating other animals. (Note that some students may recognize that mice can be omnivores.)

Students should recognize that as all of the organisms die, they become food sources for the fungi and bacteria, which are decomposers.

Have students consider what would happen if the mice were removed from the community. Their charts should reflect the following responses:

- *These animals have lost a source of food:* coyote, bobcat, and hawk.
- *These organisms (or parts of organisms) are less likely to be eaten:* seeds and grass.
- *These are the remaining food sources for each of the animals in the first square:* grasshoppers, deer, and squirrels for the coyote; squirrels and deer for the bobcat; squirrels for the hawk.
- *These primary consumers are left:* squirrels, deer, and grasshoppers.

A Visit to a National Park

You may wish to display a map of the United States and have a volunteer locate the Great Smoky Mountains National Park. Students may find it interesting to know that the first people to live in the region were the Cherokee Indians. The area was made into a park by an act of the United States Congress in 1926.

Provide students with time to read the lists of organisms and their food sources, and to identify the pictures. Then have them construct their food webs, as instructed on page 25. When students finish, involve them in a discussion of their diagrams and ideas. Ask a few volunteers to draw their food webs on the chalkboard. Involve students in a discussion of the questions on page 25 to summarize the lesson.

Answers to
In-Text Questions

1.
   - producers: trees (apple, poplar, spruce, pine), leaves, grass, grain, blueberries
   - herbivores: deer, red squirrels, porcupines, rabbits
   - carnivores: weasels, foxes, snakes, frogs, hawks, centipedes, salamanders
   - omnivores: bears, insects, mice
2.
   - Deer, mice, and bears are in competition for wild apples.
   - Deer and rabbits are in competition for twigs.
   - Weasels, foxes, bears, snakes, and hawks are in competition for mice.
   - Weasels, foxes, and snakes are in competition for frogs.
   - Weasels and foxes are in competition for mice, frogs, and rabbits.
   - Insects, frogs, salamanders, and centipedes are in competition for insects.
   - Frogs and salamanders are in competition for earthworms.

A Visit to a National Park

The Great Smoky Mountains National Park in Tennessee and North Carolina has a wide variety of animals. Here is a list of some of them and a few of the foods they eat:

Animal	Food
deer	wild apples, poplar twigs
porcupine	inner bark of spruce and pine trees
mouse	grain, seeds of wild apples
rabbit	grass, twigs
red squirrel	seeds from pine and spruce trees
weasel	mice, frogs, rabbits
fox	mice, frogs, rabbits
bear	wild apples, blueberries, mice
insect	grass, leaves, smaller insects
snake	frogs, salamanders, mice
frog	insects
hawk	snakes, weasels, mice, red squirrels
centipede	insects, worms
sowbug	dead materials from all organisms
earthworm	dead materials from all organisms
salamander	worms, insects

- Sowbugs and earthworms are in competition for dead materials from all organisms.

You may wish to point out that even though the porcupines and red squirrels both depend on pine and spruce trees for food, they eat different parts of the trees and, therefore, are not in competition.

3. Answers may vary; possible food chains might include:
   - grass → rabbit → weasel → hawk
     The hawk is the top carnivore.
   - grain → mouse → snake → hawk
     The hawk is the top carnivore.
   - seeds of wild apples → mouse → bear
     The bear is the top carnivore.

Construct a food web for these animals and their food sources, and then answer the following questions.

1. Name the organisms that fill each of these niches: producer, herbivore, carnivore, omnivore.
2. Which animals are in competition for the same food? Name the animals and their food.
3. Construct a food chain consisting of four links in which the final link is the top carnivore in the food web. Which animal is the top carnivore? Does it have any enemies? Explain.
4. What would happen to your food web if no more frog eggs hatched?
5. The environment containing your food web is under attack! There is an enormous increase in the population of the spruce budworm. In no more than six lines, write a report on what immediate effect this increase will have on the food web.

Assessment

1. Have students use pictures from magazines along with their own drawings to make a mural or bulletin-board display of a forest community. They should identify producers, consumers, herbivores, carnivores, omnivores, scavengers, decomposers, predators, and prey in the community.
2. Suggest that students make mobiles to show food chains. The top layer in the mobile should represent the top carnivore. The second layer should represent other secondary consumers. The third layer should show primary consumers. The bottom layer should show producers.

Extension

Suggest that students visit a park, or some other undeveloped area, and take notes on the plants and animals that live there. They should use their notes to draw a food web of the community.

Students should recognize that top carnivores do not have enemies because there are no predators to prey on them. However, humans are always a potential threat to top carnivores.

4. If no frogs' eggs hatched, the weasel, fox, and snake would lose a source of food. There would also be fewer animals eating insects, and the insect population would increase.
5. An enormous increase in the population of the spruce budworm would upset the balance of the entire food web. For example, the spruce trees would begin to die, and red squirrels and porcupines would have less food to eat. As a result, squirrels and porcupines might die. If the population of red squirrels declined, the hawks would have less food to eat. If the hawk population declined, the populations that it preys on would begin to increase.

The Players and Their Roles

LESSON 5

✦ Getting Started

In this lesson, students explore the effects of temperature, moisture, and light on living things. Students examine ways in which animals and people adapt to these abiotic changes in their environment. The lesson concludes by having students track seasonal changes in the abiotic factors of the environment.

Main Ideas

1. Every plant and animal has a specific range of tolerance for the abiotic factors in its environment.
2. Temperature, water, and light are significant abiotic factors for living things.
3. Animals and plants have many adaptations that enable them to withstand extremes in temperature resulting from seasonal changes.
4. Humans can survive much longer without food than they can without water.
5. Light affects growth, health, color, and structure of plants.

✦ Teaching Strategies

Ask students how they feel if they are trying to study and it is too hot or too cold. *(If it is too hot, they feel sleepy. If it is too cold, they shiver. In either case, it is hard to concentrate on studying.)* Point out to students that they, like other living things, respond best when the abiotic factors in their environment are just right.

5 The Nonliving Players

Tolerance

If you were wearing a sweatshirt in your classroom, and the temperature rose above 30°C, how would you feel? What about if the temperature fell below 10°C?

There is a range of temperatures that you can tolerate comfortably. It is called your *range of tolerance for temperature*. Predict what your range of tolerance for temperature would be for the place where you are now.

Within the range of tolerance, there is a temperature that you like best—one that is most comfortable for you. That temperature is called the *optimum temperature*. What would you guess the optimum temperature is for doing science in a classroom?

You have a range of tolerance for other conditions too. What might some of these conditions be? You and all living things are affected by conditions in the environment. At the beginning of this unit, you identified the nonliving, or abiotic, parts of the environment, such as temperature, moisture, light, and wind. How do these factors affect various living things? For each of the abiotic factors, how does the range of tolerance vary for different organisms? You will investigate these questions in the next few pages.

Temperature Tolerance in Plants

How well do seeds grow at different temperatures? Try the following Exploration to find out.

LESSON ORGANIZER

Objectives

By the end of the lesson, students should be able to:
1. Identify water, light, and temperature as important abiotic factors for living things.
2. Explain what is meant by the "range of tolerance" for a given abiotic factor in the environment.
3. Provide examples of how plants and animals respond to changing abiotic factors in the environment.

Process Skills

observing, designing experiments, drawing conclusions, reading a graph

New Terms

none

26 Interactions

EXPLORATION 3

Too Hot or Too Cold?

You Will Need

- 30 radish seeds
- 3 Petri dishes
- 3 pieces of cardboard
- paper towels
- a thermometer
- water
- scissors

What to Do

1. Cut four pieces of paper towel into the shape of a Petri dish and layer them in the bottom of the dish. Do this for the other Petri dishes.

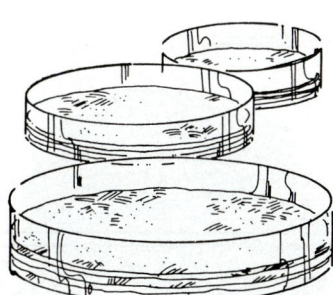

2. Now pour water into each Petri dish until the layers of paper towels are soaked. Pour off any extra water.

3. Put 10 radish seeds on the paper towels in each dish. Spread out the seeds so that they do not touch each other.
4. Cover each Petri dish with a piece of cardboard.
5. Put one Petri dish in a refrigerator; place another in a warm location (such as near a radiator or in an incubator); place the third dish in the classroom in a safe place where it won't be disturbed.
6. Measure the temperature at each place.
7. Over several days, count the number of seeds that sprout each day in each Petri dish. Add water as needed. In your Journal, make a table like the one shown below and record your observations.

Day	In Refrigerator ____°C	In Warm Place ____°C	In Classroom ____°C	No. of Seeds Sprouted
0				
1				
2				

Judging from your results, at what temperature do radish seeds sprout best? How would you redesign this experiment to find the range of temperature tolerance shown by radish seeds?

Tolerance

Have students read page 26 silently. Then involve them in a discussion of the temperature range that they find to be the most comfortable. *(A temperature range of between 20° and 25°C is comfortable for most people. A temperature of around 22° or 23°C is considered optimum.)* Help students recognize that a comfortable temperature often depends on the type of activity they are doing.

Exploration 3

Before students begin this Exploration, a refrigerator and an incubator should be available for use. Have students work in small groups to complete the activity. Observations should be made daily over a period of four or five days. Remind students that to make this a fair test, they need to keep all the variables constant except for temperature.

Answers to

In-Text Questions

- Students should observe that the radish seeds germinate best in a warm place and worst in a cold place.
- Involve students in a discussion of how they would redesign the experiment. Help them to recognize that they would have to try germinating seeds at several different temperatures in order to find the range of tolerance.

Materials

Exploration 3: 30 radish seeds, 3 Petri dishes, roll of paper towels, 3 pieces of cardboard, water, thermometer, scissors, refrigerator, incubator, Journal
Exploration 4: radish seeds, soil, containers, water, graduated cylinder or measuring cup, various kinds of seeds
Abiotic Factors and the Seasons: Journal

Teacher's Resource Binder

Resource Worksheets 1–5, 1–8, and 1–9
Activity Worksheets 1–6 and 1–7
Graphic Organizer Transparencies 1–5, 1–6, and 1–7

Science Discovery Videodisc

Disc 2, Image and Activity Bank, 1–5

Time Required

four class periods

Other Projects

Students may complete the projects individually or in pairs. Point out that they should take careful notes on their procedures and observations. Then students should use their notes to write a report that responds to the project questions.

1. Students will need to keep track of daily and nightly temperatures in order to make their charts. Areas that do not have cold winters usually have dry ones, and seasonal changes can be observed in plants. Help students adapt the activity to the climate in their area.
2. Suggest that students organize their responses to the first two questions in a chart, and then write a short paragraph in response to the third question. Temperature tolerance of different plants varies greatly.
3. Provide seed catalogs for students, or suggest that they contact a local nursery to find out the information that they need. To extend the activity, have students provide additional information about the fruit trees that will grow in their area.
4. You may wish to suggest some areas for students to use in their comparisons, such as forest, mountain, semi-tropical, temperate, coastal, and so on. Plant adaptions make plants well suited to where they live.

Temperature Tolerance in Animals

Suggest that students look in almanacs or contact the National Weather Service to find out the hottest and coldest temperature ever recorded in their area.

Help students recognize that physiological responses to temperature, such as shivering and sweating, are ways that living things adapt to extremes in temperature.

You may wish to have students work on the questions in pairs or small groups. Each group should complete all five questions.

Other Projects

To find out more about temperature tolerance in plants, select one of these research projects. Keep in mind that the first two need to be investigated at a specific time of year.

1. During the fall season, observe which plants survive the cold nights. Which plants survive frost the longest? Which plants do you think will survive the winter? Make a chart showing temperature and plant survival.
2. A spring follow-up to the fall activity: Which plants are the first to grow in the spring? Where are these plants located? What have you discovered about the temperature tolerance of different plants?
3. You are interested in growing a fruit tree in a planter or in your yard. Look in some seed catalogs to determine what kinds of fruit trees will grow successfully in your area.
4. Compare plants that are native to your area with plants that are native to an entirely different type of habitat. For example, if you live in a desert-like environment, you might want to compare native plants to the plants in a tropical rain forest. What makes plants well suited to where they live?

Temperature Tolerance in Animals

"Today's a real scorcher!"

What is the hottest temperature ever recorded for your area?

"Snowstorm warnings in effect. School's canceled!"

What is the lowest temperature you are likely to experience where you live?

What, then, is the range of temperatures your body has to endure? How do you prepare yourself for such temperature extremes? The most obvious answer is "by the clothes I wear." Can you think of some body responses that help you tolerate different temperatures? What are some examples?

How do other living things tolerate temperature extremes? To help answer this question, form small groups of three or four and choose at least two of the following questions or activities. You may need to look in books on the subject and ask informed people, such as zookeepers or veterinarians.

1. Discuss and list all the possible hardships experienced by living things (both animals and plants) in the winter. Check the ones that you think are the most stressful.

Answers to
In-Text Questions

1. Answers may vary, but possible responses include: finding food, keeping warm, finding shelter, and hiding from predators. Of these, finding food is probably the most difficult for animals. Protection may be the most difficult for plants.

(Answers continue on next page)

2. List ways in which living things cope with the difficulties you have identified.
3. In the winter, where would you expect to find the following animals: deer, bear, frog, robin, woodpecker, skunk, snake, and squirrel? How does each one survive the winter?
4. Of what value is snow in aiding temperature tolerance in plants and animals? Name some specific examples.
5. Noreen lives in Bermuda, where the temperature never goes below 15 °C. Write a letter telling her about the changes she might see in the animals and plants where you live as winter approaches.

Moisture Tolerance

Moisture is another vital abiotic factor in the lives of living things. Living things have both a range of tolerance for moisture and an optimum amount of moisture they need. Compare the range of moisture tolerance for each plant in the following pairs.

Answers continued

2. Answers may vary, but possible responses include:
   - Animals may grow heavy coats of fur to protect themselves against the cold weather.
   - Some animals add layers of fat in the fall that their bodies use in the winter when food is scarce.
   - Some animals hibernate.
   - Some animals protect themselves from predators by changing color in the winter to blend in with the snowy background.
   - Plants become dormant (alive, but not actively growing) to survive the winter.
3. Deer would be grazing in open fields and pastures. Bears would be sleeping in dens. Frogs would be hibernating in the bottoms of pools, streams, and lakes. Robins would have flown south for the winter. Woodpeckers would be looking for dormant insects in the bark of trees. Skunks would be hibernating in dens. Snakes would be hibernating underground. Squirrels would be looking for the food they stored in the fall. (Animal activities may vary depending upon which part of the country is being discussed.)
4. Snow serves as insulation for many plants and animals, and keeps their temperature above freezing. For example, mice live quite comfortably under the snow. Pond animals survive under the snow and ice where the water is not frozen. Plants are kept from freezing if they are covered with snow.
5. Students' letters will vary depending on the type of winter in their area. However, student responses should indicate an understanding of how plants and animals adapt to the season of winter.

Answers to

Moisture Tolerance

Comparisons include:
- Moss needs a very moist environment, cactus thrives in a very dry climate.
- Both sea grass and ferns need a moist environment.
- Pine trees need a moist environment, sagebrush thrives in a dry environment.

The Players and Their Roles   29

Exploration 4

This Exploration provides an excellent opportunity for students to design and perform their own experiments. This can be done at home or in the classroom. Encourage students to share their experimental designs with you before they begin. Helpful suggestions may be made at that time. If the experiment is to be done at home, write a note to the parents describing the activity so parents can cooperate.

When all the experiments have been performed and conclusions drawn, have students discuss and display their findings. Students should conclude that radish seeds have an optimum moisture need for germination even though they may germinate with varying amounts of water. Also, they should conclude that different kinds of seeds have different optimum moisture levels for germination.

Who Needs Water Most?

As students begin this writing exercise, make it clear that humor, wit, and knowledge are all acceptable in the conversations they create. Though students' compositions will vary, they should indicate a recognition that the fish would suffer the most by the water shortage, followed by the frog, earthworm, snake, and mouse.

EXPLORATION 4

Experimental Design for Moisture Tolerance

Here is a challenge to your experimental abilities. Plan experiments to answer the following questions.

- Is there an optimum amount of moisture needed for radish seeds to sprout?
- Do all seeds have the same optimum moisture level for sprouting?

Show your experimental design to your teacher to make sure that you are on the right track and to check that your plan is safe. If there is time and the materials are available, perform your experiments and find the answers to the questions.

Who Needs Water Most?

An emergency meeting of the animals was called. Present were a mouse, a frog, some earthworms, a fish, and a snake. They met to discuss the long period of drought they had been experiencing that summer. Water levels were very low. It was a crisis for most, but particularly dangerous for some.

Compose a conversation among the animals that shows the relative importance of water to each. Include the amount of tolerance that each has to the water shortage.

Water Needs—Some Questions to Discuss

1. Humans can survive for several weeks without food, but not nearly so long without water. In fact, humans need almost 3 L of water every day! In a hot desert, you would need 11 L a day because of water lost through perspiration. Why is water essential to humans?

2. Even when you eat solid food and drink liquids other than water, you are taking in part of the 3 L of water that you need daily. Suppose you have just finished this delicious dinner:

Quantity	Food	Mass of Food	Percentage of Water in Food	Mass of Water
1 small glass	orange juice	90 g	87%	
1 slice	bread	30 g	35%	
1 serving	corn kernels	85 g	79%	
1 serving	roast beef	100 g	59%	
1	baked potato	100 g	75%	
1 serving	squash	110 g	96%	
2 slices	tomato	30 g	95%	
2 leaves	lettuce	50 g	94%	
1 slice	watermelon	800 g	93%	
1 glass	milk	250 g	87%	

Can you calculate how many liters of water you ate and drank in this meal? This will require you to perform a little arithmetic. First, you will need to calculate the mass of water for each food listed above. Remember that 1 mL of water has a mass of about 1 g.

Answers to

Water Needs—Some Questions to Discuss

1. Water is essential to humans because our cells and tissues are composed largely of water. Water is the major component of blood. Also, it serves as a vehicle in which salts and wastes are eliminated from the body in the form of urine.

2.
- orange juice: 90 g × .87 = 78.3 g water
- bread: 10.5 g
- corn kernels: 67.2 g
- roast beef: 59 g
- baked potato: 75 g
- squash: 105.6 g
- tomato: 28.5 g
- lettuce: 47 g
- watermelon: 744 g
- milk: 217.5 g
  Total mass of water: 1432.6 g
  Total mass of dinner: 1645 g
  Volume of water: 1432.6 mL or 1.4326 L

3. Look at the graph below, and then answer the following questions. The amounts of water given are total starting quantities.
   (a) How long can a human live at 35 °C with 4 L of water? with no water?
   (b) How long can a human survive at 25 °C with 10 L of water?
   (c) What are the two factors on the graph that determine length of survival time?

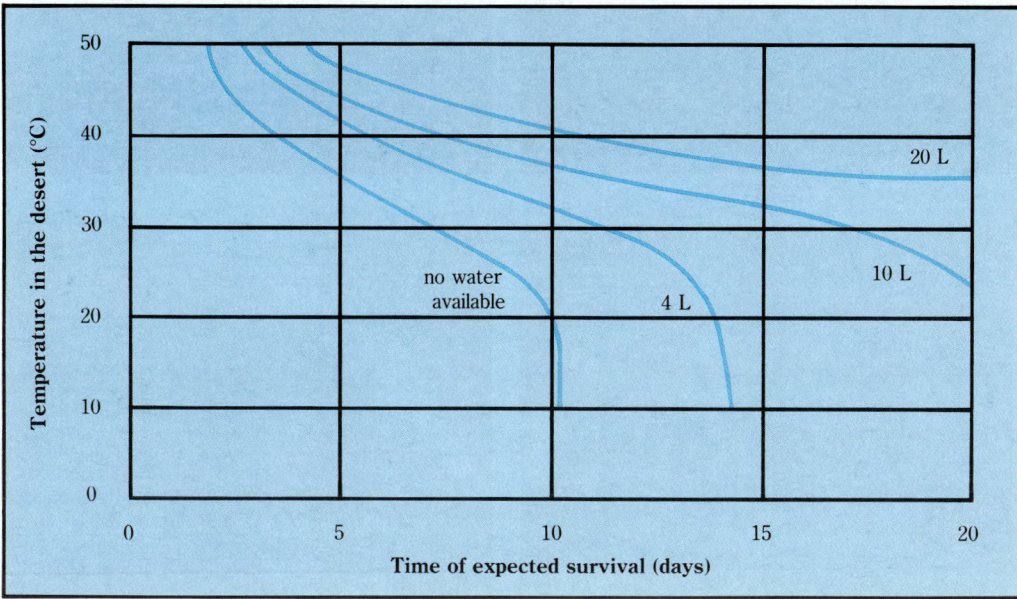

Expected Survival Time of a Human in the Desert

Problems to Ponder

1. Marine fish and sea gulls can drink salt water. Why can't you?
2. In the desert you find plants such as cacti, and animals such as the kangaroo rat, pocket mouse, and desert-dwelling lizards. How do they survive long periods without water?
3. Animals that live along shorelines are covered with water only half the time (when the tide is in). How do they survive those times when the tide is out?
4. Plants need water. What do they use it for?
5. Find out what the average monthly rainfall for your area was last year. Compare it with the rainfall in other parts of the country. On how many days did you have rain? Is there any difference in the amount and frequency of the rainfall in places near the ocean? inland? near mountains? in prairie regions?

---

3. (a) With 4 L of water at 35 °C, a human can live up to 8 days. However, without water, a human can survive for only 5 to 6 days.
   (b) With 10 L of water at 25 °C, a human can survive up to 18 or 19 days.
   (c) The length of survival time is influenced by temperature and the amount of available water.

Answers to
Problems to Ponder

Assign one question to a small group of students for further research.

1. People cannot drink salt water, such as ocean water, because they cannot tolerate high concentrations of salt. To get rid of the excess salt, humans would have to excrete so much water along with the salt that they would become dehydrated. Marine fish drink salt water and then get rid of the excess salt through special salt-secreting cells in their gills. Sea gulls are able to drink salt water because they are able to remove excess salt through salt glands just above their eyes.

2. Cacti survive, in part, because they have tough, rubbery surfaces and tiny leaves called spines that permit very little evaporation. Cacti also store water in their fleshy stems.

   Many desert animals need little water in order to survive. The water that they do need comes from the food they eat. Also, they tend to be nocturnal and search for food only during the night. This minimizes the water loss that would occur during the heat of the day.

3. Animals in a tidal region often reside within hard coverings or shells. They keep some water within their shell, and thereby protect themselves from drying out.

4. Plants use water in several ways. Minerals in the soil need to be dissolved in water before they can be absorbed into the roots. Food and minerals must be dissolved in water to get transported throughout the plant. Plants use water during the process of photosynthesis.

5. An almanac will contain some of the statistics students will need to know. Students' responses will vary depending upon where they live. However, students should discover that differences in the amount of rainfall will occur in all of the areas mentioned in the question.

Light Tolerance

Involve students in a discussion of what they already know about how plants respond to light. For example, most students have probably observed that plants bend in the direction of light. Most will know that plants need light in order to grow. Some students may remember from earlier studies that plants need sunlight in order to make food.

Direct students' attention to the pictures on page 32. Call on a volunteer to identify each one, and involve the class in a discussion of the questions.

Answers to
In-Text Questions

In response to the first question, students should recognize that the sunflowers (a) and the orange trees (d) need the most light. Ferns (b) and mushrooms (c) thrive in the shade.

If the plants were arranged in order from those that need the most sunlight to those that need the least sunlight, this order would be: sunflowers, orange trees, ferns, and mushrooms.

In response to the third question, students should conclude that (e) shows the plant at dawn because its leaves are folded in a protective fashion, and (f) shows the plant at noon because its leaves are opened to receive the sunlight.

(Answers continue on next page)

Light Tolerance

Which of the organisms shown in (a) through (d) need the most light? the least light?

Which of the pictures below was taken at dawn, and which was taken at noon? Explain.

32   Interactions

Light, especially sunlight, clearly plays a major role in the growth of plants. Can you explain the following situations?

- Geranium plants stored in the basement over the winter develop very long stems.
- The grass under a garbage can becomes yellowish.
- Most woodland flowers grow quickly and bloom very early each spring.
- Plants that grow in a forest under the branches of trees are small and scarce compared to those in open spaces.
- Moss tends to grow more abundantly on the north side of trees than on the other sides.

What conclusions can you draw about plants and light from your explanations for these interesting facts?

Some Light Problems to Investigate

The following questions can help you learn even more interesting facts about living things and light. Work in groups to find the answers.

1. Some animals are described as "nocturnal." What does this mean? What would you say about their range of light tolerance? Make a list of all the nocturnal animals you can think of. Then discover what the opposite of nocturnal is. Name some animals of this kind.
2. Find out how the amount of light affects each of the following:
   (a) Bats
   (b) Houseplants
   (c) Migrating animals and birds
   (d) Seed germination (sprouting)
   (e) Poinsettias
   (f) Human beings
3. Observe the changes that occur in the leaves, stems, and flowers of plants at different times of the day, evening, and night. Record your observations.
4. Are there animals that can get along without any light at all? Research information about animals that live in the ground and in the depths of the sea.
5. How does the amount of light affect the activity of different kinds of insects? If possible, observe the activities of some insects at different times of the day.

Answers to

Some Light Problems to Investigate

1. *Nocturnal* animals are active at night. They have a very limited tolerance for light. Some nocturnal animals include: bats, skunks, raccoons, owls, and moles.
   Animals active during the day are called *diurnal*. Diurnal animals include: squirrels, groundhogs, and most small birds.
2. (a) Bats are nocturnal. They can see, but they do not have good eyesight. They prefer darkness and rely on other senses for obtaining their food.
   (b) House plants develop pale, elongated stems if they are kept away from light. Also, they turn their leaves toward light.
   (c) It is possible, but not confirmed, that shorter days with fewer hours of daylight help stimulate animals to migrate.
   (d) No, light or the absence of light is a requirement for the sprouting of many kinds of seeds.
   (e) When the hours of daylight are restricted, poinsettia leaves begin to turn red.
   (f) Current research indicates that extended periods of low light can cause depression.
3. Students' responses will depend on the kinds of plants they observe. For example, some plants fold their leaves at night and open them in the morning. Many plants close their blossoms at night, and open them in the morning. Some, like the morning glory, close them during the day.
4. Students should discover that animals that live in the deepest parts of the oceans, in caves, and underground survive without any light at all. Encourage students to discover how these animals have adapted to living in a lightless environment.
5. Students' responses will depend on the insects that they observe, but they should conclude that light does affect insects in many ways. For example, many insects, such as crickets, begin their calls at dusk. Fireflies come out after dark. Bees, on the other hand, depend on daylight to locate flowers with nectar.

Answers continued

- Geranium plants in a basement respond to the lack of light by making longer stems and spreading out. This helps them gather the maximum amount of light.
- Grass requires light to remain green. Once the garbage can cuts off the light, the leaves become yellowish-white due to a loss of chlorophyll.
- Trilliums, violets, and other spring woodland plants develop and bloom early because sunlight is readily available on the forest floor. Once the trees produce leaves, the forest floor becomes shaded. As a result, the growth and development of many of the smaller plants that live on the forest floor slows down.
- Very little light penetrates beneath the branches of forest trees. As a result, very few types of plants survive on the forest floor.
- The north side of a tree receives less sunlight than the south side, so plants that require less light and more moisture grow there. Moss is an example of such a plant.

After discussing the questions, students should conclude that light is very important in the way plants grow and respond to their environment.

The Players and Their Roles

Abiotic Factors and the Seasons

Involve students in a discussion of the seasonal changes that occur in their area. List key words and phrases on the chalkboard, such as cold, snowy winters; warm, rainy summers; cool, dry falls. Help students to recognize that these represent abiotic factors that affect living things in their environment. Have students read the page and review the chart. When students have completed their charts, involve them in a discussion of their ideas and responses. Help resolve any differences of opinion.

Follow Up

Assessment

1. Have students work in small groups to make bulletin board displays to illustrate how abiotic factors can affect living things. Suggest they use pictures from magazines, drawings of their own, original poetry, and writings to show these abiotic factors.
2. Have students make cartoons to illustrate how temperature, moisture, and light affect plants and animals. Suggest that they use the cartoon on the first page of the lesson as a model. Have students share their work by displaying it in the classroom or organizing it into a booklet for others to read and enjoy.

Extension

1. Point out to students that naturally occurring phenomena, such as volcanoes, and events triggered by human actions, such as oil well fires, can affect the amount of sunlight that reaches the earth. Suggest that they do some research to find out what other kinds of events, both natural and human-made can affect abiotic factors in the environment. Have them share what they learn.
2. Introduce students to the term *phototropism* (a plant's reaction to light). Suggest that they do some research to learn more about phototropism, and other kinds of tropisms that are exhibited by plants and animals. Have them share what they learn by making fact sheets.

34 Interactions

Abiotic Factors and the Seasons

Although you may not have been thinking about it, you've been reading about some of the effects that seasons have on abiotic factors. For example, the temperature decreases in winter and increases in summer. There is more light in summer than in winter. The seasons also affect the amount of moisture available.

To help you track seasonal changes in abiotic factors, create a chart like the one below and complete it in your Journal. Note that two abiotic factors that you haven't discussed have been added to the chart. As you fill out the chart, also think about the biotic factors that change as the seasons change.

Place (your choice):				
Abiotic Factors	Spring	Summer	Autumn	Winter
Temperature				
Moisture				
Light				
Wind				
Soil				

Challenge Your Thinking

1. Going in Circles
   (a) What abiotic or biotic part of the environment might "?" represent?
   (b) Tell the story of the dotted line.
   (c) Six lines originate at "tree." What might the interactions be?
   (d) "Grass" has two lines leaving it. How many more lines could you put in?
   (e) Suppose a line joined "Mike" and "tapeworm." How would you describe this interaction?
   (f) Locate the following:
      • A predator–prey interaction
      • A relationship involving a carnivore
      • A relationship involving a herbivore
      • An interaction involving a scavenger

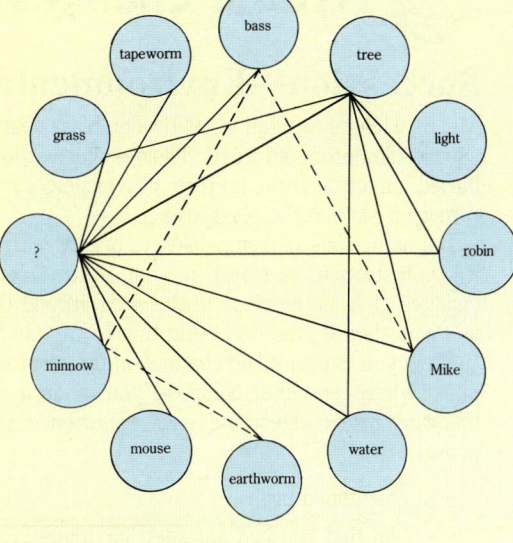

2. Imagination Please!
   Write a haiku describing any interactions that occur in nature. A haiku is a Japanese poem that has three lines of 5, 7, and 5 syllables, respectively. Ask a classmate to read your haiku to discover what type of interaction you've described.
   Here are some examples of haiku.

Hummingbirds hover
To drink the orchids' nectar   (Do you recognize
And to pollinate.   mutualism?)

Crows like it, owls don't.
Plants cannot live without it,   (What is it? Hint:
Though bats will shun it.   It's an abiotic factor.)

Answers to Challenge Your Thinking

1. The answers to *Going in Circles* are listed below.
   (a) The question mark might represent air. The large number of interrelationships that it has indicates that the missing part is critical to all other items on the chart.
   (b) The dotted line tells the story of a food chain—earthworm, minnow, bass, Mike.
   (c) Interactions may be the following:
      • Tree and grass—the shade of the tree casts a shadow on the grass and limits its sunlight.
      • Tree and water—water is absorbed through the roots along with dissolved minerals.
      • Tree and air—leaves use carbon dioxide in the air to produce food.
      • Tree and Mike—Mike may use the tree for recreation, wood, or shade.
      • Tree and robin—the tree provides a habitat for the bird.
      • Tree and light—the leaves use sunlight to make food.
   (d) Lines could be drawn from grass to:
      • Air—grass helps produce oxygen.
      • Mouse—grass provides protection from predators.
      • Earthworm—grass decays and makes food for the worm.
      • Water—grass helps absorb water and prevent erosion.
      • Light—grass requires light for photosynthesis.
   (e) parasitism
   (f) • predator-prey: robin and earthworm
      • carnivorous relationship: bass and minnow
      • herbivorous relationship: mouse and trees (seeds)
      • scavenger: earthworm and grass
2. Answers will vary. Students should emphasize interactions between living organisms and their environment. The answer to the second riddle is *sunlight*.

The Players and Their Roles

LESSON 6

✷ Getting Started

This lesson explores some of the changes that occur within a biological community. It begins by introducing students to the concept of biological succession. In *Exploration 5,* students identify changes within a community by examining a rotting log. The lesson continues by looking at some of the causes, both natural and human-made, of population changes within communities.

Main Ideas

1. Changes are constantly occurring within biological communities.
2. Nature can restore itself to its original state through succession.
3. Human actions may adversely affect the organisms that make up biological communities.
4. Changes in the population of one organism are often reflected in the populations of other organisms.

✷ Teaching Strategies

Succession— Environmental Change

Before students begin reading, direct their attention to the photograph and have them identify what it shows. *(Mt. St. Helens erupting in 1980.)* Involve students in a discussion of how a volcanic eruption would affect the communities of living things around it.

Have students write brief descriptions (poems or prose) about one of the locations listed. Invite several volunteers to read their descriptions to the class. Help students recognize that over **(Continues on next page)**

PUTTING IT ALL TOGETHER

6 Natural Changes

Succession—Environmental Change

Mt. St. Helens erupted in 1980. The blast destroyed hundreds of square kilometers of forest. Hillsides were stripped clean and then buried under several meters of volcanic ash. Almost nothing remained alive in the blast area.

But nature has a strong ability to preserve its original state and, if it is disturbed, to return to that state. Over the years, many changes have taken place in the area around the volcano. Plants and animals are gradually returning to the charred land.

Have you ever noticed changes in the plant and animal communities where you live? Suppose you made a visit to one of the following places. Describe your experience in either poetry or prose.

- an abandoned farm
- a pond that you had not seen for many years
- a place where loggers had cleared trees a few years before
- a garden left unattended for several weeks
- a field left unplowed for several years
- an abandoned lot

Changes in which one group of organisms replaces another may take days, weeks, years, or centuries, depending on the type of environment that is changing. All of the changes that occur in an area are collectively known as **succession.** The photos below illustrate succession in a pond community. During succession, one group of organisms follows, or *succeeds,* another. Succession is a slow, natural process. If possible, study an area where succession is taking place. Make regular trips to the area on your own, and record any changes you find in the plant and animal life there.

LESSON 6 ORGANIZER

Objectives
By the end of the lesson, students should be able to:
1. Explain what is meant by biological succession.
2. Describe changes that take place in biological communities.
3. Explain how changes in the population of one organism may affect the populations of other organisms.
4. Describe the characteristics of a specific biological community.

Process Skills
observing, interpreting graphs, collecting and organizing data, drawing conclusions

New Terms
Succession—changes in an area that cause one group of organisms to be replaced by another.

36 Interactions

a b

c d

Can you observe succession occurring in these scenes?

(Succession continued)
a period of time, changes occur in biological communities. If students have not had the opportunity to visit one of the places listed on page 36, involve them in a discussion of the changes they think might occur in each location. For example, an abandoned farm might become overgrown with weeds, shrubs, and fast growing trees. Review the definition for the term *succession,* and have students name areas where they have seen succession occurring. As suggested on the bottom of page 36, invite students to make a trip to an area where succession is taking place, and to record the changes that they observe.

Answer to caption: Direct student's attention to the pictures on page 37. Help students recognize that successive communities of plants help restore an area to its original condition. In the example pictured, the badly burned forest is replaced by grasses, weeds, and small shrubs. Eventually, these will be replaced by larger shrubs and fast growing trees. Finally, those will give way to a dense forest similar to the original one.

Materials

Exploration 5: field guides for identifying organisms, paring knife, Journal, alcohol thermometer, magnifying glass, trowel, white plastic tray

Exploration 6

A Woodland Community: field guides for identifying organisms, Journal, metric ruler or meter stick

A Forest-Floor Community: field guide for identifying organisms, metric ruler or meter stick, unlined paper, Journal, newspaper or white plastic tray, magnifying glass

A Pond Community: field guides for identifying organisms; plastic bottles, containers, and plastic bags for taking samples; Journal; long-handled "muck scoop" (kitchen strainer); white plastic tray; glass jar or aquarium (optional); pump (optional)

Science Discovery Videodisc

Disc 2, Image and Activity Bank, 1–6

Time Required

five to six class periods

Putting It All Together

Exploration 5

This discovery activity provides students with an opportunity to examine the biotic and abiotic changes occurring in a rotting log, how these changes are brought about, and their effect on the organisms that live there. You may wish to visit the area before the field trip in order to locate suitable logs for students to study and to determine the length of time that should be allotted for the activity. (Be sure to get the necessary permission to use the land you have selected for the field trip, and make the necessary arrangements with your school administration.) If a field trip is not possible, bring one or two rotting logs into the classroom. Students can complete the activities for Zones A, B, and C.

The day before the field trip is to take place, provide some classroom time for students to discuss what they will be doing. Ask them to make some predictions about what they think they will observe. Then divide the class into four groups, and assign each group one zone to study. Provide time for each group to read and discuss the instructions. You may wish to have students copy the instructions in the textbook into their Journals so they can use their notes as a reference during the field trip.

Once students have been given a log to study, allow them to work on their own. Be available to answer any questions and to make procedural suggestions. For each zone, students should record abiotic factors such as temperature, light, and moisture. Students can then use these factors to determine the tolerance ranges of the organisms found in each zone. Remind students that they should keep careful notes and make sketches of their observations in their Journals.

Zone A

Caution students not to sit on the log. Suggest that they determine if the kinds of organisms living on the top of the log are the same as those living near the bottom of the log. Encourage students to identify how the micro-environment of the bark changes from place to place. For example, the bark on top of the log may be drier than the bark near the ground; there may be more dirt in the bark near the ground than in the bark on the top of the log.

38   Interactions

EXPLORATION 5

Field Trip to a Changing Community

Where?
A rotting-log community in a nearby wooded area.

When?
Before the ground freezes. It may take quite a few hours to complete your visit.

Why?
You are going to discover what goes on in a fallen tree—who lives there, what work they do, and what they eat.

You Will Need
For each group:
- a field guide
- a paring knife
- your Journal and a pencil
- a thermometer
- a magnifying glass
- a trowel
- a white plastic tray

What to Do

1. Choose one or two well-rotted logs to examine.
2. Approach the log slowly and quietly. You might not be the only visitor! Look carefully for signs of animals that may be using the log.
3. Divide into four groups, with each group examining just one zone of the rotting-log community.
   Zone A: Bark
   Zone B: Wood beneath bark
   Zone C: Bottom of log
   Zone D: Soil beneath log
4. For each zone, describe and record the abiotic factors of temperature, light intensity, and dampness. For Zone D, investigate the top 6 cm of soil beneath the log.
5. For each zone, consider the tolerance ranges of the organisms for the abiotic factors listed in step 4.

Zone A
Bark

Describe the size, shape, and color of any fungi on the log.

Describe any green plants growing on the bark. The plants could be mosses, ferns, or tree seedlings.

Examine the bark closely for lichens. With a magnifying glass, observe the lichens, and make a drawing of what you see. Remember, this organism has two parts: a fungus (consumer) and an alga (producer). Since these are not visible without a microscope, you may wish to review the diagram of the lichen on page 13.

Look carefully for animal life or evidence of life (such as spider webs, animal droppings, and seed shells) on the bark. Are there holes in the bark to suggest animal life beneath?

Zone B
Wood Beneath Bark

Carefully remove some of the outer bark. Use a paring knife to break apart the wood underneath. With your magnifying glass, search it carefully for any small animals that make their home in the log. Try to identify as many as you can. Draw those you cannot identify. Do plants grow in this zone? Why or why not?

Zone C
Bottom of Log

Examine the bark and wood here carefully, as suggested for Zones A and B. Record all plant and animal life that you see.

Does life on the bottom of the log differ from that in Zones A and B? Why? Carefully roll the log away, if possible, and examine any life on the ground under it. You might find snakes, salamanders, toads, insects, or evidence of other animals (tunnels, pathways, food).

Zone D
Soil Beneath Log

How does the vegetation growing next to the log differ from that growing on the log? Push a trowel down 6 cm into the soil beneath the log. Place some of this soil on a white plastic tray and carefully observe any life. List the different kinds of organisms and types of movement you see.

Zone B

Have students remove the bark very slowly so as not to disturb any organisms living under it. Instruct students to watch carefully as the bark is removed so that they may catch a glimpse of any organisms that quickly scurry away. Caution students to use the paring knife carefully. Students should realize that green plants cannot live under the bark because of the absence of sunlight.

Zone C

Students should carefully roll the log over in order to examine the bottom of it. Suggest that they move the log as little as possible. Instruct students to be on the watch for any organisms that may quickly dart away as the log is moved. Students should recognize that many of the organisms that live here are adapted to high levels of moisture, no light, and prefer to live in the soil. They may, however, find some organisms like those in Zones A and B, such as fungi and certain types of insects.

Zone D

Have students carefully examine the ground beneath the log before they disturb it in any way. Then suggest that they take their soil samples from beneath the log, place them on white plastic trays or sheets of paper, and carefully examine them with a magnifying glass.

Students should observe that most of the plants growing next to the log have roots that extend down into the soil. Most of the plants growing on the log do not have similar root systems. They should conclude that the plants growing on the log get their nourishment from the log itself. The plants growing next to the log get their nourishment from the soil beneath the log.

Putting It All Together

Some Things to Think About

1. A tree that falls may have a mass of many hundred kilograms. The tree will eventually disappear, leaving no visible trace. What will happen to the organisms that once lived on or in the tree? the log?
2. It takes many years for a tree to rot away completely. What changes in plant and animal life might occur during this time? Can you identify fallen trees in different stages of decay?
3. If snow falls and covers the log, what will happen to the life in and on the log? What effect do you think winter has on the decaying process?
4. Compare the tolerance ranges of the organisms living in the different zones of a rotting log. Do all the organisms on the log need to tolerate the same range of conditions? Explain.
5. In nature, destructive processes such as the rotting away of trees are balanced by constructive processes. In other words, forces that renew life are constantly at work in the rotting log. How many plants and animals can you identify that have benefited from the rotting away of the log?
6. Write an article for a class newspaper explaining the niche of an animal that is associated with the rotting-log community.

In your study of the rotting-log community, you saw that the log was home to many organisms. These plants and animals used the log for shelter as well as for food. Other organisms that you didn't see—decomposers—were also present. Think back to what you read on page 19 about compost piles. Decomposers were present there, too, using the compost for food and breaking it down so that other organisms could use it for food. The decomposers in the rotting log were doing the same thing.

When the remains of living things are used in this way to help new life develop, the materials are being *recycled*. People, too, are learning to recycle many of the things they need and use. You can see many examples of recycling going on constantly in the natural world.

The matter from this decomposing animal is being recycled by the action of decomposers.

Answers to
Some Things to Think About

1. Organisms that lived in the tree will try to find a new place to live; some of these will die from exposure to predators, no location to reproduce, or lack of food. Organisms on the rotting log will have a similar experience.
2. As a tree rots, food supplies change. The bark and the wood may loosen, allowing beetles, ants, slugs, and worms to make their homes there. Plant life could send down roots into the newly formed spaces. The following is an example of a succession that could take place in a decaying tree:
   - At first, the bark and wood remain solid. Lichens may grow on the bark and small rodents may still live in the tree.
   - Next, the bark loosens. Bacterial action begins to soften the wood under the bark. Small insects burrow into the bark and wood. Lichens and mosses cover much of the bark.
   - The wood begins to break down and decay. The evidence of burrowing organisms becomes more apparent. The roots of grasses and small trees penetrate the wood, causing it to break apart.
   - Finally, the log becomes noticeably smaller. The wood becomes very soft. Numbers of insects, spiders, and centipedes increase. A variety of plant life grows on the log.
3. In winter, most plant life and fungi become dormant, or produce seeds or spores that will germinate in spring. Most animals either hibernate, migrate, or die. Many leave eggs to hatch in the spring. The decaying process slows in the winter due to the cold temperature, reduced availability of moisture, and dormant plant-life.
4. Ranges of tolerance for temperature and moisture are difficult to determine. Students might hypothesize, however, that there are more extremes of temperature and moisture in Zone A than in the other zones. For example, the moisture and temperature levels in the soil under the log would not change as much as that on the surface of the log.
5. Answers may vary, but possible responses include: lichens, mosses, small plants, ferns, snails, ants, salamanders, slugs, rodents, snakes, beetles, earthworms, spiders, moles, and centipedes.
6. Content will vary depending on the animal selected by the student. However, articles should demonstrate an understanding of how the animal fills the niche in a rotting-log community. Consider having students publish their articles in a classroom or school newspaper.

Population Changes

You know that a natural community contains a large number of niches for producers and consumers. Environmental changes that upset the producer-consumer relationship often change the natural balance within a community. Sometimes this can create serious problems.

Many settlers who came to the United States wanted to farm the land. Before they could plant crops, though, they had to cut down tracts of forest. What do you think happened to the many animals and other living things that were dependent on the forest? What would have happened if all the forests had been cut down? Do you think the settlers' fields provided a home for any living things?

Today, in many parts of the United States, forests and fields are being destroyed to make way for large real estate developments. As human communities grow larger, the communities of other living things get smaller.

However, some changes in a community occur because of what seem to be unrelated events. Consider this statement:

The number of mice in a community indirectly affects the number of rabbits in the same community.

(Population Changes continued)

Have students read the paragraphs at the top of the page, pausing at the population graph. Involve them in a discussion of what they have just read. You may find the following analysis helpful in guiding the discussion:

- Because of the increased food supply, the mice will thrive and reproduce. The immediate result will be an increase in their population.
- Because the conditions affecting the rabbits have not changed, there will be no initial effect on the rabbit population.
- Because of the increasing mouse population, it is easier for their predators, the hawks and foxes, to see and catch them.
- Because the mice now make up the major part of the hawk and fox diet, the rabbit population begins to increase. Their predators, also the hawks and foxes, are having an easier time catching mice.
- Therefore, changes in the mouse population indirectly affect the rabbit population.
- When the population of rabbits begins to increase, they are more easily seen and caught by predators than the mice. Now the pressure is off the mouse population, which had been in decline because of excessive predation and lack of food. The mouse population begins to return to its original size. Because of continued predation, the rabbit population begins to return to its original size.

Direct students' attention to the graph. Ask them to explain what it shows *(the changes in rabbit and mouse population over a period of time)*. Have students respond to and discuss each of the questions that follow the graph.

How can this be? Do rabbits eat mice, or vice versa? Mice and rabbits are both herbivores. That is, they eat green plants. Suppose that the type of food that mice like to eat starts growing faster than usual in the area. Soon there is more than plenty of food for the mice. There is no increase in the type of food that rabbits prefer.

What immediate effect will this have on the mouse and rabbit populations?

Foxes and hawks eat both mice and rabbits. Mice are the easier prey, so with the increase in the mouse population, the foxes and hawks will eat more mice than rabbits. What is the reason for this change of diet? What will happen to the rabbit population now that the foxes and hawks have changed their diets? Judging from what you've learned, does the mouse population in this community affect the rabbit population?

What do you think will eventually happen in this community? Why do think this will happen?

The changes in population of mice (—) and rabbits (- -) are shown in the graph.

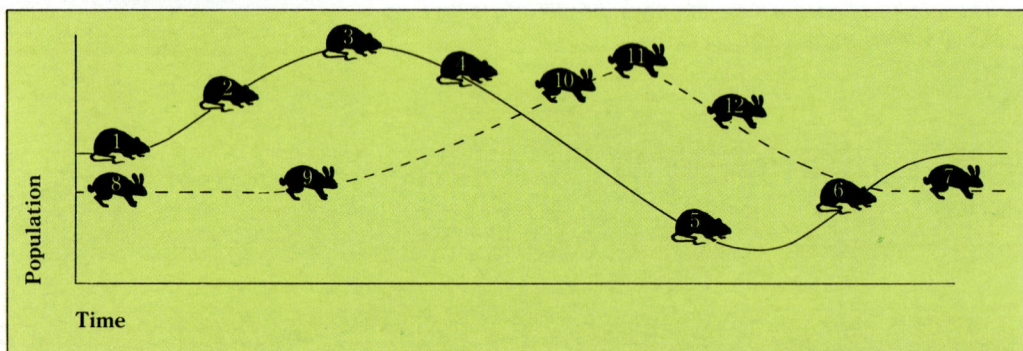

Each of the following events is represented by a number on the graph. Can you match the number to the event?

(a) The original population of rabbits; the largest population of rabbits; the largest population of mice; the smallest population of mice

(b) Mouse and rabbit populations when the increase in plants that mice like is first noticed

(c) The mouse population when they have exhausted their food supply and when they are still being preyed on by foxes and hawks more than rabbits are

(d) The rabbit population when the foxes and hawks begin preying on mice more than on rabbits

---

Answers to

In-Text Questions

a) 8—the original population of rabbits; 11—the largest population of rabbits; 3—the largest population of mice; 5—the smallest population of mice.

b) 2, 9—The mouse population begins to increase much more rapidly than the rabbit population.

c) 4—The mouse population begins to decline.

d) 10—The rabbit population increases dramatically.

e) 12—The rabbit population begins to decline.

f) 6—The mouse population begins to return to normal.

g) 6, 7—Both populations return to where they were at the beginning.

(e) The rabbit population when the mice run out of food, and foxes and hawks are preying on rabbits more than on mice

(f) The mouse population when foxes and hawks are preying more on the rabbits

(g) A return to original population counts of mice and rabbits

Population Changes to Think About and Research

1. The lemming is a small rodent that lives in cold, northern parts of the world. Every few years, according to legend, lemmings leave their habitats and drown themselves. Scientists, however, now know that they do not deliberately kill themselves. What might cause them to leave their homes and die in large numbers?

2. Study these three changes brought about by humans, and then answer the question that follows.

   In 1859, 24 European rabbits were introduced into one area of Australia. These rabbits had no natural enemies in their new environment: there were no predators, parasites, or diseases to affect them. The rabbit quickly became a pest, destroying all plant life in the area. Within a mere six years, over 20,000 of the rabbits had to be killed in an effort to control their growth.

   In 1890 an Englishman brought starlings to North America from Europe. Now the starling, a noisy and destructive bird, is found all over North America. It has displaced many of the native songbirds from their niches.

   In 1869 gypsy moth caterpillars were brought to North America from France by a man who wanted to start a cloth-making industry by breeding the gypsy moth with the silkworm. These caterpillars now kill shade trees by eating their leaves.

   How do some animals get out of control and become pests? Search for other examples of "out-of-place" animals or plants that have become pests.

3. In 1534 Jacques Cartier landed on the east coast of North America and found thousands of flightless birds that were clumsy on land but were powerful swimmers. These great auks soon became regarded as a source of profit. They were used for food; their feathers were used for making featherbeds; and they were even boiled for their oil! As a result, no one has seen a great auk alive since about 1844. Find out more about why and how this happened.

   What other animals have become extinct in the past century? Which are in danger of extinction? Are there also plants that are in danger?

A lemming

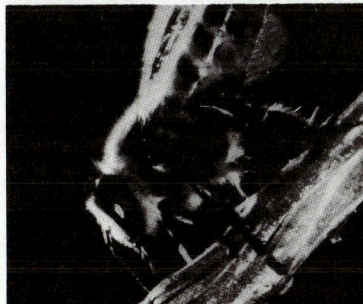

How is a "killer" bee different from a regular bee?

The last of these birds was seen in 1844.

lemmings is too great for the food supply. Thousands of lemmings leave in search of food. Their large numbers cause them to fall victim to exposure and predation.

2. When a new species is introduced into a location, where it has no natural enemies, its population increases unchecked. The resulting effect on native plants and animals may be disastrous.

3. The great auk was hunted to extinction. Eventually, so many auks were killed that there were not enough left to maintain the population—that is, more auks died each year than were born. As a result, they became extinct.

   Some animals that have become extinct in the last century include: the passenger pigeon, Carolina parakeet, Labrador duck, moa, Stellar's sea cow, and a kind of zebra called a quagga.

   Some protection agencies have lists of the plants and animals that are *endangered* in the students area. Well-known endangered species include: the grizzly bear, Florida panther, Carolina northern flying squirrel, red wolf, bobcat, cheetah, Asian elephant, gorilla, leopard, Asiatic lion, giant panda, black rhinoceros, tiger, hooded crane, West African ostrich, California condor, whooping crane, bald eagle, American peregrine falcon, and American crocodile. The many plants in the rain forests of the world are endangered due to habitat destruction. Have students do research to find out what plants native to their own area are endangered.

Answer to caption: Killer bees (Africanized honeybees) are more aggressive, hardier, and reproduce more rapidly than do European honeybees.

Answers to
Population Changes to Think About and Research

Provide time for students to research one or more of the population changes listed.

1. Lemmings are small rodents that inhabit the mountain slopes of Canada and Scandinavia. At certain times, their food supply increases abundantly. As a result, there is a rapid increase in the size and frequency of lemming litters. After several years, the population of

Exploration 6

Ideally, every student should have the opportunity to examine one of the three communities listed in the Exploration. One way to facilitate this is to divide the class into several groups. Assign each group one of the communities to study. After the field trip, provide each group with time to prepare an oral report for the class. Encourage class members to ask questions and to discuss the information that has been presented to them by each group.

The three communities listed in the Exploration are only a representation of the many biological communities that exist. If you live in an area that does not support these communities, substitute others that are more appropriate. For example, you may be in an area where desert, beach, or grassland communities are more dominant than woodland or forest communities. If your school is in an urban area, suggest that students study a park or vacant-lot community.

A Woodland Community

Have students read the material and collect the items they will need before they take the field trip. Suggest that they write down some questions that they might find the answers to. Encourage each group to thoroughly discuss procedures ahead of time.

A Tree Study

Students should easily recognize the difference between coniferous and deciduous trees. Reference books and field guides will help students identify trees by leaf shape and bark characteristics.

Encourage students to draw sketches of leaves, bark, flowers, cones, and seeds for the specific tree they have selected to study. Leaf or bark rubbings are an excellent way of recording information for later research. Students should take notes to accompany any sketches or rubbings that they make.

When examining seeds, have students consider how the seeds are adapted for dispersal. Encourage them to think about the role animals play, especially squirrels and birds, in the dispersal of seeds.
(Continues on next page)

EXPLORATION 6

A Community Study

You have already examined a rotting-log community. What other communities exist in your area? Is there a nearby woodland, pond, or stream? These examples and others offer interesting communities for study. With your classmates, choose a community that you wish to explore. Here are three possible communities for study:

- A woodland community
- A forest-floor community
- A pond community

You Will Need

- field guides for identifying organisms
- your Journal and a pencil
- a ruler or meter stick for making measurements

A Woodland Community

Woodlands provide people with many things: quiet hiking trails, beautiful spots for camping and picnicking, lakes for fishing, and so on. But too often we overlook much of the life a forest contains.

One visit is not enough to investigate all of a forest's mysteries, so this activity might make you want to return for another look.

Always keep in mind that the area you study should be left in at least the same condition as when you arrived, if not better. (Carry out those aluminum cans you find among the bushes!) If you turn over stones, be sure to return them to their original positions. Instead of taking samples of organisms back with you, try sketching them as best you can. If this isn't possible, take one organism, not a handful. Try to obtain samples in a way that won't harm the organism. For example, instead of pulling a leaf from a tree, take a leaf that has already fallen from the tree.

A Tree Study

Decide which group of trees makes up the main population of the area. If the trees have needle-like leaves, produce cones, and keep their leaves all winter, then they are *coniferous* trees. If they have flat, broad leaves, produce flowers, and shed their leaves in autumn, then the trees are *deciduous*. Choose a tree you'd like to study.

Look up the group and the name of the tree you chose in a field guide. When using a reference, pay particular attention to leaf size and shape, to how the leaves are arranged on a stem, and to how many leaves grow from a single stem. The bark can also be useful for identification purposes. Sketch a leaf of your tree, and if you can find fallen leaves, take one with you to help you learn more about the tree later.

Each type of tree within the two main groups has its own bark color and markings. You are probably familiar with the white, papery bark of paper birch trees, for instance. In your Journal, describe and sketch the bark of the tree you chose to study.

If you do your study in spring and are studying deciduous trees, check for tree flowers. Some are colorful and obvious, like cherry or apple blossoms. Others, such as the flowers of the birch and oak, are simple, small, and not at all showy. Sketch the flowers of the tree you've been studying, if they are present. Next look for its seeds on the tree or on the ground beneath. Sketch these too.

If you are studying a conifer, check for cones on the tree or on the ground nearby. Cones are made up of scales, and seeds are attached to these scales. Remove some scales and try to find the seeds or the scars showing where the seeds were once attached. Sketch a cone, a scale, and a seed or scar. Why are squirrels helpful to coniferous trees?

Observe the tree for other kinds of life. Check the bark, the leaves, and the area at the roots. List the organisms you find. Try to discover the relationship between the tree and each organism. Then construct a food chain or food web that shows how the organisms interact.

What influence do you think your tree has on other plants and animals nearby? What other organisms and abiotic conditions does it depend on? Consider its range of tolerance for each of the abiotic factors you studied earlier. Can you discover any examples of mutualism or commensalism between your tree and its immediate environment?

(A Tree Study continued)
Point out to students that a forest can be divided into layers. The upper layer, or canopy, receives the most sunlight and moisture. The amount of growth in the middle layer and in the lowest layer, or undergrowth, is dependent on the thickness of the canopy. The undergrowth is also affected by the thickness of the middle layer, the amount of moisture that penetrates the middle layer, and the soil of the forest floor. Have students examine different areas of the forest floor where the amount of undergrowth varies. They should be able to suggest reasons for the differences they observe.

Students may have difficulty constructing food chains because of an absence of observable wildlife. However, they may be able to infer food chains from the insects and other small animals that they observe on tree trunks and under rocks and logs. Encourage students to look not only for the actual animals, but for the evidence of animals, such as nests, webs, footprints, or droppings. Based on these observations, students may be able to construct a large food web of the woodland area rather than individual food chains.

Answers to
In-Text Questions

- The effect of abiotic factors and tolerance ranges will be difficult for students to measure on a single trip. However, they can observe the effect of moisture and sunlight on the growth of trees. The amount of sunlight also affects the temperature, humidity, and the types of animals and plants that will thrive in a given area.
- Students may observe examples of birds' nests in trees *(commensalism)*, lichens on tree trunks *(mutualism)*, spiders and cobwebs on bushes *(commensalism)*, and bracket fungi on trees *(parasitism)*.

Putting It All Together

A Forest-Floor Community

Have students read the material and collect the items they will need ahead of time. Suggest that they make notes on what they expect to observe. The procedures for this study are somewhat more complicated than the previous one. It is important, therefore, that each group thoroughly discusses procedures before they begin the field trip.

Smaller Plants

Encourage students to be as specific as possible when they complete their maps. Have them begin by noting that the litter that covers the forest floor consists of leaves, needles, twigs, and animal droppings. All of this constitutes the beginning of the decomposition process for the formation of new soil. Students will observe more in the litter if they use magnifying glasses.

There is a tremendous diversity of small plants, fungi, and lichens on the forest floor. For the purposes of this activity, however, naming the plants is not as important as being aware of their presence and recognizing their importance as soil builders and providers of food and shelter.

What's Under the Forest Litter?

Caution students not to injure any animals that they find and examine. Use this as an opportunity to instill respect for all living things. Also, encourage students to handle any animals very carefully since some, such as centipedes, might retaliate with small bites. For this reason, it might be a good idea if students wore gloves for this part of the activity.

If a group finds a salamander, instruct them not to pick it up by the tail. Many students have been left with only the tail in their hand, while the rest of the salamander escapes and later regenerates its lost appendage.

(Continues on next page)

EXPLORATION 6—CONTINUED

A Forest-Floor Community

A whole world exists on the forest floor. Small flowering plants, ferns, mosses, lichens, and fungi can be seen on the ground. Underneath the litter of the forest floor lies a multitude of organisms.

You Will Need

- a meter stick
- your Journal and a pencil
- a newspaper or a white plastic tray
- a magnifying glass

Smaller Plants

Use a meter stick to mark off a 1- × 1-m plot. Examine the area closely and make a map of the plant and animal life you find. Use a scale of 1:5 to draw your map. In other words, your map will be 20 × 20 cm. You can use symbols to indicate each plant species, but be sure to give a key for your symbols. Also note the soil conditions of your sample area.

If the area you examined did not contain any lichens, make a special effort to search for some nearby. Some lichens that are easy to recognize include "British soldiers," with their red tips, or "old man's beard," a pale green, stringy lichen that hangs from tree branches. Other lichens, which grow on rocks, are scaly-looking and gray or orange-red in color. Sketch them, and use a field guide to identify them when you return.

Also look for the many interesting types of mosses. One type has brownish spore cases growing out of the top.

What's Under the Forest Litter?

Mark off a 50- × 50-cm plot on the forest floor. Carefully move and look under the rocks, stones, twigs, leaves, and other debris in the area. You will be surprised by what you find. Record all traces of plant and animal life. Try to construct food chains and food webs for the list of organisms you discover.

Gather some decomposing leaves and spread them on a piece of newspaper or on a white plastic tray. You might find animal life such as millipedes, spiders, centipedes, and salamanders.

Look around you for examples of commensalism, mutualism, and parasitism. Can you see signs of different types of living things co-operating with each other? Keep a record of your observations.

Sit down and look up. What do the plants and animals that live in the top layers of a forest add to the forest floor? What do they take away? How do the large trees affect the abiotic conditions on the forest floor?

Describe the tolerance ranges of the organisms that live in the leaf litter. How do these compare with the tolerance ranges of the organisms that live in the upper levels of the forest? Explain your answer.

(What's Under the Forest Litter? continued)

Have students construct food webs rather than specific food chains. The source of energy, the producers, are still green plants. However, in the forest litter and on the forest floor, the plants are mainly represented by decaying leaves and twigs.

Consumers will primarily be arthropods, such as insects, millipedes, centipedes, sowbugs, and spiders. Millipedes and some insects are the main herbivores. Centipedes, most spiders, and some of the insects are the main carnivores. Larger animals such as salamanders are also carnivores.

Many of the organisms that live in the forest litter and on the forest floor are scavengers, such as sowbugs, slugs, and snails. Most of the parasites are microorganisms such as bacteria. Some mites may attach to other animals in the litter and be parasitic. But, these will be extremely difficult to observe, even with a magnifying glass.

Answers to
In-Text Questions

- Commensalism is easy to observe on the forest floor. For example, many of the animals that live in the litter use leaves and other debris for shelter and protection. Some animals find shelter in the roots of growing plants. Mutualism may be harder to observe. There are examples of insects—often beetles—that cooperate with one another in food gathering and storage, but the likelihood of observing such behavior is remote.
- Plants and animals from the top layers of the forest supply the forest-floor community with leaves, seeds, nuts, and berries, as well as decaying organisms. The organisms from the top layers take away light from the forest floor.
- Sunlight, temperature, and moisture on the forest floor are greatly affected by the surrounding trees.
- The tolerance of organisms for these abiotic factors varies widely in the upper levels of the forest. On the forest floor, light, moisture, and temperature tend to be constant, so that organisms require less of a tolerance range in order to live there.

Putting It All Together

A Pond Community

Have students read the page and collect the items they will need ahead of time. Suggest that they make some notes on what they expect to observe. Have each group discuss the procedures they will use to explore the pond community before the field trip begins. **CAUTION:** Scout the pond community prior to the Exploration to be sure it is safe. Be aware of the bank and pond-bottom conditions and of the location of deep water.

There are many activities that can be done in and around a pond. Looking for animal life may be the most exciting part of the field trip. Encourage students to look for organisms near any decaying organic matter in the pond. Have them try to classify the animals they find as either carnivores or herbivores.

If possible, have a field guide available for each group of students, or develop fact sheets illustrating some of the common plants and animals found in a pond community. This information enables students to identify organisms quickly and to recognize some of their specific features.

You may wish to suggest that students compare the organisms that live in the three different zones of the pond—surface dwellers, such as water striders, boatmen, and floating plants; those organisms that live in the water, such as fish, turtles, and frogs; and bottom dwellers, such as insect larvae. Each zone has slightly different environmental conditions, and, therefore, is inhabited by different kinds of organisms.

Students can use their observations to suggest which characteristics organisms need to survive in each zone. Some students may observe that a few organisms are able to move from zone to zone, but students should realize that these moves are usually temporary.

Setting up a pond aquarium in the classroom allows students to observe the organisms over a long period of time. Be careful not to overload the aquarium! A small number of organisms is often more successful than a large number. Proper aeration is essential for the organisms living in the aquarium.

Be sure that students recognize that the pond is visited by many animals, such as deer and hawks, searching for food and water. These should be considered part of the pond's community and included in the food webs constructed by the students.

48 Interactions

EXPLORATION 6—CONTINUED

A Pond Community

You Will Need

- field guides for identifying organisms
- plastic bottles, containers, and bags for taking samples
- your Journal and a pencil
- a longhandled "muck scoop" (kitchen strainer)
- a white plastic tray
- a glass jar or an aquarium (optional)
- a pump (optional)

What to Do

Wear suitable clothing—preferably rubber boots. **Be Careful:** *Use caution around the pond.* Before you start looking for pond organisms, look at the pond itself. Is it a natural or an artificial pond? Where does the water come from? Where does it go? Make a sketch of the pond. Show the location of any trees or bushes that surround it. Do any streams flow into or away from the pond? Are there any rocks? Also be sure to include in your sketch any human-made features such as buildings or roads. All of these can influence the life in the pond.

One of the most interesting features of a pond is its great variety of animal life. The illustration at the right shows some of the pond life you might see.

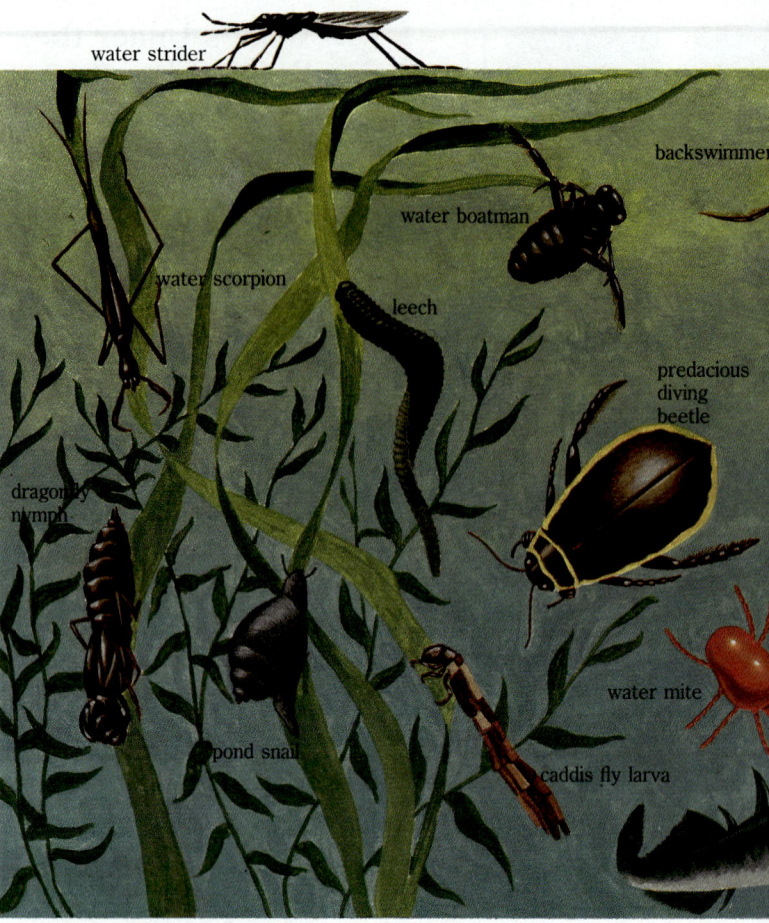

The hardest animals to find are the ones that live in the muck at the bottom. For these you will need a "muck scoop." A kitchen strainer with a long handle should work. Use the scoop to bring up some of the decaying material on the bottom. Place a scoopful on a white plastic tray and gently poke through it, looking for anything that moves. You will likely find larvae of the dragonfly, damselfly, stonefly, and other insects. There may also be different types of worms.

As you find these animals, carefully move them to a container that has only water in it. This will allow you to look at them more closely. Use a field guide to identify some of them. Can you determine which of the animals are herbivores? carnivores? scavengers? What do you think the mouth of a carnivore would be like?

Answers to

In-Text Questions

- Carnivores tend to have mouthparts and other appendages adapted for catching and holding prey. The mouthparts of herbivores are less obvious.
- As with a forest community, commensalism is easy to observe in a pond community. For example, fish and frogs use the debris of water and land plants for shelter and protection. Mutualism is harder to observe. There are examples of aquatic organisms that eat parasites and dead tissue from other organisms such as fish, but the likelihood of observing such behavior is remote. Leeches that live on the blood of other organisms are a good example of parasitism.
- The tolerance ranges of organisms living in the bottom muck are quite small because the abiotic conditions in this area of a pond remain fairly constant throughout the year.
- Abiotic conditions for organisms surrounding the pond will differ in light level, availability of oxygen, amount of water, and temperature.

whirligig beetle
mosquito larva
killifish
tadpoles
salamander larva
dragonfly nymph
mayfly nymph
stickleback
bullhead

There may be some larger animals around your pond. You may see the animals themselves or perhaps just signs of them. In spring, frogs' eggs may be found. Can you see signs of muskrats or beavers? Make a list of any animals or birds you see.

Look for examples of commensalism, mutualism, and parasitism. Can you find signs of different organisms living or working together?

Describe the tolerance ranges of the organisms that live in the bottom muck. How are the abiotic factors different for the organisms that live in the area surrounding the pond?

When you finish your study of the pond, leave behind as few human signs as possible. Your footprints will remain, but take your plastic bottles, containers, bags, and other equipment with you. They are not a natural part of the pond.

Before you leave the pond, take some of the muck from the bottom, some of the plants, a variety of animals, and pond water. You can use these materials to set up a small pond aquarium at home or in your classroom. Use a large glass jar or an aquarium. Whichever you use, try to make it as natural as possible. The water in your aquarium will have to be *aerated* (have oxygen added). You can do this with a small pump.

Watch the animals closely. Describe how each one moves. How does each feed? How do they react to each other?

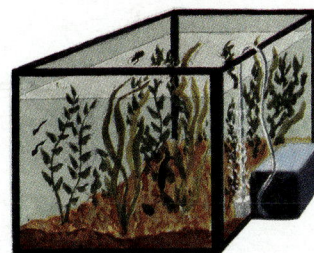

Construct one large food web for the pond community, including all the organisms you either found or saw signs of. Are there some that you didn't see but that you think might visit the pond occasionally? How do they fit into your web? Classify the organisms as herbivores, carnivores, omnivores, or producers. Did you find any decomposers in or around the pond?

Assessment

1. Have students imagine that they are preparing land to make a garden. The land is plowed, raked, and ready to go. But something happens and the garden never gets planted. Ask students to draw three illustrations that show what happens to the garden at the end of the summer, after one year, and after several years. Invite volunteers to use their illustrations in an oral report about succession.

2. Have students complete the following chart for a biological community, such as a woodland, forest-floor, or pond community.

Role or Relationship	Example
herbivore	
predator—prey	
carnivore	
producer	
decomposer	
scavenger	
omnivore	
commensalism	
mutualism	
parasitism	

Display the charts where other students may study and discuss them.

Extension

1. Succession from a lake community to a forest community goes through specific stages. Suggest that students discover how a lake becomes a forest. They can share what they learn in a presentation to the class.

2. Challenge students to research how specific plants and animals have adapted to the communities in which they live. For example, frogs are adapted to a pond community by having webbed feet for swimming. Deer are adapted to living in a forest community by being camouflaged to blend in with their environment.

Putting It All Together 49

LESSON 7

✸ Getting Started

In this lesson, students investigate the interactions between people and the environment. They also examine how human activity has impacted the delicate balance of nature. Students are asked to consider whether people are a part of the environment or masters who control it. In considering these two divergent points of view, students investigate the problem of acid rain.

Main Ideas

1. The impact of humans on the environment has often upset the delicate balance of nature.
2. Acid rain is one example of the serious consequences that human activity can have on the environment.
3. People can have a positive influence on improving the environment and correcting environmental mistakes of the past.

✸ Teaching Strategies

Ask students to consider how people interact with the environment. Write some of their ideas on the chalkboard. *(People use natural resources, dispose of wastes, grow food, and create pollution.)* Use the list to point out that people can either work with or work against nature.

Students should understand that the upper diagram illustrates that people are part of the environment. They should recognize that the bottom diagram represents the idea that people are in complete control of the environment.

 Humans: Boss or Co-Worker?

Which of the diagrams below do you think better shows the interactions between humans and their environment? In your Journal, write down all the reasons for your choice.

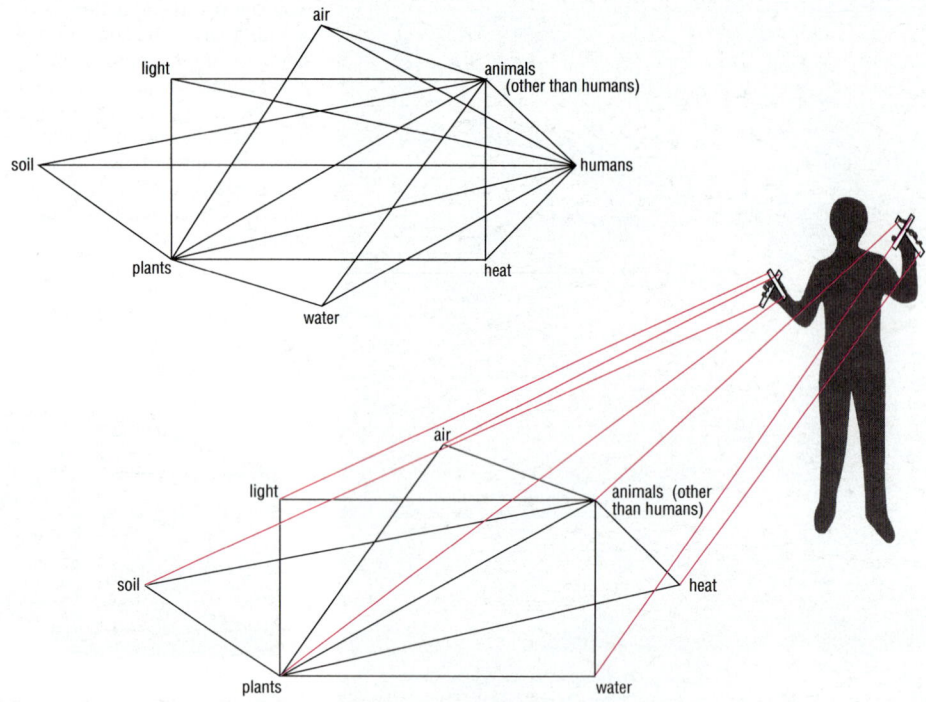

Some people consider themselves to be like other animals. They know that humans influence both the living and the nonliving environments. But they believe that humans are, in turn, equally influenced by the environment and all that it contains. Take a look at the two webs above. Which one supports this viewpoint?

Other people see themselves as separate from animals. They feel that humans have a special power or authority over both living and nonliving things in the environment. They know they influence their environment, but they often behave as if the environment has little influence on them. What you read about the great auks earlier in the book is a good example of this viewpoint. As you can see, this attitude has created many serious problems. Which of the two webs above fits this viewpoint?

Humans can manipulate and use the environment in ways that no other living thing on earth can. This ability carries with it the duty to consider the effects of our actions.

LESSON 7 ORGANIZER

Objectives

By the end of the lesson, students should be able to:
1. Identify some positive and negative influences people can have on the environment.
2. Explain the consequences of acid rain on the environment.
3. Identify some of the sources of acid rain.
4. Suggest ways of preventing acid rain and reversing its effects on the environment.
5. Identify ways in which people can have a positive effect on the environment and correct past environmental mistakes.

50   Interactions

Facing the Problems

When humans think they have absolute control over nature, they often find that they have created dangerous situations. Here are some issues to consider. Think about the issues, research them, and discuss them with your fellow students.

1. Can humans bring harm upon themselves in their unwise use of the environment? How? Can you give any local examples?
2. What kind of harm can result from the careless use of pesticides? Is there a problem in the area where you live? What evidence of a problem is there? What is being done about it?
3. What are the problems created by smog in areas such as southern California? Is there air pollution where you live? What is being done about it?
4. What have industrial wastes done to the rivers and lakes? What are "hazardous wastes"? How do these kinds of wastes affect the way humans live?
5. What happens to forests and farmland when human communities, including industries and highways, expand? Is this a problem in your area?

We have come to realize that when we damage living things or thoughtlessly change nonliving things, we hurt ourselves. Because we are part of the web of life, we are affected by changes in our surroundings. Are we smart enough to survive? Survival means thoughtful, careful interactions between human beings and all parts of the environment—living, once living, and nonliving.

Can Individuals Make a Difference?

In this lesson, you have thought about and researched some environmental problems. Government and industrial agencies exist to help solve such problems. There are also many concerned individuals who encourage awareness of environmental issues. They often take an active role in solving and preventing environmental problems.

One group of students, called Kids Against Pollution (KAP), has testified before local and state governments, served as witnesses before the Environmental Protection Agency, proposed an amendment to the Constitution, and even made presentations at the United Nations. What prompted these students to do so much? One eighth-grade KAP member gave this answer: "I want my generation to be known as the one that did something about the environment." Here is the story of a group of students who decided to get involved with solving one environmental problem—acid rain.

Spraying pesticides

How can smog be reduced?

Members of Kids Against Pollution have testified before the Environmental Protection Agency about their environmental concerns.

Answers to
Facing the Problems

You may wish to assign each of the problems to a group of students to research. Have each group report their findings to the class. The following are general responses to the questions.

1. Unwise use of the environment can include: toxic waste dumps, air pollution, and pesticide pollution. These harm people by poisoning food and air, causing birth defects, and causing disease.
2. Pesticides can contaminate the food we eat, kill wildlife, and pollute water.
3. Smog can cause serious illnesses, such as lung cancer, emphysema, and heart disease.
4. Industrial wastes have polluted rivers and lakes. Hazardous wastes include chemical and nuclear waste.
5. Forests get cut down and are never replaced. Farmland can be lost to development.

Answer to caption: A reduction in pollutants, especially those from cars and industry, would reduce smog.

Can Individuals Make a Difference?

Have students read the material silently. Then involve them in a discussion of whether or not they believe people can stop pollution.

Process Skills
observing, analyzing, inferring, drawing conclusions

New Terms
Acid rain—occurs when pollution from burning gas, oil, and coal mixes with water vapor in the air to form acid; this acid then falls to earth with snow and rain.

pH scale—a scale, ranging from 0 to 14, used to measure the acid or alkaline content of a substance.

Materials
The Story of Mr. Ripley's Class: Journal
What the Students Found Out in Their Research: Journal

Teacher's Resource Binder
Graphic Organizer Transparency 1–8

Science Discovery Videodisc
Disc 2, Image and Activity Bank, 1–7

Time Required
four class periods

Putting It All Together 51

The Story of Mr. Ripley's Class

It would be useful to have a variety of resources about acid rain available for students. The value of this lesson is that it deals with a current environmental problem; provides an example of how students contribute to the public awareness on a pressing social issue; and illustrates how students can effectively contribute to the solution of a serious environmental problem. The lesson also provides considerable information about acid rain and examines its harmful effects, causes, and possible solutions.

Have students read the page silently. Then involve them in a discussion of each of the questions. Accept all reasonable responses without comment. Suggest that students record their ideas in their Journals. Point out that as they read more about Mr. Ripley's class, they will discover more about acid rain and the answers to these questions.

To extend the activity, have students bring in articles about acid rain that they find in magazines and newspapers. Use the articles to set up a reading center. Have students add to the center throughout the following weeks.

The Story of Mr. Ripley's Class

Mr. Ripley's class in environmental science is no ordinary class. They study environmental problems that affect their community and beyond. They have expressed their concerns in such practical ways as writing a page on environmental issues in the local newspaper and participating in energy conservation activities. They have also conducted debates before the whole school on controversial environmental problems.

The students are quick to take the opportunity to initiate action. So when the city council announced the formation of a committee to study acid rain, Mr. Ripley's class decided to get involved. Here are some of the questions they investigated:

- What is acid rain?
- What causes rain to become acidic?
- Is it a natural problem, or is it caused by people?
- What harm does it do?
- Is it a widespread problem?
- Does it do any harm locally?
- If acid rain is harmful, what can we do as a group of people and as a government to control the problem?

How well informed are you? How many of these questions can you answer? What other questions would you have asked?

52   Interactions

What the Students Found Out in Their Research

Air pollution has led to the creation of acid rain. **Acid rain** occurs when pollution from burning gas, oil, and coal mixes with water vapor in the air to form *acid*. This acid then falls to earth with snow and rain. Acid rain kills trees and makes ponds and lakes unable to support life. Some of the lakes in the eastern United States are now so polluted that they can barely support life.

The acid content of a substance is measured on a **pH scale** of 0 to 14. Look at the scale at the right. It shows the acid content of many different liquids and foods. A pH of 0 indicates the most acidic solution, and a pH of 14 indicates the most basic. Anything that has a pH value below 7 is acidic. The lower the number, the greater the acid concentration. The middle point, pH = 7, is neutral.

A pH value greater than 7 indicates the presence of a *base*. Basic solutions are often referred to as *alkaline*. Alkaline substances can *counteract*, or neutralize, acids; that is, they lower the acid content of a substance.

Check the scale for yourself.

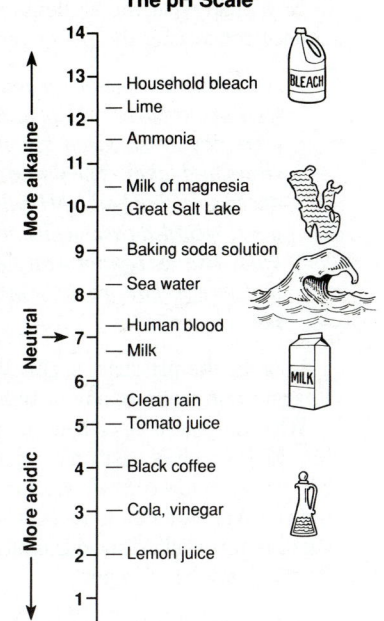

1. Is ordinary rainwater acidic?
2. At what pH does ordinary rain become acid rain?
3. Which has more acid in it—tomato juice or lemon juice?
4. Have you heard of people taking milk of magnesia for an acid stomach? Why would they do this?
5. Homeowners often put lime on their lawns in the spring. Why?
6. When the pH drops by one point, the acid content increases by 10 times. For example, a solution with a pH of 4 is ten times as acidic as a solution with a pH of 5. If the pH dropped by two points, what would the increase in acid content be?

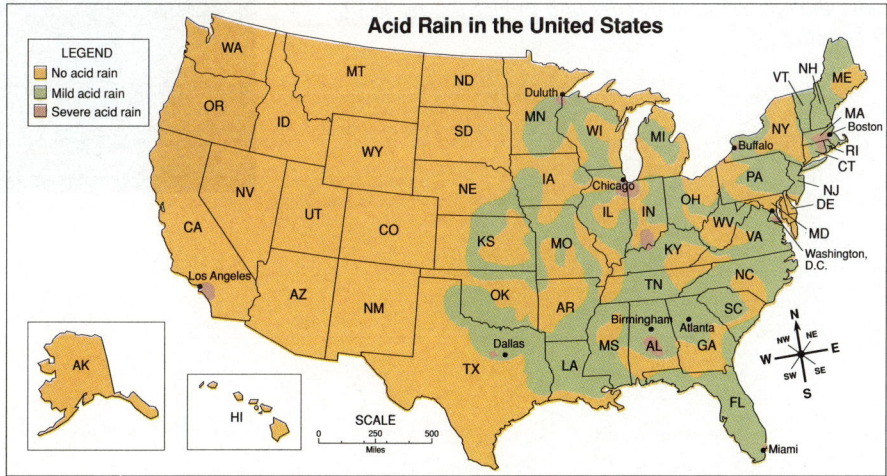

What the Students Found Out in Their Research

Have students read the page silently. Then call on a volunteer to describe how acid rain results from air pollution. If students seem to be having difficulty grasping the concept, draw a simple diagram on the chalkboard. Then have students write brief explanations in their Journals describing how acid rain is formed. Involve the class in a discussion of their ideas until they seem comfortable with the concept.

Direct students' attention to the pH scale. Be sure that they understand which end of the scale represents bases and which represents acids. Point out that the number 7.0 on the scale is neutral because it lies exactly in the middle of the acid/base range.

Answers to In-Text Questions

1. yes
2. at a pH below 5.6
3. lemon juice
4. Milk of magnesia has a pH of 10.5. It is a base, and bases neutralize acids or reduce the acid content of a substance or solution.
5. Soils are often acidic. Lime is a very basic compound having a pH of 12.4. Spreading lime on acidic soil neutralizes, or reduces, the acidic content of the soil.
6. 100 times

Direct students' attention to the map and have them describe what it shows *(areas in the United States that are affected by acid rain)*. Have students use the map to identify specific locations that are affected by acid rain. *(Continues on next page)*

Putting It All Together 53

(What the Students Found Out continued)

Have students begin reading the page, pausing after the first part of the report to the city council. Have students turn back to the pH scale on the previous page to find out what liquid has a pH near 2.4. *(vinegar)*

Encourage students to speculate as to why acid rain kills fish and trees. Accept all reasonable responses. *(Acid rain finds its way into lakes, rivers, and streams and raises the acid levels. This upsets the delicate abiotic balance in those communities and causes many fish and other aquatic organisms to die. Acid rain falls on forests and kills the plants living in the forest community that cannot tolerate high acid levels.)* Help students recognize that acid rain upsets the existing abiotic balance in a biological community, which results in serious consequences for the organisms that live there.

Have students read the rest of the page. Ask them to respond with a show of hands if they are surprised by the amount of sulfur and nitrogen oxides that are poured into the atmosphere every day. If possible, display pictures that show dramatic examples of pollution from car exhaust and factory emissions.

Now read what the students wrote for their report to the city council concerning the pH of rainwater in their city.

Acid rain has become a major problem in our city. The trees in our parks are not as healthy as they used to be, and every day we see dead fish wash up on the shores of our rivers and lakes. The thunderstorm we had last week produced rain that had a pH value of 4.0. This is far from the worst recorded case, which was a pH value of 2.4 in Scotland. But we cannot wait for our pH values to worsen to that level before we do something about it. We must act now.

Look at the pH chart again. What liquids have a pH near 2.4? Imagine rain made of any of these liquids!

What do you think causes the pH of some rainwater to be low? Mr. Ripley's class easily discovered the answer. Industries, cars, and trucks release gases (sulfur oxides and nitrogen oxides) into the air. When rain or snow falls, these gases dissolve in the water vapor to form sulfuric acid and nitric acid. Here is more information from the students' report.

In North America, air pollution is a major concern. Every day, at least 165,000 metric tons of sulfur and nitrogen oxides are released into the atmosphere. This pollution remains in the air for up to 10 days, traveling thousands of kilometers from its source, depending on the wind currents. In other words, sulfuric acid clouds rising from our factories in and around the Pittsburgh area might be over the Great Lakes region in two days. So our pollution affects more than this community. Likewise, another region's pollution is making our situation worse. If every community reduced the amount of sulfur and nitrogen oxides that are released by cars and industries, we would all be better off.

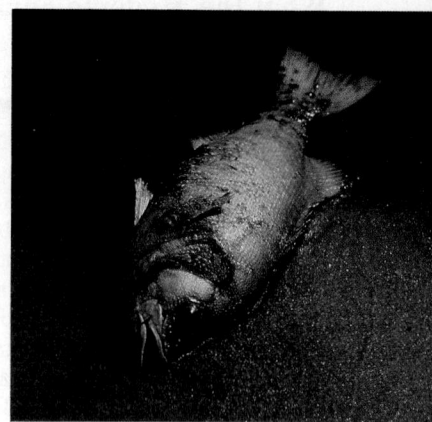

54   Interactions

The Effects of Acid Rain

Mr. Ripley's class found plenty of information about the effects of acid rain on the environment. They learned that acid rain lowers fish populations and seriously harms aquatic life. Acid rain also destroys soil and trees. The students realized that when the soil is affected, crops will not grow as well as they would in soil with normal acidity. When acid rain kills trees, the economy suffers. That's because forests supply us with lumber for construction, paper, and pulp.

Do some research yourself and find out how acid rain actually affects the life processes in lake fish. Why do lakes become more acidic over time?

Some Controls

With your classmates, discuss what can be done about the acid rain problem. Make a class list of possible solutions.

Here are three of the eight specific recommendations that Mr. Ripley's class made in their report to the city council. Did you include any of these among your own suggestions?

1. Place more regulatory controls on factories and industries that release sulfur and nitrogen oxides into the air.
2. Penalize companies or individuals who do not follow the regulations.
3. Provide more funding for research on the effects of acid rain.

What could be done to reduce this source of pollution?

The Effects of Acid Rain

Have students read the material silently. Point out that acid rain not only kills fish, but also other aquatic organisms, such as frogs, salamanders, and insects. Involve students in a discussion on how acid rain may affect the food webs of aquatic communities. Help them recognize that when one organism is affected by acid rain, the chances are that many others will be affected, too. For example, if there are no more frogs, then wading birds, such as herons, cranes, and egrets, will stop coming to ponds and lakes to look for food. If the fish die, eagles will also stop coming in search of food.

Encourage students to research and find the answers to the question at the end of the selection. The high acid level of the water raises the acid levels of the blood in the fish and poisons them. Lakes become more acidic over time because acid rain continues to fall on them.

Some Controls

Before students read the selection, involve them in a discussion of what might be done about the acid rain problem. List their suggestions on the chalkboard. Then have students read the material to see if they included the recommendations made by Mr. Ripley's class. *(Additional suggestions include: manufacture non-polluting automobiles, provide more funding for research on non-polluting energy sources, increase public awareness of the problem, and support political candidates who favor strong pollution control measures.)*

Answer to caption: Increased public transportation, non-pollutant engines, and cleaner burning fuels would help eliminate pollution from automobiles.

Putting It All Together 55

Answers to
More Things to Do

Encourage students to select one or more of the activities to complete.

1. Encourage students to work creatively on their posters. Suggest that the messages might have more impact if they touch on local concerns about acid rain.
2. Acid rain's impact strongly affects the fishing, lumber, and tourist industries. As the public becomes more aware of the problem, it is likely to exert pressure on public officials to find a solution. The public itself contributes to the problem by heavily relying on automobiles for transportation.
3. One possible ranking is: industry, cars and trucks, uninformed public, government preoccupied with other matters, inappropriate education, weather conditions.
4. The pH of local rain is likely to be lower (more acidic) in the spring than in the fall since people drive more often in better weather conditions.
5. The burning oil wells caused a rapid increase in the acid rain in the region. The smoke from the burning oil wells also caused a drop in temperature in the region by blocking out some of the sunlight. Prevailing wind patterns were altered, affecting the important rains of the monsoon season.

More Things to Do

1. Prepare posters for the school and community alerting people to the acid rain problem. Use the information on acid rain provided in these pages.
2. What occupations in your state are likely to be affected by acid rain? What effect do you think the increasing problem of acid rain will have on the general public? How does the public contribute to the problem?
3. Rank these factors in terms of how they contribute to the acid rain problem today: weather conditions, industry, cars and trucks, an uninformed public, a government that is preoccupied with other matters, and inappropriate education.
4. Call your local weather service and ask if they monitor the pH level of local rain. If so, chart the pH values during the next couple of months to see how they fluctuate. Would you expect the same values in the fall as in the spring? Why or why not? What local factors might influence the pH values?
5. In 1991, during the war in the Persian Gulf, many oil wells in Kuwait were set on fire. Smoke from burning oil typically contains high concentrations of sulfur dioxide. With your classmates, research the effects the burning oil wells may have on the region and on surrounding countries.

✸ Follow Up

Assessment

1. What is the meaning of the title of this lesson: *Humans: Boss or Co-Worker?* Which one of these two designations for humans do you consider to be true? Explain why.
2. Describe an interaction that humans have had with your local environment and that you consider to be harmful. Make some suggestions about how to correct this harmful interaction.

Extension

1. Point out to students that, in addition to acid rain, there are other environmental problems facing the world. They include such things as the greenhouse effect, the deterioration of the ozone layer, and deforestation of the tropical rain forest. Suggest that students choose one of these topics to research. Then they should present their findings to the class in an oral report.
2. Point out to students that acid rain is a global problem. Suggest that they research to discover other countries that are being affected by it, and what they are doing to solve it. Have them share what they learn by presenting an acid rain "world news report" to the class.

56   Interactions

Challenge Your Thinking

Sketch a graph of the population changes in John's community to show what happened. For an example of a population graph, remember the mice and rabbits on page 42.

Answers to
Challenge Your Thinking

The graph should illustrate that the rodents' increase in population is due to an absence of predators (the hawks). Killing the chicken hawks saved the chickens and the rats that consumed the grain.

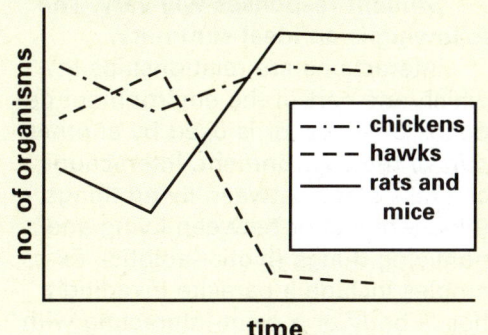

Putting It All Together 57

UNIT 1
Making Connections

Summary for
The Big Ideas

Student responses will vary. The following is an ideal summary.

Interactions are relationships in which one part of the environment depends on, uses, or is used by another part of the environment. Interactions can take place between living things, (biotic–biotic) or between living and nonliving things (biotic–abiotic). Examples include a parasite invading a host's body or a plant interacting with sunlight to make food.

The biotic parts include plants, animals, fungi and bacteria, or the living parts of the environment. The abiotic, or nonliving, parts include rocks, light, air, water, chemicals and heat.

The place that an organism lives is its habitat, and its way of life is its niche.

Commensalism is the dependence of one organism on another organism that is neither harmed nor helped by the relationship. In parasitism one organism depends on another for habitat and food; the second organism is harmed or destroyed by the relationship. Mutualism is a sharing of benefits between two organisms.

Producers are plants that make food for the community from the sun's energy, carbon dioxide, and water. Consumers are generally animals. They do not make their food as plants do. Consumers that eat plants are called herbivores. Consumers that eat only animals are called carnivores. Scavengers eat and dispose of the bodies of dead animals. Decomposers reduce the remains of plants and animals by causing organic material to decay.

A food chain shows how a series of organisms depends on one another for food. Food chains move from a producer to a primary consumer to a secondary consumer, and up to the top carnivore. A food web consists of several interconnected chains. It shows how a given organism has links with many other members of the community.

Unit 1 Making Connections

The Big Ideas

In your Journal, write a summary of this unit, using the following questions as a guide.

- What are "interactions"? Give examples.
- What do the biotic and abiotic parts of the environment include?
- How is an organism's "niche" different from its "habitat"?
- Can you illustrate the differences between commensalism, mutualism, and parasitism?
- How would you describe the roles of producers, consumers, scavengers, and decomposers in a community?
- What is the difference between a food chain and a food web?
- How do each of the abiotic parts of the environment contribute to the welfare of a community?
- What kinds of changes take place in a community over time?
- Why are there continuous changes in plant and animal populations?

Checking Your Understanding

1. The mystery word tells what it's all about!

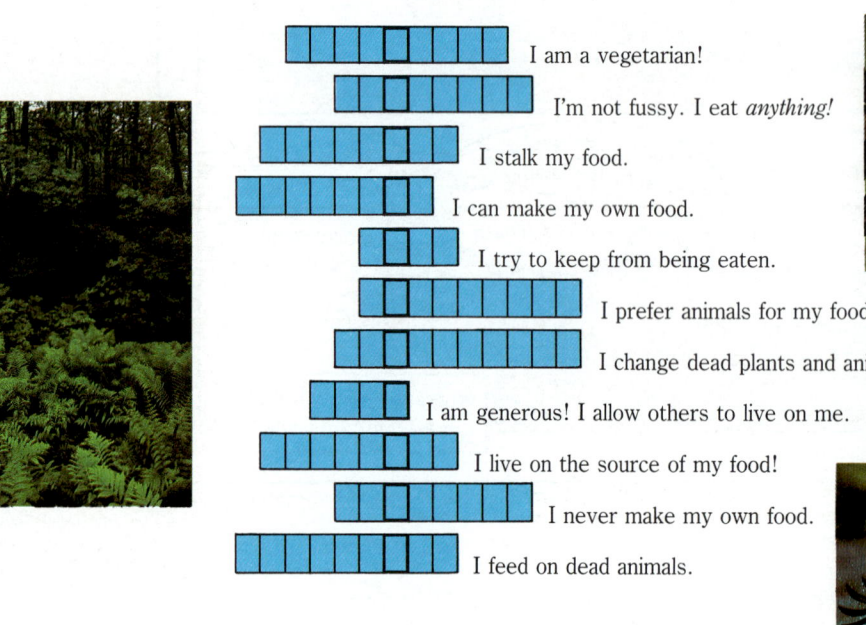

- I am a vegetarian!
- I'm not fussy. I eat *anything!*
- I stalk my food.
- I can make my own food.
- I try to keep from being eaten.
- I prefer animals for my food.
- I change dead plants and animals into soil.
- I am generous! I allow others to live on me.
- I live on the source of my food!
- I never make my own food.
- I feed on dead animals.

Plants and animals depend on an optimum temperature for growth, and they exhibit a range of tolerance for temperatures. Similarly, living organisms require moisture, and they vary in moisture tolerance range. Light is required by all plants for the production of food. Plants tend to turn toward light, and animals have a varied response to light.

Disrupted communities tend to return to their original state. After an environment is destroyed, over time, one group of organisms can begin to grow again. This is followed by another group that succeeds the first group. In turn, more complex plants succeed the early plants until the original state is reached. This process is called succession.

Changes occur in a community because of the organisms' dependence on each other for food. As a certain plant increases in number, so will the herbivores that use the plant as a primary food source. This increases predation on herbivores, and the population of predators. Eventually, the predators will drastically reduce the herbivore population.

2. Arrange these organisms into a food chain: grass, grasshopper, bear, fish, and frog.
3. Suppose that during the drought last summer, a forest fire burned a large portion of a forest near your town or city. Predict the kinds of changes that will occur in the plant and animal communities there over the next 100 years.
4. Draw a concept map that shows the energy relationships between the following items: *producer, sunlight, carnivore,* and *herbivore.*

Reading *Plus*

You have been introduced to interactions among organisms and their environment. By reading pages S1–S22 in the *SourceBook,* you will take a closer look at the role of organisms in ecosystems. You will learn more about how energy and materials are used in these ecosystems and how pollution affects our biosphere.

Updating Your Journal

Reread the paragraphs you wrote for this unit's "For Your Journal" questions. Then rewrite them to reflect what you have learned in studying this unit.

Answers to
Checking Your Understanding

1. The answers in order are: herbivore, omnivore, predator, producer, prey, carnivore, decomposer, host, parasite, consumer, scavenger. The mystery word is *interaction.*
2. The order of the food chain is grass, grasshopper, frog, fish, bear.
3. Over the years, the land where the forest stood will undergo succession. Succession is a natural series of changes where one type of population is replaced by another. The first population to appear will be the pioneer plants (lichens, mosses, and grasses). Next, small seed plants like shrubs and bushes will appear. Finally, larger plants and trees appear. When the community returns to its original state, called the climax community, succession is completed.
4.
```
        Sunlight
           |
    provides energy for
           |
        Producers
           |
     which produce
   food for themselves
         and for
           |
       Herbivores
           |
   which are consumed by
           |
       Carnivores
```

About Updating Your Journal

The following are sample ideal answers.
1. Changes in the environment include abiotic changes brought about by natural or human-made destruction and population changes brought about by changes in local resources. **Patterns of Change** are reflected in a community by the orderly progression of changes, known as succession.
2. People influence the environment by using natural materials, eating plant and animal life, disposing of waste products, and maintaining large populations. People help the environment by recycling, conservation, and wildlife preservation. Humans harm the environment by pollution, destruction of habitats, and endangering various species.
3. The bee and the flower have a relationship characterized by mutualism. The bee gets food from the flower. The flower uses the bee to transport pollen. Both rely on abiotic factors such as water, temperature, and light.
4. No. The plant is making food from the sunlight; both organisms are affected by temperature and moisture.

Making Connections 59

Science in Action

♦ Background

A veterinarian is to animals what a physician is to people. Even the educational requirements are similar. To be a veterinarian, a person needs four years of preveterinary study and four years of study at a college of veterinary medicine. The courses of veterinary school are similar to those at a medical school, and they include such subjects as anatomy, surgery, and chemistry.

Once a veterinarian has received a degree and the necessary licenses, he or she may go into private practice. However, some veterinarians go into government service with such agencies as the U.S. Public Health Service or the U.S. Department of Agriculture. Others serve in programs sponsored by such agencies as the Peace Corps and the World Health Organization. Some of the more interesting places where you might find a veterinarian include race tracks, circuses, and zoos.

Science in Action

Spotlight on Veterinary Science

The healing and care of animals is the job of veterinarians like James Love. Most veterinarians work in clinics, but James is in charge of an animal care station for a medical research center. We talked with James in the primate section of the station.

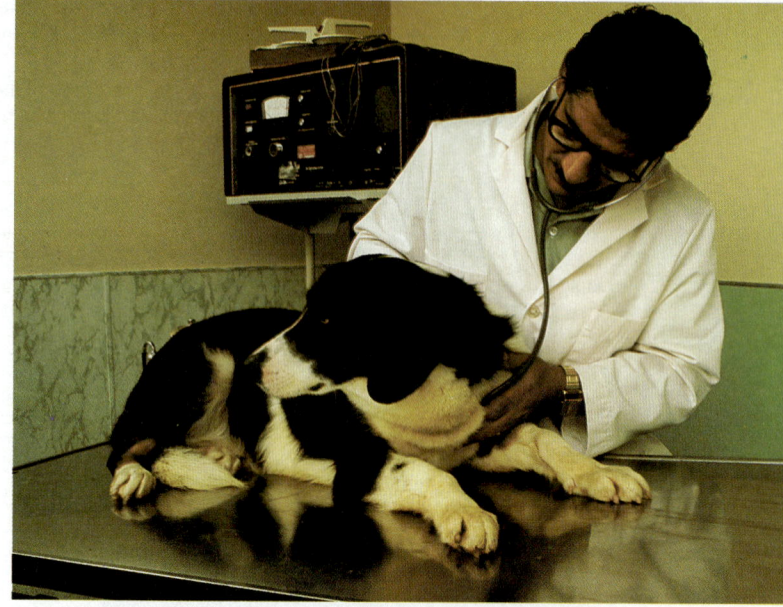

James: We have a small colony of baboons here that are used for research purposes. However, since they have become an endangered species, the use of primates like baboons in research has become limited. We consider them to be important for research since they are very similar to human beings in many ways. Rats, mice, and guinea pigs are not enough like humans to be used for certain types of research.

One of the most exciting events we've had was the first birth of a baboon here. Immediately after birth, a newborn baboon is able to cling to its mother with all four limbs. This makes it possible for the mother to walk, run, and move about in trees without holding on to her young.

Q: What else does a veterinarian do at a medical research center?

James: Scientists come to the animal care station when they want to do experiments using animals. They discuss with me the kind of care the animal should have, the special equipment that might be needed, and any special diets that might be required. We also advise them about surgical procedures.

Q: Are there alternatives to using animals for medical research?

James: There is a movement all over the world to reduce or abolish the use of animals in research. Research on tissue cultures instead will develop very rapidly, I think, and will replace many of the animals being used in research at present. Also, future technological advances may be able to simulate the responses of animals without actually using the animals in the tests.

Q: Is the main emphasis in veterinary medicine on prevention, or on cure?

James: In caring for small farm animals, such as poultry and pigs, it is becoming uneconomical to treat individual animals for disease. It is just too expensive to call for the vet. The emphasis now is on preventing disease by vaccination or by adding antibiotics to feed or drinking water. Similar preventive medicine programs have been developed in the cattle industry.

♦ Using the Interview

Before students begin reading, invite them to share any "veterinarian" stories they know. Encourage them to identify the kind of pet they took to the veterinarian and to describe what happened. Invite students to make comparisons between what a veterinarian does for animals and what a physician does for people.

When students finish reading, you may wish to evaluate their understanding of the interview by asking questions similar to the following:
- In what kind of a facility does James Love work? What is his position? *(He is in charge of an animal care station for a medical research center.)*
- What is one alternative to using animals for medical research mentioned by James Love? *(Some research can be done on tissue cultures instead of on animals.)*
- Why does James Love believe the emphasis in treatment of farm animals has shifted from individual treatment to prevention? *(It has become too expensive to call a doctor every time an animal gets sick. Treatment for sick farm animals is becoming uneconomical.)*

After students have finished discussing the interview, ask them if they think they would like to be a veterinarian. Encourage them to explain their responses.

Q: Do you have any thoughts about junior high and high school students doing research using animals?

James: Animals should not be harmed unnecessarily. Nevertheless, there is much research involving the observation of animals that can be done by young people. Those interested should first write to the Humane Society for information on the kinds of research projects that can be done. (2100 L St., N.W., Washington, D.C. 20037).

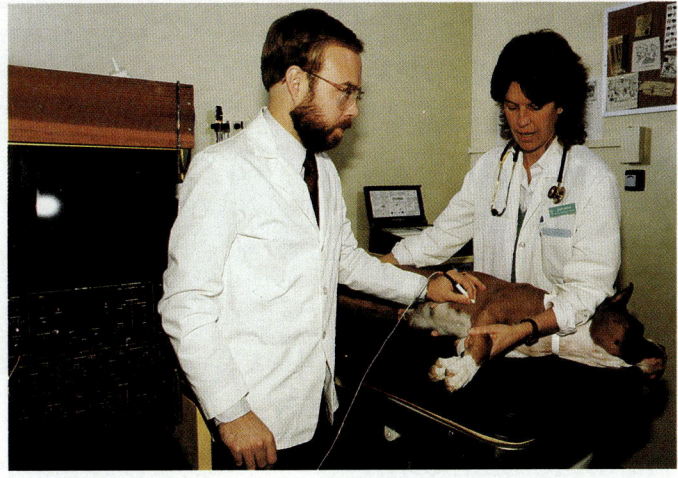

Q: Can you suggest some research projects that students might do?

James: One project that might be interesting would be to place two rats in separate cages and give only one an exercise wheel. Then over a period of time, you could compare their food intake, water intake, and weight change.

There are any number of projects of this sort, but you should write to the Humane Society or consult a veterinarian before you start your project.

A Project Idea

Some groups of people oppose the use of animals for research or medical testing. Others feel that it is acceptable to use animals for this purpose. This could be the topic of a class debate. To prepare for the debate, investigate this issue, being sure to consider both sides fairly and carefully. Why are animals used for testing? What alternatives are there to using animals in medical research?

◆ Using the Project Idea

Provide students with time to research and become familiar with the issue of medical testing on animals. Determine which class members support medical testing on animals and which are against it. Select a debating team from each group of students. Set aside a particular day and time for the debate to take place. After the debate has been presented to the class, involve students in a discussion about which side they felt had the best arguments.

If a debate is not feasible, you may wish to have students research the issue and suggest that they share what they learn with their classmates.

◆ Going Further

Ask students to investigate the following questions:
- What is the history of veterinary medicine in the United States?
- What are some areas of specialization in veterinary medicine?
- Set up an interview with a veterinarian. Discover what kinds of animals he or she treats. What advice about pet care does the veterinarian suggest?
- How do people feel about medical research on animals? Take a survey of family members, neighbors, and friends.

Science in Action

Science and Technology

◆ Background

The field of oceanography began in 1872, when the British Navy vessel, *H.M.S. Challenger,* set sail for a three and a half year expedition of the world's oceans. The ship was outfitted with the most modern scientific instruments of the time. Despite numerous difficulties, the 110,224-km voyage amassed enough notes, journals, and logbooks to compile the unprecedented *Challenger Reports,* which took 20 years to complete.

Today, there are more than 5000 oceanographers in the United States alone. Most of these scientists specialize in one of several fields of oceanography. For example, physical oceanographers study the wave, current, and tidal movements of the ocean. Chemical oceanographers study the chemical composition of ocean water. Marine geologists explore and map the landforms beneath the sea. Marine biologists study marine organisms and their changing environment.

Science and Technology

Oceanography

It has been said that the ocean is the final frontier on the earth. While many people are excited about going *up*, into space, some people have turned their attention *downward*, into the ocean. One such person is Dr. Sylvia Earle, an oceanographer who, in 1979, made the deepest untethered solo dive in history—381 meters. In her record-breaking dive off the coast of Hawaii, she planted a flag on the ocean floor in imitation of the first astronauts on the moon.

Dr. Sylvia Earle on ocean floor

The Blue Unknown

The ocean is largely unknown. Maps of the earth show different kinds of geographical regions, like deserts and rain forests. Such regions support their own kinds of **ecosystems,** or communities of plants, animals, and bacteria. The ocean, however, is still represented on many maps by a uniform blue color. This suggests that it is all the same under the surface—which certainly isn't true. There are many different geographic regions and ecosystems in the ocean waiting to be explored and studied.

Freedom Underwater

Going down to study the ocean depths has always been difficult. But this is changing. New underwater vehicles, or *submersibles,* are making it possible for single individuals to "fly" under the sea with a freedom that has never been possible before—to do "barrel rolls with the whales," according to Dr. Earle. Older submersibles were like dirigibles, with very little mobility. The new vehicles are more like airplanes.

Touching Down

New submersibles may soon make it easy for individuals to go down to the very deepest part of the ocean, a depth of nearly 11 kilometers. This trip has been made only once before, in 1960. The trip was made in the *Trieste,* a Swiss-built bathyscaph consisting of a steel observation compartment attached to a submarine-shaped float. This particular vehicle is no longer in use.

Dr. Earle in the *Deep Flight* submersible.

◆ Discussion

When students have finished reading the feature, involve them in a discussion of what they have read. You may wish to use questions similar to the following:

1. Why do we call the ocean "the final frontier on earth?" *(The ocean is considered the final frontier because it is the last, largely unexplored area of the earth.)*
2. How are modern submersibles different from older ones? *(Modern submersibles have more mobility than older ones. They make it possible for single individuals to "fly" under the sea with a freedom that was impossible with older submersibles.)*
3. What are divers likely to see as they explore greater depths in the ocean? *(The deeper they dive, the dimmer the light from above. Below 100 m, there is not enough light for green plants to grow. Only animals and bacteria live below this level. Divers who explore the very deep regions of the ocean will see a variety of beautiful and bizarre creatures, such as sightless fish, luminescent animals that glow green or blue, and fish with enormous mouths and exaggerated teeth.)*

A Sunless World

The new technology may even open up the depths of the ocean to recreational divers. If so, the world the divers see will be unlike anything they have encountered before. The deeper down you go, the dimmer the light from above becomes. As you dive below 100 meters there is not enough light for plants to produce food by photosynthesis. Only animals and bacteria live below this level.

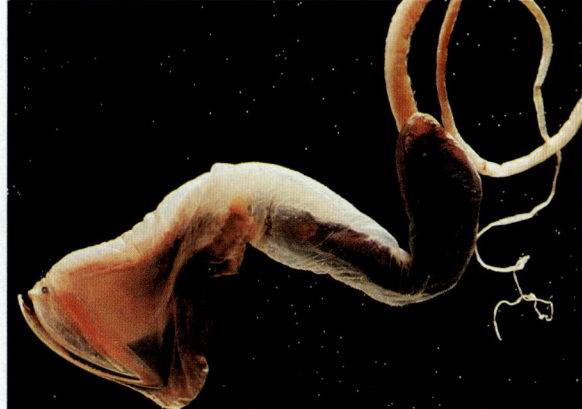
Deep-sea swallower

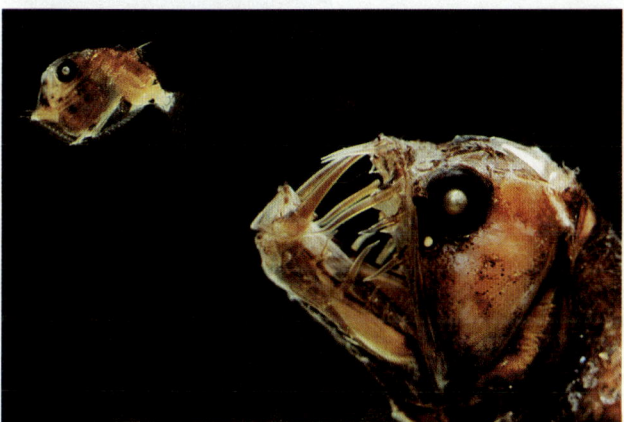
Viperfish chasing a hatchetfish

People who go down to the sunless depths bring back reports of a world that is at once beautiful and bizarre. Sightless fish, luminescent creatures that glow green or blue, and fish with enormous mouths and exaggerated teeth—these are some of the fascinating sights that await the underwater explorer, perhaps even the "tourist" of the future.

Protecting the Ocean

One of the things that people see when they go down into the ocean is trash. Things like tangled fishing lines, for example, may do significant damage to life undersea. Do you think that a greater public awareness of such dangers would help bring about reforms? Do you think that undersea exploration is itself a danger to the environment?

Find Out for Yourself

Do some research to find out about the different classifications of life in the ocean. What are some of the animals that live at great depths? Where do they get their food? How are they adapted to their world? Why do you think that so many of them have such large mouths and teeth? What is the function of the *bioluminescence* of certain deep-water organisms?

✦ Critical Thinking

To promote logical-thinking skills, you may wish to ask questions similar to the following:

1. Why do you think more attention has been paid to space exploration than to ocean exploration? *(Some students may feel that space exploration is more exciting. Others may feel that more benefits are likely to come from space exploration. Still others may recognize the political dimension of space exploration.)*

2. What are some of the obstacles that scientists must overcome in order to explore the ocean depths? *(Overcoming the tremendous pressure created by the water; providing divers with air, food, shelter, and protection; providing an easy and mobile way of moving through the water; providing habitats in which to live on the ocean floor.)*

3. How are ocean and space exploration similar? *(Both involve exploring unknown regions; both involve unknown dangers as well as unknown rewards; in both cases, explorers must provide a safe way of living in a hostile environment.)*

✦ Going Further

1. Some people feel that too much money, time, and research are spent on space exploration and not enough are spent on ocean exploration. Others feel that space exploration is more important. Do some research to find out how you feel about this issue, using the following questions as a guide:
   - How important is it to learn all we can about the earth before exploring beyond it?
   - What benefits are likely to result from both space and ocean exploration?
   - Can we afford not to explore both outer space and the ocean? Why or why not?

2. The development of submersibles and other types of underwater diving gear has made great strides in the past 10 years. Do some research to find out what kinds of submersibles are currently being used and what kinds are being developed for future use.

UNIT 2
Diversity of Living Things

✳ Teaching the Unit

☀ Unit Overview

This unit focuses on the diversity of living things, the reasons for diversity, and how scientists make sense of this diversity. In the section *Such Variety!*, students consider the great diversity of living things and how plants and animals are adapted to their habitats and niches. In the section *Beginnings and Endings*, students are introduced to Darwin's theory of natural selection. In the final section, *Keeping Track*, students develop their own classification systems and learn how scientists classify plants and animals.

☀ Using the Themes

The unifying themes emphasized in this unit are **Evolution, Scale and Structure,** and **Patterns of Change.** The following information will help you weave these themes into your teaching plan. A focus question is provided with each theme as a discussion tool to help you tie the information in the unit together.

Evolution can be emphasized throughout this unit. The concepts in Lessons 1 and 2 are outcomes of evolution. Lesson 3 presents the theory of evolution by natural selection to explain how living things change and diversify. In Lesson 4 students explore how the interactions of organisms and chance events affect the survival, and thus the evolution, of living things.

Focus question: *Why do organisms change over time?*

Change occurs as species adapt to new environmental conditions. Ancestors of porpoises, for example, lived on land but began to inhabit the ocean. A gradual change in structure occurred over the subsequent generations because, in the ocean, certain variations were more successful than others.

Scale and Structure is emphasized in Lessons 1 and 2, which examine the diversity of living things and their adaptive structures. In Lesson 3, the study of the finches on the Galápagos Islands illustrates the interplay of structure and function. In Lessons 5 and 6, students learn that the classification of living things is based on common structures.

Focus question: *How does an organism's structure affect its ability to survive?*

An organism's structure enables it to gather food and/or nutrients, tolerate the climate of its habitat, and avoid predators. Discussions of Darwin's finches and their bill adaptations are excellent examples of how structure affects organisms' chances for survival.

Patterns of Change is reinforced as students study the life cycles of frogs, mosquitoes, and grasshoppers in Lesson 7. Evolution and extinction, discussed in Lessons 3 and 4, are also patterns of change.

Focus question: *What patterns of change can be seen among the life cycles of organisms?*

Most organisms undergo a series of changes from birth to adulthood. Grasshoppers go through a pattern of changes called incomplete metamorphosis, in which "in-between" stages have the same general appearance as the adult. Frogs undergo complete metamorphosis, in which "in-between" stages are different in appearance and structure from the adult.

☀ Using the *Science Discovery* Videodiscs

Disc 1 *Science Sleuths,* Neo-Cassava: The Tropical Miracle

A foundation has transplanted the neo-cassava from a tropical rainforest. It hopes to grow this inexpensive, nutritionally balanced plant on experimental farms as a remedy for world hunger. After an initial promising harvest, the crop is dying. The Science Sleuths must analyze the evidence for themselves to determine the cause of the problem.

Disc 2 *Image and Activity Bank*

A variety of still images, short videos, and activities are available for you to use as you teach this unit. See the *Videodisc Resources* section of the **Teacher's Resource Binder** for detailed instructions.

☀ Using the *SciencePlus SourceBook*

In Unit 2, students are introduced to organisms from each of the five kingdoms and learn how the history of life on earth correlates to the development of complex organisms. The unit examines the theory of evolution by natural selection and evidence for the common ancestry of all life on earth.

PLANNING CHART

SECTION AND LESSON	PG.	TIME*	PROCESS SKILLS	EXPLORATION AND ASSESSMENT	PG.	RESOURCES AND FEATURES
Unit Opener	64		observing, discussing	For Your Journal	65	Science Sleuths: *Neo-Cassava: The Tropical Miracle* Videodisc Activity Sheets TRB: Home Connection
SUCH VARIETY!	66			Challenge Your Thinking	85	
1 Living Things in All Shapes and Sizes	66	2 to 3	observing, comparing, describing, writing	Exploration 1 Exploration 2	67 70	Image and Activity Bank 2–1 TRB: Resource Worksheets 2–1, 2–2, and 2–3
2 Why Is There Diversity?	74	2 to 3	observing, describing, inferring, illustrating	Exploration 3	84	Image and Activity Bank 2–2
BEGINNINGS AND ENDINGS	86			Challenge Your Thinking	93	
3 The Origins of Diversity	86	2	observing, predicting, formulating hypotheses, drawing conclusions			Image and Activity Bank 2–3
4 The Loss of Diversity	90	2	analyzing, explaining cause and effect relationships, researching			Image and Activity Bank 2–4 TRB: Resource Worksheet 2–4
KEEPING TRACK	94			Challenge Your Thinking	123	TRB: Resource Worksheet 2–9
5 Making Sense of Diversity	94	5	observing, comparing, classifying, designing a classification system			Image and Activity Bank 2–5 TRB: Resource Worksheets 2–5 and 2–6 Resource Transparency 2–1 Graphic Organizer Transparencies 2–2, 2–3, and 2–4
6 Ordering Diversity	112	2	observing, comparing, taking a survey, analyzing data	Exploration 4	114	Image and Activity Bank 2–6 TRB: Resource Worksheet 2–7 Activity Worksheet 2–8 Graphic Organizer Transparencies 2–5 and 2–6
7 Life Cycles: Diverse Stages	116	2	observing, sequencing, illustrating, comparing	Exploration 5	116	Image and Activity Bank 2–7
End of Unit	124		applying, analyzing, evaluating, summarizing	Making Connections TRB: Sample Assessment Items	124	TRB: Activity Worksheets 2–10 and 2–11 Science in Action, p. 126 Science and Technology, p. 128 *SourceBook,* pp. S23–S44

*Time given in number of class periods.

Meeting Individual Needs

☀ Gifted Students

Explain that there are many direct and indirect causes for the endangerment and extinction of certain species. Direct causes are those in which humans intentionally exploit species for a particular purpose, such as for the pet trade (parrots, tropical snakes); for food (dodos, whales, sea turtles, bison); for sport hunting (polar bears, hawks); for bounty hunting (wolves, coyotes, mountain lions); for the fashion industry (elephants, skins from various animals); for superstitious uses (rhinoceros horns); and for commercial uses (bison, whales).

Indirect causes are those in which humans unintentionally endanger a species. Examples include habitat destruction (deforestation, construction, farming); pollution (pesticides, oil spills); and exotic introductions (goats in the Galápagos Islands, starlings throughout the United States, rabbits in Australia). Have students research one of these causes of endangerment. Encourage them to illustrate the problem on posters and provide a list of suggestions of solutions to the problem. Display the posters.

☀ LEP Students

1. To help students understand the relationship between an animal's structure and way of life, have them design an animal for a particular habitat. Tell them to consider these questions when designing their animals: What are the size and shape of its feet, legs, and neck? How many legs does it have? Is the body covered by fur, feathers, scales, skin, or shell? Is the body long and thin, or is it bulky? How does the animal defend itself? What is the shape of its head, eyes, and ears? What food does it eat? Be sure that students have reasons for each of the structures they draw. Have students label the structures they have chosen in both English and their native language. Then ask students to write a short paragraph explaining how the structures are used in the organism's environment. Correct their answers only for science content, not for grammar.
2. To help students understand the life cycles of insects, have them capture a caterpillar and raise it to adulthood. They will need a small bottle of water, a 2-L milk carton, scissors, glue, a small piece of screening, clear plastic wrap, and masking tape. Cut large windows on two adjoining sides of the carton. Glue the screening over one window opening and plastic wrap over the other. Find a caterpillar eating foliage. Insert the stems of the foliage into a bottle of water. Put the caterpillar and the bottle of foliage into the carton. Close the top, seal it with masking tape, and open it only when the food supply needs replenishing. Have students make drawings and keep detailed records to record what happens. The caterpillar will pupate and undergo metamorphosis to the adult stage. Students should label their work in both English and their native language.
3. At the beginning of each unit, give Spanish-speaking students a copy of the *Spanish Glossary* from the *Teacher's Resource Binder*. Also, let Spanish-speaking students listen to the *English/Spanish Audiocassettes*.

☀ At-Risk Students

Unit 2 offers many motivating activities for students to gain firsthand knowledge of the diversity of living things. For example, *Exploration 3* is a creative exercise for reinforcing the connection between structure and function. Encourage students to make models of their creature and its environment.

The Extinction Game provides a concrete, hands-on approach to learning about endangerment and extinction. Students will be able to see how the forces behind endangerment and extinction affect the population of different animals.

To show how an animal's form is related to function, have students prepare permanent casts of footprints found in soft soil or mud along a stream. They will need some cardboard, paper clips, a small sack of plaster of Paris, water, a pan, and a mixing spoon. To make a cast, students should form a cylinder with the cardboard, securing it with a paper clip. Place it on top of a track, and pour a mixture of plaster into the cardboard cylinder. As the plaster hardens, a model of the track forms. Students should make casts of at least two different kinds of footprints. Then they can compare the structure and how the feet are used (e.g., for wading, walking, swimming, perching, or carrying objects).

Some students may have difficulty classifying organisms. Give them practice in classifying other things that are more familiar to them, such as clothes or foods. After they have practiced developing criteria for classifying these things, they may find it easier to understand why living things are classified the way they are.

Cross-Disciplinary Focus

Drama

Have students research how human exploitation is causing the rhinoceros, or some other organism, to become endangered. Ask them to create a play that tells the story. For example, the basic plot could deal with the hunting and killing of a rhinoceros, the removal of its horn, the selling of the horn, the use of the horn for ceremonial daggers in Arabia, or the buying of some powdered horn in pharmacies in the Orient.

Art

Have students make a display that illustrates the great diversity of forms found in nature. For example, they could collect and mount a variety of leaves or sketch the different branching patterns of trees. They could collect shells of various shapes and sizes. They may also do some research on the function of natural forms (e.g., the spiral pattern of a nautilus shell).

Cross-Cultural Focus

Plants from the Tropics

A great variety of plants can be found in tropical countries. Many houseplants, such as philodendrons, snake plants, African violets, and orchids, originally came from tropical rain forests as did many fruits, such as citrus fruits and bananas. Many other fruits, such as mangoes, passion fruits, papayas, and guavas, are native to tropical countries. Have interested students find out more about houseplants or fruits from the tropics. If possible, bring in some unfamiliar tropical fruits for students to sample. If there are students in your class who come from a tropical country, have them discuss the fruits that are available in their native countries.

Have students research the effects that the plants grown in an area have on the culture of the people living there. For example, in the rain forests of South America, many plants are used for medicinal purposes, for building materials, and for food.

Resources

 ## Bibliography for Teachers

Lawrence Hall of Science, Center for Multisensory Learning. *Environments.* Berkeley, CA: Lawrence Hall of Science, 1983.

_____, Outdoor Biology Instructional Strategies. *Adaptations.* Berkeley, CA: Lawrence Hall of Science, 1980.

_____, Outdoor Biology Instructional Strategies. *Lawns and Fields.* Berkeley, CA: Lawrence Hall of Science, 1982.

National Wildlife Federation. *Amazing Mammals, Part I and II.* Ranger Rick's NatureScope. Washington, DC: National Wildlife Federation, 1987.

_____. *Endangered Species: Wild and Rare.* Ranger Rick's NatureScope. Washington, DC: National Wildlife Federation, 1986.

 ## Bibliography for Students

Herberman, Ethan. *The City Kid's Field Guide.* New York, NY: Simon and Schuster, 1989.

Lerner, Carl. *Plant Families.* New York, NY: Morrow, 1989.

National Wildlife Federation. *Endangered Animals.* Washington, DC: National Wildlife Federation, 1989.

Zimm, Herbert. *Golden Nature Guides Series.* New York, NY: Golden Press. Publication years vary.

 ## Films, Videotapes, Software, and Other Media

Animal Life Databases.
Software.
For Apple II family.
Sunburst Communications
39 Washington Avenue
Pleasantville, NY 10570

Animals with Backbones.
Animals Without Backbones.
Film and Videotape.
Coronet Film & Video
108 Wilmot Road
Deerfield, IL 60015

Classification.
Software.
For Apple II family.
Prentice-Hall Allyn & Bacon
Sylvan Avenue
Englewood Cliffs, NJ 07632

Classifying Animals with Backbones.
Classifying Animals Without Backbones.
Software.
For Apple II family.
D.C. Heath Software
2700 N. Richardt Ave.
Indianapolis, IN 46219

Classifying Plants and Animals.
Film and Videotape.
Coronet Film & Video
108 Wilmot Road
Deerfield, IL 60015

Family Identification.
Software.
For Apple II family.
Conduit
University of Iowa, Oakdale Campus
Iowa City, IA 52242

UNIT 2

Unit 2: Diversity of Living Things

✹ Unit Focus

Ask students to take out their Journals and list as many living things as they can in one minute. Then have them share their lists with each other. Tell them that biologists believe there are as many as 30 million different kinds of living things on earth. About 3 million have been named and described. Ask students if they have ever thought about the great diversity of living things in the world around them. Ask them why they think such diversity exists. Accept all reasonable answers at this point.

✹ About the Photograph

This photograph is of a panther chameleon, which is a native of Madagascar. Chameleons are perhaps the most specialized group of tree-living lizards. Their adaptations include: the ability to change skin colors; a sticky tongue that can shoot out and capture prey; keen, protruding eyes that move independently to better locate prey; feet that are divided, with three toes on one side and two on the other side for gripping branches; and a prehensile tail that can be wrapped around branches to keep the lizard still while it is waiting for prey.

SUCH VARIETY!
1. Living Things in All Shapes and Sizes, p. 66
2. Why Is There Diversity? p. 74

BEGINNINGS AND ENDINGS
3. The Origins of Diversity, p. 86
4. The Loss of Diversity, p. 90

KEEPING TRACK
5. Making Sense of Diversity, p. 94
6. Ordering Diversity, p. 112
7. Life Cycles: Diverse Stages, p. 116

Using the Photograph

Ask students to speculate about the possible advantages or disadvantages the chameleon may have in changing its skin color. (Answers will vary. The changes are a response to heat, light, and the emotional state of the animal.)

List other adaptations of chameleons on the chalkboard. Point out that the structure and behavior of an organism is usually adapted to its way of life. Ask students to infer from the characteristics mentioned where these lizards might live and what they might eat. (Their grasping toes and prehensile tail allow them to live in the trees. Their powerful eyes that can move independently are able to spot small moving insects, and they can catch these insects with their long, sticky tongue.)

About Your Journal

Students should answer the Journal questions to the best of their abilities.

These questions are designed to serve two functions: to help students recognize that they do indeed have prior knowledge about the topic, and to help them identify any misconceptions they may have. In the course of studying the unit, these misconceptions should be dispelled.

For Your Journal

Write a paragraph summarizing your thoughts about each of the following:
1. Why are there so many different kinds of plants and animals on earth?
2. Why can't a fish live both on land and in water like a frog does?
3. What are some of the problems that develop when species become extinct?
4. What features of the panther chameleon shown here do you think have enabled it to survive in its habitat?

Diversity of Living Things

LESSON 1

✳ Getting Started

In this lesson, students observe and describe the diversity of living things around them. First, they observe the diversity of living things in one square meter of grass. Then they observe the diversity that exists in a park or large area. Finally, they solve some riddles that describe the diverse characteristics of living things.

Main Ideas

1. Living things exemplify diversity in their sizes, shapes, and physical structures.
2. A diversity of living things can be found in all environments.

✳ Teaching Strategies

Before class, collect some specimens of plants, small insects, and other living things. Place the plant specimens in jars of water so that they won't wilt. Place the animal specimens in small jars. To provide ample oxygen for the animals, cover the tops of the jars with pieces of cheesecloth secured with rubber bands.

Set up some work stations at which you have placed a few different specimens. Have students work in small groups of three or four to examine the diversity among the organisms. Ask them to compare and contrast the size, structure, and shape of the specimens. In this way, they will be able to observe firsthand the diversity among structures such as roots, stems, leaves, legs, and skin coverings.

(Continues on next page)

66 Diversity of Living Things

SUCH VARIETY!

Living Things in All Shapes and Sizes

There are more than 30 million kinds of living things in the world. These living things come in all shapes, sizes, and structures. **Diversity** is the term that biologists use to describe differences in living things. In this unit you will look at the diversity of living things.

On this page and the next are illustrations of four living things, or **organisms.** Look at each one carefully. How do they differ from one another?

lizard

mushroom

LESSON  ORGANIZER

Objectives

By the end of the lesson, students should be able to:
1. Observe and describe the diversity of living things, including differences in size, shape, and structure.
2. Compare the diverse characteristics of living things.

Process Skills

observing, comparing, describing, writing

New Terms

Diversity—a term biologists use to describe different characteristics in all living things.
Organisms—living things.

With a classmate, discuss the diversity you see. Here are some questions to help you begin.

1. What is the relative size of each organism?
2. What features does each organism have? In other words, does it have roots, stems, leaves, flowers, eyes, legs, hair, scales, skin, or any other notable features?
3. What one characteristic does each organism have that distinguishes it from the other three?

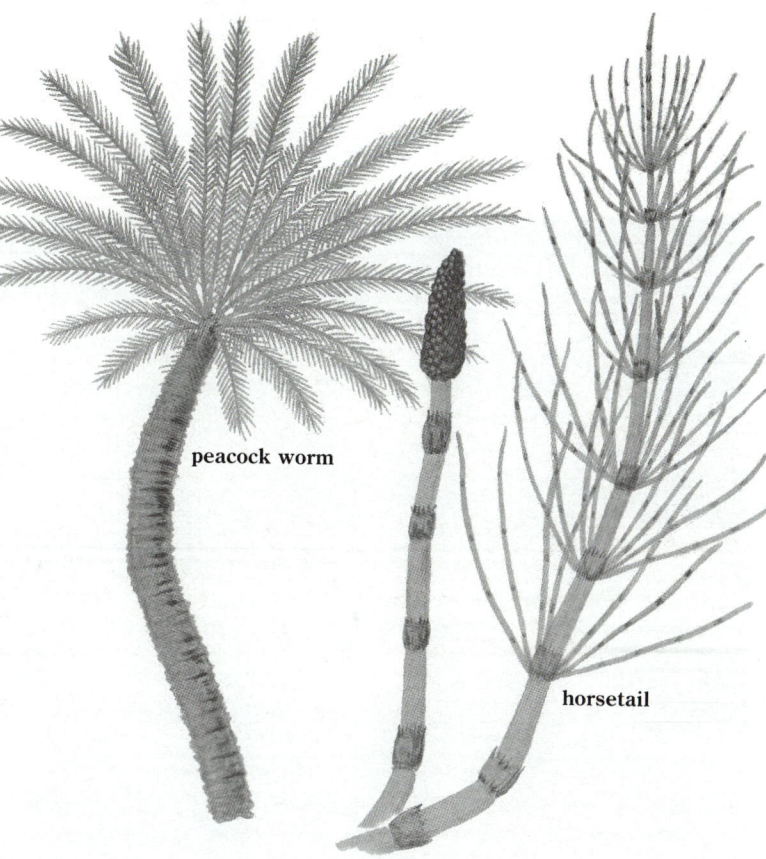

peacock worm

horsetail

You have just described four living things that illustrate the great diversity of organisms. Living things can differ in size, habitat, appearance, eating habits, and methods of self-protection, to name just a few characteristics. Now you will investigate diversity a little further. Do either Activity 1 or Activity 2 of Exploration 1.

EXPLORATION 1

Looking for Diversity

Activity 1

Diversity in a Lawn

While playing ball on the grass or stretching out in your lawn chair to enjoy the sunshine, have you ever wondered about what is happening beneath you? There may be a food-hunting expedition or a ferocious battle going on there. You can learn a lot by getting down on your knees and carefully observing a small area of lawn. In this activity you will mark off one square meter of ground, preferably the day before the Exploration. Then you will study the diversity of animal and plant life in this square meter of lawn.

You Will Need

- sticks (4 per group)
- measuring tape
- heavy string and scissors
- a magnifying glass
- your Journal and a pencil

What to Do

Find a convenient grassy area. Using the measuring tape, measure out 1 m of ground and place a stick at both ends of the tape.

Place one end of the tape at a 90-degree angle from one of the sticks. Measure out 1 m and push a stick into the ground. Repeat this step twice to make a square. Then tie string to the sticks to outline your square.

Materials

Exploration 1
Activity 1: Journal, 4 craft sticks (about 15–20 cm long), measuring tape, heavy string, scissors, magnifying glass, pencil
Activity 2: Journal, pencil
Exploration 2: Journal

Teacher's Resource Binder

Resource Worksheets 2–1, 2–2, and 2–3

Science Discovery Videodisc

Disc 2, Image and Activity Bank, 2–1

Time Required

two to three class periods

(Teaching Strategies continued)
Have students examine the four organisms pictured on pages 66 and 67. Ask students to work in pairs to read and answer the questions asked on these pages. You might want to make reference materials on these organisms available so that students' answers will be more detailed. Then reassemble the class and involve them in a discussion of their responses to the questions. You may wish to have them answer these same questions for the specimens they observed at their work stations. At this point, you may wish to make a composite list of all the differences noted. Encourage the use of the words *diversity* and *organism* in all class discussions.

Exploration 1
Activity 1

Have students read pages 67–69 before doing the Exploration. Tell them that on the first day, they will set up their lawn areas, observe the plant life there, and fill in the table for plants. On the second day, they will explore animal life and fill in the table for animals. Note that the tables require a count of the number of kinds, or *species,* of plants and animals. Students need only a general notion of *species* as a *kind* of organism to complete this activity.

Divide the class into small groups of two or three students each. Before they go outside, remind them not to disturb the area they'll be working in any more than is necessary. Let students "scout out" a good area of lawn to examine. Encourage them to find an area that has not been mowed for a while so that they can observe whole plants.

Students should first get a general feel for the area by noting any large plants or animals that they see. Then they can get down on their hands and knees to take a closer look at the organisms in their marked-off area.
(Continues on next page)

Such Variety! 67

(Activity 1 continued)

It may be impossible to count the exact number of certain kinds of plants or animals, so discuss some guidelines for recording their observations. Suggest that they estimate the number by sampling (counting the number in a representative 10- × 10-cm square and then multiplying that number by 100).

At first, students may think that all the plants look the same. However, on closer examination with a magnifying glass, they should be able to see that there are differences in leaf shape and size. In addition to drawing the plants, students could pull up a specimen of each different kind of plant they see. Later, students could place these specimens between wax paper and press them using heavy books. Each one could be identified and placed on a separate page in a booklet.

(Continues on next page)

EXPLORATION 1—CONTINUED

Observe the plant life in your study site. In your Journal, fill in a table like the one shown here. Then analyze the information you collected by considering the questions below.

Quick Sketch of Each Plant	Number of Same Kinds (Species) of Plants	Description of Each Plant (including size, appearance, color, and any other features you observe)	Common Name

Questions

1. How many types of plants did you find?
2. Which is the smallest plant you found? The biggest?
3. Which plant did you find in the greatest numbers? The fewest?
4. Do all the plants you found have some feature in common? If so, what is it?
5. How does the plant life affect the animal life in your study site?

The next day, carefully approach your staked area to see whether any larger animals, such as butterflies, are present. If you come to the area too quickly or noisily, you may scare away some animals. Using a magnifying glass, look closely for any smaller animals that might be hidden among the plants. In your Journal, fill in a table similar to the one shown here. When you have completed the table, analyze this information by considering the questions below the table. Then share your findings with your classmates.

Quick Sketch of Each Animal	Number of Same Kinds (Species) of Animals	Description of Each Animal (including size, appearance, color, and any other features you observe)	Common Name

Questions

1. How many types of animals did you find?
2. Did you hear animal sounds but not actually see the animal?
3. Did you find any evidence of animal life but not see the animals themselves?
4. Which is the smallest animal you found? The biggest?
5. Which animal did you find in the greatest numbers? The fewest?
6. How does animal life affect plant life in your study site?

Activity 2
Diversity Around You

What to Do

For your Journal, prepare two tables like those in Activity 1. Next, walk around your school, your neighborhood, or a park and fill in your tables. Then, answer the questions that follow each table.

(Activity 1 continued)

Also instruct students to look for evidence of animal life, even if the animals themselves are not present. For example, they may find plants that have been partially eaten.

Students may know the common names for some of the organisms they see. If not, as they draw it, allow them to give a descriptive name to each organism that they find. At this point, it is not important to learn the correct names of the organisms.

When the activity has been completed, involve the class in a discussion of the results. Ask: Did all of you find the same kinds of plants and animals? Did you find these plants and animals in the same numbers? Students may have different names for the same organism. This may be a good time to introduce the concept of standardization of names. Explain that scientists standardize plant and animal names so that they can communicate with each other about different organisms.

Activity 2

You may wish to have students complete this activity in class or as a homework assignment. Tell students this activity is similar to *Activity 1*, except that they are asked to observe a much larger area from a greater distance. Explain that rather than observing small plants and animals found in a tiny patch of ground, they will be observing and describing larger organisms such as trees, shrubs, mammals, and birds.

After the activity has been completed, involve the class in a discussion of their results. It will be interesting for students to compare their tables. If they observed different areas, they will have observed different kinds of plant and animal life.

Such Variety!

Exploration 2

As an introduction to this Exploration, you may wish to read the following poem to your students. Before beginning, ask students to list several words that they associate with bats. After they have heard the poem, ask students if their ideas about bats have changed at all.

Bats

*A bat is born
Naked and blind and pale.
His mother makes a pocket of her tail
And catches him. He clings to her long
   fur
By his thumbs and toes and teeth.
And then the mother dances through
   the night.
Doubling and looping, soaring,
   somersaulting—
Her baby hangs on underneath.
All night, in happiness, she hunts and
   flies.
Her high sharp cries
Like shining needlepoints of sound
Go out in the night and, echoing back,
Tell her what they have touched.
She hears how far it is, how big it is,
Which way it's going:
She lives by hearing.
The mother eats the moths and gnats
   she catches
In full flight; in full flight
The mother drinks the water of the
   pond
She skims across. Her baby hangs on
   tight.
Her baby drinks the milk she makes
   him
In moonlight or starlight, in mid-air.
Their single shadow, printed on the
   moon
Or fluttering across the stars,
Whirls on all night; at daybreak
The tired mother flaps home to her
   rafter.
The others all are there.
They hang themselves up by their toes.
They wrap themselves in their brown
   wings.
Bunched upside-down, they sleep in air.
Their sharp ears, their sharp teeth, their
   quick sharp faces
Are dull and slow and mild.
All the bright day, as the mother sleeps,
She folds her wings about her sleeping
   child.*

—Randall Jarrell

In Exploration 1 you discovered the diversity that can be found in the living things around you. Now here are some riddles showing how much more diversity exists. See how many you can solve!

EXPLORATION 2

The Puzzling Diversity of Living Things

Activity 1

Who Am I?

In the following riddles, you'll find information about the structures, habits, and habitats of some organisms. Read each riddle carefully and think about what it tells you. Which living thing does each riddle describe? If you're having trouble figuring out what a riddle is describing, look at the pictures for some hints. (The pictures are not in order.) If you are really stuck, the answers follow in code. All you have to do is break the code!

Riddle 1

I move slowly when I am young but very quickly when I'm an adult. I eat flying insects, which I hunt near water. I have to be a strong flier to catch my food. When I stretch my four wings, I look like a helicopter. I have two more legs than a dog, and I have very large eyes. I am cold-blooded and have an external (outside) skeleton. Sometimes I'm very colorful. Who am I?

Riddle 2

I can walk, run, and swim. I can see well, but my sense of smell is not very sharp. I am warm-blooded. I am very adaptable and can live in many different environments. I really enjoy changing my environment. I care for my young for many years. I stand upright. Who am I?

Riddle 3

I must live in damp or wet places, avoiding the dry heat of summer and the cold of winter. If I'm living in a cold climate, I become dormant in the winter. If I'm a female, I produce young by laying eggs in water. I survive by eating anything that moves and that I can swallow. I can sing very well. Some of my close relatives can secrete a sticky white poison that can kill or paralyze dogs or other enemies who may try to eat them. Who am I?

Riddle 4

I live in a lake, marsh, salt bay, or on a beach. I eat mostly fish and crustaceans. Although I can fly, I catch fish only by swimming. My great throat pouch is handy for scooping up fish. I fly by alternating several flaps of my wings with a glide. I always fly with my head hunched behind my shoulders. I nest on the ground in colonies. I have a wingspread of 2.5 to 3.0 m. My close cousins live only by the ocean, while I can venture inland. I am happy to report that these cousins are growing in number, even though they suffered from DDT poisoning a few years ago. Who am I?

Riddle 5

I have a very high body temperature. It is usually 7 degrees warmer than a human's. My feet are well adapted for grasping things. I have four toes on each foot: two point forward, and two point backward. I have stiff, spiny tail feathers that act as a prop when I hunt food. I eat tree-boring insects, ants, acorns, flying insects, berries, and sap. My home, which I make myself, is a hole in a tree. I use my bill to chisel the wood. Who am I?

Riddle 6

I have pointed green stalks above the ground and a rounded brown bulb below. People must pull me out of the soil before I can be useful to them. Cooks use me to improve the taste of food. If people bite me, I can bite back, making their eyes water. Who am I? (This riddle is based on an Anglo-Saxon riddle that's over a thousand years old!)

Riddle 7

I live in cold, well-oxygenated water, and I'm a fast, strong swimmer. I am slim, sleek, and colorful. I'm a carnivore; I eat mostly insects and smaller members of my own kind. I spawn my eggs during the spring in small, clear streams. I'm cold-blooded. Who am I?

Riddle 8

I undergo wondrous changes during my life. At the beginning, I am a sweet-smelling pink and white blossom. Later I'm a hard, green ball that makes your eyes and mouth pucker if you try to eat me. Finally, I become a sweet, juicy red or yellowish fruit. People say I keep physicians away. Who am I?

Riddle 9

I am a big animal. My mass is about 225 kg, but my tail is only about 15 cm long. I am dark in color. Generally, I live on forest floors and in thickets. When it begins to get really cold, I enter my shelter for the winter. I don't have very good sight, but my senses of hearing and smell are keen. Using these senses, I find lots of food: small animals, insects, any flesh, garbage, leaves, grasses, berries, nuts, and fruits. I am a good climber, and if I'm disturbed, I may retreat to the upper branches of a tree. Who am I?

Riddle 10

I am warmblooded and hairy. I feed milk to my young. I have no upper teeth, but my lower teeth tell you what kind of food I eat. I chew my cud, and I have a complex stomach. The males of my kind have huge, branching antlers. I have a heavily maned neck. Humans, wolves, and mountain lions are my only enemies, but mountain lions usually won't attack me when I'm fully grown. My young are not camouflaged from these enemies until their winter hair grows out. Sometimes you can hear the males of my kind give a high-pitched bugle call. If this call is answered by another male, a battle may follow. Who am I?

Activity 1

Have students work individually, at home or in class, to solve each of the riddles. There are enough details provided in each riddle to lead students to the realization that every species has unique features. Ask students to list in their Journals all the information given and to research any information that they do not understand. If they cannot solve the riddles from the clues given, then the pictures and coded answers should help.

Answers to

Who Am I?

1. dragonfly
2. human
3. toad
4. white pelican
5. woodpecker
6. onion
7. rainbow trout
8. apple
9. brown bear
10. elk

Such Variety!

Activity 2

Have students work on this activity individually. Make sure that students have enough information to write their riddles. Instruct them to include the same amount of detail in their riddles as the riddles in their texts did.

To help students with their riddles, ask them to find answers to the following questions about the organism: Where does it live? What does it eat? How does it move? What appendages does it have? How does it produce young? What structures does it have? Is it cold-blooded or warm-blooded? Make sure students realize that these terms refer not to the actual temperature of the blood but to the animal's ability (or lack thereof) to adapt its blood temperature to that of its environment. The technical terms are *ectothermic* (cold-blooded) and *endothermic* (warm-blooded).

Students should be encouraged to select animals or plants that are unfamiliar to them, so that they can practice their research skills as well as their writing skills.

You may wish to supply students with the names of some organisms that they can use to write riddles about. You might even write the organism names on separate slips of paper, and let students take turns to choose the one they will write about. Include species of plants and animals that are common in your area. This will give students an opportunity to become more familiar with the living things around them.

In addition to this activity, students could play *20 Questions*. A volunteer could begin by saying: "I'm thinking of an animal" or "I'm thinking of a plant." Other students would be allowed to ask 20 questions about the animal or plant. Each question must have a *yes* or *no* answer. Encourage students to think about the differences between various living organisms.

EXPLORATION 2—CONTINUED

Here are the coded answers to the riddles:
1. CQZFNMEKX
2. GTLZM
3. SNZC
4. VGHSD ODKHBZM
5. VNNCODBJDQ
6. NMHNM
7. QZHMANV SQNTS
8. ZOOKD
9. AQNVM ADZQ
10. DKJ

Hint for decoding:
ZMS = ANT

Activity 2
You Be the Writer!

Read the riddles again carefully, this time noticing which characteristics of the organisms are used to describe their diversity.

Then try writing your own riddles for some of the living things pictured here. Before you start, do some research to find out about the ones you chose. After writing your riddles, see whether your friends can solve them.

Answers to
In-Text Questions

The organisms pictured on page 72, from left to right, are: owl, pine branch, fungi, narcissus, bat, and leopard.

The organisms pictured on page 73, from left to right, are: kangaroo, sea slug (nudibranch) and sea anemone, platypus, jellyfish, octopus, and shark.

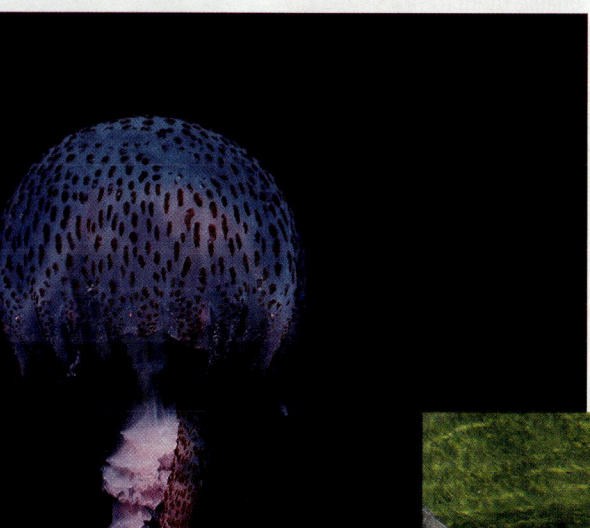

✶ Follow Up

Assessment

1. Give each student a plant or animal specimen, or a picture of a plant or animal. Have each student write a riddle noting the characteristics of that organism. Each riddle should contain enough information to enable another student to identify the organism.
2. Ask students to exchange riddles with their classmates and identify each organism.

Extension

1. Have students use sweep nets to sample and compare the insects living in two different grassy areas. To make a sweep net, bend a wire coat hanger into the shape of a circle. Put an old pillow case on the wire. Use wire or strong tape to fasten the net to an old broom or mop handle. To take a sample, sweep the grass back and forth with the net. Then hold the net closed until the contents are emptied into a jar.
2. Take students on a trip to a zoo to observe the diversity of animals. Assign an animal for each student to research in more depth. They should find out such information as the native environment for the animal, what it eats, and how it produces its young. Reports and illustrations could be displayed in the classroom.

Such Variety! 73

LESSON 2

✳ Getting Started

In this lesson, students learn how plants and animals are adapted to their particular habitats. First, they observe and discuss some adaptations that animals have for obtaining food, for protection, and for locomotion. Then they consider how plants are adapted to their particular habitats.

Main Ideas

1. There are many unique environments on earth, with many habitats for living things in each of these environments.
2. Adaptations enable organisms to survive in their habitats.
3. Plants and animals have a variety of structures and behaviors that help them adapt to their habitats.

✳ Teaching Strategies

Have students read page 74 and study the photographs on pages 74 and 75. Ask them to list the plant and animal adaptations that they observe. Have students categorize the adaptations in terms of whether they are used for obtaining food, for protection, or for locomotion.

Ask students the following questions. How are a tree frog's feet adapted for climbing trees? *(It has suction cups on its feet.)* How are a camel's feet adapted for living in the desert? *(It has wide padded feet that prevent it from sinking into the sand.)* How is a giraffe's body shape adapted for obtaining food? *(It has a long neck for reaching tree leaves.)* How is a pelican's beak adapted for catching food? *(It has a large beak and pouch for catching and holding fish.)*

(Continues on next page)

❷ Why Is There Diversity?

Why are there so many different types of living things? Why isn't there a single "all-purpose" organism or at least one all-purpose plant and one all-purpose animal? These are not easy questions. This lesson will help you find some answers.

You already know that there are many different environments on our planet. There are deserts and oceans, grasslands and mountains, forests and tundra. In each of these environments, there are many different habitats for living things.

Since there is such a diversity of habitats, no single organism can survive in all of them. Obviously an organism that lives on the ocean floor would have great difficulty surviving in a desert!

You learned in Unit 1 that each organism plays a particular role in its habitat. Just as no single organism can survive in all habitats, no single organism can play all roles in one habitat.

All this leads to another question. How is each organism suited to play a certain role in a certain habitat? Keep on reading!

The Fine Art of Survival

Adaptations are the features organisms have that enable them to survive and produce young. Both animals and plants have many different kinds of structures and behaviors that help them survive in their environments. Below are three categories of adaptations to discuss: adaptations for obtaining food, for protection, and for locomotion. The organisms on this page and the next will give you some hints.

1. Think about the great variety of structures and behaviors (adaptations) that animals and plants have for obtaining their food.
2. Can you think of animal and plant adaptations that could be used for protection? Give some examples.
3. Consider an animal's locomotion—its movement from place to place. Name some adaptations that various animals have for locomotion. Is each adaptation you thought of related in some way to the organism's habitat? Explain.

LESSON ❷ ORGANIZER

Objectives

By the end of the lesson, students should be able to:
1. Observe and describe animal and plant adaptations.
2. Infer how animal and plant adaptations help organisms to survive in their particular habitats.

Process Skills

observing, describing, inferring, illustrating

New Terms

Adaptations—features that enable organisms to survive and produce young.

74 Diversity of Living Things

(Teaching Strategies continued)

Have students select an animal of their choice and draw that organism in its correct habitat. Display the students' pictures in the classroom in three groups—adaptations for obtaining food, for protection, and for locomotion.

Answers to

The Fine Art of Survival

1. **Structures and adaptations for obtaining food:**
   - The Venus' flytrap has leaves adapted for catching insects.
   - The spider has glands that secrete silk, which it uses to build the webs that catch its prey.
   - The hummingbird has a long, thin bill for reaching the nectar in flowers.
   - The plant has large leaves adapted for catching sunlight to make food.
   - The lion has sharp claws for catching its prey and sharp teeth for eating it.
   - The mosquito has a sucking tube to obtain blood from its victims.
2. **Structures and adaptations for protection:**
   - The cactus has sharp spines to protect it from being eaten.
   - The blowfish has spines and can puff itself up as a warning to potential predators.
3. **Adaptations for locomotion:**
   - The long arms, legs, and tail of the gibbon enable it to swing from tree branches.
   - The wings of the hummingbird and mosquito enable them to fly.
   - The fins and tail of the blowfish are for swimming.

Appendages—protruding parts of an organism's body.
Camouflage—adaptation that enables an organism to blend in with its environment.
Mimicry—adaptation in which one organism gains protection by looking like another type of organism that predators avoid because of its undesirable smell, taste, or sting.

Materials

Exploration 3: Journal

Science Discovery Videodisc

Disc 2, Image and Activity Bank, 2–2

Time Required

two to three class periods

Such Variety! 75

Disappearing Acts

The photographs on pages 76 and 77 show various animals that use camouflage to hide from their predators. Have students read the text and observe the photographs.

If time permits, ask students to go outside to look for examples of animals that use camouflage (e.g., rabbits or deer in fields, or moths on tree bark). Encourage students to draw the camouflaged animals and their backgrounds. Display the pictures in the classroom.

As another extension, have students draw, color, and cut out objects or animals that they believe would be hard to see if placed in the school yard. Have half the students place their animals in plain sight. Then have the other half see how many of the animals they can find. Repeat this activity having the two groups switch roles. Ask: Which animals were the easiest to find? Which were the most difficult to find? Why? **(Continues on next page)**

Disappearing Acts

Some animals in this world have spectacular survival techniques. But you have to be a very good observer to see them. Like escape artists or magicians, some animal species have developed an amazing variety of illusions. To escape the ferocious jaws, beaks, and fangs of hungry predators, they simply "disappear" into their environment! Other animals do the same trick, not to avoid being prey, but in order to catch their own prey.

Living things pull off their "disappearing acts" by a method called **camouflage** that allows them to blend in with their environment. Some organisms resemble twigs or leaves. Others resemble nonliving things like stones. Many organisms blend in perfectly with the color of their surroundings. The following examples illustrate how the shape and color of organisms can protect them in their environments.

The Katydid

Some types of katydids look like green leaves. If caterpillars accidentally nibble at the edges of the katydid's wings, spots of brown surrounded by yellow rings appear. These markings are just like the spots of decay that would appear on a leaf.

The Chameleon

Chameleons usually appear green or brown to blend in with their surroundings. However, when attracting a mate, defending their territory, or responding to light or temperature changes, chameleons may change to black, red, yellow, white, or orange.

The Stick Insect

Stick insects look like twigs. Their shape and color protect them from being eaten by birds.

76 Diversity of Living Things

The Polar Bear

Polar bears blend in with the snow and ice that are usually part of their surroundings.

The Flower Mantis

The flower mantis looks like a flower. Its appearance attracts insects to feed on what they think is flower nectar. The unsuspecting insects are then devoured by the deceptive mantis.

The Grasshopper

Some young grasshoppers can change their color to match their environment. For instance, nymphs that eat green leaves become green.

The Snowshoe Hare

Snowshoe hares are white in winter to blend in with the snow, and brown in summer to blend in with trees, grasses, and weeds.

Who Am I?

*We each have our special color and shape;
From our enemies it's a way to escape.*

"I'm an extremely changeable fellow;
Any color will do—black, white, or yellow!"

"As for me, I specialize in white—
So I can match my habitat site."

"I am white too, for part of the year;
But when summer comes, then brown will appear."

"Colors—green and brown—are important for me too
But it's a shape like a leaf that really sees me through."

"Yes, shape is important—I look like a flower.
But that's to bring insects for me to devour."

"I could change color when I was small.
Then I was a tasty treat for all."

"I have the best trick that you've ever heard
I look like a twig—it'll fool any old bird."

(Disappearing Acts continued)

If there is time, have students try another activity to demonstrate how camouflage can help an animal to survive. Scatter an equal number of differently colored toothpicks on some green grass. Tell students that the toothpicks represent insects. Allow each student 10 seconds to pick up as many toothpicks as possible. Count the number of toothpicks of each color found. Ask: Which colors were easiest to find? *(the colors most different from green)* Which colors were most difficult to find? *(the colors most similar to green)* Why? *(Green toothpicks are camouflaged better than the other toothpicks.)*

Answers to
Who Am I?

- chameleon
- polar bear
- snowshoe hare
- katydid
- flower mantis
- grasshopper
- stick insect

Such Variety! 77

Attention, Please!

The photographs on page 78 show examples of animals that announce their presence with bright colors, rather than using their coloring as camouflage. The bright colors may attract a mate or warn predators to stay away. Often, brightly colored animals have another form of defense, such as a sting or a distasteful smell or taste.

Have students work in small groups of three or four to observe the pictures and answer the questions. Then reassemble the class and involve them in a discussion of their answers.

Answers to
In-Text Questions

- The male peacock and cardinal use their bright colors to attract mates. The females, who tend the eggs and hatchlings, are drab in color.
- The tiger swallowtail butterfly uses its colors to attract mates.
- The stripes on the skunk warn potential predators to stay away or get sprayed.
- The stripes on the wasp warn predators to stay away or get stung.
- The lionfish has long, fan-like fins and a brightly striped body. These colors warn potential enemies to stay away or risk contact with the potent venom contained within the fish's grooved spines.

Attention, Please!

Some animals are adapted to attract attention. Which animals shown here have an adaptation that

- Attracts members of their own kind for mating purposes?
- Warns possible enemies (predators) of their bitter or smelly secretions or powerful stings?

Mimicry: Looking Dangerous

Some animals, such as the Mexican milk snake, have warning features that protect them from being eaten by predators. The Mexican milk snake is not poisonous, but it looks like the Texas coral snake, which predators avoid because a coral snake's bite can be serious. The Mexican milk snake *mimics* the Texas coral snake. Through **mimicry,** the harmless Mexican milk snake escapes being eaten.

Mexican milk snake

Texas coral snake

There are some very strange examples of mimicry in the animal kingdom. The larvae of the hawkmoth resemble certain poisonous snakes. Another type of moth mimics hornets. Its wings and coloring look like a hornet's, and this moth even has what looks like a stinger on its abdomen. When threatened, the moth bends and twists like a hornet that is searching for a place to inject its venom.

Another interesting form of mimicry is found in insects that have evolved large eyespots that resemble a vertebrate's eyes. These are often on the upper side of the wings, and they are usually covered when the insect is at rest. When the insect is disturbed, however, it spreads its wings, exposing the eyespots. Through experiments, scientists have shown that many birds are frightened by these eyespots.

A bird would avoid these 'eyes'.

Something to Research

Find out about the mimicry of the drone fly. What does it mimic? From which animals is the drone fly protected by its mimicry?

Plants Also Adapt to Their Habitats

Point out that like animals, plants show a variety of adaptive structures. This variety is necessary for survival in different habitats.

You may wish to arrange a field trip for students to observe plant adaptations. If there are different habitats near the school (e.g., a grassy field, a marsh, or a forest), take students to each habitat and observe and compare the different plants in each area. Then ask students to explain how the plants are adapted to each different habitat.

While on the field trip, suggest that students participate in a scavenger hunt to find plant adaptations. Make a list of plant adaptations they could find. Possible examples include a plant that an animal might avoid (a plant with thorns, spines, or stinging hairs), a plant that can live in the shade, a plant that can climb, a plant that can live in water, a plant that has a strong odor, a plant with flowers or fruits, a plant growing on another plant, and a plant that can keep its leaves all winter. Have students look for these adaptations, record the names of the plants in their Journals, and make a quick sketch of each one. To evaluate the hunt, ask volunteers to show their sketches to the rest of the class.

Ask students to read pages 80 and 81 and look at the illustrations. There are 13 plants shown. Divide the class into small groups of two or three students. Ask each group to use the list of plant names on page 81 to identify the plants.

Then have the groups match the 13 varieties of plants with the eight habitat adaptations listed on page 80.

Plants Also Adapt to Their Habitats

Plants, like animals, are uniquely suited to their habitats. Although plants cannot move about like animals, they do have certain adaptions that allow them to survive and reproduce in particular habitats.

Observe the plants shown on these two pages. Then decide which plant best matches each habitat adaptation listed below.

1. A plant that is able to hold on to rocks in swift rivers and streams
2. A plant that can store water
3. A plant that can catch insects
4. A plant that can compete with other plants for sunlight
5. A lawn mower-proof plant
6. A tree that can withstand high winds
7. A plant that grazing animals would not eat because of the plant's protective surface features
8. A plant that is capable of living on the surface of a pond

80 Diversity of Living Things

Perhaps you were able to identify some of these plants because you have seen them before. Their common names are:

Barnyard grass
Barrel cactus
Birch tree
Broadleaf plantain
Coconut palm
Dandelion
Duckweed
Pitcher plant
Pondweed
Scotch thistle
Sundew
White waterlily
Wild rose

Match the names with the pictures.

Answers to

In-Text Questions

The plants that correspond to the habitats described are:
1. d
2. b
3. j, l
4. c, i
5. a, f, g
6. i
7. b, h
8. k, m

The names of the plants are:
(a) broadleaf plantain
(b) barrel cactus
(c) birch tree
(d) pond weed
(e) barnyard grass
(f) dandelion
(g) scotch thistle
(h) wild rose
(i) coconut palm
(j) sundew
(k) white waterlily
(l) pitcher plant
(m) duckweed

Such Variety!

Answers to
Adaptations for Seed Dispersal

Have students make sketches to show how the pea seed could be adapted to meet each of the needs listed. Drawing the seed adaptations will help students see the relationship between form and function. Encourage students to be creative. Possible solutions:

- Add waterproof air sacs so that the seed could float in water.
- The pod could be modified so that it would pop open and eject its seeds, like a touch-me-not pod. Also, if the pods were positioned high on a tall, flexible stalk, they would be thrown farther from the parent plant.
- Animals are attracted to pea seeds, but they eat and destroy them. If the seeds were surrounded by a tough coat that was resistant to digestion and then surrounded by tasty flesh, animals would eat the flesh and eliminate the seeds in different locations.
- Having a sticky substance or burrs on its surface would enable the seed to be picked up by furry animals.
- Having wings or tufts would enable the seeds to be carried away by the wind.

A Picture Puzzle

As an extension, have students go on a seed hunt. Encourage them to use their observational and deductive skills to find out how each kind of seed is dispersed.

- Parachute seeds (e.g., dandelion, sycamore, and milkweed) can be blown to see how they are carried away by the wind.
- Helicopter seeds (e.g., maple, ash, elm, tree of heaven, and linden seeds) can be tossed into the air to see how they descend.
- Sling-shot seeds (e.g., jewelweed, also called touch-me-nots) can be touched to see how they burst and fling their seeds.
- Hitchhiker seeds (e.g., burdocks and beggar's-ticks) will stick to clothing, hair, and fur.
- Indigestible seeds in fruits can be seen when the fruit is cut open.
- Boat seeds (e.g., coconuts, sea beans, and cranberries) can be seen floating in water.

Adaptations for Seed Dispersal

Most plants produce seeds that grow into new plants. But seeds cannot always survive if they simply drop directly beneath the plant that produced them. They may need to be moved to another area. Many types of seeds, therefore, are adapted to travel. This helps ensure that new plants will grow and survive.

Look at a pea seed. Do you think it could travel very far on its own? How might you change it so that it has a greater chance to travel and survive? Working with some classmates, design changes in the pea seed that will enable it to do the following:

- Float on water
- Be thrown a distance by the parent plant
- Attract an animal that would carry it to another location
- Hitchhike on an animal for some distance
- Be carried by the wind

A Picture Puzzle

Examine each seed closely. Identify the adaptation that is being used to help each seed travel away from the parent plant. Do these adaptations resemble the suggestions you made for the pea seed?

Answers to
In-Text Questions

This exercise enables students to check the ideas they thought of in the previous exercise. Have them make tables in their Journals with the headings: *Plant* and *Adaptation for travel*.

- **a.** acorn—attractive as a food source for animals; spread and/or buried by these animals
- **b.** squash—eaten by animals; seeds dispersed in feces
- **c.** cucumber—eaten by animals; seeds dispersed in feces
- **d.** pod—drying, bursting, expelling seeds
- **e.** pod—drying, bursting, expelling seeds
- **f.** burdock—seed has hooks that stick to clothing or fur
- **g.** blueberries—eaten by animals; seeds dispersed in feces
- **h.** maple seed—wings allow the seed to drift in the air
- **i.** milkweed—pod breaks open and silky hair on seed helps it to be airborne, even in a gentle breeze
- **j.** hay—small hooks on the seeds attach to clothing and fur
- **k.** cones—washed away by heavy rains
- **l.** apple tree—fruit eaten by animals; seeds dispersed in feces
- **m.** coconut—floats in water
- **n.** poppy—pod throws out seeds
- **o.** mountain ash—berries eaten by birds; seeds dispersed in feces
- **p.** dandelion—airborne by silky tufts of hair

Such Variety! 83

Exploration 3

Encourage students to draw and then describe the structures of an animal that meets the specifications outlined in the activity. Do not tell students at this point that these are actually the features of a lobster or crayfish. (Lobsters have two pairs of feelers for testing food; appendages around the mouth for creating water currents; two large pincers for holding food, for breaking open shells, and for defense; four pairs of legs for walking and holding food; smaller appendages under the tail for swimming and holding eggs; and tail appendages for quickly swimming away.)

Follow Up

Assessment

1. Ask students to design a plant that could survive in a particular habitat, such as a shady spot, a pond, a swiftly flowing stream, or a desert. Then ask them to write a short paragraph explaining how its adaptations help it to survive.
2. Provide students with pictures of animals. Ask them to observe the animals' structures, to describe the animals' adaptations, and to deduce where the animals could live.

Extension

1. Take students on a field trip to do some bird-watching. Ask them to pay particular attention to the adaptations of the birds' beaks and feet. Have students use their observational and deductive skills to discover some facts about each bird's way of life. Use photographs of birds if a field trip is not possible. *(For example, ducks and geese have webbed feet for swimming and broad, flat beaks for straining food from water. Woodpeckers have two toes in front and two in back with sharp claws for climbing trees. Most songbirds have three toes in front and one behind for perching on limbs of trees. Ground birds, such as pheasants and partridges, have feet useful for running and scratching the ground. Ground birds also have short, stout beaks for ground feeding. Birds of prey have sharp, curved talons for grasping prey and strong, hooked beaks for tearing apart flesh. Seed-eating birds have short, thick bills for crushing seeds. Insect-eaters have slender, pointed beaks useful for picking up insects. Herons have beaks like long spears for catching fish.)*
2. Students could make a bulletin-board display about seed dispersal. Seeds could be collected and arranged into groups according to their mode of dispersal.

EXPLORATION 3

Designing an Animal

Here's another riddle for you: How is your thumb like a monkey's tail? Give up? They're both **appendages,** or protruding parts of the body. In fact, your entire hand and arm is an appendage. So are an insect's antennae, or feelers.

Appendages are adaptations designed to help an animal perform various specialized tasks. Appendages help an animal live successfully in its environment.

What to Do

Design an animal with appendages that will enable it to have the following characteristics:

1. The animal lives in water.
2. Its heavy body needs a lot of support as the animal walks.
3. It can walk on the bottom of a body of water for kilometers without stopping.
4. It can dart away suddenly from its enemies by swimming.
5. It can create water currents to bring food in the water to its mouth.
6. It tests its food before eating it.
7. It has appendages to hold larger pieces of food.
8. It can break apart hard bits of food.
9. Its diet includes shelled animals.
10. It has appendages to hold its young.
11. It has formidable defensive weapons.

In your Journal, draw the creature you designed. Does it look like any animal you have seen before?

---

84   Diversity of Living Things

Challenge Your Thinking

1. In letters from Tim Lawson, your pen pal in Australia, you've learned about the Australian "bush," which includes giant grain fields and tropical forests. You've also learned about the "outback," an area of great space, scattered inhabitants, and desert regions where beautiful wild flowers grow when it rains. Tim has told you about vegetation such as the gum tree (grown nowhere else) and the golden-flowered wattle. And he's written about strange animals like kangaroos, koalas, and wombats.

   Write Tim a letter telling him about the many kinds of plants and animals he would see during a visit to your area of the United States. Tell him about how these organisms are adapted for survival in your environment.

2. Sketch an adaptation that
   (a) Would enable an animal to cut down a tree.
   (b) A water plant might use to float on the surface.
   (c) Might be used as a defense against predators.
   (d) Would enable a plant to live with very little water.
   (e) Would enable an organism to blend in with its environment.

Answers to Challenge Your Thinking

1. Answers will vary according to your area of the country. Students might consult reference books written about your state or field guides for your region of the country. Add more variety to this assignment by allowing students from other states or countries to write about their native regions.

2. Answers will vary, but possible solutions are listed below. Encourage creativity.
   (a) sharp, serrated beak or teeth
   (b) broad, waxy leaves that prevent saturation; air-filled, sac-like leaves or flowers
   (c) hard shell; offensive smell or spray; sharp appendages; various forms of mimicry
   (d) large internal storage areas for water; leaves with limited surface area from which water could escape through evaporation
   (e) camouflage

LESSON 3

Getting Started

In this lesson, students are introduced to the theory of natural selection. First, they analyze an experiment about light- and dark-colored peppered moths. Then they observe the differences between Darwin's finches and speculate about what the differences mean.

Main Ideas

1. Natural selection explains how the different features of a species change over generations in response to changing environmental conditions.
2. Darwin developed his theory of natural selection by observing the differences in the shapes and sizes of the beaks of finches that inhabited the Galápagos Islands.

Teaching Strategies

Natural Selection

Have students read page 86 to learn about Darwin and his theory of natural selection. Ask volunteers to summarize Darwin's ideas and think of examples that support it.

Make sure that students understand that variations occur naturally in populations of organisms. Sometimes pressures from the environment, such as a reduced food supply, allow individuals with certain variations to better survive in the environment. Those organisms that survive are able to reproduce, and their characteristics are then passed on to the next generation.

BEGINNINGS AND ENDINGS

3 The Origins of Diversity

Natural Selection

Sometimes, over many generations, the characteristics of plants and animals change. The changes enable the plants and animals to live successfully in a new habitat. For example, the porpoise is an animal that lives, eats, and reproduces in the water. But unlike other aquatic organisms, the porpoise breathes air through lungs, just like land animals do. The porpoise's ancestors were land animals that, over time, changed their habitat from land to water. They probably adapted to the oceans because they found more food or more easily escaped predators there than when they lived on land. The modern porpoise, which is a strong and fast swimmer, shows just how successful these changes were.

Changes can also permit a species to live more successfully in its own habitat. Polar bears, for example, have not always been white. But the polar bears that were white were more successful in finding food. Blending in with their environment enabled them to sneak up on their prey. Hence white polar bears survived to reproduce. Over many years, the white polar bears grew in number, while the dark ones gradually disappeared.

In the 1800s, Charles Darwin developed a theory that organisms had *evolved,* or changed over many generations, by a process he called **natural selection.** Darwin called the process natural selection because the environment favors, or selects, organisms that are fit for survival. Darwin thought that organisms with traits that are well suited to the environment survive and reproduce at a greater rate than organisms that are less suited to the environment. The well-suited organisms thus pass desirable traits on to their offspring. Natural selection explains how dark polar bears were replaced by white ones and how porpoises became aquatic animals.

Sometimes the habitat of a species will change in some way. This change may force the species to move to a new habitat, adapt over many generations to the changed habitat, or become extinct. One example of a species that adapted to a changed area is the peppered moth, which you are about to examine in detail. Study the research data given on page 87. Do these facts support the theory of natural selection?

Which polar bear is harder to find?

LESSON ORGANIZER

Objectives

By the end of the lesson, students should be able to:
1. Predict how a species will adapt to changes in its habitat.
2. Observe Darwin's finches and use the theory of natural selection to analyze their differences.

Process Skills

observing, predicting, formulating hypotheses, drawing conclusions

New Terms

Natural selection—theory developed by Charles Darwin explaining how the characteristics of a species can change over many generations such that the species better suits the environment.

---

86 Diversity of Living Things

The Case of the Peppered Moth

Date:
The 1800s, during the Industrial Revolution

Place:
England, near the industrial city of Birmingham

Researcher:
Professor Kettlewell and his assistants

White moth on a light trunk

Dark moth on a dark trunk

Research Subject:
The peppered moth. Peppered moths have lived in the forests of England for thousands of years. They rest on the trunks of trees during the day and are the diet of many birds. Peppered moths vary in color, from light-colored to dark-colored.

Research Problem:
Did air pollution, which covered tree trunks with black soot from the many new industries, affect the survival of the peppered moth?

Conditions:
Before the Industrial Revolution, the tree trunks were light-colored. The trunks and branches were also covered with silvery-white lichen.

As the Industrial Revolution progressed, pollution killed the lichen and blackened the tree bark.

Hypothesis:
Kettlewell formed a hypothesis about the peppered moth. Form your own hypothesis. It should include the effect that you think the Industrial Revolution had on the peppered moth's survival.

Procedure:
With his assistants, Kettlewell then experimented to test his hypothesis. They used the following procedure:

1. They located two areas. One was a wooded area with lichen-covered oak trees. The other was a wooded area that had been subjected to pollution for many years.
2. They released a known number of light-colored and dark-colored peppered moths into each area.
3. After a given amount of time, they recaptured as many moths as they could.

Results:
In the unpolluted area, more light-colored moths survived. In the polluted area, more dark-colored moths survived.

Discussing the Results:

1. Do these results support your hypothesis? (They supported Kettlewell's hypothesis.)
2. How can you explain the results?
3. Why didn't Kettlewell release the moths only into the polluted area?
4. Do Kettlewell's findings support the theory of natural selection? Explain your answer.

Materials
none

Science Discovery Videodisc
Disc 2, Image and Activity Bank, 2–3

Time Required
two class periods

The Case of the Peppered Moth

This activity is a good illustration of natural selection. Students are asked to formulate their own hypotheses and to compare their ideas with the historical facts.

Have students work in small discussion groups to read the page and answer the questions. Suggest that they read up to, but not including, *Procedure*. Then ask each group to formulate its own hypothesis before reading the one that Kettlewell formed.

Answers to
Discussing the Results

1. If students' hypotheses were incorrect or untestable, allow them to restate a hypothesis in light of the results from Kettlewell's experiments.
2. When the light-colored bark was covered with black soot and the lichens were destroyed, the light moths lost their camouflage. They then became easy prey for predators (birds). Under the same circumstances, however, the dark moths had the advantage of camouflage.
3. The moths released into the unpolluted area were used as the control group in the experiment.
4. Kettlewell's findings support the theory of natural selection because more of the dark-colored moths survived to carry on the population in the polluted area; by contrast, the light-colored moth population dropped.

Darwin's Finches

Point out to students that the Galápagos Islands are a living laboratory of natural selection. This is because they are volcanic islands isolated from the South American mainland by some 960 kilometers of ocean. Ancestors of the plants and animals that inhabit the islands today were brought from the mainland by wind currents or floating debris. The organisms that arrived there first had virtually no competition from other species since there were no species already there.

Darwin's Finches

In 1835 Charles Darwin spent five weeks visiting the Galápagos Islands as the naturalist on H.M.S. *Beagle*. He observed, recorded, collected, and preserved everything he could of the islands' natural history. There were many strange and colorful animals. But what excited him most were drab little birds that made unmusical sounds—finches. These birds all resembled one another closely, except for one set of features—the size and shape of their beaks. Separated for thousands of years on the different islands of the Galápagos, the finches of each island had adapted in a unique way to their own environment. The differences among the finches of the various islands are shown in this table.

Charles Darwin

Differences Among Some of Darwin's Finches

Name of the Finch	Feeding Habit	Form of Beak
small tree finch	Uses delicate bill to eat aphids and small berries.	
large tree finch	Grinds fruit and insects with parrot-like bill.	
small ground finch	Uses pointed bill to eat tiny seeds and pick ticks from iguanas.	
large ground finch	Conical bill enables it to eat large, hard seeds.	
cactus finch	Long bill probes for nectar in cactus flowers.	

Answers to
In-Text Questions (page 89)

1. The beak of the small tree finch is broad and short, enabling it to eat aphids and small berries. The beak of the large tree finch is similar in structure to that of the small tree finch, but it is larger. Its size enables the bird to obtain and grind larger fruit and insects. The beak of the small ground finch is long and pointed, enabling it to obtain tiny seeds from ground surfaces and ticks from iguanas. The large ground finch has a strong conical beak that enables it to obtain and crush hard seeds. The cactus finch has a long beak that it uses to probe for nectar in cactus flowers.
2. Yes. Each island had a slightly different food source. The finches that had the appropriate beak for feeding on each particular food source had the greatest chance of surviving and passing on that characteristic to their offspring.
3. (a) Accept all reasonable responses. Scientists hypothesize that the original finches were probably seed eaters, like those found on the South American mainland.

1. How is the structure of the beak well suited to the diet of each group of finches?
2. Do Darwin's finches support the theory of natural selection? Give reasons for your answer.
3. It has been speculated that Darwin's finches reached the Galápagos from the mainland of South America as a single flock perhaps a million or more years ago. Think about the following questions and explain your answers.
   (a) What do you think the original finches looked like? Why?
   (b) Is it possible that the original birds were various species that arrived on the islands at different times?
   (c) Assume that one flock of finches gave rise to the 14 different species now existing on the islands. For this to occur, would it be significant that the Galápagos consist of many small islands rather than one large one?
   (d) What advantages would the finches have had in arriving on the islands under the following conditions?
      (i) When no other species with exactly the same diet existed
      (ii) When there were no predators
      (iii) When there were no parasites to live on the finches and weaken them
4. How has diversity helped Darwin's finches to survive?

Adaptation Adventure

Would you like to live in a new, vastly different environment? How about the moon? Underwater? A planet in outer space? What adaptations (including technological ones) would help you survive? Think about the adaptations that were required for the *Apollo* astronauts to live and travel on the moon for several days.

Create a tale about how you and a group of your friends survived in a very different environment. Describe the conditions of the new environment, the difficulties you would encounter, and the adaptations you would need to survive. You might be able to develop technological solutions (adaptations) to these new problems quite quickly. But changes in the structure or function of organisms' bodies (natural adaptations) often take many, many generations to occur—or they may never occur at all. What might happen to your species if the necessary technology is not available?

H.M.S. *Beagle* arrives in South America. Can you find the Galápagos Islands on a map?

Adaptation Adventure

Encourage students to read the paragraphs and then write some creative tales about a strange new habitat and the adaptations they would need for survival there. Suggest to students that they make illustrations of the adaptations to accompany their stories.

Answer to
In-Text Question

The species may become extinct if the necessary technology is not available.

Follow Up

Assessment

Ask students to describe and analyze how different species of finches came to inhabit the Galápagos Islands and how their feeding habits became specialized.

Extension

1. Encourage students to do some library research on Darwin's theory of evolution by natural selection. Have students find out about the fossil remains that he found in Argentina and the inferences he made about them.
2. Darwin was greatly influenced by the theories of the geologists Charles Lyell and James Hutton. Have students find out what these theories were and how they influenced Darwin.

(b) No, because the finches have so many similarities other than the shapes of their beaks.
(c) Yes. The birds' beaks had to adapt to the food sources available, which varied from island to island.
(d) (i) There would have been no competition from other species for the same food.
   (ii) The finch population could feed and reproduce with no threat to its existence from predators.
   (iii) The finch population could remain healthy and strong without any damaging effects from parasites.
4. Because of the diversity of their beaks, the finches were able to successfully inhabit many different habitats on the islands. Diversity also reduced competition for limited food sources.

LESSON 4

★ Getting Started

In this lesson, students play *The Extinction Game,* which should help them to better understand the kinds of changes that can affect a species' survival. This lesson also helps students identify the positive and negative effects that humans can have on populations in the environment.

Main Ideas

1. Extinction results from a number of natural and human-made causes.
2. In an environment with little diversity, there is a greater chance of living things becoming extinct.
3. Human actions can prevent animals from becoming extinct.

★ Teaching Strategies

Ask students to bring in newspaper and magazine articles describing endangered species. Ask students to read the articles and think about the various reasons why plants and animals become extinct.
(Answers to In-Text Questions follow on next page)

4 The Loss of Diversity

Have you ever thought about what would happen if there were very little or no diversity among living things? Imagine what the results would be if:

- All bears were black.
- All rabbits were white.
- All plants were 5 cm tall.
- No insects could fly.
- The only living things in the oceans were seals.

Getting Started

Make 20 cards each for the following animals: whooping crane, bowhead whale, wood bison, sea otter, Peary caribou, eastern cougar, white pelican, Eskimo curlew. Have 20 blank animal cards on hand in case some of the animal populations grow beyond 20 members. Make two copies of each extinction card shown along the bottom of this page and the next.

Also make one cardboard token for each animal to travel on the gameboard pathway.

The Extinction Game

You have seen that animals may become *extinct* when their habitat changes significantly. The Extinction Game will help you discover the kinds of changes that may affect the survival of certain animals. Two to eight people can play at one time.

To play the game, choose a certain animal from the list below. Begin with a population of 20 of your animal. To be a winner in the game, you need to finish before your animal becomes extinct. You will be using two sets of cards for this game: "animal cards," which represent each of the eight animals, and "extinction cards," which tell you how many members to add or subtract from your animal's population.

Place extinction cards face down in a pile on the table. (After drawing an extinction card, place it on the bottom of the pile.) Roll the die to determine who begins—the highest goes first. Then, in turn, each player throws the die and makes a move. Remove or add animal cards from your pile as required by the extinction cards. If your animal becomes extinct, you are out of the game.

Extinction is a process that generally happens over a long period of time. This period may be represented by traveling the path a second time with your surviving population.

Animal Cards

Extinction Cards

Make two copies of each extinction card.

LESSON 4 ORGANIZER

Objectives

By the end of the lesson, students should be able to:
1. Explain some of the natural and human-made pressures that can cause extinction.
2. Describe how humans can cause, but also help prevent, extinction of an organism.

Process Skills

analyzing, explaining cause and effect relationships, researching

New Terms

Endangered species—a species that is near extinction and may not survive in the wild unless it is protected.
Extinction—the dying out of a species.

90   Diversity of Living Things

Materials

The Extinction Game: dice, Blackline Master of animal and extinction cards, scissors (or cardboard to make cards and tokens, scissors, and markers or crayons)

Teacher's Resource Binder

Resource Worksheet 2–4

Science Discovery Videodisc

Disc 2, Image and Activity Bank, 2–4

Time Required

two class periods

Answers to
In-Text Questions

Have students read the introduction to the lesson and speculate about the results for each given situation. A few suggestions:

- If all bears were black, those in the Arctic would have no camouflage.
- If all rabbits were white, those that live in woods and fields would easily be seen by predators.
- If all plants were 5 cm tall, overcrowding would result. Plants would not be able to take advantage of the resources available in the higher and lower layers of a forest ecosystem.
- If no insects could fly, some insect-eating birds and bats would starve, and many flowers would not be pollinated.
- If the only living things in the oceans were seals, they would have nothing to eat.

The Extinction Game

Divide the class into small groups of between two and eight students. Ask each group to make a set of 20 of each of the animal cards and two of each of the extinction cards. Invite each group to play the game. *Note:* The *Teacher's Resource Book* contains pre-made animal and extinction cards that can be photocopied instead of having students make the cards themselves. There are five more extinction cards in the *Teacher's Resource Book* material than in the Pupil's Edition.

You might want to make reference material available for students who want to learn more about the animals on the extinction cards.

The extinction cards contain realistic information about the dangers and possible causes of extinction. You may wish to give students a simple pre-test and post-test to see what they have learned as a result of playing the game. Questions could include "What types of natural changes influence the survival of a species?" and "What are some human-made changes that can influence the survival of a species?"

Beginnings and Endings

Back from the Brink

Have students read about the brown pelican on page 92. Ask: Why should we be concerned about the brown pelican? What difference would it make if it became extinct? *(The arguments for saving the brown pelican, or any other species, can be categorized in terms of:*
ethical values—*respect for all forms of life*
scientific values—*contributions that a species can make to the study of science*
ecological values—*an organism's role in balancing an ecosystem.)*

Answers to
In-Text Questions

The sequence students choose should indicate their understanding that DDT caused the decline of the brown pelican on West Anacapa Island, but that the pelican population there was restored to viable numbers in the 1980s. Suggested sequence: f, h, d, a, e, g, c, b

Answers to
Another Close Call

Students' answers will vary but should indicate that in Texas and Louisiana, a pesticide poisoned the fish that brown pelicans ate. When the fish died, the pelicans' food supply was cut off, and the population of pelicans dropped almost to zero. The use of the pesticide has since decreased, but the numbers of pelicans in Texas and Louisiana will probably remain low because of decreased food supply, oil spills, and the loss of habitat.

Back from the Brink

The brown pelican is a fish-eating coastal bird that nests along the Atlantic, Pacific, and Gulf shores of the Americas. Below are some facts about brown pelicans on West Anacapa Island, a major breeding colony for brown pelicans in California. The facts trace the brown pelican's story from the days when it lived in large numbers on West Anacapa Island, through its decline, to its comeback from the brink of extinction. But the facts are not placed in the right sequence. Can you put them in a better order?

(a) Investigators discovered that a chemical company had been dumping waste DDT into the Los Angeles sewer system for some time.

(b) An average of 5000 pairs of brown pelicans nested on West Anacapa Island from 1985 to 1989.

(c) In 1973 the brown pelican was placed on the United States' list of *endangered species*—those species that may not survive in the wild unless they are protected.

(d) When brown pelicans ate DDT-contaminated fish, the DDT accumulated in their bodies. This caused the shells of their eggs to be so thin and fragile that they often broke while the eggs were being laid or incubated.

(e) In 1970 only one brown pelican hatched on West Anacapa Island.

(f) Brown pelicans nested in large numbers on West Anacapa Island.

(g) In 1972 the use of DDT was banned.

(h) DDT worked well in killing mosquitoes and other insects, but biologists discovered that DDT had contaminated the Pacific Ocean and the fish in it. This caused considerable harm to the food chain.

Another Close Call

Examine the following diagram closely. What does it tell you about the survival of the brown pelican in Texas and Louisiana?

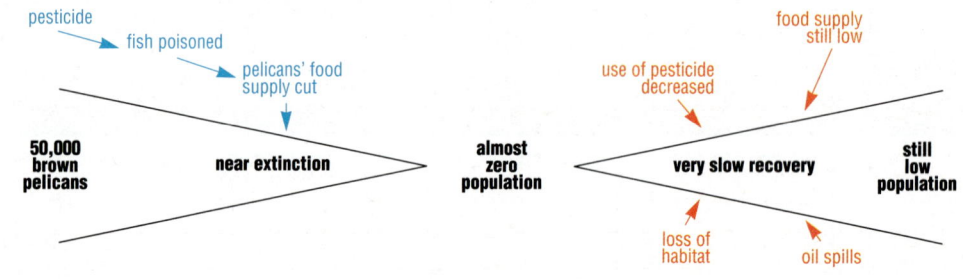

✷ Follow Up

Assessment

1. Ask students to select one or two important reasons for saving wildlife. Encourage them to write a newspaper editorial on the question, "Why save wildlife?"
2. Ask students to write a short story told from the point of view of an endangered species. The story should incorporate some of the natural and human-made pressures that have threatened the species and some that have helped it.

Extension

Show students pictures of species that became extinct due to natural causes (e.g., dinosaurs and saber-toothed tigers). Then show them pictures of animals that are now extinct due to human interference (e.g., the dodo bird and the passenger pigeon). Ask them to do some research on one of these animals to find out more details about what happened.

Suggest that students find out about the protection offered to endangered species (especially those in their area) by wildlife departments; zoos; and local, state, and national parks.

Challenge Your Thinking

1. Natural selection is sometimes referred to as *survival of the fittest*. Do the examples of the peppered moth and Darwin's finches illustrate the idea of survival of the fittest? If so, how? Think of some other examples that either support or refute this idea.

2. It is believed that the ancestors of modern giraffes did not all have long necks. Some had long necks, while some had short ones. Use the idea of *survival of the fittest* to explain why giraffes no longer have short necks.

3. An article in the local newspaper stated: "Everything that humans do to the environment causes animals to become extinct." Write a letter to the editor of the newspaper to disagree with the statement. In your letter, describe some of the ways that humans help prevent animals from becoming extinct.

Answers to Challenge Your Thinking

1. Yes. The peppered moths and finches that were best suited to their environment (the fittest) were the ones that survived and reproduced. Other examples supporting the idea of *survival of the fittest* might include the porpoise and the polar bear. Make sure students understand that "fitness" is a measure of both survival *and* reproductive success. (A mule may be very "fit" and strong, but it has no reproductive success because it is sterile.)

2. The short-necked giraffes probably had to compete with deer and other animals for the leaves on the lower parts of trees. Faced with a scarcity of leaves, the short-necked giraffes might have eventually been in danger of starving. Able to reach the leaves on the upper parts of trees, giraffes born with longer necks survived in greater numbers and thus had a greater reproductive advantage. As a result, the trait for long necks was passed on to future generations.

3. Students' letters could mention the protection that wildlife departments, zoos, and parks offer to endangered species. They could mention recycling and pollution-reduction efforts, which help prevent the destruction of habitats. Encourage students to include in their letters their own innovative ways to prevent extinction.

LESSON 5

✶ Getting Started

In this lesson, students are introduced to the concept of classification. They learn that most living things are classified into one of two kingdoms—plants or animals. Then they learn about the subgroups in each kingdom.

Main Ideas

1. Diversity creates the need to classify living things into groups and subgroups.
2. Most living things can be classified into two kingdoms—the *plant kingdom* and the *animal kingdom*.
3. The animal kingdom is divided into *vertebrates* and *invertebrates*.
4. The plant kingdom is divided into two groups—*mosses* and *plants with tubes for conducting food and water*.

✶ Teaching Strategies

Ask students to explain how they group things in their everyday lives. Ask: Do you arrange your clothes in a certain way? Do you arrange record, cassette, or compact disc collections in a particular way?

Discuss the problems that would arise if there were no classification systems in libraries, telephone books, or the mail system. Students should conclude that things are classified for the convenience of people who study or use them. Explain to students that living things are classified for the same reason.

KEEPING TRACK

5 Making Sense of Diversity

Grouping Living Things

As you may well imagine, it's difficult to keep track of over 30 million kinds of living things! The easiest way to do this is to group the organisms. You already have lots of experience with grouping things. For example, in your kitchen, how is the silverware grouped? Are the dishes kept in one place and the dish towels somewhere else? Are the pots and pans separated from the drinking glasses? Here are some more situations to think about in which items are sorted into groups, or *classified:*

- How are the books classified in the school library?
- How does the telephone company use grouping to keep track of everyone with a phone?
- There are over 240 million people in the United States. How does the postal service group all these people so that mail can be delivered to every person?

Name some other situations in which items are divided into groups.

Now think about some ways in which you might classify living things.

LESSON 5 ORGANIZER

Objectives

By the end of the lesson, students should be able to:
1. Classify living things into two kingdoms—*plants* and *animals*.
2. Classify animals into two groups—*vertebrates* and *invertebrates*.
3. Classify plants into two groups—*mosses* and *plants with conducting tubes*.

Process Skills

observing, comparing, classifying, designing a classification system

New Terms

Classify—to sort into groups.
Kingdoms—the major divisions of living things.

94   Diversity of Living Things

Try to place all the living things in the illustration on page 94 into two groups. Then further divide each group into two more subgroups. Invent a name for the living things in each group and subgroup. Your classification system should look like this:

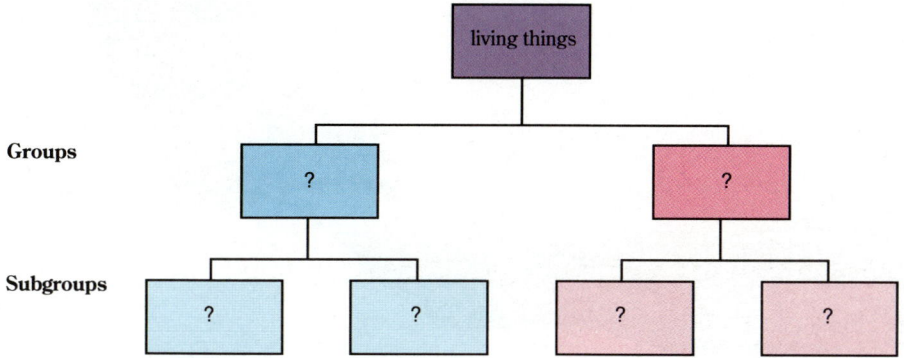

Paulette devised this way to classify the organisms:

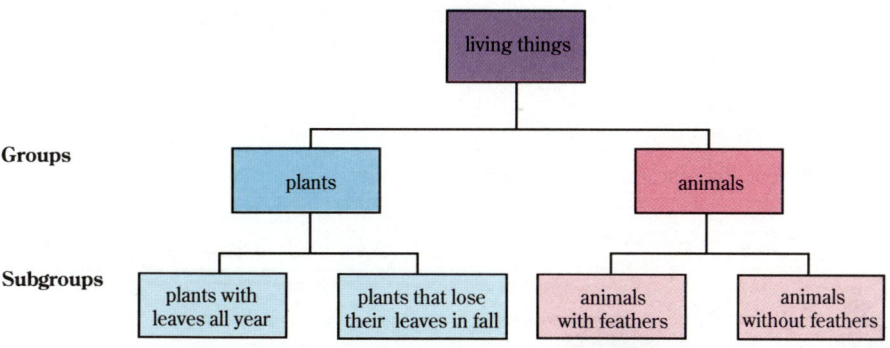

Paulette's system is only one among many ways of classifying organisms. Compare your system with Paulette's. Try classifying the living things in the picture according to Paulette's system. How useful do you think her system is?

Now let's look more closely at the differences between organisms—differences that help you make a classification.

1. Choose one of the living things in the illustration on page 94. In your Journal, list the distinguishing features that you could use to identify it. Read your list to some of your classmates. Ask them to suggest other organisms that would also match the features on your list. How many different organisms did your classmates suggest? Do you always need to talk about *each single living thing*, or can you talk about *a group of living things*?

2. Try doing the same thing again with another organism.

Vertebrates—animals that have internal (inside) skeletons with which to support themselves.
Vertebral column—the backbone.
Invertebrates—animals without backbones.

Materials

Putting It All Together: A variety of preserved or live specimens of plants and animals (optional), Journal
A Simplified Classification System: Journal

Teacher's Resource Binder

Resource Worksheets 2–5 and 2–6
Resource Transparency 2–1
Graphic Organizer Transparencies 2–2, 2–3, and 2–4

Science Discovery Videodisc

Disc 2, Image and Activity Bank, 2–5

Time Required

five class periods

Grouping Living Things

Have students read the introduction to the lesson on page 94. It lists examples of classification in our everyday lives.

Then divide the class into small groups of three or four students and have the groups work through the classification activities on pages 95 to 97. Tell them that there is more than one way to classify the organisms in the picture on page 94. Ask them to figure out their own classification system before looking at Paulette's system in the middle of page 95.

Answers to

In-Text Questions (page 94)

- Library books are classified by author, topic, and title.
- Area codes indicate a specific region of each state, and the first 3 digits of a telephone number indicate which part of the city the number will access. Telephone books are alphabetized.
- The mail system uses zip codes.

Answers to

In-Text Questions (page 95)

1. & 2. Answers will vary, but the descriptions should include distinguishing features that can be used to identify the animals or plants. Students should be able to guess each other's organisms from the descriptions and to name other related organisms that have the same features. Students will realize as they do the activity that a group of living things—not just a single living thing—may match the descriptions they have written.

(Answers continue on next page)

Keeping Track 95

Answers continued

3. Diagrams may vary. One possible diagram could identify *Living Things* as the main group, with subgroups of *Plants* and *Animals*. Plants could be divided further into land plants and water plants. Animals could be divided further into those that live only in the water and those that can live out of the water.

4. Diagrams may vary. One possible diagram could divide all plants into two groups—*Trees* and *Shrubs and other smaller plants.* Trees could be further divided into *Trees that have leaves year-round* and *Trees that lose their leaves in the fall.* Shrubs and other smaller plants could be divided into *Those with woody stems* and *Those with soft, green stems.*

(Answers continue on next page)

3. Look at the living things in this illustration. Make a group and subgroup diagram that can be used to identify them.

4. Try a slightly different classification system for these four plants.

96 Diversity of Living Things

Paulette divided the living things in the picture into plants and animals. Could you use her grouping for the following organisms?

Answers continued

The eagle and worm can be classified as animals. The flowers, palm tree, ferns, and cactus can be classified as plants. The mushroom, spirogyra, and amoeba cannot be classified into either group, since there are distinct differences between them and plants or animals. Therefore, Paulette's system is not inclusive enough for all of the organisms shown.

The five-kingdom system of classification is discussed on page 98 of the Pupil's Edition. Detailed information about each of the five kingdoms also is provided in the *SourceBook,* pp. S24–S34. Ask students to answer this question again after the five-kingdom system has been discussed.

Keeping Track

The Animal Kingdom

As a class, devise several ways in which the organisms in the illustrations on pages 98 and 99 could be classified into two groups. Consider all possibilities (e.g., with or without legs, with or without shells, with or without fur, live in the water or on the land, do or do not fly, do or do not have tails). Have students classify all the animals using each suggested classification system.

After students have presented their own classification systems, have them read page 100, which presents the animal classification system that biologists use. Ask students to classify the animals pictured on pages 98 and 99 as either vertebrates or invertebrates.

Emphasize that the major feature distinguishing vertebrates from invertebrates is the presence or absence of a backbone and an internal skeleton.

Answers to
In-Text Questions

The organisms pictured can be divided into two groups: those with backbones (vertebrates) and those without backbones (invertebrates). The invertebrates in the pictures include: snail, mussel, grasshopper, lobster, octopus, and spider. The snail, mussel, and octopus are mollusks. Most mollusks have hard shells to protect their soft bodies, although some mollusks, such as the octopus, lack an external shell. The grasshopper, lobster, and spider are arthropods. They have hard external skeletons that protect their bodies. Their muscles are attached to these exoskeletons.

The vertebrates include: moray eel, penguins, snake, giraffe, tree frog, dog, fish, chick, salamander, tiger, and crocodile.

moray eel

Biologists, the scientists who study living things, realized the need for a consistent system of classification. For this reason, about one hundred years ago, they divided all living things into two kingdoms: the animal kingdom and the plant kingdom. However, throughout the past century, scientists have been gaining more knowledge about the differences among living things. This has created the need to devise more precise systems for grouping all organisms. Most biologists now group living things into five kingdoms, and many biologists have even proposed a six-kingdom system. Either way, there are many living things that are regarded as neither plants nor animals.

Here is how a modern biologist would group the living things shown in the photos on page 97. Notice that the organisms that are neither plant nor animal have been placed in the "Other Kingdoms" category.

penguins

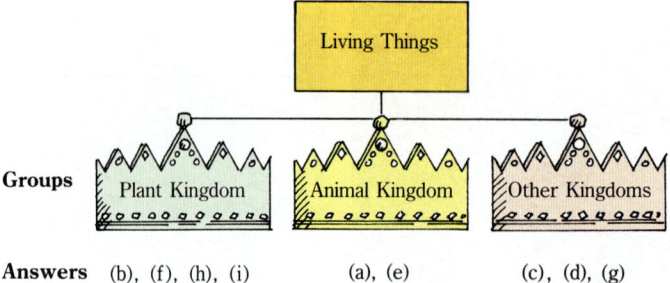

Groups	Plant Kingdom	Animal Kingdom	Other Kingdoms
Answers	(b), (f), (h), (i)	(a), (e)	(c), (d), (g)

snail

The Animal Kingdom

Pictured here are some members of the animal kingdom. Imagine that you are a biologist, and try dividing these animals into two groups. (Hint: Think about what supports their bodies.)

snake

mussel

Answers to
Invertebrates

Features that could be used to classify the organisms include number of legs, presence of antennae, location of habitat (on land or in water), and presence of eyes. The names used to label the subgroups could be based on the characteristics chosen.

The names of the organisms are:
- (a) millipede
- (b) jellyfish
- (c) spiny sun sea star
- (d) coquina clams
- (e) black and yellow argiope spider
- (f) earthworm
- (g) sand dollar
- (h) purple-veined anemone
- (i) Iceland scallop
- (j) painted lady butterfly
- (k) leech
- (l) crayfish
- (m) centipede
- (n) tarantula
- (o) oyster
- (p) brittle star with purple sea urchin below
- (q) horned flatworm
- (r) harvestman (daddy longlegs)
- (s) dogbane leaf beetle
- (t) common octopus
- (u) grasshopper
- (v) jonah crab
- (w) rotifer

The more you know about different animals, the easier it is to make useful groupings. To classify, you need to carefully observe the animals' features—not only their outside features, but their inside features too.

What characteristics did you use to classify the animals in the photos on pages 98 and 99? As you probably realize, your job was difficult because you had to depend mostly on what you could observe in the pictures. Biologists, though, study the actual animals, examining them on both the outside and the inside. The type of body support an animal has is one feature that biologists use when classifying animals. Many animals have internal (inside) skeletons to support themselves. These animals are placed in a group called **vertebrates.** Vertebrates are animals with a *vertebral column,* or backbone.

Other animals depend on a hard, external support system. For example, clams, lobsters, and grasshoppers each have a hard shell. Earthworms have tough muscles that give them support. Other animals don't really need a means of support. Their surroundings are their support. This is true for many of the animals that live and drift about in the water. Clams, lobsters, grasshoppers, earthworms, and animals that drift about in the water all can be placed in a group called **invertebrates**—animals without backbones. Is this the way you divided the animals on pages 98 and 99 into two groups? If not, do so now.

Invertebrates

Here are pictures of some invertebrates. Remember that none of these animals has a backbone for support. How might you group these animals? What features would you use as a basis for making subgroups? What names might you use to label your groups?

Keeping Track 101

The Classification System Used by Biologists

Ask students to read the material on page 102. Call on volunteers to read aloud the material in the tables.

If there is a beach nearby, take students on a field trip to collect, observe, and classify the animals or animal remains that they find on the beach or in shallow water. You could also ask students to bring in shells and other animal remains that they find on the beach. Set up a classroom display of the different phyla of invertebrates, to which students can add their specimens.

If there is no beach nearby, choose a seashore area to research as a class. Assign a specific organism to each student. Students should find out as much as possible about the organism, and how its features allow it to survive in its habitat. Students can illustrate their organisms, and the illustrations can be organized into a display.

The Classification System Used by Biologists

By studying invertebrates in detail, biologists have developed a classification system for them. To help you understand their system, study these two tables. The first table identifies the features that biologists use to classify invertebrates into different subgroups. The second table divides one subgroup, the arthropods, into even more subgroups.

After studying the tables, see whether you can place all the animals pictured on pages 100 and 101 into the classification scheme on page 103. Your task will not be easy; using pictures of animals to observe identifying features is more difficult than using the actual animals.

The Identifying Features of Invertebrates

Distinguishing Features Biologists Use to Group Animals	Name of the Subgroup	Examples of Animals in the Subgroups
worm-like, round bodies with many segments	annelids (means "arranged in rings")	earthworm
very soft bodies, which in most cases are protected by a shell	mollusks (means "soft-bodied")	clam (example with a shell) octopus (example without a shell)
covered with a spiny skin	echinoderms (means "spiny-skinned")	starfish
having many jointed appendages (attachments such as legs, feelers, tail)	arthropods (means "jointed legs")	lobster

The Identifying Features of Arthropods

Distinguishing Features Biologists Use to Make Subgroups of Arthropods	Name of the Subgroup	Examples of Arthropods in the Subgroup
six legs, three main body parts, wings	insects	grasshopper
eight legs, two main body parts	arachnids	spider
more than eight legs, many other kinds of appendages, crusty covering	crustaceans	lobster
numerous legs	millipedes centipedes	millipede centipede

102 Diversity of Living Things

A Simplified Classification System for Invertebrates

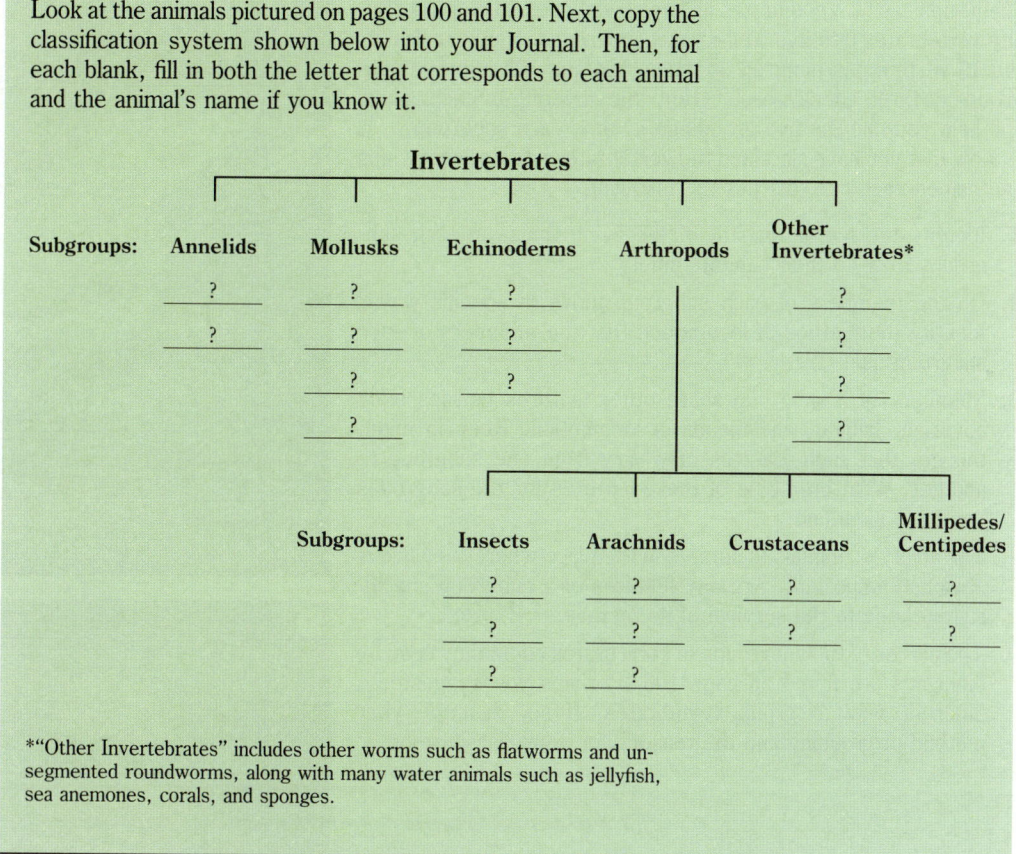

Look at the animals pictured on pages 100 and 101. Next, copy the classification system shown below into your Journal. Then, for each blank, fill in both the letter that corresponds to each animal and the animal's name if you know it.

*"Other Invertebrates" includes other worms such as flatworms and unsegmented roundworms, along with many water animals such as jellyfish, sea anemones, corals, and sponges.

Now think some more about the subgroups.

1. Did you have any problems deciding which subgroup each invertebrate belonged to?
2. It's interesting to think about where invertebrates live. How many are found in water? in moist places? on dry land?
3. Does it appear that their structure enables them to live successfully in various places? Why do you think this is the case?

Answers to
A Simplified Classification System for Invertebrates

The invertebrates should be classified as follows:
- *Annelids*—earthworm and leech
- *Mollusks*—clam, oyster, scallop, and octopus
- *Echinoderms*—sea star, sand dollar, brittle star, and sea urchin
- *Other Invertebrates*—jellyfish, flatworm, anemone, and rotifer
- *Insects*—beetle, butterfly, and grasshopper
- *Arachnids*—harvestman (daddy longlegs), argiope spider, and tarantula
- *Crustaceans*—crayfish and crab
- *Millipedes/Centipedes*—millipede and centipede

Answers to
In-Text Questions

1. Students will probably answer in the affirmative. It is difficult to classify invertebrates using only photographs, especially if you are not familiar with their structures.
2. Most of these animals live in the water. Exceptions include insects and arachnids that can live on dry land, and earthworms, snails, and millipedes that prefer moist environments.
3. Yes. Because of the diversity of structures among invertebrates, they can live successfully in many different places.

Keeping Track 103

Vertebrates

Students should find it easier to classify vertebrates into subgroups because students are more familiar with vertebrates. To help students remember how vertebrates are classified, have them summarize the characteristics of each subgroup in their Journals. Encourage them to include simple drawings of animals that belong in each subgroup.

Answers to
In-Text Questions

The five subgroups are as follows:
- *Fish*—scales and/or fins
  Includes: (b) lionfish, (i) angel fish, and (o) shark
- *Amphibians*—no scales
  Includes: (e) frog, (g) toad, and (m) salamander
- *Reptiles*—scales, no fins
  Includes: (d) snake, (j) alligator, and (n) turtle
- *Birds*—feathers
  Includes: (c) loon, (k) duck, and (l) owl
- *Mammals*—fur or hair covering
  Includes: (a) bat, (f) human, and (h) fur seal

(Answers continue on next page)

Vertebrates

Because you are so well acquainted with the vertebrates around you (including yourself!), you will be able to generate vertebrate subgroups easily. Vertebrates all have backbones. Biologists classify vertebrates into five well-known subgroups. Can you classify the 15 vertebrates pictured on pages 104 and 105 into five subgroups of three animals each? (Hint: the covering of each animal will help you find the five groupings.) Name each subgroup.

Now that you have classified the vertebrates into five subgroups and named each one, discuss the following:

1. Identify the normal habitat of the living things in each subgroup. How do their habitats differ?
2. Do the members of each subgroup move in specific ways? Identify the method of locomotion that the members of each subgroup use.
3. Members of a subgroup share other features besides body covering, habitat, and means of locomotion. Keep in mind, though, that such features may vary from one subgroup to another. What are some of these features for the subgroups you have identified?
4. The ability of animals to survive in the place where they live is obviously important. Are any features of a subgroup specifically related to the survival of its members? Explain.
5. The names of the subgroups of vertebrates commonly used by biologists are found on page 106. Perhaps you suggested a different way to group vertebrates. If so, describe your method of grouping and the reasons for your classification.

104 Diversity of Living Things

Answers continued

1. Fish live in water. Amphibians live in water and on land. Reptiles live mostly on land. Birds live mostly on land and in the air. Mammals live mainly on land, although some live in the air or in the water.
2. Fish swim by using their fins, tails, and body movements. Amphibians walk or run by using their legs. Some also swim or jump using their large hind legs. Reptiles (except for snakes) walk or run on two pairs of legs. Snakes use their internal muscles to crawl. Birds use their wings to fly and their legs to walk, run, or hop. Most mammals use their two pairs of legs to walk or run.
3. Answers may vary. Students may suggest methods of reproduction and respiration that subgroups have in common. Fish and amphibians lay their eggs in water. Reptiles also lay eggs, but they are fertilized inside the female's body. The young of mammals develop in the mother's body until they are born. Fish use gills to breathe. Amphibians breathe through their moist skin and lungs. Reptiles, birds, and mammals breathe through lungs.
4. The features described in answers 2 and 3 help the animals in each subgroup to survive in their specific habitats. For example, fish have fins, tails, and streamlined bodies that enable them to swim. Gills enable them to breathe in water.
5. This is an open-ended question that gives students practice in classification. Respect every attempt that students make to classify according to consistent criteria. They might suggest classification by habitat, size, color, form of locomotion, reproduction, or type of appendages, to name just a few.

Keeping Track 105

Putting It All Together

This activity gives students an opportunity to consolidate all the information they have learned so far about classification. Have them copy the tables on page 106 into their Journals and then fill them in using the illustrations on pages 107 to 109. They can also include any preserved or live specimens available in the classroom.

To help students remember how to classify vertebrates and invertebrates, set up a chart on a bulletin board entitled "The Animal Kingdom." The chart could have the following headings: *Group, Subgroup, Examples,* and *Major Characteristics.* Divide students into small groups and have each group fill in part of the chart. Students could cut out pictures from old magazines or draw pictures of animals and place them around the bulletin board.

Find an area near the school that is an interesting place for students to search for animals. Have students take their Journals, a pencil, and a spoon for digging. Have them look under rocks and logs, and dig up some soil. They should record the names of any animals they recognize and make sketches of those that are unfamiliar. You may wish to take some natural history field guides that students can use to identify the animals they do not know.

Instruct students not to disturb the animals they find. Remind them to put back any logs, rocks, or soil that they have moved. After the activity is completed, have students classify all of the animals they found into their correct groups. Arthropods can also be classified into their correct classes.

Putting It All Together

Many animals are illustrated on the next three pages. Your task is to classify them in your Journal using a classification table like the one below. Fill in both the letter that corresponds to each animal and the name of the animal. This table is an overall classification system for animals. It brings together everything you have learned about classification for the animal kingdom.

If your classroom contains any living or preserved specimens that are not represented in the pictures, classify them as well. Remember to write only in your Journal.

Animal Kingdom Classification System

Invertebrate Subgroups		Examples
Annelids		
Mollusks		
Echinoderms		
Arthropods	Insects	
	Crustaceans	
	Arachnids	
	Millipedes/Centipedes	
Other Invertebrates		

Vertebrate Subgroups	Examples
Fish	
Amphibians	
Reptiles	
Birds	
Mammals	

Sometimes the outward appearance of an animal can fool you, so be sure to examine each animal carefully. For example, the turtle does have a hard covering, but it is not a mollusk. Why not?

Answer to
In-Text Question

Even though the turtle has a shell, it is not a mollusk. Mollusks have soft bodies covered by a hard shell made of calcium compounds. Turtles do not have soft bodies. They have all the characteristics of reptiles, including a backbone, a scale-covered body, and lungs. Their shells are made of a hard, horn-like material that lies over a deeper bony layer. Students should understand that many characteristics combine to determine the classification of an organism.

Keeping Track 107

Answers to
In-Text Questions (pages 107–109)

The animals in the illustrations should be classified as shown below. Note that *(n) amoeba* does not fit into any of the categories.

INVERTEBRATES
- *Annelids*—(m) bristle worm, (o) leech
- *Mollusks*—(d) snail, (h) clam
- *Echinoderms*—(i) sea star, (p) brittle star
- *Arthropods:*
  Insects—(b) beetle
  Crustaceans—(c) crab
  Arachnids—none shown
  Millipedes/Centipedes—(f) millipede

VERTEBRATES
- *Fish*—none shown
- *Amphibians*—(j) toad, (k) salamander
- *Reptiles*—(l) sea turtle, (r) snake
- *Birds*—(s) duck, (t) penguins, (u) ostrich
- *Mammals*—(a) manatee, (e) horse, (g) flying squirrel, (q) dolphin

108 Diversity of Living Things

Keeping Track 109

The Plant Kingdom

Have students use the classification diagram on page 110 to classify the plants in the illustrations on pages 110 and 111. If possible, bring in some sample specimens of plants from each subgroup. You may wish to ask students to bring in small twig cuttings, leaves, flowers, and seeds or fruits from cone-bearing and flower-bearing plants.

Students may not realize that all trees, other than conifers, have flowers that develop into pods or fruits. Deciduous trees (those that lose their leaves in autumn and winter) usually flower in the spring, before the leaves develop.

Take students on a scavenger hunt to discover how many examples of both flowering and nonflowering plants they can find in the schoolyard or a park. (Algae, lichens, mosses, ferns, and conifers are all examples of nonflowering plants.) Encourage students to observe the various places these plants grow. Ask: Where do they grow? How are they adapted to their habitats? *(One example is a lichen, which is actually a fungus and an alga living together. Lichens can be found growing on rocks and on the north side of trees. They do not need the nutrients that soil provides. Ferns and mosses grow in damp and shady places. Conifers live in colder places and can be found on north-facing slopes.)*

The Plant Kingdom

The great variety of plants in the world led scientists to develop a classification scheme for them, just as was done for animals. At this stage, it may not be clear to you why biologists use the particular groupings shown below. Nonetheless, you can use them to classify plants around you. Practice by using the pictures of plants on this page and the next.

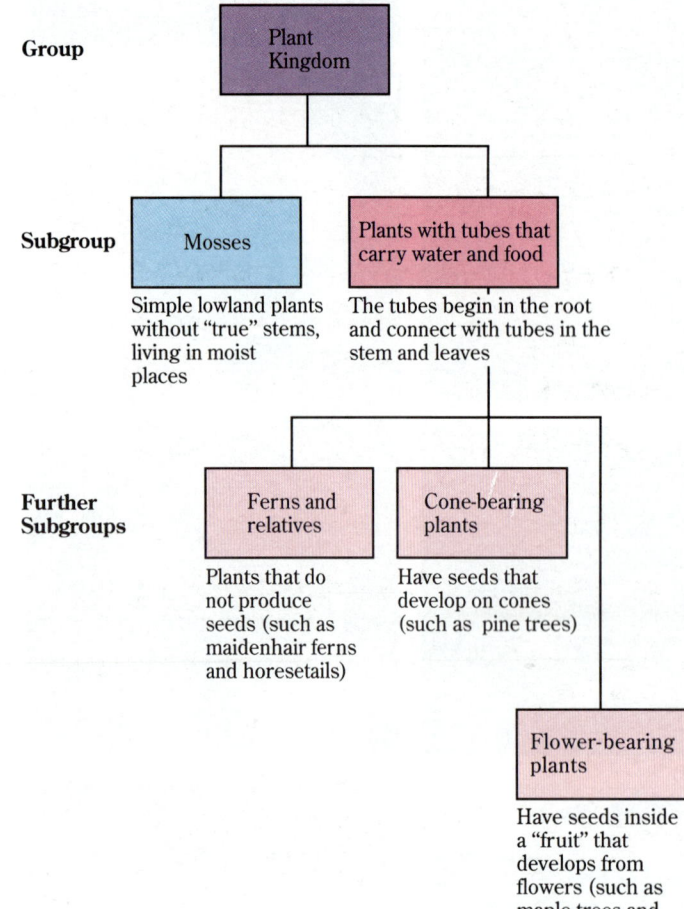

Group: Plant Kingdom

Subgroup:
- Mosses — Simple lowland plants without "true" stems, living in moist places
- Plants with tubes that carry water and food — The tubes begin in the root and connect with tubes in the stem and leaves

Further Subgroups:
- Ferns and relatives — Plants that do not produce seeds (such as maidenhair ferns and horsetails)
- Cone-bearing plants — Have seeds that develop on cones (such as pine trees)
- Flower-bearing plants — Have seeds inside a "fruit" that develops from flowers (such as maple trees and violets)

cedar

roses

110 Diversity of Living Things

fiddlehead fern

oak

moss

Answers to
In-Text Questions

The plants in the illustrations should be classified as follows:
- *Mosses*—moss
- *Ferns and relatives*—fiddlehead fern
- *Cone-bearing plants*—cedar
- *Flower-bearing plants*—roses, oak

✹ Follow Up

Assessment

1. Bring in samples of several different kinds of plants, including mosses, fern leaves, branches from deciduous trees, cones and branches from conifers, and seeds and fruits from flowering plants. Ask students to classify these specimens into the correct groups and subgroups. Have students explain their reasons for classifying the specimens the way that they did.
2. Bring in pictures of vertebrates and invertebrates. Ask students to classify the animals into the correct groups and subgroups and explain their reasoning.

Extension

1. If possible in your area, suggest a "forest study." Students could take a field trip to a forested area and classify the many different types of plants and animals that they see. Encourage them to draw pictures of the different organisms and identify them if they can. They should then classify the organisms that they can name.
2. Remind students that microscopic organisms cannot be classified as plants or animals. For example, protozoans, which belong to the kingdom Protista, are one-celled, animal-like organisms that can be seen only with a microscope. Ask students to fill a jar half-full of mud and water from the bottom of a pond or stream so that they can observe these one-celled creatures. Suggest that they add about a dozen grains of cooked rice to the jar and then leave it undisturbed for a week, away from direct sunlight. Then ask them to use a microscope to observe the variety of protozoans in one drop of the water.

Keeping Track 111

LESSON 6

✷ Getting Started

This lesson introduces students to the classification system of Carolus Linnaeus. Students learn to categorize living things into kingdoms, phyla, classes, orders, families, genera, and species.

Main Ideas

1. Scientists classify living things into progressively smaller and more specific groups, according to their similarities.
2. The scientific name of a living thing is made up of two words. The first is the genus name. The second is the species name.
3. There is diversity within a species.

✷ Teaching Strategies

Have students read the introductory material on Linnaeus. Tell them that Linnaeus' idea to classify plants and animals by a system was not a new one. Aristotle developed a system for classifying plants and animals over 2000 years ago. His system divided all organisms into two groups—plants and animals. He grouped animals according to *where they lived*—in the air, on land, or in the water.

By contrast, Linnaeus' system classified organisms on the basis of *structure*. After Darwin's theory gained acceptance, biologists decided that classifying organisms according to their evolutionary relationships would be more useful. Structural characteristics usually reflect evolutionary relationships, so biologists have retained most of Linnaeus' groupings.

6 Ordering Diversity

The Swedish scientist Carolus Linnaeus, who lived in the eighteenth century, can be called the "master of order." He devised a scientific system for classifying living things according to their similarities. It is essentially his system that we still use today. You can see Linnaeus's classification system at the right.

You might be wondering why some of the groupings in his classification system have Latin names. In Linnaeus' time, Latin was the language used by all educated people. In fact, "Carolus Linnaeus" is a latinized form of "Karl von Linné"—his real name!

It was Linnaeus' ambition to name all the living things in the world. He did not quite achieve this goal, but he did name over 770 plants and 4400 animals.

Carolus Linnaeus

A Name You Can Claim

Where do you fit into Linnaeus' system? First, you are clearly a member of All Living Things. Next, you belong to the animal kingdom. Because you are a vertebrate (you have a backbone), you belong to phylum Chordata. Then, because you are warm-blooded, have hair (not feathers), and were born alive (instead of hatching from an egg), you fit into class Mammalia. There are certain similarities between you and animals such as gorillas and chimpanzees, so the next grouping you belong to is order Primates. You belong to the family Hominidae and the genus *Homo*. The most specific grouping is species. Your specific name is *sapiens*. Combine the names for our genus and our species, and you get *Homo sapiens*—the name for all human beings. (You might find it interesting to look in a dictionary to see what our name means.)

You have more in common with other humans *(Homo sapiens)* than you do with other primates, such as chimpanzees. But you have more in common with chimpanzees than you do with other mammals, such as cats, for instance. However, there are still more similarities between you and cats than between you and maple trees, which are also living things.

Every single type of living thing, from a fungus to a frog, has its own **scientific name**. A scientific name is a combination of its genus and its species names. It's important to remember that a scientific name contains both of these parts. You wouldn't refer to a wolf, for instance, simply as *lupus*. A wolf's scientific name is *Canis lupus*. Don't get him confused with *Canis familiaris*—your pet dog!

Now do some research. What are some organisms that belong to the animal kingdom? to phylum Chordata? to class Mammalia? to order Primates? to family Hominidae? to genus *Homo*?

	All Living Things
(Group)	Kingdoms
(Subgroup)	Phyla
	(singular: Phylum)
(Further	Classes
Subgroupings)	Orders
	Families
	Genera
	(singular: Genus)
	Species

LESSON 6 ORGANIZER

Objectives

By the end of the lesson, students should be able to:
1. Explain the system that biologists use to classify living things.
2. Observe and describe diversity within the species *Homo sapiens*.

Process Skills

observing, comparing, taking a survey, analyzing data

New Terms

Scientific name—the genus and species names of a living thing.

112 Diversity of Living Things

A Legend: Wisakedjak Names All Creatures

A Cree legend says that the fox was the first animal to be given a name. This happened because one animal outsmarted Wisakedjak, the great spirit of the Western Indians, who said the animal was very "foxy." Wisakedjak named all the animals, fishes, birds, and insects, and he told each of them where and how to live.

He told the fish they must live in the water and swim. He told the ducks that they could live on the water, on land, and even fly in the air to escape their enemies. Wisakedjak gave the deer a white tail and said he could run fast and jump far. Squirrels could climb trees and hide in the tallest branches. To the slow porcupines, he gave a special suit of quills, which all animals learned to respect. Rabbits were trained to sit very still so they could not be seen.

Dr. Ivan H. Crowell

Do you recall any animals' names that are quite descriptive of what the animals are like? What are they?

There are many stories about how animals and plants got their characteristics. This painting illustrates the Ojibwa Indian folktale of how they received their colors.

Materials

Exploration 4: Journal

Teacher's Resource Binder

Resource Worksheet 2–7
Activity Worksheet 2–8
Graphic Organizer Transparencies 2–5 and 2–6

Science Discovery Videodisc

Disc 2, Image and Activity Bank, 2–6

Time Required

two class periods

Answers to

A Name You Can Claim

- The name *Homo sapiens* comes from the Latin words *homo,* meaning "man," and *sapere,* meaning "to know."

Ask students to do some research and list some animals that belong in each grouping. In doing this research, they will discover that each subgroup is more specific than the preceding one. For example, they could list the following animals:
- Animal kingdom—worms, insects, clams, frogs, lizards, birds, cats, humans
- Phylum Chordata—frogs, lizards, birds, cats, humans
- Class Mammalia—cats, dogs, cows, humans
- Order Primates—gorillas, chimpanzees, humans
- Family Hominidae—all forms of humans, living and extinct
- Genus *Homo habilis* and *Homo erectus* (both extinct), and *Homo sapiens* (modern humans)

A Legend: Wisakedjak Names All Creatures

Point out that every culture in the world has stories of how plants and animals came to be and of how they acquired their characteristics. Many of the legends are based on keen observations of the natural world. This Cree legend describes characteristics that enable certain animals to survive (e.g., deer can run fast, squirrels can climb trees, and porcupines have quills for protection).

Often, the names given to animals reflect their outstanding features (e.g., crossbill, woodpecker, and grasshopper). Challenge the students to think of as many descriptive names for animals as they can.

Primary Source:
Reprinted from calendar by Ivan H. Crowell, with minor alterations to correct punctuation.

Keeping Track 113

The Same But Different—Diversity Within Species

Ask students to read the introductory material on page 114 and then ask: What examples of the diversity of features in *Homo sapiens* can you identify? *(Answers will vary, but possible responses include eye, hair, and skin color; height and body build; shape of eyes, nose, and mouth; and hair texture.)*

Exploration 4

Have students work in groups of four to complete this activity. Ask them to copy the table on page 115 into their Journals and fill in the appropriate information for group and family members.

You may need to clarify some of the characteristics listed on the table. Point out the difference between attached and free earlobes and between pointed and straight hairlines (i.e., widow's peak or not). Demonstrate the two thumb positions when the hands are folded together, and have a student demonstrate tongue-rolling.

The Same But Different—Diversity Within Species

Do you recall your scientific name? Right, you're a *Homo sapiens*. This is the name of the genus and species to which you belong. Every person you know is also a *Homo sapiens*. All the organisms within a species are alike. Thus you share certain characteristics with all human beings—characteristics such as the general arrangement of your facial features, the number of your fingers and toes, and the ability to reason, to name just a few. People belonging to the same family tend to resemble each other even more closely. This is known as a family resemblance. It occurs because every person inherits certain characteristics from his or her parents, grandparents, and so on.

Despite family resemblances, however, everyone is unique. You are "one of a kind." Not one of your friends or acquaintances is exactly the same as you. Nor are you a duplicate of your parents or grandparents or brothers or sisters (even if you are a twin).

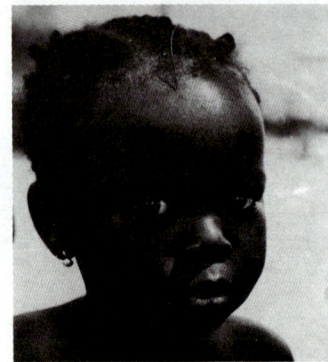

Every human is unique. How many examples of the diversity of features in *Homo sapiens* can you identify?

EXPLORATION 4

Tracing Similarities and Differences

In this Exploration, you will investigate some of your own inherited characteristics to determine how much you are like your family. You will do the same for some of your classmates.

What to Do

In the classroom, work in groups of four. Examine six inherited characteristics for each member of the group. In your Journal, construct a table like the one on page 115 and record each member's characteristics. Repeat this procedure with family members at home.

114 Diversity of Living Things

Characteristic	Student				Family	
	You	(2)	(3)	(4)	(1)	(2)
Hand-folding, thumb position						
(1) Left over right						
(2) Right over left						
Earlobes						
(3) Attached						
(4) Free						
Hairline						
(5) Pointed hairline						
(6) Straight hairline						
Tongue-rolling						
(7) Can roll						
(8) Can't roll						
Digit next to little finger						
(9) Hair						
(10) No hair						
Toe next to big toe						
(11) Same length or longer than big toe						
(12) Shorter than big toe						

Analyze Your Data

1. Study your table and then compare your results with the other groups in your class.
2. Calculate the percentage of individuals in your group who have each characteristic. What does this tell you?
3. Does anyone have all of characteristics 1, 3, 5, 7, 9, and 11? Does anyone have all of characteristics 2, 4, 6, 8, 10, and 12? Are there any two people in the class who have identical characteristics? Did you know that there are 64 possible combinations of these six characteristics?
4. You have looked at only six pairs of characteristics. There are hundreds of other characteristics you might have considered. Do you agree or disagree with these statements?

   "You're *just* like (name) ."

   "Why, you are just the spitting image of your mother/father!"

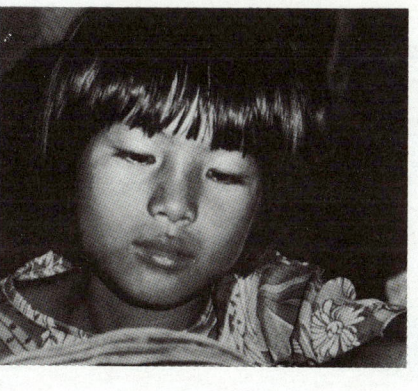
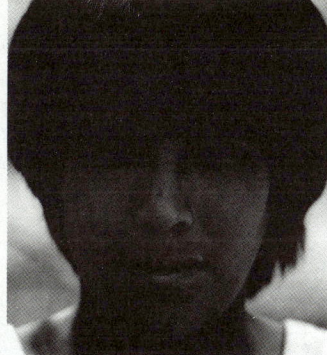

Answers to
Analyze Your Data

As students complete the table and answer the questions, they will realize that there are hundreds of characteristics within the human species that can vary. Students should conclude that no two people are exactly alike.

Assessment

Have pictures of various vertebrates and invertebrates available. Ask students to:
(a) Identify the animals that belong to the same kingdom as humans. *(all animals)*
(b) Identify the ones that belong to the same phylum as humans. *(all those with backbones)*
(c) Identify the ones that belong to the same class as humans. *(all mammals)*
(d) Identify the ones that belong to the same order as humans. *(primates)*
(e) Explain how the animals in each group or subgroup are similar to humans.

Extension

1. Have students research the scientific names for various organisms that are familiar to them, such as plants and animals that live in their area. Have students find the derivations of these names in the dictionary. They will discover that most scientific names are Latin.
2. There are many other inherited traits that students can use in a survey about classmates and family members. Examples include the ability to form a "V" by spreading the middle and ring fingers apart while keeping the other fingers together; the ability to spread apart the toes; the ability to wiggle the little toe while keeping the other toes still; and color blindness. Ask students to tabulate their results and make an oral report to the class to share their findings.

Keeping Track 115

LESSON 7

✸ Getting Started

This lesson focuses on the diversity of features among organisms in different stages of their life cycles. In *Exploration 5,* students observe and record the changes that a frog undergoes as it grows from egg to adult. Then students examine and label diagrams of the life cycles of two types of insects. Students conclude the lesson by researching the diversity that exists in plant life cycles and in the behavior patterns of insects.

Main Ideas

1. There is diversity within the different stages of an organism's life.
2. There is diversity in animal behavior and structure.

✸ Teaching Strategies

Have students read the introductory material on life cycles. You might want to bring in actual specimens or photos of plants and animals in various stages of development. Challenge students to identify the plant or animal.

From Eggs to Frogs

Read the poem aloud to your students. Then ask students to describe the changes the frog undergoes. *(When the frog hatches from an egg, it has a tail and no legs. The back legs and then the front legs gradually appear, and the tail grows shorter.)* Ask students if they know of any other animals that change in appearance during their life cycles. *(Insects such as flies, butterflies, moths, bees, and beetles have four distinct stages in their life cycles. Most amphibians undergo a process of metamorphosis in which the adults look very different from the larvae.)*

116   Diversity of Living Things

7 Life Cycles: Diverse Stages

You have seen how much diversity there is among different groups of living things. And you know that there is diversity even within a single group. There is also diversity within the different stages of an organism's life. Most organisms go through a series of stages from birth to adulthood to the birth of offspring in the next generation. This series of stages in an organism's life is called its **life cycle.** In the life cycles of many animals, the young animals live freely (on their own) and look quite different from the adults they will become. How many animals like this can you think of? Let's look at the life cycles of three living things.

From Eggs to Frogs

A small, round frog egg looks very different from an adult frog. Between these two stages in the life cycle of a frog, diverse forms can be observed.

In the early spring, down by the marsh,
The croaking of frogs is deep and harsh.
When the male sends out his mating song,
The female responds—and before long,
She deposits her eggs in a shallow pond.
Then the male secretes sperm to complete the bond.
The egg divides into millions of cells—
"It's a tadpole!" the young boy yells.
But watch for a few weeks, and what do you see?
No longer a streamlined shape, swimming so free.
Its gills vanish and its tail disappears—
"It's a frog!" the young boy cheers.

LESSON 7 ORGANIZER

Objectives

By the end of the lesson, students should be able to:
1. Describe, illustrate, and label the stages in the life cycle of a frog, a mosquito, and a grasshopper.
2. Sequence the stages in the life cycles of organisms undergoing complete metamorphosis and those undergoing incomplete metamorphosis.

Process Skills

observing, sequencing, illustrating, comparing

New Terms

Life cycle—the series of developmental stages in an organism's life.
Metamorphosis—changes that occur from the egg to the adult stage in insects and in many other animals, such as the frog.

EXPLORATION 5

Observing the Changes

You Will Need

- frog eggs
- a large jar of pond water
- algae or cornmeal
- gravel
- a pond plant
- a tall stick
- a terrarium (or a cardboard box fitted with clear plastic windows)
- plastic wrap for windows
- scissors and tape
- soil
- a small log
- a green plant
- a dish of water
- flies, worms, and small insects
- a magnifying glass
- a spoon

What to Do

To study the diverse forms in the life cycle of a frog, you must construct two homes. The tadpole "nursery" is simply a large jar with a layer of gravel on the bottom. Add a pond plant, a tall stick, and pond water to complete the home.

A small terrarium or a cardboard box fitted with clear plastic windows is the beginning of an ideal home for a frog. Soil, a small log, a green plant, and a dish of water are all you need to meet the frog's needs.

Be sure to construct the home so that the frog can obtain plenty of air. Remember to feed it flies, bits of worm, and small insects.

After you have completed the homes, the tenants may move in. At this point, keep accurate notes and diagrams in your Journal so that you can complete your study with a detailed account of the life cycle of the frog.

Some Suggested Procedures

1. Obtain some frog eggs (or *spawn*, as they are called) from your teacher.
2. Put the eggs in the tadpole nursery. Using a spoon, take a few eggs out of the nursery each day and examine them with a magnifying glass.
3. Write and draw any changes you observe taking place in the jelly covering.
4. When the eggs have hatched, carefully examine a tadpole to find the outside gills it uses to take in oxygen.
5. Watch the growth of the leg buds and legs. Do the back or the front legs appear first?
6. Observe the tadpoles' tails daily. What happens to the tails?
7. Look for signs of lung development. The tadpoles will begin to use their nostrils to breathe at the surface of the water. When the tadpoles climb up the stick and remain out of water, it is time to change residences.
8. Observe the frog(s) carefully for several weeks. Make daily records and drawings of limbs, size, color, breathing, and locomotion.
9. To summarize your findings, make a poster-size diagram showing the life cycle of a frog. Include your detailed observations on the diagram or in an accompanying table.

Materials

Exploration 5: Journal; frog eggs; large jar; pond water; gravel; pond plant; tall stick; terrarium (or cardboard box, scissors, tape, and plastic wrap for windows); soil; small log; green plant; dish of water; algae or cornmeal; flies, worms, and small insects; magnifying glass; spoon; poster board; markers

If frog or toad eggs are not available, you may wish to have students culture mealworms or fruit flies instead. (See teaching strategies on page 117 for directions.)

Science Discovery Videodisc

Disc 2, Image and Activity Bank, 2–7

Time Required

two class periods

Exploration 5

Frog eggs are not available at all times of the year. In most temperate areas, the eggs are laid in the spring. Look in ponds for a jellylike mass of frog eggs or a long, thin string of toad eggs. Take only a few eggs. If overcrowding occurs after the eggs hatch, return most of the tadpoles to the water where the eggs were found. Return only tadpoles or frogs collected locally. Frog eggs can be bought from bait shops or biological supply companies.

Feed tadpoles algae, or sprinkle small amounts of cornmeal on the surface of the water.

If frog or toad eggs are not available, mealworm or fruit fly cultures are easy to keep in the classroom. Both undergo complete metamorphosis. The four stages in their life cycles can be observed easily.

To culture fruit flies, attract some fruit flies to a container using over-ripened fruit. After trapping, transfer the fruit flies to small jars containing chunks of bananas. Place six to eight fruit flies in a jar, and plug the mouth with loose cotton. Soon, eggs will be deposited. In two to three days, the larvae will hatch. Place a piece of paper in each jar so that when the larvae are ready to pupate, they have something to crawl on.

To culture mealworms, purchase about 10 larvae at a fish-bait store or aquarium shop. Place the mealworms in a large jar with some oatmeal and a slice of apple or potato for moisture. Put a screen cover on the jar to prevent the adult beetles from escaping. Adults may be fed small amounts of raw carrots.

Ask students to record and illustrate the changes that the insects undergo. Have students record the length of time each stage in the life cycle lasts.

Answers to

Some Suggested Procedures

4. The external gills will become internal gills after a few days.
5. The hind legs appear first.
6. The tails become shorter and shorter until they eventually disappear.

From Eggs to Adult Insects

Have students read the introductory material on page 118. After they have looked at the two life cycles, they can copy and label them in their Journals. Refer students to the *Label Hints* on page 120 for help in labeling.

The mosquito's life cycle is an example of *complete metamorphosis,* in which the stages between egg and adult are quite different from the adult in appearance and structure.

In contrast, the grasshopper's life cycle is an example of *incomplete metamorphosis,* in which the stages between egg and adult each look very much like the adult. Through the process of molting, the young nymph grows larger.

Answers to In-Text Questions

The correct labels for the life cycle diagram are:
The Life Cycle of a *Mosquito*
An example of *complete metamorphosis*

1. eggs
2. larva
3. pupa
4. adult

(Answers continue on next page)

From Eggs to Adult Insects

Insects are more abundant than any other group of animals. They inhabit every part of the earth except mid-ocean. They have been found in the atmosphere several kilometers above the earth. The life cycles of different species vary, but all insects undergo changes as they develop from young forms to adult forms. These changes are easy to identify because each stage generally looks quite different from the one before it.

Here are the life cycles of two insects. Carefully examine the two cycles and the "Label Hints" on page 120. Then copy the cycles into your Journal and use the "Label Hints" to label the diagrams.

The Life Cycle of a ___?___

An example of $\underline{?}$ $\underline{?}$
(this word means "total") (this word means "change")

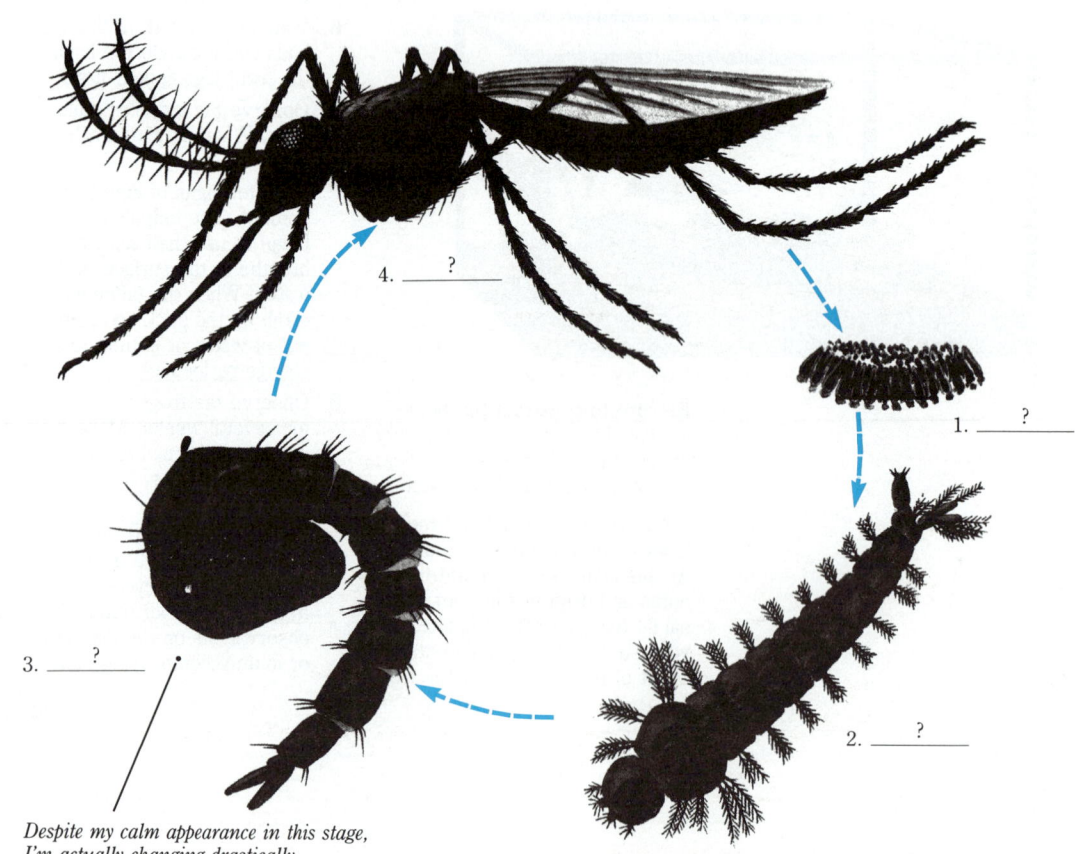

Despite my calm appearance in this stage, I'm actually changing drastically.

118 Diversity of Living Things

The Life Cycle of a ___?___

An example of ___?___ ___?___
(this word means "partial") (this word means "change")

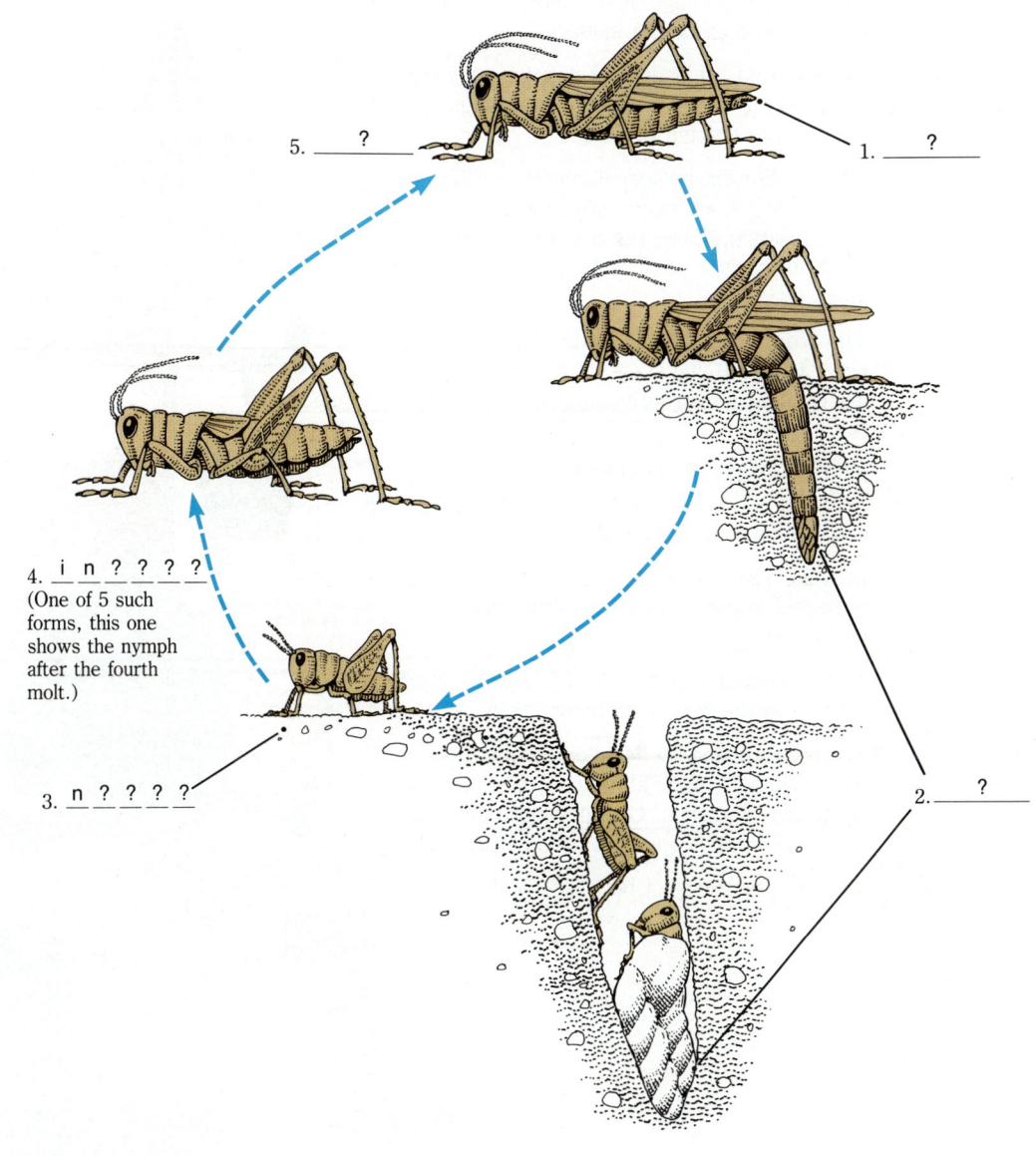

5. ___?___

1. ___?___

4. i n ? ? ? ?
(One of 5 such forms, this one shows the nymph after the fourth molt.)

3. n ? ? ? ?

2. ___?___

Answers continued

The correct labels for the life cycle diagram are:
The Life Cycle of a *Grasshopper*
An example of *incomplete metamorphosis*

1. ovipositor
2. eggs
3. nymph
4. instar
5. adult

Keeping Track 119

Label Hints

Adult	The last stage in the life cycle of an insect (and other animals).
Ovipositor	The pointed structure at the end of the body of a female grasshopper. The female uses it to lay her eggs.
Egg	The beginning stage in the life cycle of an insect (and many other animals).
Larva	In some kinds of insects, this is the stage that looks like a worm, moves a great deal, and eats a large amount of food.
Pupa	Little motion can be seen during this stage, but great changes are taking place inside an outer case. Inside the case, the larva changes into its adult form. Not all insects go through this stage.
Metamorphosis	The word that describes the changes which occur from the egg stage to the adult stage in insects (and in many other animals, such as the frog).
Complete metamorphosis	In this kind of metamorphosis, the "in-between" stages are quite different in appearance and structure from the adult.
Incomplete metamorphosis	In this kind of metamorphosis, the "in-between" stages have the general appearance of the adult.
Nymph	A grasshopper's young, having a large head relative to the rest of its body.
Molting	The breaking and shedding of the outer skeleton of animals, such as insects, as their soft body grows inside. Grasshoppers molt six times. The wings appear on the fourth molt.
Instar	The form of an insect between moltings.

What Is It Really Like?

Your class has been invited to explain life cycles to a group of 10-year-olds who really like puppets. Write the script for a skit in which one puppet is an elderly mosquito and another is an elderly grasshopper. They're talking about their youth and the changes they went through when they grew up. Toward the end of the skit, another puppet appears—a frog. The frog compares his life cycle with the other two. What would the frog say?

Diversity in Plant Cycles

Do you think there is diversity among the stages of plant life cycles? Could there be "young plants" that live and look different from the adults they will become? Research the life cycle of a fern, and you may make some valuable discoveries.

Answers to
What Is It Really Like?

This activity reinforces what students have learned about life cycles. Encourage students to be both accurate and creative. Their scripts should include the differences in the life cycles of the mosquito, grasshopper, and frog.
mosquito: egg, larva, pupa, adult
grasshopper: egg, nymph, instar, adult
frog: egg, tadpole, adult

Answers to
Diversity in Plant Cycles

In this research exercise, students learn that plants also have life cycles with distinctly different stages. Young plants can look quite different from the adult plants they will become.
Life cycle of a fern:
A spore, which comes from the brown spots called *sori* found on the underside of the fern leaf, germinates to form a small, heart-shaped structure called the *prothallus.* Egg and sperm cells are produced by the prothallus. When fertilization occurs, the fertilized cell develops into a *sporophyte*—a fern plant, with roots and curled-up fern leaves called *fiddleheads.* The fern leaves eventually become mature and produce sori, and the cycle continues.
Life cycle of a seed plant:
seed, seedling, adult plant, flower, fruit

Keeping Track

Answers to
Help Wanted

Harvester ants: clean and excavate small nests in sand or fine gravel; forage and collect seeds all day long; are ferocious when disturbed, and fearlessly bite intruders
Digger bees: dig cells in the ground and provision cells with nectar and pollen
Leaf-cutting ants: harvest bits of leaves and other bits of plants on which to grow fungus in the nests
Milking ants: milk honeydew from aphids
Potter (mason) wasps: build cells of mud; fasten cells to twigs and other objects; bring butterfly or beetle larvae to the nest; seal the cells

✹ Follow Up

Assessment

1. Give students modeling clay and ask them to make models of an insect that undergoes complete metamorphosis. *(Insect should have four stages—egg, larva, pupa, and adult.)* Also have students do this for an insect that undergoes incomplete metamorphosis. *(Insect should have an egg stage, several nymph stages in which it gradually gets bigger, and an adult stage.)*
2. Ask students to draw an ideal home for a tadpole and an ideal home for an adult frog. Then ask them to explain why the two homes are different. *(The home for the tadpole should be a large jar of pond water because the tadpole has gills and swims with a tail. The home for the adult frog should contain fresh air, soil, a dish of water, and green plants because the frog has lungs and legs.)*

Extension

1. Ask students to research the life cycles of some insects that cause damage to crops or property in their area. Have students find out the stages in which the insects are most harmful. Also ask students to identify the stages at which the insects can best be controlled. *(Less than 1 percent of insect species are harmful. Examples of insects that destroy crops include boll weevils, Hessian flies, corn earworms, chinch bugs, and Colorado beetles. Examples of insects that destroy property include clothes moths, carpet beetles, silverfish, and termites. The stages in which the insects are most harmful will vary according to species.)*
2. Have students research and write reports on 17-year cicadas or lodgepole pines. Both have interesting and unusual life cycles. *(In the case of the 17-year cicada, eggs hatch in the twigs of trees and shrubs. Once hatched, the nymphs fall to the ground, dig into the soil, and feed on roots. The nymphs remain in the ground until they are full-grown. Finally, they dig up to the surface of the soil, where they shed their skin and emerge as large, winged adults. The entire process takes 17 years. Lodgepole pine cones release their seeds only when there has been a fire. During a fire, the cone protects the seeds inside. Afterward, the scales open and let out the seeds.)*

One Form with Many Functions

Help Wanted

Insects are versatile creatures. They perform many tasks, just as humans do. However, insects have practiced their jobs for millions of years. In this time, insects have perfected their tasks so well that individuals now have specific jobs to perform for the benefit of their community.

The following want ads are taken from *The Worker's Journal* of Insectville. Study each one closely and then write your own ads for the insects shown below. Do some research to discover what each insect can do. Work in groups of two or three. Each group should prepare one ad for a class edition of *The Worker's Journal*.

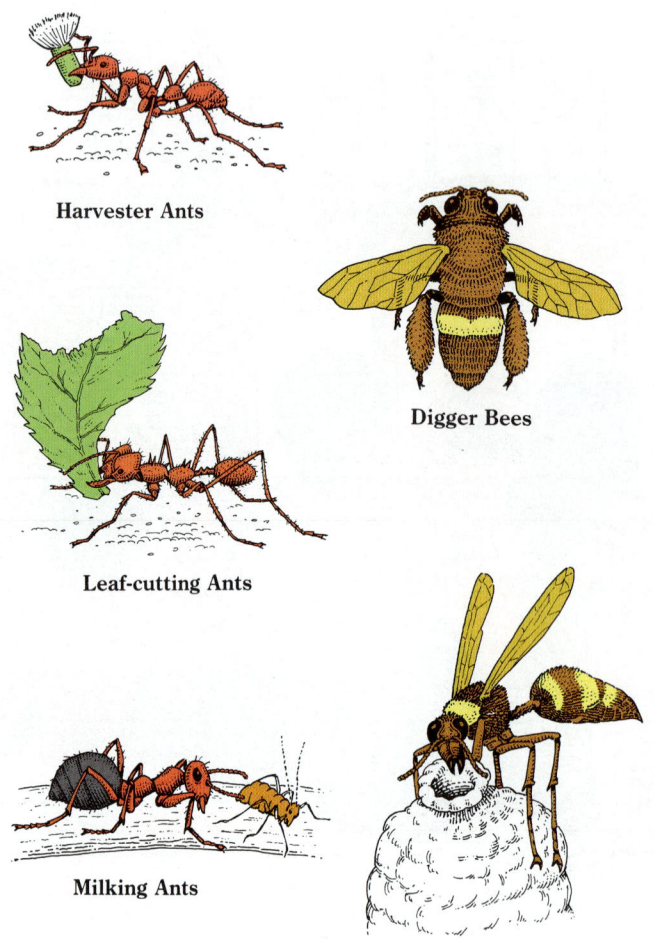

Harvester Ants

Digger Bees

Leaf-cutting Ants

Milking Ants

Potter (Mason) Wasps

The Worker's Journal
CLASSIFIED

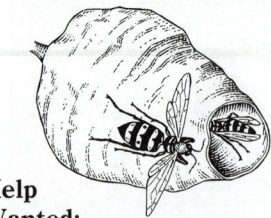

Help Wanted: Papermaking Wasps
Two papermaking wasps needed to construct 10 nests of paper. Workers must obtain wood fibers from local dead trees. Must have good chewing ability to break down the fiber and mix it with their saliva to form pulp. Ability to shape the pulp into six-sided cells is also essential. **Apply at Hives, Inc., Box 999**

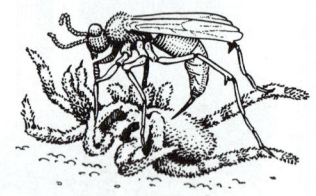

Help Wanted: 5 Female Pepsis Wasps
Must have armored bodies to hunt tarantulas and a lethal sting to paralyze the tarantula's body. Must have strength to drag the huge tarantulas to evacuated tunnels. Need strong mandibles and legs to dig new tunnels. Must be able to scrape the hair off a section of the tarantula's abdomen and deposit an egg on the cleared spot. Must be able to fill in the tunnel with sand and soil before leaving. **Apply at SPPW Tunnel 28 (Society for the Preservation of Pepsis Wasps, Inc.)**

Challenge Your Thinking

1. **The Animal Kingdom Pie**
   Can you explain the meaning of this pie chart? Which animals would you place in the missing piece of pie?

2. According to the classification systems used by biologists, why is:
   (a) a turtle more closely related to an alligator than to an eel?
   (b) an earthworm more closely related to a grasshopper than to a snake?
   (c) an octopus more closely related to a clam than to a lobster?
   (d) a whale more closely related to a tiger than to a shark?
   (e) a maple tree more closely related to a rosebush than to a pine tree?

3.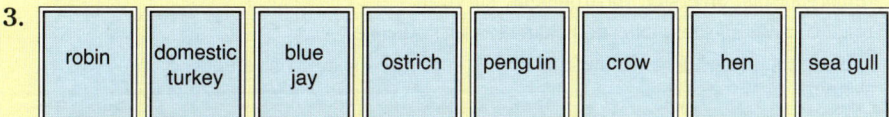

   Classify the birds above according to their characteristics. Divide them into two groups and write the characteristic each group shares in the boxes in the first row. Then divide each of these groups into two subgroups and write the characteristic each subgroup shares in the boxes in the second row. Under each of the four boxes in the second row, write the names of the birds that fit into that subgroup. Remember to write only in your Journal.

   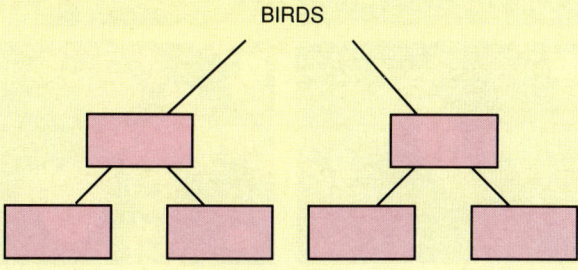

Answers to Challenge Your Thinking

1. The whole pie represents all of the animal kingdom, three-quarters of which are arthropods. Of the arthropods, the insects represent the largest section. The missing portion contains some of the invertebrates, such as worms, but it primarily contains the animals with backbones.

2. (a) A turtle is more closely related to an alligator than to an eel because both the turtle and the alligator are reptiles, while the eel is a fish.
   (b) An earthworm is more closely related to a grasshopper than to a snake because both the earthworm and the grasshopper are invertebrates, while the snake is a vertebrate.
   (c) An octopus is more closely related to a clam than to a lobster because both the octopus and the clam are mollusks, while the lobster is a crustacean.
   (d) A whale is more closely related to a tiger than to a shark because both the whale and the tiger are mammals, while the shark is a fish.
   (e) A maple tree is more closely related to a rosebush than to a pine tree because both the maple tree and the rosebush are flower-bearing plants, while the pine tree is a cone-bearing plant.

3. See chart at left. Other categories students might consider: migratory/non-migratory; nests in trees/nests on land; fast flyer/slow flyer; or fast runner/slow runner.

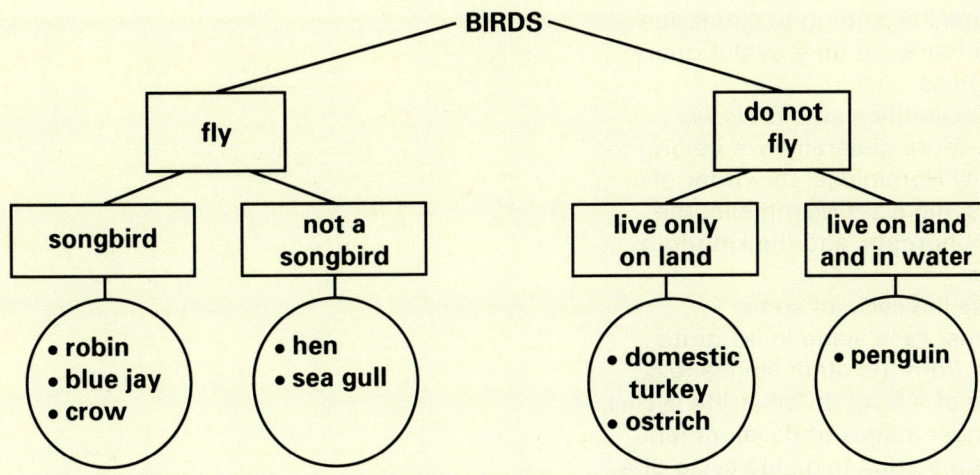

Keeping Track 123

UNIT 2
Making Connections

Summary for
The Big Ideas

Student responses will vary. The following is an ideal summary.

Diversity describes the many differences in size, shape, and structure that exist among living things. It arises as organisms adapt to a great variety of habitats and conditions over time.

Animals are adapted for *protection* by devices such as coloring, shape, mimicry, odors, stingers, claws, and teeth; for *obtaining food* by different types of teeth, beaks, and sensing structures; for *attracting a mate* by bright color patterns and by communication signals; and for *locomotion* by structures such as wings, legs, and fins.

Plants have root structures to anchor them and to obtain water and other nutrients. They have leaves that collect the sun's energy to make food. Some have sharp points to ward off animals. Some have flowers to attract pollinators. They can spread seeds by air, wind, animals, water, or mechanical propulsion.

Natural selection suggests that organisms with favorable traits are the most likely to survive, reproduce, and pass on those traits to their young.

Darwin examined the differences between the beaks of finches on the Galápagos Islands. He concluded that after being separated for thousands of years on the different islands of the Galápagos, the finches of each island had adapted in a unique way to their environment.

Species can become extinct when they cannot adapt to changes in environmental conditions. These changes can occur naturally or through human manipulation of the environment.

We classify things because dividing things into groups makes it easier for us to locate items or information quickly and efficiently. Grouping things also helps us recognize similarities and differences in a systematic way.

Living things are grouped and subgrouped according to similarities and differences in their evolutionary relationships.

Our scientific name is *Homo sapiens*. More generally, we belong to the family Hominidae, the order of Primates, the class Mammalia, the phylum Chordata, and the kingdom Animalia.

In the life cycle of some organisms, each stage looks quite different from the other stages (e.g., life cycle of a frog). In other life cycles, the younger stages of development look like the adult (e.g., life cycle of a grasshopper).

Unit 2 Making Connections

The Big Ideas

In your Journal, write a summary of this unit, using the following questions as a guide.

- What is diversity? Why is there diversity among living things?
- What are some adaptations animals have for protection? for obtaining food? for attracting a mate? for locomotion?
- How are plants adapted to survive and reproduce?
- What is meant by natural selection?
- What evidence did Darwin use to develop his theory of how life evolved?
- What kinds of conditions cause species to become extinct?
- Why do we classify things?
- How, in a general way, are living things classified?
- How would you classify yourself, according to Linnaeus's system? What is your scientific name?
- How is diversity shown in the life cycles of living things?

Checking Your Understanding

1. For each group of words below, write one or two sentences that show how the words are related to each other.
   (a) adaptation, predator, peppered moths, habitat
   (b) organisms, diversity, environment, survival
   (c) diversity, species, inherited characteristics, unique
2. Classify the living things in these photos according to some consistent system.

a

b

c

d

e

f

g

h

124   Diversity of Living Things

3. Draw a concept map that shows how the following words are related to each other: *Texas coral snake; camouflage; mimicry; snowshoe hare; adaptations; snow; grasses, trees, and weeds; brown in summer; white in winter;* and *Mexican milk snake.*

Reading *Plus*

You have been introduced to the great diversity of life and to how this diversity arises. You also learned how to classify animals and plants into groups and subgroups. By reading pages S23–S44 in the *SourceBook*, you will take a closer look at classification, at some of the ways diversity came about, and at the origin of life.

Updating Your Journal

Reread the paragraphs you wrote for this unit's "For Your Journal" questions. Then rewrite them to reflect what you have learned in studying this unit.

Answers to Checking Your Understanding

1. **(a)** *Peppered moths* use camouflage as an *adaptation* that protects them from *predators* in their *habitat.*
   (b) Adaptations help *organisms* in their struggle for *survival* in a given *environment*. Diversity arises from variations in how each species adapts over time to its habitat.
   (c) While *inherited characteristics* make members of a *species* alike in many ways, there is also great *diversity* within a species. People belonging to the same family, for instance, may look similar, but despite family resemblances, every human is *unique.*
2. See the classification diagram on page S160.
3. See concept map on page S160.

About Updating Your Journal

The following are sample ideal answers:
1. Plants and animals adapt to a variety of habitats in a number of different ways, resulting in structural and behavioral diversity. Such evolutionary changes enable organisms to live more successfully in their given habitats. *(The formation of these adaptive structures demonstrates the **Evolution** of living systems.)*
2. Fish have adapted only to an aquatic environment; frogs breathe through gills during the aquatic part of their life cycle and then develop lungs so that they can survive on land.
3. The extinction of a species affects the organisms that depend on that species for food, for habitat, or for other life needs. Organisms that depend on the consumers of the extinct species will also be affected, and so on throughout the food chain.
4. The panther chameleon can change the color of its skin. Chameleons have excellent eyesight, and their eyes move independently to locate prey.

Making Connections 125

Science in Action

◆ Background

The appeal of a career as a medical doctor is widely known. As a result, the competition to get into medical school is intense. There are about 270,000 applications written each year by 28,000 students. However, only 17,000 of those students are accepted by medical schools. Medical school generally takes four years, with the first two spent primarily in classrooms and laboratories, and the latter two in clinics. This is followed by three or four years of residency in a hospital. Doctors concentrating in surgery, internal medicine, and other specialties spend three to seven years combined with their residency in further study. Finally, doctors must pass a medical-board examination.

Science in Action

Spotlight on Medicine

Medical Practice

Harold Still has been a family physician for nearly 40 years. Physicians diagnose and treat diseases and disorders of the human body. If you wish to enter this profession, you must spend four years in undergraduate study at a university. After this, four years of training in medical school is required. Upon graduating from medical school, you would receive an MD (Doctor of Medicine) degree. Then you would spend between three and seven years as a resident in a hospital, preparing for a specialty. Your specialty might be in family medicine, neurology, or surgery, just to name a few.

Q: Why did you become a medical doctor, and what do you find most interesting about your work?

Harold: Perhaps it was my mother who encouraged me in this direction. In any case, I wanted to be a physician from the time I was about 12 years old.

I really enjoy patient contact. I guess that's why I like being a family doctor. I like taking care of people. In particular, I like looking into the whole life of a person, not just an illness. I am interested in the relationship between the illness and the person's social environment, including work and family.

Over the past 14 years, I have also enjoyed teaching family medicine to future doctors.

Q: What kind of background and education should a person who wants to enter medicine have?

Harold: Concern for people, confidence, and the ability to interview and relate to people are all necessary personal characteristics. An understanding of the scientific basis of medicine is also very important for a family physician.

Doctors should have extensive knowledge of medicine, of course, but they should also have a good understanding of English, history, and nonscientific subjects. I feel strongly that students should not cut themselves off from the humanities and social studies when planning to become a doctor. I like to see the development of a well-rounded person. I think such physicians are the most effective in dealing with patients.

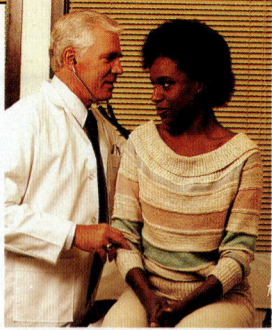

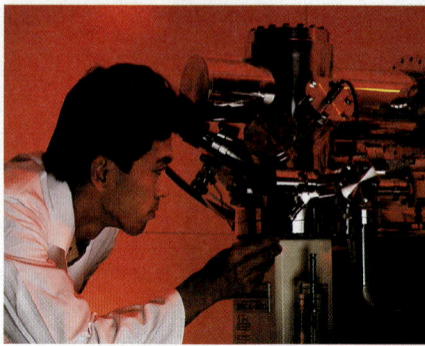

Medical Research

Barbara Pope is a medical researcher who studies the immune system of the body and the causes of cancer. Barbara has a medical degree, but instead of seeing patients, she prefers to do research.

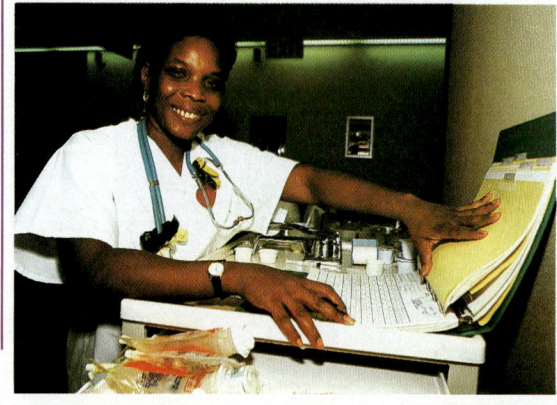

◆ Using the Interview

To check students' understanding of the article, ask questions such as the following:

- What are the major differences between Dr. Harold Still's profession and Dr. Barbara Pope's profession? *(Harold works with individual patients on a day-to-day basis, whereas Barbara works as a researcher doing laboratory work and writing scientific articles.)*
- What characteristics are important in a family doctor? *(Besides an understanding of the scientific basis of medicine, a family doctor should have concern for people and the ability to interview and relate to people.)*
- What characteristics are important in a medical researcher? *(A medical researcher should enjoy analyzing experimental data and scientific papers. Like the family practitioner, a medical researcher must understand the scientific basis of medicine.)*
- Do you think you would like a career as a medical doctor? If so, do you think you would prefer being a family doctor or a medical researcher? Why?

Q: What do you find most stimulating about your work?

Barbara: The joy of analyzing experimental data is one very stimulating part of scientific work. It's exciting when you bring the data together, come up with a pattern of what you think is happening in a given field, and make your own hypothesis. There's a lot of work, but it's very rewarding work.

Q: Will there ever be a "grand" cure for all forms of cancer?

Barbara: Cancer is a very complex problem, and there is probably no single answer to it. There are a lot of partial answers, however, and eventually all the partial answers will cure cancer. But it's going to be a while yet before we understand enough basic biology. There are many systems of the body involved, including the immune system. Only when we understand how all of these systems interact will we be close to solving the problem of cancer.

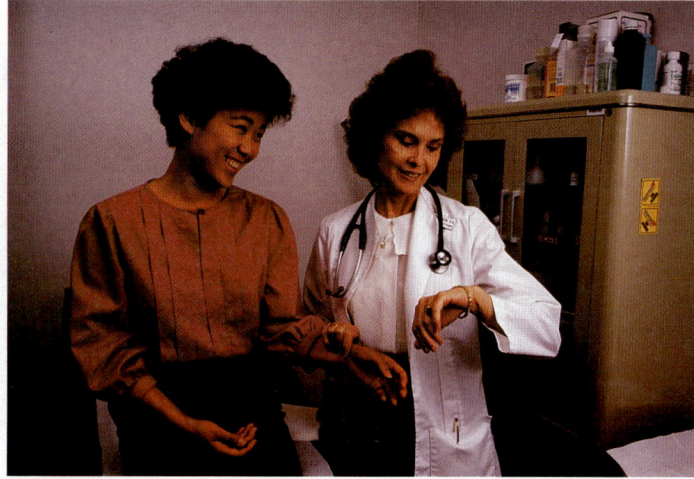

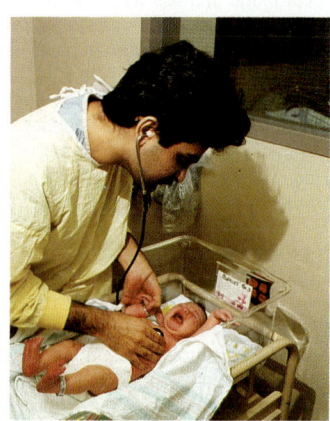

Some Project Ideas

1. In many countries, medical care is basically free to all citizens, regardless of their level of income. In such countries, the government either owns and operates the health-care system, or it finances health care through national health insurance.

    Do some research to learn more about health-care systems. Do you think our private health-care system is adequate? What are its advantages and disadvantages? What are the advantages and disadvantages of a national health-care system?

2. Do some research to find out how AIDS has changed the practice of medicine. Check references such as *Magazine Index* or *Reader's Guide to Periodical Literature* for magazine articles on the subject. In addition, AIDS service organizations, your city or county health department, and local hospitals can direct you to further information. Share the information you gather with your classmates.

3. Various forms of cancer are associated with smoking. What is the connection between smoking and cancer? Present your information in a dramatic and convincing manner to your classmates.

    For assistance, contact your family physician and/or the American Cancer Society in your area. Your local library will also have valuable information on this topic.

✦ Going Further

Ask students to investigate the following questions:
- What are the most challenging problems in medical research today?
- What are some of the different areas in which medical doctors can specialize? What would the typical workday look like for each of the specialists you choose?
- Should health-care professionals who are infected with HIV (the virus that causes AIDS) be required to inform their patients about their condition? Why or why not?

✦ Using the Project Ideas

Divide students into groups of four or five. Assign a project idea to each group. Ask students to design a plan for carrying out the project.

1. The private health-care system in the United States usually works for people who have health insurance. However, many people do not have health insurance, leaving them with limited access to medical attention. A national health-care system provides equal treatment for everyone, regardless of income level. However, national health care is sometimes substandard. Students should find out the reasons behind the difference in quality of care between private and public health care.

2. In their research, students can focus on the following topics:
    - As a result of AIDS, medical personnel have paid closer attention to infection control with all patients, not just those with AIDS. With tougher infection-control standards, fewer infections are now spread in hospitals and other medical care facilities.
    - As a result of AIDS research, many unexpected medical discoveries have been made that may yield cures for cancer and other diseases.
    - The need for AIDS treatment has forced a reduction in the number of years the FDA requires for the testing of certain drugs. Changes in FDA drug-testing policies may benefit many people who have life-threatening conditions.
    - Also as a result of AIDS, alternatives to keeping patients in the hospital have been devised, such as hospices and home health-care arrangements.

3. Smoking can cause lung cancer, emphysema, and heart disease. There is also a connection between pregnant women who smoke and low birth-weight babies.

    To provide students with up-to-date information, you might want to request brochures from the American Cancer Society. [1599 Clifton Rd., NE, Atlanta, GA 30329. (404) 320-3333.]

Science in Action

Science and Technology

♦ Background

Advanced technology has made large-scale hunting of animals easy and profitable. This has caused many species to be hunted to extinction. Others, such as sea turtles, whales, and many mammals are still being hunted for their hides, despite the fact that they are endangered.

Even more threatening are the indirect human assaults on natural ecosystems. Many species are becoming extinct due to the habitat destruction caused by deforestation, wetland drainage, construction, farming, dam building, and strip mining.

Climatic changes due to greenhouse warming, acid rain, and ozone depletion will force many species to become extinct. Those species are integral members of ecosystems. When these members are destroyed, humanity is attacking its own life-support systems.

Today, many humans are working to save wildlife by protecting animal habitats and breeding species in captivity. The American bison, the Arabian oryx, the nene goose, the Andean condor, and the sea otter are a few animals that people have helped save from extinction.

Science and Technology

Extinction in the Modern World

In the year 1990 alone, it is estimated that *10,000 different species* of living things became completely extinct! In the year 2000, the studies say, the number of species that become extinct will be 20,000. These numbers may seem even more astonishing when you realize that in 1975 the number of extinctions was "only" 100.

It probably won't surprise you that humans are the cause of this massive destruction of life. The human population of the earth is rapidly increasing, and people are encroaching on life all around them.

Of course, there have always been extinctions from purely natural causes. Dinosaurs died off long before humans appeared on the earth. But in the past, species that died off were replaced by new species. Today the rate of extinction is so great that there is not enough time for new species to evolve to replace the ones that are lost.

The Technology Factor

As the human population increases, people move into new territory. They clear forests, drain swamps, and build dams. Technology, with innovations such as chain saws and bulldozers, has made humans effective at altering the environment very quickly. As a result, the habitats of many plants and animals have been destroyed.

Technology has also made humans effective at killing animals. Inventions such as guns and explosive harpoons have made it possible for people to kill animals on very large scales. The passenger pigeon, now extinct, was once so numerous that clouds of them would darken the sky for hours. It was killed off completely, mostly for food and fertilizer. And the American Bison would be extinct today if conservationists had not stopped their slaughter.

Certainly not all human destruction of life is deliberate. A lot of it comes as a result of the pollution that technology so often brings. Chemical wastes, pesticides, accidental oil spills, and the like, have poisoned the natural habitats of many plants and animals. Often there is a *chain reaction*. For example, poisonous wastes in the ocean might kill a plant or fish that is the food supply of a certain bird. So, because of the pollution, the bird dies too.

The American Bison has been saved from extinction.

♦ Discussion

When students have finished reading the feature, involve them in a discussion of what they have read. You may wish to use questions similar to the following:

1. What reasons can you name to explain the increasing number of species that are becoming extinct? *(The human population on earth is rapidly increasing, and people are encroaching on all forms of life around them.)*

2. What is different now regarding how species die off and new ones evolve? *(The rate of extinction is now so great that there is not enough time for new species to evolve.)*

3. What is the difference between an endangered species and an extinct species? *(An endangered species is in danger of becoming extinct. An extinct species no longer exists and will never exist again.)*

Conservationists are working to save the California condor.

Can Technology Help?

The work of conservationists has many sides, including proposing laws to protect endangered species. Often it involves working directly with the plants or animals. At times modern technology proves helpful. For example, researchers may attach radio-tracking devices to animals in order to learn the animals' habitats and migratory routes. In this way, they can determine if a species is being threatened by environmental hazards.

Saving the Condor

In some cases, animals are caught and raised in captivity in order to keep them safe from dangers in the environment. For example, conservationists had captured all living California condors by the end of 1987. It is hoped that they can be released back into nature when sufficient numbers have been bred, and when the environment has been made safe. By the time you read this, some of the great birds may be back in the wild.

Controversy

Money is very often the reason that people deliberately kill living things in large numbers. There is a lot of money to be made from, say, killing elephants just for their tusks. But it may not be just a matter of greed. For some people, such activities might be their only source of income. Therefore, their lives would seem to depend on doing such things.

Do you think there should be laws protecting endangered species? How important is it, after all, to save an animal like the blue whale or the whooping crane? How would you feel if your personal livelihood were threatened by a law that protected some endangered species, perhaps a "minor" one that you had never even heard of before?

African elephants are being killed in large numbers.

Find Out for Yourself

Do some research to find out why the California condor was in danger of becoming extinct. What laws have been enacted by the state of California to make the environment safe for these birds? What are some other animals or plants that have become extinct in recent years? What were the reasons for their extinction?

◆ Critical Thinking

To promote logical-thinking skills, you may wish to ask questions similar to the following:

1. Many conservationists felt that the California condor should not have been captured for captive breeding. They felt that it was better for the condor to vanish in the wild rather than to die in a cage. How do you feel about this issue? *(Opinions will vary. The birds were taken into captivity to protect them from whatever was harming them in the environment. Others argued that a condor killed by starvation, poisoning, or poaching could hardly be called "death with dignity.")*

2. How does the fashion industry contribute to endangerment of organisms? *(In the early part of this century, feathers were the rage in the hat industry. Hundreds of different species of birds were killed so that their feathers could be used to decorate hats. Only a timely change in fashion stopped this slaughter. Currently, crocodile skins are made into wallets, belts, shoes, and handbags. Furred mammal skins are also in great demand. Ivory from the African elephants' tusks is used to make jewelry.)*

◆ Going Further

1. Have interested students find out how the introduction of exotic species can lead to the endangerment of native species (e.g., the introduction of goats in the Galápagos Islands or rabbits in Australia).
2. Have students send for information about endangered species from organizations such as: The Whale Protection Fund, The World Wildlife Fund, and Greenpeace.
3. Students may be interested in becoming personally involved in helping to save an endangered species. If so, have them do some research to find out what species are endangered in their area, and what is causing that endangerment. This information might be available from your state's Department of Fish and Game. Students could launch a publicity and/or fund-raising campaign to help save these species. For example, they could write editorials to the local paper, hand out fliers, make posters to display in public places, and collect money to aid organizations helping to save the species.

Science and Technology

UNIT 3 Solutions

✦ Teaching the Unit

Unit Overview

In this unit, students are introduced to the properties of solutions by drawing from their prior knowledge of solutions in everyday life. In the section *A Solution or Not?*, students use this prior knowledge to formulate a definition of solutions, solutes, and solvents. Then students apply their understanding of solutions to the familiar characteristics of hard and soft water. In the section *Separating Solute and Solvent,* students investigate distillation, evaporation, and chromatography as methods used to separate solutes from solvents. In the final section, *Concentration of Solutions,* students explore the characteristics of solutions that have different concentrations.

Using the Themes

The unifying themes emphasized in this unit are **Scale and Structure, Systems and Interactions,** and **Patterns of Change**. The following information will help you weave these themes into your teaching plan. A focus question is provided with each theme as a discussion tool to help you tie the information in the unit together.

Scale and Structure is the predominant theme in the first section. As students compare the properties of mixtures to those of true solutions, they make inferences about the structure of solutions. The concept of scale can be related to the size of the particles (molecules and ions) present in a true solution.

Focus question: *How are the size of particles related to the formation of solutions?*

Students should understand that solutions form when particles of the solute fill in the empty spaces between the molecules of a solvent. Because of the small size of the particles and because these particles are uniformly mixed, the individual particles of the components of a solution are not visible.

Systems and Interactions is an integral part of Sections 2 and 3. For example, a solution represents an interaction between two or more substances. Evaporation and distillation of a solution are interactions that separate solute from solvent. Both processes require the input of energy in the form of heat or air in motion.

Focus question: *How does water interact with the earth and its organisms to produce a solution?*

Students can draw on their investigation of the water cycle to conclude that as water strikes the earth in the form of precipitation, it aids in the process of erosion, putting dissolved pieces of the landscape into solution. Dissolved plant and animal matter also become part of the solution, helping to cycle nutrients through the environment.

Patterns of Change is the major focus of Section 3. Changes that can be discussed in connection with solutions include boiling-point elevation, freezing-point depression, and an increase in buoyant force.

Focus question: *What factors bring about changes within a solution?*

Adding more solute to a solution increases the boiling point and lowers the freezing point. The temperature of a solution can determine the solubility of the solute, and the physical and chemical properties of the solute itself determine its interaction with the solvent.

Using the *Science Discovery* Videodiscs

Disc 1 *Science Sleuths, Green Thumb Plant Rentals #1* Plants rented by MicroDiscovery (MDY) from Green Thumb Incorporated (GTI) are dying. The owner suspects a rival company of sabotage. The Science Sleuths must analyze the evidence for themselves to determine the true cause of the plant failures.

Disc 2 *Image and Activity Bank*
A variety of still images, short videos, and activities are available for you to use as you teach this unit. See the *Videodisc Resources* section of the **Teacher's Resource Binder** for detailed instructions.

Using the *SciencePlus SourceBook*

Unit 3 reviews the important terms discussed in the text and introduces the new terms colloids and suspensions. Special classes of solutions known as acids, bases, and salts are then discussed, with emphasis on their properties and important practical uses.

129A

PLANNING CHART

SECTION AND LESSON	PG.	TIME*	PROCESS SKILLS	EXPLORATION AND ASSESSMENT	PG.	RESOURCES AND FEATURES
Unit Opener	130		observing, discussing	For Your Journal	131	Science Sleuths: *Green Thumb Plant Rentals #1* Videodisc Activity Sheets TRB: Home Connection
A SOLUTION OR NOT?	132			Challenge Your Thinking	151	
1 Solutions and Non-Solutions	132	2 to 3	investigating, observing, predicting, writing	Exploration 1	134	Image and Activity Bank 3–1
2 Parts of a Solution— Solutes and Solvents	140	4	analyzing, classifying, investigating, inferring	Exploration 2 Exploration 3 Exploration 4	145 146 147	Image and Activity Bank 3–2 TRB: Resource Worksheets 3–1, 3–2, and 3–3 Graphic Organizer Transparencies 3–1, and 3–2
3 Soluble, Not Very Soluble, or Insoluble?	148	2	observing, investigating, recording, comparing	Exploration 5 Exploration 6	148 150	Image and Activity Bank 3–3 TRB: Resource Worksheet 3–4 Graphic Organizer Transparency 3–3
SEPARATING SOLUTE AND SOLVENT	152			Challenge Your Thinking	167	
4 The Salt Industry	152	1 to 2	observing, investigating, comparing, designing an experiment	Exploration 7	155	Image and Activity Bank 3–4
5 Desalination and Distillation	156	1 to 2	observing, analyzing, comparing, illustrating			Image and Activity Bank 3–5
6 Another Solution	160	2	observing, inferring, predicting, formulating hypotheses	Exploration 8	161	Image and Activity Bank 3–6
7 Sugar Solutions	164	1 to 2	observing, analyzing, inferring, predicting	Exploration 9	166	Image and Activity Bank 3–7
CONCENTRATION OF SOLUTIONS	168			Challenge Your Thinking	185	
8 Getting Colder	168	2	observing, formulating hypotheses, analyzing, drawing conclusions	Exploration 10 Exploration 11 Exploration 12	168 169 169	Image and Activity Bank 3–8
9 How Much Solute?	170	2	observing, investigating, comparing, formulating hypotheses	Exploration 13 Exploration 14	172 174	Image and Activity Bank 3–9
10 Saturation	176	2	investigating, measuring, interpreting data, making a graph	Exploration 15	176	Image and Activity Bank 3–10 TRB: Resource Worksheet 3–5 Resource Transparency 3–4 Graphic Organizer Transparency 3–5
11 Solutions in the Environment	179	2	writing, researching, calculating			Image and Activity Bank 3–11 TRB: Activity Worksheet 3–7 Resource Worksheet 3–6 Graphic Organizer Transparency 3–6
12 Growing Crystals	183	2	investigating, measuring, observing, inferring	Exploration 16	183	Image and Activity Bank 3–12 TRB: Activity Worksheets 3–8 and 3–9 Graphic Organizer Transparency 3–7
End of Unit	186		applying, analyzing, evaluating, summarizing	Making Connections TRB: Sample Assessment Items	186	TRB: Activity Worksheet 3–10 Graphic Organizer Transparency 3–8 Science in Action, p. 188 Science and the Arts, p. 190 *SourceBook,* pp. S45–S60

*Time given in number of class periods.

Meeting Individual Needs

Gifted Students

1. Pose one or more of the following "what if" questions for students:
   - What would daily life be like if solutions did not exist?
   - What would the oceans be like if salt did not dissolve in water? What would beaches be like?
   - What would the world be like if water did not act as a "universal solvent"? How would this effect the human body?
2. Explain to students that the amount of oxygen dissolved in water varies greatly depending on the temperature. Warm water cannot hold as much dissolved oxygen as can cold water. Have students study a map or globe, and make predictions about the water temperature and the oxygen concentration in different ocean areas. Have them do library research to check their predictions. Suggest that they research the importance of dissolved oxygen as a factor in determining the abundance and variety of marine life in given locations.

LEP Students

1. Using magazine and newspaper pictures, have students create a scrapbook of examples of solutions and mixtures. Suggest that they include examples from publications in their native language, if available. Students can annotate their scrapbooks in English or their native language. Students could explain how they classified each example as a solution or a mixture and name the solvents.
2. The words "weak" and "strong" are often used in English to describe solutions that are dilute and concentrated, respectively. Set up solutions of "weak" and "strong" tea. Have students work in mixed-language, cooperative-learning groups to compare the solutions using English and native language vocabulary.
3. At the beginning of each unit, give Spanish-speaking students a copy of the *Spanish Glossary* from the *Teacher's Resource Binder.* Also, let Spanish-speaking students listen to the *English/Spanish Audiocassettes.*

At-Risk Students

This unit affords many experimental opportunities for students to gain insight into the underlying properties of solutions. *Exploration 6* allows students to practice controlling variables while exploring the factors that put a solute into solution. In *Exploration 8,* students act as both scientist and detective as they analyze inks using paper chromatography. This exercise also provides a good opportunity to review the terms associated with experimental procedure.

As an extension activity, ask students to identify appliances used to make solutions in their own kitchens. For each appliance they identify, have them explain its effect on the solution-making process. For example, electric mixers and food processors speed the solution process by agitation or increasing the surface area available to the solvent. Heat sources speed the solution process by increasing the temperature of the solvent and enhancing solubility.

Cross-Disciplinary Focus

Creative Writing

Hydroponics is a technique for cultivating plants without using soil. Instead of soil, the plants are grown in a solution containing all of the essential plant nutrients. Either artificial or natural light sources are used. Discuss hydroponics with the class, highlighting the possible advantages and disadvantages of this technique. Suggest that students write a futuristic short story in which hydroponics plays an important role. Possible story settings could include space colonies, floating cities on the open ocean, and so on.

Design and Metal Alloys

Metal alloys provide materials with particular physical properties essential to the design or function of an object. For example, silver and gold used in jewelry are combined with other metals to make the jewelry stronger and more resistant to abrasion. Sterling silver is an alloy of 925 parts of silver to 75 parts of copper. Gold alloys used in jewelry are described by the karat system. Pure gold is described as 24-karat because all 24 parts are gold. The gold commonly used in jewelry is 14-karat, which has 14 parts gold and 10 parts of copper. Using information on metal prices from the daily newspaper, suggest that students calculate the actual worth of the metals used to manufacture a given piece of jewelry or other manufactured item. Have students research the uses and properties of alloys. To share their work, they could develop classroom displays or bulletin boards illustrating their research findings.

Cross-Cultural Focus
Plant Dyes and Medicines

Cultures from all over the world have learned to use solutions made from local plant materials for medicines and dyes. Asian, African, North and South American, European, Australian, and Pacific Island indigenous peoples have all developed medical treatments based on plant extracts. Extracts from local plants were also used for dying cloth and other materials. Until the invention of synthetic dyes in the early twentieth century, dye plants, such as indigo, were widely cultivated and had great commercial importance. Have students research local cultivated or wild plants used for medicine or dyestuffs. (Local natural history museums may be another source of information about uses of plants native to your area.) They will find that some of these plants are familiar kitchen herbs used today, such as sage and mint. A simple extract can be made by immersing these plant materials into boiling water. The water-soluble components will dissolve in the hot water. This solution is known as a "tea," or infusion. Students can then use the techniques of paper chromatography described in *Exploration 8* on page 161 to investigate the components of the plant extract.

Resources

 ## Bibliography for Teachers

Barber, Jacqueline. *Crime Lab Chemistry.* Great Explorations in Math and Science (GEMS) Series, ed. by Lincoln Bergman and Kay Fairwell. Berkeley, CA: Lawrence Hall of Science, 1990.

Center for Multisensory Learning. *Mixtures and Solutions.* Berkeley, CA: Lawrence Hall of Science, 1990.

Chemical Education for Public Understanding Program (CEPUP), Lawrence Hall of Science. *Groundwater Contamination, The Fruitvale Story.* Menlo Park, CA: Addison Wesley Publishing Company, Inc., 1990. Includes optional kit.

Sumrall, William J., and Fred W. Brown. "Consumer Chemistry in the Classroom: Science from the Supermarket." *The Science Teacher,* 58 (April 1991): 29–31.

Windholz, Marcia, ed. *Merck Index: An Encyclopedia of Chemicals, Drugs, and Biologicals.* Rahway, NJ: Merck & Company, Inc., 1983.

 ## Bibliography for Students

Asimov, Isaac. *How Did We Find Out About the Atmosphere?* New York, NY: Walker Publishing Company, Inc., 1985.

Fialkov, Yu. *Extraordinary Properties of Ordinary Solutions.* Chicago, IL: Imported Publications, 1985.

Gardner, Robert. *Kitchen Chemistry: Science Experiments to Do at Home.* Englewood Cliffs, NJ: Simon & Schuster, Inc., 1989.

Lewis, James. *Measure, Pour, and Mix: Kitchen Science Tricks.* New York, NY: Meadowbrook Press, 1990.

Pringle, Lawrence. *Rain of Troubles: the Science and Politics of Acid Rain.* New York, NY: Macmillan Publishing Company, Inc., 1988.

 ## Films, Videotapes, Software, and Other Media

Alterations in the Atmosphere.
Videotape.
Films for the Humanities, Inc.
P. O. Box 2053
Princeton, NJ 08543

Mixtures and Solutions.
Videotape.
Eureka!
Lawrence Hall of Science
Univ. of California
Berkeley, CA 94720

Solubility.
Software.
For Apple II Family.
Educational Materials and Equipment Company
Old Mill Plain Road
P. O. Box 2805
Danbury, CT 06813-2805

Solutions: Ionic and Molecular.
Videotape and Film.
Coronet Film & Video
108 Wilmot Road
Deerfield, IL 60015

Solutions.
Software.
For Apple II Family.
Prentice Hall Allyn & Bacon
Sylvan Avenue
Englewood Cliffs, NJ 07632

Water Pollution.
Software.
For Apple II Family, IBM PC, PC jr.
Educational Materials and Equipment Co.
Old Mill Plain Road
P. O. Box 2805
Danbury, CT 06813-2805

UNIT 3

Unit Focus

To introduce the unit, pose the following questions:
- What is a mixture? *(a substance made up of two or more identifiable components)*
- What is a solution? *(Students may suggest that a solution is something dissolved in water.)*
- Can there be a solution with no water? *(yes)*
- What might such a solution be like? *(Students may suggest other liquid solvents.)*
- Can gases form solutions? If so, name one. *(Yes. Soda water is a solution of carbon dioxide gas dissolved in water.)*

Now, suggest that students develop a general definition of a solution. *(A possible response: a mixture of substances that is transparent and appears to be of the same composition throughout.)*

About the Photograph

This photograph shows ocean waves near the shore of the Pacific Ocean. Examples of solutions and mixtures are visible. Ocean water is a solution containing 10 major dissolved ions and 61 other elements. Other bodies of water also contain quantities of particulate matter. The surrounding air is also a solution. It contains about 79 percent nitrogen, 20 percent oxygen, and 1 percent other gases.

Unit 3 Solutions

A SOLUTION OR NOT?
1. Solutions and Non-Solutions, p. 132
2. Parts of a Solution—Solutes and Solvents, p. 140
3. Soluble, Not Very Soluble, or Insoluble? p. 148

SEPARATING SOLUTE AND SOLVENT
4. The Salt Industry, p. 152
5. Desalination and Distillation, p. 156
6. Another Solution, p. 160
7. Sugar Solutions, p. 164

CONCENTRATION OF SOLUTIONS
8. Getting Colder, p. 168
9. How Much Solute? p. 170
10. Saturation, p. 176
11. Solutions in the Environment, p. 179
12. Growing Crystals, p. 183

130 Solutions

For Your Journal

Write a paragraph summarizing your thoughts about each of the following:
1. Why can't you see a sugar cube after it has been stirred in water?
2. How could you separate the sugar from the water later?
3. What difference would it make if water were not such a good solvent?
4. Is sea water a solution or a non-solution? Explain your answer.

✷ Using the Photograph

Have students study the photograph. To promote discussion, ask them questions similar to the following:
- Are there any solutions in this photograph? *(the ocean, the air)*
- What characteristics do solutions have in common with the ocean? *(The particles of different substances in solutions cannot be seen.)*

Involve students in a discussion of the ocean and the atmosphere as solutions. Ask them to brainstorm a list of substances that could be dissolved in these solutions. *(Possible responses include: salt dissolved in the ocean; oxygen dissolved in the air.)* Discussion topics that could lead to such inferences include: needs of organisms, the water cycle, the carbon-oxygen cycle, erosion, air and water pollution, sewage disposal, combustion, international shipping, air travel, and so on.

✷ About Your Journal

Students should answer the Journal questions to the best of their abilities.
These questions are designed to serve two functions: to help students recognize that they do indeed have prior knowledge about the topic, and to help them identify any misconceptions they may have. In the course of studying the unit, these misconceptions should be dispelled.

Solutions   131

LESSON 1

✹ Getting Started

This lesson introduces the term *solution* and examines the various meanings of the word as it applies to solving mysteries, problem solving, and chemistry. Students are introduced to the Tyndall Effect and how it can be used to identify solutions. The lesson concludes with a bulletin-board activity in which students write facts about solutions on index cards.

Main Ideas

1. The term *solution* has multiple meanings.
2. Most solutions are liquid, transparent mixtures.
3. The parts of a solution are spread uniformly throughout the mixture.
4. The Tyndall Effect is a method of identifying a solution; no true solution shows the Tyndall Effect.

✹ Teaching Strategies

Direct students' attention to the three pictures at the top of page 132. Call on a volunteer to read the captions. Discuss the different ways the word *solution* is used. Ask students to make up several sentences to practice using the word.

Have students silently read the questions at the bottom of the page. They are intended to stimulate interest in the unit. Accept all reasonable responses at this point. The correct answers are given on the following page as a discussion aid.

A SOLUTION OR NOT?

1 Solutions and Non-Solutions

The detective solves the crime. That's one kind of solution.

The math student solves the problem. That's another kind of solution.

The chemist does these things and more. The chemist also mixes solutions!

In this unit you are going to be a chemist most of the time, but you will also be a mathematician and a bit of a detective! So after much detecting, calculating, and mixing, you will be able to tell or show anyone the answers to these questions:

- How is a solution different from a non-solution?
- What is a solute? a solvent?
- What solutes are very soluble in water? not very soluble? insoluble?
- How can you separate a solute from a solvent?
- How does a solute affect the boiling and freezing points of a solvent?
- What is a "dilute" solution? a "concentrated" solution?
- How can you test a solution to see whether it is dilute or concentrated?
- What is meant by "saturated" and "solubility"?
- Does temperature affect the amount of solute that can be dissolved?

LESSON  ORGANIZER

Objectives

By the end of the lesson, students should be able to:
1. Distinguish between mixtures that are solutions and those that are not.
2. Recognize the Tyndall Effect and describe it in relation to solutions.

Process Skills

investigating, observing, predicting, writing

New Terms

none

Materials

Exploration 1
Part 1: 6 beakers; 100-mL graduated cylinder; 10-mL graduated cylinder; 1 mL powdered juice mix; 10 mL rubbing alcohol; milk; 1 mL sugar; 1 mL motor oil; 10 mL sand; 10 mL salt; 500 mL water; eyedropper;

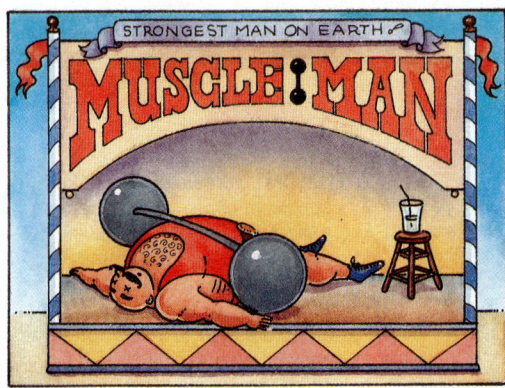

A Case of Foul Play

At first, it appeared that the carnival's Muscle Man had passed out from trying to lift too much weight, or possibly from the heat.

Binti Sirba, a paramedic, examined the Muscle Man and told Sergeant Malikia Noble, the carnival's security guard, "You're not going to believe it, but this man isn't unconscious. He's just fast asleep!"

Sergeant Noble frowned. "But that's ridiculous. No one falls asleep in the middle of lifting weights. Something's fishy here." Looking out at the audience, she asked, "Would someone tell me what the Muscle Man did right before lifting the weights?"

A boy spoke up. "Sure. He bragged that he was going to lift more weight than we'd ever seen anyone lift before. He gulped down a bunch of whatever's in that cup there, took a few deep breaths, lifted the weights to shoulder level, and then boom! Down he went." Others nodded. "Yep. Out like a light."

After hearing this, Ms. Sirba carefully examined the Muscle Man's drink. It appeared faintly yellow and transparent (clear), and it smelled like lemons.

"That's lemonade from the stand across the way," Sergeant Noble told Ms. Sirba. "He drinks it all the time, although it's not very tasty. It's that powdered lemonade mix you add water to."

"Hmm. That helps, but I still need to make one more test." Ms. Sirba put some of the lemonade into a clear glass, shook it, and held it up to the light. "This isn't a pure lemonade solution," she concluded. She held it up for the security guard to see. "My guess is that someone's added sleeping powder to the lemonade in that cup."

Chemical tests at a forensic lab showed that a potent dose of insoluble sleeping powder had been added to the Muscle Man's lemonade. Thanks to the paramedic's astute observations, further investigations were carried out, and the man who operated the lemonade stand was later convicted of the crime.

stirring rod; bright light or flashlight; piece of cardboard; pencil or knife; metric ruler; safety goggles
Part 2: various mixtures such as instant coffee and water, tea and water, chalk dust and water, soft drinks, salt and water, index cards

Science Discovery Videodisc

Disc 2, Image and Activity Bank, 3–1

Time Required

two to three class periods

Answers to

In-Text Questions

- A solution is a homogeneous mixture. A non-solution is not homogeneous.
- A solute is the substance in lesser quantity in a solution. The solvent is the substance in greater quantity in a solution.
- A material is soluble if it "disappears" into the solvent. It is not very soluble if it only partially disappears. If all the material remains visible, it is insoluble.
- Evaporating the solvent is one way to separate a solute from solution.
- Solutes raise the boiling point and lower the freezing point of a solvent.
- A dilute solution has less solute than a concentrated solution.
- Some ways to test whether a solution is dilute or concentrated: measure its mass/unit volume; perform glass-tube test (the more concentrated solution is more dense and thus sinks to the bottom of the tube); compare the intensity of the solution's color; and perform float tests (the more concentrated the solution, the better able an object is to float in the solution).
- Saturated means that the maximum amount of solute is dissolved. Solubility measures the number of grams of solute it will take to saturate 100 g of solvent.
- At higher temperatures, more solute can be dissolved.

A Case of Foul Play

The story on page 133 sets the stage for *Exploration 1.* Call on a volunteer to read it aloud. Ask students to make preliminary evaluations of the situation described based on the facts given.

Exploration 1

Part 1

Divide the class into small groups and distribute the materials. Encourage each group member to prepare at least one mixture. Keep the graduated cylinder and stirring rod clean following the preparation of each mixture. Ask groups to examine all the mixtures and list the differences between the solutions and non-solutions.

All examples in *Part 1* are mixtures. However, only the powdered juice mix and water, rubbing alcohol and water, and sugar and water mixtures are solutions. Emphasize that in contrast to solutions, the other mixtures are formed by two or more substances that each remain identifiable when blended together. The most obvious example of this is the sand and salt mixture.

It may be useful at this point to explore the distinctions among the terms *transparent, translucent,* and *opaque.* Involve the class in a discussion of the meanings of these words. Have students brainstorm a list of familiar liquids that are transparent, translucent, and opaque. *(Transparent: water, pure vegetable oils, and filtered apple juice. Translucent: unused motor oil and many fruit juices. Opaque: milk, blood, and used motor oil.)*

In the discussion, note that many liquid solutions (for example, the vegetable oils or filtered apple juice) may be colored but are still transparent.

Ask students to name as many characteristics of solutions as they can. Put their responses on the board. The following is a sample list.

Characteristics of a liquid solution:
- It is a mixture.
- Light can pass through it.
- Two or more substances are distributed uniformly throughout one another.
- The particles that the substances are made of are indistinguishable.
- It may have color or be colorless. (Many solutions can be both colored and transparent.)

(Continued on next page)

EXPLORATION 1

Solution or Non-Solution?

Part 1

How was the paramedic, Ms. Sirba, able to distinguish the non-solution in the Muscle Man's cup from a pure solution? To tell a solution from a non-solution, try this experiment.

You Will Need

- 6 beakers
- a 100-mL graduated cylinder
- a 10-mL graduated cylinder
- the substances shown below
- water
- a stirring rod

What to Do

In separate beakers, prepare the six mixtures listed at the right. Stir each one well to ensure that it is completely mixed.

Mixture #1
Add 1 mL of powdered juice mix to 100 mL of water.
Mixture #2
Add 10 mL of rubbing alcohol to 100 mL of water.
Mixture #3
Add 2 or 3 drops of milk to 100 mL of water.
Mixture #4
Add 1 mL of sugar to 100 mL of water.
Mixture #5
Add 1 mL of oil to 100 mL of water.
Mixture #6
Add 10 mL of sand to 10 mL of salt.

Now separate the mixtures into two groups: those mixtures that are solutions (Mixtures #1, #2, and #4) and those that are not solutions (Mixtures #3, #5, and #6). Examine the two groups carefully. What differences do you see? Based on your observations, how will you know a solution when you see one?

Suppose you have been asked to appear in court as the scientific expert for the trial of the lemonade stand man. In giving evidence before the court, you are asked how to tell a solution from a non-solution. To illustrate your answer, you decide to show the court the six mixtures and explain the differences between the three solutions and the three non-solutions. Prepare your explanation in writing.

On the next page is a memorandum from Sergeant Noble to the forensic laboratory. How does your description of what a solution is compare with Sergeant Noble's?

134 Solutions

Police Department
Criminal Investigation Branch

Memorandum

To: ACME Forensic Laboratory
From: Sergeant M. Noble
Re: Carnival Lemonade

After a preliminary examination of the scene of the crime, it appeared that the victim had merely fallen asleep on the job. However, further investigation showed that the lemonade the carnival employee had consumed was not a pure solution of lemonade powder and water. This led us to suspect criminal activity.

Anyone with even a little background in chemistry knows that a solution is a mixture of two or more things, but that not all mixtures are solutions. In a solution, one of the materials becomes hidden, so that the mixture appears to be only one thing. You might say that one part becomes camouflaged. Sometimes, as in the case of food coloring and water, a solution can be colored, but it is still transparent. Any liquid mixture that is transparent must be a solution.

When we held the lemonade up to the light, it showed the Tyndall Effect. This test provided further evidence that the lemonade was not a pure solution. I suspect foul play and recommend that you test the enclosed sample to find out what was added to the lemonade drunk by the carnival employee.

M. Noble

(Exploration 1 continued)

In response to the idea that students appear as scientific experts in the trial of the lemonade man, suggest that they prepare in writing the differences between a solution and a non-solution. Call on volunteers to share their explanations.

As an alternative, you may want to organize a classroom dramatization. Students could work in groups to study the written expert testimony on page 135, and then develop a set of questions to ask the witnesses in courtroom-style proceedings. You may choose a student to serve as a judge or moderator. Other students could play the roles of Sergeant Noble, paramedic Binti Sirba, and the attorney for each side.

In addition to reinforcing the students' knowledge of solutions, a courtroom dramatization offers the opportunity to discuss the differences and similarities between scientific evidence and legal evidence. For example, in the hypothetical case of Muscle Man and the sleeping powder, courtroom evidence could include nonscientific testimony about the character of the chief suspect, the lemonade stand operator. The scientific testimony would then be that which could be backed up by experimental evidence; for example, the use of the Tyndall Effect.

The use of scientific expert testimony is increasing as courts and legislatures throughout the United States are becoming involved in making decisions on environmental and health issues. The courtroom dramatization activity could be extended by having students locate and discuss newspaper articles in which scientific expert testimony has played a role.

Memorandum

Ask students to read the memorandum from Sergeant M. Noble. Have them compare what they learned about the characteristics of solutions in *Part 1* to those described in the memorandum.
(Continues on next page)

A Solution or Not?

(Exploration 1 continued)

Have students read file card #17 on page 136. This card describes and explains a phenomenon known as the Tyndall Effect. That is, a liquid that contains suspended particles large enough to scatter or reflect the path of a beam of light passing through it exhibits the Tyndall Effect. As a result, the boundaries of the light path become visible. An example of the Tyndall Effect is shown in the illustration at the bottom of page 136. True solutions do not exhibit the Tyndall Effect (shown in the illustration on the bottom of page 137) since the light travels through the mixture without reflection or scattering.
(Continues on next page)

EXPLORATION 1—CONTINUED

Laboratory File W/P 106 (#17 and #18) in the forensic lab deals with the Tyndall Effect. Card #17 describes and explains it, and Card #18 tells how to test for it. Study these cards and then test Mixture #3—milk and water—with the solution tester shown on the bottom of the page.

File W/P 106   **#17**

Topic
TYNDALL EFFECT

Description:

Some mixtures that first appear to be solutions may prove not to be solutions after all. If the path of a bright light shining through a mixture can easily be seen, then that mixture is not a solution. In other words, the path of light passing through solutions is not visible. No matter what color they are, true solutions made of liquids are transparent.

Explanation:

In a non-solution, there are particles large enough to scatter or reflect light, allowing us to see the path of the light as it passes through. This scattering is called the Tyndall Effect. True solutions do not exhibit the Tyndall Effect.

A Non-Solution

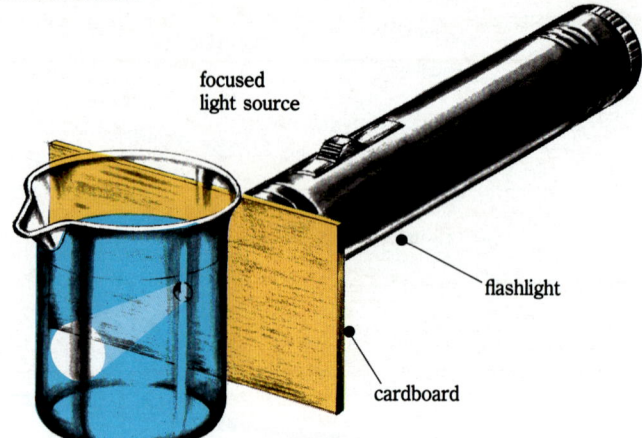

focused light source

flashlight

cardboard

136   Solutions

File W/P 106   #18

Topic
TESTING FOR THE TYNDALL EFFECT

Materials
- a bright light, lamp, or flashlight
- a piece of cardboard
- mixture(s) to be tested
- a pencil or knife
- a metric ruler

Making the Solution Tester
Using a pencil or knife, make a small, neat hole in the center of the piece of cardboard. The hole should be no more than 0.5 cm in diameter. Place the cardboard over the light source so that the only light you can see comes through the hole.

The Test
Press the light and the cardboard tightly against the beaker holding the mixture. If you can see the path of the light through the mixture, then it is not a true solution. The test should be done in a place with limited light.

A True Solution

(Exploration 1 continued)
Have students read file card #18 on page 137. Divide the class into small groups and distribute the materials needed to make solution testers. These will be used in *Part 2* on page 138. Direct students' attention to the illustration at the bottom of page 137. It shows the results of testing a true solution. Ask: What does the illustration show? *(A light source passing through a transparent liquid without any visible traces of its path.)* Then have students turn back to the illustration on page 136 that shows the results of testing a non-solution. Ask: Why is the path of the light visible in the non-solution? *(The non-solution has larger particles that scatter and reflect the light.)*

Before continuing to page 138, have students review their data from *Part 1* and predict the results of testing for the Tyndall Effect on the liquid mixtures from *Part 1*. *(The powdered juice mix and water, the rubbing alcohol and water, and sugar and water mixtures should show results similar to the true solution on page 137. The milk and water and the oil and water mixture should show results similar to the non-solution on page 136.)*

Relate the information on these pages to the memorandum from Sergeant Noble by asking the following question:
- Why did Sergeant Noble test for the Tyndall Effect? *(to find out whether the liquid in the cup was a pure powdered lemonade solution)*
- If the results of the Tyndall-Effect test had been negative, would that have proven that the liquid was a pure powdered lemonade solution? Why or why not? *(No, because the substance added to the lemonade could have been soluble and could have given a negative result for the Tyndall Effect.)*

A Solution or Not?

Part 2

Divide the class into small groups and distribute the materials. Have groups read page 138 and perform *Part 2*. To obtain the best results, here are some hints:

- A darkened room is necessary to test for the Tyndall Effect.
- All mixtures should be diluted so that they are nearly transparent and colorless. Otherwise, the test may not work.
- Use the purest water possible.
- Get glassware as clean as possible.

Be sure that students understand that the Tyndall Effect is the actual path of the light through the particles in the mixture. The reason the path of light can be seen through a non-solution is that the particles are large enough to scatter the light; the particles of a true solution are too small to scatter light.

Answers to
In-Text Questions

Have the groups list all their observations. The results should be as follows:

- solutions—sugar and water, coffee and water, tea and water, salt and water, and the soft drinks only if they're flat or in a sealed bottle
- non-solutions—freshly opened soft drinks; milk and water; and chalk dust and water

  Note that freshly opened soda water is not a true solution because where the bubbles are formed, the mixture is heterogeneous (the bubbles are pockets of pure gas). Freshly opened soda water might display the Tyndall Effect because the pockets of gas could scatter the light. Soda in a sealed bottle and flat soda have an equilibrium concentration of gas in solution. (This can be shown by heating the soda and forcing out more gas bubbles.)

Detective Work

Set up the three beakers as shown in the illustration on page 138. Place water in one; food coloring and water in the second; and food coloring, water, and a drop of milk in the third. Explain that all are mixtures. Ask: Are all of these mixtures solutions? How can you find out? *(You could use the solution tester.)*

138 Solutions

EXPLORATION 1—CONTINUED

Part 2
What to Do

Test Mixture #3 (milk and water) with the solution tester. Can you see the light path through it? Now test Mixture #4 (sugar and water).

Next, repeat the test with other mixtures. Some that you can use are instant coffee and water, tea and water, chalk dust and water, soft drinks, and salt and water. Which ones are true solutions?

Detective Work

Are the mixtures pictured below all solutions? Discuss with a classmate. (Hint: There is water present in all beakers.)

Solution Facts

You will come across many interesting facts about solutions and non-solutions. Perhaps your class could set aside some space on a bulletin board for "Solution Facts." Use index cards to record any interesting facts about solutions or non-solutions, and put them up on the bulletin board.

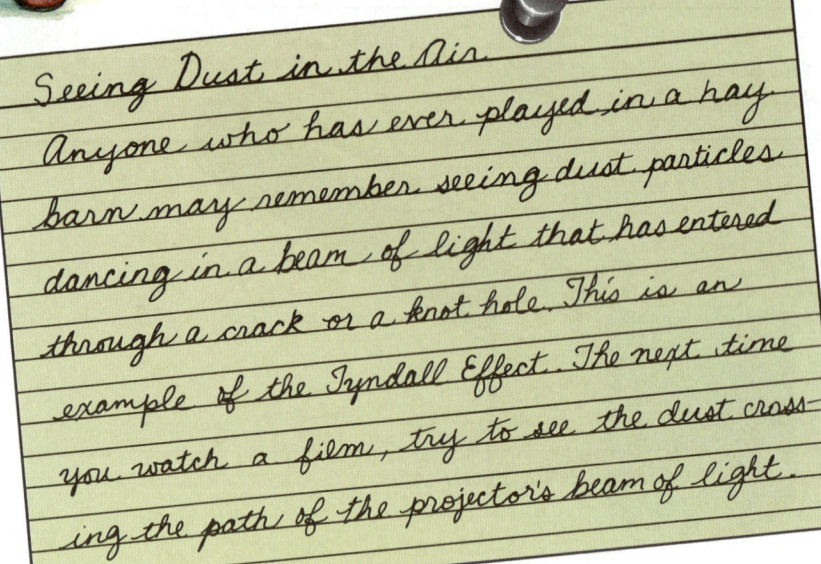

*Seeing Dust in the Air.
Anyone who has ever played in a hay barn may remember seeing dust particles dancing in a beam of light that has entered through a crack or a knot hole. This is an example of the Tyndall Effect. The next time you watch a film, try to see the dust crossing the path of the projector's beam of light.*

Point out that a clear liquid cannot automatically be classified as a solution; it may be a single substance, such as water or alcohol. If students identify the clear liquids as solutions, ask them how they can tell a single substance from a solution. This question will lead into the next lesson, *Parts of a Solution—Solutes and Solvents.*

Why Is the Sky Blue?
Did you know that air without the presence of dust, soot, and other such particles is a true solution? It is light from the sun being scattered by these particles that causes the sky to appear blue. If there were no particles in the air, the sky would be black.

Can Solutions Be Solid?
Yes they can! The Liberty Bell in Philadelphia, Pennsylvania is made of bronze — a solution of copper and tin. Solid solutions are called alloys. Copper and tin ore are melted and then mixed together. When this solution cools, it forms the alloy bronze. Unlike liquid or gaseous solutions, solid solutions aren't transparent.

Do you have a Solution Fact to contribute?

Follow Up

Assessment

1. Present students with the following scenario: You are describing to someone who is several years younger why vinegar is a solution. What would you say? *(Students should include a basic definition of a solution, indicate that vinegar is a transparent mixture of several components, and describe how vinegar could be tested to ascertain that it is a solution and not just a mixture.)*

2. Many substances in the students' everyday lives are solutions. Ask them to make a list of five such substances and explain why they think each is a solution. *(Examples could include cleaning solutions such as bleach and detergents, clear beverages such as carbonated water or juices, and cosmetic solutions such as perfumes. Students should explain that they are all transparent mixtures.)*

Extension

Have students investigate alloys: what they are, how they are made, and why they are useful. Some examples include bronze, stainless steel, and sterling silver.

Solution Facts

The purpose of this exercise is to introduce new information to expand students' understanding of what a solution is. Solutions can be combinations of gases and gases, liquids and gases, liquids and solids, liquids and liquids, or solids and solids. Examples of these combinations are given in the table on page 140 of the student text. All solutions have a homogeneous composition and all solutions except solid and solid mixtures have the property of transparency to light.

Have students create a bulletin board display of their own solution facts. Encourage them to include drawings or magazine photographs showing examples of solutions to make their displays more interesting.

A Solution or Not?

Lesson 2

✴ Getting Started

This lesson begins by introducing the terms *dissolving, solvent,* and *solute.* Students apply these terms as they analyze solutions from their everyday world. Then students learn about solvents other than water. Environmental concerns are introduced as students study the effects of oil spills on wildlife. The lesson concludes with students investigating the differences between hard and soft water and how substances dissolve in each.

Main Ideas

1. Solutions can be made from different combinations of gases, liquids, and solids.
2. In solutions, the substance that is in greater quantity is the solvent and the substance that is in lesser quantity is the solute.
3. Many substances, both soluble and insoluble, are added to water. This can be harmful to the environment.
4. Hard water is a solution of compounds containing iron, calcium, and magnesium.

✴ Teaching Strategies

Hold up a soft-drink bottle, an object made of brass, and an empty beaker and ask students to identify the solution(s).
(All three are actually solutions: soft drinks are solutions of sugar, carbon dioxide gas, flavoring, and water; brass is an alloy consisting of the solids copper and zinc; and air is a solution made up mainly of nitrogen and oxygen.)
(Continues on next page)

---

2 Parts of a Solution— Solutes and Solvents

Earlier you prepared a solution of sugar and water. Sugar is a solid, and water is a liquid. You also prepared a solution of rubbing alcohol and water—both liquids. Other possible combinations are listed in the data table below. For each combination, an example is given. Can you provide another example of each combination?

Some Combinations	Example	Your Example
solid and liquid	salt water	?
gas and liquid	soft drinks	?
liquid and liquid	rubbing alcohol and water	?
solid and solid	bronze (a solution of copper and tin)	?
gas and gas	air	?

Naming the Parts of a Solution

What happens when you mix things to make a solution? When a solid and liquid are mixed together, the solid separates into particles that spread evenly throughout the liquid. This process is called **dissolving**. The particles of both parts of the solution, solid and liquid, are too small to be seen. This explains why the solution is transparent. Since the particles of the solid spread evenly to every part of the liquid, the solution is the *same* throughout.

Dissolving in Action

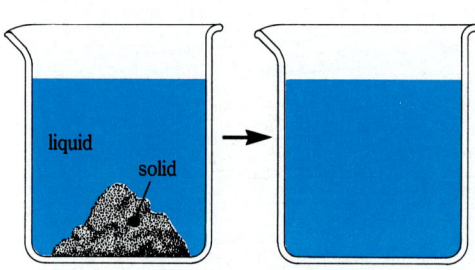

This is what you might observe as a solid dissolves into a liquid. Where did the solid go?

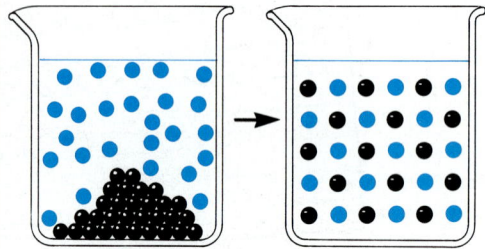

This is what you might observe if you could actually see the particles. Check to see if the solution is the same throughout.

LESSON ORGANIZER

Objectives

By the end of the lesson, students should be able to:
1. Identify the parts of a solution as the solvent and solute.
2. Name examples of solutions in which the solvent and solute are various combinations of gas, liquid, and solid.
3. Identify the solute and solvent in various everyday solutions.
4. Name examples of useful and harmful solutions.
5. Identify and test for hard water.

Process Skills

analyzing, classifying, investigating, inferring

New Terms

Dissolving—the process in which a solute separates into particles that spread evenly throughout a solvent.
Solvent—a substance that dissolves a solute to make a solution.

---

140 Solutions

If a solid or a gas is dissolved in a liquid, the liquid is usually the **solvent**. The solid or gas is the **solute**. What happens when you have other combinations, such as liquid–liquid, gas–gas, or solid–solid? In these cases, the substance of which there is a *greater* quantity is the solvent. The substance of which there is a *lesser* quantity is the solute.

> The *solvent* dissolves the *solute*.
>
> The *solute* dissolves *in* the *solvent*.

In these examples, which part is the *solute,* and which part is the *solvent?*

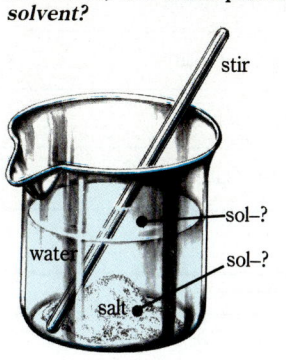

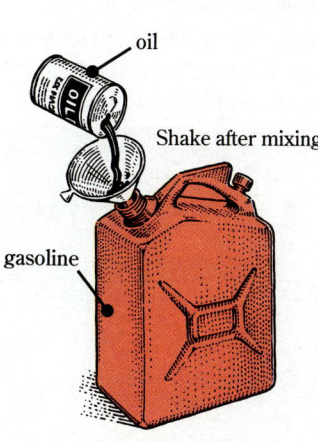

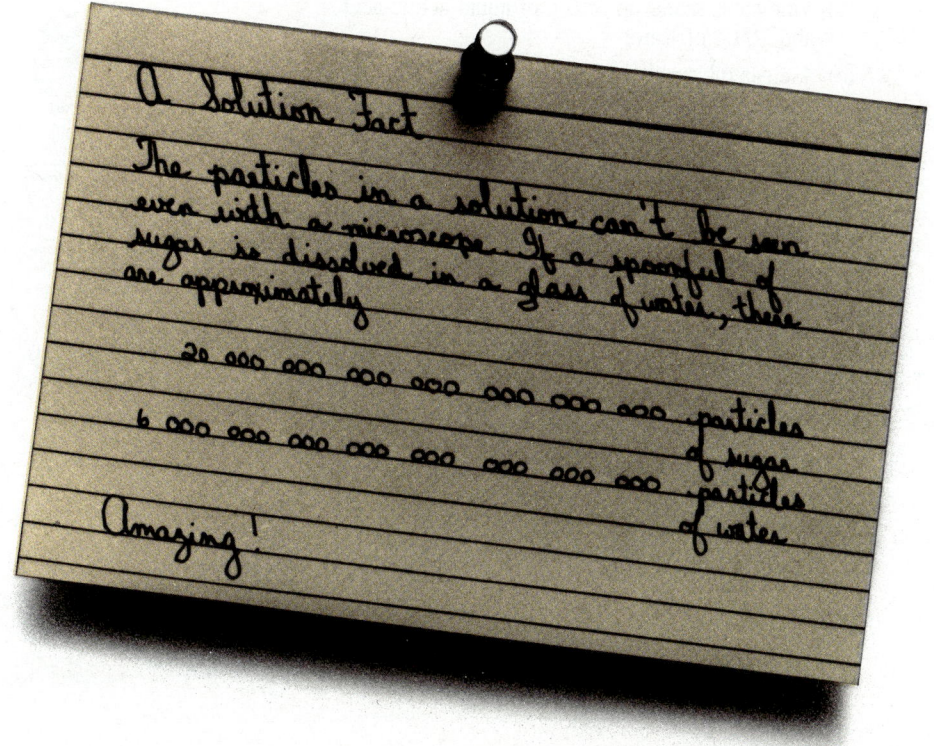

(Teaching Strategies continued)
Call on a volunteer to read aloud the top of page 140. Involve students in a discussion of the table. Ask them to suggest other examples of each combination. (Examples of solutions: solid and liquid—sugar and water; gas and liquid—air and water; liquid and liquid—antifreeze and water; solid and solid—iron and carbon [steel]; gas and gas—nitrous oxide and oxygen [laughing gas])

Naming the Parts of a Solution

After reading this section, ask a student to review the meaning of the term *dissolving* in his or her own words. Then direct students' attention to the diagrams at the bottom of the page. Involve students in a discussion of how these diagrams illustrate the concept of dissolving. Students may need help with the idea that a liquid is a collection of particles that are separate from one another, and so small that they cannot be seen with a microscope. To illustrate this, use colored solutes as an example. The individual particles cannot be seen, but their presence is indicated by the color. The color is also evidence of the homogeneous nature of a solution. This can be demonstrated by dissolving a single crystal of potassium permanganate in a beaker of water.

Answer to caption:
- In the first beaker: the water is the solvent, the salt is the solute.
- In the second beaker: the mineral water is the solvent, the carbon dioxide is the solute.
- In the oil can: the gasoline is the solvent, the oil is the solute.

Solute—a substance that is dissolved in a solution.
Hard water—water that contains certain types of solutes; it will not form as many suds when mixed with soap.
Soft water—water that is mostly free of dissolved matter.

Materials

Solutions from A to Z: Journal
Exploration 3: 25-mL graduated cylinder; 35 mL soft water (distilled water); 10 mL tap water; beaker; basic, unscented hand soap; 2 test tubes; test-tube rack; knife; eyedropper; tap water from different sources; stirring rod; safety goggles; Journal

(Organizer continues on next page)

Solutions from A to Z

Ask students to make a list in their Journals of any unfamiliar words and to look up their meanings. For example: alloy, molten, pulp, mineral spirits, DDT, or sparkling water.

Answers to
Table 1

Type of Solution
solid and liquid:
- (c) baking soda and water
- (e) mineral spirits and dried paint
- (h) sugar and water
- (i) rubbing alcohol and ice
- (n) minerals and rainwater (acid rain)
- (p) limestone and weak acid
- (r) sodium chloride and water, potassium chloride and water
- (t) salt and water
- (u) seltzer tablet and water
- (v) insecticide powder and water
- (w) DDT powder and water
- (y) some vitamins and water

gas and liquid:
- (a) oxygen and water
- (j) carbon dioxide and water
- (m) sulfur dioxide and water
- (o) carbon dioxide and water
- (u) seltzer tablet and water

liquid and liquid:
- (b) acetic acid and water
- (d) molten copper and molten zinc
- (e) mineral spirits and liquid paint
- (f) ethylene glycol and water
- (h) lemon juice and water
- (k) food coloring and water
- (v) liquid insecticide and water

solid and solid:
- (z) copper and gold

gas and gas:
- (g) water vapor and air
- (l) discharge gas and air
- (q) water vapor and air
- (s) nitrogen gas and oxygen gas
- (x) iodine vapor and air

(Answers continue on next page)

Solutions from A to Z

Here are situations involving real solutions. Read through them with a classmate. For each situation:

1. *Identify the type of solution* (solid in a liquid, liquid in a liquid, etc.). In your Journal, enter this information in a table similar to Table 1.
2. *Identify the solvent and solute in each solution.* Enter this information in your Journal in a table similar to Table 2 on page 143. Again, remember to write only in your Journal.

(a) Fish breathe oxygen, which is part of water.
(b) Vinegar is made up of 5 g of liquid acetic acid and 100 g of water.
(c) A clear mixture of baking soda and water often helps an upset stomach.
(d) Brass is an alloy that can be made by mixing 65 parts of molten copper with 35 parts of molten zinc.
(e) Mineral spirits will easily remove paint from your hands.
(f) Antifreeze used in a car radiator is a mixture of the liquid ethylene glycol and water. The ethylene glycol is present in the greater quantity.
(g) On a windy day, snow and ice disappear into thin air without melting.
(h) Sugar, lemon juice, and water make up lemonade.

Table 1

Type of Solution	Situations (by letter)
solid and liquid	c, e, etc.
gas and liquid	
liquid and liquid	
solid and solid	
gas and gas	
other (specify)	

(Organizer continued)

Exploration 4: 95 mL distilled water; beaker; 25-mL graduated cylinder; sugar; sodium chloride; magnesium chloride; calcium chloride; iron chloride; calcium nitrate; magnesium nitrate; 7 test tubes; test-tube racks; basic, unscented hand soap; knife; stirring rod; safety goggles; eyedropper

Teacher's Resource Binder
Resource Worksheets 3–1, 3–2, and 3–3
Graphic Organizer Transparencies 3–1 and 3–2

Science Discovery Videodisc
Disc 2, Image and Activity Bank, 3–2

Time Required
four class periods

(i) Rubbing alcohol is a good lock de-icer because it can dissolve water drops frozen in the keyhole.
(j) When you open a bottle of sparkling water, gas suddenly appears everywhere and rises to the top.
(k) A few drops of food coloring can change the color of water.
(l) Pulp mills discharge a gas that spreads through the air. The gas can be smelled kilometers away.
(m) Sulfur dioxide in the air mixes with water droplets in the atmosphere to form acid rain.
(n) Minerals in the ground can be dissolved by acid rain.
(o) Rain falling on the earth mixes with carbon dioxide from decaying plants and the roots of plants. The result is the formation of a weak acid.
(p) In some places, this weak acid dissolves limestone to make limestone caves.
(q) Water from lakes and streams is continually evaporating into the air.
(r) Clear ocean water contains a considerable amount of sodium chloride and potassium chloride (both are types of salt).
(s) Air is made up of almost four parts nitrogen gas to one part oxygen gas.
(t) A bit of table salt in water helps heal cuts.
(u) A seltzer tablet added to water fizzes and then becomes clear. The solution tastes a little salty.
(v) Runoff water from some farms and forests contains insecticides.
(w) There are traces of DDT in ocean water, even in the Arctic and Antarctic.
(x) If iodine crystals are dropped into a dry, warm container, a purple vapor suddenly appears and spreads evenly through the air in the container.
(y) When foods are boiled, some of their valuable vitamins are lost in the process.
(z) Gold jewelry usually has some copper in it to make it stronger and less expensive.

Table 2

Situation	Solute	Solvent
a		
b		
c	baking soda	water
d		
e		
f		
g		
h		
i		
j		
k		
l		
m		
n		
o		
p		
q		
r		
s		
t		
u		
v		
w		
x		
y		
z		

Answers to
Table 2

See answers to Table 2 below. In some examples, such as (o), (p), and (u), a chemical change occurs. However, the resulting mixture is still a solution.

Situation	Solute	Solvent
a	oxygen	water
b	acetic acid	water
c	baking soda	water
d	zinc	copper
e	paint	mineral spirits
f	water	ethylene glycol
g	water vapor (from snow and ice)	air
h	sugar and lemon juice	water
i	ice	rubbing alcohol
j	carbon dioxide	water
k	food coloring	water
l	discharge gas	air
m	sulfur dioxide	rainwater
n	ground minerals	acid rain
o	carbon dioxide	rainwater
p	limestone	weak acid
q	water vapor	air
r	sodium chloride and potassium chloride	ocean water
s	oxygen gas	nitrogen gas
t	table salt	water
u	carbon dioxide	water
v	insecticides	runoff water
w	DDT	ocean water
x	iodine vapor	air
y	vitamins	water
z	copper	gold

Answers continued

Note that (e), (h), (u) and (v) have each been listed twice. In (e), the answer depends on whether the paint is in liquid or solid form when it is removed from the hands. In (h), both sugar (a solid) and lemon juice (a liquid) are combined with water to make lemonade. In (u), the seltzer tablet forms both a gas (carbon dioxide) and a solid that dissolve in water. In (v), some insecticides are in solid form when dissolved in water, others are in liquid form. There might also be some confusion when a phase change precedes the solution formation. For example, in (g) and (x), the solid first changes into a gas, and then the two gases form a solution.

A Solution or Not? 143

Waste Not, Want Not

Call on a volunteer to read aloud the top half of page 144. Involve the class in a discussion of water and its many uses. Help students to realize what a precious resource water is. Ask them to think of substances that pollute the water supply in their area. Then ask them to identify which of these substances dissolve in water. Students can use a table similar to the one on the right side of page 144 to record their findings.

As an extension to this discussion, ask students to do some research to discover where waste water goes in their community. Then they can find out if any of the wastes from the water are recycled. Some sewage treatment plants can process waste into a sludge suitable for agricultural and industrial uses, but this is not widely done at the present time.

Other Solvents Besides Water

Suggest that students copy the table at the bottom of page 144 into their Journals, and then add to the list of solvents. Ask them to follow up by doing library research to discover which of the solvents in their table are dangerous.

The following could be added to the table of solvents and their uses.
- Dry cleaning fluid dissolves grease and other dirt on clothing.
- Mineral spirits dissolve oil on axles and bicycle chains.
- Silver dissolves mercury in order to make fillings for teeth.
- Molten cryolite dissolves bauxite during the process of extracting aluminum.
- Gasoline dissolves lead compounds in the making of leaded gasoline.
- Wax dissolves colored dyes in the making of colored candles.

One interesting demonstration is to place two watch glasses on an overhead projector. Put some iodine crystals on each watch glass and add a few drops of water to one and alcohol to the other. (The iodine will be very soluble in alcohol and not very soluble in water.)

Waste Not, Want Not

Did you notice that water was involved in many of the solutions on pages 142 and 143? In the Middle Ages, people who were interested in science searched for a "universal solvent"—a liquid that would dissolve everything. Water comes closest to being this universal solvent.

Think of the many things water is used for each day, such as cooking, washing, drinking, cleaning, swimming, and industry. And after it is used, water often contains many wastes. Some waste materials are easily seen when we inspect waste water. But many other wastes dissolve in water and are therefore invisible.

Make a list of about 10 or 12 substances that can end up polluting the water supply. Look around your school, home, and neighborhood. Decide which of the materials can be dissolved in water.

Much of the used water ends up at a sewage plant, where it is partially cleaned before being released into the environment. Where does waste water go in your community? If there is a sewage treatment facility nearby, visit it and find out how the waste is removed from the sewage. Can any of these wastes be *recycled*?

Substance	Does It Dissolve in Water?
grease	No
sugar	Yes

Other Solvents Besides Water

Water is a great solvent, but as you have just seen, it does not dissolve everything. For example, it does not do a good job on cleaning grass stains from clothes. Nor does it work very well on greasy hands.

There are other solvents besides water. Many of them are useful for doing things that water cannot. The table below lists only a few of these solvents. Perhaps you know of others to add to the table. You can also ask a chemist, a pharmacist, or a hardware store clerk for further suggestions. Add about six more items to the table—more if you can!

Solvent	Solute	Use
alcohol	iodine	tincture of iodine
methyl alcohol	grease	windshield wiper fluid
toluene	rubber	rubber cement
turpentine	oil paint	paint thinner
ethyl acetate	nail polish	nail polish remover
gasoline	engine oil	fuel for two-cycle engines
alcohol	ice	de-icer for keyholes

Water is a safe solvent. It does not burn and is not poisonous. However, many other solvents are dangerous. Use your library to find out which solvents in your table can be dangerous.

Waterproof Birds

Water does roll off a duck's back! When a duck or other water bird preens itself, it is actually replacing natural oils that cover its feathers. Since oil does not dissolve in water, the natural oils act as a raincoat that makes the bird waterproof. Without this coating of oil, its wings would become waterlogged, and the bird would not be able to fly or even float. The water bird would be in danger of drowning.

The grebe in the photograph is covered with too much oil. It was caught in an oil slick—a danger to wildlife that is becoming more common every year. How could the grebe be cleaned? What will happen to the bird's natural oils in the cleaning process?

Grebe covered by oil

EXPLORATION 2

A Project: Oil Spills

Our coastlines and wildlife are sometimes harmed by oil slicks. The oil is released into the water when oil tankers clean their bilges or when these ships are involved in accidents, such as the one in Prince William Sound, Alaska, in 1989.

Find out how environmental agencies get rid of oil slicks on water surfaces and on beaches. How are birds and other animals that have been in contact with oil treated?

Oil slick

Waterproof Birds

Call on a volunteer to read this section aloud. Then direct students' attention to the photograph. Involve the class in a discussion of the cleanup methods used on the grebe. This will illustrate the effects of oil spills on wildlife. Students might suggest methods of cleaning the oil-soaked grebe that include washing it in soapy water or carefully wiping off its feathers with a solvent such as mineral spirits. Soap works by helping the grease and oil mix with the water. With either treatment, the bird's natural feather oils will also be removed.

Explain to students that oil spills can affect wildlife in many ways. For example, birds and mammals produce natural oils and waxes that coat their feathers or fur. The coating of natural oils helps keep the animal dry and warm. When the natural oils are removed, the feathers (or fur) lose their insulating properties and the animal can easily develop hypothermia and die. Birds that have been cleaned by humans cannot be released back into the environment until they have regained their natural oily coating.

The petroleum from an oil spill can also cause hypothermia directly by soaking and matting fur or feathers. The insulating quality of these body coverings comes from their ability to trap air. For example, the fluffy, tightly packed breast feathers of water birds hold a layer of warmed air next to the bird's skin. If these feathers become matted with petroleum, they will no longer protect the bird.

Animals can also ingest petroleum as they try to clean or preen themselves. Toxic compounds can damage the liver, kidneys, and other internal organs. Volatile compounds can cause respiratory problems. Animals that have ingested petroleum can also develop a form of anemia.

Exploration 2

This activity provides students with an opportunity to explore a serious, real-life problem that relates to their study of solutions.

Since *Exploration 2* is a research problem, suggest that students do this as an at-home assignment. Encourage them to use their school or local library to research the necessary information. Students' research should include information on the methods of cleaning up oil spills and the treatment of affected wildlife. Another possible resource would be to have students use telephone directories to contact local and state environmental agencies to discover how these agencies are involved in clean-up activities. If possible, have a representative from such an organization visit your classroom to answer questions.

A Solution or Not? 145

Exploration 3

Divide the class into small groups and distribute the materials.

Some procedural hints:

- To be sure the soft water samples are free of solutes, use clear rainwater or distilled water.
- On the day of the activity, prepare a quantity of soapy water by adding scrapings of pure soap to 250 mL of soft water. (This amount makes enough for 10 groups to each have 25 mL.) Be sure to use pure soap and not a detergent. Most detergents contain foaming agents that will interfere with the results.
- Mix the soap solution thoroughly without excessive shaking. Make sure the solution is very soapy.
- Test the soap solution by adding 10 drops to 10 mL of soft water. This mixture should make soap suds when shaken. If it does not, modify the procedure by having students use 20 drops (instead of 10) in 10 mL of soft water.

The amount of soap suds formed in the different samples of water is used as a rough indicator of how hard the various water samples are. Results may vary depending on the local tap water. Generally, municipal tap water contains many solutes. If the tap water in your area comes from an underground source, it will probably contain many minerals dissolved from rocks.

Students can take the following measures to ensure more reliable results:

- test equal amounts of soft water and tap water
- add equal amounts of soapy water to the soft water and to the tap water
- shake both test tubes the same number of times and with the same force
- use water samples of the same temperature

Students can also compare water from the hot water tap with water from the cold water tap. Water taken from the hot water tap, if used at room temperature, will generally test "harder" than water taken from the cold water tap. This is because it has been heated in a holding tank containing a high concentration of dissolved minerals.

Ask volunteers to collect water from different sources in your city. These samples can then be brought to the classroom and tested.

Hard Water

Hard water is *not* frozen water! It is water that contains certain types of solutes. **Soft water,** such as rainwater or melted snow, is mostly free of dissolved matter. Hard water will not form as many suds when mixed with soap as soft water will. Does your drinking water contain solutes? Here is an easy test to determine whether your drinking water is hard or soft.

EXPLORATION 3

Tap Water— Hard or Soft?

You Will Need

- a 25-mL graduated cylinder
- 10 mL soft water (distilled water works best)
- 10 mL tap water
- 25 mL soapy water, made by adding flakes of basic, unscented hand soap to 25 mL soft water. Mix well, but do not make suds.
- 2 test tubes
- a test-tube rack
- an eyedropper

What to Do

In one test tube, place 10 mL of drinking water from the tap. Into the second test tube, pour 10 mL of soft water. Add 10 drops of soapy water to each, and shake the test tubes an equal number of times. Compare the amount of suds formed in each test tube. Is the tap water harder than the soft water? Do you think this is a reliable test?

Make a list of at least three things that you did in this experiment to ensure that your results would be reliable.

Do you think the results would have been the same if you had used drinking water from the hot-water tap?

Collect samples of water from different sources in your city and test each one for hardness.

146 Solutions

What Not to Do!

Will this procedure give reliable results?

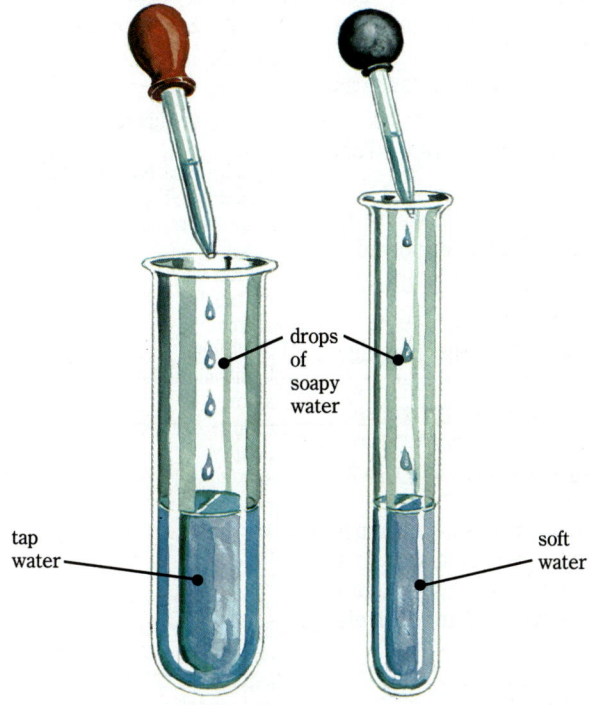

EXPLORATION 4

Solutes in Hard Water

You Will Need

- sugar
- sodium chloride
- magnesium chloride
- calcium chloride
- iron chloride
- calcium nitrate
- magnesium nitrate
- a 25-mL graduated cylinder
- distilled water
- 7 test tubes
- test-tube racks
- soapy water, made according to instructions in Exploration 3
- an eyedropper

What to Do

In test tubes, dissolve a few crystals of each of the solutes listed above in separate 10-mL amounts of distilled water. Test the solutions for hardness, as you did in Exploration 3. Make a table that records which of the solutes is able to make suds.

What are some of the solutes that may be responsible for making water hard?

Something to Solve

Does the amount of calcium chloride added to the 10-mL portion of distilled water affect the number of drops of soapy water required to make suds? What do you predict? Try adding different amounts of calcium chloride to test your prediction.

Something to Solve

Suggest that students design a mini-experiment to solve this question. Ask them to record step by step what they do, what they predict, and what happens. Point out that they should control variables, except for the amount of calcium chloride, throughout their experiment. Then reassemble the class and involve them in a discussion of their results.

Assessment

1. Ask students to make a list of all of the solutions they come into contact with from the time they get up in the morning until the time they go to bed at night. Name the solute and solvent for each one. *(Their lists will most likely contain cosmetic, household, and food and drink solutions.)*

2. Present students with the following scenario: At home, your family complains about how difficult it is to make lots of suds when taking a shower. How could you apply what you have learned from this lesson to explain why this is so? *(The home water supply must contain a large quantity of solutes which keeps the soap from making suds.)*

Extension

1. Have students research how household water softeners work. *(Various methods exist based on precipitation or ion exchange.)*

2. Have students research the kinds of restrictions that have been placed on three different hazardous solvents. They should describe the properties of the solvents, their toxicities, proper handling, and disposal techniques.

What Not to Do!

Help students to define the word *reliable*. Have them equate it with *a fair test*. That is a concept they should be familiar with. The procedure shown in the illustration will not give reliable results because different volumes of water are involved.

Exploration 4

Divide the class into small groups and distribute the materials. Again, have students make soapy water using soap-bar shavings. Students will find that magnesium chloride, calcium chloride, calcium nitrate, magnesium nitrate, and iron chloride produce hard water. Since sodium chloride did not affect the hardness of water, students should infer that it is not the chloride that makes water hard, but rather the metallic component of each chloride.

A Solution or Not? 147

LESSON 3

★ Getting Started

This lesson helps students describe the process of dissolving by classifying the solute as: **soluble, not very soluble, or insoluble.** In *Exploration 5,* students investigate a number of substances and classify them according to how well they dissolve in water.

Main Ideas

1. Substances vary in their capacity to dissolve in water.
2. The rate of dissolving may be increased by grinding the solute, by stirring the mixture, and by increasing the temperature of the solvent.

★ Teaching Strategies

Have students compare and contrast the differences among the three new terms *soluble, not very soluble,* and *insoluble.* Involve them in a discussion of variables and how these affect the accuracy of a test, by using the terms *fair test, variable,* and *controlled experiment.*

Exploration 5

Divide the class into small groups and distribute the materials. Have groups read the instructions on page 148. Direct their attention to the materials list at the top of the page. Ask: What are the variables in this Exploration? *(Possible variables include type of substance, amount of substance to be tested, quantity of water, amount and force of stirring, particle size, and water temperature.)* Ask: Which of these variables should be held constant? *(All variables should be kept the same, except for the substance being tested.)*

Reminder: each substance should be allowed to dissolve for 5 minutes.

148 Solutions

3 Soluble, Not Very Soluble, or Insoluble?

As you have seen, water comes closest to being a universal solvent because it dissolves a great many substances. However, you know that there are some substances that it cannot dissolve, such as grease and oil. Are there still other substances that water can dissolve a little but not completely?

Here are some terms that describe how well a solvent (such as water) dissolves a solute.

- A substance is **soluble** if it disappears in the solvent. This shows that it has dissolved completely.
- A substance is *not very soluble* if some of it is still visible in the solvent. This shows that only some of it has dissolved.
- A substance is **insoluble** if none of it disappears in the solvent. This shows that none of it has dissolved.

There are certain factors that affect how well a substance dissolves. One of them is time. When testing how soluble a substance is, you should allow enough time to make sure you get reliable results. Another factor, as you will see later, is temperature. There are others besides. These factors are also known as **variables.**

Whenever you are comparing different substances, you must remember to keep all the variables except one in every test *constant*—in other words, the same. When you do this, you are *controlling* the variables, and your test results will more likely be fair and reliable.

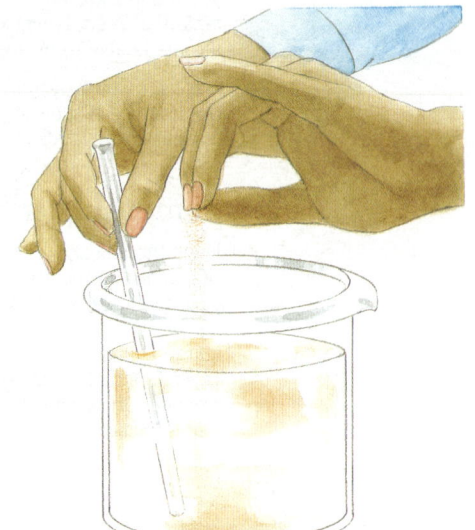

EXPLORATION 5

How Soluble Is It?

You Will Need

- a 10-mL graduated cylinder
- 21 test tubes or beakers
- test tube racks
- a stirring rod
- a variety of substances to test:
  flour
  sodium bicarbonate (baking soda)
  magnesium sulfate (Epsom salts)
  candle wax
  mothballs
  starch
  chalk dust
  bouillon cubes
  alum
  orange drink crystals
  pepper
  salt
  gelatin
  sugar
  coffee
  tea
  instant coffee
  instant tea
  chili powder
  aspirin
  seltzer tablet
  solid watercolor paint

What to Do

1. To 10 mL of water, add a pinch or a piece of one of the solutes listed above and stir. Does the substance dissolve? If it does, add a second pinch and stir again. If you observe no change, leave it for 5 minutes, stirring occasionally. Why is it necessary to leave it for 5 minutes?
2. Decide whether the substance is
   (a) soluble,
   (b) not very soluble, or
   (c) insoluble.

LESSON ORGANIZER

Objectives

By the end of the lesson, students should be able to:

1. Distinguish between solutes or substances that are soluble, not very soluble, or insoluble.
2. Identify and test for variables that will increase the rate of dissolving.

Process Skills

observing, investigating, recording, comparing

New Terms

Soluble—able to dissolve or pass into solution.
Insoluble—unable to dissolve or pass into solution.
Variables—factors that affect a process.

3. Repeat the test on the other materials you have gathered. Write down any variable(s) in your experiment. Can you think of anything you could do to make this a more controlled test?
4. In your Journal, record your findings in a table like the one below.

Substance	Soluble	Not Very Soluble	Insoluble	Was Time A Factor?	Any Other Variable(s)	Special Observations

An Experiment Out of Control

John had results that did not compare well with those of his classmates. Here's how he performed each test:

Test 1

He half-filled a test tube from the cold-water tap, dumped in some crystals, and shook the test tube until his arm got tired.

Test 2

He added 5 mL of water to a beaker, added two crystals of another solute and stirred the mixture once or twice with a stirring rod.

What would you suggest to John to help him improve his experimentation technique?

Detective Work

Soluble or not soluble— that is the question!

Dyani and Mario were having a friendly argument over the experiment they had just finished. Mario thought that baking soda should be entered in their data table under the "Not Very Soluble" heading. Dyani insisted that baking soda was insoluble. Suddenly Mario snapped his fingers and said, "I know a way we can figure out who's right!" What do you think Mario's idea was?

Answers to
In-Text Questions

2. **(a)** *soluble:* magnesium sulfate, alum, orange drink crystals, salt, sugar, instant coffee, instant tea, and aspirin
   (b) *not very soluble:* gelatin, bouillon cubes, coffee, tea, seltzer tablet, and solid paint
   (c) *insoluble:* flour, candle wax, mothballs, starch, chalk dust, pepper, and chili powder
3. To make this a more controlled test, a precise amount of solute could be added instead of a pinch or piece of the substance.

Answers to
An Experiment Out of Control

- Keep the amount of water constant.
- Keep the source of water constant.
- Keep the amount of crystals constant.
- Keep the amount of stirring constant.
- Keep water temperature constant.

Answers to
Detective Work

One possibility is to stir the baking soda in distilled water, filter it, and let the filtrate evaporate. Any residue that is left is evidence of partial solubility of the baking soda.

Materials

Exploration 5: 10-mL graduated cylinder, 21 test tubes or 50-mL beakers, test-tube racks, stirring rod, water, variety of substances (flour, baking soda, Epsom salts, candle wax, mothballs, starch, chalk dust, bouillon cubes, alum, orange drink crystals, pepper, salt, gelatin, sugar, coffee, tea, instant coffee, instant tea, chili powder, aspirin, seltzer tablet, solid watercolor paint), clock or watch, safety goggles, Journal
Exploration 6: sugar cubes, water, 150 mL-graduated cylinder, 2 beakers, stopwatch or watch with second hand, Journal

Teacher's Resource Binder

Resource Worksheet 3–4
Graphic Organizer Transparency 3–3

Science Discovery Videodisc

Disc 2, Image and Activity Bank, 3–3

Time Required

two class periods

Exploration 6

Have students predict how long an unstirred sugar cube will take to dissolve. Divide the class into teams and distribute the materials. Ask each team to read the instructions carefully and to follow the rules. Have students begin by recording their ideas in their Journals for speeding up the dissolving time. Remind them that they can do only *one* thing to speed up the rate.

In the Exploration, the amount of solute and solvent are held constant. The variables students can change include the amount of agitation, the water temperature, and the size of the sugar particles. Whichever method is chosen, have them record their data such as number of shakes, temperature, and so on in their Journals.

After students have completed the activity, reassemble the class so that they can compare their predictions with their actual findings. To reinforce understanding, be sure that the terms *variable, constant,* and *control of variables* are used in the discussion.

Answers to
Detective Work

Some possible hypotheses:
- The water in one of the beakers was stirred.
- One of the beakers was heated.
- One of the candies was crushed.
- Someone stole the candy.
- A combination of all of the above.

EXPLORATION 6

Let's Have a Sugar Cube Race!

If you drop a sugar cube into a glass of water, how long will it take to dissolve? Will it disappear in 5 minutes, 10 minutes, or even longer? What could you do to speed up the rate at which it dissolves? Before reading further, make a list of things you could try. Record the list in your Journal.

Object of the Race
To dissolve one sugar cube in 150 mL of water as rapidly as possible.

You Will Need
- 2 sugar cubes
- a 250-mL graduated cylinder
- 2 beakers
- a stopwatch

Rules

1. Each team will be given a sugar cube to add to 150 mL of water. Watch it dissolve without disturbing it. Make a note of the time it takes.
2. Each team must then decide what single thing they will do to the sugar cube or the water to speed up the rate at which the cube dissolves.
3. Each team will get a new sugar cube and another 150 mL of water.
4. Taking care to note the time, start your procedure for speeding up the dissolving of the cube. Record the time it takes the cube to dissolve completely.

What was the single variable you changed to speed up the rate of dissolving? What were the variables of the other teams in the class? Which of these variables is the most important?

Detective Work
The Case of the Missing Candy

Two identical candies are placed in beakers of water. Fifteen minutes later the experimenter returns to the room and sees that one piece of candy has disappeared.

What do you think could have happened while the experimenter was out of the room? Can you explain why one piece of candy disappeared and one did not?

Assessment

Present students with the following scenario: Cassie wants to find out how long it takes to dissolve a sugar cube at different temperatures. Design an experiment to help her. What variables should she keep the same? *(volume of water, time allowed to dissolve, and mass of sugar cube)* What do you think Cassie will discover when she completes her experiment? *(She should discover that the sugar dissolves faster at higher temperatures.)*

Extension

Have students investigate how stains are removed from clothing, furniture, and other items. Have them make a poster illustrating stain removal techniques and relating the process to the principles of solutions. *(Generally, stain removal techniques take advantage of the solubility properties of the stain substance. In most cases, the stain substance is not soluble in water, but very soluble in some other liquid. Soaking the stain in such a liquid allows it to dissolve.)*

Challenge Your Thinking

1. Based on your knowledge of hard water, explain the "bathtub ring." What solvents might you use to get rid of this ring?

2. Sarah keeps confusing *solute* with *solvent*. How are you able to remember the difference between these similar-sounding terms?

3. Van is trying to determine if water containing detergent will dissolve salt faster than water alone does. When you ask how he plans to perform this experiment, he tells you:

   I'm going to add a pinch of salt to 25 mL of water and time how long it will take to dissolve while stirring. Then I'll add salt to some water containing liquid dishwashing detergent, stir, and check the time again.

   What variables must you remind Van to control so that his test results will be reliable?

4. The illustration below shows two beakers of water. One beaker contains a sugar cube and the other an ice cube. After 15 minutes, both cubes will have disappeared. Will both beakers now contain a solution? Justify your answer.

Answers to Challenge Your Thinking

1. Because some tap water often contains many dissolved solutes, some of these solutes may come out of solution and combine with the dirt and oil from our bodies, along with the bath soap, forming a residue around the tub. Solvents you would use to get rid of the ring should dissolve grease as well as whatever dissolved matter is in your water supply. For example, if your water contains dissolved calcium carbonate, you could dissolve it with an acid such as vinegar.

2. Encourage students to be creative. One possible answer would be to note that *solvent* is a longer word, and the solvent is the substance that is in greater quantity in a solution.

3. Variables that Van must control include the amount of salt added, amount of water, source of water, and force with which the mixture is stirred.

4. Provided that the ice cube consists of pure water, the ice cube melted in water yields pure water, which is not a solution. Only the beaker with the sugar and the water contains a solution.

A Solution or Not?

LESSON 4

Getting Started

This lesson introduces students to the salt industry. Through the examination of two different methods of salt extraction, students learn about the importance of evaporation and boiling. The lesson concludes with students designing a process of their own to extract salt from a salt and sand mixture.

Main Ideas

1. Evaporation is the process by which a liquid changes gradually to a gas.
2. Boiling is the process by which a liquid changes rapidly into a gas. Each liquid has its own boiling point.

Teaching Strategies

Call on a volunteer to read aloud the introduction to this lesson. Ask students to cite examples of processes that involve evaporation and boiling that they are aware of in their everyday world. Then instruct them to silently read the essays about salt preparation on pages 153 and 154 in order to discover the similarities and differences of the two processes. In Zambia, salt is extracted by dissolving it from the ashes of dried grasses. In Nova Scotia, salt is extracted from underground deposits by dissolving it in water.

Explain to students that in addition to food uses, salt and salt solutions are widely used in the chemical industry. Salt is used in the manufacture of sodium bicarbonate, sodium hydroxide, hydrochloric acid, chlorine, soap, and other chemical products.

SEPARATING SOLUTE AND SOLVENT

4 The Salt Industry

Around the world, salt is obtained in several different ways. Many methods use an *evaporation* or *boiling* process to separate the solute (salt) from the solvent (water).

Evaporation is the gradual process by which a liquid changes into a gas or vapor. For example, wet clothes will dry over time by taking heat from the surrounding air. The drying, or evaporation, process will go faster if you add heat, as in a clothes dryer.

Boiling is the process by which a liquid rapidly changes into a gas. This occurs when the temperature of the liquid is raised to its boiling point. Water boils at 100 °C at sea level. At higher elevations, the boiling point of water is a bit lower. Other liquids have their own characteristic boiling points.

Read the two reports that follow, written by students from different parts of the world. They describe how salt is produced in their countries. The first report is written by students in Zambia, a country in south-central Africa. The second report is by a student in Nova Scotia, a province of Canada.

As you read the reports, count the number of times the evaporation process and the boiling process are referred to. Think about the ways in which the two processes are similar.

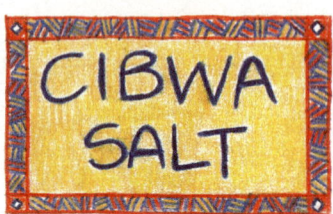

By:
Michael Nosenge
and
John Bavalya

LESSON  ORGANIZER

Objectives

By the end of the lesson, students should be able to:
1. Define evaporation and boiling and explain how these methods can be used to separate a solute from a solvent.
2. Give examples of evaporation and boiling.

Process Skills

observing, investigating, comparing, designing an experiment

152 Solutions

Cibwa salt is found mainly in the country of Zambia. This kind of salt is found in a sort of grass.

In August, the grass is collected using hoes and then spread on the ground to dry. This grass may remain there for about two months. When it is dry, it turns brown in color.

In September and October, the Cibwa salt is extracted from the grass. People do this by burning the dried grass, which turns it into ash. The salt is in the ash.

This ash is put on a filter made of another kind of grass, and water is poured over the ash into a clay pot. The water dissolves the salt and leaves behind the insoluble impurities of the ash on the filter.

The salt solution in the clay pot is placed over a fire. During the boiling, the water evaporates away. When the water has all evaporated, the salt is in the bottom of the pot in solid form.

The people in our part of Zambia still make salt the same way.

Grass

Collecting the grass

Drying the grass

Boiling away the water, leaving the salt behind

Pouring water over the ash

Burning the dried grass

Making Cibwa Salt

You might want to explain to students that the preparation of Cibwa salt from Zambian grass takes advantage of the fact that plants absorb salt from the soil. Plants, such as the Zambian grass, absorb sodium and other ions from the soil. Many important plant nutrients, such as nitrogen, potassium, phosphorus, calcium, and magnesium occur in the form of soil ions. These and other ions are released from weathered rocks and become part of the *soil solution.* The soil solution consists of all the water contained in the soil and its dissolved solids, liquids, and gases. Materials that become a part of the soil solution are available for absorption by plant roots.

The carbon compounds in the grass are converted into carbon dioxide and water when the grass is burned. The remaining residue contains a relatively high amount of salt, which is then water-extracted. The extract is filtered, and the insoluble material is left behind. Evaporation of the filtrate recovers the dissolved salt.

New Terms

Evaporation—the gradual process by which a liquid changes into a gas or vapor.

Boiling—the process in which liquid rapidly changes into a gas at its boiling point (the maximum temperature at which the liquid can exist as a liquid).

Materials

Exploration 7: 15 g sand and salt mixture, heat source, beaker, funnel, filter paper, stirring rod, paper towels, tweezers, magnifying glass, stand, safety goggles, water

Science Discovery Videodisc

Disc 2, Image and Activity Bank, 3–4

Time Required

one to two class periods

Separating Solute and Solvent

The Salt Plant at Nappan, Nova Scotia

The following is background information which you can share with your students. The Nova Scotian salt deposits were formed by evaporation of water in a partially enclosed bay over a long period of time. The ancient sea that formed these salt beds existed in the Mississippian period, approximately 350 million years ago. At that time, hot and dry climatic conditions made it possible for extensive salt beds to be deposited. Over millions of years, the salt beds were buried beneath sediments. Later, tectonic forces in the Nova Scotia area created a fold in the rock. This raised the ancient salt beds close to the earth's surface—making it possible for the salt beds to be mined today. The salt-making process described on page 154 involves making an underground saturated salt solution, which is then pumped out and processed.

The Nova Scotian salt beds represent only a small part of world salt resources. Extensive beds of rock salt also exist in Pakistan, Iran, Germany, eastern Europe, the former Soviet Union, the southern United States, China, and West Africa. Natural brines are found in the Dead Sea region, Austria, France, Germany, India, the United States, and Great Britain. Ocean water is a salt resource all over the world. Twenty liters (5.2 gal) of ocean water contain approximately 2.2 kg (1 lb) of sodium chloride.

As a research activity, have students investigate the source of salt commonly sold in their local area. *(In the northeastern United States, salt is made from natural brine pumped from the ground at Syracuse, New York. In the southeast and midwest states, salt is mined from underground salt domes. In the Pacific coast states, salt is evaporated from ocean water.)*

BY JILL OICKLE

Much of our area has large salt deposits deep underground. It is believed that these salt beds were formed millions of years ago when salt seas evaporated, leaving the salt behind. In some areas of Nova Scotia, the salt is mined using tunnels. But in Nappan, where I live, a different method is used.

First a hole is drilled down to the salt bed. Then water is pumped from a storage tank down through a pipe. This water dissolves some of the salt, making an underground cavern full of salt water. This salt water is then forced back to the surface where the water is evaporated, leaving the salt behind.

154 Solutions

EXPLORATION 7

It's Your Turn to Design!

You Will Need

- sand–salt mixture
- a heat source
- beakers
- a funnel and filter paper
- stirring rods
- paper towels
- tweezers
- a magnifying glass
- a stand
- water

Ask for any other apparatus you need.

What to Do

Here's the problem! You have just found a deposit of salt, but it is quite impure. In fact, it's 50 percent sand. Work as a team with several other people to devise a scheme to extract the pure salt. Check your plan with your teacher to make sure it's safe, and then carry out your experiment. Report on the success of your design.

Exploration 7

Divide the class into groups (teams) and distribute the materials. Each team should formulate a step-by-step plan to carry out in their investigation. Check their plans for safety in advance.

One method of extracting the pure salt from the sand and salt mixture is as follows:

1. Add about 25 mL of water to the mixture. Stir it.
2. Pour the mixture through a filter and collect the clear solution, or filtrate, that passes through the filter paper.
3. Heat the filtrate gently to evaporate the water.
4. Closely examine the substance that remains to determine what it is. *(It is salt.)*

An alternative approach is to have students pour 15 to 20 mL of the filtrate into a jar and let it evaporate naturally.

Point out to the class that filtration does not remove dissolved substances; it is not possible to separate solute from solvent by means of filtration.

✳ Follow Up

Assessment

Propose the following situation to students: Imagine that you live far from civilization. You have accidentally mixed your salt supply for the next six months with flour. How might you recover the salt? *(Answers will vary. The mixture could be added to water. The flour will not dissolve in water, but the salt will. The water in the salt solution can then be boiled off or allowed to evaporate, leaving the salt behind.)*

Extension

1. It has been proposed that highly hazardous waste, such as that from nuclear power plants, will be buried in vast salt deposits that are found throughout the United States. Find out why these deposits have been proposed for such a purpose.
2. Along the Gulf Coast of North America are numerous, unusual deposits called salt domes. Find out what these are and how they form. How is the buoyancy principle related to their formation?

Comparing Methods

Compare the method of salt extraction used in Zambia with the method used in Nova Scotia. What similarities are there in the methods of gathering the salt? How was water used in each process? How was the salt separated from the water? Where did the water go?

Answers to

Comparing Methods

The Zambian and Nova Scotian methods of gathering salt are quite different. In Zambia, the salt is taken out of the soil by grass plants and does not need to be mined. Only extraction, filtration, and evaporation steps are necessary. Both methods are, however, similar in that they require water for the extraction and refining of the salt. In Zambia, water is used to separate the salt from the ash by dissolving the salt. In Nova Scotia, the salt is dissolved in water and the salt solution is then pumped to the surface.

In Zambia, desalination is achieved by boiling the salt water in clay pots. The water evaporates, leaving the salt behind. In Nova Scotia, the salt is removed without boiling. In both cases, the water escapes into the atmosphere.

Separating Solute and Solvent 155

LESSON 5

✷ Getting Started

In this lesson, students are introduced to the processes of desalination and distillation. The lesson continues with a discussion of the earth's water cycle, and a creative writing exercise.

Main Ideas

1. Distillation is a process that combines boiling, which creates a gas, with condensation by cooling. The solute and the solvent are separated from each other, and both are retained during the process.
2. In nature, evaporation, condensation, and precipitation (the water cycle) can be compared to the process of distillation.

✷ Teaching Strategies

Call on a volunteer to read aloud the introduction on page 156. Involve the class in a discussion of the new terms, *desalination* and *distillation*. Point out that in the process of distillation, both solute and solvent are saved. Distillation differs from simple evaporation and boiling in that it includes the additional step of cooling to change the vapor back into liquid form.

The Future of Desalination

Call on a volunteer to read aloud the material on desalination. Involve students in a discussion of desalination and the role it might play as a future source of fresh water. Elicit their responses to the open-ended questions at **(Continues on next page)**

Desalination and Distillation

In certain parts of the United States and the world, fresh drinking water is in short supply. One method of meeting the ever-increasing demand for fresh water is to remove the salt from ocean water. This process is called **desalination.** In this case, it is the solvent (water), not the solute (salt), that is being saved.

In some situations, it may be useful to save both the solvent and the solute. To save the solvent, as is done in desalination, or to save both the solute and the solvent, a process called **distillation** is used. Let's look more closely at both desalination and distillation.

The Future of Desalination

Should we be using desalination of the ocean as a source of fresh water? Here are some facts about desalination:

- Desalination can relieve water shortages in dry regions along seacoasts.
- One large desalination plant alone can produce about 280 million liters of fresh water daily.
- The desalination process requires large amounts of energy, regardless of how that energy is produced.
- Providing desalted water for dry regions far from seacoasts requires a transportation system.
- The world's desalination plants today produce only a small fraction of the world's daily demand for fresh water.
- Nuclear-powered desalination plants might someday produce desalted water and electricity at the same time.

Form small groups and discuss the following questions. Do you think it would be wise for us to rely on the oceans for our supply of fresh water? Why or why not? How do you think the cost of getting water from desalination compares with the cost of fresh water from other sources? What are some ways in which we could make better use of the fresh water already available to us from rain, lakes, and rivers?

LESSON ORGANIZER

Objectives

By the end of the lesson students should be able to:
1. Describe the processes of desalination and distillation.
2. Illustrate the process of distillation.
3. Describe the process of the water cycle and analyze its importance.

Process Skills

observing, analyzing, comparing, illustrating

New Terms

Desalination—removing the salts from a substance, usually ocean water, in order to make it drinkable.

156   Solutions

Distillation

Distillation involves two steps. The first step is heating the solution to change one part of the solution into a gas or vapor. The second step involves cooling the vapor back into liquid form and collecting it.

Here is a design that Karen used for collecting both solvent and solute by distillation. What are some weaknesses in her design? Is she taking the necessary safety precautions?

With other students, discuss how Karen's design might be improved, using similar equipment. Check your suggestion with your teacher (for safety), and then test your own design. Compare your results with those of other groups.

Something to Write About!

Here's a chance to apply what you have learned about evaporation, boiling, and distillation.

During an expedition to a remote mountainous area, you learn that a person living there urgently needs distilled water to help treat a rare medical condition. You have a teakettle, a camp stove, and a few bottles. Draw a diagram to show how you could use them to obtain distilled water. Explain how your method works. See how many different ways your class can devise for distilling water with just the equipment listed.

- **Distillation**—a process in which a solution is heated until it evaporates, leaving behind dissolved materials. The second step involves cooling the vapor back into liquid form and collecting it.
- **Water cycle**—the continuous movement of water from the atmosphere to the earth and back again. This distillation process purifies the earth's water supply.

Materials

Distillation: materials will vary with student designs.

Science Discovery Videodisc

Disc 2, Image and Activity Bank, 3–5

Time Required

one to two class periods

(The Future of Desalination continued) the bottom of the page. The drawbacks to dependence on desalinated ocean water include transportation problems and the expense of desalination. In the United States, it costs about 1 dollar to produce 3800 L of fresh water from sea water. Water taken from fresh water sources such as lakes and rivers costs about 45 cents. Students may suggest water conservation as a better way to use the fresh water that is already available.

Distillation

Have students study the distillation apparatus in the illustration on page 157. They can work individually or in cooperative-learning groups to locate possible flaws in the design and suggest improvements. After students have completed the exercise, reassemble the class and involve them in a discussion of the pros and cons of the different groups' various designs. Then choose one to test as a demonstration. (The test criterion should be volume of solvent recovered.)

Answers to
In-Text Questions

The main flaw of Karen's design is that it permits most of the distilled solvent to escape from the beaker in the form of steam or vapor. Also, it appears that the solution in the flask is being heated very rapidly. This could create a safety hazard if the stopper was blown out by the force of the vapor pressure.

A possible improvement would be to lengthen the condensing tube so that the liquid would have more time to cool before reaching the receiving beaker. The lengthened condensing tube could be coiled or possibly jacketed in such a way that the condensing distillate would be cooled.

Something to Write About!

Have students draw their distillation set-ups, and discuss alternatives to the laboratory-type equipment used in the classroom. *(Students could boil the water, condense the vapor, and collect the condensate in the bottles.)*

Separating Solute and Solvent 157

What Is the World's Largest Distillery?

Call on a volunteer to read aloud pages 158 and 159. Direct students' attention to the illustration of the water cycle. Then involve them in a discussion of how the water cycle is similar to the distillation process they just studied. *(Both are two-step processes involving a warming-up and a cooling-down stage. Also, when water evaporates, it leaves behind dissolved solutes.)*

Using the illustration, ask students to trace the path of a water molecule from the ocean to the atmosphere, onto the land, and back into the ocean again. *(Water evaporates from the oceans, lakes, and streams and from the soil. Water also enters the air by transpiration, a process in which plants give off water vapor into the atmosphere. When water vapor rises in the atmosphere, it expands and cools. As the vapor is cooled, some of it condenses, or changes into liquid water, and forms clouds. Water then falls from the clouds as rain, snow, sleet, and hail.)*

This might be a good time to discuss why the ocean is salty. Students should recognize that water can pick up and dissolve mineral salts as it flows across surface rocks on its way back to the ocean. The salts are then concentrated in the ocean basin by continued distillation of water.

(Continues on next page)

What Is the World's Largest Distillery?

Surprisingly enough, it is the earth itself! The earth's water is constantly involved in a purification process similar to that of a giant distillery. This process is called the **water cycle.**

Each day, vast quantities of water evaporate from the oceans, lakes, rivers, and land. Eventually, this moisture condenses and falls as rain or snow. This is nature's way of purifying the world's water supply. How is nature's way similar to the method you developed to purify water?

The Water Cycle

158 Solutions

Because of the water cycle, all the water on the earth today was here when the earth was formed. The next time you have a drink of water, stop and ask yourself where the water may have been before. Perhaps that very water, or part of it, had been in a royal bath in ancient Egypt. Or perhaps George Washington Carver might have drunk some of it. Take the two cartoon characters that appear on this page and develop a cartoon strip showing their adventures through time. How many different and unusual places can you picture? Let your imagination roam, but there is one rule: at some point, you must include the water cycle in your cartoon series.

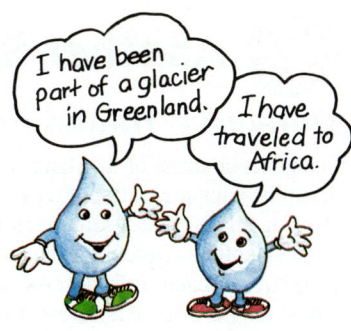

(What Is the World's Largest Distillery? continued)

As preparation for the writing and illustrating exercise on page 159, you may want to discuss detours that fresh water can make on its way back to the ocean. For example, a water molecule could be frozen for thousands of years in a glacier before returning to the ocean as meltwater or part of an iceberg. A water molecule could also seep into the ground and become part of a reservoir of groundwater that becomes sealed off somehow for many centuries. Such deposits are sometimes known as fossil water. A water molecule could also spend varying amounts of time in the body of a plant or animal.

The cartoon strip using the two water drops as characters encourages students to be creative and imaginative. Display their cartoons on the bulletin board. (Each cartoon should clearly indicate the water cycle.)

 **Follow Up**

Assessment

Have students draw a distillation apparatus and label the function of each part in relation to the distillation process.

Extension

1. Have students research the importance of water in regulating the earth's temperature.
2. Have students research the importance of water in bodily functions. For example, how water makes digestion, circulation, and regulation of body temperature possible. Students can also find out what role the process of dissolving plays in the transportation of vitamins and minerals in the body.

Separating Solute and Solvent   159

Lesson 6

Getting Started

This lesson shows students how the knowledge of solutions can be applied to everyday situations. Detective work is used as an example. Terms related to scientific methodology, such as *observation, prediction, inference,* and *hypothesis* are reviewed.

Main Ideas

1. An inference is a statement that attempts to explain an observation.
2. A hypothesis is a possible explanation of a question or problem, based on observations. It is a testable statement.
3. A prediction states the expected outcome of a future event.
4. Paper chromatography is a separation method in which the components of a solution are absorbed at different locations along a piece of filter paper.

Teaching Strategies

Detective Hanamoto's Forgery Case

Call on a volunteer to read aloud the introduction on page 160, pausing before Detective Hanamoto's summary. Involve the class in a discussion of the things both scientists and detectives do. Highlight making observations, making inferences, making predictions, formulating hypotheses, making educated guesses, and performing experiments.

Call on another volunteer to read the detective's summary of the forgery case. Ask students how they would go about solving the case. Accept all reasonable responses.

6 Another Solution

Detective Hanamoto's Forgery Case

Scientists do many of the things detectives do. This includes using scientific equipment, detectives' methods, and detectives' ways of thinking about clues. Detectives must be good observers, and so must scientists. Detectives search for clues, just as scientists do. Detectives make **inferences** and draw conclusions from the clues; so do scientists. Detectives prepare an explanation or a **hypothesis** to explain why certain things happen the way they do. Scientists do the same. Both detectives and scientists often make educated guesses about something that has happened or that might happen. Detectives working in a forensic lab often conduct experiments and make measurements; so do scientists. In this lesson, you will be asked to do many of the things detectives and scientists do: observe, make inferences, form hypotheses, make predictions, and conduct experiments.

Detective Yori Hanamoto made the following summary of a forgery case he investigated. As you read his report, try to pick out the similarities between detective work and scientific investigation.

Detective Hanamoto's summary:

The accused was charged after a complaint was made about an unauthorized withdrawal of money from another person's bank account. The accused claimed he was given a check for the amount. The police investigation suggested that the accused had altered the check: instead of "$7000," the check was now made out for "$70,000"!

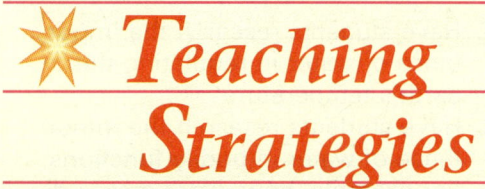

LESSON 6 ORGANIZER

Objectives

By the end of the lesson, students should be able to:
1. Identify and analyze the methods that scientists use for solving problems.
2. Explain the process of paper chromatography.
3. Separate solutions by means of paper chromatography.

Process Skills

observing, inferring, predicting, formulating hypotheses

New Terms

Inferences—conclusions, based on observations or facts, that attempt to make sense of the observations.
Hypothesis—an explanation of why certain things happen. In an experiment, a hypothesis is tested.

160 Solutions

On close examination with a magnifying glass, the check still looked genuine. The only way to settle the issue was to perform a test on the ink. If the check was a forgery, there would be a difference in the inks where there shouldn't be; that is, there would be a difference between the "Seven" and the "ty." There would also be a difference between the comma in "70,000" and the numbers themselves. After I made the test, there was no doubt: the check was a forgery.

What kind of test could Detective Hanamoto have used? Play detective by trying Exploration 8.

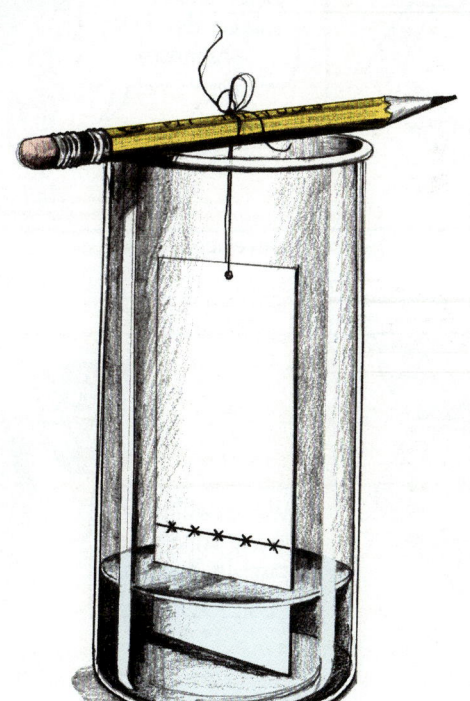

EXPLORATION 8

Analysis of Inks

You Will Need

- a metric ruler
- a strip of filter paper about 10 cm long and 6 cm wide (You can also use ordinary newspaper—but use only the edge so that there is no ink on it to confuse your results!)
- a beaker or other wide-mouthed container
- distilled water
- equipment to permit the paper to be held upright in the container
- several samples of washable (water-soluble) inks. Certain felt-tipped pens can be used.

What to Do

1. Pour water into the container to a depth of 2 cm.
2. Draw a line in pencil at a distance of 3 cm from the bottom of the filter paper or newspaper.
3. Put small crosses in pencil at least 1 cm apart on the line you have drawn.
4. Place a drop of one of the inks on one of the pencil crosses. Put one drop of each of the other inks on each of the other crosses.
5. Let the inks dry, then suspend the strip by some means so that the bottom 2 cm of the strip is immersed in the water inside the container.
6. After 15 minutes, remove the filter paper or newspaper and let it dry.

What do you observe?

Exploration 8

Divide the class into small groups and distribute the materials. Make certain that students test only water-soluble inks. The best ink to use is from water-soluble transparency markers. Capillary tubes or elongated eyedroppers are excellent for transferring drops of liquid onto the filter paper. Alternatives to filter paper are paper towels or coffee filters.

Instruct students to touch the end of the tube or eyedropper to the ink and then to an "X" on the filter paper. Just the right amount of liquid will be transferred. If using a pen, make a small dot on the filter paper.

After students have completed their investigations, you can share the following analysis of the activity.

The solvent (water) is pulled up the paper by capillary action. The solutes in the solution are carried along at different rates, depending upon the solubility of each component and how much each component adsorbs (adheres) to the paper. A component that is very soluble tends to move farther up the paper. Another component that is less soluble, or more strongly adsorbed by the paper, moves a smaller distance. In this manner, separation of components occurs. Two identical colors that travel up the paper at the same rate and the same distance are probably the same substance. However, if one red spot spot travels up the paper 1 cm and the other red spot travels up the paper 3 cm, then they are different substances.

Materials

Exploration 8: metric ruler, scissors, strip of filter paper (10 cm × 6 cm), beaker or wide-mouthed container, thin rod or pencil, thread, several samples of washable, water-soluble inks (or felt-tipped pens), distilled water, clock or watch, capillary tubes or elongated eyedroppers, safety goggles, Journal

Science Discovery Videodisc

Disc 2, Image and Activity Bank, 3–6

Time Required

two class periods

Separating Solute and Solvent

Answers to Questions

1. Students should apply their observations from *Exploration 8* to answer this question. Detective Hanamoto must have dissolved some of the ink from the check in a very small amount of water, and then used a chromatography technique. If the two inks were different, Detective Hanamoto should have observed that the inks traveled up the paper different distances and at different rates.
2. Inks are a true solution. Students should suggest testing the solution for the Tyndall Effect. They will find that a very dilute solution of ink and water is transparent, and does not show the Tyndall Effect.
3. *Observation:* information gathered using one's senses
   Inference: a statement that attempts to explain an observation
   Hypothesis: an explanation of an expected outcome of an experiment
   Prediction: a statement of an expected outcome of a future event

(Answers continue on next page)

EXPLORATION 8—CONTINUED

Questions

1. Now explain why Detective Hanamoto *inferred* that the check was written with two different inks.
2. Do you think that inks are a true solution? Devise an experiment to find out.
3. Read the following cartoon, which shows Keenan being a detective. Then explain the difference between observations, inferences, hypotheses, and predictions.

Observation Inference

Hypothesis Prediction

162　Solutions

4. Did you notice the various ways in which Detective Hanamoto was doing the types of things that scientists do? Complete the table below in your Journal, using specific examples from Detective Hanamoto's summary.

What Scientists Do	What Detective Hanamoto Did
observes	?
hypothesizes	?
infers	?
predicts	?
experiments	?

5. "Being scientific" means doing the things scientists do. In learning about solutions, you are being scientific. The list below suggests a number of ways in which scientists learn about our world. For each one, suggest in your Journal an example of a situation in this unit in which *you* have been scientific.

Being Scientific Means	I Was Being Scientific by
observing	?
measuring	?
classifying	?
inferring	?
predicting	?
experimenting	?

6. *Detective Work at Home.* The method you used to separate the colors of ink is called "paper chromatography" (*chroma* means "color"). You can use the same method to determine whether food coloring is made up of only one color or is, instead, a solution of different colors. Try it, and report back to your class.

6. This activity can be done at school or at home. Reinforce the need for parental permission before performing this activity at home. Also, warn students that nontoxic food colorings can stain skin and clothing. If they do this activity at school, make up solutions of colors and have students "play detective" to discover the colors that make up each solution.

Assessment

1. Have students write their own detective story explaining how chromatography helped them to solve a crime. (*The process of chromatography should be described.*)
2. Have students write a paragraph in which they relate the concept of solubility to the chromatography process. (*The solutes in the solution are carried along at different rates depending upon the solubility of each component.*)

Extension

1. Ask students to find the word "forensics" in the dictionary. Then have them research the kinds of work a forensic scientist does. Ask them to explain how a knowledge of solutions is important when doing forensic activities.
2. Have students research other pigment solutions, such as fabric dyes. This would include comparing the characteristics of plant dyes and synthetic dyes.

Answers continued

4. Detective Hanamoto acted like a scientist by doing the following:
   - He examined *(observed)* the check carefully. He also made careful observations of the results of his experiments.
   - He *hypothesized* that the accused had used a different ink to make the changes on the check. Therefore, testing the ink would reveal differences.
   - From his results, he *inferred* that the accused was guilty.
   - He *predicted* that a test would show whether or not the inks were different.
   - He performed an *experiment* to test the inks.

5. Some possibilities include:
   - Observing the Tyndall Effect
   - Measuring liquid volumes in investigations
   - Classifying solutes and solvents
   - Inferring the characteristics of solutions and nonsolutions
   - Predicting the results of evaporation of a solvent
   - Experimenting with the composition of inks

Separating Solute and Solvent

LESSON 7

✳ Getting Started

This lesson examines the effect of solutes on the boiling point of a solvent. Students explore this concept by reading about the process for making maple sugar. Then students make some artificial maple syrup in *Exploration 9*. Students examine how honey is made and how bees make use of the evaporation process.

Main Ideas

1. Solutes raise the boiling point of solvents.
2. The greater amount of solute in a given amount of solvent, the higher the solution's boiling point.
3. In the making of honey in nature, water is evaporated from the honey; in the making of maple products, water is boiled away.

✳ Teaching Strategies

Call on a volunteer to read aloud the introduction and Beth's report on the sugarbush on page 164. One of the purposes of this story is to illustrate that solutions boil at a higher temperature than do pure solvents. Point out that as the solution becomes more concentrated, its boiling point goes up. Some of the important points in Beth's report follow:

- Sap is a solution made up of 98 percent water and 2 percent sugar.
- Maple syrup is bottled at 104°C.
- Increasing the sugar content of a solution results in raising its boiling point.
- Maple candy is boiled at 110°C.

7 Sugar Solutions

Maple syrup and maple sugar are products made from the sap of the rock maple tree, often called the sugar maple. The sap of the sugar maple is a solution of sugar and water. Much can be learned about solutions by visiting a grove of sugar maples (a sugarbush) and seeing how maple syrup is made. That is just what Beth's class did. She visited a sugarbush in Vermont, which is the largest producer of maple syrup in the United States.

Read the report Beth presented to her English class the day after she and her class made the trip. As you read the report, write down three important discoveries Beth made about solutions. Compare your list with that of a classmate, giving reasons for your choices.

Mr. Lowther, the owner of the sugarbush I visited, starts tapping his sugar maple trees about the second week of March. If it is a warm spring, he begins sooner. If it's a large tree, Mr. Lowther can bore several holes into it. A small spout, called a spile, is pounded into each hole. In this sugarbush, all the spiles are connected to a network of plastic tubing that carries the sap to the sugar house. There the sap is boiled down to make maple syrup. Mr. Lowther said that a single tree can supply between 40 L and 100 L of sap in a season.

The sugar house is an exciting place, full of steam and delicious odors from the boiling sap. I found out that sap is only 2 percent sugar and 98 percent water! The sap is boiled in the evaporator until the water content is decreased to 34 percent and the sugar content increased to 66 percent. Mr. Lowther said that it takes 60 L of sap to make 1 L of maple syrup. Only when the syrup has the proper water and sugar content is it drawn off, strained, and bottled. I asked Mr. Lowther how he knew when the syrup was 66 percent sugar. He said he uses a thermometer! That's right—when the temperature reaches 104°C, the solution is right for bottling.

Mr. Lowther makes other maple products as well. He prepared a whole batch of maple candy for the class by boiling some syrup in a big pot until the thermometer reached 110°C. Then he poured the syrup onto a patch of clean snow where it hardened. It was great!

A trip to the sugarbush is an interesting way of investigating solutions. What a great time we had!

Here is a chart of the other maple products Mr. Lowther makes, and their boiling points.

Maple butter	113°C
Maple cream	114°C–115°C
Maple wax	119°C
Hard maple sugar	123°C

LESSON 7 ORGANIZER

Objectives

By the end of the lesson, students should be able to:
1. Explain the effect of solutes on the boiling point of a solvent.
2. Investigate the process used for making artificial maple syrup.

Process Skills

observing, analyzing, inferring, predicting

New Terms

none

164 Solutions

Questions to Ponder

With a classmate, discuss the answers to the following questions.

1. What method of removing a solute from a solvent is used in making maple products?
2. What effect does sugar have on the boiling point of water? (Pure water boils at 100 °C.)
3. What is the relationship between the number of degrees of change in the boiling point of the solution and the percentage of sugar in the solution?
4. How did Beth know that maple wax contains less water than maple cream?
5. At most, how many liters of maple syrup could one tree produce in a season?

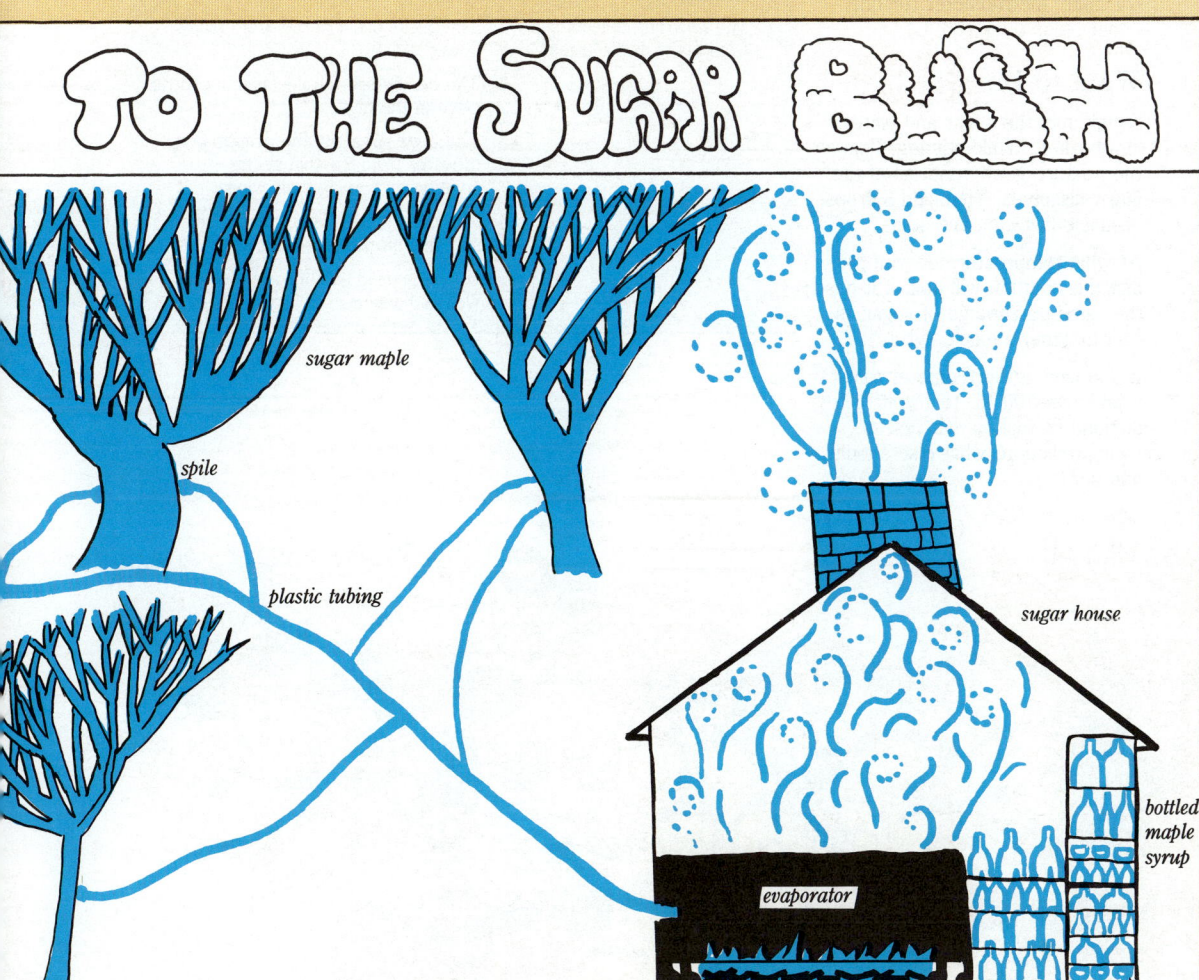

Materials

Exploration 9: beaker, balance, 65 g white sugar, 50-mL graduated cylinder, 35 mL water, 3 drops of maple flavoring, eyedropper, hot plate, stirring rod, candy thermometer (or one that reads over 100 °C), safety goggles
The Honey Factory: Journal

Science Discovery Videodisc

Disc 2, Image and Activity Bank, 3–7

Time Required

one to two class periods

Answers to
Questions to Ponder

Have students work in pairs to discuss and formulate answers to the questions. Have them begin by identifying the steps in the process of making maple syrup. These are shown in the illustration on pages 164 and 165.

1. The method of removing a solute from a solvent in making maple products involves boiling off the solvent (water) and leaving the solute (sugar) behind.
2. The sugar (solute) has the effect of raising the boiling point of water (solvent).
3. The higher the boiling point of the solution, the higher the percentage of sugar in the solution.
4. Beth knew that maple wax contains less water than maple cream because the boiling point of maple wax is higher.
5. According to Beth's report, a single tree can produce between 40 and 100 L of sap in a season. Since about 60 L of sap is required to make 1 L of maple syrup, a tree that produces the maximum of 100 L of sap will yield about 1.7 L of maple syrup.

Separating Solute and Solvent

Exploration 9

For safety reasons, this Exploration should probably be done as a class demonstration. If no artificial maple flavoring is available, check the effect of sugar on the boiling point of water. A solution similar in concentration to Mr. Lowther's maple syrup should boil at 104°C. To ensure a fair test, measure the boiling point of pure water first. (Results may not be exactly 100°C.) Discuss the possible reasons for the discrepancy:

- Most thermometers are not completely accurate.
- The location of the thermometer in the beaker might influence the results.
- Boiling temperature is affected by elevation.

Stress that there is a real difference in temperature between the boiling point of pure water and that of a sugar solution. The specific temperatures are not as important.

Answers to
The Honey Factory

Suggest that students work in pairs to complete the table by filling in the steps of Mr. Lowther's operation.

Mr. Lowther's Business
1. Mr. Lowther collects sap from many maple trees.
2. All the sap from Mr. Lowther's trees is carried by plastic tubing to the sugar house.
3. Mr. Lowther speeds up the evaporation of water from the sap by boiling the sap.
4. Mr. Lowther stores the product in bottles.
5. Maple syrup and other maple products are used by humans as food.

EXPLORATION 9

Making Artificial Maple Syrup

You Will Need
- a beaker
- a balance
- 65 g white sugar
- a 50-mL graduated cylinder
- 35 mL water
- 3 drops maple flavoring
- a hot plate
- a stirring rod
- a candy thermometer (or one that reads over 100 °C)

What to Do

Simply mix the sugar and water in a beaker. While stirring, heat the mixture gently until all of the sugar dissolves. Then add 3 drops of maple flavoring and stir.

At what temperature do you predict this solution will boil, if it has the same percentage of sugar as Mr. Lowther's maple syrup?

If you have a thermometer that reads over 100 °C, test your prediction. To make a fair test of your prediction, what else should you do?

The Honey Factory

The honey made by bees comes from *nectar*, a solution that is mainly sugar and water. The bees gather nectar from flowers and then store it in open cells in the beehive. The bees fan their wings in front of the cells to help evaporate the excess water in the nectar. When the solution has reached the right thickness, they close off the open cells with wax. The result is honey. Bees can do what Mr. Lowther does without boiling.

Actually, there are many similarities between making maple syrup from sap and making honey from nectar. One is given in the table below. In your Journal, fill in the rest of the steps from Mr. Lowther's operation that correspond to the descriptions of the steps performed by the honeybees.

Mr. Lowther's Business	The Honeybees' Business
1. Mr. Lowther collects sap from many maple trees.	1. The bees collect nectar from many flowers.
2. ?	2. The bees store the nectar in a central place in the hive.
3. ?	3. The bees speed up the evaporation of water from the thin nectar solution by fanning their wings.
4. ?	4. The bees store the product in cells and seal off the cells with wax.
5. ?	5. Honey is used as a food by bees as well as humans.

Follow Up

Assessment

1. Explain the following situation to students. Jeff was boiling a beaker of sea water in order to find how much salt would be left behind. He noticed that the temperature of the boiling liquid kept increasing. Ask students to explain why. (*As the concentration of solute increased, the boiling temperature increased.*)
2. Ask students to suggest ways of making syrup from maple sap without heating. (*Students may suggest using fans or some other means to move the air and speed up the evaporation process.*)

Extension

Have students get two identical plastic containers. Into one, ask them to pour some of the artificial maple syrup they made in *Exploration 9*. Into the other container, ask them to pour pure water. Have students place both containers in a freezer. Ask them to predict the results. (They may be surprised when they discover that the concentrated sugar solution will not freeze. The reason for this is discussed in Lesson 8.)

Challenge Your Thinking

1. You are given a clear liquid. It does not show the Tyndall Effect when a light is shined through it. Without tasting the liquid, how would you tell whether it is just one substance or a solution? Is there any case where your method might not work?

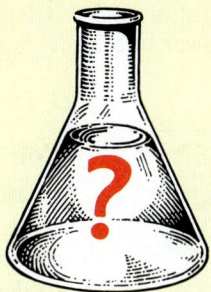

2. Sometimes you find spots on your dishes after they have been washed. How can you explain this?

3. Li is boiling a beaker of sea water to find out how much salt remains.

    (a) Describe all of the things Li might observe.

    (b) What will happen to the temperature of the boiling liquid as time passes? Why does this happen?

4. Salim wondered what would happen to the artificial maple syrup he made if he put it in the freezer. To his surprise, it would not freeze. Why wouldn't it freeze?

Answers to Challenge Your Thinking

1. You could use paper chromatography. Different components of the solution will travel different distances up the filter paper, at different rates. This method would not work if the components of the solution were colorless. You could also try boiling off all of the solvent to see if any dissolved substances remained behind. Liquids can also be separated by differences in their boiling points.

2. Spots on your dishes are an indication that there are dissolved solutes in your water. As the water on the dishes evaporates, the solutes remain behind as a residue.

3. (a) Li would observe that the water undergoes a phase change and becomes a gas. When all of the water has evaporated, the dissolved salts will be left behind in solid form.

    (b) As time passes, more of the water will evaporate, resulting in a higher concentration of salt in the remaining liquid. The higher the percentage of dissolved solutes, the higher the boiling point.

4. Just as solutes raise the boiling point, they also lower the freezing point.

Separating Solute and Solvent

Lesson 8

Getting Started

The activities in this lesson provide students with the opportunity to **study the effect of a solute on the freezing point of a solution**. Students then investigate the freezing of water with and without a solute.

Main Ideas

1. Solutes lower the freezing point of solvents.
2. The greater amount of solute in a given amount of solvent, the lower the solvent's freezing point.
3. The melting point of ice and the freezing point of water is 0°C.
4. Adding salt to ice lowers the melting point, often to below the temperature of the surroundings, and therefore causes the ice to melt.

Teaching Strategies

Have students study the cartoons and explain the message in each. The message is this: the presence of a solute lowers the freezing point of water.

Exploration 10

Have students work in pairs. Caution them to avoid banging or breaking the thermometers.

When the temperature remains constant (about 0°C), any remaining ice should be removed from the cup. Point out to students that the temperature of the ice and water will go no lower than 0°C. The new water volume will equal the original volume plus the water from the melted ice. Since the mass of 1 mL of water is 1 g, the mass of ice used to lower the water temperature can be determined.

CONCENTRATION OF SOLUTIONS

Getting Colder

Beth discovered that there is a relationship between the boiling point of a solution and the amount of solute (sugar) dissolved in the solvent (water). Would a solute affect the freezing point of a solvent as well? The following drawings suggest an answer to this question.

What's the message of these cartoons? Here are three experiments to help you find out: one for the laboratory (Exploration 10), one for home (Exploration 11), and one already completed for you to analyze (Exploration 12).

EXPLORATION 10

What Is the Temperature of Water When It Freezes?

You Will Need

- a 250-mL graduated cylinder
- a plastic-foam cup
- ice cubes
- a thermometer

What to Do

1. Add 150 mL of water to a plastic-foam cup.
2. Now add an ice cube and stir with a thermometer. **Be Careful:** *Stir slowly and carefully so that the thermometer does not break.* Observe the changes in temperature.
3. Keep adding ice cubes until the temperature will not go any lower. What is the lowest temperature the water can reach?
4. As soon as the temperature is constant, remove any ice that is left. How much ice did you use to cool the water to its lowest temperature? There is a way you can find out, without finding the mass of the ice.

LESSON ORGANIZER

Objectives

By the end of the lesson, students should be able to:
1. Explain the effect of solutes on the freezing or melting point of a solvent.
2. Analyze an investigation to determine its purpose and its results.

Process Skills

observing, formulating hypotheses, analyzing, drawing conclusions

New Terms

none

---

168 Solutions

EXPLORATION 11

Effect of a Solute on the Freezing of Water

You Will Need
- a 15-mL measuring spoon (1 Tbsp.)
- 3 juice glasses
- 45 mL of sugar
- 15 mL of salt
- water

What to Do

Pour 75 mL of water into a glass. Put 45 mL of sugar in the second glass and 15 mL of salt in the third glass. Add enough water to the second and third glasses to make their solutions even with the water level in the first glass. Stir. Put the glasses in the freezer. Examine them every half hour. In what order do they freeze? Why do you think this is so?

Water

Sugar solution

Salt solution

EXPLORATION 12

Mr. Acosta's Home Investigation

What to Do

Analyze the results of Mr. Acosta's experiment and draw some conclusions from his findings.

What He Did

Mr. Acosta surprised his family by taking over the kitchen one evening to do some experimenting. He had brought home several thermometers. He took the salt, measuring cups, and glasses out of the cupboard. If you analyze the table of results below, you will learn what he did and the information he found. What conclusions can you make about the effect of the solute on the solvent?

Trial	Solutions Mixed	Temperature When a Crust Forms on Top	Time to Freeze Completely
1	5 mL of salt in 100 mL of water	−3.6 °C	1 hour 5 minutes
2	10 mL of salt in 100 mL of water	−7.0 °C	1 hour 30 minutes
3	15 mL of salt in 100 mL of water	−10 °C	1 hour 45 minutes
4	20 mL of salt in 100 mL of water	−14 °C	2 hours
5	100 mL of water (no salt)	0 °C	48 minutes

Look again at the drawings on p. 168. For each drawing, write a two-line caption that explains in scientific terms what the drawing illustrates.

Answers to Exploration 11

The plastic cup filled with water will freeze first, the sugar solution second, and the salt solution third. Students may be confused by this because there are more milliliters of sugar than of salt. However, the sugar particles are so large that there are fewer in 45 mL than there are in 15 mL of salt. Therefore, the sugar freezes first because there are fewer particles dissolved in the water.

Answers to Exploration 12

Possible conclusions:
- Adding salt to water lowers its freezing point.
- Adding a solute to a solvent lowers the freezing point of the solvent.
- The more solute in a given amount of solvent, the greater the lowering of the freezing point of the solvent.

Scientific captions for the cartoons on page 168:
- Making Ice Cream: Adding more salt lowers the melting point of the ice so that the temperature of the liquid water can go below 0°C.
- Antifreeze in the Car: Adding antifreeze to the water in the radiator keeps it from freezing by lowering its freezing point.
- The Salt Truck: Putting salt on ice makes the ice melt at a lower temperature.

Materials

Exploration 10: 250-mL graduated cylinder, plastic-foam cup, ice cubes, alcohol thermometer, 150 mL water

Science Discovery Videodisc

Disc 2, Image and Activity Bank, 3–8

Time Required

two class periods

✷ Follow Up

Assessment

Ask students to summarize the important concepts in this lesson.

Extension

Have students design an experiment to determine the best ratio of antifreeze to water for a car driven in weather −5° to −10°C.

Concentration of Solutions

LESSON 9

★ Getting Started

Through the activities in this lesson, students will investigate the concentration of solutions.

Main Ideas

1. A solution becomes more concentrated when more solute is added to it and more dilute when more solvent is added to it.
2. Concentration can be expressed as number of grams of solute dissolved in 100 g of solvent.
3. Ways to compare the concentration of two solutions include: measuring the masses of a given volume, testing the buoyancy of objects, layering the solutions, and comparing freezing points.

★ Teaching Strategies

Call on a volunteer to read aloud page 170. Involve students in a discussion of the new terms *concentration, concentrated,* and *dilute.*

Explain to students that if water were added to solution A, the color would become less intense as the solution became more dilute. Adding more solvent would make solution B more dilute. Adding more solute would make solution B more concentrated.

At this point, you might want to introduce the concept of density. Density is mass per unit volume. This is directly related to the concentration of solutions. The more concentrated solution will have more particles of solute per unit volume, and will therefore have greater density.
(Answers to Concentrate on These! follow on next page)

9 How Much Solute?

Sap is a *dilute* sugar solution. Maple syrup is a more *concentrated* sugar solution. What do these terms mean?

Both beakers in the illustration contain the same type of solution. But how might you determine which one contains more solute?

The beakers show an example of two different **concentrations** of the solute and solvent. The one with *more* solute in the solution is the **concentrated** solution. The one with *less* solute in the solution is the **dilute** solution.

What would happen to the color of solution A if you added a lot of water to it? How would adding water to solution A change its concentration? Would solution A become more concentrated or more dilute? Now suggest two ways of changing the concentration of solution B. Would each change make solution B more concentrated or more dilute?

By now you may be noticing that the concentration of a solution is a *relative* thing. It applies only when you compare solutions made of the same solute and solvent. You can compare the concentration of one solution with another having the same solute and solvent by using the terms *more concentrated* or *more dilute.*

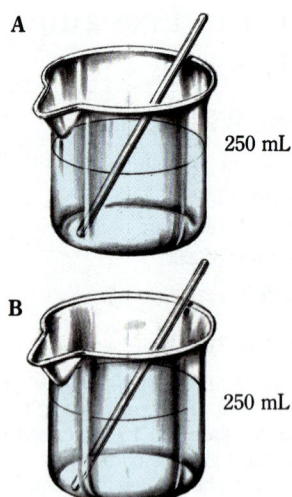

Concentrate on These!

1. Beakers C and D contain the same solute and solvent. The amount of solute is the same in both beakers.
   (a) What is the difference between solutions C and D?
   (b) After stirring, which will be more dilute? more concentrated?

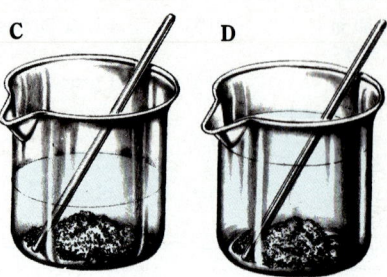

Stir to dissolve.

2. Beakers E and F contain the same solute and solvent. The amount of solvent is the same in both beakers.
   (a) What is the difference between solutions E and F?
   (b) Which will be more dilute? more concentrated?

Stir to dissolve.

LESSON 9 ORGANIZER

Objectives
By the end of the lesson, students should be able to:
1. Use the terms *concentrated* and *dilute* correctly.
2. Change a dilute solution into a concentrated one and vice versa.
3. Demonstrate ways of comparing the concentrations of two solutions made up of the same solvent and solute.

Process Skills
observing, investigating, comparing, formulating hypotheses

New Terms
Concentration—the amount of solute dissolved in a solvent.
Concentrated solution—a solution with a relatively large amount of solute.
Dilute solution—a solution with a relatively small amount of solute.

170 Solutions

3. Stir to dissolve completely. Then add more solvent to solution G. The result is a more __?__ solution.

4. Stir to dissolve completely. Then boil off some of the solvent in solution H to make a more __?__ solution.

G

H

Stir to dissolve.

Numbers for Concentration

If you and your classmates were told to prepare a dilute solution of salt and water, would all of the results be identical? Probably not. The concentration of each solution would likely differ. In order for everyone to make an identical solution, you would have to be told the amounts of both the solute and the solvent. Another way of stating this is to say that you would need to know the concentration of the solution.

The concentration of a solution is often expressed in "x grams of solute" dissolved in "100 g of solvent." Here are the concentrations of a few familiar solutions.

Solution	Solute	Solvent
maple sap	2 g sugar	100 g water
maple syrup	194 g sugar	100 g water
vinegar	5 g acetic acid	100 g water
sea water	3.5 g salt	100 g water
apple juice	not less than 0.035 g vitamin C	100 g water

Mystery Solutions

It is not always possible to identify the more concentrated solution just by looking at it. However, there are other properties or characteristics that can be used to distinguish a concentrated solution from a dilute one. For example, what property or characteristic of maple sap is used to determine the right concentration of maple syrup? Exploration 13 suggests some other ways to discover the relative concentrations of solutions.

Answers to

Concentrate on These!

1. (a) D contains more solvent than does C.
   (b) D will be more dilute; C will be more concentrated.
2. (a) F contains more solute than does E.
   (b) E will be more dilute; F will be more concentrated.
3. dilute
4. concentrated

Numbers for Concentration

Have students read the material and study the data table. They should realize that the hypothetical salt solutions discussed in the text will vary. Evaluate student understanding of the data table by asking questions similar to the following:

- Which solution contains the least solute? *(The apple juice contains only 0.035 g of vitamin C in 100 g of water.)*
- Which has the highest concentration? *(The maple syrup contains 194 g of sugar in 100 g of water.)*
- Which is more concentrated, maple syrup or maple sap? *(maple syrup)*

Mystery Solutions

Discuss the questions about maple syrup with students. Students should recall from Lesson 7 that the property of maple sap that is used to determine the right concentration of maple syrup is the boiling point of the solution. The more concentrated the sugar solution, the higher its boiling point.

Materials

Exploration 13: 2 salt solutions (300 g salt and 1 L water, 75 g salt and 1 L water), 2 beakers, 10-mL graduated cylinder, 2 hard-boiled eggs, 2 test tubes, test-tube rack, food coloring, clear plastic straw or glass tube, plastic-foam cup, ice cubes, metric ruler, thermometer, balance, standard masses

Exploration 14: 3 salt solutions (300 g salt and 1 L water, 75 g salt and 1 L water, and pure water), 3 test tubes, test-tube rack, 3 used pencils about 8 cm long each (or 2 straws), 6 thumbtacks (or 6 nails and modeling clay), 25-mL graduated cylinder, ruler, scissors, balance, standard masses

Science Discovery Videodisc

Disc 2, Image and Activity Bank, 3–9

Time Required

two class periods

Concentration of Solutions 171

Exploration 13

Divide the class into small groups. Set up three stations around the classroom, one for each test. If supplies are plentiful, have a number of set-ups at each station.

The salt solutions for this Exploration are prepared as follows:
- Concentrated salt solution—In a 2-L container, dissolve 300 g of salt in 1 L of water.
- Dilute salt solution—In another 2-L container, dissolve 75 g of salt in 1 L of water.

Stir each solution, allowing time for the salt to dissolve completely. Label the containers "A" and "B", respectively.

The materials for each station can be set up as follows:
- Station 1—*Test 1* requires 2 salt solutions, 2 beakers, and a balance.
- Station 2—*Test 2* requires 2 salt solutions and 2 hard-boiled eggs.
- Station 3—*Test 3* requires 2 salt solutions, 2 test tubes, a test-tube rack, food coloring, a clear plastic straw or glass tube, and a metric ruler.
- Station 4—*Test 4* requires 2 salt solutions, 2 plastic-foam cups, several ice cubes, and a thermometer.

Answers to
In-Text Questions

Test 1
The mass of the more concentrated salt solution will be greater because the same volume of solvent contains a greater mass of the solute.

The mass of 10 mL of a concentrated salt solution prepared as directed above (300 g of salt in 1 L water) is 13 g. The mass of 10 mL of a dilute salt solution prepared as directed above (75 g of salt in 1 L water) is 10.75 g. The mass of 10 mL of pure water is 10 g.

Test 2
The egg will float better in the more concentrated salt solution. Concentrated solutions, because of their greater density, generally demonstrate greater buoyancy than do dilute solutions or pure solvents. Buoyancy is a force that opposes the force of gravity.

Explain to students that tests based on buoyancy are only useful when distinguishing between solutions of different concentrations made with the same solute and solvent.
(Answers continue on next page)

EXPLORATION 13

Concentrated or Dilute?

You Will Need
- 2 salt solutions of different concentrations—solution A and solution B
- 2 beakers
- a balance
- a 10-mL graduated cylinder
- 2 hard-boiled eggs
- 2 test tubes and test-tube rack
- food coloring
- a clear plastic straw
- a plastic-foam cup and ice
- a ruler
- a thermometer

What to Do
Form small groups for this Exploration. Your teacher will give each group two salt solutions—solution A and solution B. One is concentrated; the other is dilute. Perform the following tests on each solution. Explain how each test enables you to distinguish the dilute solution from the concentrated one.

Test 1
Find the mass of a beaker. Add 10 mL of solution A to the beaker and find its new mass. What is the mass of 10 mL of solution A? Repeat with 10 mL of solution B. How does the mass of solution A compare with the mass of solution B? Which is more concentrated—solution A or solution B? Why? Return each solution to its original container.

Test 2
Drop an egg into both solutions. What happens to the egg in solution A? in solution B? Which is more concentrated, solution A or solution B? Why?

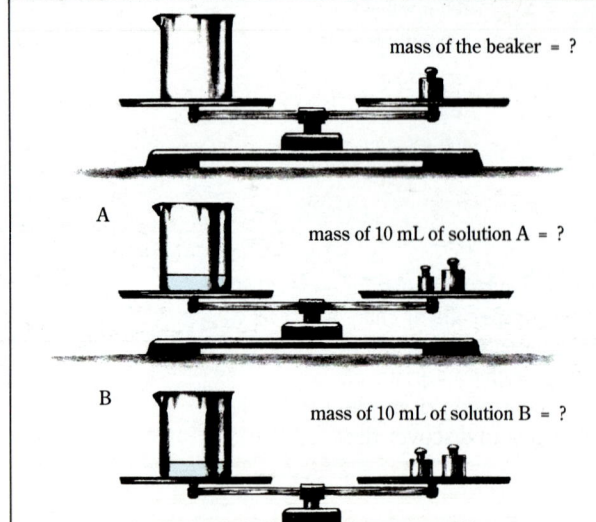
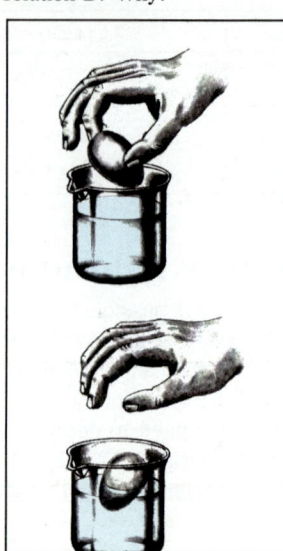

Test 3

Fill two test tubes halfway with each solution. Add several drops of food coloring to test tube B. Using a clear plastic straw or glass tube, form two separate layers of the solution inside the straw.

Step 1
Lower the straw into Solution A by about 1 cm. Let solution A enter the straw. Now cover the top of the straw securely with your finger.

Step 2
Transfer the straw with the solution from test tube A to test tube B, and lower the straw by 3 cm or more.

Step 3
Lift your finger off the top of the straw to let solution B enter. Cover the top of the straw again. Remove from test tube B. Which solution is on top, solution A or solution B? Or did the two mix together?

Reverse the whole procedure, this time putting the straw into test tube B first. Can you form layers with the colored solution on the bottom? on the top? Which appears more concentrated, solution A or solution B? How did you decide?

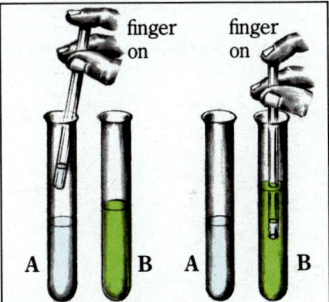
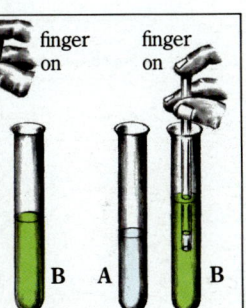
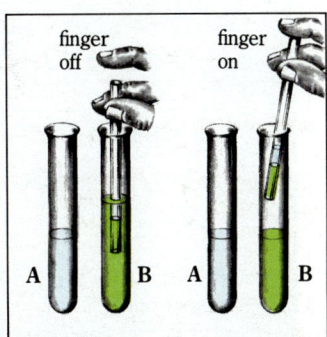

Test 4

Place 25 mL of solution A in a plastic-foam cup. Add ice to the cup and stir carefully with a thermometer. What is the lowest temperature you obtain? Repeat the procedure with solution B. Which solution has the lowest freezing point? Which solution is more concentrated, solution A or solution B? Why?

Can you suggest other tests that might let you distinguish solutions of different concentrations?

Answers continued

Test 3
If students are careful, it is possible to layer (without mixing) the two salt solutions inside the straw or the tube. The more concentrated salt solution will form the bottom layer because it contains more solute and is therefore more dense. Students will most likely be able to infer that the "heavier" (more concentrated) solution will sink to the bottom.

Test 4
Remind students to be careful not to bang or break the thermometers.

Students will find that they can achieve a lower temperature reading with the more concentrated salt solution. This is because the freezing point of a solution decreases proportionally with the amount of solute it contains.

Other tests to distinguish solutions of different concentrations include:
- Comparing boiling points of the solutions
- Evaporating equal volumes of the solutions and comparing the mass of any solids that remain

Concentration of Solutions

Solution Fact

Direct students' attention to the photograph on page 174. Some may have visited the Dead Sea or other bodies of water with high salt concentrations, such as Utah's Great Salt Lake. Ask students to share their experiences concerning floating in salty bodies of water. *(Higher salt concentrations cause greater buoyancy.)*

Answer to

In-Text Question

The salt solution in the Dead Sea is very concentrated. It exerts a greater buoyant force on the human body than does fresh water.

Exploration 14

This investigation explores the relationship between buoyancy and concentration. Use the salt solutions prepared for *Exploration 13* and a sample of pure water as the three solutions to be tested.

Once students have completed this Exploration, you might ask them to layer the salt solutions in a clear plastic straw or tube as they did for *Test 3* in *Exploration 13*.

Solution Fact

The following is an advertisement from the window of a travel agency.

THE DEAD SEA

Enjoy life at the Dead Sea! Spend your next vacation on the shores of the Dead Sea, where you can enjoy the sunshine ... and the water. We guarantee that you will float. A great vacation spot for all you sinkers!

What property of the Dead Sea makes it easier to stay afloat in its waters?

EXPLORATION 14

Make a Concentration Tester for Salt Water

You Will Need

- 3 different concentrations of salt solutions
- 3 test tubes
- test-tube rack
- 3 used pencils about 8 cm long each (or 2 straws, each cut in half)
- 6 thumbtacks (or 6 nails and modeling clay)
- a balance
- a 25-mL graduated cylinder
- a ruler
- scissors

174 Solutions

What to Do

1. Make a tester using one of the methods illustrated.

2. Find the mass of 25 mL of each solution. Then pour 25 mL of each solution into its own test tube.
3. Place one tester in each test tube as shown.

In which solution does the pencil or the straw stand the highest? the lowest? in between?

Compare the masses of the solutions. Can you see any relationship between the mass of a solution and its concentration? between the concentration of a solution and how well an object floats in it?

Do these solutions have the same concentration?

Follow Up

Assessment

1. Have students describe how and why the concentration tester they built in *Exploration 14* works. In their description, ask them to include such words as *dilute, concentrated,* and *concentration.*
2. Have students explore other tests for concentration:
   (a) Measure and record the mass of two equal volumes of solutions. Next, place the solutions in separate dishes and let them evaporate. Then, find the mass of the solute remaining in each dish.
   (b) Float a small ruler vertically in each of two solutions. Record the depth to which it sinks in each. (This works best in a tall narrow container such as a graduated cylinder.)

Extension

Have students research bodies of water with high salt concentrations. Encourage them to write a report on how these bodies of water were formed.

Answers to

In-Text Questions

- The pencil (or straw) tester will stand highest in the concentrated salt solution (300 g of salt in 1 L of water) and lower in the dilute salt solution (75 g of salt in 1 L of water). The tester will stand lowest in the pure water sample. The solution with the greater mass has the greater amount of solute, and is thus more concentrated. Students should realize that objects have greater buoyancy in the more highly concentrated solutions.

- When comparing the masses of the solutions, students will find that the most concentrated solution has the greatest mass because it contains the greatest amount of solute. The mass of 25 mL of the concentrated salt solution is 32.5 g. The mass of 25 mL of the dilute salt solution is 26.9 g. The mass of 25 mL of pure water is 25 g.

Answer to caption: None of the solutions pictured have the same concentration because the testers each float at different heights.

Concentration of Solutions 175

LESSON 10

✷ Getting Started

In this lesson, students examine the concepts of saturation and solubility. They also investigate the effect of temperature on solubility.

Main Ideas

1. A solution is saturated when no more solute can be dissolved in it.
2. Solubility is expressed as the number of grams of solute that can be dissolved in 100 g of solvent.
3. Solubility changes with temperature, and, in most cases, solubility increases as temperature increases.

✷ Teaching Strategies

Exploration 15

You may wish to prepare the salt samples a day in advance. Divide the class into small groups and distribute the materials. Once students have their solubility values, review their calculations with them. Involve students in a discussion of why their results might vary from the value given in the table on page 177. *(The main source of error probably occurred in determining when the solution was saturated.)*

Answers to

Comparing Solubility Figures

- Approximately 3.6 g of salt will saturate 10 mL of water at room temperature.
- Approximately 36 g of salt will dissolve in 100 mL of water at room temperature.

(Answers continue on next page)

---

10 Saturation

A sponge soaks up water easily, but there comes a point after which it is unable to soak up any more water. At that point a sponge is said to be **saturated**. When no more solute can dissolve into the solvent, a solution is said to be saturated. Earlier, you learned that water does not dissolve all things equally well. For instance, much more solute is needed to saturate the solution if the solute is sugar than if the solute is baking soda. The amount of solute needed to saturate a solution can be measured.

How Much Solute for Saturation?

You Will Need

- a salt sample
- a 10-mL graduated cylinder
- water
- a beaker
- a stirring rod

What to Do

1. Using a dry graduated cylinder, measure the volume of salt in your sample.
2. Add salt gradually to a beaker containing 10 mL of water until no more salt dissolves. Make sure you stir well after each addition of salt.
3. Measure what is left of your sample of salt. How many milliliters of salt did you use?

Comparing Solubility Figures

One milliliter of salt has a mass of about 1.2 g. What mass of salt did you use in 10 mL of water? What mass would dissolve in 100 mL of water? What mass of salt is needed to saturate 100 g of water? (Remember, 100 mL of water is about the same as 100 g of water.) How does this figure compare with the figure for salt given in the Table of Solubilities on p. 177?

LESSON 10 ORGANIZER

Objectives

By the end of the lesson, students should be able to:
1. Explain saturation and solubility.
2. Analyze the effect of temperature on solubility.
3. Read a table of solubilities and make a graph based on that information.

Process Skills

investigating, measuring, interpreting data, making a graph

New Terms

Saturated—the point at which no more solute can dissolve into the solvent in a solution.

Solubility—the amount of solute needed to saturate a solution; the number of grams of solute that can dissolve in 100 g of solvent.

176 Solutions

Saturation and Temperature

The amount of solute needed to saturate a solution is a measure of the solution's **solubility**. Solubility is often expressed as the number of grams of solute that can dissolve in 100 g of solvent. Do you think the temperature of a solvent affects its solubility? The following solubility table may provide the answer. It shows the maximum amount of different solutes (in grams) that can be dissolved in 100 g of water at different temperatures.

Table of Solubilities

Temperature (°C)	Salt (sodium chloride) (g)	Sugar (sucrose) (g)	Baking Soda (sodium bicarbonate) (g)	Alum (potassium aluminum sulfate) (g)
0	35.7	179	6.9	4.83
10	35.8	191	8.2	6.90
20	36.0	204	9.6	9.63
30	36.2	220	11.1	14.30
40	36.5	238	12.7	22.20
50	36.8	260	14.5	30.30
60	37.3	287	16.4	46.90
70	37.6	320	—	74.50
80	38.1	362	—	138.80
90	38.6	414	—	—
100	39.2	487	—	—

Making Sense of the Data

1. How does temperature affect the amount of solute that can be dissolved?
2. What substance is most soluble at 0 °C? least soluble at 0 °C?
3. What is the order of solubility of the given solutes at 0 °C? at 60 °C?
4. What is the solubility of sugar at 0 °C? At what temperature is the solubility of sugar approximately doubled?
5. Look at the solubility of other solutes at 0 °C. At what temperature would the solubilities of each be approximately doubled?
6. What substance shows the least increase in solubility as the temperature rises?
7. How much sodium chloride must you add to 100 g of water at 50 °C in order to saturate the solution?
8. Suppose you added 25 g of alum to 100 g of water and heated the solution to 80 °C. Would all of the alum dissolve? What would happen if you now cooled this solution of alum to 20 °C?

Materials

Exploration 15: salt sample (5 mL of salt), 10 mL water, 10-mL graduated cylinder, beaker, stirring rod
Picturing the Data: graph paper

Teacher's Resource Binder

Resource Worksheet 3–5
Resource Transparency 3–4
Graphic Organizer Transparency 3–5

Science Discovery Videodisc

Disc 2, Image and Activity Bank, 3–10

Time Required

two class periods

Answers continued

- 36 g of salt
- The maximum amount of solute to dissolve in 100 g of water is 36 g. Student results may vary, but should approximate this number.

Saturation and Temperature

Call on a volunteer to read aloud the top of page 177. Once the meaning of solubility is understood, introduce the question of how temperature affects solubility. Direct students' attention to the *Table of Solubilities.* Ask: As the temperature in column 1 increases, what do you notice about the quantity of each solute? *(The number of grams of each solute increases as the temperature increases.)* Point out to students that this is not true for all solutes.

Answers to

Making Sense of the Data

1. The amount of solute that can be dissolved increases as the temperature increases.
2. Sugar is most soluble at 0 °C; alum is least soluble at 0 °C.
3. Least to most soluble:
   - at 0 °C—alum, baking soda, salt, sugar
   - at 60 °C—baking soda, salt, alum, sugar

   There is a dramatic increase in the solubility of alum, greater than that of baking soda and salt. Thus, it can be inferred that temperature has a greater effect on the solubility of some substances than on others.
4. The solubility of sugar at 0 °C is 179 g per 100 g of water. Solubility of sugar doubles at 80 °C.
5. Solubilities double:
   - salt: it does not double
   - baking soda: between 40° and 50 °C
   - alum: at about 20 °C
6. Salt shows the least increase in solubility as the temperature rises.
7. 36.8 g
8. Yes, all the alum would dissolve. If cooled to 20 °C, the solution would be able to hold only 9.63 g. Therefore, 15.37 g (25 g − 9.63 g = 15.37 g) would come out of the solution.

Concentration of Solutions

Answers to

Picturing the Data

1. The starting points show the solubilities of the solutes at 0°C. Therefore, you know that all the solutes have different solubilities at 0°C.
2. The line for salt is almost horizontal. This shows that temperature has little effect on its solubility.
3. The lines for sugar and alum are very curved. This shows that temperature has a great effect on their solubilities.
4. The steep curve tells you that the solubility of that substance greatly increases with temperature.

Answer to caption: The graph on page 178 shows the effect of temperature on the solubility of sugar in water.

Follow Up

Assessment

Have students write a paragraph comparing and contrasting the effects of temperature on the solubilities of salt, sugar, baking soda, and alum in water. (They may refer to the table in their text.)

Extension

Have students research industrial or household processes that take advantage of temperature effects on solubility. Have them make posters or flow charts to illustrate these processes.

Picturing the Data

Another way of representing the data in the table is to show it in a graph such as the one below. The vertical axis of the graph is a scale of the solution's solubility. The horizontal axis shows the temperature. Use the information in the Table of Solubilities to graph each solute. Use different symbols for each solute (for example, X for salt, * for sugar, O for baking soda, and + for alum).

1. All the curves start at a different place on the vertical axis. What does this tell you?
2. Which line is almost horizontal? What does this tell you?
3. Certain lines are very curved. What does this tell you?
4. One curve is very steep. What does this tell you?

This is the graph of one solute. Which solute is it?

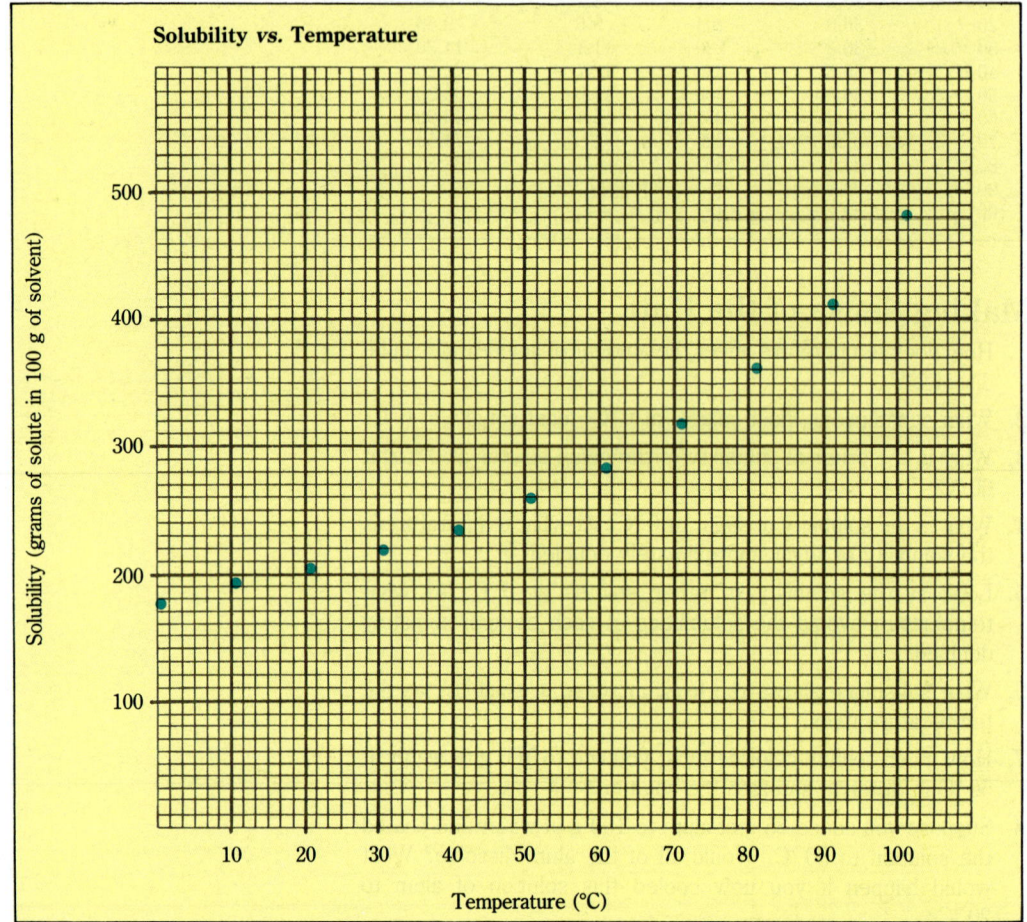

178 Solutions

11 Solutions in the Environment

While reading the following section about two important solutions, make your own list of any words referring to solutions. The list at the right will help you get started. Your list will be useful later on when you will be working on a crossword puzzle.

concentrated, concentration, dilute, dissolve, distillation, evaporate, mixture, saturated, solute, solution, solvent, transparent

The Atmosphere

The air around you and around the earth is called the **atmosphere**. It is a solution. How do you know that it's a solution? Here's how:

(a) The atmosphere is a mixture.
(b) Like liquid solutions, the atmosphere is transparent.
(c) The atmosphere can be separated into its parts.
(d) The concentrations of the parts need not always be the same.

The atmosphere is like a huge blanket around the earth. It is several thousand meters thick. Human actions, nevertheless, can change the concentrations of some of its parts. When we drive a car, the gasoline burns. This produces a gas called *carbon dioxide* that becomes one of the solutes in the atmosphere. Burning wood, oil, and other fuels also produces carbon dioxide. Some people think that putting so much carbon dioxide into the air will warm up the "blanket" of the atmosphere, slowly warming up our climate. What effect would this have on us?

Carbon dioxide isn't the only gas produced when we burn fuel. Factories that burn coal send many metric tons of *sulfur dioxide* into the air every day. Sulfur dioxide is a very poisonous gas, and it causes changes faster than carbon dioxide. You already know from studying Unit 1 that sulfur dioxide is soluble in water, and when the two are mixed, the solution they create is acidic. Rain that falls through air with sulfur dioxide in it is a solution called "acid rain." Many lakes and rivers are becoming more and more acidic because of the acid rain falling into them. Some are so acidic that fish cannot live in them any more.

What can you do to stop the atmosphere from being "saturated" with pollution?

LESSON 11 ORGANIZER

Objectives

By the end of the lesson, students should be able to:
1. Recognize that the atmosphere and the ocean are solutions.
2. Name some solutes that, in excess, can pollute the atmosphere.
3. Use the vocabulary of solutions correctly.

Process Skills

writing, researching, calculating

New Terms

Atmosphere—the solution of air surrounding the earth.

(Organizer continues on next page)

LESSON 11

Getting Started

In this lesson, students expand their vocabulary of solution words and then practice using these words. They learn that the atmosphere and the ocean are considered to be solutions. Air pollution is examined in regard to the solutes in the atmosphere. Then students use their "solutions" vocabulary to complete a crossword puzzle and write a short paragraph. The lesson concludes with a discussion of the amount of gold found in sea water.

Main Ideas

1. The atmosphere is a solution.
2. One pollutant in the atmosphere is sulfur dioxide. One result of sulfur dioxide pollution is acid rain.
3. The ocean is the largest liquid solution on earth.

Teaching Strategies

Have students review the list of solution words shown at the top of page 179. If necessary, suggest that they look back through the unit for the definitions of any words they are unsure of. As a Journal exercise, encourage students to define these terms in their own words.

The Atmosphere

Call on a volunteer to read aloud the material on page 179. Then involve students in a discussion of the new term *atmosphere*. Review the reasons the atmosphere is considered a solution.

Answer to caption: Discuss with students causes of atmospheric pollution and possible solutions to the problem. Accept all reasonable responses.

Concentration of Solutions 179

The Ocean

Call on a volunteer to read page 180 aloud. Involve students in a discussion of the solutes in sea water and the importance of the two dissolved gases for water animals and plants.

Sea water is a complex solution containing many dissolved substances. You might want to explain to students that a measure of the overall amount of all of these salts present in a sample of sea water is known as *salinity*. The salinity of a sample of sea water can vary depending on where the sample is taken. Observed variations in salinity result from such factors as incoming fresh water from river systems and rate of evaporation.

To illustrate the principles discussed on page 180, have students evaporate some sea water. If sea water is not available in your location, use the steps given below to make a solution of artificial sea water of standard salinity as defined by the International Association of Physical Sciences of the Ocean.

Making Artificial Sea Water:
1. To 500 mL of distilled water, add 12.03 g anhydrous $MgSO_4$ and stir until completely dissolved.
2. In another 500 mL of distilled water, dissolve 2.19 g of anhydrous $CaCl_2$.
3. In a large container, mix the solutions prepared in steps 1 and 2 above.
4. To the mixed solution prepared in step 3, add 35.1 g NaCl and stir until dissolved.
5. Add 1.5 g KCl and stir solution thoroughly.

Adaptations of marine organisms to life in salt water is a possible topic for student research or class discussion.

The Ocean

The atmosphere is not the only large and important solution in the world. The ocean is another. Indeed, the ocean is the largest liquid solution on earth. Parts of it are deep enough to swallow up Mount Everest—the world's highest mountain!

The solvent of this huge solution is water. The ocean contains many different solutes. What are they? This is the sort of question that can be answered by a chemist. Here is what a chemist would find when separating the water from the solutes in some sea water.

Parts of the ocean are deep enough to swallow up Mount Everest.

Mt. Everest
8,850 m

Marianas Trench
(Pacific Ocean)
11,036 m

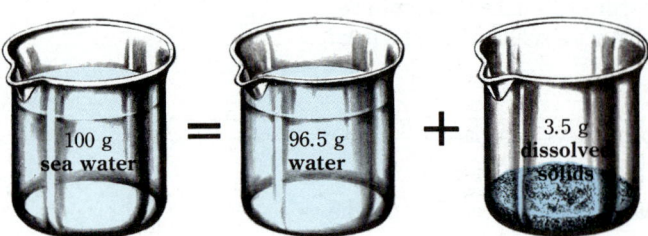

If 100 g of sea water is evaporated to dryness, 3.5 g of dissolved solids will be left behind. Most of the dissolved solids, about 2.4 g, is sodium chloride—table salt. The remainder consists mainly of compounds of magnesium, sulfur, calcium, potassium, bromine, and iodine. Someday, the ocean may be a valuable source of these materials, but so far only sodium chloride, magnesium, and bromine are being extracted on a commercial basis.

Solids are not the only solutes found in sea water. Two important gases that are dissolved in the ocean are oxygen and carbon dioxide. Without dissolved oxygen, most water animals could not exist; without dissolved carbon dioxide, plant life in the ocean would die.

(Organizer continued)

Materials

Word Hunt: Journal

Teacher's Resource Binder

Activity Worksheet 3–7
Resource Worksheet 3–6
Graphic Organizer Transparency 3–6

Science Discovery Videodisc

Disc 2, Image and Activity Bank, 3–11

Time Required

two class periods

Word Hunt

The answers to the clues for this crossword puzzle are probably in the list on page 179. Copy the puzzle into your Journal and complete it there.

Clues

Down
1. The largest liquid solution on earth is the _____.
2. All solutions are _____, except a solid and a solid.
3. In acid rain, sulfur dioxide is the _____.
5. The most abundant solute in the sea is _____.
9. Solutes _____ in water.

Across
4. Water is a good _____.
6. Sea water, oil and water, milk, air, and muddy water are all examples of _____.
7. When a liquid changes to a gas, it _____.
8. The atmosphere and ocean are both _____.
10. The amount of solute in a solvent is its _____.

Answers to
Word Hunt

Down
1. The largest liquid solution on earth is the *ocean*.
2. All solutions are *transparent*, except a solid and a solid.
3. In acid rain, sulfur dioxide is the *solute*.
5. The most abundant solute in the sea is *salt*.
9. Solutes *dissolve* in water. (Some students may take exception to this item, as not all solutes dissolve in water.)

Across
4. Water is a good *solvent*.
6. Sea water, oil and water, milk, air, and muddy water are all examples of *mixtures*.
7. When a liquid changes to a gas, it *evaporates*.
8. The atmosphere and the ocean are both *solutions*.
10. The amount of solute in a solvent is its *concentration*.

At this point, you may wish to use scrambled words, anagrams, or other word games to focus students' attention on unit vocabulary words not included in the *Word Hunt* puzzle. Computer software programs are available that will construct crossword and other puzzles using words of your choice.

Concentration of Solutions 181

Scientific Talk

Tell students that they can change the endings of the solution words when they are constructing their sentences. You may wish to give students the choice of using a story approach in writing their paragraph or a purely factual one.

Answers to

Gold from the Sea

Involve students in a discussion of the difficulty of communicating ideas without using the proper vocabulary. You may wish to have a thesaurus in the classroom to help students with their search for synonyms.

1. A sample answer for the first sentence follows:
   "Water is sometimes called the universal (material in which other materials are mixed in such a way that they become invisible), since so many (dissolved materials) are found in it."
2. Have students check their local newspaper for a report on the current price of gold, which fluctuates daily. For example, gold was recently reported at a price of $350.00 per troy ounce.
   To calculate the worth of gold in one cubic kilometer of sea water, students will need the following information:
   - There are 12 troy ounces in one pound.
   - One troy ounce equals 31.103 g.
   - If 1 troy ounce of gold costs $350, then 1 gram costs $350 divided by 31.1 = $11.25.
   - Each cubic kilometer of sea water contains 0.004 metric tons of gold; that is, 4 kg or 4000 g of gold. Therefore, the value of the gold in one cubic kilometer of sea water would be 4000 g × $11.25/g = $45,000.

Scientific Talk

Take another look at the terms in the list on page 179. Write a short paragraph that contains all of these words.

Gold from the Sea

Water is sometimes called the universal *solvent*, since so many *solutes* are found in it. Even gold is found *dissolved* in sea water. If just one cubic kilometer of this *solution* is *evaporated*, we will recover 0.004 metric tons of gold. That's a big gold mine!

1. To find out how difficult it is to talk about solutions without using any of the words listed on page 179, rewrite the paragraph above, replacing the italicized words with words or phrases that mean the same thing.
2. Listen to the national news or look in a newspaper for today's gold prices. Can you work out how many dollars' worth of gold is found in 1 cubic kilometer of sea water?

✱ Follow Up

Assessment

Locate an article that concerns the atmosphere or the oceans. Have students read the article and briefly summarize it using "solution" words from the unit.

Extension

Have students collect magazine and newspaper articles regarding air and water pollution. Then suggest that they use their articles in a bulletin-board display.

12 Growing Crystals

Growing your own crystals is fun, and all it requires is a suitable solute and some patience. Many solutes are suitable for crystal growing. A few of these are copper sulfate, magnesium sulfate, and alum. Alum will be used in Exploration 16. You may try others if you wish.

Tourmaline

Aragonite

Sugar crystals (magnified 40 times)

Quartz

Gypsum

EXPLORATION 16

Growing a Crystal

You Will Need

- a 100-mL graduated cylinder
- water
- alum
- a beaker
- a stirring rod or a straw
- an evaporating dish
- a 25-mL graduated cylinder
- plastic wrap
- a small jar
- thread
- a ruler
- a thermometer

LESSON 12

✶ Getting Started

The activity in this lesson provides students with an opportunity to explore and investigate the technique of growing crystals.

Main Ideas

1. When solutions evaporate, the solute left behind is often in crystal form.
2. The rate of evaporation of the solvent determines the rate at which the crystal grows.

✶ Teaching Strategies

Exploration 16

Divide the class into small groups and distribute the materials. You may wish to speed up step 1 of this Exploration by measuring out the 30 g of alum for each group in advance. Keep in mind that maximum results are achieved when the crystallization is allowed to continue for several weeks.

As the alum crystals grow, have students clean the beakers containing the growing solution every few days. This is done by pouring the liquid into another beaker, washing out the crystals that have formed on the bottom, and then returning the solution to its original beaker. The thread should also be cleared (carefully) of excess crystals. In their Journals, have students record the shape and size of their crystals.

LESSON 12 ORGANIZER

Objectives

By the end of the lesson, students should be able to:
1. Infer the crystalline nature of many solids.
2. Demonstrate a technique for growing crystals.

Process Skills

investigating, measuring, observing, inferring

New Terms

none

(Organizer continues on next page)

Concentration of Solutions 183

Answers to
A Final Challenge

Possible answers to this section follow:
- One way of finding out the mass added to the crystal each day would be to measure its mass on a laboratory balance on a daily basis. Subtracting the previous day's mass measurement would give the mass of one day's growth.
- After collecting and calculating daily growth measurements over a period of time, the average daily growth could be determined by totaling all the data and dividing by the number of days involved.

Assessment

1. Ask students to explain the relationship between the evaporation of the solvent and the size of the crystal. *(As the solution evaporates, more alum comes out of solution and adds to the size of the crystal.)*
2. Have students make a series of drawings to explain the procedure for growing crystals.

Extension

1. Have students grow crystals using other solutes: sugar, salt, and copper sulfate.
2. Have students research the crystallization process that took place within the earth to form each of the mineral crystals shown on page 183.

EXPLORATION 16—CONTINUED

What to Do

1. **Prepare a saturated solution**
   (a) From the Table of Solubilities on page 177, determine how much alum is needed to make a saturated solution in 200 mL (or 200 g) of water at 20 °C. Measure out this amount of alum and then add about 10 g.
   (b) Dissolve the alum in a beaker by adding 200 mL of water at about 50 °C. (Water from the hot-water tap works fine.)
   (c) Let the solution cool to room temperature (2–3 hours). After cooling, there should be some undissolved solute on the bottom of the beaker. If there is, the solution is saturated.

2. **Grow a seed crystal**
   (a) Carefully pour about 20 mL of this saturated solution into an evaporating dish. *Be careful not to transfer any undissolved solute.* Cover your beaker with plastic wrap to keep the rest of your solution from evaporating.
   (b) Place the evaporating dish containing the 20 mL of solution in a quiet spot where it will not be disturbed. In a couple of days, you should have a supply of seed crystals of sufficient size to use in step 3.

3. **Suspend your seed crystal**
   (a) Remove the plastic wrap from the container holding your supply of saturated solution. Slowly pour this liquid into a clean jar. *Be careful that no undissolved alum enters the jar.* The jar should be small enough so that it will be nearly full.
   (b) Choose a well-formed crystal from your supply of seed crystals and tie a thread around it.
   (c) Attach the thread to a stirring rod or straw so that when the straw or rod is placed across the mouth of the jar, the seed will be suspended in the center of the solution. Place the jar in a quiet spot and observe what happens over the next week or two.

A Final Challenge

In millimeters (mm), measure the dimensions of your crystal every few days. Keep a record of its growth. Also make a diagram that shows its shape and size. On the average, how much mass is added to your crystal each day? How could you find out?

(Organizer continued)

Materials

Exploration 16: 30 g alum, 25-mL graduated cylinder, 100-mL graduated cylinder, beaker, stirring rod or straw, evaporating dish, plastic wrap, small jar, thread, 200 mL of water, ruler, thermometer, Journal

Teacher's Resource Binder

Activity Worksheets 3–8 and 3–9
Graphic Organizer Transparency 3–7

Science Discovery Videodisc

Disc 2, Image and Activity Bank, 3–12

Time Required

two class periods plus observation time during a two-week period

Challenge Your Thinking

1. Maya was given five solutions of sugar in water, in different concentrations. Four of the solutions had been dyed the following colors by adding food coloring: red, yellow, green, and blue. The fifth solution was clear. She used the layering test to determine their concentrations. The illustration below shows her results.

   Which solution is the most dilute? Which is the most concentrated? Arrange the five liquids from the least concentrated to the most concentrated.

2. Josh wants to check a claim by Brand X that their canned fruit is packed in very light syrup (dilute sugar solution). Suggest a way that he can compare the "lightness" of the syrup used by Brand X with that of the syrup used by Brand Y.

3. Salt is mixed with sand and placed on roads in the winter after snowstorms in many states. This practice does not work as well in Alaska as it does in Washington. Why not?

4. The graph shows how temperature affects the solubility of gases and most solids in water.

   (a) When can a lake hold more dissolved oxygen—in the winter or in the summer? Why?

   (b) Liter for liter, which ocean can hold more salt—the Caribbean or the North Atlantic? Why?

   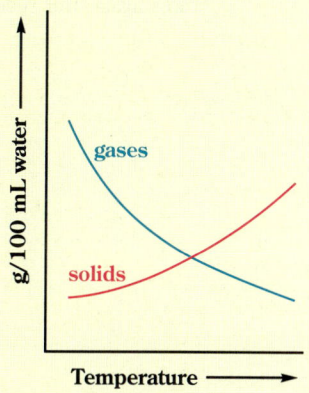

Answers to Challenge Your Thinking

1. The order from least to most concentrated is *red, green, blue, yellow,* and *clear.*
2. One answer would be to boil the two solutions. The solution with the most dissolved solutes will boil at a higher temperature. Another method would be to compare the freezing points of the two solutions. The solution with more dissolved solutes would have a lower freezing point.
3. The temperatures are frequently much lower in Alaska than in Washington. Therefore, the melting point of ice is not lowered enough to cause melting.
4. (a) According to the graph, gases become less soluble at higher temperatures. Therefore, a lake would hold more dissolved oxygen in the winter.
   (b) The Caribbean can hold more dissolved salt because solids are more soluble at higher temperatures.

UNIT 3
Making Connections

Summary for
The Big Ideas

Student responses will vary. The following is an ideal summary.

A solution is a homogeneous mixture; a non-solution is not homogeneous because its component parts are still visible. The parts of a solution are the solute and the solvent. The solute is no longer visible because it is uniformly mixed throughout the solution.

Solubility refers to the maximum quantity of solute that will dissolve in a solvent. To determine whether a substance is soluble, you would attempt to dissolve it in a solvent. A substance is soluble if it disappears in the solvent. This shows that it has dissolved completely. A substance is not very soluble if some of it is still visible in the solvent. A substance is insoluble if none of it disappears in the solvent, showing that none of it has dissolved.

A solid solute can be separated from a solvent by boiling off the solvent or by allowing the solvent to evaporate.

A solute might dissolve faster at higher temperatures and with agitation of the particles.

Evaporation is the gradual change from a liquid into a gas. In the water cycle, water evaporates from the oceans, rivers, lakes, and land. Honeybees also use evaporation to get rid of the excess water in their nectar solutions. Boiling occurs when a liquid rapidly changes into a gas. To make maple syrup, the excess water is removed by boiling.

The presence of a solute in a solvent can lower its freezing point and raise its boiling point. This property of solutions can be useful when it is necessary to lower the freezing point of a substance, such as ice on a road, or raise the boiling point of a substance, such as the boiling point of water in a radiator.

A concentrated solution, because it contains more particles of dissolved solute, will have a higher boiling point and a lower freezing point. A solution of greater concentration will also exert greater buoyancy upon an object floating in it.

The *solubility* of a substance refers to the maximum quantity of a solute that can be dissolved in a solvent. When this maximum point is reached, the solution is said to be *saturated*. *Temperature* has a direct effect on the solubility of substances. In general, gases become less soluble at higher temperatures, while solids become more soluble.

Unit 3 Making Connections

The Big Ideas

In your Journal, write a summary of this unit, using the following questions as a guide.

- How is a solution different from a non-solution?
- What are the parts of a solution? Why are they not visible?
- How would you determine whether a substance is soluble, not very soluble, or insoluble?
- How would you separate a solute from a solvent?
- How would you make a solute dissolve faster?
- How is evaporation different from boiling? Which process occurs in the water cycle? in the making of maple syrup? in the making of honey?
- How does the presence of a solute affect the boiling point and freezing point of water? Of what practical value is this?
- How would you distinguish by experiment the difference between a concentrated solution and a dilute solution of a given solute in a given solvent?
- How are the terms *saturated*, *solubility*, and *temperature* related?

Checking Your Understanding

1. Suppose 6 g of salt is mixed with 50 g of water.

   (a) What is the mass of the solvent?
   (b) What is the mass of the solute?
   (c) What is the mass of the solution?

2. Jennifer discovered that when she put a little washing soda (sodium carbonate) in water, the solution became warm. She wondered if the amount of washing soda mattered. So she found 5 bottles, filled each one with 100 mL of water, and took the temperature of the water in each. Jennifer then added different amounts of soda to each and measured the temperatures again. Her results are on the next page.

Bottle	Number of Teaspoons* of Soda Added	Amount of Water (mL)	Temperature Before (°C)	Temperature After (°C)
A	1	100	22	26
B	2	100	22	30
C	3	100	22	34
D	4	100	22	37
E	5	100	22	39

*1 teaspoon of washing soda has a mass of about 7.1 g.

(a) What variables were held constant?
(b) What variable changed?
(c) What variable changed in response?
(d) What did Jennifer find out?
(e) Which solution was most dilute?
(f) If washing soda happened to make a blue solution, which solution would be the deepest blue?

3. Draw a concept map that shows how the following words are related: *solute, antifreeze, solvent, boiling point, salt,* and *freezing point.*

Reading Plus

Now that you have been introduced to solutions and non-solutions, let's take a closer look at these mixtures and at some of the special properties they display. By reading pages S45–S60 in the *SourceBook,* you will learn more about different solutions and about the importance of water in making many of them. You will also learn more about acids, bases, and salts.

Updating Your Journal

Reread the paragraphs you wrote for this unit's "For Your Journal" questions. Then rewrite them to reflect what you have learned in studying this unit.

Answers to
Checking Your Understanding

1. (a) The mass of the solvent is 50 g.
   (b) The mass of the solute is 6 g.
   (c) The mass of the solution is 56 g.
2. (a) Variables held constant are the size of the bottle used, the starting temperature, and the amount of water added.
   (b) The variable that changed was the amount of washing soda added.
   (c) The temperature after adding the washing soda was the variable that changed in response.
   (d) Jennifer found out that the addition of washing soda raised the temperature of the solution.
   (e) The solution with only 1 teaspoon added was the most dilute.
   (f) The solution with 5 teaspoons would be the most concentrated, and would therefore be the deepest blue.
3.

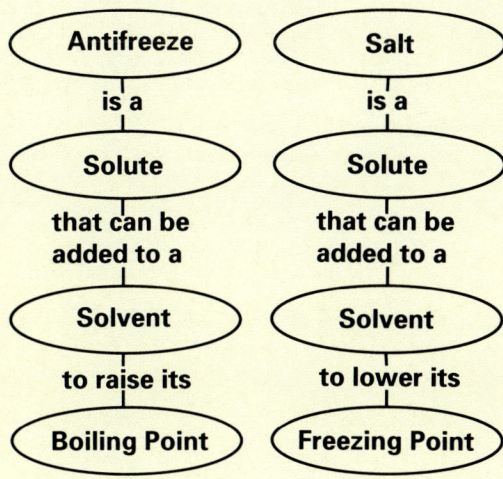

About Updating Your Journal

The following are sample ideal answers.
1. After a sugar cube is dissolved in water, it can no longer be seen because the particles of sugar have become mixed uniformly with the particles of water. *(The theme of **Scale and Structure** relates to the size of particles in relation to the formation of a solution.)*
2. The sugar can be separated from the water by allowing the water to evaporate or by boiling off the water until only the sugar remains.
3. If water were not such a good solvent, many things would be different. For example, water helps to break down substances in the soil, allowing organisms to then make use of these nutrients. The solvency of water is also a major factor in the erosion of the earth. Water also helps organisms carry out bodily functions.
4. Sea water is solution of water and many dissolved solutes including salts, oxygen, and carbon dioxide. It is a solution because any liquid mixture that is transparent must be a solution.

Making Connections 187

Science in Action

✦ Background

Biochemists and chemists at major universities conduct specialized, original laboratory research. They direct laboratory research teams composed of graduate students and postdoctoral fellows, and they may also teach undergraduate courses. Educational requirements include a Ph.D. degree in Chemistry or Biochemistry, as well as additional years of postdoctoral research experience. Excellent written and oral communication skills are needed to prepare grant proposals, research papers, seminars, and lectures.

Science in Action

Spotlight on Chemistry

Biochemistry

Kathleen Mailer is a biochemical researcher and university teacher with a Ph.D. in chemistry. Her research concerns the biochemistry of living tissues, mainly the heart. Kathleen says she loves research and greatly enjoys teaching. The most stimulating part of her job, she says, is the freedom to wonder whether an idea is true and then to try to find out.

Q: Why did you choose to do heart research?

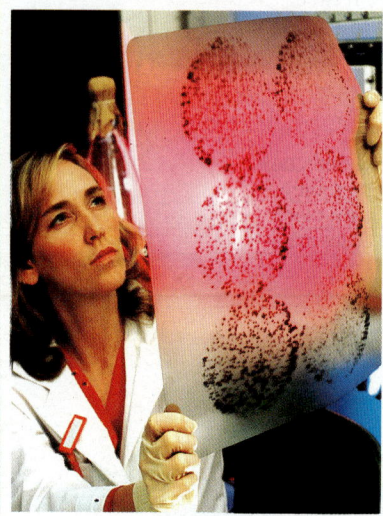

Kathleen: To me, the heart is the most fascinating part of the body. It's the one organ that must never stop. Your muscles, your liver, and other organs can stop temporarily, but not your heart.

The heart is like a super-efficient factory. Even during the sprint at the end of a marathon, only a part of its capacity is being used. It is said to be "over-engineered."

Nevertheless, the human heart can be afflicted with heart disease. Heart disease also occurs in rabbits, but very rarely in rats. It's curious that the same things that predispose human beings to heart attacks, such as absence of exercise or bad diets, also predispose rabbits to heart attacks—but not rats. It is important, therefore, to study the differences between rat hearts and rabbit hearts.

Chemistry Teaching and Research

Enrique Diaz has devoted years to teaching chemistry in high school and college. In addition, he did chemical research before becoming a teacher. The basic qualification for his field of work is a Ph.D. in chemistry. Enrique obtained his doctorate after earning a Bachelor of Arts in languages and a Master's degree in science. His many and varied responsibilities as a researcher, teacher, and administrator make his life busy and rewarding.

✦ Using the Interview

After students have read the interviews, involve them in a discussion of the many areas within chemistry that can be studied. To develop your discussion, you may wish to use questions similar to the following:
- What are some ways you could find out the answers to questions that no one else has ever researched? *(You could start by doing library research and formulating a hypothesis about something that was known to happen. Next, you could design an experiment to test that hypothesis. Then, based on your data, you could draw some conclusions.)*
- What research topics might be of interest to a biochemist? *(Biochemical topics include: the structure and function of the human body, food chemistry, or the chemistry of plants, animals, protists, bacteria, or viruses. Generally, biochemists research the structure and function of biological molecules that occur in the cells of plants, animals, and other organisms.)*
- What research topics might be of interest to a chemist? *(Chemists investigate the structure of compounds and their chemical properties. They isolate and characterize existing compounds, and study how new compounds can be synthesized.)*
- Why is a knowledge of basic math, chemistry, and physics important for everyone in today's world? *(Answers will vary, but students should recognize that the increasing role of technology in today's world makes it necessary for people to have an understanding of science and math in order to make informed decisions about their lives.)*

Q: How did you get interested in science?

Enrique: As a child, I was curious and always wanted to know how things worked and why things were as they were. My interest in chemistry was sparked by very enthusiastic high school and college chemistry teachers. Their love of chemistry and teaching helped tremendously when the studies became more difficult.

Q: What do you enjoy most about your work?

Enrique: I enjoy the opportunity to work with students and to pass on some of the enthusiasm I have for chemistry and research. Making a new compound that no one has made before is quite a thrill. Then, determining the structure of that compound makes all the effort worthwhile.

Q: What subject areas are important for students who are interested in becoming chemists?

Enrique: First of all, they must have a knowledge of basic math and chemistry and physics. Of course, this is important for everyone—not just for chemists. Scientific literacy is a must for today's world.

Students interested in chemistry must also be able to write clearly, because scientists share their research discoveries with other scientists by writing articles that describe their work.

Q: Are there any particular abilities that are most important to a chemical researcher?

Enrique: I think that an analytical problem-solving approach, logical thinking, attention to detail, and careful library research habits are all important abilities. One hour in the library recording what others have already discovered can save many hours in the laboratory.

A Project Idea

You've probably seen labels on laundry detergents that say "Phosphate-Free." Find out why these detergents are better for the environment than detergents containing phosphate.

Chemists have played a key role in replacing the phosphates in laundry detergents with safer substances. Find out what these safer substances are and why they won't damage the environment as much as phosphates do.

◆ Using the Project Idea

Explain to students that phosphorus is an element essential for the survival of all organisms. It is only when phosphates are in the environment in excessive quantities that they contribute to pollution. In order to fully understand the relationship that phosphorus has with the environment, students can also research the following related topics:

- How phosphorus cycles through the environment
- How the amount of phosphorous in the soil determines the type of plants present (for example, the native plants of Australia)
- How plants have evolved structures that retain phosphorus in nutrient-poor soils
- The effect of using phosphates on agricultural lands
- How mycorrhizal fungi help absorb phosphorus from the soil

The library, chemists at local universities, and the water quality division of the Environmental Protection Agency would all be excellent sources of information.

◆ Going Further

Ask students to investigate the following questions:

- How do chemists and biochemists use chromatography in their work?
- Research chemists are involved in the development of new drug treatments for diseases. As a safety precaution, these drugs are only available after extensive testing. However, some patients who are very ill have requested the right to take medicines that have not been thoroughly tested. Do you think these patients should be allowed to take these medicines? Explain.

Science and the Arts

✦ Background

The science of perfume making relates directly to the principles of solubility discussed throughout Unit 3. In order to extract and concentrate fragrant substances from natural materials, solvents are used. Generally, essential oils from plants are not very soluble in water. Therefore, alcohol or petroleum-derived solvents are more effective. Steam distillation is used because it takes advantage of variable solubility in regard to temperature.

The ancient Egyptians used oils and animal fats as solvents to extract fragrant compounds from the resin of the north African plant, *Boswellia.* This resin, known as frankincense, contains large amounts of the compounds pinene and limonene. Pure pinene has a strong aroma of turpentine, and limonene has a lemon scent. Pinene has insecticidal properties and may be produced by plants as a defense against insects.

Science and the Arts

The Perfume Maker's Art

Making perfumes is an ancient art. It was practiced, for example, by the ancient Egyptians, who rubbed their bodies with a substance made from soaking fragrant woods and resins in water and oil. From certain references and formulas in the Bible, we know that the ancient Israelites also practiced the art of perfume making. Other sources indicate that the art was known to the early Chinese, Hindus, Arabs, Greeks, and Romans.

A Complex Art

Have you ever wondered why some perfumes are traditionally so expensive? To begin with, a fine perfume may contain over 100 different ingredients! And it's not just a matter of mixing things together that have pleasant odors, as you will see.

Plant, Animal, or ?

The perfume ingredients you are probably most familiar with come from fragrant plants or flowers like sandalwood, jasmine, or patchouli. These plants get their pleasant odor from their *essential oils,* which are stored by the plants in tiny baglike parts called *sacs.* The parts of the plants used for perfumes may include their flowers, roots, or leaves.

Another kind of ingredient that is used in making perfume comes from animals. These substances are useful as *fixatives,* to make other fragrances long-lasting. In some cases, the substances as they occur full-strength in nature have distinctly unpleasant odors! For example, *civet musk* is actually a foul-smelling liquid that the

Jasmine flowers

African civet

civet, a small cat-like animal, sprays on its enemies. Other examples of fixatives are *castor,* from the beaver, and *ambergris,* from the sperm whale.

A third class of ingredients is human-made, or synthetic. In some cases, these ingredients are chemically the same as naturally occurring materials. In others, however, they are chemicals unknown in nature. These extend the creative options of the perfume maker.

✦ Promoting Observational Skills

Ask students to carefully study their surroundings in order to answer the following questions.
- Sample various scents from your environment and make a list of those that you think could be used in a perfume. Explain what you like about each scent in your list. Your list could include flowering plants, fruits, herbs and spices, and flavorings such as vanilla extract.
- Have you smelled any of the scents listed in the table on page 191? Which ones do you prefer? Which ones do you dislike? Why do you like some scents more than others? *(Student answers will vary.)*

✦ Discussion

Promote discussion in the classroom by posing the following questions.
- Why do you think perfumes have so many ingredients? *(Blending ingredients can produce different, more pleasing aromas.)*
- Do you think that one perfume could smell good to everyone? Why or why not? *(Probably not. Smell perception varies from one individual to another.)*
- How do you think synthetic ingredients extend the creative options of the perfume maker? *(Synthetic ingredients help to create new aromas by making ingredients not found in nature. Synthetic ingredients also make it possible to manufacture naturally occurring ingredients that are difficult to obtain.)*

Taking "Notes"

When you smell a perfume, the first odor you detect is called the *top note*. It is a very fragrant and refreshing odor that evaporates rather quickly. The *middle note*, or *modifier*, gives a more substantial character to the impression made by the top note. The *base note*, or *end note*, is the odor that lasts the longest.

A Catalog of Fragrances

Perfume fragrances can be grouped according to their most dominant odors. The table below shows some of the major groups, with examples.

Group	Examples
Floral	Jasmine, rose, gardenia
Spicy	Clove, cinnamon, nutmeg
Woody	Sandalwood, cedarwood
Mossy	Oak Moss
Orientals	Combinations with vanilla, balsam, etc.
Herbal	Clover, sweetgrass
Leather/tobacco	Leather, tobacco, birch tar
Aldehydic	Synthesized from aldehydes (these generally have a fruity odor)

Roses have long been admired for their fragrance.

Find Out for Yourself

Test a number of different perfumes and colognes to see if you can distinguish three different "notes" of each. Write a report on your findings. Do some research on the history of perfume making.

Perfumes have other uses than as personal fragrances. Make a list of as many of these uses as you can.

◆ Find Out for Yourself

Perfumes and other scented products can sometimes trigger reactions in sensitive individuals. Therefore, this activity is not recommended for students with asthma or other allergic conditions. Other students who choose this activity should work in a well-ventilated area and test only one or two perfumes each. Students may discover the phenomenon of *sensory accommodation*, in which an aroma becomes less perceptible after a period of exposure. Students can extend this activity further by reporting on this phenomenon.

Perfumes are widely used to scent a variety of soaps, detergents, cosmetics, and other consumer products. Perfumes have also been developed for industrial uses, either to cover up an unpleasant aroma, or to enhance an existing one. As an extension activity, have students investigate the use of scents in consumer products. They could design experiments to compare the characteristics of scented and unscented versions of the same product.

◆ Going Further

- The history of perfume making offers an opportunity to infuse multicultural perspectives into the science curriculum. Students can report on the development of perfume-making techniques in various cultures around the world.
- Have interested students investigate the mechanism of smell in humans or in other organisms. Students should discover how solutions relate to the chemical reactions that occur when using the sense of smell.

Science and the Arts

UNIT 4 Force and Motion

✷ Teaching the Unit

☀ Unit Overview

In this unit, the concepts of force and motion are developed in ways that allow students to draw from personal experiences, observations, and previous knowledge. In the section *Understanding Forces*, students identify the forces described in a story about a sailboat race. Next, magnetic, electrical, gravitational, and elastic forces are introduced. In *Measuring Forces*, students learn how force can be measured with force meters. In the section *Frictional Force*, students consider the causes and the effects of friction, and learn about ways to reduce it. In *Going Through the Motions*, students do thought experiments to develop an understanding of Newton's first law of motion. Then students consider concepts of inertia and forces in pairs.

☀ Using the Themes

The unifying themes emphasized in this unit are **Systems and Interactions, Stability,** and **Energy.** The following information will help you weave these themes into your teaching plan. A focus question is provided with each theme as a discussion tool to help you tie the information in the unit together.

Systems and Interactions is a theme that is evident throughout the unit. Students learn that all forces are characterized by an agent, a receiver, and an effect. The interaction between the agent and the receiver results in a force that can be observed or inferred and that can be measured. Students find evidence of a system of interactions in the concept of motion—every action is found to have an equal but opposite reaction.

Focus question: *Describe the system of forces interacting in a game of pool.*

In a game of pool, the cue stick is the agent of force; the cue ball that it strikes is the receiver. The effect is that the cue ball begins to move. In order to move, the force must be great enough to overcome the inertia of the cue ball and the friction between the ball and the table. When the cue ball hits another ball, the system of interaction is the same, but now the cue ball is the agent and the second ball is the receiver.

Stability is another theme emphasized in this unit. Students learn about Newton's Laws of Motion and the stability of objects due to inertia and balanced forces. Change in motion occurs only when unbalanced forces or a new force alters the movement of an object.

Focus question: *How does the motion of objects demonstrate stability?*

Stability results when all the forces in a system are balanced. Using Newton's Laws, students should understand that an object remains stable in its motion or position until it is acted upon by an unbalanced force. Objects such as the moon and the planets, for example, are relatively stable in their motion; they are subject to balanced forces and inertia, which keep them moving in a steady path.

Energy is apparent throughout this unit; most notably in Section 1, which describes how forces affect matter. Energy is a necessary component of force. Motion that results from an applied force is evidence of this energy.

Focus question: *How is the force you exert in doing a push-up the end result of the sun's energy?*

Students should realize that a person doing a push-up must use energy to apply the force necessary to move his or her body to overcome the force of gravity. The ultimate source of the energy used by the person is the sun. Plants use the sun's energy to make their own food; humans then eat plants or other organisms that have eaten plants to obtain energy. Energy from food is stored as chemical energy until it is needed.

☀ Using the *Science Discovery* Videodiscs

Disc 1 *Science Sleuths, A Day at the Races*
The Zahlers, a father and daughter competing in the Pine Block Derby, are suspicious when another entry scores an unusual number of points. Mr. Zahler suspects another entrant of cheating. The Science Sleuths must consider all of the evidence to determine if and how someone cheated during the race.

Disc 2 *Image and Activity Bank*
A variety of still images, short videos, and activities are available for you to use as you teach this unit. See the *Videodisc Resources* section of the **Teacher's Resource Binder** for detailed instructions.

☀ Using the *SciencePlus* SourceBook

Unit 4 extends the study of motion by emphasizing observation and description of different types of motion. Newton's Laws of Motion and Gravitation are introduced, along with a discussion of momentum.

191A

PLANNING CHART

SECTION AND LESSON	PG.	TIME*	PROCESS SKILLS	EXPLORATION AND ASSESSMENT	PG.	RESOURCES AND FEATURES
Unit Opener	192		observing, discussing	For Your Journal	193	Science Sleuths: *A Day at the Races* Videodisc Activity Sheets TRB: Home Connection
UNDERSTANDING FORCES	194			Challenge Your Thinking	209	
1 Forces at Work	194	2	observing, inferring, recognizing relationships, analyzing	Exploration 1	198	Image and Activity Bank 4–1
2 Different Kinds of Forces	200	2	observing, analyzing, inferring, classifying	Exploration 2	200	Image and Activity bank 4–2
3 What Makes Things Fall?	204	2	recognizing relationships, problem solving, creative writing, analyzing			Image and Activity Bank 4–3 TRB: Activity Worksheet 4–1 Graphic Organizer Transparency 4–1
MEASURING FORCES	210			Challenge Your Thinking	224	
4 The Tools to Use	210	3	measuring, comparing, predicting, extrapolating	Exploration 3 Exploration 4 Exploration 5	210 212 214	Image and Activity Bank 4–4 TRB: Resource Worksheets 4–2 and 4–3 Graphic Organizer Transparency 4–2
5 Estimating the Size of a Force	216	2	estimating, measuring, making a table, predicting	Exploration 6 Exploration 7	217 218	Image and Activity Bank 4–5 TRB: Resource Worksheet 4–4 Graphic Organizer Transparency 4–3
6 How Strong Are You?	220	2 to 3	estimating, predicting, measuring, recording			Image and Activity Bank 4–6
FRICTIONAL FORCE	225			Challenge Your Thinking	235	
7 Take a Look at Friction	225	2 to 3	designing an experiment, controlling variables, formulating hypotheses, interpreting data	Exploration 8	229	Image and Activity Bank 4–7 Resource Transparency 4–4
8 Causes and Consequences of Friction	231	2	observing, experimenting, drawing conclusions, creative writing	Exploration 9 Exploration 10	231 233	Image and Activity Bank 4–8 TRB: Activity Worksheet 4–5 Graphic Organizer Transparency 4–5
GOING THROUGH THE MOTIONS	236			Challenge Your Thinking	255	
9 Thought Experiments	236	2	observing, inferring, analyzing, drawing conclusions			Image and Activity Bank 4–9
10 Puffball: A Game of Force and Motion	242	1 to 2	playing a game, predicting, analyzing, drawing conclusions			
11 What Is Inertia?	246	2	predicting, experimenting, observing, analyzing	Exploration 11 Exploration 12	248 250	Image and Activity Bank 4–11
12 Forces in Pairs	251	2 to 3	observing, inferring, predicting, analyzing	Exploration 13	254	Image and Activity Bank 4–12 TRB: Activity Worksheet 4–6 Graphic Organizer Transparency 4–6
End of Unit	256		applying, analyzing, evaluating, summarizing	Making Connections TRB: Sample Assessment Items	256	Science in Action, p. 258 Science and Technology, p. 260 *SourceBook,* pp. S61–S76

*Time given in number of class periods.

Meeting Individual Needs

Gifted Students

Have students make a model satellite launcher. Materials they will need include: a bucket, a basketball, a coat hanger, a sinker or other mass, a 30-cm piece of string, masking tape, and a test tube. Give them the following instructions: Place the ball in the bucket. Bend the hanger so that 30 cm of it is straight and the rest is curved into a circular shape. Using masking tape, secure the circular part of the hanger on the ball so that the straight, 30-cm length stands upright. Attach the sinker to the string. Tape the other end of the string to the bottom of the test tube. Place the test tube upside down on the top of the upright hanger. The ball represents the earth, and the sinker represents the satellite. Find out what happens when the satellite is launched in the following ways:

- With a slight tap, push the sinker up and away from the surface of the ball. *(This is what happens when an object is projected at low speed straight up from the earth. The object falls back to its starting point.)*
- With a slight tap, push the sinker off the surface of the ball at an angle. *(The mass moves away from the ball and falls back at some distance from the starting point.)*
- With a strong tap, push the sinker off the surface of the ball at an angle. *(The mass circles the ball and lands.)*

Conclusion: This shows three ways to launch a satellite. The third way is the only one that gives it enough force to orbit.

LEP Students

1. Give students the opportunity to experiment further with noncontact forces. Provide the following materials: magnets, metal washers, various objects that magnets will and will not attract, scissors, bits of paper, wool, rubber balloons, and string. Tell students to find as many different ways as they can to make the objects move without touching them. Have students summarize their findings in posters.
2. Supply students with a list of new terms from the unit. These could include: force, noncontact forces, weight, mass, gravitational force, newton (N), inertia, action force, and reaction force. Suggest that they make a bilingual dictionary by writing one English term on each page. They should illustrate each term, and then students should translate the term into their native language.
3. At the beginning of each unit, give Spanish-speaking students a copy of the *Spanish Glossary* from the *Teacher's Resource Binder*. Also, let Spanish-speaking students listen to the *English/Spanish Audiocassettes.*

At-Risk Students

1. Use Lesson 6 from the section *Measuring Forces* as a basis for a hands-on testing of the strength of forces. Students should work in pairs or small groups and measure the forces involved in a force-meter tug-of-war.
2. In order to illustrate the effect of forces on motion, students can have a car race with a partner using the following materials: small toy cars, oil, foil, some cloth, wax paper, sandpaper, and 2 wooden ramps. Ramps should be propped up with books so that they are at the same angle. Have students find out how slow they can make the cars go—the slowest car wins. Students should discuss strategies with their partners. Then they should list the things that made the cars go slower. In a second race, the students should attempt to make their cars go as fast as possible. Students should make a list of what causes the car to move quickly.

Cross-Disciplinary Focus

Technology in Space

Travel in outer space, in the absence of gravity, poses many technical problems. For example, how can you eat or drink without the food and drink escaping and floating around in the space capsule? How do you sleep at night without floating out of bed? How do you keep your muscles from weakening? Have interested students find out how scientists have solved these and other problems created by the absence of gravity. Encourage them to think of some of their own designs for making space travel more convenient and comfortable.

Creative Writing

Have interested students use what they have learned about force and motion to write a science fiction story about a trip to another planet. The concepts that could be included in the story are:

- the mass of an object always stays the same, even though its weight changes
- friction and gravity depend on your location in space
- even in space, every action force has a reaction force

 ### Cross-Cultural Focus
Student-Based Vocabulary

Have students generate synonyms for the words "force" and "motion." Slang terms, words in foreign languages, phrases from songs, and other popular sources are appropriate responses. For each word given, the student should use the term in a complete sentence. Compile a list of these words, and use them when introducing new concepts in this unit. This can be a fun activity and may help relate the new concepts to the students' own vocabulary.

A Japanese Train that Travels on Air

Have interested students find out about the magnetically levitated train that is being built in Sakaigawa, Japan. The train, which will run from Tokyo to Osaka, will travel at record speeds. The only friction that will slow it down will be air friction. Students should do research to find out about transportation technologies in other countries. Have students find out how these transportation systems suit the needs of the countries they investigate.

 ### Bibliography for Teachers

Gunstone, R.F. and M. Watts. *Force and Motion: Children's Ideas in Science.* London, England: Open University Press, 1985.

Lawrenz, F. "Misconceptions of Physical Science Concepts Among Elementary School Teachers." *School Science and Math,* 86, (1986): 654-660.

McCloskey, M. "Intuitive Physics." *Scientific American,* 248, (1983): 122-130.

 ### Bibliography for Students

Darling, David. *Could You Ever Travel to the Stars?* Minneapolis, MN: Dillon Press, 1990.

Hancock, Ralph. *Understanding Movement.* Morristown, NJ: Silver Burdett & Ginn, Inc., 1984.

Stannard, Russell. *The Time and Space of Uncle Albert.* New York, NY: Holt, Rinehart & Winston, 1990.

Zubrowski, Bernie. *Building and Experimenting with Spinning Toys.* New York, NY: William Morrow & Co., Inc., 1989.

 ### Films, Videotapes, Software, and Other Media

Airplanes and How They Fly.
Film and Video.
Coronet Film and Video
108 Wilmot Road
Deerfield, IL 60015

Buoyancy.
Film and Video.
Coronet Film & Video
108 Wilmot Road
Deerfield, IL 60015

Introductory Mechanics.
Software.
For Apple II.
Conduit
University of Iowa
Oakdale Campus
Iowa City, IA 52242

Isaac Newton & Full Graphic Newton
Software.
Krell Software
Flowerfield Bldg. #7
Suite 1D
St. James, NY 11780

Newton's First Law, Newton's Second Law, Newton's Third Law.
Software.
For Apple II and IBM PC.
Prentice-Hall Allyn and Bacon
Sylvan Avenue
Englewood Cliffs, NJ 07632

UNIT 4

Force and Motion

✹ Unit Focus

Before beginning the unit, involve students in a discussion of their ideas about force and motion. Present them with some representational diagrams. For example, draw a diagram of a girl on a bicycle. Tell students that she is not pedaling or pressing on the brakes, yet the bicycle is slowing down. Ask: Are there forces acting on the bicycle? Why do you think so? *(There is the force of air friction and rolling friction opposing the forward motion of the bicycle. There is the force of gravity pulling downward on the bicycle, and the force of the road pushing up on the bicycle. There is the force of the girl pushing down on the bicycle, and the upward push of the bicycle on the girl.)*

✹ About the Photograph

This photograph shows the exact moment of impact between a ball and a pane of glass. At this instant, the force of the fast-moving ball shatters the glass.

Special photographic techniques are required to capture events such as this—high-speed film, special shutters, high-intensity light. Until recently, taking such a photo would not have been possible.

UNDERSTANDING FORCES
1. Forces at Work, p. 194
2. Different Kinds of Forces, p. 200
3. What Makes Things Fall? p. 204

MEASURING FORCES
4. The Tools to Use, p. 210
5. Estimating the Size of a Force, p. 216
6. How Strong Are You? p. 220

FRICTIONAL FORCE
7. Take a Look at Friction, p. 225
8. Causes and Consequences of Friction, p. 231

GOING THROUGH THE MOTIONS
9. Thought Experiments, p. 236
10. Puffball: A Game of Force and Motion, p. 242
11. What is Inertia? p. 246
12. Forces in Pairs, p. 251

For Your Journal

Write a paragraph summarizing your thoughts about each of the following:
1. How can you recognize a force?
2. You know that the surface of the earth changes constantly—trees fall in a forest, mountains erode, and rivers change course. What forces are responsible?
3. How are forces at work in this picture?

✸ Using the Photograph

Ask students to imagine the event taking place in this photograph in slow motion. Then ask them to describe the scene, step-by-step, in terms of the interaction of forces. *(The students' answers may start with how the ball was given the force in the first place—presumably with the swing of a bat. However, the forces that can be identified in this photo include the action force of the ball and the reaction force of the glass. You would be able to see that when the ball hit the glass, the glass actually pushed back against the ball before breaking. The force on the glass was great enough to start the pane moving, while the force on the ball was not great enough to stop the motion of the ball. Therefore, the pane shattered and the ball continued to travel forward after passing through the glass.)*

✸ About Your Journal

Students should answer the Journal questions to the best of their abilities.
These questions are designed to serve two functions: to help students recognize that they do indeed have prior knowledge about the topic, and to help them identify any misconceptions they may have. In the course of studying this unit, these misconceptions should be dispelled.

Force and Motion

LESSON 1

✵ Getting Started

In this introductory lesson, students read an action-filled story about a sailboat race. Then they identify the forces that are described in the story, the objects that are exerting the forces, the objects receiving the forces, and the effects of these forces. In *Exploration 1*, they observe photographs of forces at work and identify the agents, receivers, and effects of these forces.

Main Ideas

1. A force is a push or a pull exerted by one object on another.
2. A force can alter the shape or the motion of an object.
3. A force may be represented by an arrow since it has both direction and size.

✵ Teaching Strategies

The Race

Call on a volunteer to read aloud the sailing adventure on page 194. Allow students some time to look at and enjoy the photographs on pages 194 and 195.

Ask students if they have ever been sailing. If there are students who have sailed, ask them to describe their experiences and to explain how the sails must be manipulated to catch the wind.
(Continues on page 196)

UNDERSTANDING FORCES

1 Forces at Work

The Race

Today was the day of the race! Kathleen and Kimiko were excited! With difficulty, they dragged their sailboat across the sand into the water.

A favorable west wind was blowing. The craft was ready to go. Their hearts beating hard, they pulled on the halyards with nimble fingers to raise the sails.

With one hand, Kimiko took control of the tiller. With the other, she tugged on the sheet for the mainsail. Kathleen, sitting mid-ship, secured the sheet of the jib.

As a gentle breeze pushed on the sails, the craft moved slowly and steadily out beyond the jetty. There the full force of the wind stretched the sails. The craft suddenly sped up. They were off!

The boat shuddered as the choppy waves slapped and sprayed the hull. With the experience of a summer's sailing, the girls sailed toward the starting line. Just as they reached the line, the starter's flag was lowered. The race was on! Kimiko pushed on the tiller. The rudder twisted against the water. Their course was set. The sails strained and the mast creaked as the boat skimmed and broke the water's surface. Kathleen pushed downward on the centerboard to offset the leeway. Passing the second red buoy on the starboard, Kimiko decided to make a quick turn. She drew the tiller toward herself, hauled in the sheet at the same time, and hollered "Coming About!" The boat swung around until the wind was at the stern. The boom of the mainsail swung suddenly from one side of the craft to the other. In an instant, the boat changed its course.

Kathleen, who was watching another boat coming up very rapidly on the leeward side, responded too late to the warning. The boom smacked her arm. She fell over the side and with a splash hit the water.

Quickly, almost automatically, Kimiko brought the boat up into the wind. The sails flapped harmlessly; the boat slowed down. . .

LESSON 1 ORGANIZER

Objectives

By the end of the lesson, students should be able to:

1. Identify forces as pushes or pulls exerted by one object (the agent) upon another object (the receiver).
2. Infer what the effect of a force will be on the receiver, such as changing its shape or motion.

Process Skills

observing, inferring, recognizing relationships, analyzing

New Terms

Force—a push or pull exerted by one object upon another.

194 Force and Motion

Materials

Exploration 1: Journal

Science Discovery Videodisc

Disc 2, Image and Activity Bank, 4–1

Time Required

two class periods

Understanding Forces

(The Race continued)

Definitions of the sailing terms:

- Halyards—ropes used to haul up the sails.
- Tiller—lever at the back of a craft used for steering.
- Mainsail—the principal sail on the mast of the boat.
- Sheet—a rope for holding and controlling a sail's position.
- Jib—forward sail of a boat.
- Hull—the frame or body of a boat.
- Rudder—a vane in the water to which the tiller is attached; it enables a boat to be steered.
- Centerboard—a straight, adjustable board that extends beneath the boat into the water; it increases stability when a boat tips or slips sideways.
- Leeway—the drift of a boat toward the leeward (away from the wind) side.
- Buoy—a marker in the water that indicates the port side of a channel.
- Starboard—the right side of a boat.
- Stern—the back end of the boat.
- Boom—the horizontal support that stretches the sail; it is attached to one end of the mast.
- Leeward—toward the side of the boat turned away from the wind.
- Jetty—a piece of land jutting out into the water.
- Tack—to sail into the wind first in one direction and then in another to make forward progress in the direction of the wind.

Recognizing Forces

Action words (verbs) may refer to forces (pushes or pulls) or may describe motion, which is caused by a force. Ask students to distinguish between force words and motion words as they write their list of words.
Action words referring to forces: dragged, blowing, beating, pulled, hit, tugged, pushed, stretched, twisted.
Action words referring to motion: moved, sped up, skimmed, broke, fell, swung (around).

The sailing terms used in this story about Kathleen and Kimiko may be new to you. Look up the definitions for the terms below, or ask someone who sails. Then read the story again.

- halyards
- tiller
- mainsail
- sheet
- jib
- hull
- rudder
- centerboard
- leeway
- buoy
- starboard
- stern
- boom
- leeward
- jetty

Recognizing Forces

Sailing is an exciting, fast-paced activity. Many forces are at work as a sailboat skims across the water. Sailors put out tremendous effort and action to keep their boats sailing. Notice how many action words appear in the story above: pushed, hauled, smacked, and slapped, just to name a few. Write a list of all the action words in the story and all the forces that you recognize.

Questions

1. Can you have a force without some object (someone or something) exerting the force and some other object receiving the force?
2. Name some effects that forces have on objects.
3. Is it possible for one object to exert a force on another without ever touching it? Where are there examples of this in the story?
4. Examine the story for pairs of opposing forces, which are forces that work against each other. How many did you find?
5. How would you define or describe a force? Make sure your definition includes all the things you have discovered about forces.
6. Write an exciting conclusion to the story. Illustrate the events of your story that involve forces. Then identify the forces found in your drawing.

Answers to Questions

1. No. A force is a push or a pull that one body exerts on another. Sometimes, the object exerting the force and/or the object receiving the force are only inferred. In the phrase, "over the side she fell," students should infer that the earth's gravity pulled her down.
2. Forces are able to change the shape of an object (the sail), give motion to an object at rest (move the sailboat), and alter the speed or direction of an object in motion (when the craft suddenly sped up).
3. Yes. One object can exert a force on another without direct contact. In the story, Kathleen lost her balance and fell over the side. The earth exerted a gravitational force on Kathleen. This is an example of force at a distance.

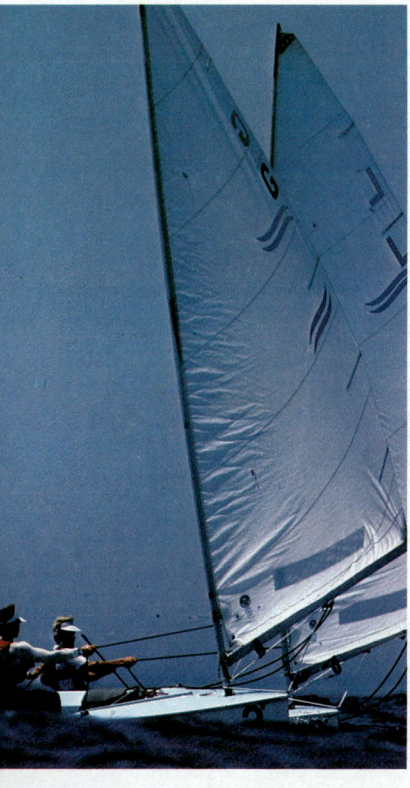

Describing Forces

The passages below are taken from the story of the boat race. Pay careful attention to the underlined words and the words in italics. Then answer the questions, and see what you can discover about forces.

1. What is similar about the forces described in these sentences?
   (a) "With one hand, Kimiko took control of the tiller. With the other, <u>she</u> *tugged* on the <u>sheet</u> . . ."
   (b) "As a gentle <u>breeze</u> *pushed* on the <u>sails</u> . . ."
   (c) "The <u>boom</u> *smacked* her <u>arm</u>."
2. What do these sentences tell you about the effects of forces?
   (a) ". . . the full force of the wind *stretched* the sails."
   (b) "The craft suddenly *sped* up."
   (c) "In an instant the boat *changed* its course."
   (d) "She *fell* over the side . . ."
   (e) ". . . the boat *slowed down* . . ."
3. Think of other examples of forces that could cause an object to do the following:
   (a) change its shape   (c) speed up or slow down
   (b) begin to move   (d) change direction
4. What aspect of force does this quotation describe? "Kathleen pushed *downward* on the center board . . ."

The Direction of Forces

In the story of the race, the directions of most of the forces are not described. When scientists make diagrams of forces, they often use an arrow to indicate the direction of a force. The length of the arrow gives you an idea of the size of the force. The tail of the arrow indicates the point where the force is exerted.

Which way do you think the table will move?

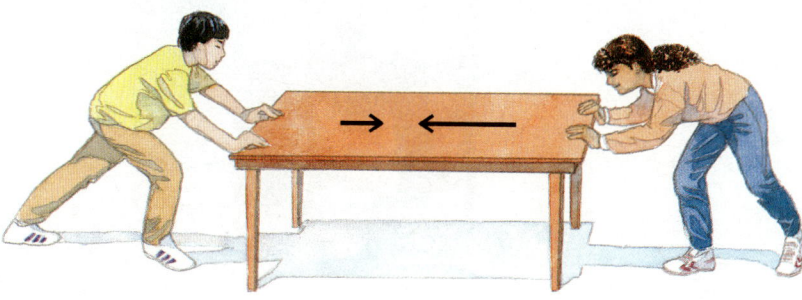

4. Opposing forces are implied in the story. For example, the force of the wind on the boat is opposed by the frictional force of the boat against the water. When the two forces are balanced, the boat moves at a constant velocity. If the wind's force is less than the opposing water resistance, the boat will slow down.

5. A force is a push or pull that one object exerts on another.
6. Be sure students have identified the forces that they have included in their stories. Make a bulletin board display with the stories and illustrations.

Answers to
Describing Forces

1. In these sentences, one object (Kimiko, a breeze, the boom) exerted a push or pull on another object (the sheet, the sails, her arm).
2. The effects of forces can:
   (a) change the shape of an object
   (b) increase the speed of an object
   (c) change the direction of the motion of an object
   (d) cause an object at rest to fall
   (e) decrease the speed of an object
3. Examples of forces that could cause an object to
   (a) change its shape: an air pump inflating a football, wind blowing against a tree
   (b) begin to move: a person opening a door, a baseball pitcher hurling a ball
   (c) speed up or slow down: a person pushing another on a swing, a cyclist pushing harder on the bike pedals, a driver pushing the brake pedal of a car, a person turning off the motor of a moving boat
   (d) change direction: a wind blowing a paper airplane from its course, a hockey puck striking a skate blade during a goal shot
4. The quotation shows that forces have a direction.

The Direction of Forces

The table in the illustration would move to the left since the force arrow pointing to the left is longer. Have students demonstrate opposing forces on a desk or book. Ask them to make a sketch of the demonstration and to use arrows to illustrate the forces that were exerted.

Understanding Forces

Answers to
Exploration 1

Counterclockwise from top left:
- **(a)** *Agent:* the tackle; *receiver:* the player with the ball; *effect:* the forward motion of the ball-carrier is stopped.
- **(b)** *Agent:* the hammer; *receiver:* the wedge; *effect:* the wedge is driven downward into the wood, splitting it.
- **(c)** *Agent:* the balloons; *receiver:* the girl's arm; *effect:* the downward pull of gravity on the girl's arm is slightly offset by the upward pull of the balloons.
- **(d)** *Agent:* the parachute; *receiver:* the drag-racer; *effect:* the forward motion of the drag racer is slowed.
- **(e)** *Agent:* the motorist; *receiver:* the car (also the jack); *effect:* the car is lifted upward by the jack.
- **(f)** *Agent:* the girl; *receiver:* the boy; *effect:* the girl is hurled backward and the boy rolls forward.
- **(g)** *Agent:* the archer; *receiver:* the bow; *effect:* the bowstring stretches and the bow resists by deforming (bending).

In many photos, the forces of gravity and/or friction are at work. For example, the combined forces of the tackle and gravity cause the ball-carrier to fall. Students should recognize that most receivers have several different forces acting on them at the same time.

EXPLORATION 1

A Picture Puzzle

Several forces are illustrated in these pictures. For each picture, pick out the object (the agent) that is having a noticeable effect on another object (the receiver). In your Journal, sketch each situation, and place arrows indicating the direction of each force and the point where each force is being applied. Write the name of the agent and the receiver, and describe the effect of the force.

Are there cases where two forces are acting on the same object? What is their combined effect on the object?

(a)

(b)

(c)

(d)

198   Force and Motion

(e)

(g)

(f)

🟊 Follow Up

Assessment
Perform a series of demonstrations of forces at work. For example, pound a nail into a piece of wood, blow on a piece of tissue paper, pull a wagon, smash a piece of clay, pick up a book, and push a door. For each demonstration, ask students to identify the agent, the receiver, and the effect of the force.

Extension
Take students on a walk around the school. While they are on the walk, ask them to observe and list as many agents, receivers, and effects of forces as possible.

Understanding Forces 199

LESSON 2

✷ Getting Started

In this lesson, students are introduced to a variety of different forces. These include buoyant, magnetic, electrical, gravitational, elastic, and frictional forces.

Main Ideas

1. There are many types of forces, including buoyant, magnetic, electrical, gravitational, elastic, and frictional.
2. Some forces, including buoyant, elastic, and frictional forces, are contact forces. Other forces, including magnetic, electrical, and gravitational, are noncontact forces.

✷ Teaching Strategies

Tape paper clips to the bottom of a toy boat. Place the boat in a shallow pan of water. Ask students if they can think of a way to get the boat to move. They might suggest pushing or pulling it, blowing on it, and so on. Move a magnet around underneath the pan. Explain that magnets exert a noncontact force. Explain that in this lesson, they will be learning about many different kinds of contact and noncontact forces.

Exploration 2

Set up one experiment at each station around the classroom. (For large classes, two setups of each station would be preferable.) Divide the class into small groups. Instruct them to move from one station to another at 6-minute intervals.

2 Different Kinds of Forces

EXPLORATION 2

You're familiar with the many different game booths at a carnival. For this Exploration, you will test your skill at a carnival of science.

Experiencing Forces

The seven mini-experiments in this Exploration provide examples of specific forces at work. Each one is set up at its own station in the classroom. At each station you will find materials arranged as in the illustrations on pages 201 to 203. Your purpose at each station is to discover which forces are at work. As you work through the mini-experiments, think about these questions:

- What is the agent, receiver, and effect of each force?
- Are there any new types of forces?
- How are they different?
- Are there any *noncontact forces*—forces exerted by an agent that does not touch the receiver?

Record your observations and conclusions in your Journal.

You Will Need

- an empty jar
- a pail
- water
- a glass
- paper clips and rubber bands
- a magnet
- a plastic bread bag, cut in strips
- a sock
- hooked masses
- string
- books
- drinking straws
- a bottle
- newspapers and a cloth towel

LESSON 2 ORGANIZER

Objectives

By the end of the lesson, students should be able to:
1. Identify and classify different types of forces.
2. Distinguish between contact and noncontact forces.
3. Analyze the forces at work in several new situations.

Process Skills

observing, analyzing, inferring, classifying

New Terms

Noncontact forces—forces exerted by an agent that does not touch the receiver.

Force and Motion

Station 1

What to Do

1. Push the empty jar slowly into a pail of water. Keep its open end facing up.
2. Submerge it; let the water fill the jar.
3. Now take the jar out of the water.

What forces do you feel at each step?

Station 2

What to Do

1. Bring a strong magnet near the glass container.
2. Try to lift the paper clips to the top of the glass.
3. Lift the container off the table.

Think about the forces involved.

Station 3

What to Do

1. With one hand, hold up two long strips cut from a plastic bread bag. Observe them as they hang freely.
2. Now rub both sides of the strips with a wool sock. Allow the strips to hang freely again. Observe any forces at work.
3. Bring one of the strips near a table or a wall. What happens?

Identify the forces acting on the plastic strips.

Answers to

In-Text Questions

Station 1
- agent: person's hand and the water; receiver: the jar; the effect: as the jar is pushed into the water, the water exerts an opposing upward buoyant force upon the jar. If the jar is released, it pops upward. As the filled jar is lifted upward, the opposing downward force (the weight of the jar and water) is apparent.
- new force: buoyant force
- noncontact force: none

Station 2
- agent: the magnet; receiver: the paper clips; effect: the magnet, if it is strong enough, attracts the paper clips even though the glass is between it and the paper clips. These forces are magnetic.
- new force: magnetic force
- noncontact force: magnetic force

Station 3
- agent: electrons (static electricity); receiver: the plastic strips; effect: when the plastic strips are rubbed with a wool sock or piece of fur, they move apart from each other. Also, the strips tend to be attracted to whatever else is close by. These pulling and pushing forces, which are exerted at a distance, are electrical.
- new force: electrical force
- noncontact force: electrical force

(Answers continue on next page)

Materials

Exploration 2
Station 1: empty jar, pail, water
Station 2: magnet, drinking glass, paper clip
Station 3: scissors, plastic bread bag, wool sock
Station 4: several sheets of paper
Station 5: rubber bands, hooked masses, soft object
Station 6: rubber bands, scissors, paper clips, 2 books, string, drinking straws, metric ruler
Station 7: bottle, newspaper, cloth towel, board, book, metric ruler

Science Discovery Videodisc

Disc 2, Image and Activity Bank, 4–2

Time Required

two class periods

Understanding Forces 201

Answers continued

Station 4
- agents: gravity (the earth) and air resistance; receiver: paper; effect: there is a downward force (gravity) pulling on the pieces of paper. There is also an upward force (air resistance) pushing on the paper, especially on the uncrumpled piece. The single sheet of paper falls more slowly because of air resistance.
- new force: none
- noncontact force: gravity

Station 5
- agents: the rubber band and the earth; receiver: the masses; effect: the rubber band is exerting an upward force on the masses, and the masses are exerting a downward force on the rubber band.
  new force: elastic force
- noncontact force: gravity

Station 6
- agent: hand pulling rubber band and rubber band pulling the book; receiver: the book; effect: the rubber band applies a pulling force on the books. The books apply a pulling force on the rubber band. It takes more force to start two books moving than it did for one book. The friction between the books and the table opposes the forward motion of the books. The opposing force of friction is lessened by the use of the drinking straws.
- new force: none
- noncontact force: none

Station 7
- agent: gravity (the earth) and the surface on which the bottle rolls; receiver: the bottle; effect: the force of friction opposes the motion of the bottle. The amount of frictional force depends upon the type of materials over which the bottle rolls.
- new force: none
- noncontact force: gravity

EXPLORATION 2—CONTINUED

Station 4
What to Do
1. How fast does paper fall? Why does it fall? Try dropping the following:
   - (a) a single sheet of paper
   - (b) a piece folded into quarters
   - (c) a piece crumpled into a ball
2. Try two different pieces at the same time. Do some fall faster than others?

What are the forces involved here?

Station 5
What to Do
1. Hang a hooked mass on the rubber band.
2. Add more masses. How many are needed to break the band? Let the masses drop onto something soft.
3. Replace the broken rubber band with a new one so the next person can try the experiment.

Identify all forces.

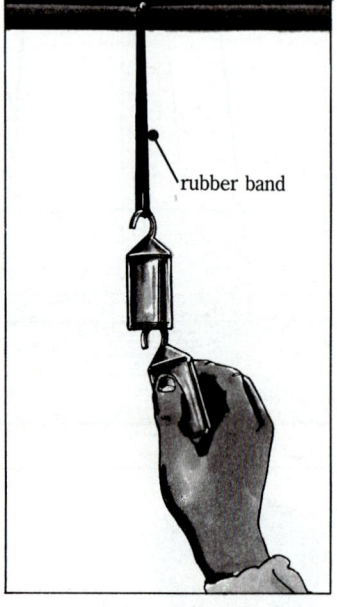

rubber band

Station 6
What to Do
1. Cut a rubber band and tie it securely to a paper clip. Attach the clip to a string tied around a book.

   How long must the band stretch before the book starts moving?
2. Try it again with another book placed on top of the first one. What do you notice?
3. Place the books on drinking straws and repeat the experiment. How has the opposing force changed?

What tends to prevent the books from moving?

---

202 Force and Motion

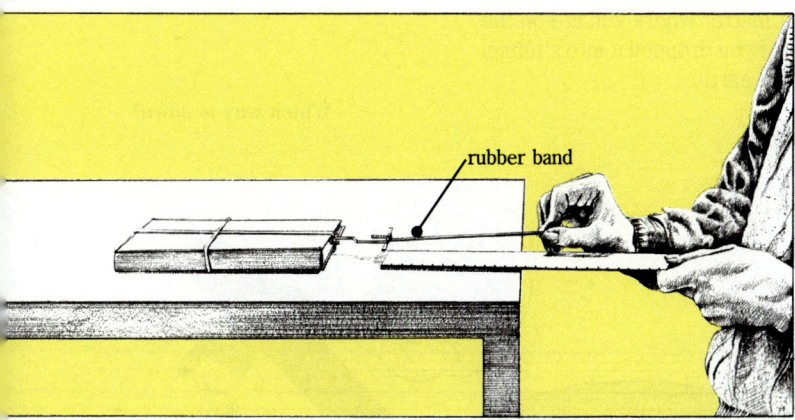

rubber band

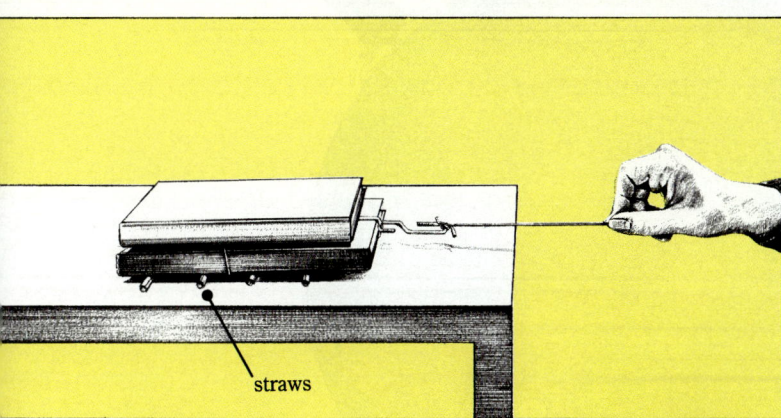

straws

Station 7
What to Do

1. Find out how far a bottle will roll on the floor before stopping.
2. Place several sheets of newspaper on the floor and try again.
3. Repeat the experiment by rolling the bottle over other surfaces, such as a cloth towel spread on the floor.

What is slowing down the bottle?

Which Forces Did You Meet in the "Carnival of Science"?

1. Water exerted an upward **buoyant force** on the empty bottle at Station 1. Where did you see this force before?
2. A **magnetic force** was present at Station 2. Did the magnet need to touch the paper clips to exert an attractive force on them?
3. An **electrical force** acted upon the charged plastic strips at Station 3. Each strip exerted a force on the other. Is this a noncontact force like the magnetic force?
4. A **gravitational force** acted upon the sheets of paper at Station 4. Is this a non-contact force? What agent is exerting the force? Why did the uncrumpled sheets fall more slowly?
5. The rubber bands at Stations 5 and 6 exerted an **elastic force**. Where else could you find elastic forces?
6. **Frictional forces** were exerted on the bottle and book at Stations 6 and 7. As an object moves in one direction, frictional forces oppose the motion. In what direction was the frictional force at Stations 6 and 7 acting?

starting point
board
floor

Answers to
Which Forces Did You Meet in the "Carnival of Science"?

1. Floating or swimming are examples of a buoyant force. The sailboat in the story was kept afloat because of the buoyant force of the water.
2. The magnet does not have to touch the paper clips to attract them. This is an example of a noncontact force.
3. Each strip exerts a noncontact force on the other. That is, the forces act in pairs.
4. The earth exerts a gravitational force that pulls the paper to the ground. It is a noncontact force, since the earth does not have to touch the paper to exert a pulling force on it. The uncrumpled sheets fall more slowly due to their larger surface areas. The air puts more frictional force on them.
5. Other examples of elastic forces are: the ropes and sail being stretched by the wind in the story and the elastic force of the bow acting upon an arrow.
6. Friction acts in a direction opposite to the direction of the motion of an object.

✱ Follow Up

Assessment

1. Ask students to name an example of a buoyant force, an electrical force, a frictional force, and an elastic force. In each case, ask them to identify the agent and the receiver of the force, and to state what effect the force has or might have on the receiver.
2. What is the most obvious type of force (electrical, frictional, and so on) in each situation below? Explain your choice.
   a) a jack in the box *(elastic)*
   b) a gymnast on a trampoline *(elastic)*
   c) deep-sea diving *(buoyant)*
   d) parachuting *(frictional and gravitational)*
   e) a skier heading down a mountain *(gravitational)*

Extension

1. Have students adjust a faucet so that a very thin stream of water flows from it. Give a comb a charge by running it through your hair several times. Hold the comb 2 or 3 cm from the stream of water. (*The water should be strongly attracted by the charge of the comb.*)
2. Have students cut a strip of thin cardboard about 2 cm by 10 cm. Fold it in half lengthwise and balance it on a pencil point. The pencil point should indent but not puncture the cardboard, so that the cardboard can turn easily. Put a charge on a comb by running it through your hair. Hold it near one end of the cardboard strip. (*The cardboard turns on the pencil point and moves toward the comb.*)

Understanding Forces 203

LESSON 3

✳ Getting Started

In this lesson, students focus on the concept of gravity. In the discussion of Newton's laws, the concepts of mass and weight are developed.

Main Ideas

1. Every object in the universe exerts an attractive force on every other object.
2. The size of the gravitational force that one object exerts on another depends upon the masses of the objects and the distance between them.
3. The weight of an object changes with its location in the universe; the mass of an object remains constant.

✳ Teaching Strategies

Demonstrate dropping a ball to the students. Release it so that it simply falls to the floor. Then throw the ball a few feet across the room. Ask students to name the different forces that occur in each activity. *(In dropping the ball, gravity and air friction act on the ball. In throwing the ball, gravity, air friction, and the force exerted by your arm are all acting on the ball.)* Throw the ball again and ask students to trace the path that the ball takes until it hits the ground. Point out to students that gravity is acting on the ball. Otherwise the ball would continue moving out into space.

Call on a volunteer to read aloud page 204. Involve the class in a discussion of the questions. Sample answers follow on the next page.

3 What Makes Things Fall?

You are holding a ball in your hand. You let it go. It falls. What is the direction of its motion? Is there a force acting on the ball? If so, what agent exerts this force? What is the direction of the force in relation to you? to the earth? Does it matter where you are on the earth? What would happen to the ball if you dropped it into a tunnel that passed through the center of the earth?

Which way is down?

LESSON 3 ORGANIZER

Objectives

By the end of the lesson, students should be able to:
1. Understand that all objects exert an attractive force on one another.
2. Recognize how the masses of two objects and their distance from each other affects the gravitational force between them.
3. Distinguish between weight and mass.

Process Skills

recognizing relationships, problem solving, creative writing, analyzing

New Terms

Gravitational force—this force causes objects to move toward each other.
Mass—the amount of matter present in an object.

204 Force and Motion

Blow up a round balloon to use as a model of the earth. Rub the whole surface of the balloon with a cloth or against your hair. Now hold little pieces of paper near its surface at different places. Release the paper.

Does the paper fall toward the balloon? What kind of force is acting on the paper? What is the agent of this force?

The balloon (or earth model) exerted an *attractive* force on the bits of paper. It "pulled" the paper toward its surface. The earth appears to do something similar to the ball. The earth exerts an attractive force on the ball and pulls the ball toward its surface.

The forces exerted by the model and the earth itself are similar in two ways: both are attractive, and both are noncontact forces. However, there is an important difference between the earth and the model.

The attractive force exerted by the balloon is *electrical*. You gave the balloon an electrical charge by rubbing it against your hair or the cloth. The force exerted by the earth, however, is not electrical. It is an attractive force that all objects made of matter exert on each other. Because the earth is made of matter, it is able to attract other things made of matter.

The force exerted by the earth is called **gravitational force**. This force causes objects to move toward each other. The gravitational force the earth places on an object is called the **weight** of the object. To find out where the theory of gravitational force came from, read the following story.

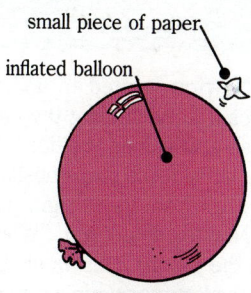
small piece of paper
inflated balloon
electrical force

gravitational force

Weight—the gravitational force the earth places on an object.

Materials

What Makes Things Fall? ball, round balloon, piece of cloth, bits of paper

Teacher's Resource Binder

Activity Worksheet 4–1
Graphic Organizer Transparency, 4–1

Science Discovery Videodisc

Disc 2, Image and Activity Bank, 4–3

Time Required

two class periods

Answers to

In-Text Questions (page 204)

- The ball moves downward when it is released because the force of gravity is acting on the ball.
- The earth exerts this force.
- The direction of the force is downward in relation to a person standing on the earth.
- The direction of the force is toward the center of the earth in relation to the earth.
- No matter where you are on the earth, the direction of the force of gravity is always toward the center of the earth.
- If a ball were dropped into a tunnel that passed through the center of the earth, it would oscillate back and forth from one side of the earth's surface to the other. Air friction would oppose its motion, shortening its path at each oscillation, until it would come to rest at the center of the earth.

Answer to Caption: "Down" in this picture means toward the center of the earth.

Divide the class into small groups. Distribute balloons, pieces of cloth, and pieces of paper to each group. Ask them to read the top two paragraphs on page 205 and try the balloon-earth model.

Reassemble the groups and ask: How is the model similar/dissimilar to the real earth? *(Both the balloon and the earth exert a pulling, noncontact force directly toward the surface. The balloon exerts an electrical force, while the earth exerts a gravitational force.)*

Call on a volunteer to read aloud the remainder of page 205. Be sure all students are comfortable with the meanings of the new terms *gravitational force* and *weight* before proceeding to page 206.

Understanding Forces

A Striking Discovery

Use a bowling ball and a rubber mallet to demonstrate how a moving ball changes direction when a force is applied to it. First, ask students to predict the direction the moving ball will take after it is hit with the mallet. Many will think that it will move in the direction of the applied force. However, the ball is already moving, so it will take a direction in between its initial direction of motion and the direction of the applied force.

Explain to students Newton's thought experiment that describes how satellites orbit the earth. Begin by drawing successive pictures of stones being thrown from the top of a mountain with greater and greater force. Tell students that each stone is thrown with more force than the one before it. The first stone curves in an arc for 1 kilometer before falling to the ground. The second stone curves 10 kilometers, the third curves 100 kilometers, and so on. Explain that if a stone were thrown with enough force, it would continue to fall in an arc that follows the curvature of the earth; that is, it would begin to orbit the earth.

Answers to
In-Text Questions

Earth is the agent of gravitational force, all matter is the receiver of the force, and the effect of the force is that objects fall toward (are attracted to) the earth.

A Striking Discovery

Why do objects fall? If an object starts to move, isn't there a force causing this movement? The scientist Isaac Newton pondered these questions. The old story says that he was sitting under a tree when an apple bounced off his head and set him thinking!

Newton wanted to understand the motion of the moon. Why does it keep circling the earth? Newton already knew that if you push a ball along the floor, it moves in a straight line—even after the initial force stops pushing it. It takes a second force to change the ball from its straight path.

Newton wondered, "What if the earth exerts an attractive force on the moon? This would explain why the moon orbits the earth instead of moving in a straight line." The earth's force causes the moon to change its path just as a force exerted on a moving ball causes the ball to change its path. Then Newton thought, "If the earth exerts an attractive force on an object far away, it must also exert an attractive force on objects near its surface. It is this force that causes objects to fall toward the earth."

Thus, Newton had an explanation for both the apple falling and the moon curving from its straight path. The earth exerts a gravitational force on every object. What are the agent, receiver, direction and effect of this gravitational force?

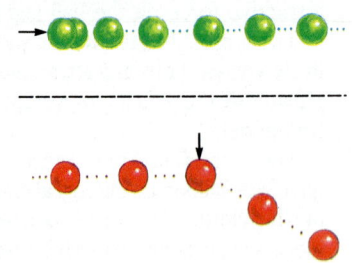

Changing the direction of a moving ball

206 Force and Motion

Newton's Laws

Newton's laws of motion and gravitation were extremely important discoveries. They explained many fundamental properties of our universe. For example, Newton concluded that every object in the universe exerts an attractive force on every other object in the universe. This law applies to huge bodies of matter like stars and planets: the sun attracts the earth; the earth attracts the sun. It also applies to bodies of very different size: the earth attracts the ball; the ball attracts the earth. It is true of small bodies as well: the table attracts the chair; the chair attracts the table. With small objects, the gravitational force is extremely small. With large objects, the force is much greater. The size of the attractive force depends on the size of the object. In other words, the strength of the force depends on how much matter there is in the object. The amount of matter is what scientists call **mass.**

Did Newton ever test his theory? No. That came much later when Lord Cavendish devised an experiment that proved that Newton's theory was correct. He showed that if you doubled the quantity of matter (mass) in one of the bodies, the attractive force also doubled. Cavendish also demonstrated that when bodies were separated farther apart from each other, the force of attraction became smaller.

Dr. X

Write a science fiction story about the future. Here is the scenario: Dr. X has just invented an anti-gravity spray. The harmless-looking liquid can cancel out the gravitational force exerted by the earth.

Now he wants to try it out. He chooses your school and sprays it all over the cafeteria and gym. Your principal rushes out of the office just as . . .

Newton's Laws

Call on a volunteer to read aloud page 207. Students may have difficulty accepting Newton's law of gravitation. It is difficult to imagine that as the earth exerts a pulling force on a rubber ball, the ball also exerts a pulling force on the earth. Ask: If this is so, why does the ball move toward the earth, yet the earth does not move toward the ball? (*The earth has a much greater mass; thus, the pulling force that the ball exerts on the earth is not great enough to make it move a measurable amount.*)

Explain to students that in Cavendish's experiment, large lead balls were brought close to small lead balls. The mutual gravitational attraction between the small and large balls caused a vertical string to twist by a measurable amount.

Answers to
Dr. X

Encourage students to be creative. These stories will help you gauge how well the students have understood the idea of "gravitational force." Students should understand that gravity would no longer hold objects in place. The motion of the earth revolving would exert a force on the objects, causing them to move tangentially away from the earth.

If students do not have time to complete this activity in class, it can be done as a homework assignment. The results can be displayed on a bulletin board.

Understanding Forces 207

A Massive (Or Weighty) Problem

Call on a volunteer to read aloud page 208. Be sure that students understand the difference between mass and weight. *(The mass of an object remains constant throughout the universe. An object's weight depends on the gravitational forces acting on it. An object can become weightless if the earth's gravitational force is negligible, but it cannot become massless.)* Ask students to name examples of situations where it would be appropriate to use one term or the other.

✸ Follow Up

Assessment

1. Ask students to explain why each of the following statements is true:
   (a) You weigh slightly less at the top of a mountain than you do at its foot. *(You weigh less at the top of a mountain because you are farther from the center of the earth. Weight depends on your mass and your distance from the center of the earth.)*
   (b) The moon moves in an elliptical path around the earth. *(The moon revolves around the earth because the earth's gravity is constantly pulling the moon toward it and away from a straight-line path.)*
2. Do the following demonstrations: hang one, then two, then three washers from a rubber band; use an equal-arm balance to measure the mass of three washers; measure your weight with a bathroom (spring) scale. Ask: Which method of weighing objects would give the same results anywhere in the universe? *(The equal-arm balance would give the same result because it is measuring mass, not weight. On the equal-arm balance, the mass of the washers is balanced against the mass of standard masses. The stretch of the rubber band and the spring scale measure the gravitational pull on an object, which is its weight. Therefore, the measurements would change, depending on the object's distance from the center of the earth.)*

A Massive (Or Weighty) Problem

You know that the earth exerts a gravitational force on objects. This force is commonly called the weight of the object. If the earth were to disappear, an object would have no reason to fall. The object would become "weightless." This does not mean that it would no longer contain any matter. The object contains the same amount of matter with or without the earth's pull. That is, an object has the same mass. Obviously, mass and weight are really quite different ideas.

How much do you weigh? Suppose you were to take off in a spaceship and go far above the earth's atmosphere. What would happen to your weight? Remember that Cavendish showed that the attractive force of one object for another decreases as the distance between them increases.

At 6400 km above the earth (the distance from Washington D.C. to Paris, France), you would weigh only ¼ of your weight on earth. Of course, you would still have the same mass.

Earth
object has weight

No Earth
no weight

Extension

1. Have students write a story about a trip to the moon. Their stories should include descriptions of how gravity changes as they travel from the earth's surface to the moon's surface and back again. Encourage them to describe what it would feel like to eat, to move around, and to do simple tasks on the moon's surface. Mention that the moon has one-sixth of the earth's gravity. Have students read their stories to the class. You might want to display their stories on a bulletin board for others to read and enjoy.
2. Have students research and write a report on Newton's original thought experiment in which he hypothesized that the same force that causes an apple to fall also keeps the moon and the planets in orbit.

Force and Motion

Challenge Your Thinking

1. Your weight on the moon is the measure of the attractive force that the moon exerts on you. On the moon, you would weigh only ⅙ of your weight on earth. Why?
   Would you have any less mass on the moon?
   Why could you jump so much farther on the moon than on the earth?

2. If there were no gravitational force at a baseball game, what would happen if:
   (a) a ball were hit by a batter toward left field?
   (b) the left fielder made a jump to catch the ball?
   (c) the batter started to run for first base?
   (d) the surprised coach jumped out of the dugout?

3. How do you win a game of tug-of-war? What kinds of force must you overcome? Why must you play on even ground for the game to be fair?

4. The spring scale below is being used to find the weight of a jar with bees resting on the bottom of it. All of a sudden, the bees start flying around. What happens to the pointer on the scale? What does this tell you about weight? about force?

Earth 6400 km **Moon** 1700 km

The mass of earth is 80 times greater than the mass of moon.

Answers to Challenge Your Thinking

1. Weight is a measure of gravitational pull on an object. Therefore, if the gravitational pull of the moon is ⅙ that of the earth, you will weigh on the moon ⅙ of what you weigh on earth.

   Your mass will remain constant whether you are on the earth or the moon.

   The gravitational force acting on you when you jump on the moon is less than the force acting on you when you jump on the earth. Therefore, it takes less force to overcome gravity on the moon, and you can jump farther.

2. (a) The ball would continue on indefinitely in its path until another force stopped it. There would be no downward movement due to the pull of gravity.
   (b) The left fielder would continue upward until another force acted upon him.
   (c) In the absence of gravity, the batter could hardly start running. Since he is weightless, there would be little friction between his shoes and the ground. The upward motion of the first step would cause the runner to leave the ground, and without gravity, he would continue upward.
   (d) The coach would continue upward and not return to earth unless another force acted upon him.

3. The force of one side must overcome the force on the other side. The friction of the player's feet against the ground acts against the force that the other team is exerting. If the ground is uneven, this could give one team an unfair advantage because friction would be increased. Also, gravity would give the downhill-team an added advantage.

4. When the bees start flying, the pointer on the scale remains in the same place. Even when the bees are flying, they are exerting a force on the air in the bottle, which pushes down on the scale. This force is the same as when the bees are resting on the bottom of the bottle.

Understanding Forces 209

LESSON

✸ Getting Started

In this lesson, students discover how to measure the size of a force. They begin by looking at examples of applied force. Then they make their own force meters. The lesson concludes with students using a spring to measure gravitational force.

Main Ideas

1. Elastic materials that stretch uniformly when masses are added to them can be used to measure force.
2. The unit of force, the newton, is approximately the weight of 100 g on the earth's surface.

✸ Teaching Strategies

Pull and push an object across a table. Ask students if they can think of a way to measure the amount of force used to move the object. (*Accept all reasonable answers.*) You may wish to let students try some of their own suggestions before beginning this lesson.

Answers to
Exploration 3

(a) A mass stretches the rubber band, and the added mass stretches it farther. The ruler measures how far the rubber band stretches. The length of the stretch could be used as a measure of force, but it may not be directly proportional to the mass added.

(Answers continue on next page)

MEASURING FORCES

4 The Tools to Use

Forces cannot be seen. But there is a way to discover the size of a force—measure the size of its effect. In this section, you will follow a series of Explorations showing you how to make *force meters* and how to use them to measure forces.

EXPLORATION 3

Force Meters

Carefully study the following examples. In each case, what effect is brought about by the applied force? If the force were increased, how would the effect change? How could you use the effect of the force as a means of measuring the force? The first example has been completed for you. For the remaining examples, draw a diagram in your Journal showing what would happen if there was an increase in force. Your drawing should also show how you would use the effect to measure the size of the force. If you prefer, you may describe in words the increase in force, its effect, and how to measure the size of the force.

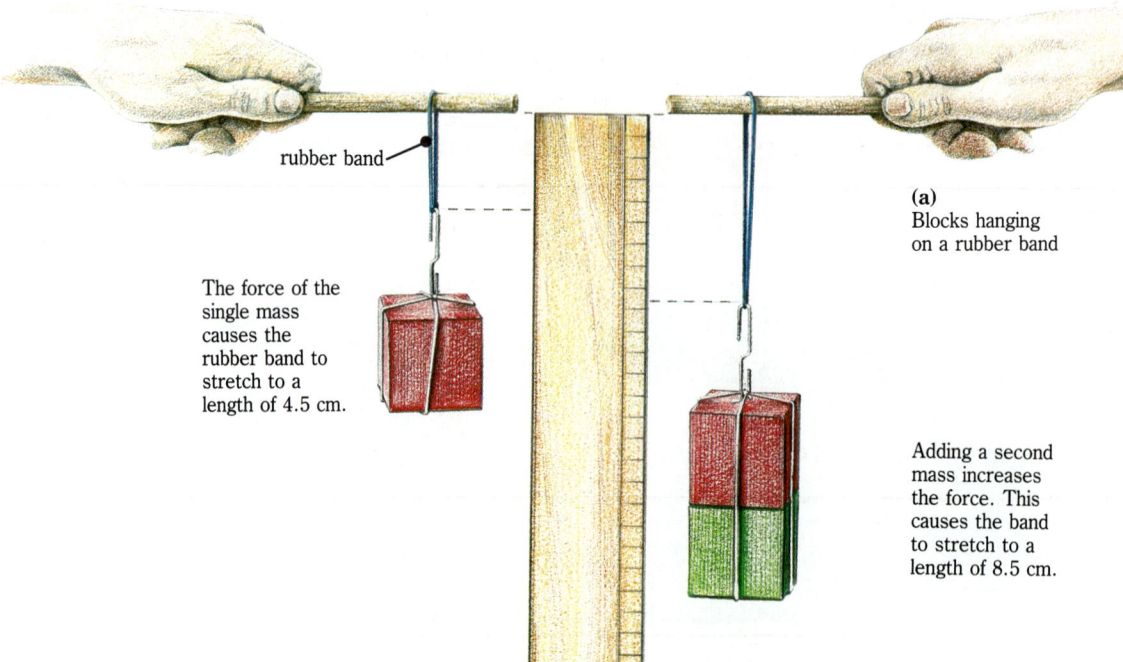

rubber band

The force of the single mass causes the rubber band to stretch to a length of 4.5 cm.

(a) Blocks hanging on a rubber band

Adding a second mass increases the force. This causes the band to stretch to a length of 8.5 cm.

LESSON ORGANIZER

Objectives

By the end of the lesson, students should be able to:
1. Suggest several ways to measure gravitational force.
2. Construct and calibrate a force meter.

Process Skills

measuring, comparing, predicting, extrapolating

New Terms

Newton (N)—the SI unit used to measure force; one newton is equal to the gravitational force that the earth exerts on a 100-g mass on the earth's surface.

210   Force and Motion

EXPLORATION 3—CONTINUED

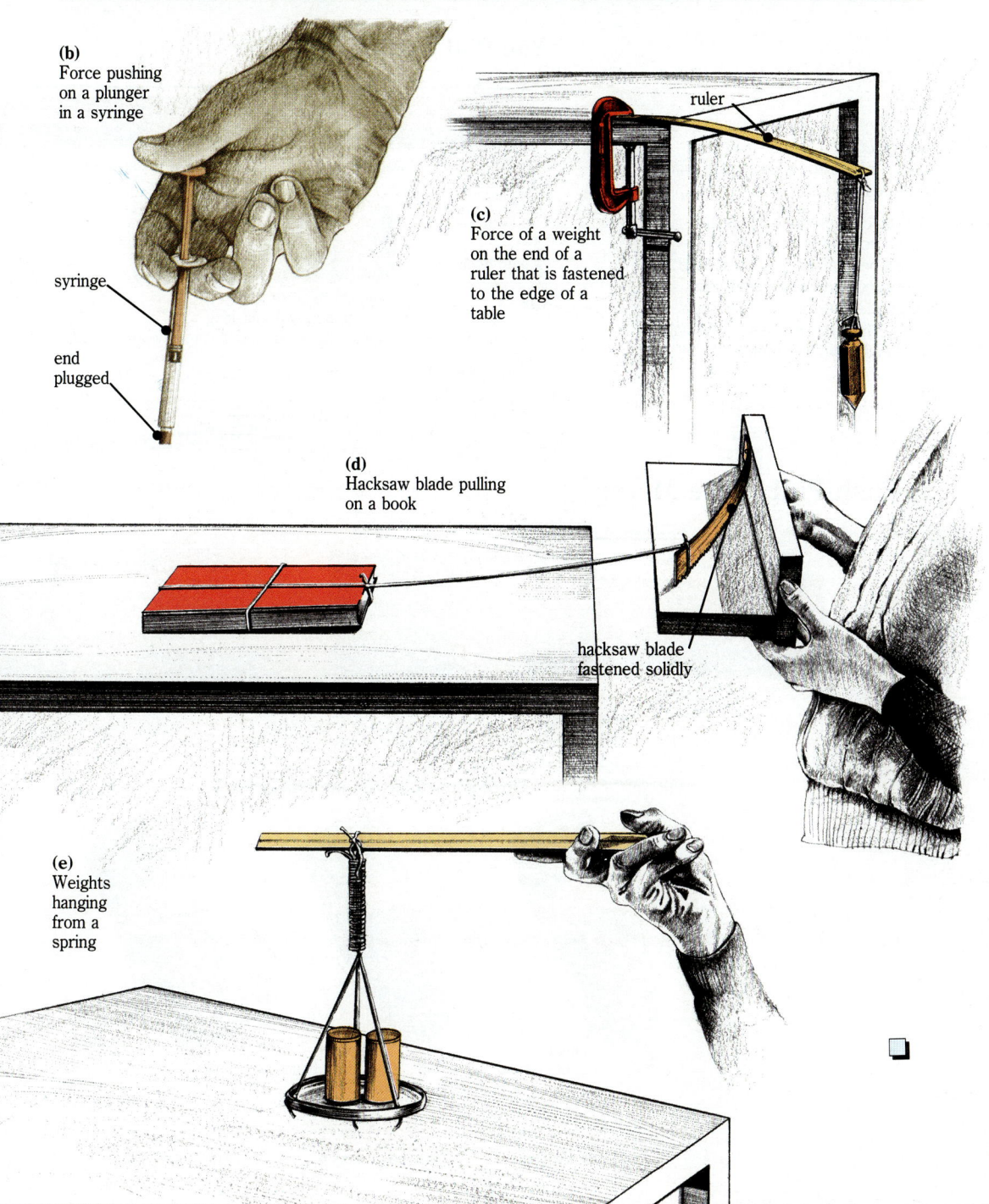

(b) Force pushing on a plunger in a syringe
- syringe
- end plugged

(c) Force of a weight on the end of a ruler that is fastened to the edge of a table
- ruler

(d) Hacksaw blade pulling on a book
- hacksaw blade fastened solidly

(e) Weights hanging from a spring

Answers continued

(b) A force pushes down the plunger of the syringe. The greater the applied force, the farther the plunger moves down. The distance it moves down could be used as a measure of the force, although the distance moved may not be proportional to the force applied.

(c) A mass hung from the end of the ruler bends the ruler. An increase in force would result from hanging another mass on the string. This increase in force would bend the stick farther. The size of the arc that the stick bends could be used as a measure of force.

(d) The increase in force could come from placing another book on top of the first book. The effect is that the hacksaw blade bends more. The amount the blade bends could be used as a measure of the force.

(e) The increase in force could come from adding more weights. The effect would be to stretch the length of the spring. The length of the spring increases as the applied force increases. The increase in the stretch of the spring is directly proportional to the increase in the force applied. This increase could be used to measure the size of the force.

Materials

Exploration 3: Journal
Exploration 4: 30-cm wooden dowel, 1.5 to 2 cm in diameter; cup hook; rigid cardboard tube; paper clips; rubber bands; masking tape; three 100-g masses
Exploration 5: coil spring, washers, 100-g, 500-g, and 1000-g hooked masses, meter stick, force meter, Journal

Teacher's Resource Binder

Resource Worksheets 4–2 and 4–3
Graphic Organizer Transparency 4–2

Science Discovery Videodisc

Disc 2, Image and Activity Bank, 4–4

Time Required

three class periods

Measuring Forces 211

Exploration 4

Divide the class into small groups and distribute the necessary materials. Each student should make a force meter. The procedures to follow are illustrated in the text. Invite members from each group to make force meters that measure the three different degrees of forces. (Have students refer to the diagrams at the bottom of page 212.)

Other materials students may want to use to design their own meter: wooden blocks, hammer and nails, hacksaw blades, springs, cardboard, protractors, and hooked masses. **(Continues on next page)**

EXPLORATION 4

Making Your Own Force Meter

Here are plans for constructing a force meter. This method uses rubber bands to measure the size of forces.

Construct the meter or design one of your own. Instead of using rubber bands, you could use a spring or some other elastic material. After completing your design, find the materials you will need and build the meter.

You Will Need

wooden dowel 1.5 cm to 2.0 cm in diameter

 cup hook

rigid cardboard tube (strengthen with masking tape, with holes on opposite sides near end)

 bent paper clips

 rubber bands

A Push-Pull Force Meter

Three ways to fasten the rubber bands

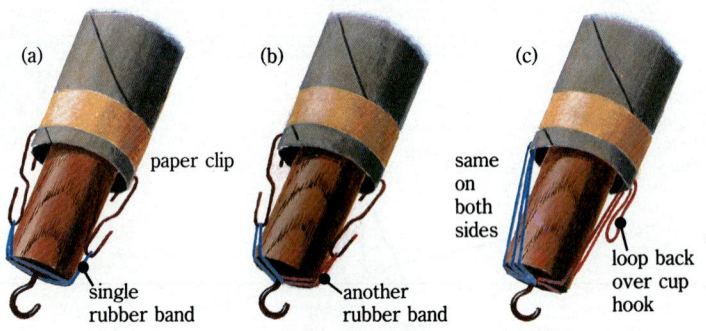

(a) paper clip — single rubber band
(b) another rubber band
(c) same on both sides — loop back over cup hook

1. Smaller Forces   2. Medium Forces   3. Larger Forces

---

212   Force and Motion

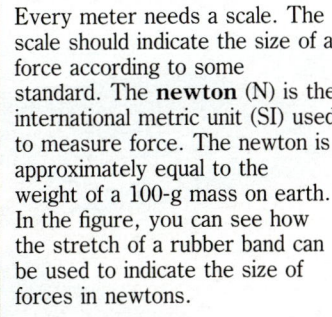

Every meter needs a scale. The scale should indicate the size of a force according to some standard. The **newton** (N) is the international metric unit (SI) used to measure force. The newton is approximately equal to the weight of a 100-g mass on earth. In the figure, you can see how the stretch of a rubber band can be used to indicate the size of forces in newtons.

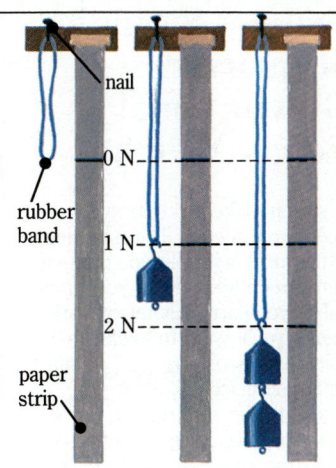

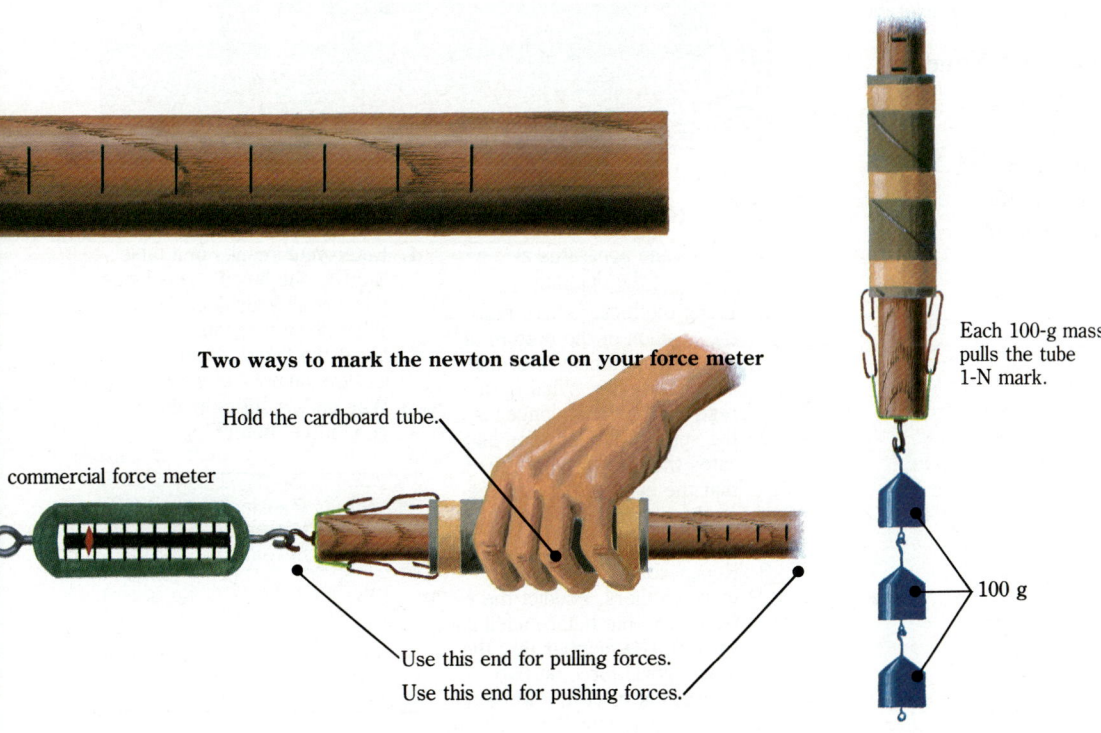

Two ways to mark the newton scale on your force meter

Hold the cardboard tube.

commercial force meter

Use this end for pulling forces.
Use this end for pushing forces.

Each 100-g mass pulls the tube 1-N mark.

100 g

(Exploration 4 continued)
Reassemble the groups and call on a volunteer to read aloud the top of page 213. Involve students in a discussion of the meaning of the new term *newton.* Direct their attention to the diagram at the top of page 213. The diagram should help students review the idea of calibration. To calibrate their force meters, students will need several 100-gram masses or a commercial force meter that measures force in newtons. To illustrate how students can calibrate their meters, refer them to the diagrams at the bottom of page 213.

You may wish to have students complete *Exploration 7* immediately after constructing their force meters so that they can test how well their meters work.

Measuring Forces

Exploration 5

Call on a volunteer to read aloud the first paragraph of the Exploration. In each Exploration, students are encouraged to make a formal laboratory report of their experiment. Involve students in a discussion of what is expected of them in terms of writing their reports. Ask students to enter in their Journals the title, purpose, predictions, and the table shown on page 214. Instruct students to perform the Exploration and enter their results, answers to questions, graphs, and any conclusions they have into their Journals. Students should discover that the amount the spring stretches is directly proportional to the number of washers added to it. Thus, their graphs should approximate a straight line. Also, springs stretch out in a linear fashion, which makes measuring the force easier.

EXPLORATION 5

Using a Spring to Measure Gravitational Force (Weight)

A spring makes an extremely good force meter. The following experiment will help you discover why. Keep a record in your Journal as you carry out this experiment. You'll want to keep precise records of what you do and what you discover. After you complete the experiment, write a formal report using the information from your Journal. In your report include a title, purpose, procedures, predictions, results, discoveries, and answers to the questions in this Exploration.

You Will Need

- a coil spring
- washers
- hanging masses
- a meter stick
- a force meter

What to Do

1. Set up the apparatus as shown in the diagram.
2. Using the meter stick, read the position of the bottom of the spring. Repeat after adding one washer. What is the result? How much longer is the spring? The position indicates the size of the force that the earth exerts on one washer; that is, the weight of one washer.
3. Repeat for 2, 3, 4, 5, or more washers. Predict the effect each time before adding a washer. (Remember that the gravitational force on two washers is twice the force on one washer. How large would the size of the force be on 3, 4, or 5 washers?)
4. Enter your results in a table like the one below. This helps you see all your data clearly. What does your table tell you about the length of stretch and the number of washers? Why does a spring make a good force meter?

Number of Washers	Position of Bottom of Spring	Stretch of Spring
0		
1		
2		
etc.		

Answers to
Questions (page 215)

1. Students should look at their graphs to extrapolate the weight of the object in number of washers based on the amount the spring stretches.
2. Again, students can look at their graphs to extrapolate the weight of the object in washers based on the amount the spring stretches.
3. If the book is not so heavy that it distorts the spring, the spring scale can be used to measure the force needed to move the book across the table.

5. Another useful way to study your data is to make a graph of it like the one to the right. In your Journal plot the amount of stretch against the number of washers.

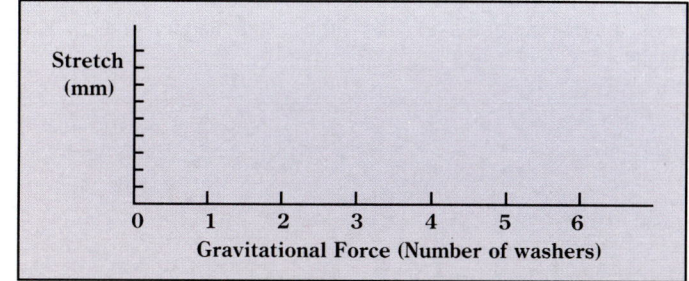

Questions

1. Suppose that an object hung on the spring causes it to stretch 22 mm. What would be the size of the gravitational force on the object (its weight)? Use your graph to find the answer.
2. Choose an object that is not too heavy to distort the spring. Predict how many washers it will weigh. Now hang it on the spring and find out if your prediction was correct.
3. Could you use the spring to measure the force needed to pull a book across a table?

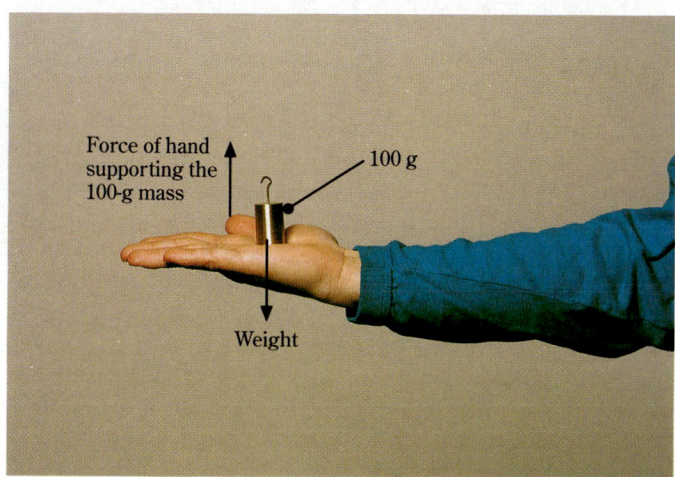

Converting Washers into Newtons (N)

1. Hold a 100-g mass in your hand. How much force are you exerting to hold it up? Find out by placing the mass on a commercial force meter marked in newtons.
2. Repeat for a 500-g mass and a 1000-g (or 1-kg) mass.
3. How many "washers of force" are equal to 1-N of force? Use the commercial force meter to find out.
4. For the original spring you used, how many millimeters would it have to stretch to indicate a 1-N force?

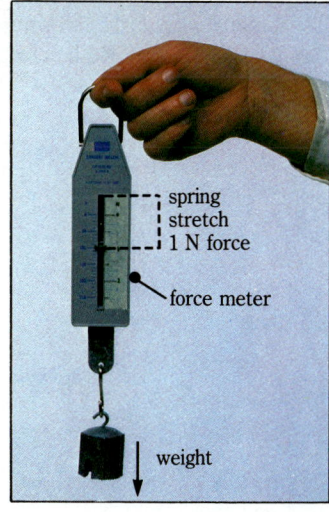

Answers to

Converting Washers into Newtons (N)

1. It takes 1 N of force to hold a 100-g mass at the surface of the earth.
2. It takes 5 N to hold a 500-g mass and 10 N to hold a 1000-g mass.
3. To find out how many "washers of force" are equivalent to 1 N of force, students should hang washers on the force meter until the meter registers 1 N.
4. Again, students can look at their graphs to find out how far the spring stretches when the equivalent of 1 N of force in washers is hung on the spring. They could also hang approximately 100 g of washers on their springs to get their estimate.

Follow Up

Assessment

1. Provide students with a spring, three 100-gram masses, and several objects. Ask them to use the spring to measure in newtons the gravitational force on these objects. *(The 100-gram masses could be placed on the spring to see how far they make the spring stretch. A graph of weight in newtons versus centimeters of stretch could be drawn. Then the objects could be placed on the spring to measure how far they make the spring stretch. Their weight in newtons could then be estimated from the graph.)*
2. Have students compare the elasticity of a spring to that of a rubber band. Suggest that they measure the position of the ends of the spring and rubber band as additional masses are added to them. Then they could remove the additional masses one at a time, again checking the positions of the ends of the spring and rubber band. Have them repeat the experiment a second time. Ask: Does the spring/rubber band stretch the same amount for each added mass? Which is a better force meter? Why? *(The spring makes a better force meter because it stretches nearly the same amount with each added mass, and it does not lose its elasticity. The rubber band stretches different amounts with each added mass, and it does lose its elasticity.)*

Extension

Have students find out about different kinds of scales, including spring scales used to weigh produce, bathroom scales, and equal-arm balances. Ask them to explain how each scale works by using diagrams and explanatory paragraphs.

Measuring Forces

Lesson 5

✷ Getting Started

In this lesson, students estimate how many newtons of force are needed to lift different objects. In *Exploration 6,* they feel different sizes of forces with their fingers. In *Exploration 7,* they test their ability to first estimate and then measure different sizes of forces.

Main Ideas

1. The size of a force needed to lift something can be estimated.
2. The size of a force can be measured in newtons with a force meter.

✷ Teaching Strategies

Estimating the Size of a Force

Set up a table filled with objects similar to those shown in the photograph. Encourage students to actually lift these objects; let them feel the forces that need to be applied to overcome the objects' weights. From this experience, they should be able to estimate the weights of the objects listed.

Copy the list of items from page 216 onto the chalkboard. Record a few of the students' estimates for each item. Then have the class agree on one answer for each item. Look in the answer box to see how close the estimates were.

5 Estimating the Size of a Force

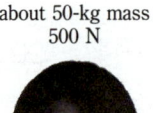

You are now familiar with the use of the newton (N). But what does a newton feel like? Try lifting and holding objects like these:

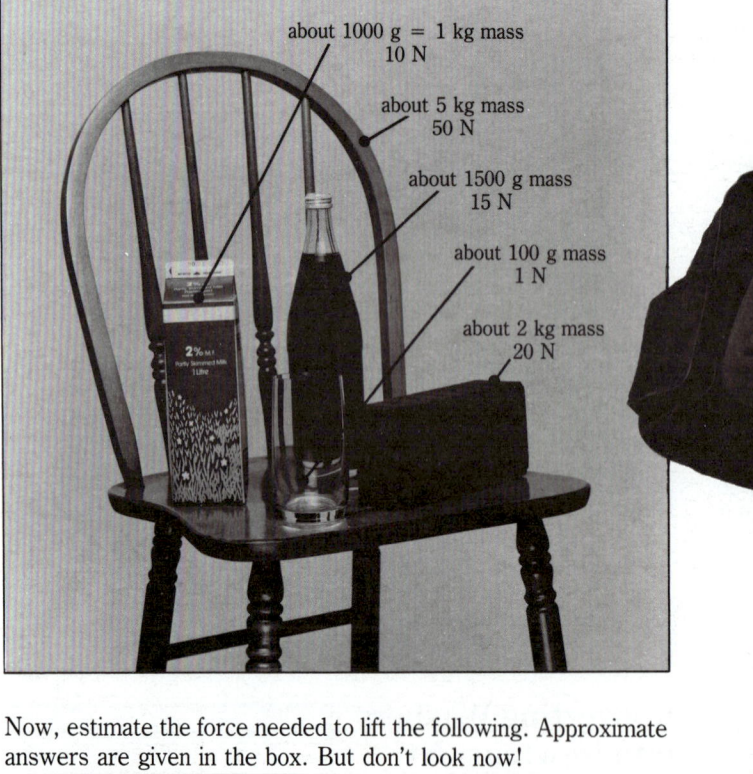

Now, estimate the force needed to lift the following. Approximate answers are given in the box. But don't look now!

(a) Your science textbook
(b) Your teacher
(c) A pail of water
(d) A metal wastepaper basket
(e) A dozen eggs
(f) A basketball
(g) A hammer
(h) An iron shovel
(i) A concrete block
(j) $5 in quarters
(k) A newborn baby
(l) A compact car

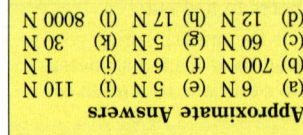

Approximate Answers
(a) 6 N (e) 5 N (i) 110 N
(b) 700 N (f) 6 N (j) 1 N
(c) 60 N (g) 5 N (k) 30 N
(d) 12 N (h) 17 N (l) 8000 N

LESSON 5 ORGANIZER

Objectives

By the end of the lesson, students should be able to:
1. Estimate, in newtons, the force needed to lift an object.
2. Measure the size of a force in newtons.

Process Skills

estimating, measuring, making a table, predicting

New Terms

none

216 Force and Motion

EXPLORATION 6

A Finger Exercise

Here's how to feel different sizes of forces with your finger.

You Will Need

- a meter stick
- a 1-kg mass or equivalent

What to Do

1. Balance a meter stick over the edge of the table (at the 50 cm mark).
2. Place a 1-kg mass on the ruler so that it is centered at the 45-cm mark.
3. Press down at 100 cm to raise the mass. You are using a 1-N force.
4. Press down at 60 cm for a 5-N force.
5. Now shift the 1-kg mass to 0 cm. Press at 100 cm for 10 N, at 75 cm for 20 N, at 60 cm for 50 N, and at 55 cm for 100 N.

With the 1-kg mass at 0 cm you can calculate the force in newtons with this formula:

$$\frac{500}{(P - 50)}$$

where P = pressing point.

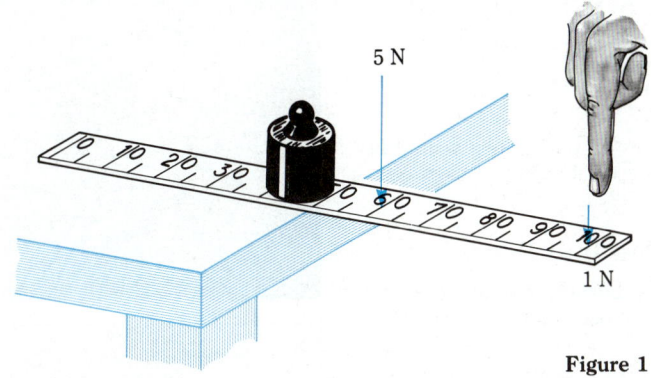

Figure 1

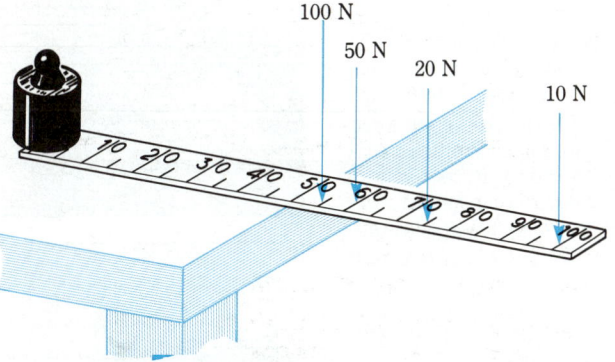

Think About It

Can you exert a 100-N force with one finger? with two fingers? What is the maximum size of force each finger can exert? Are all your fingers equal in strength? How does your left hand compare to your right hand?

Exploration 6

Divide the class into small groups and distribute the materials. Encourage all group members to try the activity so that they have the first-hand experience of feeling different sizes of force.

The meter stick and table edge form a lever. The edge of the table serves as the fulcrum. The distance of the force from the fulcrum times the amount of force on each side of the balance should be equal. For example, 10 N placed 5 cm from one side of the fulcrum would balance 1 N placed at 50 cm on the other side of the fulcrum (10 N x 5 cm = 1 N x 50 cm).

Answers to
Think About It

Students will discover that:
- They cannot exert 100 N of force with only one finger.
- They cannot exert 100 N of force with two fingers.
- The maximum size of force each finger can exert will vary with the fingers used.
- Their thumbs and index fingers are probably stronger than their other fingers.
- This will depend on whether students are left or right-handed.

Materials

Estimating the Size of a Force: brick, drinking glass, bottled soft drink, carton of milk
Exploration 6: meter stick, 1-kg mass, table
Exploration 7: Journal, force meter, masking tape, string or cord, scissors, sandbag, rope, pulley, 3 books, brick, board

Teacher's Resource Binder

Resource Worksheet 4–4
Graphic Organizer Transparency 4–3

Science Discovery Videodisc

Disc 2, Image and Activity Bank, 4–5

Time Required

two class periods

Exploration 7

Set up this activity so that students move from one location to another to measure the 12 situations. Letter the locations (a–l) as on the chart on page 219. Suggest that students work in pairs to measure the forces. Encourage students to copy the chart on page 219 into their Journals, and then to predict the size of the force needed in each situation. Suggest that students use masking tape, string, or cord to attach their force meters to the various objects shown in the pictures. As they measure the actual force that must be exerted in each situation, they should record their measurements.

As they work through the measuring activities, students may make the following discoveries:

- It takes more force to start an object moving than to keep it moving.
- It takes more force to lift an object vertically than to drag it horizontally.
- It takes less force to move a brick up an inclined plane than it does to lift it vertically the same distance.
- It takes less force to lift an object with a movable pulley than with a fixed pulley.

EXPLORATION 7

Estimating and Measuring the Size of Forces

Now it's time to test your ability to estimate different sizes of forces.

You Will Need

- a commercial force meter or the one you made yourself

What to Do

1. Construct a table in your Journal like the one on page 219. As you perform this Exploration, fill in the information to complete your table.
2. Predict the size of the force exerted in each of the situations shown. Record your estimates in your Journal.
3. Select a force meter which enables you to make an accurate measurement of the force required in each case. Record these measurements in your Journal.

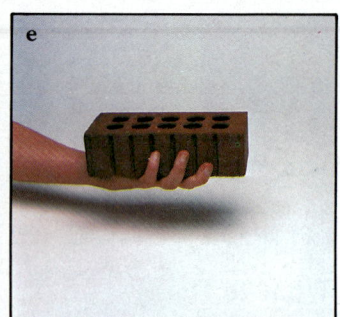
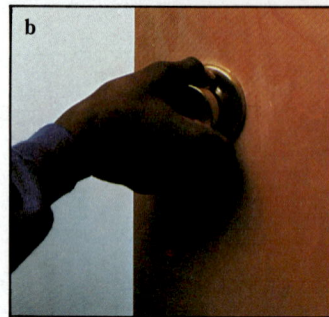
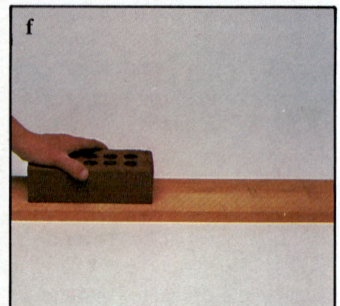
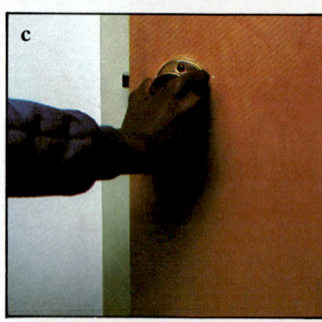
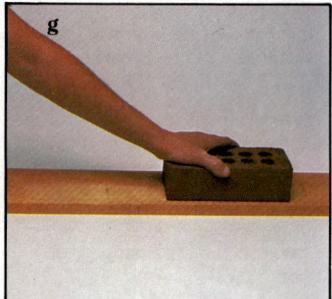
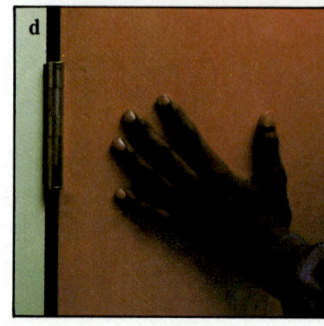
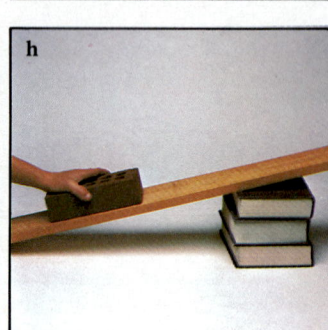

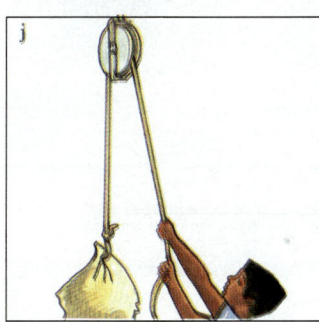
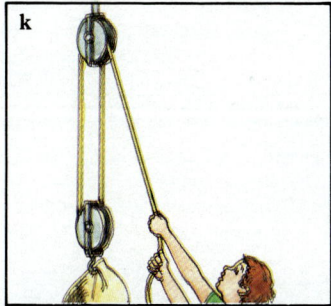
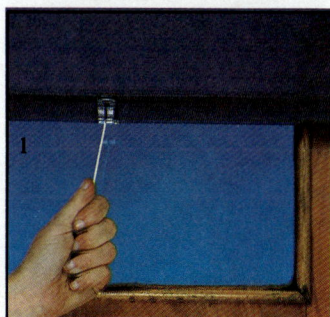

Situation	Estimated Force (N)	Measured Force (N)
(a) Opening a drawer		
(b) Pulling a door knob		
(c) Closing a door using the knob		
(d) Closing a door by pushing near the hinge		
(e) Lifting a brick		
(f) Causing a brick to start moving on a horizontal board		
(g) Keeping a brick moving on a horizontal board		
(h) Pushing a brick up an inclined board		
(i) Pushing a table		
(j) Lifting a sandbag using a pulley		
(k) Lifting a sandbag using two pulleys		
(l) Pulling down a screen or blind		
(m) Your choice		

Assessment

1. Show students pictures of various objects. Ask them to estimate the force in newtons needed to lift the objects.
2. Provide students with a force meter and ask them to measure the force needed to open a door, close a door, lift a book, and pull a book across a table.

Extension

1. Have students research the units that are used to measure force and mass in the English system of measurement (pounds and slugs). Then ask them to convert these units to SI units of measurement.
2. Ask students to think of things they do daily that require pushing and pulling. Have them measure and record in newtons the force needed to perform these daily tasks.

Measuring Forces

LESSON 6

★ Getting Started

In this lesson, students first estimate and then measure the strength of different muscle groups in their bodies. Also, they increase their knowledge about the concepts of mass and weight.

Main Ideas

1. Muscle strength can be measured in newtons.
2. A force meter can be used to measure muscle strength.
3. Mass and weight are measured with different kinds of scales and in different units.

★ Teaching Strategies

Ask students to estimate the force in newtons that they think their arm muscles can exert. Then ask if they think their leg muscles are stronger or weaker than their arm muscles. Students should make estimates before they actually measure their strength.

How Strong Are You?

Set up four stations around the room with one activity at each station. If possible, have numerous sets of materials available at each station so that a number of students can work at the same time. It would be best to have students work in pairs.
(Continues on next page)

6 How Strong Are You?

Here's a chance to measure your own strength. First, guess how many newtons you'll be able to exert. Then see how close your guesses were to your true strength.

1. Arm-Muscle Strength

How far can you stretch the luggage cord? Measure the length that the cord stretches in cm.

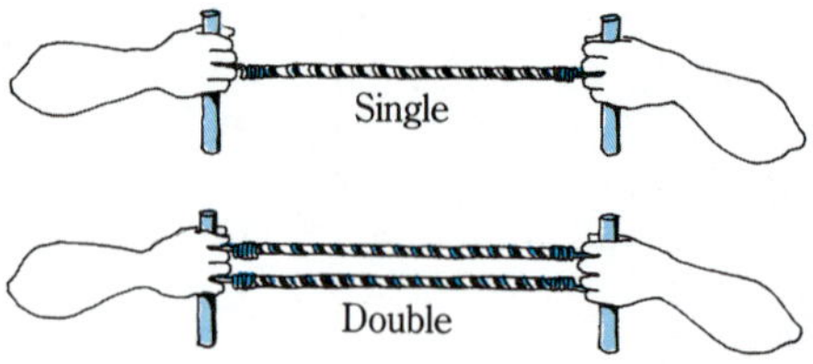

Single

Double

Find the force you used! Place the cord beside a meter stick. Pull the cord with a force meter until it matches the length that you were able to stretch the cord.

Now record the number of newtons shown on the force meter.

LESSON ORGANIZER

Objectives

By the end of the lesson, students should be able to:
1. Estimate the strength of different muscles in newtons.
2. Measure the strength of different muscles in newtons.
3. Distinguish between measuring mass and weight.

Process Skills

estimating, predicting, measuring, recording

New Terms

none

220 Force and Motion

2. Forearm Pull

Stretch the luggage cord using only your forearm muscles.

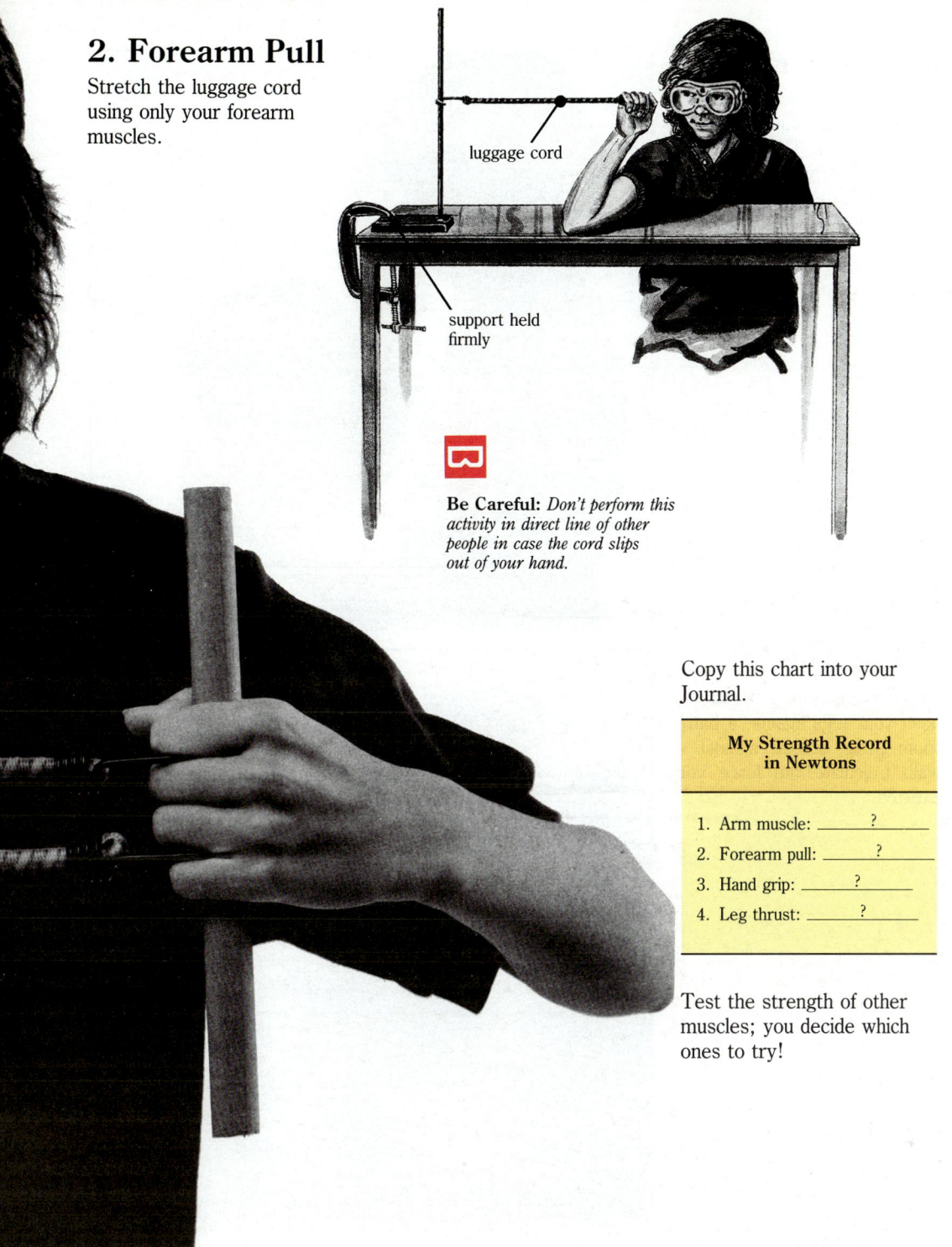

Be Careful: *Don't perform this activity in direct line of other people in case the cord slips out of your hand.*

Copy this chart into your Journal.

My Strength Record in Newtons
1. Arm muscle: _____?_____
2. Forearm pull: _____?_____
3. Hand grip: _____?_____
4. Leg thrust: _____?_____

Test the strength of other muscles; you decide which ones to try!

(How Strong Are You? continued)

It is inevitable that students will turn these activities into a competition. However, caution them to take the competition lightly in order to minimize the risk of muscle strain and to prevent the equipment from being broken.

Before students begin, have them copy the table on page 221 into their Journals so that they can keep accurate records of their results. Suggest that they carefully study the illustrations on pages 220 through 222 so that they will know how to set up the equipment and perform each activity.

1. Arm-Muscle Strength

Caution students not to do this activity in the direct line of other people in case they let go of the cord or the cord breaks.

2. Forearm Pull

Students need to position themselves so that their arms are securely and firmly resting on the table. They should wear safety goggles in case the cord slips during the activity.

Materials

1. **Arm-Muscle Strength:** 2 luggage cords, 2 dowels, force meter, meter stick, nail, hammer, tape, piece of wood
2. **Forearm Pull:** luggage cord, table, vice or C-clamp, metal stand, safety goggles
3. **Hand-Grip Strength:** bathroom scale
4. **Leg Thrust:** chair, bathroom scale

Science Discovery Videodisc

Disc 2, Image and Activity Bank, 4–6

Time Required

two to three class periods

Measuring Forces

3. Hand Grip

Remind students to multiply the reading on the scale by 10 to find the force in newtons. The scale must have metric measurements.

4. Leg Thrust

Suggest that one student sit in the chair and another student push against the chair from the back so that it doesn't move. Again, remind students to multiply the reading on the scale by 10 to find the force in newtons.

Encourage students to measure the strength of other muscles. To measure a shoulder push, have students push their shoulders against a bathroom scale that is braced against the wall. To measure the strength of a forefinger press, have them press their forefingers against a bathroom scale. The strength of a little finger can be measured by pulling directly on a force meter. Students should suggest an additional strength-measuring activity and perform it.

3. Hand Grip

Grip a bathroom scale with both hands. Be sure to use a scale that shows kg. The amount in kg multiplied by 10 gives you the force in newtons.

4. Leg Thrust

Using a chair as a backrest, push your legs against a bathroom scale braced against a wall. Calculate the force you exerted.

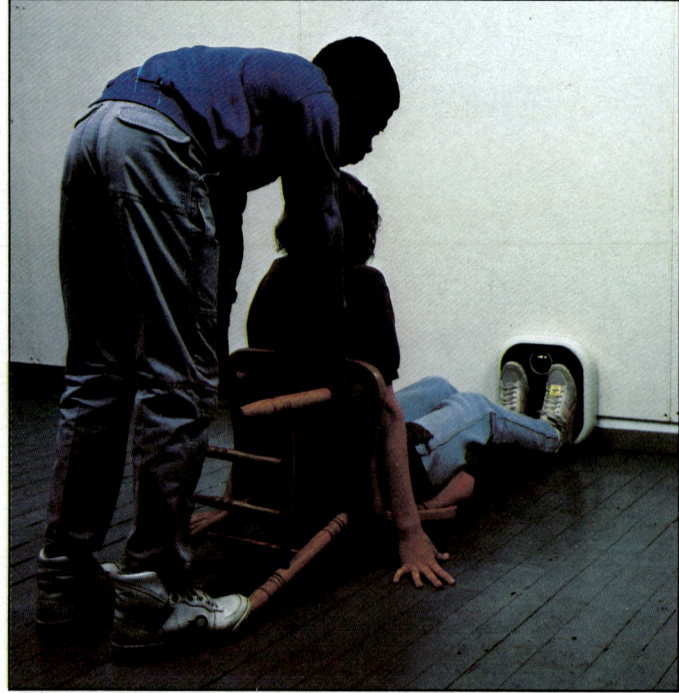

Measuring Mass and Weight

As you have already learned, the quantity of matter in an object is called its mass. Whether you move an object to the top of Mount Everest or to the moon, its mass remains the same. The object still has the same quantity of matter in it. Mass is usually measured in kilograms (kg).

Weight, on the other hand, is a measure of the gravitational force exerted on an object. The newton is the unit used to measure gravitational force (weight) and all other forces. A spring scale is the device used most often to measure weight.

Commonly, the terms "mass" and "weight" are interchanged and used incorrectly. This can cause some confusion—even for students of science. Carefully study the table and picture below. Account for the difference in the weight (N) of the spring scales shown in the figure.

Mass is . . .	Weight is . . .
the quantity of matter in an object.	a measure of the gravitational force exerted by the earth on an object.
constant for an object taken anywhere in the universe.	varied as you move away from the earth's surface.
measured with an equal arm balance or a pan balance.	measured with a spring scale.
measured in grams (g) and kilograms (kg).	measured in newtons (N).

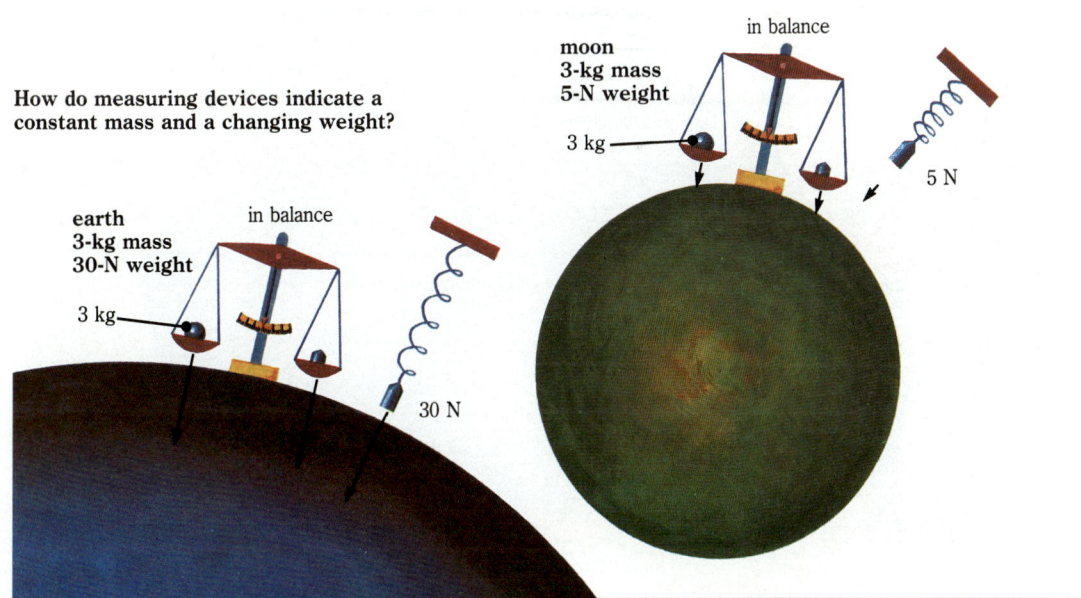

How do measuring devices indicate a constant mass and a changing weight?

 **Follow Up**

Assessment

1. Photocopy a diagram of the major muscle groups in the human body. Draw lines to several muscle groups, for example, those in the forearm, fingers, and thighs. Ask students to estimate the force in newtons that these muscle groups can exert.
2. Present the following scenario to students: You are in a life-threatening situation: you need to push or pull something with either your arm or leg muscles. Which muscles would you choose to use? Why?

Extension

1. Have students find some world records for weight lifting. If the data is given in kilograms, have them convert the measurements to newtons. If the school has weight-lifting equipment, they can compare their own strength to the strength of those who have broken world records.
2. With the assistance of a physical education teacher, have interested students participate in a body-building program. Suggest that they keep track of their improvement by recording how much force their different muscles can exert.

Measuring Mass and Weight

Ask students to explain what they already know about mass and weight from reading this unit. (Refer to Lesson 3.)

Call on a volunteer to read the top of page 223. Pause after the three paragraphs have been read to be sure students are comfortable with the differences between measuring mass and weight. Point out that in everyday life, these terms are often used interchangeably.

Then direct students' attention to the chart in the middle of the page. It summarizes the differences between the concepts of measuring mass and weight.

Refer students to the diagrams at the bottom of the page and have them answer the question. The equal arm balance measures the same mass for a given object despite the pull of gravity. An object measured on a spring balance changes weight as gravitational force changes.

Measuring Forces

Answers to
Challenge Your Thinking

1. Corrected sentences are:
   (a) If I go far enough away from the earth's surface, I will get to a place where my *weight* will be almost zero.
   (b) An object of 200 N on the surface of the earth *will measure less than 200 N* in the orbiting space station.
   (c) In the space station, I would use an *equal-arm balance* to find the mass of an object.
   (d) Rachel would get the same results on the moon; mass remains constant.
   (e) Food is sold in the grocery store by *mass*—for example, $2.50 for a kg of sugar.
2. Allow students to present creative or unusual answers. The focus of this question is on the process of finding the force needed to perform the specified activity. Possible answers include:
   (a) Calculate your own mass in kg. Multiply by 10 to determine the force in newtons.
   (b) Attach a force meter to the doorknob, and measure the force needed to pull the door closed.
   (c) Students should realize that pushing or pulling an object requires the same force. Push a force meter against the back of the car and read the measurement when the car starts to move. Or, attach a force meter to the front bumper of the car and measure the force needed to pull the car into motion.
   (d) The boulder could be suspended from an enormous force meter; its weight could be estimated; or a lever could be used to measure the force needed to lift it.

Challenge Your Thinking

1. Take the mass and weight quiz! People often confuse the meaning and use of the terms mass and weight. Rewrite these incorrect sentences and correct the confusion.
   (a) If I go far enough away from earth's surface, I will get to a place where my mass will be almost zero.
   (b) An object of 200 N on the surface of the earth should still measure 200 N in the orbiting space station.
   (c) In the space station, I would use a spring scale to find the mass of an object.
   (d) Rachel placed a football on one pan of an equal arm balance and got a perfect balance with a 250-g mass on the other pan. Naturally, she couldn't expect the same result on the moon.
   (e) Food is sold in the supermarket by weight—for example, $2.50 for a kg of sugar.

2. Design an innovative way to measure the force needed to:
   (a) perform a chin up
   (b) close a door
   (c) push a sports car
   (d) lift an enormous boulder

224 Force and Motion

FRICTIONAL FORCE

7 Take a Look at Friction

You already saw some examples of friction in Exploration 1. Take another look at those examples. What did they tell you about friction? Now read the three case studies below and answer the questions.

Case Study A

As Jennifer reaches the level field, she gradually slows down.

1. What slows down the skis?
2. Is there a force involved?
3. What is the direction of the force?
4. What name do we give to this kind of force?

direction of motion

hard-packed snow

Case Study B

Driving east on a level stretch at 100 km/h, Andy shifts into neutral. The car coasts for 300 m before coming to a stop.

1. Identify the frictional force and its direction.

direction of motion

road surface

LESSON 7 ORGANIZER

Objectives

By the end of the lesson, students should be able to:
1. Observe the direction and effects of friction in a variety of situations.
2. Identify factors that influence the magnitude of frictional force.

Process Skills

designing an experiment, controlling variables, formulating hypotheses, interpreting data

New Terms

none

(Organizer continues on next page)

LESSON 7

✸ Getting Started

In this lesson, students explore case studies about the effects of friction. They are presented with hypotheses about friction and are asked to determine what kinds of experiments would test these hypotheses.

Main Ideas

1. Friction opposes a starting or continuing motion.
2. Friction is greater when motion is starting than when it is continuing.
3. Friction varies with the kinds of surfaces opposed to one another.

Answers to
Case Study A

1. The frictional force between the skis and the snow slows down the skis.
2. yes
3. The direction of the force is opposite to the direction of motion.
4. frictional force

Answer to
Case Study B

1. The frictional force is the force the road exerts against the tires. Since the car is traveling eastward, the frictional force must be acting westward.

Frictional Force   225

Answers to
Case Study C

1. The frictional force is the force the floor surface exerts against the bottom of the refrigerator.
2. In this case study, the object (the refrigerator) is not moving, but friction is exerted on it by the floor.
3. The force of Don and Susan together is slightly larger than the frictional force exerted by the floor since together they are able to move the refrigerator.

Answers to
Questions

Divide the class into small groups to discuss the statements. Reassemble them and involve the groups in a discussion of their responses.
1. Possible responses include: frictional forces that slow down skateboard wheels, ice skate blades, doors on hinges.
2. Examples include: sand against the bottom of the boat that stops the boat from moving easily as it is dragged; water against the bottom of the boat that slows down its motion; air friction against the boat that slows down its motion.
3. If there was no frictional force opposing the motion of the skis, they would keep going forever. However, if air friction is considered, then the skier would eventually slow down due to friction between the skier and the air.
4. If Andy had been driving on gravel, the car would have stopped sooner. If he had been driving on ice, he would have coasted farther before stopping.
5. Don could move the refrigerator himself by placing it on rollers or on a smoother surface. Both solutions reduce the friction between the floor and the refrigerator.

Case Study C

Don dropped his book behind the large refrigerator. He pulled on one edge of the refrigerator to move it away from the wall. It did not budge. His sister, Susan, joined him. Together they just managed to move it.

1. What is the frictional force?
2. What is it doing here that is different from the frictional force in the other case studies?
3. How do the combined forces of Don and Susan compare with the size of the frictional force?

Case Study Conclusions

The following general conclusions about frictional forces can be drawn from the case studies:

1. When one surface moves against another, a frictional force results. The frictional force works against the direction of motion.
2. If one body moves over a stationary body, the stationary body tends to slow down the moving body.
3. If a force tries to move a body at rest on a surface, the surface resists this motion with an opposing force of friction.

Questions

Discuss these questions with a group of classmates. Try to reach an agreement about each one.

1. Name some other examples in which frictional forces slow down motion.
2. In the sailboat story at the beginning of this unit, were there any examples of friction? If so, identify them.
3. In case study A, what if there were no frictional forces opposing the motion of the skis?
4. For case study B, what would have happened if Andy had been driving on gravel? on ice?
5. In case study C, how could you help Don move the refrigerator by himself? In your solution have you reduced, increased, or maintained the size of the force of friction?

(Organizer continued)

Materials

Exploration 8: bricks, force meter, board, sheet of glass, plastic drinking straws

Teacher's Resource Binder

Resource Transparency 4–4

Science Discovery Videodisc

Disc 2, Image and Activity Bank, 4–7

Time Required

two to three class periods

Measuring Friction

How large or small are frictional forces? How can they be increased or decreased? You measured friction in Exploration 7. Review these measurements to see whether they can help you discover some new ideas about friction.

Janell had some ideas (hypotheses) about friction. She decided to check her new ideas by conducting a series of experiments.

The results of her experiments are summarized on page 228. Look at the results, then try to solve the Friction Hypothesis Puzzle below. In this puzzle, you will match Janell's experimental results with the hypotheses she began with. Do the experimental results confirm all of her hypotheses?

Friction Hypothesis Puzzle	
Experiments designed to test the hypotheses:	**Janell's hypotheses:**
(a) Experiments 1 and 4	1. The size of the frictional force depends on the kind of surfaces that are rubbing together.
(b) Experiments 2 and 3	2. The heavier the object, the greater the force of friction.
(c) Experiments 3 and 4	3. The force of "starting" friction is greater than the force of "sliding" friction.
(d) Experiments 3 and 5	4. The force of "rolling" friction is greater than the force of "sliding" friction.
(e) Experiments 3 and 6	5. The force needed to lift an object is greater than the force needed to drag a body along a level surface.
(f) Experiments 4 and 7	6. Frictional force is constant for an object on a surface, no matter what the area of the surface contact is.

✸ Teaching Strategies

Measuring Friction

Review the results of *Exploration 2,* Stations 6 and 7, on pages 202 and 203. Ask: What did you discover about friction in this Exploration? *(Friction was decreased when straws were placed between the table and the book. Friction was the greatest between the towel and the bottle, and the smallest between the floor and the bottle.)* Then refer students to *Exploration 7,* situations a, f, and g on page 218. Ask: What did you learn about friction in this Exploration? *(It took more force to lift a brick than it did to drag it horizontally. The frictional force on the brick was less than its weight. It took more force to start the brick moving than it did to keep it moving.)*

Call on a volunteer to read the top half of page 227. Then direct students' attention to page 228. You may wish to perform each of the experiments. Ask students to read the description of each experiment and to look at the diagram. Pause after each one to be sure students understand what the experiment was about and how the result was determined.

Make sure students understand that the size of frictional force can be inferred from the force needed to move an object or to keep it moving. The minimum force needed to start an object moving is a measurement of the force of the starting friction of the surface against the object. The force that causes an object to slide on a surface at a steady rate equals the force of the sliding friction that the surface exerts on the object.

Friction Hypothesis Puzzle

Students will figure out which hypothesis in column 2 matches up with the groups of experiments in column 1. Ask students if each of the six hypotheses was confirmed. *(All of the hypotheses were confirmed except number 4. The force of sliding friction is greater than the force of rolling friction.)*
(Answers to In-Text Questions follow on next page)

Answers to
In-Text Questions

(a) 5
(b) 1
(c) 3
(d) 2
(e) 6
(f) 4

Follow up with a brief discussion of student answers.

Expt. No.	Janell's Experiment	Diagram	Results
1	Janell measured the weight of a brick.		Weight = 20 N
2	Janell measured the force needed to start the brick moving on a glass surface.		Force = 3.5 N
3	Janell measured the force needed to start the brick moving on a wooden surface.		Force = 6.3 N
4	Janell measured the force needed to keep the brick moving on wood once it had started to move.		Force = 5.0 N
5	Janell measured the force needed to start two bricks moving on a wooden surface.		Force = 12.5 N
6	Janell turned the brick on its narrow side and measured the force needed to start it moving on wood.		Force = 6.3 N
7	Janell placed straws between the brick and the wooden surface and measured the force needed to keep the brick moving.		Force = 0.70 N

Force and Motion

EXPLORATION 8

Your Turn to Test Friction

Do you still have doubts about Janell's hypotheses? After all, she compared the results of only two or three experiments for each of the hypotheses. Perhaps more proof is needed. It might be valuable to conduct more experiments.

Here is your opportunity to confirm or correct the results of Janell's experiments and to find out whether or not her hypotheses hold true.

What to Do

1. Choose one of Janell's hypotheses to test.
2. Design your own experiment.
3. Decide what equipment you will need.
4. Collect the equipment.
5. Perform the experiment.
6. Record your data.
7. Check your results. Did you perform the test enough times? Were the variables controlled?
8. Decide whether the hypothesis you tested is right or wrong.

More About Friction

By now, you realize that friction is present all around us. From the cars on the highway to the sneakers on a basketball court, friction is a familiar force. Summarized here are some important scientific points about friction.

Friction . . .

- opposes a starting or continuing motion.
- is greater when motion is started than when motion is continued.

Exploration 8

Divide the class into groups of three or four students. Have the groups design and perform experiments to test Janell's hypotheses. Make sure that each of the hypotheses is chosen by at least one group.

This Exploration can be used to emphasize the fact that science advances through formulating and testing hypotheses. Experiments may prove or refute hypotheses, or they may cause them to be modified. The fact that some hypotheses will be tested several times could be used to emphasize the need for replication of results—if the same conditions are used in a series of experiments, the same results are expected. Explain the idea of experimental error, which can result from imperfect procedures and uncertainties in measurement.

Involve students in a discussion of experimental design, paying particular attention to the control of variables. For example, ask: Would doing Janell's Experiments 2 and 4 be a fair test of the third hypothesis? *(No, since the type of surface should have been the same in both experiments.)*

Before students begin the Exploration, you may wish to choose one of the hypotheses and, as a class, design the complete experiment that would test it. This "class design" could provide a model for students for the remaining five hypotheses. Then have students design and perform their experiments and record their results. Students should make tables or charts to record their data. When all groups have finished their experiments, involve them in a discussion of their results. Class results could be compiled on the chalkboard.

More About Friction

Call on a volunteer to read aloud page 229 and 230. Then divide the class into small groups to discuss the five questions at the bottom of page 230. These questions are designed to prepare students for the next lesson. Expect variety in students' answers.

Frictional Force

Answers to
In-Text Questions

1. two surfaces in contact with each other
2. oiling, using lubricants, using bearings
3. wearing down or damaging surfaces; excess heat
4. as traction between wheels and the ground for riding in a car or on a bike; allows us to walk
5. water or air friction

✷ Follow Up

Assessment

Present students with the following hypothesis—friction is greater between snow and rubber than between smooth ice and rubber. Ask them to design an experiment to test this hypothesis.

Extension

Have students use the following method to test the frictional force between different surfaces. Use a thumbtack to secure a 2-m length of string to the end of a block of wood. Tie a bent paper clip to the other end of the string. Dangle this end of the string over the edge of a table. When washers are hooked onto the paper clip, the force of gravity on the washers will pull the block across the table. Tape different kinds of materials to the block and to the table, and record the force needed to move the block for each run. Try wax paper, aluminum foil, newspaper, cardboard, sandpaper, and strips of cloth. Find out which combinations produce the least and the most friction.

Friction . . .
- is greater for heavier weights.

- varies in size with the kinds of surfaces opposed to one another.
- is reduced when the opposing surfaces are separated by something round.

Now, answer the following questions about friction.

1. What causes friction?
2. What are some ways to reduce friction?
3. What harm can friction cause? Why?
4. Where is friction useful? Why?
5. What substances besides solids can cause friction?

 # Causes and Consequences of Friction

EXPLORATION 9

The Cause of Friction

Is it easy to move one piece of sandpaper over another? No—it's rough going! Turn over the two pieces of sandpaper and move their smooth surfaces over each other. Is it easier? Why is there a difference? The rougher the surfaces, the harder it is to move one over the other. The bumps and irregularities on the surfaces interlock with one another and disrupt motion. This causes friction.

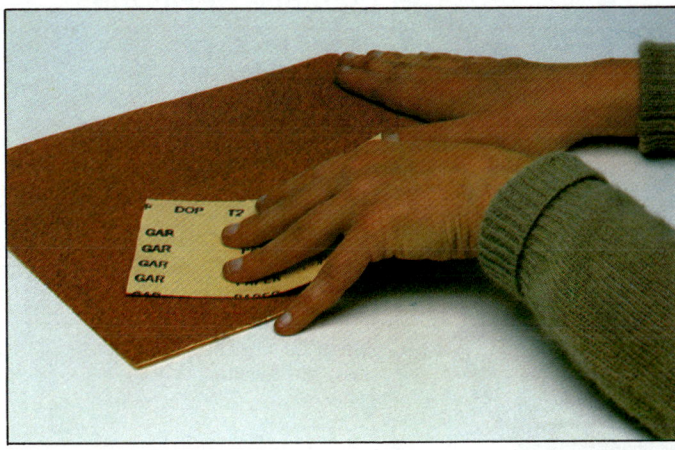

But there is friction even when the smooth sides are rubbed together. If you examine the smoother side with a magnifying glass, you see that it is not perfectly smooth. Test the smoothness of various surfaces listed below with your finger.

- piece of metal
- ordinary paper
- book cover
- window glass
- table top
- floor tile
- wall
- sandpaper

Reducing Friction

If we make rough surfaces smoother, we reduce friction. Cover the rough side of two pieces of sandpaper with margarine or grease. Now rub the surfaces together. Not all the friction is gone, but it is significantly less.

Try rubbing your finger on glass. Then place a drop of oil between your finger and the glass. Glass without oil seems quite smooth. Is there any difference with oil?

Roller bearings

Which surfaces are roughest? smoothest?

Even the smoothest surfaces appear rough when viewed with a microscope. Can you identify the objects shown in the magnified photograph below?

(ballpoint pen on paper)

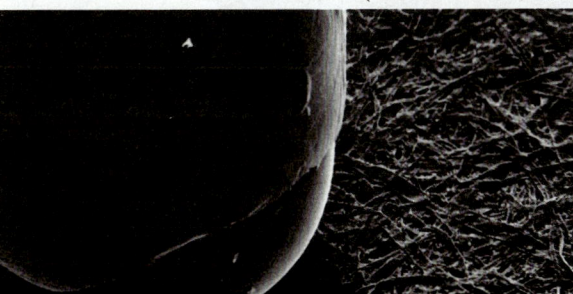

LESSON ORGANIZER

Objectives

By the end of the lesson, students should be able to:
1. Describe the cause of friction.
2. Explain ways in which friction may be helpful or harmful, and suggest ways to reduce or increase it.

Process Skills

observing, experimenting, drawing conclusions, creative writing

New Terms

none

(Organizer continues on next page)

LESSON

✸ Getting Started

In this lesson, students are introduced to the causes of friction and the ways in which friction acts on matter. Also, they learn about some ways to reduce friction. Then students perform some friction experiments and observe how ball bearings work.

Main Ideas

1. Friction results from the interlocking of opposing surfaces as they move past each other.
2. Friction may be reduced by smoothing opposing surfaces, by lubrication, or by separating opposing surfaces with ball bearings.

✸ Teaching Strategies

Have students observe the surfaces of different materials under a compound microscope. Ask: Which surfaces appear the smoothest? Which appear the roughest? Between which surfaces is there the most friction? the least friction?

Exploration 9

Divide the class into small groups and distribute the needed materials. Allow groups to work through all the parts of *Exploration 9*, except *Science "Friction"* on page 232.

At that point, reassemble the class and involve them in a discussion of what they discovered.

Reducing Friction

Students should find that when they put oil on the glass, friction is decreased.

Frictional Force 231

Science "Friction"

In this creative writing exercise, students should identify the sources of friction, and describe the effects of the lack of friction in everyday situations. Some examples:
- You cannot walk.
- You cannot turn a doorknob.
- Everything you touch begins to slide.
- If you try to open a door, it will crash into the wall and the force against the frame will cause the nails, no longer held by friction, to pop out, and everything will fall on the floor.
- If you try opening a drawer, it comes out in your hands and the contents fall out and move smoothly along the floor, bouncing off the walls without slowing down.

EXPLORATION 9—CONTINUED

It is very important to oil or grease the moving metal parts of machines. Can you suggest some examples where this is done? Often, metal surfaces which must move or turn against each other are separated by metal balls or rollers. This reduces the friction between the moving parts. Recall that rolling friction is less than sliding friction. Wheels of roller skates, bicycles, motorcycles, and cars turn on metal axles that are separated from the wheels by ball or roller bearings. Adding oil to the bearings further reduces the friction.

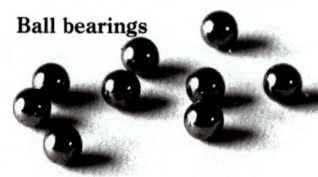
Ball bearings

Ball bearings in a bicycle wheel

The Harm of Friction

What harm does friction do? Suppose you rub two pieces of sandpaper together for 30 seconds. What do you observe? Feel how warm the surfaces are. You may also notice that the surfaces have been slightly worn down.

> **Friction**
> - opposes motion
> - produces heat
> - causes surfaces to wear away

Science "Friction"

In the following pages you will read about many examples of friction—some harmful, some helpful. Before you look at these pages, try writing a little science "friction."

Your home has been invaded by a UA (unfriendly alien) armed with an AF (anti-friction) spray gun. You open the front door to enter your house and . . .

Describe what happens in the next few minutes.

(Organizer continued)

Materials

Exploration 9
The Cause of Friction: sandpaper, magnifying glass, ordinary paper, book cover, microscope, piece of metal
Reducing Friction: margarine or grease, oil, drinking glass, sandpaper
The Harm of Friction: sandpaper, watch or clock with second hand
Exploration 10
Transportation by Brick: brick, metric ruler, newspaper or cloth, wooden dowels, margarine, grease, or motor oil, roller skate, large plastic bowl, water, large container to hold water
Comparing the Friction of Different Types of Paper: paper towel, wax paper, 2 jars, scissors, water, wooden dowel, string, hole-punch, various kinds of paper, cardboard, different grades of sandpaper, paper reinforcements
Reducing Friction: book, jar lid, glass marbles, 2 paint cans, motor oil, roller bearings, ball bearings, sleeve bearings

Teacher's Resource Binder

Activity Worksheet 4–5
Graphic Organizer Transparency 4–5

Science Discovery Videodisc

Disc 2, Image and Activity Bank, 4–8

Time Required

two class periods

EXPLORATION 10

Friction Projects

Transportation by Brick

What to Do

1. Move a brick a distance of 10 cm in various ways: dragging only the brick; wrapping it up and dragging it; dragging it on rollers; dragging it on a lubricated surface; dragging it on a roller skate; placing the brick in a large plastic bowl and dragging it over water. Compare the force needed in each of the various methods.
2. Draw conclusions about how friction is decreased in vehicles used for transportation.

Comparing the Friction of Different Types of Paper

What to Do

1. Set up your equipment as shown in the illustration. Remember, paper and jars must be of equal size.
2. Attach a string to the center of the wooden dowel. Pull gently on the string. Do the two jars move along evenly? Or does one lag behind, causing the dowel to rotate? Why? Where is the friction greater?
3. Add water to one of the jars until the two move along evenly. What does adding water do? How do the frictional forces compare now?
4. How can you use your data to compare the friction between waxed paper and the table, and between the paper towel and the table?
5. Compare the friction between other surfaces. Use various kinds of paper, cardboard, and sandpaper of different grades.

Reducing Friction

What to Do

1. Try rotating a book on a table.

2. Place a jar lid over enough marbles to fill the lid. Place the book on top and rotate it.

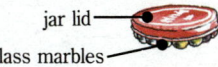
jar lid
glass marbles

3. Try rotating one paint can over another. Separate them with glass marbles in the outer ridge of the top of one can. Add oil and observe what happens.

4. Go to a garage, auto salvage shop, bicycle shop, or ball bearing company. Get samples of roller bearings, ball bearings, and sleeve bearings. Find out how each type works. Make a list of devices that use the different types of bearings.

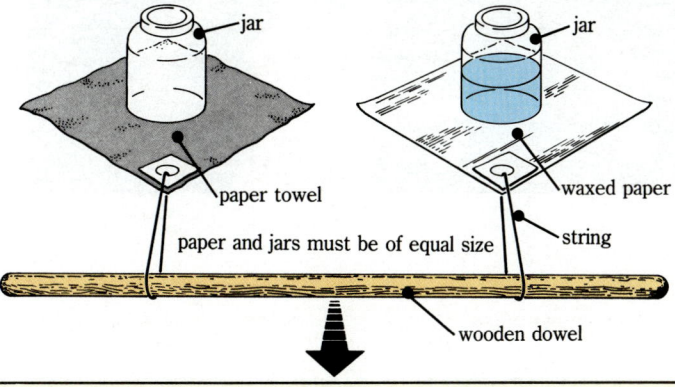

Exploration 10

Set up five stations around the classroom. Divide the class into small groups and ask them to travel from one station to the next performing the activity at each one.

Answers to In-Text Questions

Transportation by Brick

1. Dragging the brick on wooden dowels requires the least force; dragging it by itself requires the most force.
2. On land, wheels are used to reduce the friction between transportation vehicles and the surfaces on which they travel. Making the surface of the brick smoother or lubricating the surface on which it is dragged also reduces friction. The outside surfaces of vehicles are made as smooth as possible in order to reduce friction.

Comparing the Friction of Different Types of Paper

2. No, the jars do not move evenly. The jar on the paper towel lags behind because the friction between the paper towel and the table beneath it is greater than the friction between the wax paper and the table beneath it.
3. Adding water increases the friction between the wax paper and the table beneath it. The added weight of the water causes the downward force to increase. This increased force causes the irregularities of the wax-paper surfaces to press down on the irregularities in the table surface. This makes it more difficult for the two surfaces to slide past one another.
4. Most students should hypothesize that there is more friction between the table and the paper towel than between the wax paper and the table.

Reducing Friction

The marbles, which serve as bearings, make it much easier to turn the book and rotate the paint cans. Adding oil would make movement even easier. If students do not have access to different types of bearings, obtain several of each type so that students can observe them and experiment with how they work. The following information will help you describe the functions of each.

Ball bearings are placed in a groove between two surfaces. They roll back and forth so that the two surfaces do not rub each other.
They are found in wheels, roller skates, ball point pens, and light equipment.

Roller bearings are long, solid cylinders that roll; they are placed so that two surfaces roll over each other rather than rub together.
They are found in light and heavy equipment and in wheels.

Sleeve bearings look like thin tubing; they surround an inner, solid cylinder.
They are found in light equipment and in bushings in automatic transmissions.

Answers to
Friction: Friend or Foe?

The surfaces rubbing together to cause friction are the following:
Track shoes: bottom of the left shoe and the surface of the track; the bottom of the right shoe and the starting block
Parachute: air and the parachute; air and the parachutist
Race car: tires and the pavement; moving parts of the engine rubbing against each other; air and the car body
Skier: snow and bottom of the ski; skier's foot and the ski; ski pole and the ground
Chain saw: blade and log; hands and handle of the saw
Slide trombone: pieces of the slide; lips and mouthpiece

1. slide trombone, skier
2. parachute, chain saw
3. inside the car engine, the slide trombone
4. track shoe, parachute, tires on race car
5. *Track shoes:* have track changed to ice
   Parachute: reduce the size of the parachute
   Race car: make the car more aerodynamic; improve lubrication qualities of engine oil
   Skier: change the snow to ice
   Chain saw: lubricate the blade of the saw
   Slide trombone: add oil to the slide
6. *Track shoes:* put cleats on track shoes
   Parachute: add weight to the parachutist; increase the size of the parachute
   Race car: roughen the road surface; do not put oil in engine
   Skier: make bottom of skis rough
   Chain saw: push on saw with more force
   Slide trombone: remove oil from the slide

Friction: Friend or Foe?

Picture Puzzle

In the pictures on this page, identify the surfaces that are rubbing together to cause friction.

1. In which situations is there little friction?
2. In which situations is there a lot of friction?
3. Where is friction harmful?
4. Where is friction helpful?
5. How could you reduce friction in each of the pictures?
6. How could you increase friction in each of the pictures?

✷ Follow Up

Assessment

During the course of a day, when is friction useful? when is it harmful? *(Useful: for walking, running, skipping, riding a bike, riding in a car, riding in a bus, writing on the chalkboard, when pulling on a rope
Harmful: wears out shoes, tires, wheels, machine parts)*

Extension

Have students investigate the parts of a car that depend on friction for safe operation and the parts of a car in which friction is undesirable. *(Tires, belts, and brakes depend on friction for safe operation. The clutch, treads, pistons, transmission, and bearings are worn down by friction.)*

Challenge Your Thinking

Do you know the following friction facts? Read through them, and answer the questions about them. Afterward, find two friction facts of your own.

1. A hovercraft travels faster than a boat. Why is this so?
2. Chalk marks differently when its writing end is pushed rather than pulled across the blackboard. In what way are the marks different? Why are they different?
3. Archeological discoveries reveal that the wheel was first made in Assyria in 3000 B.C. How does the wheel reduce problems with friction?
4. A crucial part of a space shuttle is the 33,000 thermal panels for re-entry into the earth's atmosphere. Why are the panels so important?
5. Sliding quickly down a rope may burn your hands. Why?
6. An ancient method of lighting fires was to rub two pieces of wood together. How could this start a fire?
7. Shown here is a road sign used internationally. What does it mean?

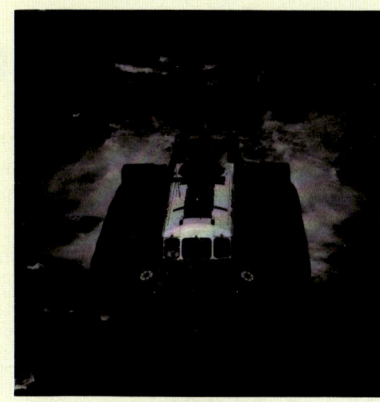

Answers to Challenge Your Thinking

1. A hovercraft travels faster than a boat because it hovers in the air above the water. By reducing its contact with the surface of the water, friction is also reduced. Friction with air is less than the friction with water.
2. When chalk is pushed across a blackboard, it experiences a greater downward force. This increase in force pushes the surface of the chalk farther into the surface of the board, leaving a heavier mark.
3. The wheel reduces friction because rolling objects experience much less friction than sliding objects. The points of contact are reduced. In fact, were the wheels not somewhat roughened at the point of contact with the surface, they would spin in one spot instead of moving across the ground.
4. Thermal panels prevent space shuttles from burning up from the intense heat caused by friction between the earth's atmosphere and the space shuttle.
5. Sliding down a rope while gripping causes excessive heat due to friction.
6. When two pieces of wood are rubbed together, the friction causes enough heat to be generated so that the wood ignites.
7. The road sign means "slippery when wet." Water on the road may cause the tires to slip, greatly decreasing the friction between the road and the tires.

Frictional Forces 235

LESSON 9

★ Getting Started

In this lesson, students take a look at the motion of objects and how forces affect this motion. They are then presented with five thought experiments that lead them to a conceptual understanding of Newton's first law of motion.

Main Ideas

1. If the forces acting on an object are balanced, the object's motion will not change.
2. An unbalanced force causes an object to speed up, slow down, or change direction.

★ Teaching Strategies

Explain that it is not always possible for scientists to perform experiments in the real world. For example, on earth it is difficult to carry out experiments to show how objects move in the absence of gravity. Instead, scientists perform thought experiments.

Call on a volunteer to read aloud the first two paragraphs on page 236. Divide the class into small discussion groups. Let them debate the answers to these questions within their groups. The thought experiments on pages 237 to 241 will allow them to test their ideas and change them if necessary.

Answers to

A Survey

1. Yes, a force is necessary to start a ball rolling on a level surface.

(Answers continue on next page)

236 Force and Motion

GOING THROUGH THE MOTIONS

 **Thought Experiments**

Everyone is familiar with the motion of objects, but not everyone has the same explanation for how objects move through space, or how forces are related to their motions. See what you can discover by conducting a survey.

Look at the questions in the survey below, and answer them on your own. Then ask the views of at least two other people. How do your views differ? How are they the same?

A Survey

1. Is a force necessary to start a ball rolling on a level surface?
2. Once the ball is moving, must a force be exerted on the ball to keep it moving?
3. Is a force required to slow the ball down?
4. In the 1500s, Galileo wondered what kept the moon circling about the earth. What explanation for the moon's orbit do you have?
5. Once the Apollo spacecraft got away from the earth on its path toward the moon, the astronauts shut off the rocket engines. Is a force necessary to keep the spacecraft going at over 3000 km/h?
6. The path of the Apollo spacecraft was straight; the path of the moon is circular. How would you explain this difference?

To find out whether or not your answers are correct, think about the following five thought experiments. Remember, scientists regularly perform thought experiments as they sort out their ideas and develop their hypotheses.

LESSON ORGANIZER

Objectives

By the end of the lesson, students should be able to:
1. Predict what happens to an object at rest or an object in motion when unbalanced forces act upon it.
2. Identify objects with balanced or unbalanced forces acting upon them by observing their motion.

Process Skills

observing, inferring, analyzing, drawing conclusions

New Terms

none

Thought Experiment 1

Over 350 years ago, Galileo performed an experiment similar to the one below. For the purpose of his thought experiment, Galileo regarded the surfaces involved in his experiment as being so smooth that there was no friction between them. Of course, there are no surfaces like that in the real world.

(a) When a ball rolls down the slope on the left, we know that it rolls up the slope on the right to the same height—2 m.

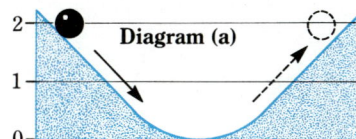

(b) The ball does the same thing here in diagram (b), but note how much farther it goes.

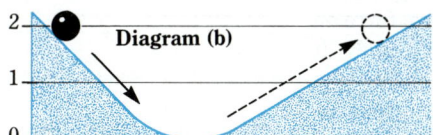

(c) How far would you predict the ball would go in diagram (c)? diagram (d)? diagram (e)? Is any force necessary to keep the ball moving along the horizontal plane (e)?

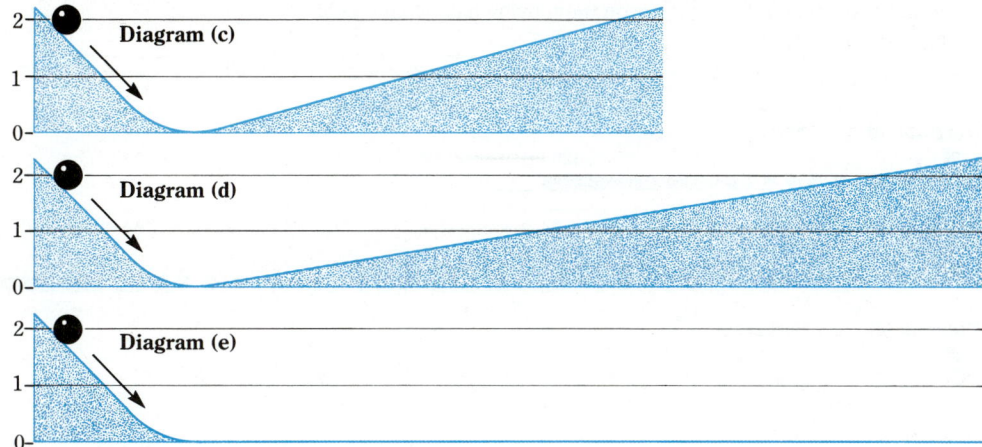

Galileo said that once an object is moving, it tends to keep on moving in a straight line without slowing down or speeding up. But, if a force were placed on this moving object, the force could slow it down, speed it up, or change its direction.

Materials

Thought Experiments: Journal

Science Discovery Videodisc

Disc 2, Image and Activity Bank, 4–9

Time Required

two class periods

Answers continued

2. No. However, the force of friction (and gravity as well, if it is moving upward) will eventually bring the ball to a stop. A steady force equal to the force of friction must be exerted on the ball to keep it moving at uniform speed.
3. Yes. On earth, the force of friction slows down a moving ball. The force of gravity slows down a ball that is moving upward. In outer space, another force would be required to slow down the ball.
4. If the moon were not being pulled in by the earth, it would travel in a straight line at a steady speed.
5. Once the spacecraft got far enough away from the earth, there was no friction and only minimal gravitational force opposing its motion. The force of the original thrust of the engines kept it going.
6. The path of the Apollo spacecraft was straight because it was moving directly away from the earth—opposite to the pull of the earth's gravity—and because the moon's gravity had little effect on the spacecraft until it began to orbit the moon. The path of the moon is approximately circular because the earth's gravity is acting on the moon as it is moving past the earth. The earth's gravity is great enough to cause the moon to orbit the earth.

Thought Experiment 1

Call on a volunteer to read aloud the first paragraph. Then direct students' attention to each of the lettered diagrams. Draw the diagrams on the board and ask volunteers to indicate where they think the ball will stop rolling in each case.

In diagram (a), in the absence of friction, the ball will move up to the same vertical height at which it started. In diagrams (b) to (d), as the slope becomes less steep, the ball will move farther before it is pulled back by gravitational force. In diagram (e), the ball will move forever on a level surface, since there is neither friction nor gravitational force to oppose its motion.

Hence, in Galileo's *Thought Experiment 1,* in addition to an object having a natural state of rest, a moving object has a natural state of steady motion when moving in a straight line.

Going Through the Motions

Thought Experiment 2

Call on a volunteer to read aloud the first paragraph on page 238. Suggest that students work individually to answer questions 1 and 2 in their Journals. After a specified amount of time, involve the class in a discussion of their diagrams and what can be inferred from them.

Answers to
In-Text Questions

1. The surfaces represented by the diagrams are:
   (a) grass
   (b) living room carpet
   (c) concrete road
   (d) bowling alley
   (e) a frictionless surface.
2. If there were no opposing force on the moving ball, it would keep moving in a straight line at the same speed forever.

Thought Experiment 2

Suppose Galileo lived in the twentieth century. He might think about the game of bowling. When a player places an initial force on the ball, the ball leaves his or her hand and moves away quickly. It meets an opposing force of friction. Diagrams (a) through (e) show what is likely to happen. Of course, how far the ball goes depends upon the surface on which it is moving.

1. In your Journal, match each diagram with the appropriate surface: grass, a concrete road, a frictionless surface, a bowling alley of indefinite length, living room carpet

2. If there were no opposing force on the moving ball, what would happen to its motion?

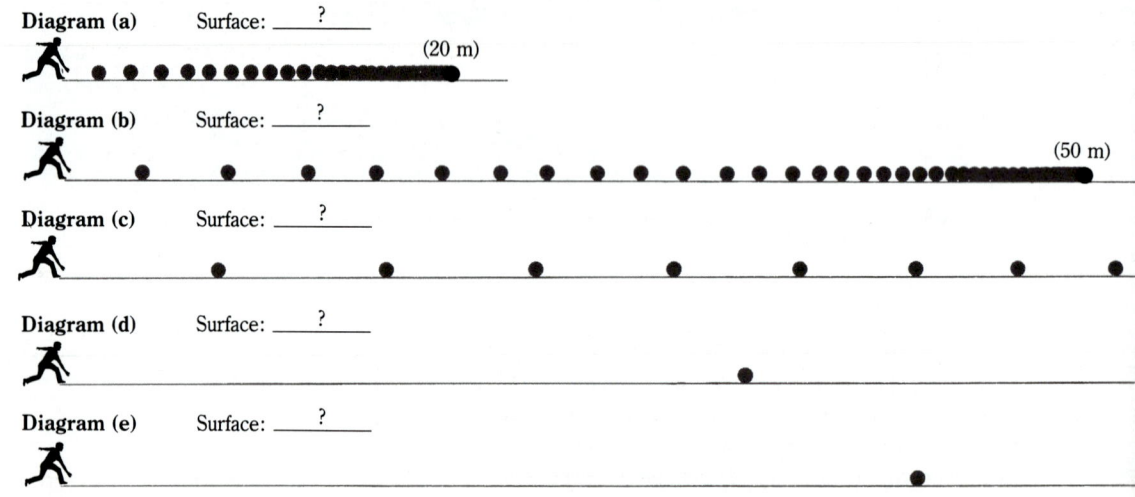

238 Force and Motion

Thought Experiment 3

Recall what you learned earlier about gravitational force.

1. In diagram (a), what stops Bob from throwing the ball higher?
2. Why can he throw the ball so much higher in diagram (b)?
3. What would happen to the ball if Bob threw it off the spacecraft in diagram (c)?

(a) Earth's surface 10 m

Questions, Part 1

If you have been working through these Thought Experiments carefully, some conclusions should already be clear. Record your answers in your Journal.

Conclusions	Questions
A force is needed to start an object moving.	What force starts the object moving in each Thought Experiment?
An object may keep moving when this force is no longer exerted on the object.	In each Thought Experiment, when was the force that started the object in motion removed from the object?
If there are no forces on the moving object, it will keep moving in a straight line with steady motion.	Which drawings in each Thought Experiment show that a force is needed to change the direction or the speed of motion?
Objects in motion may be slowed down by opposing forces.	What are the opposing forces in each of the Thought Experiments?

(b) Moon's surface 60 m

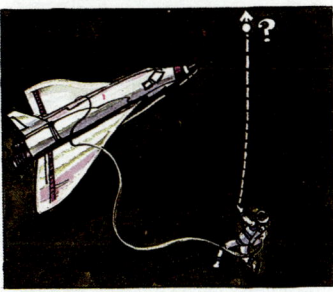

(c) Spacecraft

Answers to
Thought Experiment 3

Direct students' attention to the three diagrams at the side of page 239. Call on volunteers to answer the three questions by referring to the diagrams.

1. In diagram (a), the earth's gravity stops the ball from going higher.
2. In diagram (b), the ball can be thrown higher because the gravitational force on the moon is only one-sixth of that on earth.
3. A ball thrown off a spacecraft would respond to the gravitational pull of the earth and would go into an orbit similar to that of the spacecraft.

Answers to
Questions, Part 1

- The force that starts an object moving is: gravitational force in *Thought Experiment 1;* a push by the bowler's hand in *Thought Experiment 2;* and a push by Bob's hand in *Thought Experiment 3.*
- The force that started the object in motion was removed when: the ball reached a level surface in *Thought Experiment 1;* the ball left the bowler's hand in *Thought Experiment 2;* and the ball left Bob's hand in *Thought Experiment 3.*
- A force is needed to change the direction or the speed of motion: diagram (e) in *Thought Experiment 1;* diagram (e) in *Thought Experiment 2;* and diagram (c) in *Thought Experiment 3.*
- The opposing forces are: gravitational force in *Thought Experiment 1;* frictional force in *Thought Experiment 2;* and gravitational force in *Thought Experiment 3.*

Going Through the Motions

Thought Experiment 4

Ask students to silently read page 240. Suggest that they look carefully at the diagrams to answer the questions. After a specified amount of time, involve the class in a discussion of their answers. Be sure they understand the meaning of the term *unbalanced*.

Answers to
In-Text Questions

1. The weight does not fall because the upward force exerted by the string balances the downward gravitational force.
2. In diagram (b), the upward force exerted on *A* by the string attached to *B* balances the downward gravitational force on *A*. *A* and *B* are equal in weight; therefore, *A* does not move.
3. In diagram (c), *A* moves downward because the gravitational force acting on it (*A*'s weight) is greater than the upward force acting on it (*B*'s weight).
4. In diagram (d), *A* moves upward because the force pulling *A* up (*B*'s weight) is greater than the downward gravitational force acting on *A* (*A*'s weight).

Answers to
Questions, Part 2

- Don was unable to move the refrigerator alone because the force he exerted on it was less than or equal to the force of friction. Since the refrigerator was at rest, it stayed at rest.
- In *Thought Experiment 4,* there is an unbalanced force in diagrams (c) and (d). In each case, *A* moved due to this unbalanced force.

Thought Experiment 4

1. You are holding a weight, as in diagram (a). Why doesn't the weight fall?
2. In diagram (b), the added weight, called weight B, takes the place of your hand. Weight A is still not moving. Why?
3. What happens to A in diagram (c)? Why?
4. What happens to A in diagram (d)? Why?

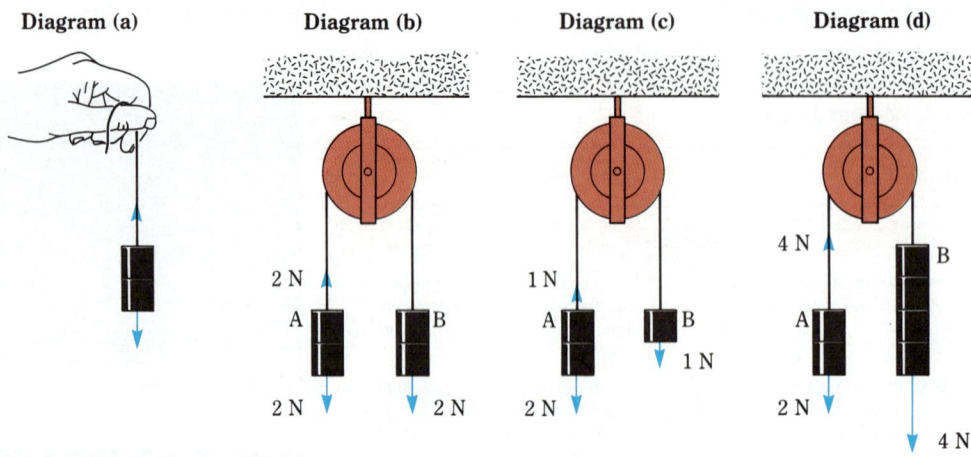

Diagram (a) Diagram (b) Diagram (c) Diagram (d)

Questions, Part 2

Draw some more conclusions from these Thought Experiments and record them in your Journal.

Conclusions	Questions
If forces on an object at rest are balanced, the object stays at rest. It is just as if there were no forces on the object.	Do you remember Don's attempt to move the refrigerator? Why couldn't Don move it?
An unbalanced force causes an object at rest to start moving.	In which parts of Thought Experiment 4 is there an unbalanced force?

240   Force and Motion

Thought Experiment 5

1. A car is moving at 70 km/h. If the car is allowed to coast freely in a straight line until it stops, what would slow it down?

Diagram (a) (The solid arrows refer to forces. The dotted arrows refer to the speed of the car.)

2. Now, instead of allowing the car to coast, the motor is kept running. The gas pedal is pressed just enough to maintain the car's speed at 70 km/h. What is the size of the force that the car's motor is providing to move the wheels along the road?

Diagram (b)

3. Suppose that the speed of the car is steadily increasing. How do you think the force of friction and the force provided by the motor compare? Are the forces on the car balanced or unbalanced?

Diagram (c)

Questions, Part 3

Here are the final conclusions to be drawn from the Thought Experiments.

Conclusions	Questions
If the forces on an object in motion are balanced, the object will stay at rest or in steady motion in a straight line.	Which diagram in Thought Experiment 5 shows balanced forces?
An unbalanced force causes an object to speed up or slow down.	Which diagrams in Thought Experiment 5 show this? What else can an unbalanced force do to an object in motion?

Answers to
Questions, Part 3

- In *Thought Experiment 5*, diagram (b) shows balanced forces.
- Diagrams (a) and (c) show that an unbalanced force causes an object to change speed. In (a), there is no force being exerted in the forward direction after the car begins to coast, so the force of friction slows down the car. In (c), there is a greater force provided by the car's motor in the forward direction. This forward force is greater than the force of friction. Therefore, the car speeds up.
- An unbalanced force can also cause a moving object to change direction.

Assessment

Ask students to use diagrams and to explain in writing how Galileo came to the conclusion that a ball will roll forever if there is no force to cause it to speed up or slow down. *(Galileo said that if a ball rolls down an incline on the left, it tends to roll up an equivalent height on the right. The ball rolls a greater distance as the angle of the incline on the right is reduced. If the plane on the right is horizontal, it will roll on forever.)*

Extension

Stand on a bathroom scale inside an elevator. Watch what happens to your weight as the elevator:
- starts up *(weight increases)*
- moves at a constant speed *(weight stays constant)*
- doesn't move at all *(weight stays constant)*
- moves up and stops *(weight decreases)*
- starts down *(weight decreases)*
- moves down and stops *(weight increases)*

Ask students to explain what happens in writing. They can also draw diagrams if they need to.

Answers to
Thought Experiment 5

1. The frictional forces of the road surface on the tires and the air against the body of the car would slow down the car.
2. The size of the force provided by the car's motor to keep the wheels moving is the same as the size of the opposing frictional forces. Therefore, in regard to the direction of motion, all forces on the car are balanced, and the car moves with constant speed in a straight line.
3. If the speed of the car is steadily increasing, the force provided by the motor is greater than the opposing force of friction. The forces on the car are unbalanced—the motor is exerting the greater force. Therefore, the car moves with increasing speed.

Going Through the Motions

Lesson 10

✷ Getting Started

In this lesson, students play a game of Puffball in order to apply their knowledge of forces and motion. They analyze the direction of motion that results when balanced and unbalanced forces act on a puffball in several different game situations.

Main Ideas

1. An object at rest with balanced forces acting on it will remain at rest; an object in motion with balanced forces acting on it will continue in a steady, straight motion.
2. An unbalanced force will cause an object at rest to start moving and an object in motion to speed up, slow down, or change direction.

✷ Teaching Strategies

Tell students that as they play a game of Puffball, they should analyze the forces that are changing the motion of the ball.

A game is always a fun way to learn about science. To begin, call on a volunteer to read page 242 so that students can learn how to play Puffball. However, before the game begins, have the class establish some rules. Some possibilities include: straws cannot touch the ball, keep your hands behind your back, do not touch the table with your body, and so on.

Divide the class into small groups and distribute the materials. Have as many setups of the game as possible.
(Continues on next page)

10 Puffball: A Game of Force and Motion

Here is a game you can play in class or on your own. As in any game, there are forces at hand. The purpose of Puffball is to blow a light ball into your opponent's goal. The straws cannot touch the ball, and your hands cannot touch the table.

You Will Need

- a playing area of approximately 1 m × 1.5 m
- a barrier about 4 cm high around the edges of the area
- 2 goal areas
- a plastic-foam or table tennis ball
- drinking straws
- 2 players (or teams of 2 or 3 players)

Play for 15 minutes and then write down all you have observed about the effects of two or more forces on the motion of a light ball.

LESSON 10 ORGANIZER

Objectives

By the end of the lesson, students should be able to:
1. Predict what happens to an object at rest when balanced and unbalanced forces act on it.
2. Predict what happens to an object in motion when balanced and unbalanced forces act on it.

Process Skills

playing a game, predicting, analyzing, drawing conclusions

New Terms

none

242   Force and Motion

Next, work through the situations on the following pages. A dotted arrow beside the ball indicates the direction in which the ball is moving before Magda or Pauline begins to blow. No dotted arrow means that the ball is at rest. The direction of the straw in Magda's or Pauline's mouth indicates the direction of the force.

Your task is to show the size and direction of the forces. Copy the diagrams into your Journal. Use solid arrows on your diagrams as shown in the examples below. Also, predict whether a point is going to be scored, and if so, who scores it. In a difficult case, *you* may set the conditions and decide the outcome. Two examples have been given to help get you started.

Example 1

Ball is at rest. Pauline is blowing the ball toward Magda's goal.

It will go in. Pauline gets a point.

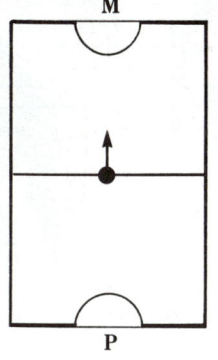

Example 2

Ball is at rest. Magda and Pauline are both exerting a force. Who is exerting the greater force? Where does the ball go? Who scores the point?

You decide that: Magda exerts a greater force. The ball goes in Pauline's goal. Magda scores.

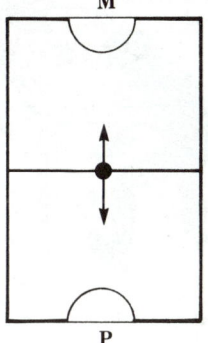

(Teaching Strategies continued)

After students play the game for a while, ask them to record everything they have observed about the effects of the forces on the balls. Then involve students in a discussion of what they have discovered about the ball's motion when one or more forces act upon it. The game and your discussion set the stage for the second part of the exercise: the analysis of forces being exerted on the ball and its resultant motion in different game situations.

Review the two sample game situations on page 243. Make sure students understand that they should copy the diagrams on page 244 and 245 into their Journals. Instruct them to draw arrows to show the size and direction of the forces acting on the ball in each diagram. Remind students that they need to predict who will score; when more than one outcome exists, students may choose the winner. Point out to students that the dotted arrows in the diagrams indicate that the ball is already moving before the forces described are exerted.

Materials

Puffball: playing area 1 m × 1.5 m, barrier about 4 cm high, 2 goal areas, plastic-foam or table-tennis ball, plastic drinking straws, masking tape, meter stick or metric ruler, watch or clock, Journal

Time Required

one to two class periods

Going Through the Motions

Answers to

The Games

(a) With a single force in the direction of Pauline, the ball moves into Pauline's goal. Magda scores a point.

(b) Pauline's greater force causes the ball to move in the direction of Magda, so the ball moves into Magda's goal. Pauline scores a point.

(c) The girls are exerting equal and opposite forces, but the ball is already moving toward Magda, so it continues to move in that direction at the same speed. Pauline scores a point.

(d) With a single force in the direction of Magda, the ball moves into Magda's goal. Pauline scores a point.

(e) With a single force in the direction of Pauline, the ball, which is moving toward Magda, slows down. The ball may either go into Magda's goal anyway, slow down, or, if Magda blows hard enough, the ball will stop and then move in the opposite direction into Pauline's goal.

(f) Both girls are exerting a force, but Magda's is greater. However, the ball is already moving toward Magda's goal. Magda's force may not be great enough to reverse the ball's direction, and Pauline may score. If Magda's force is great enough to stop the ball and turn it around, it will go into Pauline's goal.

(g) Both girls are exerting an equal force, so the ball remains at rest.

(h) With a single force in the direction of Magda, the ball, which is already moving toward Magda, speeds up and goes into Magda's goal. Pauline scores a point.

(i) Both girls are exerting a force, but Magda's is greater. The ball is already moving toward Pauline. Therefore, the ball speeds up as it moves toward Pauline and goes into Pauline's goal. Magda scores a point.

(Answers continue on next page)

The Games

(a) Only Magda is exerting a force.

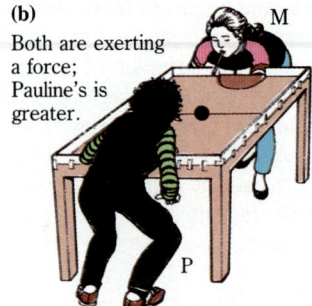
(b) Both are exerting a force; Pauline's is greater.

(c) Both are exerting an equal force.

(d) Only Pauline is exerting a force.

(e) Only Magda is exerting a force.

(f) Both are exerting a force; Magda's is greater.

(g) Both are exerting an equal force.

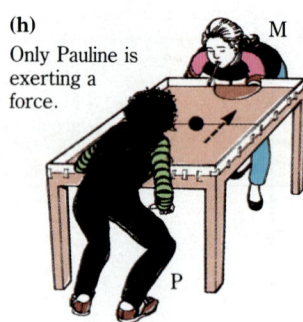
(h) Only Pauline is exerting a force.

(i) Both are exerting a force; Magda's is greater.

(j) Both are exerting an equal force.

(k) Neither is exerting a force.

(l) Both are exerting an equal force.

(m) Only Magda is exerting a force.

(n) Neither is exerting a force.

Questions

Check how well you have understood the forces and motion in Puffball by answering the following questions.

In which of situations (a) to (n) . . .

1. are there essentially no forces?
2. are there single forces?
3. are there balanced forces?
4. are there unbalanced forces?
5. is the ball at rest set into motion? Why?
6. does the ball remain in constant motion? Why?
7. does the ball, already in motion, slow down? Why?
8. does the ball, already in motion, speed up? Why?
9. does the ball, already in motion, change direction? Why?
10. does the ball at rest stay at rest? Why?

(never stops)

Answers continued

(j) Even though the girls are exerting equal forces, they are not opposite each other at the table. They are at right angles. The ball moves diagonally across the table. No one scores a point.

(k) Since the ball is moving steadily toward Pauline and no additional forces are being exerted, the ball goes into Pauline's goal. Magda scores a point.

(l) The forces being exerted by the girls are equal and opposite, so the ball continues to move to the left at a steady speed. No one scores a point.

(m) The ball is moving across the table when Magda blows on it, so it would move diagonally toward Pauline's corner. No one scores a point.

(n) The ball remains at rest since there are no forces acting on it. No one scores a point.

Answers to
Questions

1. n
2. a, d, e, h, m
3. c, g, l, j
4. b, f, i
5. a, b, d, j (A single or unbalanced force acts on the ball.)
6. c, k, l (There are no forces on the ball or the forces on the ball are balanced.)
7. e, f (There is a single force, or unbalanced forces oppose the motion.)
8. h, i (There is a single force or an unbalanced force in the direction of motion.)
9. m (There is a single force not in the direction of the motion.) The ball could also change direction in situations e and f.
10. g, n (No forces or balanced forces act on it.)

✸ Follow Up

Assessment

Have two volunteers play a game of Puffball. As they do, call on various students to give a running commentary on the forces that are acting on the ball to make it move.

Extension

Provide students with a bowling ball and a rubber mallet. Ask them to predict which direction the moving ball will take after it is hit with the mallet. *(It does not travel in the direction it is hit, but at an angle between its original direction of motion and the angle at which it was hit.)*

Going Through the Motions

LESSON 11

✳ Getting Started

With a series of simple experiments, this lesson illustrates what is meant by an object's inertia.

Main Ideas

1. The inertia of an object is its tendency to remain at rest when already at rest and to remain in motion, in the same direction and at the same speed, when already in motion.
2. The amount of inertia an object has depends on its mass.

✳ Teaching Strategies

Call on a volunteer to read the introduction to the new term *inertia*. Refer students to the two cartoons at the top of page 246. Explain that the cartoons show the mental state of *inertia*, as well as *inertia* in the scientific sense of the word. In the psychological sense of the word, it reflects an unwillingness or lack of desire to move or change. In the scientific sense of the word, the boy at rest tends to stay at rest and the girl in motion tends to stay in motion unless an outside force is applied. Point out that objects must have mass to have inertia. Also, explain that inertia is *not* a force.

Where Have You Met Inertia Before?

Call on another volunteer to read aloud the bottom of page 246. Involve students in a discussion of where they have encountered inertia before.
(Continues on next page)

11 What Is Inertia?

It takes a force to move a body.

When a body is moving, it takes a force to alter its motion.

You have seen that objects at rest tend to stay at rest. Objects in motion tend to continue in motion, maintaining a constant speed in a straight line. This tendency is called **inertia**. Inertia is not a force. It is the state in which an object remains as it is. A measure of an object's inertia is its mass. The greater the mass, the greater the inertia—that is, the greater the force needed to move the object.

Where Have You Met Inertia Before?

You are standing unsupported in a parked bus. The bus starts moving forward. What happens to you? What is inertia doing to you? Would holding onto something help?

At rest

Here you go!

LESSON 11 ORGANIZER

Objectives

By the end of the lesson, students should be able to:
1. Analyze the effects of inertia in everyday experiences.
2. Relate the inertia of an object to its mass.

Process Skills

predicting, experimenting, observing, analyzing

New Terms

Inertia—the state in which an object remains as it is. Objects at rest tend to remain at rest, while objects in motion tend to stay in motion at a constant speed in a straight line.

246 Force and Motion

You were at rest, so you tended to stay at rest. Therefore, as the bus moved forward, you tended to stay where you were. The result was that you fell backward.

Now the bus is moving along. You are standing in the aisle unsupported and are keeping your balance quite easily. Suddenly, the bus driver applies the brakes, slowing down and stopping the bus. What effect does inertia have on you?

Stop!

You were in motion. When the bus slowed down, you tended to continue forward. As a result, you fell forward.

Suppose you are sitting in the back seat of a car being driven at 80 km/h. Without warning and with tires squealing, the driver swerves around a sharp corner. What happens to you? In what direction do you tend to go in relation to the car? in relation to the road?

Can you think of other situations in which you have experienced the effects of inertia?

What does inertia feel like? Try devising some experiments to help you "feel" inertia. You may use such devices as a four-wheeled cart, a bicycle, a skateboard, or a see-saw. For example, try sitting on a cart blindfolded. Have a couple of friends pull the cart. Can your inertia tell you when they are starting, speeding up, slowing down, or stopping?

(Where Have You Met Inertia Before? continued)
Some possibilities include:
- When an airplane takes off, passengers are pressed back into their seats.
- When hurtling around in a roller coaster, your body is thrown from side to side when the roller coaster makes sudden turns, and you are pushed up when the roller coaster suddenly plunges downward.
- When a washing machine spins the water out of clothes, the water flies off in a straight line rather than taking the circular path that the clothes are forced to take.

Call on another volunteer to read page 247 aloud. Direct students' attention to the diagrams of the buses on pages 246 and 247. Ask students to relate some of their experiences standing up in buses, trains, or subways. Then involve the class in a discussion of the questions that are asked on both pages. If a car makes a sharp right hand turn, you will tend to keep moving in a straight line in relation to the road, and to the left in relation to the car.

Have students work in groups to design some of their own demonstrations of inertia. If you have a cart, skateboard, or small wagon available, one group could try the experiment suggested in the text. (Warn students to be careful not to hurt anyone with abrupt changes in motion.) When all the groups have finished their experiments, have them perform a demonstration for the class.

Materials

Exploration 11
Station 1: metric ruler, cardboard barrier, dominoes
Station 2: coin, piece of thick paper
Station 3: raw egg, hard-boiled egg
Station 4: newspaper, 2 flat sticks
Station 5: toy cart (plastic), thread, modeling clay, 2 bricks, long wooden board, 5 books, scissors

Science Discovery Videodisc

Disc 2, Image and Activity Bank, 4–11

Time Required

two class periods

Going Through the Motions 247

Exploration 11

Set up five stations around the room. If possible, have two sets of equipment at each station. Divide the class into groups of three or four. Ask each of the groups to rotate from one station to the next. When all the groups have finished making their observations, reassemble the students to discuss their observations.

Answers to
In-Text Questions

Station 1
When a strong force is applied, the frictional bonds between the bottom domino and the others are broken. The bottom domino moves, while the others remain where they are because of inertia. When a smaller force is applied, the frictional forces are not overcome, and more than one domino moves. Inertia is resistance to force. The greater the mass, the greater the inertia, or resistance to force. Therefore, when there are fewer dominoes, their inertia is more easily overcome.

Station 2
If the paper is pulled out quickly from under the coin, the coin should stay in place due to its inertia.

Station 3
The raw egg does not revolve easily because of the inertia of its semiliquid contents that are not stuck to the shell. The contents tend to remain at rest while the shell tends to rotate. As a result, the rotational movement is impeded. This does not occur with the hard-boiled egg because the inside of the egg is attached to the outside, and the force is applied to the entire egg as a unit.

(Answers continue on next page)

EXPLORATION 11

Experiencing Inertia

The five mini-experiments in this Exploration provide other examples of inertia. At each station you will find materials arranged as shown in the illustrations below.

Spend as much time as you need at each station. Then explain how inertia is working in each case. Record your observations and explanations of inertia in your Journal.

Station 1
Use a ruler to hit the bottom domino sharply toward the barrier. What happens if you hit it gently? Why? Can you reduce the pile completely, one domino at a time?

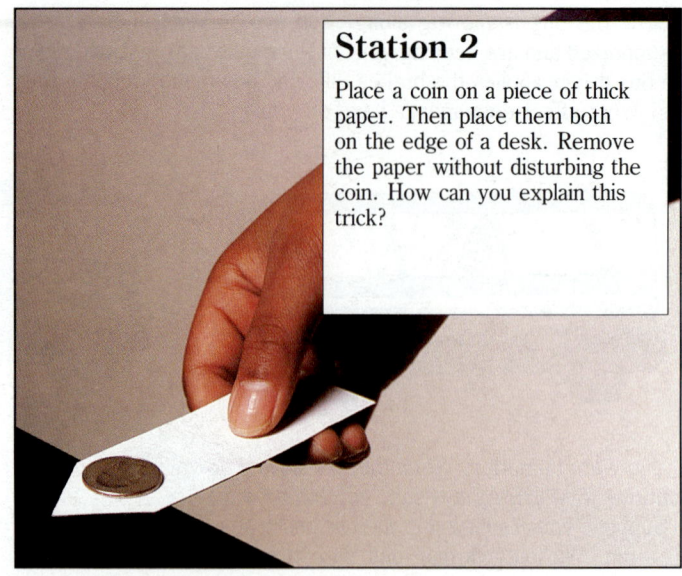

Station 2
Place a coin on a piece of thick paper. Then place them both on the edge of a desk. Remove the paper without disturbing the coin. How can you explain this trick?

Station 3
Which egg is hard-boiled? Spin each egg on a smooth surface. How does inertia help you decide which egg is boiled and which is not?

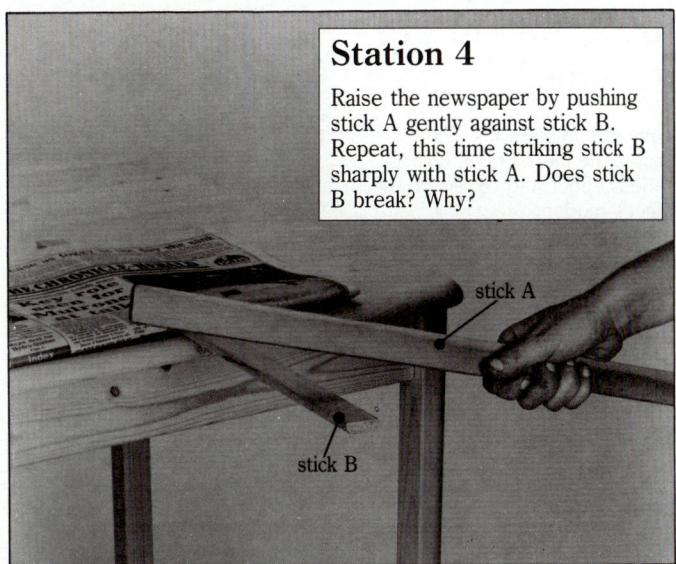

Station 4
Raise the newspaper by pushing stick A gently against stick B. Repeat, this time striking stick B sharply with stick A. Does stick B break? Why?

Station 5
Let the cart crash into the bricks. What happens to the clay passenger? Why? How far does it go? Does the height of the ramp make any difference? Use a thread to tie the clay figure to its seat. Try another crash. What happens? Could you devise a better seat belt?

four-wheeled cart with an improvised stationary seat

Answers continued

Station 4
Stick B will break if it is hit with enough force. The inertia of the newspaper and the air above it holds the stick in place. Therefore, it will break rather than move when it is hit.

Station 5
The clay passenger continues to move forward when the cart hits the bricks. The higher the ramp, the faster the cart will be moving when it hits the bricks, and the farther the clay passenger will be thrown. A seat belt made from a wider material, such as a piece of ribbon, would be better than using a piece of thread. The restraining force of the ribbon on the passenger would be spread over a wider area, so the passenger would be less likely to be thrown out or cut. Use this experiment to reinforce the value of wearing seat belts in cars.

Exploration 12

Remind students that mass and inertia are related. The greater the mass, the greater the inertia. As an introduction to the Exploration, ask the following questions. Students should answer these questions in their Journals prior to the Exploration and then again following it. Ask students if their answers changed.

- For bodies with greater inertia and greater mass, is a greater force needed to set the body into motion or alter its existing state of motion? *(yes)*
- If the same size force is placed on two bodies with different levels of inertia and thus different masses, would the motion of the two bodies be altered differently? How? *(Yes, the force would have a greater effect on the body with the lesser mass.)*

Answers to Questions

1. Students will observe that the smaller (less massive) pencil will move much farther than the heavier (more massive) one.
2. The forces applied by the two prongs of the clothespin are the same. Since the forces on both pencils are the same, and the heavier pencil travels a shorter distance, one could logically conclude that a given force has less effect on a body of greater mass.
3. Some practical examples of this principle are: a bicycle is much easier to stop than a train; a puff of air can move a puffball, but not a baseball; it is easier to stop a truck that is empty than a truck that is full.

EXPLORATION 12

Something to Try at Home

Prove this hypothesis: The greater the inertia of an object, the less effect a force has on the object's motion.

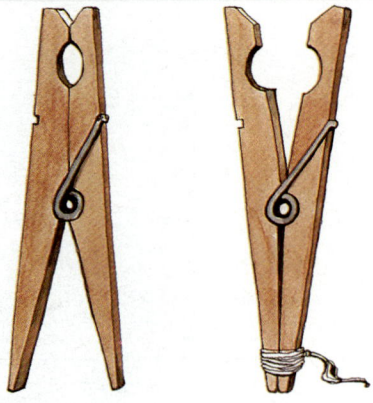

You Will Need

- a clothespin
- two pencils of different size
- string
- matches
- a cookie sheet

What to Do

1. You should do this activity with an adult present. Be sure no young children are playing nearby.
2. Tie the ends of the clothespin together with a string.
3. Set up the clothespin and the pencils on an overturned cookie sheet as shown in the drawing.
4. When the pencils are in place and everyone is safely away, burn the thread.
5. Try other objects of differing masses.

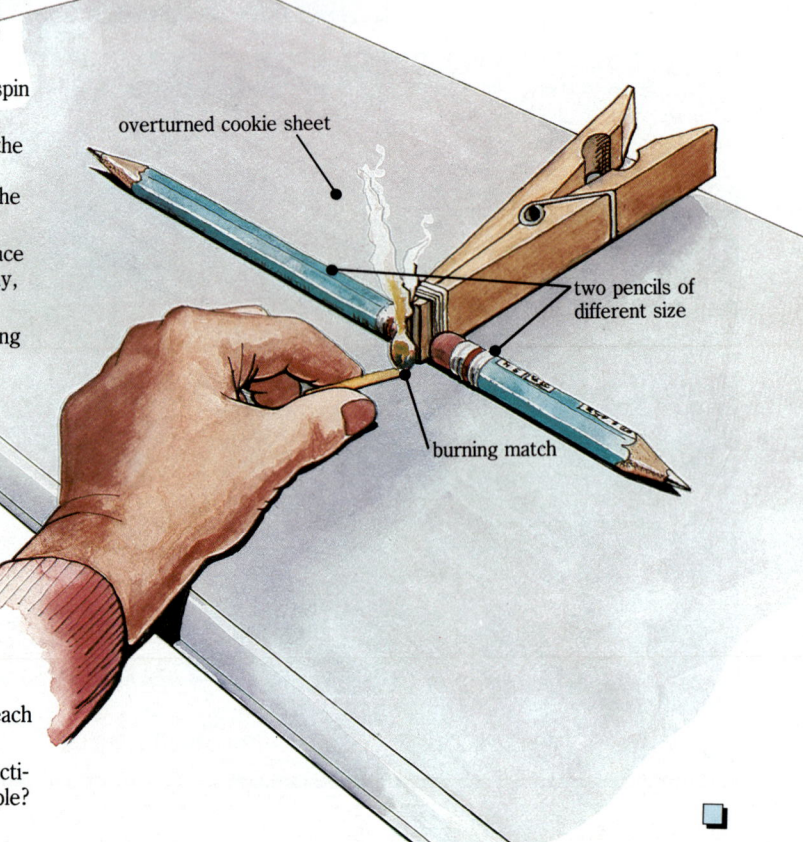

Questions

1. What do you observe?
2. Is the force the same on each pencil? Explain.
3. Can you think of some practical examples of this principle?

Follow Up

Assessment

1. Using ropes, suspend two cans with ropes from the ceiling or a doorway. One should be empty and the other full of sand. Both cans should have lids. Ask students to push each can in turn to find out the force required to swing each can the same amount. They should also try stopping them when they are moving. Ask: Which can has more mass? How can you tell based on what you know about inertia? *(Students should understand that the can with the larger mass [greater inertia] is more difficult to accelerate.)*
2. The concept of inertia could easily be called the "concept of lazy objects." Name some examples of "lazy objects." *("Lazy objects" would be those that have great mass and therefore great inertia.)*

Extension

Provide students with a roll of paper towels. Ask them to tear off a sheet with a quick jerk and then with a slow pull. Ask them to explain why a quick jerk is more effective.

12 Forces in Pairs

Until now, you have been looking at forces acting on the same object in balanced or unbalanced states. But it is also interesting to consider two forces acting on two different objects at once. Study Contests A, B, and C, and then work through the equations.

Contest A
Compare the size and direction of Deven's and Jill's forces. Do you observe that

The force of Deven on Jill	=	The force of Jill on Deven
(to the right)		(to the left)

Contest B
Mark and Tomás are having a force meter tug-of-war. Compare the size and direction of their forces. Do you notice that

The force of Mark on Tomás	=	The force of Tomás on Mark
(to the ___?___)		(to the ___?___)

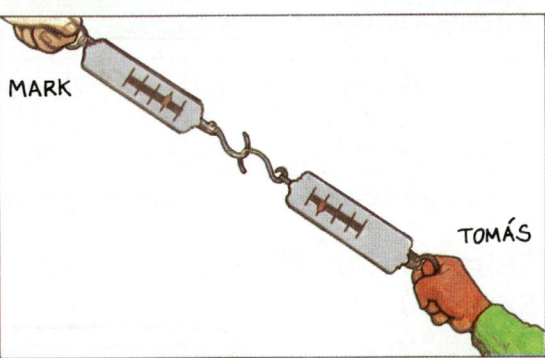

Contest C
Suppose Team Red pulls on Team Blue with a force of 800 N. With what force does Team Blue pull on Team Red?

The force of Team Red on Team Blue	=	The force of Team Blue on Team Red
(to the ___?___)		(to the ___?___)

LESSON 12 ORGANIZER

Objectives
By the end of the lesson, students should be able to:
1. Identify the reaction force for any action force.
2. Describe how action and reaction forces affect the motion of the two objects involved.

Process Skills
observing, inferring, predicting, analyzing

New Terms
Action—a force that acts in one direction.
Reaction—a force that acts in the direction opposite to an action.
(Organizer continues on next page)

LESSON 12

Getting Started

In this lesson, students investigate **Newton's Law of Action and Reaction**. First, they observe and analyze several illustrations of action and reaction forces. Then they participate in hands-on experiments involving action and reaction forces.

Main Ideas
1. If object A places a force on object B (an action), then object B places a force back on object A (a reaction), which is equal in size but opposite in direction.
2. Both action and reaction forces can produce or affect the motion of the objects upon which the forces are applied.

Teaching Strategies

Contests A and B
Using two bathroom scales, have two students demonstrate the action and reaction forces illustrated in the first contest. Contests A and B verify that when one object places a force on another object, the second object puts an equal but opposite force on the first object. In the illustration for Contest A, the force of 40 N that Deven is pushing with is equal to the force that Jill is pushing with.

Use two force meters to demonstrate what happens in Contest B. The same pulling force should be read on the two meters.

As long as the force meters are not moving, the force of Mark on Tomás (upward and to the left) is equal to the force of Tomás on Mark (downward and to the right).

(Contest C follows on next page)

Going Through the Motions   251

Do All Forces Act in Pairs?

You may recall that in Exploration 2, two strips of plastic were rubbed with a wool sock. They pushed each other apart. P and Q are both equally distant from the dotted vertical line. What does this suggest about the size of the two forces?

Now, use what you learned from Contests A, B, and C to write the "force-pair" equation for P and Q.

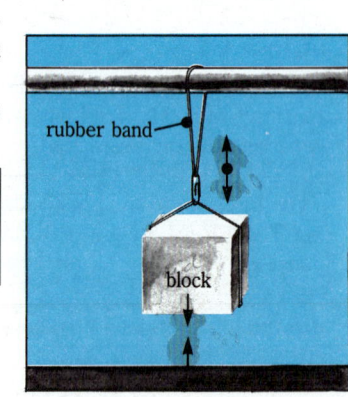

Isaac Newton's Reaction

In addition to his other important discoveries, Newton noticed that forces always occur in pairs. The forces are always equal in size to each other but opposite in direction. He called the forces **action** and **reaction**. Newton's formula states: For every action, there is an equal but opposite reaction.

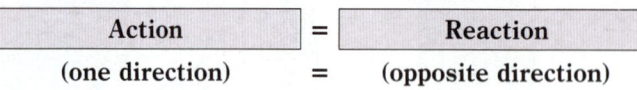

| Action | = | Reaction |

(one direction) = (opposite direction)

There is a clearer way to write this:

If A puts a force on B, then

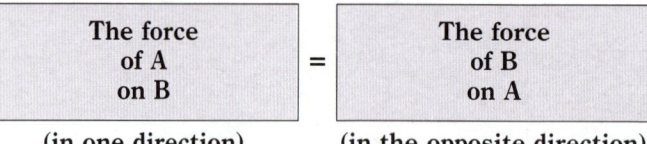

| The force of A on B | = | The force of B on A |

(in one direction) (in the opposite direction)

More Action-Reaction Pairs
Example 1

The block puts a force on the rubber band, stretching it downward. The rubber band puts a force on the block, holding it upward. Draw the force-pair equation.

Another pair of (noncontact) forces is also at work in this case:

| The gravitational force of the earth on the block | = | The gravitational force of the block on the earth |

(earth pulling the block downward) (the block pulling the earth upward)

Contest C

As long as neither team is moving, the force of Team Red on Team Blue (to the left) is equal to the force of the Team Blue on Team Red (to the right). Therefore, Team Blue pulls on Team Red with a force of 800 N.

Answers to
Do All Forces Act in Pairs?

The fact that P and Q are equidistant from the dotted vertical line suggests that the two forces are equal. The "force-pair" equation for this is: *the force of P on Q (to the right) = the force of Q on P (to the left)*

Answers to
More Action-Reaction Pairs

Example 1

Force-pair equation: *the force of the block on the rubber band (downward) = the force of the rubber band on the block (upward).*

Although the forces on the block and the earth are equal, the resultant motions are not equal. A small force on the block causes it to move, but such a small force exerted on the earth causes it to move an unmeasurable amount.

(Answers continue on next page)

(Organizer continued)

Materials

Exploration 13
Station 1: test tube with cork to fit, stand to suspend the test tube, candle, matches, thread, tape, scissors, safety goggles
Station 2: water bottle with cork to fit, seltzer tablet, paper towels, straws or wooden dowels, water, safety goggles
Station 3: roller skates, bag of sand
Station 4: balloon

Teacher's Resource Binder

Activity Worksheet 4–6
Graphic Organizer Transparency 4–6

Science Discovery Videodisc

Disc 2, Image and Activity Bank, 4–12

Time Required

two to three class periods

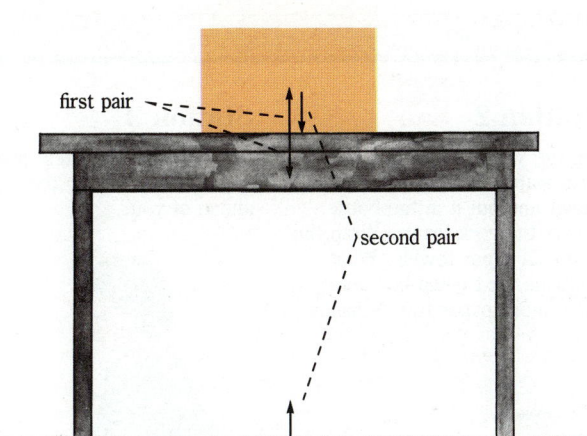

Example 2

What is the force-pair equation for each pair?

First pair: the block on the table

Second pair: the table on the earth

Example 3

Pat asked his friends to answer this question.

Here are their results. What is the force-pair equation?

Answers continued

Example 2

Force-pair equation for the first pair: *the force of the block on the table (downward) = the force of the table on the block (upward).* Force-pair equation for the second pair: *the force of the earth on the table (upward) = the force of the table on the earth (downward)*

Example 3

Force-pair equation: *the force of Pat on the wall = the force of the wall on Pat.*

To convince students that the wall is pushing back with an equal and opposite force, have them try this activity.

Exploration 13

Set up four stations, one experiment per station. If possible, have two sets of the necessary materials at each station. Divide the class into groups that can rotate from station to station. Then reassemble the class to discuss their observations.

Station 1: For more uniform and dramatic results, have students add 3 mL of water to the test tube and use a Bunsen burner instead of a candle.

Stations 1 and 2:
CAUTION: Students should not stand in front of the cork; it may become a projectile. Students should use extreme caution around these stations.

Answers to
Exploration 13

Station 1
Action: force of heated air inside test tube pushing against cork.
Reaction: force of cork pushing against heated air in test tube.
Observation: resultant motion is that both test tube and cork move, but cork moves farther because it has less mass.
Force-pair equation: force of heated air on cork = force of cork on heated air.

Station 2
Action: the force of the carbon dioxide gas against the cork.
Reaction: the force of the cork against the carbon dioxide gas.
Observation: the resultant motion is that both the bottle and the cork move, but the cork moves farther because it has less mass.
Force-pair equation: force of cork on carbon dioxide gas = force of carbon dioxide gas on cork.

Station 3
Action: the push of the skater against the bag of sand.
Reaction: the push of the bag of sand against the skater.
Observation: both the skater and the bag of sand move, but the bag of sand moves farther because it has less mass.
Force-pair equation: force of skater on bag of sand = force of bag of sand on skater.

Station 4
Action: the force of the balloon against the escaping air.
Reaction: the force of the escaping air against the balloon.
Observation: balloon moves in a direction opposite to the escaping air.
Force-pair equation: force of the balloon on the escaping air = force of the escaping air on the balloon.

EXPLORATION 13

Making Use of Reaction Force

The four mini-experiments in this Exploration show various uses of reaction force. At each station you will find materials set up as shown in the illustrations below.

In each situation, watch what happens. Identify the action-reaction pair of forces. What is each force doing? In your Journal, make a sketch, drawing in the appropriate forces. Write the force-pair equation for each situation.

Station 1

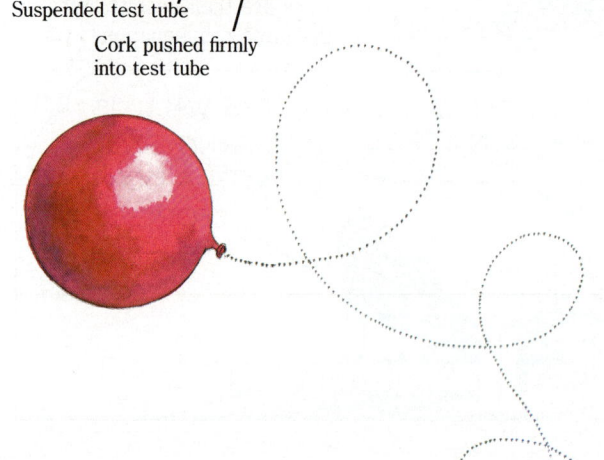

Suspended test tube
Cork pushed firmly into test tube
candle

Station 2

Fill the water bottle one-third full with water. Break up a seltzer tablet and put it in the bottle. Insert the cork firmly. Wrap the bottle in paper towels. What happens next could be messy, so have more paper towels handy.

straws or wooden dowels

Station 3

Stand on roller skates, holding a bag of sand. Throw the bag ahead of you.

Be Careful: *Use caution when standing on roller skates.*

Station 4

Blow up a balloon. Release it. What happens? How far did it go? Where did it go? Does the shape, size, or amount of air in the balloon matter? Explain.

✵ Follow Up

Assessment

R. H. Goddard was a pioneer in the field of rocketry. In 1919 he published a paper in which he proposed a way to propel rockets into space. Critics claimed that rockets would not work in space. They believed that a rocket is propelled forward by the push of exhaust gases on the atmosphere. In space there would be nothing for the gases to push against. Ask students to

Challenge Your Thinking

1. **Reaction Please!**
   Look at each situation below and identify the action and reaction forces involved.

2. Are forces, motions, and inertia found only in nonliving systems? Where do you find physical actions and reactions in your own body? Try doing a jumping-jack. Did you experience inertia? feel the force of gravity? cause an object to start moving? Draw a diagram showing direction of force in a person doing jumping-jacks.

Answers to
Challenge Your Thinking

1. • Action—the earth pulls the moon toward the earth; Reaction—the moon pulls the earth toward the moon.
   • Action—the girl puts a force on the wagon; Reaction—the wagon puts an equal and opposite force on the girl.
   • Action—the earth puts a downward force on the skydiver; Reaction—the skydiver puts an equal and opposite upward force on the earth.
   • Action—the boy on the right puts a force on the boy on the left; Reaction—the boy on the left puts an equal force in the opposite direction on the boy on the right.
2. Forces, motion and inertia are found in living as well as nonliving things. Physical actions and reactions are found in muscles and bones, the blood stream, and every organ. To perform a jumping-jack, a body must exert enough force to move despite its own inertia. A person must also temporarily overcome the force of gravity by jumping in the air. Arms, legs, head, and neck move during a jumping-jack.

explain the fallacy of this argument. *(The action force, which propels the rocket forward, is the force exerted by the expanding gases that result from fuel combustion. The reaction force is the rocket pushing against these gases.)*

Extension

Have students design a toy that is propelled by a reaction force. For example, they could attach a blown-up balloon to the back of a toy boat or car to propel it forward.

Going Through the Motions 255

UNIT 4
Making Connections

Summary for
The Big Ideas

Student responses will vary. The following is an ideal summary.

All forces are exerted by an agent that places the force on a receiver. A force always has an effect. All forces occur in pairs.

Gravitational force is the attractive force that exists between all matter. Gravitational force attracts two bodies of matter together.

Mass refers to the quantity of matter in an object. Mass is constant. Weight refers to the gravitational force that the earth places on an object. It changes with distance from earth.

The stretch of a rubber band or spring on a force meter is one way to measure force.

Mass is measured in grams or kilograms. Force and weight are measured in newtons.

Frictional force opposes the movement of an object or slows an object in motion. Friction increases with heavier masses and rougher surfaces. Roller bearings, oil, and smooth surfaces decrease friction.

An object at rest with balanced forces acting on it will not move, while a moving object with balanced forces acting on it will continue to move at the same speed, in the same direction. With unbalanced forces, an object begins to move or changes direction.

An action-reaction equation represents the concept that every action has an equal but opposite reaction.

Inertia is a characteristic of an object. It reflects the fact that a body at rest tends to stay at rest and a body in motion to stay in motion. A single force or unbalanced forces are required to change the state of motion.

Unit 4 Making Connections

The Big Ideas

In your Journal, write a summary of this unit, using the following questions as a guide.
- What characteristics do all forces share?
- What is a gravitational force and what does it do?
- How can you distinguish the mass of an object from its weight?
- What devices may be used to measure forces?
- What units are used to measure mass? force? weight?
- What does frictional force do? How can you increase it? decrease it?
- What happens to an object (at rest or in motion) if balanced forces are at work on it? unbalanced forces?
- What does an action-reaction equation represent?
- What is inertia? How does it relate to forces?

Checking Your Understanding

1. You began to learn about forces in the story of the sailboat race on page 194. Read the story again, and identify the following:
   (a) frictional forces (Name the agent, receiver, and effect.)
   (b) gravitational forces (Name the agent, receiver, and effect.)
   (c) buoyant forces (Name the agent, receiver, and effect.)
   (d) balanced forces
   (e) unbalanced forces
   (f) examples of inertia
   (g) pairs of action-reaction forces

2. One of the largest tug-of-war pulls in the world was performed by 1015 high school students in 1976.
   (a) If each student literally pulled his or her own weight, what force did the rope have to endure? (Estimate the average weight of a student.)
   (b) Can you identify an action-reaction force pair at work? What is its equation?

Answers to
Checking Your Understanding

1. Answers to questions (a) through (e) can be found in tables on page S161. Examples of inertia (f) include: the inertia of the boat opposes the force exerted by Kathleen and Kimiko to move it; the inertia of Kathleen on the boat is overcome by the force of the boom hitting her. For (g), wherever there is an action force, there is an equal and opposite reaction force.

2. (a) If each student is assumed to have an average mass of 55 kg, and thus a weight of 550 N, and each throws his or her total weight into the effort, the total force exerted on both sides of the rope would be 1015 students x 550 N/student = 558,250 N.
   (b) One action-reaction pair: the force of the students (on one side) pulling on the rope/the force of the rope pulling on those students. The force-pair equation: *the force of students (on one side) pulling on the rope = the force of the rope pulling on those students.*

(Answers continue on next page)

3. Jenny, after experiencing free-fall, opens her parachute and drifts steadily downward until her feet hit the ground. She is carried along for several meters before coming to a complete stop.
   (a) In each stage of her fall, describe what forces are acting on Jenny. Give the directions of these forces.
   (b) At what points are there unbalanced forces on Jenny? balanced forces?
   (c) When is Jenny most aware of inertia?
4. Draw a concept map showing the relationship between these words: *newtons, mass, matter, weight,* and *grams.*

Reading *Plus*

Building on what you know about force and motion, you can discover more about describing motion, the laws of motion, and the relationship of force to work and power. Read pages S61–S76 in the *SourceBook* to learn about the physical laws of force, motion, work, and power.

Updating Your Journal

Reread the paragraphs you wrote for this unit's "For Your Journal." Then rewrite them to reflect what you have learned in studying this unit.

Answers continued

3. (a) During free fall, the downward force of gravity on Jenny is much larger than the upward force of wind resistance on her body.
   (b) When she opens her parachute, she will feel like she is traveling upward at first, but she will really be feeling the force of wind resistance against her chute slightly overcoming the force of gravity (unbalanced force). Once she has reached her maximum velocity, she will continue downward at a steady speed until she reaches the ground (balanced force). At that point, the force of the ground upward overcomes the downward force of gravity, stopping her fall (unbalanced force). Wind blowing on her chute propels her several meters before friction between her feet and the ground causes her to stop (unbalanced force).
   (c) Jenny is probably most aware of inertia when she hits the ground.

4.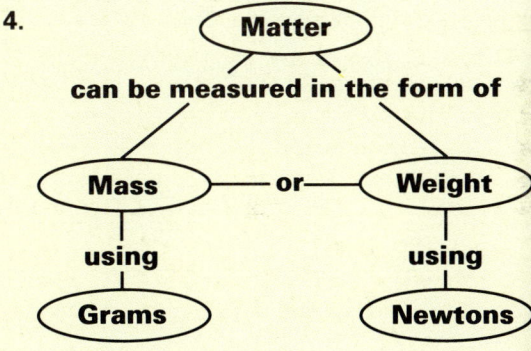

About Updating Your Journal

The following are sample ideal answers.
1. A force is a push or pull, or a reaction against a push or a pull, that affects matter.
2. The forces primarily responsible for the changes in the landscape are gravity and friction.
3. The agent of force is the ball. The ball, which is in motion, applies a force to the glass. The effect of the force on the glass is that the glass shatters. *(The relationship between the agent and the receiver illustrates the theme of **Systems and Interactions**.)*

Making Connections   257

Science in Action

◆ Background

Electrical engineers design, develop, test, and supervise the manufacture of electrical and electronic equipment. Electrical equipment includes: power generating and transmission equipment used by electric utilities, electric motors, machinery controls, automobiles, aircraft, and the lighting and wiring in buildings. Electronic equipment includes: radar, computers, communications equipment, and consumer goods such as televisions and stereo sets. Electrical engineers also design and operate facilities for generating and distributing electric power.

Science in Action

Spotlight on Electrical Engineering

Electrical Engineering Technology

Craig Gamble is an electrical engineering technologist for a company that supplies electrical energy. He earned the qualification for this position by completing a two-year program in electrical engineering technology after graduation from high school.

Q: What do you do?

Craig: I design the systems that carry electricity to homes and other buildings.

Q: What do you find most interesting about your work?

Craig: I think the best thing is seeing a job through, from the beginning stages to its completion. I follow a job through budgeting, design, and completion in the field.

Q: Is the science you studied in high school important in relation to your work now?

Craig: Most of my technical knowledge, aside from on-the-job experience, I gained while training to be a technologist. But basic ideas of mathematics and physics—particularly electricity—I learned in high school. Especially important in my present work is understanding the transmission of electrical energy and how high voltages carried by transmission lines are "stepped down" to the lower voltages required in homes and industry.

Q: Do you tend to do your work alone, or does your work involve cooperation with others?

Craig: I am very much involved with other people in my work. In fact, I would say that the ability to work with other people is very important. First of all, I usually consult colleagues in my own department about my new designs. Then having done the design work, I work with several other departments, some concerned with the technical aspects, others concerned with customer service. Then I work with the builders of the system, and then with its operators.

Research and Training

Wagih Fam is a professor of electrical engineering and the head of the Electrical Engineering Department at a university.

Q: What do you do, and what do you find most interesting about your work?

Wagih: My working life can be divided into two parts: teaching and research. As far as teaching is concerned, the joy lies in

◆ Using the Interview

Have students read the two interviews. Perhaps one student could play the role of the interviewer and two other students could play the roles of the interviewees. Use the following questions to promote discussion.

- What are the differences between the job of an electrical engineering technologist and that of an engineer? *(Electrical engineering technologists are involved with practical design and production work. They are not involved with jobs that require extensive theoretical scientific and mathematical knowledge.*

   Engineers who work in research and training apply theories and principles of science and mathematics to practical, technical problems. They are the link between scientific discovery and its useful application.)

- What traits are important for an electrical engineering technologist? *(They should be able to work as part of a team, be creative, and have an analytical mind.)*
- What traits are important for an engineer in research and training? *(They should have the same traits as those mentioned above, plus a love of teaching and a desire to do research.)*
- Do you enjoy making things like kites, windmill generators, and crystal radios?
- Do you think you have the personality and talents to pursue a career in electrical engineering?

giving knowledge to my students. It is particularly rewarding to discover that they have really understood what I have said and are able to use and apply this new knowledge.

As for my research, I find it very exciting to obtain new results and new knowledge. I spent three years on my Ph.D. project. I studied and measured the flow of electrical and magnetic forces in the air gaps of electric machines.

When the method you apply in researching or in designing a project succeeds, it is a great joy. When it fails, you experience great frustration. But the most stimulating thing, of course, is obtaining new knowledge and passing it to other people through your publications.

Q: What activities as a young person eventually led you into engineering?

Wagih: In my earlier days, I had an interest in building toys. My brother and I used to have a lot of fun making our own kites. We didn't go to a store and buy a kite and fly it. We had to make them ourselves and then make them fly. Of course, sometimes we would meet with disappointment. An accident, like punching through the paper or cutting the tail off, might happen. The kites might have to be rebuilt. We would try to make them better, always improving on the previous design.

In my final year in school, I started building my own electromagnets. With one of these, I built a doorbell. Later on, my brother and I started to wind transformers and motors, and to repair radio receivers.

Aside from these things, I was interested in painting, drawing, and drafting. That interest gave me an edge later on when I went into engineering and studied engineering drawing.

Some Project Ideas

1. Build a device that operates with power from a solar cell.
2. Construct a windmill generator that converts the energy from the wind into electrical energy. Use a small windmill directly coupled to a small DC generator. Use the electrical power output from the generator to light up a few flashlight bulbs or to drive a small electric motor.

For Assistance

Advice and assistance can be obtained from electrical engineers, electronics technologists, and electronics teachers. Most of the materials needed can be purchased at an electronics supply store or through a scientific supply house. Your teacher can give you an address for one of these. Other materials may be found at home.

◆ Using the Project Ideas

Have students read silently Some Project Ideas and choose the one that interests them the most. You may wish to have students who choose the same project work together. Solar cell and model windmill kits are available through Edmund's Scientific Supply House at (609) 547-3488. As an alternative, give students the opportunity to plan a research project based on an idea of their own.

Help students to locate professionals in the electrical engineering field who could help them with one of the project ideas. Students could write to the following address for possible contacts in their area.

Institute of Electrical and Electronics Engineers
United States Activities Board
111 19th Street NW
Washington, D.C.
20036

◆ Going Further

Ask students to investigate the following questions:
- How is electrical engineering alike and different from other engineering careers?
- What are the most challenging problems in electrical engineering today?
- Who are some important electrical engineers from the past and the present? What were their accomplishments?
- Research has shown a connection between nearness to high tension power lines and birth defects and other health problems. Do you think there could be a connection? Do some research to find out.

Science and Technology

◆ Background

Rockets have been used in warfare for hundreds of years. Today's rockets are extremely accurate. They can hit and destroy airplanes and missiles that fly faster than the speed of sound.

Scientists use rockets for exploration and research in the atmosphere and in space. Rockets carry scientific instruments to gather information about the upper atmosphere. Since 1957, rockets have shot hundreds of satellites into orbit around the earth. Rockets also carry instruments far into the solar system to explore the other planets and their moons.

The use of rockets for human space flight began in 1961. In 1969, rockets carried astronauts to the first landing on the moon. In 1981, rockets launched the first space shuttle. In the future, rockets will be used to carry people to Mars and possibly to other planets.

Science and Technology

Rocket Science

Rockets were used by Chinese armies in A.D. 1232. How effective they were as weapons we do not know for sure, but the use of rockets for military purposes spread throughout the world. In the War of 1812, for example, rockets carrying explosives were used by the British against the United States—which is what gave us the words "the rockets' red glare" in our national anthem. And certainly, we are familiar with the terrible destructive power of rocket-launched warheads in modern warfare.

The majority of rockets in the world today are still military. But the largest rockets, by far, are the ones used for space technology. It takes tremendous amounts of power to send loads into space.

The earliest rockets were fireworks.

The Saturn V launch vehicle was used to carry Apollo spacecraft to the moon.

Measuring Rocket Power

The power of rocket engines is described in terms of the amount of force, or thrust, they produce. For a rocket to get off the ground, its thrust must be greater than the weight of the rocket itself, including its fuel. Otherwise, it would just sit there.

The most powerful rocket of them all was the Saturn V, which carried astronauts to the moon. The cluster of five engines in its first stage had a combined thrust of over 33,000,000 newtons. By comparison, the engines or *boosters* that lift today's space shuttle missions off the ground produce a combined thrust of 25,800,000 newtons.

◆ Discussion

Promote discussion in the classroom by posing the following questions.

1. When were the first rockets used in military battle? *(In A.D. 1232, the Chinese armies used rockets.)*
2. What does it take for a rocket to lift off the ground? *(A rocket's thrust must be greater than the weight of the rocket.)*
3. In regard to rockets, what is the fuel dilemma? *(To propel heavier loads into space, more fuel must be used to produce more thrust. But, when more fuel is added, so is more weight. This makes it harder to lift the rocket off the ground.)*
4. In a multistage rocket, the first stage falls away from the rest of the rocket after burning its fuel supply. Then the second stage ignites and carries the payload farther out into space. How does this help to solve the fuel dilemma? *(The second stage does not need to propel the weight of the first stage. It needs only to propel its own weight.)*

The Fuel Dilemma

To get heavier loads up into space, you must use more fuel to produce more thrust. But when you add more fuel, you also add more weight to be carried—an enormous amount of it, as it turns out. A major part of the work that rocket fuel must do is to carry its own weight, as well as the weight of all the tanks and equipment that hold and transfer the fuel. In fact, the weight of the fuel in a vehicle like Saturn V is much greater than the weight of the *payload*—the object that is actually sent into space.

Getting up to Speed

For space exploration, it's not enough just to get a load off the ground. It must be brought up to a high velocity. In the case of a mission to the moon or the planets, a spacecraft must reach a velocity known as the escape velocity. For an object leaving the earth, this velocity is 40,000 km per hour. If escape velocity is not reached, a spacecraft will start to slow down the moment its engines are shut off and will eventually stop. Then it will start to fall back to Earth.

The Long Haul

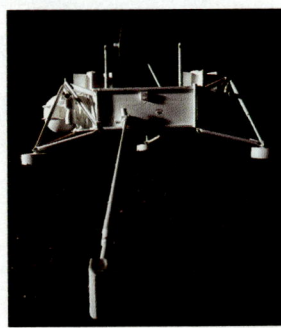

Mars lander

In designing a rocket engine, you would also have to consider how long the trip will last. In the case of a mission with a human crew, a time of 10 years might be considered entirely too long! And remember—the longer the mission, the greater the amount of supplies you would need to take along. So, for a mission to be practical, you might have to achieve a velocity even greater than the escape velocity. But greater velocity requires more fuel, and this adds more weight to the rocket. Obviously, the engineering of a rocket for a space mission requires a lot of thought and planning.

Find Out for Yourself

Suppose you are a part of a team of scientists who are designing a mission to Mars. You want to land on Mars, explore the planet, and then return to the earth. What questions would you raise about the different forces and velocities involved? Do some research and try to find answers to your own questions.

Space technology is not cheap. The cost of building a space shuttle is over 2 billion dollars, and the average cost of a flight is over 200 million dollars! How much of the cost comes from taxpayers? Do you think we as a nation are sacrificing other things in order to pay for our space program? Do space efforts pay for themselves in benefits? Would you be in favor of sending missions, or a series of them, to the moon or Mars? Why?

✦ Critical Thinking

To promote logical-thinking skills, you may wish to ask questions similar to the following:

1. How do you think the use of rockets helped to determine the outcome of the recent war in the Persian Gulf? *(The United States had guided missiles that could accurately seek out and destroy both stationary and moving targets. These included Patriot missiles that could destroy Scud missiles in flight.)*
2. How is a balloon similar to a rocket? *(In each, the reaction force due to the flowing gas propels the balloon or rocket.)*
3. Missions to Mars could be launched from a space station or from the moon. Why is this more practical than launching a rocket to Mars from Earth? *(The rocket would not have to carry fuel necessary to reach the escape velocity required to leave Earth.)*

✦ Going Further

1. Ask students to decide if there should be less research and development of rockets that launch nuclear warheads in light of the breakup of the Soviet Union. Then have students write persuasive paragraphs explaining their thinking.
2. Have students research the use of meteorological rockets for collecting information about the atmosphere.
3. Have students research the design and use of the two basic kinds of rockets: solid-propellent rockets and liquid-propellent rockets.
4. Have students research how to build and launch model rockets. With the help of a hobby store, they may be able to build and launch a model rocket. A single-stage rocket requires less skill and experience to make than a multistage or glider rocket.

UNIT 5
Structures and Design

✴ Teaching the Unit

☼ Unit Overview

In this unit, the science of structure is examined from the point of view of technology. In the section *The Science of Structure,* students explore the integrity of structures and the forces acting upon these structures. Students are introduced to some basic design elements used in engineering, including different kinds of beams, trusses, arches, and domes. Students then examine how these design elements influence the durability of various structures. In the final section, *The Art of Design,* students take an imaginary journey through history to explore the development of structure and design. Students then focus their attention on modern urban environments and, by the end of the unit, integrate what they have learned to create their own urban designs.

☼ Using the Themes

The unifying themes emphasized in this unit are **Scale and Structure, Systems and Interactions,** and **Stability.** The following information will help you weave these themes into your teaching plan. A focus question is provided with each theme as a discussion tool to help you tie the information in the unit together.

Scale and Structure is a dominant theme in each of the lessons. The unit focuses on the science of structure and the way in which the design of a structure is often determined by its size and scale in relation to its surroundings.

Focus question: *Why might a designer choose a dome to create a large, open space in a structure?*

Students should understand that designers take advantage of the way that a dome reacts to the forces upon it. The downward force of gravity is displaced laterally, so that interior supports are not necessary. The dome shape also allows for a very high ceiling without interior support.

Systems and Interactions is developed in Lessons 1, 2, 3, and 5 through observations and discussions of how the elements of a structure interact to maintain the structure's integrity.

Focus question: *In designing a bridge, what factors might a designer take into account when choosing a system of structural support?*

Designers should take into account all of the forces in the environment that could interact with the bridge structure to cause its failure. For example, the amount and type of traffic the bridge is expected to support, the distance spanned, wind, temperature changes, ice formation, and possibility of earthquake are all factors that could interact with the structure.

Stability can be discussed in connection with Lessons 1, 2, and 3. This theme can be developed by discussing how structures are designed so that they will last.

Focus question: *What design features contribute to a structure's durability?*

Students might want to refer to ancient structures depicted in this unit to answer this question. For example, in Roman architecture, the use of concrete arches combined a durable material with a design element that withstood the forces of gravity. Thus, the strength of materials, design features, and the interaction of these factors with the environment contribute to a structure's longevity.

☼ Using the *Science Discovery* Videodiscs

Disc 1 *Science Sleuths,* The Collapsing Bleachers
South Youth Organization built bleachers for Water Derby Day. One of the bleachers collapsed, and now the city is threatening to take away the organization's funding. The head of the organization thinks the bleachers were sabotaged by someone who hates the group. The Science Sleuths must analyze the evidence for themselves to determine the true cause of the structure's failure.

Disc 2 *Image and Activity Bank*
A variety of still images, short videos, and activities are available for you to use as you teach this unit. See the *Videodisc Resources* section of the **Teacher's Resource Binder** for detailed instructions.

☼ Using the *SciencePlus SourceBook*

Unit 5 takes a broad look at design by presenting ways in which other cultures through history have solved their own design problems. Other influences on design, including the environment and religion, are also discussed.

PLANNING CHART

SECTION AND LESSON	PG.	TIME*	PROCESS SKILLS	EXPLORATION AND ASSESSMENT	PG.	RESOURCES AND FEATURES
Unit Opener	262		observing, discussing	For Your Journal	263	Science Sleuths: *The Collapsing Bleachers* Videodisc Activity Sheets TRB: Home Connection
THE SCIENCE OF STRUCTURE	264			Challenge Your Thinking	292	
1 Planning the Structure	264	3	analyzing, evaluating, designing models, constructing models	Exploration 1	271	Image and Activity Bank 5–1 TRB: Resource Worksheet 5–1 Graphic Organizer Transparency 5–1
2 Response of Structures to Force	272	4	analyzing, graphing, designing models, constructing models	Exploration 2 Exploration 3 Exploration 4	273 275 279	Image and Activity Bank 5–2 TRB: Resource Worksheets 5–2, 5–4, and 5–5 Activity Worksheet 5–3 Graphic Organizer Transparencies 5–2, 5–3, and 5–4
3 Engineering Structures	282	3	analyzing, measuring, observing, constructing models	Exploration 5 Exploration 6	283 286	Image and Activity Bank 5–3 TRB: Activity Worksheet 5–6 Resource Worksheet 5–7 Graphic Organizer Transparencies 5–5, 5–6, and 5–7
THE ART OF DESIGN	293			Challenge Your Thinking	309	
4 Beauty in Structures	293	2	analyzing, researching, writing, illustrating	Exploration 7 Exploration 8	293 301	Image and Activity Bank 5–4
5 A Question of the Environment	304	2	analyzing, evaluating, designing models, constructing models	Exploration 9	308	Image and Activity Bank 5–5
End of Unit	310		applying, analyzing, evaluating, summarizing	Making Connections TRB: Sample Assessment Items	310	Science in Action, p. 312 Science and the Arts, p. 314 *SourceBook,* pp. S77–S88

*Time given in number of class periods.

Meeting Individual Needs

Gifted Students

1. Suggest to students that they design "moon shelters" for future residents of the moon. Point out that they will have to consider how their shelters will provide air, food, and water as well as protection from the moon's harsh environment. Call on several volunteers to show and explain their designs to the class. Display the remaining "moon shelters" around the classroom.

2. As an extension to *Exploration 1,* students might enjoy constructing a more elaborate bridge. They will need four ring stands (for the supports), cardboard (for the deck), and string (for the cables). Encourage students to try different arrangements of string, which can be threaded through the cardboard. When students have settled upon a design that they like, they should then test their designs against forces in the environment such as weight and wind. (A fan can be used to supply the wind.)

LEP Students

1. Invite students to make posters to show examples of the architecture typical in their native country. Suggest that they include a brief bilingual caption to explain what the structure is. Display the poster diagrams around the classroom for all students to enjoy. Encourage students to bring in photographs of structures from their homelands to share with the class.

2. Suggest that students make a "picture dictionary" using terms from their native language that identify different kinds of structures. For example, the dictionary might include words for *store, office building, home, theater, bridge, grocery store, restaurant, bakery, hardware store, apartment building, expressway,* and so on. A picture should be drawn to illustrate what each term means. Encourage students to include in their dictionaries some of the technical words given in the list on page 272.

3. At the beginning of each unit, give Spanish-speaking students a copy of the *Spanish Glossary* from the *Teacher's Resource Binder.* Also, let Spanish-speaking students listen to the *English/Spanish Audiocassettes.*

At-Risk Students

This unit affords many opportunities for students to develop their own problem-solving skills, while observing first-hand how forces affect structures. In *Explorations 1, 4,* and *6,* students construct their own model bridge, test various structural shapes and materials, and make and compare arches and beams.

To reinforce the concepts covered in the unit, suggest that students have a "house of cards" contest. The object is to see who can make the largest structure using playing cards. As students build their structures, have them identify how cards are used as beams, columns, cantilevers, trusses, and buttresses. Other categories in the contest might include the structure that uses the most cards and the structure with the best design.

Cross-Disciplinary Focus

Biology

Bird nests range from the complex to the simple. Among the many species of birds, there is an almost endless variety of design used by birds in creating their nests. Ask students to do research to find out the many ways that birds solve their design problems. Students can compare size, location, shape, and materials used. Some birds of interest include Tailorbirds, who sew leaves together; Weavers, who use elaborate knots and stitches; Edible-nest Swiftlets, who construct their nests entirely out of hardened saliva; Western Grebes, who build floating platforms for their nests; and Cliff Swallows, who create multiple-dwelling apartments out of mud. Ask students to point out any parallels that they can see with human designs.

Mathematics

Suggest to students that they make graphs to compare different kinds of structures. For example, bar graphs could be used to show the comparative sizes of the tallest buildings, the longest bridges, the largest airplanes, the biggest city parks, and so on. Encourage students to be creative when they make their graphs. For example, a graph comparing the tallest buildings could be drawn over an image of the Sears Tower. Display the graphs around the classroom or in the school library. Ask students to point out how the size of a structure influences its design elements.

Point out to students that structure and design are usually based on geometric shapes. Suggest that they do some research on a group of geometric shapes called polyhedrons. These include cubes, tetrahedrons, octahedrons, and dodecahedrons. Have students make models of these different shapes to share with their classmates.

Architecture
Point out to students that a geodesic dome is one kind of structure that supports itself, is strong, and is easy to set up and take apart. Suggest that they do some research to find out what geodesic domes look like, how they are made, what they are used for, and who invented them. Have students report what they learn to the class.

Cross-Cultural Focus
Low-Rise Housing
As the world's population has grown and land has become scarce and expensive, different solutions have been proposed to make the most of available space. In recent years the high-rise apartment was proposed as one possible answer. However, high-rise living often creates many new problems, including lack of adequate play areas and safety concerns. One possible solution is called urban low-rise housing, in which individual units are linked together. This type of urban planning is not new, however. It was used in North America before the time of European settlers by the Anasazi Indians of the Southwest and by the Dogon people of Mali in Africa. Ask students to research dwellings used by other cultures, and to evaluate whether or not these could be used successfully today.

Architectural Influence
Point out to students that many different cultures have influenced the architecture of the United States. Suggest that they do some research to discover what some of these influences have been. Encourage students to share what they discover by preparing an oral presentation for the class. If students need some suggestions, you might point out that Spanish and Native American culture has influenced the architecture of the West and Southwest. Furniture design has been influenced by the countries of Scandinavia. Architecture of Asian communities in some major cities includes designs that are often very similar to that of their native countries.

Bibliography for Teachers
Gimpel, Jean. *The Cathedral Builders.* New York, NY: Harper & Row, 1983.
Gordon, J. E. *Structures: Or Why Things Don't Fall Down.* New York: Plenum, 1978.
Huxtable, Ada Louise. *Architecture Anyone?* New York, NY: Random House, 1986.
Mark, Robert. *Light, Wind, and Structure: The Mystery of the Master Builders.* Cambridge, MA: Massachusetts Institute of Technology Press, 1990.
Rifkind, Carol. *A Field Guide to American Architecture.* New York, NY: Bonanza Books, 1984.
Rybczynski, Witold. *Home, A Short History of An Idea.* New York, NY: Penguin Books, 1987.

Bibliography for Students
Bates, Robert L. *Stone, Clay, Glass: How Building Materials Are Found and Used.* Hillside, NJ: Enslow Press, 1987.
Hamey, L. A., and J. A. Hamey. *The Roman Engineers.* Minneapolis, MN: Lerner Publications Company, 1982.
Lewis, Alun. *Super Structures.* New York, NY: The Viking Press, 1980.
MacGregor, Anne, and Scott MacGregor. *Domes.* New York, NY: Lothrop, Lee & Shepard Books, 1981.
Packard, Graham. *Bridges.* New York, NY: The Bookwright Press, 1987.
Sandak, Cass R. *Skyscrapers.* New York, NY: Franklin Watts, 1984.

Shapiro, Mary J. *How They Built the Statue of Liberty.* New York, NY: Random House, 1985.
Zubrowski, Bernie. *Messing Around with Drinking Straw Construction.* Boston, MA: Little, Brown and Company, 1981.

Films, Videotapes, Software, and Other Media
Archaeology Search.
  Software.
  For Apple II Family.
  Macmillan/McGraw-Hill
  220 East Danieldale Rd.
  De Soto, TX 75115
Architecture: The 80s and Beyond.
  Videotape and Filmstrip.
  American School Pub.
  Princeton Road
  P. O. Box 408
  Hightstown, NJ 08520
The Middle Ages: Culture of Medieval Europe.
  Film and Videotape.
  Encyclopaedia Britannica
  425 North Michigan Ave.
  Chicago, IL 60611

The Mystery of the Master Builders. Videotape.
  Coronet Film & Video
  108 Wilmot Road
  Deerfield, IL 60015

UNIT 5

Unit Focus

Display a photograph, magazine picture, or poster of a large urban area. Call on students to identify the different kinds of structures they see in the illustration. Record their responses on the chalkboard. *(Student responses might include such structures as roads, bridges, houses, stores, churches, piers, and so on.)*

Then involve students in a brief discussion of what keeps a building from falling down. *(Most students will recognize that the materials used and the underlying supports, such as beams and columns, keep a building standing.)*

About the Photograph

The Eiffel Tower was named after its designer, Alexandre Gustave Eiffel. He designed it for the Centennial Exposition of 1889, held in Paris, France. Alexandre Eiffel was a structural and aeronautical engineer. Eiffel designed many other structures, including the framework for the Statue of Liberty.

The tower is 300 m tall and rests on a base 101 m². It contains about 6400 metric tons of iron and steel. However, for its size, it is a very lightweight structure. During World War I, it was used as a military observation station. Since 1953, it has been used as a television transmission tower.

Unit 5 — Structures and Design

THE SCIENCE OF STRUCTURE
1. Planning the Structure, p. 264
2. Response of Structures to Force, p. 272
3. Engineering Structures, p. 282

THE ART OF DESIGN
4. Beauty in Structures, p. 293
5. A Question of the Environment, p. 304

For Your Journal

Write a paragraph summarizing your thoughts on each of the following:
1. What is the difference between structure and design?
2. What features does this tower have that keep it standing?
3. How do beauty, the environment, and buildings relate to one another?
4. How does the size of a building affect its structure?

✳ Using the Photograph

Have students study the photograph. To promote discussion, ask them questions similar to the following:

- How would you describe the structure? *(Students should recognize that it sits on a large base. As it soars upward, it becomes increasingly smaller and smaller.)*
- What kinds of materials do you think were used to build the tower? *(iron and steel)*
- What geometric shapes make up its construction? *(rectangles, triangles, arches, diamonds, and so on)*
- What keeps the structure standing? *(the supportive framework, the materials used, the design)*

✳ About Your Journal

Students should answer the Journal questions to the best of their abilities.
These questions are designed to serve two functions: to help students recognize that they do indeed have prior knowledge about the topic, and to help them identify any misconceptions they may have. In the course of studying the unit, these misconceptions should be dispelled.

Structures and Design 263

LESSON 1

✴ *Getting Started*

The lesson begins with a scenario in which the effect of a new development on the surrounding community is being debated. After reading and discussing the differing points of view, students are encouraged to offer their own opinions. In an effort to determine the causes of structural collapse, students then analyze structures that have failed. In a similar fashion, students examine a series of structures that have survived. They are encouraged to analyze what these structures were designed to do and what factors contributed to their reliability. The lesson concludes with students constructing a model bridge.

Main Ideas

1. Opinions may differ about the best solution to a complex problem.
2. All structures, whether natural or artificial, must be designed to withstand potentially destructive forces.
3. Problems of design and structure may have a variety of solutions.
4. Valuable information about designing structures can be learned from studying those structures that have failed and those that have survived.

THE SCIENCE OF STRUCTURE

1 Planning the Structure

Meeting of the Council

The mayor of Zenith could not be heard over the noise in the Chamber of the City Council.

"Come to order! Come to order!" Mayor Jones pounded the gavel at his desk. The citizens in the gallery and the council members seated in the chamber continued arguing.

Seldom had there been an issue that stirred up so much debate among the citizens of Zenith. ABC Developers were seeking the approval of the Zenith City Council to redevelop the old neighborhood of Riverwood across the river from the city's business center.

"Fellow council members, fellow citizens," the mayor appealed. "We will not solve this problem by shouting at one another. Please, everyone will have a chance to speak. Council Member Kowalski, you have the floor."

Council Member Kowalski: I have received many letters and phone calls from residents and businesses in Riverwood. They support the proposal. With the money ABC has offered them, they will be able to relocate in other parts of the city. The proposal will result in many new jobs involving the construction of new buildings. The city of Zenith will benefit greatly from this proposal.

Council Member Casaro: Mr. Mayor, not all the residents and businesses in Riverwood support ABC's proposal. I present to the Council a signed petition from over 2000 residents and businesses opposing this proposal. I also present a letter from the Zenith Heritage Foundation that opposes the proposal as it now stands. They write that "the proposal would mean the destruction of many historic homes and buildings. The new buildings would obstruct the view of the beautiful mountains that surround our city."

LESSON 1 ORGANIZER

Objectives

By the end of the lesson, students should be able to:
1. Analyze information in order to formulate a solution to a problem.
2. Explain the importance of design in natural and artificial structures.
3. Explain why it is important to analyze the design of structures that have failed as well as the design of structures that have survived.

Process Skills

analyzing, evaluating, designing models, constructing models

New Terms

none

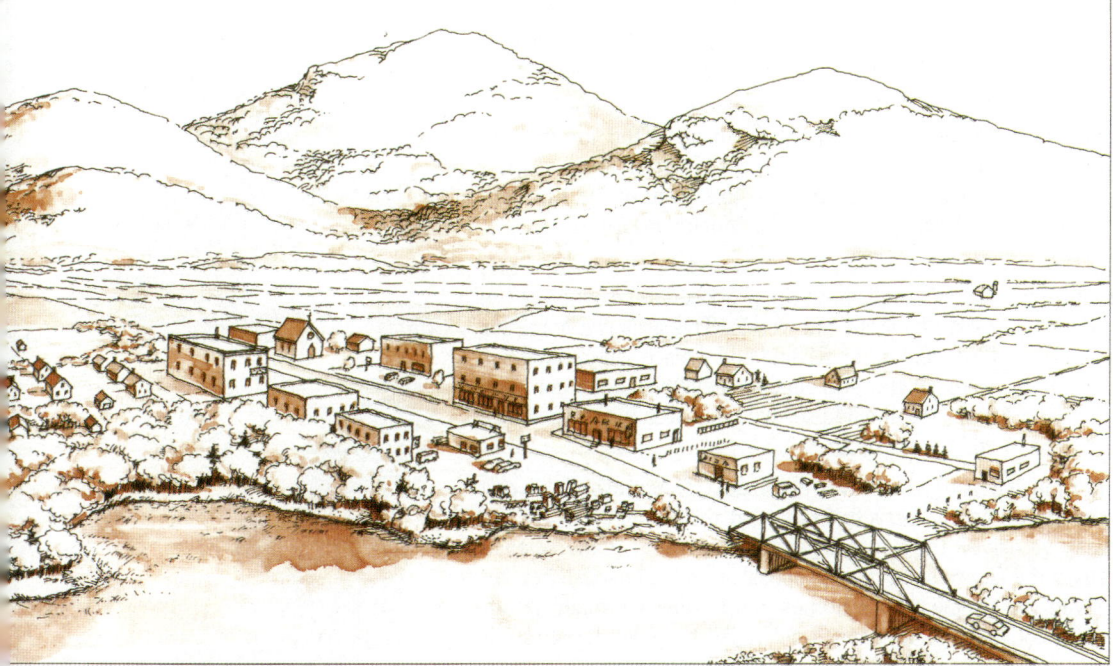

The present Riverwood

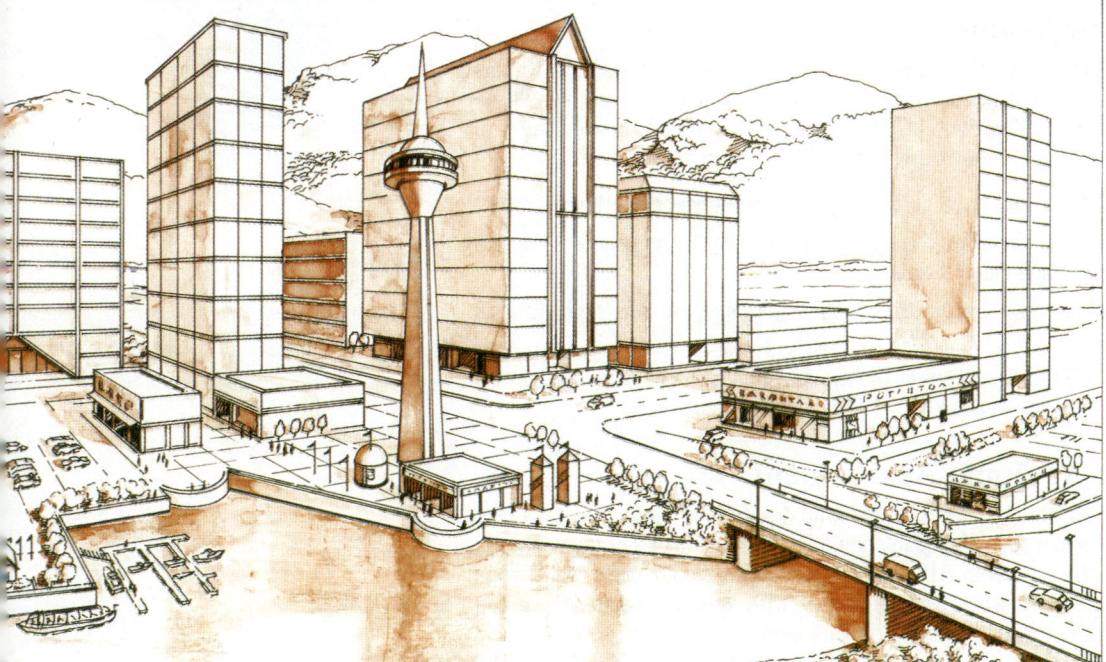

ABC Developers' proposed redevelopment of Riverwood. The original homes and businesses are gone.

✷ Teaching Strategies

Ask students to identify some of the different kinds of structures they see everyday—homes, office buildings, bridges, highways, railroads, airplanes, and so on. Keep track of their suggestions by listing them on the chalkboard. Then ask students to identify some of the ways in which these structures are designed to fulfill their functions. For example, an airplane has wings to enable it to fly, a house has beams to support its roof, a bridge has a superstructure that keeps it from collapsing. Point out that in this lesson, students will learn about the relationship that exists between design and structure.

Meeting of the Council

Have students read page 264 silently, and provide them with time to examine the pictures of present-day Riverwood and the proposed redevelopment plan. Then help students to identify and summarize the two points of view presented by council members Kowalski and Casaro. Have students discuss the merits of both points of view before continuing on to page 266. Ask them to show by a raise of hands which point of view they agree with based on the information they have so far. Record the results on the chalkboard.
(Continues on next page)

Materials

Exploration 1: 4 index cards (each approximately 7.6 cm × 12.5 cm), 20-cm length of narrow masking tape, scissors, metric ruler, 50-g mass

Teacher's Resource Binder

Resource Worksheet 5–1
Graphic Organizer Transparency 5–1

Science Discovery Videodisc

Disc 2, Image and Activity Bank, 5–1

Time Required

three class periods

The Science of Structure 265

(Meeting of the Council continued)

Have students continue reading silently the debate of the future of Riverwood. When they have finished, invite them to discuss whether or not their previous opinions have changed with this additional information.

As an alternative, have a group of students role-play the city council scenario for the class. When they have finished, involve the class in a discussion of the issues.

Your Opinion

You may wish to divide the class into groups of four or five students to discuss and debate the future of Riverwood. Suggest that each group select a leader to guide the discussion, and have one member take notes. Explain that each group should reach an agreement on a solution to the problem. When the groups have finished their discussions, have the leader from each group present a summary of their conclusions to the class.

If the issues presented so far have sparked considerable interest in the topic, you may wish to have the class organize a debate. Select debating teams and provide class time for the debate to take place. Instances of development in your own community can be discussed. Ask students to decide whether the changes made were an improvement, or whether more harm was done than good. Issues to focus on include monetary concerns, concerns of the environment, and concerns of the people living in and around the development site. Once the debate has been completed, have the class discuss which side they felt presented the most convincing arguments.

An At-Home Assignment

Encourage students to complete the home assignment and share their ideas with the class. As an alternative, you may wish to have the groups that discussed the issues earlier work together to complete the assignment.

Council Member Tufts: Mr. Mayor, Riverwood is an eyesore. The homes and businesses have fallen into disrepair. The woods and marsh are littered with garbage and debris. And the woods are becoming a hazard, especially at night. Something must be done with this area.

Mayor Jones: Many civic organizations have asked for the opportunity to speak to the Council about ABC's proposal. I call upon the President of the Environmentalists' Association, Mónica Rodriguez.

Mrs. Rodriguez: Mr. Mayor and Council Members, our organization is concerned about the proposed destruction of the existing marsh and woods. This area is a home for birds and other wildlife and a place where children from all the surrounding neighborhoods come to play. We should preserve this area for our children and for the future.

After much debate, the Mayor declared, "The future of this city belongs to the youth. Let's find out what they would like to do. I suggest we ask all the seventh, eighth, and ninth grade science classes in the city to send their proposals to the Council. Then we will invite the people whose homes and businesses might be affected to comment. The Council will make the final decision." It was agreed by the Council to proceed in this way.

Your Opinion

Suppose your class was asked to find the best solution to this problem. With your classmates, decide what you think should be done with Riverwood. Have one person record the decisions of your class. Later you will refer back to these decisions.

1. Would you have all the existing structures torn down?
2. Would you alter the environment to make more space for the new buildings? Explain why.
3. What new structures would you like to see built in this area?
4. Will the town benefit the most from a new business center or a historic neighborhood? Why do you think so?

An At-Home Assignment

As you learn about structures and design, complete your own plan for Riverwood. This plan should be in the form of a detailed sketch like the one from ABC Developers. Later your class will decide whose plan is best, and together with your classmates, you can construct a model based on this plan.

Structures That Failed

The Tacoma Narrows Bridge

Before you design new structures for Riverwood, you must ensure that your structures will not fall down. To build sturdy structures that are not likely to fall, engineers carefully study those structures that have failed. When the Tacoma Narrows Bridge collapsed in 1940, engineers around the world took notice. The engineers who investigated the collapse asked themselves:

1. Why did the bridge fail?
2. Was it poorly constructed?
3. Were the materials used inferior?
4. What were the forces that caused the bridge to collapse?
5. What changes should be made in bridge design and construction to prevent catastrophes of this kind in the future?

Answer these questions. Get together with two or three of your classmates and record your best answers in your Journal. Be sure to consider all of the forces affecting the bridge.

Answers to
In-Text Questions

Have students read and discuss the questions. Be sure that they understand that all responses are acceptable. Remind them that they will have a chance to review their answers as they work through the unit. Use these correct responses to guide students' reasoning as they complete the questions.

1. The bridge failed because the design did not provide a structurally stiff frame to resist moderate winds.
2. No, the bridge was well constructed. The design was largely at fault.
3. The materials were not inferior, but they were too light. At the time, it was believed by engineers that the structure of an expansion bridge should be kept light so that the size of the cables could be minimized.
4. Wind and the oscillations of the deck of the bridge were the main factors in causing the bridge to collapse.
5. The main change brought about by this failure was to design a stiffer and heavier deck that is resistant to twisting and to lateral deflection.

Structures That Failed

Direct students' attention to the photograph on page 267 and call on a volunteer to describe what it shows. Involve students in a discussion of why scientists might be interested in studying structures that have failed. Then have students read about the Tacoma Narrows Bridge, pausing before the questions. Ask: Does anyone know where Tacoma is located? If no one responds, explain that Tacoma is a large city in the state of Washington. You may wish to display a map of the United States and call on a volunteer to locate and identify Tacoma.

Students may be interested to know that the Tacoma Narrows Bridge spanned 853 meters across Puget Sound. It collapsed after only four months of use. The bridge had been designed to withstand winds of 195 km/h, but was destroyed by a wind of only 65.6 km/h.

The wind caused the deck of the bridge to move up-and-down in a wavelike motion. The distance between the crest and trough of these "waves" eventually reached 9 m. After several hours, the deck began to twist back and forth until some of the suspending cables snapped, plunging part of the bridge into the water. No lives were lost since the bridge was closed after it started to sway.

Picture Puzzles

Have students read the page silently, or call on a volunteer to read it aloud. Point out that the details in a concept map may vary depending on how different individuals interpret the factors contributing to a structure's failure.

Involve students in a discussion of each of the pictures, focusing on the questions presented at the top of the page. Accept all reasonable responses that students can support with logical arguments. Answers to the in-text questions may be used to get students started on their concept maps.

Answers to
In-Text Questions

Shattered House
1. The house was destroyed.
2. The destruction of the house was probably caused by an earthquake. It is possible it could also have been the result of a severe storm, such as a hurricane, typhoon, tornado, or flood.
3. The house was literally pulled and pushed apart.
4. It is possible that the house could have been saved if it had been built of stronger materials and designed to withstand the forces of nature that resulted in its destruction.

Greek Ruins
1. The beams, or lintels, shown in the picture of the ancient Greek ruins probably cracked under their own weight.
2. Other factors that could have contributed to their failure may have included the settling of the structure itself, the shifting of the earth upon which the structure was built, or earthquakes and tremors.
3. Cracks are usually the result of a structure being pulled apart.
4. The cracks might have been prevented if the structure had been designed to better withstand the forces caused by earthquakes.

Picture Puzzles

What can you learn from structures that have failed? Study each of the pictures carefully. For each situation, try to answer the following questions:

1. What was the failure?
2. What might have caused the failure? Could it have been the wind, the weight of snow, the weight of the structure itself, an earthquake, or something else?
3. Was the structure being pulled apart (stretched) or being pushed together (compressed) before the failure?
4. How could the failure have been prevented?

Emily drew a concept map to analyze the possible causes of the Tacoma Narrows Bridge collapse. Perhaps you could use a similar diagram to answer the questions above.

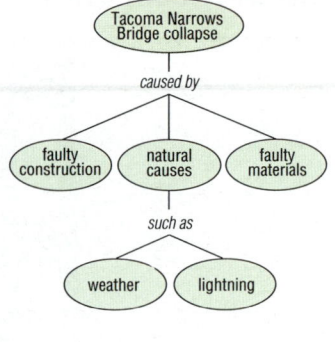

A badly-shattered house in Gonen, Turkey. What might have caused this destruction?

Stone masonry beams supported by tall columns are a common sight among the ruins of ancient Greek civilization. Notice the large cracks in the beams.

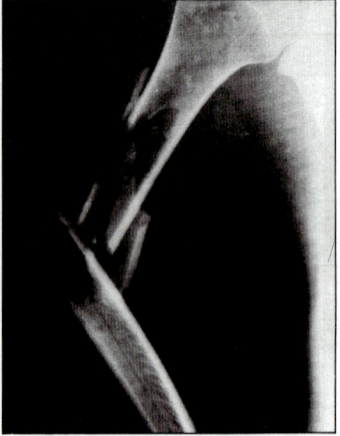

Twisted and broken bones as seen by an X ray.

X Ray
1. The bones twisted and broke.
2. The twisted and broken bones seen in the X ray could have resulted from a number of causes including a fall or a sporting or automobile accident.
3. The bones may have broken when they were violently pulled apart or pushed together.
4. There was probably no way to prevent the bones from breaking except by avoiding the situation that resulted in the break in the first place.

Structures That Haven't Failed Yet

Much can be learned about how to build safe, strong, and beautiful structures by examining those that have not failed yet. Nature and humans show their creativity in the variety and strength of the structures that they build. Study the following 10 pictures of structures that have been successful over the years.

In small groups discuss for each structure:

(a) What is it designed to do?
(b) What holds it up?
(c) What holds it together?
(d) What could cause it to fail?

Keep a record of your discussion in a table such as the one shown below.

Structure	Designed for	Held up by	Held together by	Possible causes of failure
spider's web	catching and holding insects	threads to a tree	a network of strong threads	—a very heavy insect —strong wind —heavy rain —? —?
Leaning Tower of Pisa, Italy				

A freighter

A spider's web

▲ Leaning Tower of Pisa, Italy

◀ A bird's nest with eggs

A Freighter
(a) It is designed to float on water in order to transport large amounts of cargo across the ocean.
(b) It is held up by the buoyancy, or upward force, exerted by the water on which it rests.
(c) It is held together by nuts, bolts, rivets, steel beams, and metal sheathing.
(d) Strong winds and rough seas could cause its failure, as could a collision with some other object or ship.

A Spider's Web
(a) It is designed to catch and hold insects.
(b) It is held up by silken threads attached to the branches or twigs of a tree or shrub.
(c) It is held together by a network of similar threads.
(d) Heavy objects falling into the web, strong winds, or heavy rains could cause it to fail.

Leaning Tower of Pisa
(a) It was designed as a bell tower for the Pisa Cathedral, seen at the right in the picture.
(b) The tower is held up by a system of arches and columns.
(c) It is held together by mortar, bricks, and stones.
(d) A strong earthquake or damage to the tower's foundation could cause the structure to fail. (You may wish to point out to students that the tower of Pisa leans because its foundation lies on a layer of unstable soil.)

(Answers continue on next page)

Structures That Haven't Failed Yet

Have students read the page silently. Then direct their attention to the table and discuss what it shows. You may wish to have students construct their tables in their Journals, and then discuss their ideas. Students should not be expected to recognize all of the major design and structural features of the objects shown in the photographs. The following are some ideas they may consider.

Answers to In-Text Questions

A Bird's Nest
(a) It is designed to hold and protect eggs, and to provide shelter for newly hatched birds.
(b) A bird's nest is held up by the supporting branches of a tree or shrub.
(c) It is held together by weaving grasses and twigs together.
(d) A strong wind or disturbance by a predatory animal could cause the structure to fail.

Answers continued

A Bat
(a) It is designed for flying and catching insects on the wing.
(b) It is held aloft by the forces that are created by wind resistance and pressure when it flaps its wings.
(c) It is held together by bones, muscles, skin, and connective tissue.
(d) Crashing into solid objects, severe weather, or illness could cause the structure to fail.

Hot Air Balloons
(a) They are designed to float into the air carrying passengers or cargo.
(b) They are held up by the force of heated air pushing against the inside walls of the balloon.
(c) They are held together by strong cord and thread used to sew the large panels of nylon or polyester together.
(d) Strong winds or tears in the fabric could cause the structure to fail.

A Giraffe
(a) Its structure allows it to reach the leaves in the tops of trees and shrubs.
(b) It is held up by a skeleton and strong muscles.
(c) It is held together by muscles, bones, skin, and connective tissue.
(d) A serious fall or damage sustained in a struggle could cause the structure to fail.

The "Gossamer Albatross"
(a) It is designed to fly through the air by human power.
(b) It is held up by the forces caused by the wind passing over and under its wings.
(c) It is held together by a framework of supporting structures over which is stretched a thin, fabric skin.
(d) Strong winds, severe weather, or sudden loss of power could cause the structure to fail.

A Sunflower
(a) It is designed to support a large, heavy flower and seed head.
(b) A sunflower is held up by a thick, strong stem.
(c) It is held together by fibrous connective tissues and a thick, tough, outer skin.
(d) Heavy wind and rain, insect damage, or lack of water could cause the structure to fail.

A bat

Hot air balloons

A giraffe

The "Gossamer Albatross," a human-powered aircraft

A sunflower

Hoover Dam, Colorado River

Hoover Dam
(a) It is designed to hold back and control huge amounts of water.
(b) It is held up by reinforcing rods inside its concrete walls and by the curved design of its structure.
(c) It is held together by concrete and steel reinforcing rods and beams.
(d) Earthquakes and weathering could cause the structure to fail.

Constructing on Your Own

You have examined some structures that failed and others that did not. Can you design a strong, safe structure and build a model of it?

Suppose you decide to include a bridge in your plan for Riverwood. Perhaps you want the bridge to carry automobile traffic across the river between Riverwood and the business center of the city. Or maybe you want a footbridge to cross the inlet. In either case, your bridge must be strong enough and safe enough for those who will use it.

EXPLORATION 1

Constructing a Model

You cannot design and build the actual bridge, but you can build a model. Try to build the strongest model possible.

You Will Need

- 4 standard index cards, each approximately 7.6 cm × 12.5 cm
- a 20-cm length of narrow masking tape
- scissors
- a ruler

What to Do

1. As shown in the sketch below, your model must have the following dimensions. It must cross a gap of 11 cm. To do so, it will have to be at least 12 cm long. The width of the card should be 6 cm or more.
2. If supporting structures are used, they must leave an unobstructed gap to allow for boat traffic.
3. Your model may rest on books, but it cannot be taped to them.
4. You are allowed to bend, fold, and cut the index cards.

Now comes the fun part! Can you place a mass of 50 g on your bridge without making it collapse? Congratulations! Now continue to add masses to find out which design holds the greatest load. Who had the best design? Why was it the best?

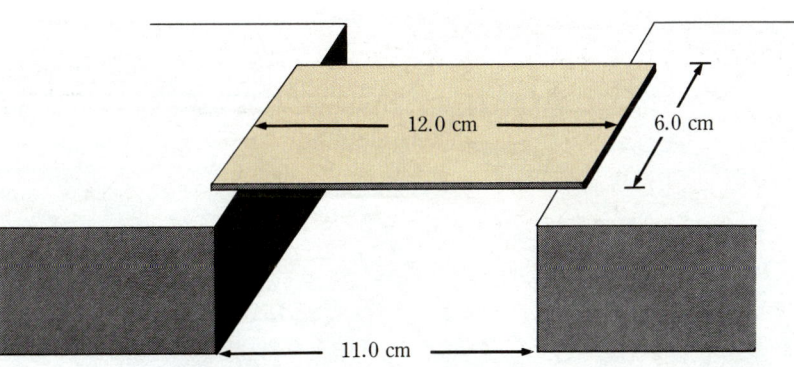

Writing About Your Model

A toy company is requesting suggestions for the construction of cheap, easy-to-make model bridges. You decide to submit your design. Write a report to Trixie's Toy Company that includes:

(a) a design that works
(b) the steps in making the toy bridge
(c) the qualities that make it a good design
(d) the pitfalls to avoid in building such toy bridges

Exploration 1

This Exploration is designed so that most students can achieve a measure of success within 30 minutes. Encourage them to compare and discuss their results as they work to complete the activity. If a hint is necessary, point out that folding the index cards will provide added strength.

When students complete the activity, help them to recognize the variety of successful solutions to the problem. Encourage them to decide which designs are the safest, which support the greatest loads, and which require the least amount of material to build.

Writing About Your Model

Call on several volunteers to present their reports to the class. Involve the class in a discussion of the ideas that are presented, and help them to reach a general consensus on the best way of constructing a model bridge.

Assessment

1. Students may enjoy collecting photographs of structures for a bulletin board display. After photographs have been collected, ask students to identify the features of the structure that have allowed it to last. *(Answers should include a discussion of materials, design, and appearance.)*
2. Have students cut pictures from magazines and newspapers that illustrate different kinds of structures. Then have them arrange the pictures on construction paper and write captions to describe the purpose of each structure.

Extension

1. Tell students to imagine that they have been asked to design a new middle school. They should work in small groups to think of several proposals. Proposals should include how the project will fit into and make use of the landscape, materials that will be used, and how the design contributes to the function. When the groups finish, have a spokesperson from each group present their ideas to the class. Have the class discuss the ideas and decide which proposal they like the best.
2. Students may enjoy doing research on some of the structures depicted on pages 269 and 270 to discover how their design is related to function. For example, how does the design of an airplane or a hot-air balloon enable it to fly? How does the design of a spider web trap insects? How does the design of a dam prevent the dam from crumbling? Have interested students write brief reports about one of these questions or a related question of their own.

The Science of Structure 271

Lesson 2

✦ Getting Started

This lesson focuses on the way in which structures respond to force. Through a series of activities, students learn to recognize the difference between tensile force, compressive force, and shear force. The lesson concludes with information on the strength of different materials and how the shape and construction of these materials can make them stronger.

Main Ideas

1. Forces applied to structures may be classified as either tensile, compressive, or shear.
2. Structures are compressed by their own weight.
3. Materials that respond to the removal of stress by returning to their original shape are said to behave elastically.
4. In an elastic response, a tensile force causes an extension that is proportional to the applied force.
5. One measure of strength is the ability of a given material or structure to withstand force without breaking.

✦ Teaching Strategies

Ask students to consider for a moment why it is important to use and understand the correct vocabulary when explaining how to do something or how something works. *(It avoids confusion and misconceptions.)* Ask students to identify occupations in which a special vocabulary is used. *(sports, science, art, medicine, space exploration, and so on)* Then have students read the lesson introduction silently. Ask them to identify any of the words on the list that they used in their descriptions to Trixie's Toy Company.

272   Structures and Design

❷ Response of Structures to Force

The designers of Riverwood will need to know whether the structures they design will stand up. Will the structures support their own weight? Can they support a load? How will the structures behave in the wind or during an earthquake? Also, the designers will have to convince the municipal authorities that the structures will be safe. To do this, they need to use the right words to communicate their ideas.

Consider your model bridge. What technical words did you use in describing your model bridge to Trixie's Toy Company?

Here is a list of words that you can use to describe the response of structures to force. Keep a list of the terms in your Journal. As you learn the meaning of each of these words, write down the definition. Remember to use what you discover in the Explorations to find the best definitions for the terms.

bending
compressing
compression
compressive force
contracting
contraction
deflection
extension

load
relaxing
response (respond)
shear
shear force
stretching
tensile force
tension

LESSON ❷ ORGANIZER

Objectives

By the end of the lesson, students should be able to:
1. Describe the behavior of structures and materials by using the following words: load, response, tension, tensile force, compression, compressive force, shear, and shear force.
2. Explain what is meant by an elastic response in a structure or material.
3. Demonstrate how to strengthen a material by shaping and joining it in various ways.

Process Skills

analyzing, graphing, designing models, constructing models

New Terms

Tensile force—a force that causes a material to stretch.

EXPLORATION 2

Three Kinds of Force: Tensile, Compressive, and Shear

You Will Need

- dull scissors
- heavy paper
- modeling clay

Part 1

1. Study the illustrations of the three structures in Diagram A: a beam, a rope, and a column. Do you observe a force on each? How does the structure respond to the force? How does each structure change in shape? For each case, describe both the force and the response of the structure to the force.

2. In Diagram B, the three structures have been redrawn. The agents of force have been replaced by arrows. The arrow indicates the direction of the force. Draw Diagram B in your Journal and correctly label the arrows with one of the following terms:

   tensile force—a force that causes stretching

   compressive force—a force that causes compression

   shear force—a force that causes bending or twisting

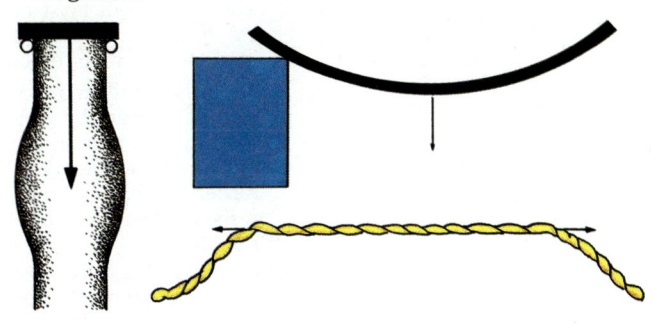

Exploration 2
Part 1

Involve students in a discussion of the illustrations. If necessary, help them to recognize that a wooden block is exerting a force on the beam, causing the beam to bend downward. The children are exerting a force on the rope, causing it to stretch. The bricks are exerting a force on the column, causing it to compress and bulge.

Answers to
In-Text Questions

1. - The weight of the block pushes down on the beam, causing it to bend.
   - Pulling on the rope causes it to stretch.
   - The weight of the blocks causes the column to compress.
2. - A *shear force* acts on the beam.
   - A *compressive force* acts on the column.
   - A *tensile force* acts on the rope.

Compressive force—a force that causes material to become more compact and pressed together.
Shear force—a force that causes bending or twisting in a material.
Deflection—the amount that a material bends when a force is applied to it.
Elastic response—the tendency of a material to return to its original shape or size after it has been stretched.

Materials

Exploration 2: dull scissors, heavy paper, modeling clay
Exploration 3
Activity 2: piece of wood (like a meter stick), metric ruler, clamp, force meter from Unit 4 or masses with hooks, string, scissors, graph paper
Exploration 4
Investigation 1: 6 index cards, masking tape, 2 books, standard masses (10-g to 20-g intervals), Journal

Investigation 2: ruler, glue or paste, 6 index cards, 6 books, Journal, standard masses (10-g or 20-g intervals), graph paper
Investigation 3: glue or paste, 6 index cards, 6 books, metric ruler, standard masses (10-g or 20-g intervals)
Investigation 4: 5 index cards, masking tape, 6 books or 2 blocks, metric ruler, standard masses (10-g or 20-g intervals), string, scissors, graph paper, force meter

(Organizer continues on next page)

The Science of Structure

Part 2

You may wish to call on a volunteer to demonstrate each of the activities for the class. Then direct students' attention to the diagrams. Help them to recognize that in each case, forces are acting in opposite directions. Students should draw the conclusion that shearing creates opposing forces.

Part 3

Have students turn to pages 267 and 268. The twisting and bending of the bridge deck is observable evidence that shear forces were largely responsible for its collapse. The house on page 268 probably collapsed from shear and tensile forces. There is evidence that parts of the house were pulled or stretched apart. There is also evidence that parts of the house were twisted out of shape. The stone masonry beams were cracked from tensile forces, or from being pulled apart. The columns were cracked from compressive forces. The bones were broken from shear forces.

Answers to
Part 4

1. When the load is removed, the beam will return to its original shape.
2. When the force is removed, the rope will contract to its original length.
3. When the load is removed, the column will return to its original shape.
4. Yes, the column experiences a load because it compresses under its own weight. If the top half of the column is removed, the bottom half will undergo a slight increase in height.

EXPLORATION 2—CONTINUED

Part 2

1. To get a better idea of shear force, take a pair of dull scissors and slowly begin to cut a piece of heavy paper. Notice how the paper bends before breaking. When a force bends or twists an object, you know that the object is experiencing a shear force. In the drawing on the right, the arrows represent the shear force of the scissors' blades.
2. Take modeling clay and form it into a small rectangular block. Grasp the top of the block between two fingers. With the other hand, use two fingers to grab the bottom of the block. Now, push the top half and the bottom half in opposite directions. You are exerting a shear force on the modeling clay.

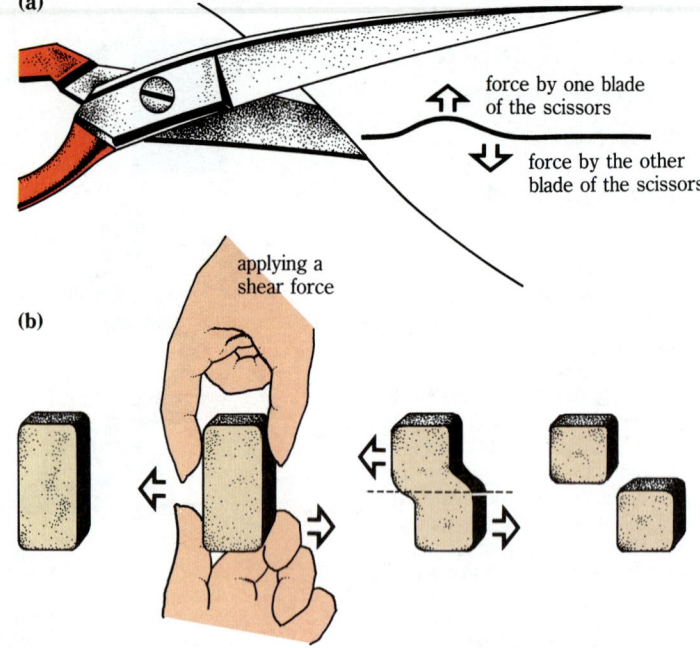

Part 3

The causes of failure in structures are forces that cause tension, compression, and shear. Look again at pages 267 and 268, and identify which of these forces were responsible for the failure of each structure. For example, the Tacoma Narrows Bridge collapsed from shear force caused by wind. Do you see where the bridge twisted and where shear force affected the roadway?

Part 4

Look at the structures in Diagram A on page 273. Read the statements that follow and answer the questions:

1. The beam is under a load, producing both tension and compression. It responds by bending. If the load is removed, how does the beam respond?
2. The rope is also under tension. It responds by stretching. If the force is removed, how does the rope respond?
3. A load is placed on the column. It responds by compressing. If the material of the column behaves like a rubber ball, how will it respond when the load is removed?

You may find that it was easy to answer these questions. However, the answer to the following question may not be so obvious.

4. If there is no weight placed on the column, does it experience any load? Imagine slicing the column in two. What happens if you remove the top half of the column? How does the bottom half respond?

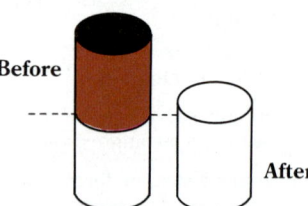

Structures do not only react to external forces; they also react to their own weight. They must be strong enough not only to withstand external forces, but to support their own weight.

(Organizer continued)

Investigation 5: 5 pieces of heavy paper, small pail, enough sand to fill the pail, hard board (approximately 10 cm × 10 cm), metric ruler, masking tape, standard masses (10-g or 20-g intervals), equal-arm balance

Teacher's Resource Binder

Resource Worksheets 5–2, 5–4, and 5–5
Activity Worksheet 5–3
Graphic Organizer Transparencies 5–2, 5–3, and 5–4

Science Discovery Videodisc

Disc 2, Image and Activity Bank, 5–2

Time Required

four class periods

---

274 **Structures and Design**

EXPLORATION 3

Elastic Responses

In the unit "Force and Motion," you made a force meter using a rubber band. As the force increased, so did the length of the rubber band. The stretching was the response of the rubber band to the tensile force on it. This stretching is said to be an *elastic response*. Why? Because when the force was removed, the rubber band relaxed and returned to its original length. Most materials used in construction respond in the same way.

Activity 1 of this Exploration is a case study, and Activity 2 is an experiment you can perform yourself.

Activity 1

Stretching Steel— A Case Study

Does steel really stretch when a strong force is applied to it? Naomi knew she could increase the tension on the strings of her guitar by turning the pegs. But did the steel strings actually get longer? Naomi went to a mechanical engineering lab where they had an apparatus to measure the change in length of steel wire. Here are Naomi's results for stretching a 1-m long steel wire with a thickness of 1.0 mm.

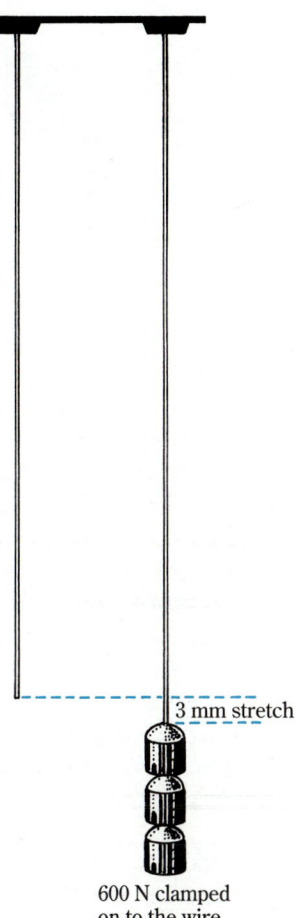

3 mm stretch

600 N clamped on to the wire

Simulation of the apparatus

Tensile (Stretching) Force (N)	Stretch (mm)
0	0
200	1
400	2
600	3
800	4
1000	5
1200	With this force, the steel string broke.

1. How much would the steel wire stretch when the force pulling it is:
   (a) 100 N? (c) 300 N?
   (b) 500 N? (d) 900 N?
2. Suppose a 10-m steel wire of the same thickness as the guitar string is used to tug a heavily loaded canoe behind a motor boat.
   (a) If the boat pulls with a steady force of 200 N,
      (i) how much will 1 m of the wire stretch? (See the results of Naomi's experiment. She used a wire that was 1 m long.)
      (ii) how much would you expect the 10-m wire to stretch?
   (b) If the motor boat suddenly yanked the wire with a force of 1500 N, what would happen?

Exploration 3

Have students read the Exploration introduction silently. To refresh their memories, you may wish to demonstrate the use of one of the force meters made in Unit 4, *Force and Motion*. Then refer students to *Part 4* of *Exploration 2*. Point out that the answers to the questions are examples of the elastic response of materials. Then have students speculate as to why it is desirable for construction materials to be elastic instead of rigid. *(Elastic materials are less likely to fail because they can adapt to changes in the forces exerted on them.)*

Activity 1

Call on a volunteer to read the introduction to the case study. Then direct students' attention to the table. Remind students that force is measured in newtons (N). To evaluate their understanding of what the table shows, ask questions similar to the following:

- Does a steel wire really stretch when a strong force is applied to it? *(yes)*
- In what increments was force added to the apparatus? *(in 200-N increments)*
- How much did the steel wire stretch for each increment of 200 N of force? *(1 mm)*
- How much force was needed to break the wire? *(1200 N)*

To extend the discussion, you may wish to graph the data presented in the table, or call on a volunteer to make a graph. In either case, involve students in a discussion of how the graph is made and what it shows.

Answers to

In-Text Questions

1. (a) 0.5 mm
   (b) 2.5 mm
   (c) 1.5 mm
   (d) 4.5 mm
2. (a) i) 1 mm
       ii) 10 × 1 mm = 10 mm
   (b) The wire would break.

The Science of Structure 275

Activity 2

You may wish to have students work in pairs or small groups to complete the activity. If space or materials are limited, consider asking two volunteers to conduct the activity as a class demonstration.

Have students read the page silently, or call on a volunteer to read it aloud. Then direct students' attention to the diagram. Be sure that they understand how to set up the apparatus and what to do. Point out to students that it is important that their measurements be as accurate as possible and that they record their results carefully. Encourage them to make a data table similar to the one in Doug's letter. Point out that Doug converted centimeters to millimeters in the "Deflection" column. Explain that this makes it easier to graph the results for step 4 on page 277.
(Continues on next page)

EXPLORATION 3—CONTINUED

Activity 2
Bending Wood

A common example of an elastic response is the motion of a diving board. The diver uses the elastic response of the board to spring into the air.

Your purpose in this Activity is to compare the amount of bending, or **deflection**, with the different forces exerted on the end of a model diving board.

You Will Need

- a piece of wood (like a meter stick)
- a meter stick for measuring
- a clamp or other means to fasten the wood to the end of a bench or desk
- masses (or use your force meter for applying known forces)
- string
- scissors

What to Do

1. Set up the apparatus as shown in the diagram.
2. Hang a mass on the free end of the stick. Measure the amount that the end deflects. Repeat for several different masses.
3. Record your results. See how Doug, an eighth grade science student, recorded his results.

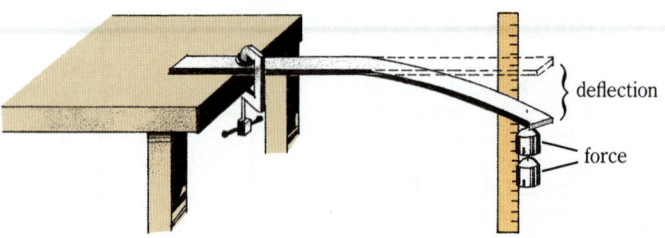
deflection
force

I used a meter stick made of spruce wood. It was 2.5 cm wide and 5 mm thick. I fastened the meter stick to the edge of a table leaving 90 cm free to bend.

Before putting weights on the end of the meter stick, I first measured the height of the meter stick above the floor. I found the height to be 78.3 cm. Then I measured the height with a weight fastened onto the end of the meter stick. The difference in the heights showed me how much the meter stick deflected. I recorded this in millimeters so it would be easier to graph.

For weights, I suspended metal blocks that had a mass of 200 g each. I remembered from studying forces and motion earlier that a mass of 100 g weighed about 1 N.

Weight (N)	Height of free end from floor before adding weight (cm)	Height after adding weight (cm)	Deflection (mm)
0	78.3	78.3	0
2	78.3	76.9	14
4	78.3	75.6	27
6	78.3	74.1	42
8	78.3	72.8	55
10	78.3	71.4	69

Structures and Design

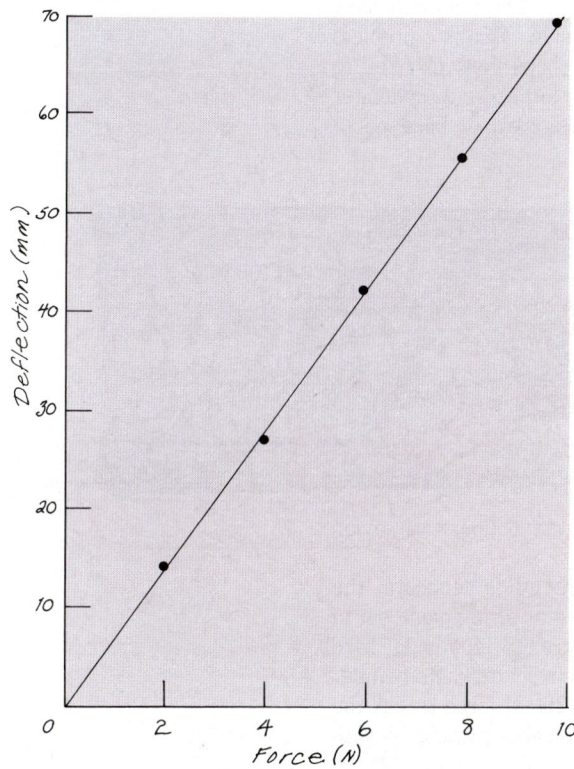

Using the Language of Tension, Compression, and Shear

Did you find definitions for most of the words at the beginning of this lesson? Use as many of those words as you can by rewriting your report to Trixie's Toy Company.

Is it easier to describe your model bridge? Using the language of engineers helps you describe structure and force precisely. By mastering scientific language, you can begin to write and think like an engineer!

4. Now graph your results. Doug let the vertical axis represent deflection and the horizontal axis represent force (weight). After plotting his points, he drew a line through them. Plot your data and draw your line. Is it a straight line or a curved one?

5. What do you conclude about the relationship between force and deflection?

6. If the length of the meter stick over the edge of the table was shorter, what difference in results would you expect? Test your prediction by repeating the experiment, but this time clamp the stick at a different position. Plot the new data on the same graph. What do you conclude?

(Activity 2 continued)

After students have performed the activity and recorded their results, have them complete step 4. Call on a volunteer to read the instructions aloud. Then discuss and review the graph on page 277. As students graph their own data, monitor their results by circulating among them. Be prepared to offer help if needed. After students have finished, have them share and discuss their graphs. Before students perform step 6, have them predict what will happen. Then have them complete the activity to test their predictions.

Answers to
In-Text Questions

4. Students should discover that the line on the graph is straight.
5. Students' graphs should show that deflection is proportional to the applied force. In other words, the amount of deflection is the same for each equal increment of force that is applied.
6. Students should discover that deflection is less when the amount of stick projecting over the edge of the table is less.

Using the Language of Tension, Compression, and Shear

At this point in the unit, students have encountered all of the words listed on page 272. Encourage them to write original sentences using each of the words, just as they were instructed to do at the beginning of this lesson. Call on volunteers to share some of their sentences with the class. Be prepared to correct any misconceptions before students continue with the lesson. Then have students rewrite their reports to Trixie's Toy Company. When they have finished, have them share their reports and discuss the merits of understanding and using correct scientific language.

Strong Materials

The designers of Riverwood need to know which materials are best for building the structures they design. The materials they select must be strong, but not too expensive. This is a problem you will face many times in your life. Perhaps you will need to select the right fabric to make your own clothes or to upholster furniture. You may want to repair a skateboard or build bookshelves.

Consider the spider's problem.

Which material do you think is the strongest: copper wire, nylon fishing line, spider's web, or wood?

What is meant by strength? Tensile strength is defined as the force needed to pull materials apart. A cord or string made out of any of these materials has its own tensile strength. If you pull hard enough on the ends of each cord, the cords will break. Remember our example of the steel wire. It took a 1200-N force to pull the wire apart. Which of the materials mentioned above will break with the least amount of force? the most?

Suppose you have long, thread-like pieces of each material. Suppose also that each "thread" has the same thickness of 1 mm. Does this change your answer? Which is the easiest to pull apart? Which is the most difficult? What is your prediction? Choose the list below that you think places the materials in order from strongest to weakest.

(a) copper, wood, nylon, spider's web

(b) wood, nylon, copper, spider's web

(c) nylon, spider's web, copper, wood

(d) copper, nylon, spider's web, wood

Here are the facts! After many tests it has been found that it takes 820 N to break the nylon line; 190 N to break the spider's web; 110 N to break the copper wire; and 81 N to break the wood. Aren't these results surprising? How do they compare with your prediction? If you thought differently, remember that the thickness of these materials plays a role in judging their strength. For example, most spider's webs that we see have threads less than 1 mm thick.

Strong Materials

Before students begin reading, ask if they have ever had to decide exactly which kind of material to use to build or make something. Call on volunteers to explain what qualities they looked for in the materials they chose. Now, have students read the first two paragraphs of the page silently. Ask them to speculate as to which of the materials—copper wire, nylon fishing line, spider's web, or wood—is the strongest. Next, have them predict the order of strength by listing the items from the strongest to the weakest. Then have students finish reading the page to discover if their predictions were correct.

Most students will be surprised to learn that spider's web, or spider's silk, is second in strength only to nylon. You may wish to explain that spider's silk is the strongest natural fiber known. Even a steel thread of an equivalent size is not as strong.

Strong Shapes

Some materials are stronger than others. But the strength of any material can be increased by changing its shape. Recall the bridges you built using index cards. What shapes had the most resistance to bending? Did you discover that a thin, flat card has comparatively little strength to support loads that cause it to bend?

The following Exploration should give you many ideas for building models of structures for Riverwood.

EXPLORATION 4

Testing for Strong Shapes

In this Exploration you will test the strength of five different shapes made out of different materials. Work with a small group, and then share your discoveries with the rest of your class.

You Will Need

- a force meter from the unit "Force and Motion," or a set of standard masses (10-g or 20-g intervals)
- a ruler
- index cards for Investigations 1 through 4
- masking tape for Investigations 1, 4, and 5
- paste or glue for Investigations 2 and 3
- heavy paper, a small pail filled with sand, and a hard board (approximately 10 cm × 10 cm) for Investigation 5

Investigation 1

Shapes Unlimited

1. Use only a single index card to make each of the five shapes shown to the right. You may use a small amount of tape to fasten the edges of the index card for shapes d and e.
2. Predict which of the five shapes you think would have the most strength to resist bending.
3. With a force meter or masses, apply force to the middle of each shape until it no longer returns to its original position when the force is removed. This happens when a permanent crease or fold is caused by the force.

(a)

(b)

(c)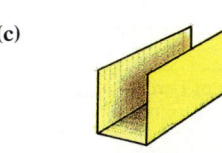

(d)

(e)

4. Make and test at least one shape of your own design.
5. Record your results in a table in your Journal. In the left column of your table, sketch each of the shapes. In the right column, give the bending force that was required to cause the shape to buckle.
6. Which of the five shapes was the strongest? Was your prediction correct? How did the shape you designed compare with these five?

Strong Shapes

Before students begin the Exploration, involve them in a discussion of the shapes they used to build their model bridges in Lesson 1. Have students arrive at a consensus about which shape was the strongest. Encourage them to speculate as to why it is important for architects and engineers to determine which shapes are the strongest before they begin building structures. *(It is important in order to avoid accidents and the waste of materials.)*

Exploration 4
Investigation 1

You may wish to have students work in pairs or small groups to complete the investigation. Before students begin, have them write their predictions about the strength of the five shapes in their Journals. Then have them test each one. Students will need to determine how to position the card shapes so that weights can be placed on them. One solution is to use the shapes to bridge the space between two books. Suggest that students gently hold the card shapes in place as masses are placed on them.

When students finish the investigation, have them discuss and compare their results. They should discover that the shapes increase in strength in the order that they are pictured, from (a) to (e), the flat shape being the weakest and the circular tube being the strongest. Have students compare their own shapes to determine who designed the strongest one.

Investigation 2

Point out to students that the term *lamination* means to bond layers of material together. It is a common way to increase the strength of the materials that are used.

You may wish to have students work in pairs or small groups to complete the investigation. Have groups read the instructions and discuss what they are to do. If space or time is limited, have volunteers perform the investigation as a class demonstration; students should make their own graphs from the results.

When students have finished, have them compare and discuss their graphs to determine if all the results were the same. Help to resolve any differences that may arise.

To extend the discussion, provide a piece of plywood for students to examine. Point out how it is made up of several layers of wood laminated together. Encourage students to discuss and speculate as to the advantages of using plywood for building projects. *(It is cheaper, stronger, and less likely to warp.)*

EXPLORATION 4—CONTINUED

Investigation 2
Lamination

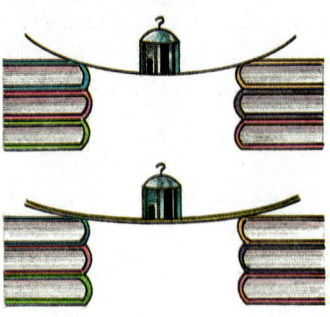

1. Make a table like the following to record your results.

Number of cards laminated together	Amount of sag (mm)

2. Arrange the books and index card as shown in the illustration above.
3. Place a standard mass on a single card so that it sags about 2 or 3 cm. Measure the sag.
4. Laminate (glue together) two cards so they have doubled in thickness. Using the same standard mass, measure the sag.
5. Repeat laminations for triple thickness and quadruple thickness. Measure the sag.
6. Make a graph similar to the one shown above and plot the sag against the number of cards laminated.
7. Extend your graph so that you can predict the amount of sag when six cards are laminated.
8. Test your prediction.

(Graph: Amount of sag (mm) vs. Number of cards laminated)

Investigation 3

Have students complete the investigation and discuss their results. They should discover that the corrugated structure is stronger than the laminated structure.

To extend the investigation, have students examine a corrugated cardboard box. Suggest that they pull the layers apart to examine them more closely. Encourage them to speculate as to why cardboard boxes are often made from corrugated material. *(Corrugated cardboard provides a great deal of strength considering the lightweight and inexpensive materials that are used in making them.)*

Investigation 3
Corrugation

1. Laminate three index cards together.

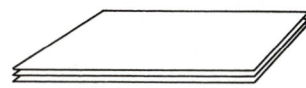

2. Find the force the laminated structure will sustain before sagging 1 cm.
3. Take another three cards and fold the middle card as shown in the diagram. Glue the cards together.

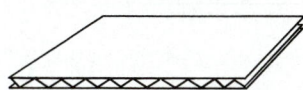

4. Find the force the corrugated structure will sustain before sagging 1 cm.
5. What do you conclude?

Investigation 4
Tubular Girders

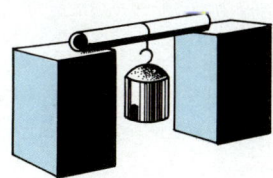

1. Roll an index card into a tubular girder. Hold it together with small pieces of masking tape.
2. Measure the diameter of the tube.
3. Find the force required to cause the girder to buckle.
4. Using the same amount of material and keeping the tube the same length, vary the diameter. Repeat for a total of five different diameters. Record your results in a table like the following:

Diameter	Maximum force

5. Make a graph of your findings (maximum force versus diameter).
6. What do you conclude?

Investigation 5
Tubular Columns

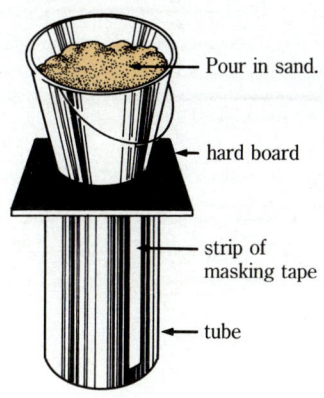

- Pour in sand.
- hard board
- strip of masking tape
- tube

1. Using heavy paper, make a tube to be used as a column.
2. Measure the diameter.
3. Construct a platform as shown. Place the pail over the center of the column as shown in the diagram.
4. Find the force required to buckle the column. You can do this by filling the pail with sand until the column buckles. Weigh the platform, pail, and sand together. This gives you the force required to buckle the column.
5. Repeat two more times, using the same amount of material for the column and keeping it the same height, but varying its diameter. Record your results in a table like the following:

Diameter	Maximum force

6. What do you conclude?
7. Instead of placing the pail over the center of the column, try placing the pail at different locations on the column to find out the effect of off-center loading.

Investigation 5

As students complete the activity, they should discover that as the diameter of the tubular column increases, more force is needed to make it buckle. Point out that this is the opposite of what happened with the tubular girder in *Investigation 4*. As students complete step 7, they should discover that off-center loading decreases the amount of force the column can hold. It also makes the column unstable and can cause it to fall over before it has a chance to buckle.

Follow Up

Assessment

1. Have students use the words listed on page 272 to make a crossword puzzle, a word "seek and find," or a matching activity between the words and definitions. Have them exchange activities and complete the one that they receive.
2. Point out to students that exercise exposes the human body to many different kinds of forces. Have them name exercises in which the body is subjected to shear, compressive, and tensile forces. Ask them to explain their choices. *(Answers could include: shear forces—snow skiing; compressive forces—weight lifting and jogging; tensile forces—stretching and bending exercises.)*

Extension

1. Suggest to students that they make a bulletin board or table display to show how laminated, corrugated, and tubular structures are used in everyday objects. Point out that they may have to do some research to find the information they need to make an interesting display.
2. Have students research one of the following topics:
    plywood
    corrugated cardboard
    particleboard
   They should find out how the material is made, what makes it strong, and what it is used for. Have students share their information by presenting oral reports to the class.

Investigation 4

Have students complete the investigation and compare and discuss their results before making their graphs. They should discover that the strength of a tubular girder decreases as its diameter increases. Some students might correctly point out that by using the same amount of material, the edges of the card will overlap, thus contributing to the strength of the girder. To account for this, the cards could be trimmed as the diameter decreases so that there is no overlap. Then have students complete their graphs. They should conclude that the strength of a tubular girder decreases proportionally with the size of its diameter.

You may wish to point out to students that a girder is a supporting structure that is used horizontally. In the next investigation they will examine the strength of a tubular column. A column is used vertically. Encourage students to speculate as to whether or not they will achieve the same results with a tubular column as they did with a tubular girder.

The Science of Structure 281

LESSON 3

✷ Getting Started

In this lesson, students examine some of the basic structures used in engineering. Students explore beams, trusses, cantilevers, arches, and domes. The lesson concludes with a discussion of how these are used in the design and construction of bridges.

Main Ideas

1. A truss framework adds strength and support to a structure.
2. For the same amount of material, an I-beam has greater rigidity and strength than does an ordinary beam.
3. A cantilever is a beam that is fixed at one end.
4. An arch provides more strength against a vertical load than does a simple beam.
5. The basic component of an arch is the wedge, which transforms vertical loads into lateral ones.
6. The dome is closely related to the arch since it is basically an arch in three dimensions.
7. Bridges are classified according to their construction: arch, cantilever, suspension, or girder beam.

✷ Teaching Strategies

Have students silently read the lesson introduction on page 282. Call on volunteers to explain what they already know about beams. *(Beams are used to span the space between two points. They are used to hold up loads. Beams may be made in different shapes.)*

Beams and Trusses

Use the questions in this section to spark discussion about the nature of beams and trusses. Most students will have seen wooden or steel beams in *(Continues on next page)*

282   Structures and Design

3 Engineering Structures

Many human-made structures are quite complex. However, they are made up of simpler components that give them their strength and shape. You can include these components in the structures you plan for Riverwood.

Beams and Trusses

Where have you seen *beams* before? What are they made of? Beams can be made of a variety of natural and artificial materials. Have you ever seen a *truss*? A truss is a framework of connected planks or steel bars that add strength and support to a structure. Trusses are frequently found in the roofs of houses and other buildings where long distances are spanned without columns.

LESSON 3 ORGANIZER

Objectives

By the end of the lesson, students should be able to:
1. Explain the advantages of using a truss or an I-beam instead of a rectangular beam to support a load.
2. Describe how a cantilever is constructed to support a load.
3. Identify that an arch and a dome are better able to support a vertical load than a simple beam is.
4. Classify bridges by their structural elements.

Process Skills

analyzing, measuring, observing, constructing models

New Terms

none

EXPLORATION 5

Activity 1
Comparing a Simple Beam with a Truss

You Will Need
- a board with one hole drilled at each end (the board can be a thin plywood strip, about 10 to 12 cm wide by 30 cm long or a similar size strip of corrugated cardboard)
- a short wooden post or piece of cardboard (10 to 15 cm long)
- standard masses
- supports (blocks or books)
- string
- 2 eye-screws

What to Do

1. Set up a beam as shown in the diagram below. Put a load of standard masses on the middle of the strip so that it sags noticeably.

2. Convert the beam into a truss, as shown in the diagram. The screws fit through the holes in the beam, and the string should be fastened to the screws. Allow the short post to sit on top of the string.

   Place the same load on it as you did before. What do you conclude?

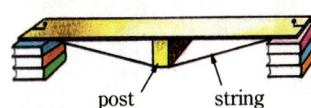

post   string

Truss

Activity 2
The I-Beam

You Will Need
- 6 boards, all the same size (boards may be strips of plywood or corrugated cardboard)
- nails
- a hammer
- tacks or glue
- standard masses
- supports (blocks or books)

What to Do

1. Nail or fasten three boards together as shown below in diagram (a).
2. Nail or fasten three boards together in the shape of an I-beam as shown in diagram (b).

(a)
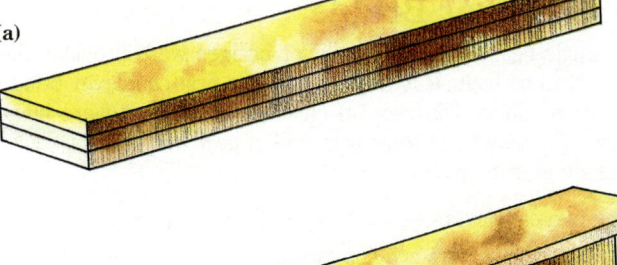

(b)
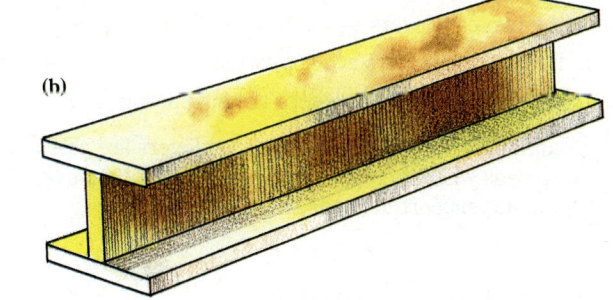

3. Put a load on the middle of the beam shown in diagram (a) so that the beam sags measurably.
4. Now place the same load on the I-beam.
5. What do you observe? Which is stronger, the I-beam or a simple beam made from the same amount of material?

Why do you think one beam is stronger than the other?

(Beams and Trusses continued) houses, malls, or gymnasiums. Then direct their attention to the photograph. Point out that the triangular structures are trusses. Ask students what the trusses are supporting. *(the roof)*

Exploration 5

Activity 1

If a wooden strip is used as a beam, it will need to be very thin in order to sag under a manageable weight. Either a 3-mm-thick piece of plywood or a wooden strip from a window shade should work well. Students should observe that the beam does not sag when it is converted into a truss. They should conclude that a truss can support more weight than can a beam.

Activity 2

Have students test the strength of the beams as they did in *Activity 1*. They should observe that the I-beam does not sag under the load. They should conclude that an I-beam can support more weight than can an ordinary beam made from the same materials. Explain to students that the I-beam is stronger because of its construction. The force of the weight on the beam is distributed farther away from the center of the beam. Thus, the same materials provide greater strength when they are in an I-beam configuration.

Materials

Exploration 5
Activity 1: board with one hole drilled at each end (the board can be a thin plywood or corrugated cardboard strip, about 10 to 12 cm wide by 30 cm long), short wooden post or piece of cardboard 10 to 15 cm long, standard masses, 2 supports (blocks or books), string, scissors, 2 eye-screws

Activity 2: 6 boards, all the same size (boards may be strips of plywood or corrugated cardboard), nails, hammer, tacks or glue, standard masses, 2 supports (blocks or books), safety goggles

Exploration 6
Activity 1: large index cards, metric ruler, standard masses (or cups filled with sand), 2 supports (blocks or books)

(Organizer continues on next page)

Cantilever

Involve students in a brief discussion of each of the pictures. Help them to recognize that:
- The tree branch is a cantilever because only one end of it is attached, or fixed, to the tree. The roots anchor the tree to the ground, offsetting the weight of the branch.
- The sign post is a cantilever because only one end of it is attached to the building. The weight of the post and the sign hanging from it are counterbalanced by the weight of the building.
- The balcony is actually made up of many cantilevers, with one end of each cantilever attached to the building. The weight of the building offsets the weight of the balcony, keeping the balcony from falling down.

Students should recognize that each of the structures have two things in common. They are each attached only at one end, and their weight is counterbalanced by the structures to which they are attached.

After students have discussed the illustrations, provide them with time to make their own drawings. When they have finished, call on several volunteers to share their work with the class. All of the drawings should show that a simple beam is supported at both ends, and that a cantilever is attached at only one end.

A Cantilever Bridge

You may wish to have students make a sketch of diagram (b), labeling the two cantilevers, the supports for the cantilevers, and the load supported by the cantilevers. Then invite a volunteer to explain his or her sketch and labels to the class. Students should recognize that the two cantilevers are the two boards held down by the cans. The cantilevers are supported by the two tables. The load that the cantilevers are supporting is represented by the weight and the board that spans the space between them.

Students should conclude that cantilevers are able to support heavy loads without collapsing because they are attached at one end to a heavier, more solid structure that counterbalances the load. The cantilever bridge in diagram (a) supports its load over a series of cantilevers, each with a greater weight on one end than on the other.

Cantilever

A *cantilever* is a simple beam that is fixed at one end. Here are some examples of cantilevers. Explain why each of them fits the definition of a cantilever.

(a)

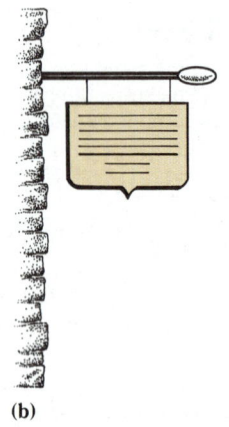
(b)

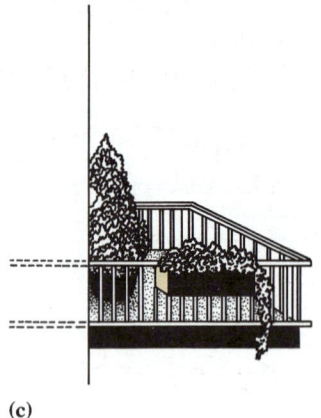
(c)

Show the difference between a cantilever and a simple beam by drawing each one supporting a person.

A Cantilever Bridge

After single-beam bridges, cantilever bridges were probably the next kind to be built. It is thought that the first cantilever bridge was built in China. Diagram (a) illustrates how they were built. Diagram (b) shows the basic principle of how they work.

Identify in diagram (b):

1. the two cantilevers
2. how the cantilevers are supported
3. the load that the cantilevers are supporting

Explain in your own words how cantilevers are able to support heavy loads without collapsing. How does the cantilever bridge in diagram (a) support its load?

Arches and Domes

By now you know there are many ways to support a load. You can support loads by using columns and beams. You can shape your beams in various ways, forming them into I-beams, for example. You can also transform a beam into a truss. And you can make a beam into a cantilever by attaching it at one end for support.

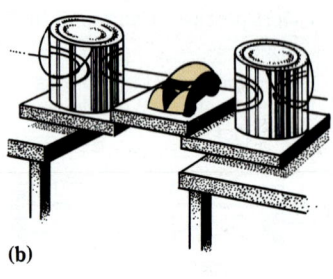
(a)

(b)

(Organizer continued)

Activity 3: modeling clay, half of a can or a large bent card, paper, pencil, protractor, standard masses (or cups filled with sand), 2 supports (blocks or books)
Bridge Identification: Journal

Teacher's Resource Binder

Activity Worksheet 5–6
Resource Worksheet 5–7
Graphic Organizer Transparencies 5–5, 5–6, and 5–7

Science Discovery Videodisc

Disc 2, Image and Activity Bank, 5–3

Time Required

three class periods

Another way to support a load is by making an *arch*. Its widespread use represented one of the most important advances made in the earliest civilizations. By the time of the Roman Empire, arch bridges and arch aqueducts (for carrying water) had become the major means of spanning distances such as valleys and rivers.

The *dome* is closely associated with the arch. A dome is basically a three-dimensional arch.

The domed cathedral Hagia Sophia in Constantinople (now Istanbul, Turkey) was the cultural center of the Byzantine Empire.

An igloo is another type of dome. Can you suggest what supports an igloo? Exploration 6 will help you answer this question.

Arches and Domes

Have students read this section silently. Then direct their attention to the photographs. Ask: What do arches and domes have in common? *(Both have a curved surface.)* Direct students' attention to the aqueduct at the top of the page. Ask: What do the arches in the aqueduct support? *(They support the top of the aqueduct. The lower arches support the upper arches.)*

Ask students to examine the photograph of Hagia Sophia and ask: What part of the cathedral is made of a dome? *(the ceiling)* Have a volunteer identify at least one arch in the picture. Then have students speculate as to what supports the structure of an igloo. Accept all reasonable responses without comment. Students will have the chance to investigate these kinds of structures further in *Exploration 6*.

To extend the discussion, point out to students that a dome is merely a series of arches arranged around a central point. To demonstrate this idea, lay several strips of paper over the curved surface of an upside-down bowl.

The Science of Structure   285

Exploration 6

Activity 1

Divide the class into small groups, and distribute the materials. The simple beam and the arch should each be constructed using a large index card. Students should use the same-sized card for each test. After students have completed the activity, reassemble the class and involve them in a discussion of their responses to the questions.

Answers to In-Text Questions

(a) The strength of an arch is greater than that of a simple beam.
(b) The higher an arch, the stronger it is. The flatter an arch, the weaker it is.
(c) The buttresses hold the arch in place, and they receive a downward as well as an outward force from the arch.
(d) The blocks support the arched card by pressing against the ends of the card. They not only receive the force of the load on the arch, but also help to hold the arch in place. The blocks support the simple beam by providing a surface on which the ends of the beam can rest.

Activity 2

Involve students in a brief discussion of the arches depicted in the photographs on page 285. Students should recognize that the arches are made of bricks, mortar, and stone. As a result, they are extremely heavy. The walls on which the arches rest must support the weight of the arches themselves and any additional weight that the arches support.

Involve students in a discussion of the two diagrams on page 286. Help them to recognize that a wall holding up an arch must exert a vertical and a lateral force on the arch because it must act as a buttress as well as a support for the weight of the arch. For this reason, a wall holding up an arch must be much stronger than a wall holding up beams and trusses where the force is acting in only one direction—straight down. One way of achieving the necessary strength in walls holding up arches is to make them thicker. You might want to point out to students that in the use of domes, the weight is transferred outward as well as downward. Because of this, architects can create a large interior space that is seemingly without support. The photograph of the interior of Hagia Sophia on page S79 will give students a sense of this kind of interior space.

286 Structures and Design

EXPLORATION 6

Comparing Beams and Arches

In this Exploration you will learn about the advantages of using an arch for spanning distances.

You Will Need
- index cards
- standard masses (or paper cups to be filled with sand)
- modeling clay
- a pencil and paper
- a protractor
- a ruler
- a can, with top and bottom removed, cut in half vertically
- supports (blocks or books)

What to Do

Activity 1
Find the maximum load that a simple beam can support before it sags 1 cm in the middle. Compare this to the maximum load an arch can support. Compare high arches with flat arches. You can hold the ends of the arch with blocks or books. These supports are called *buttresses*.

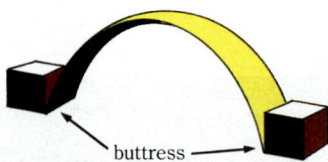
buttress

Now think about these questions:
(a) How does the strength of an arch compare to that of a simple beam?
(b) How does the shape of the arch itself affect its strength?
(c) What role do the buttresses play in an arched bridge?
(d) In what way do the blocks support the arched card differently than they do the beam?

Activity 2
Examine the photographs on page 285. Consider these things:
(a) the materials from which each arch is made
(b) the weight of the arch itself
(c) what supports each arch

Arches usually sit on the tops of walls. Each wall must not only support the weight of the arch and any load on the arch, but must also act as a buttress. In what two directions must a wall exert forces on the arch?

Study the two diagrams of structures below. One shows walls supporting an arch. The other shows walls supporting beams and trusses. Why do the walls in the arched structure need to be much thicker than those in the beam structure?

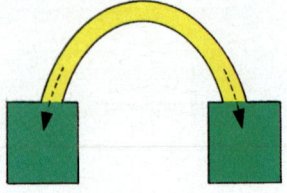
Thick walls are required to withstand force.

Thin walls are possible since the force is straight down.

Activity 3

This part of the Exploration is designed to help you understand how an arch is able to support a load. The basis of creating arches is using a wedge for support. An arch can be considered as a series of blocks, each wedged into place.

The bottom dimensions of each wedge are smaller than the top dimensions. Imagine trying to push one of the wedges downward. What would happen? You can see that an arch should have great resistance against collapsing from a downward push.

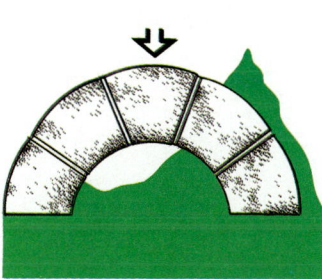

Now, make a bridge using wedges formed from modeling clay, as shown. You will need half of a can or a bent card to act as a mold for shaping the arch. With paper, pencil, and protractor, you can make a template for shaping the wedges. Make the template the same diameter as the can plus the desired thickness of the wedges.

Find the load that the clay bridge can support without collapsing. What factors would cause it to collapse?

Reshape the clay into a beam bridge. Use all of the clay. Compare the strength of the beam bridge to that of the arched bridge. For a fair comparison, what factors (variables) should be the same for both bridges?

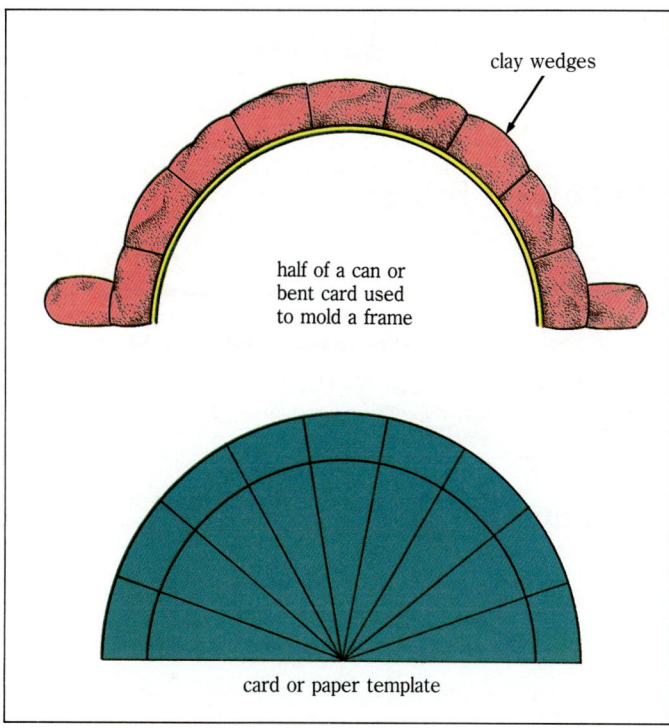

clay wedges

half of a can or bent card used to mold a frame

card or paper template

Activity 3

After students have read the explanation of how an arch is constructed, direct their attention to the diagram. Students should recognize that because of the shape of a wedge, it cannot be pushed downward, or fall through, an arch unless the entire arch gives way. Help students to recognize that the best way to protect the integrity of an arch is to make sure that the buttresses are strong enough to support it.

If there is a sufficient amount of material, divide the class into small groups and ask each group to make an arch. Each group member could make a wedge for the arch. First, the group will need to make a paper model to determine the shape of the wedges. Students should recognize that factors that could cause their arches to collapse include: weak buttresses, badly shaped wedges, wedges that are misaligned, and a load that is too heavy for the arch.

To extend the activity, have students construct arches with misaligned or misshapen wedges to determine the effect on the strength of the arch. Suggest that students try building arches with narrower wedges at the top than at the bottom to see how it affects the shape of the arch. Encourage students to "experiment" with building arches to discover what other factors may affect their strength and shape.

Students should discover that the beam bridge is much weaker than the arched bridge. For a fair comparison, both bridges should be made from the same kind and amount of material and should span the same distance.

To extend the Exploration, construct a demonstration arch using wooden wedges. Fasten sandpaper to the sides of the wedges to reduce sliding. Depending on the size of the demonstration arch and the loads placed on it, the buttresses may have to offer substantial resistance. Buttresses made from wooden blocks nailed to a platform provide one solution. However, rather than nailing the buttresses, you might have students provide the necessary resisting force.

The Science of Structure

Bridge Classification

Provide students with a few minutes to examine the four types of bridges in the illustrations on pages 288 and 289. Suggest that they write some notes in their Journals about the main characteristics of each type of bridge. Then involve students in a discussion to share their ideas. You may wish to use the following information to guide the discussion.

1. Arched Bridges

An arched bridge can be identified by the shape of its main structural feature—an arch. In this type of bridge the forces and stresses caused by vertical loads are carried along the sides of the arch to the buttresses and, eventually, into the ground.

2. Cantilever Bridge

A cantilever bridge is characterized by being fixed at one end. The illustration shows double cantilevers. The forces acting downward on one side are counterbalanced by forces acting on the other side. Each of the cantilevers is fixed to a supporting structure, which carries the downward forces to the ground.

3. Suspension Bridge

A suspension bridge is characterized by a cable or series of cables that carry the downward forces to the supports and then down to the ground.

Bridge Classification

Think of the bridges you have seen. Most bridges can be grouped as one of four types: arched, cantilever, suspension, or girder. Examples of these four types are illustrated below. What is the main characteristic of each type?

1. Arched Bridges

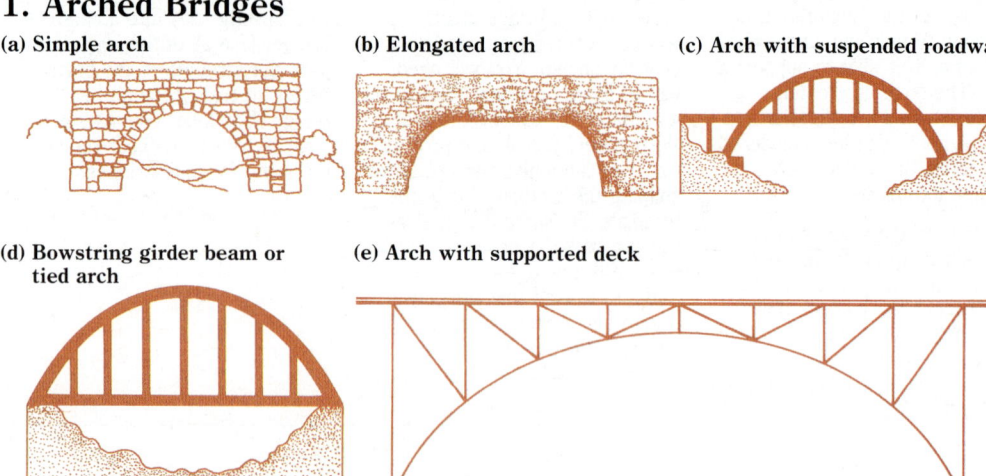

(a) Simple arch (b) Elongated arch (c) Arch with suspended roadway

(d) Bowstring girder beam or tied arch (e) Arch with supported deck

2. Cantilever Bridge

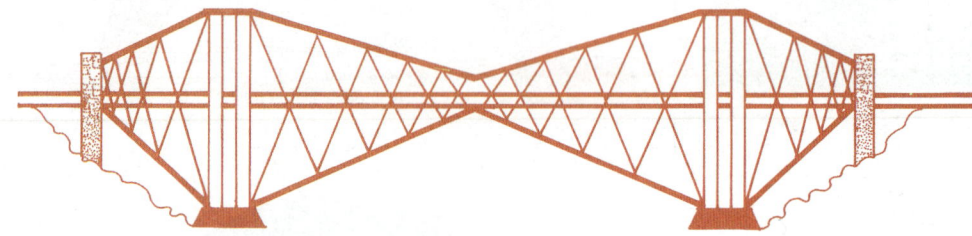

3. Suspension Bridge

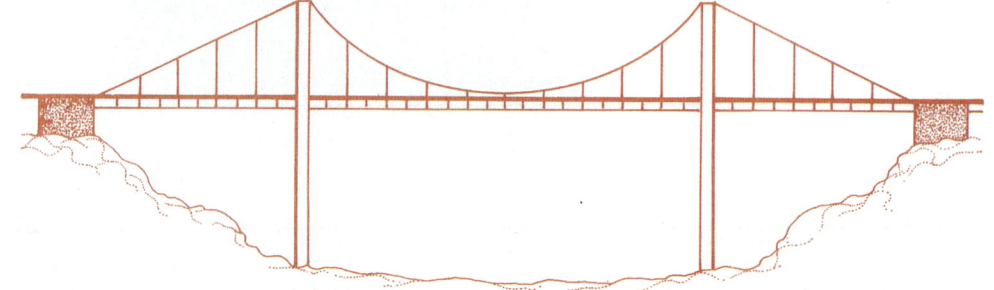

In the next section you will try your hand at classifying some bridges from around the world into these four basic types.

Famous Bridges

On page 290 there are photographs of well-known bridges from different parts of the world. Some are very old and yet are still standing. Others are quite modern. Examine each bridge and answer the following questions.

(a) Which common type does each bridge most closely resemble? (See bridge classification for help.)

(b) How is each bridge designed to resist the different forces that it might receive?

A helpful way to record this analysis is to use a table similar to this one. Copy the table into your Journal and fill in the information. Do not write in your book.

4. Girder Beam Bridges

(a) Warren girder

(b) Pratt truss

(c) Howe truss

(d) Lattice truss

(e) Stayed girder

Bridge Identification

Name	Bridge Type	Bridge Design to Resist Forces
Bridge of Sighs	Simple arch	Arch supplies strength to support the passageway and people.

4. Girder Beam Bridges

A girder beam bridge is essentially a trussed beam. The beam rests on supports that carry the downward forces to the ground. The crossed members of the girder provide strength against shear forces.

Famous Bridges

Direct students' attention to the pictures on page 290. Call on a volunteer to read the captions aloud. Then display a map of the world and ask students to locate and identify the area where each of the bridges is located.

When students have finished their analysis, involve them in a discussion of their ideas. Sample answers are given in the table below.

(Continues on next page)

Answers to Bridge Identification Table

Name	Bridge Type	Bridge Design to Resist Forces
• **Bridge of Sighs**	Simple arch bridge	The arch supplies strength to support the passageway and the pedestrians.
• **Thousand Islands Bridge**	Suspension bridge	The cables support the deck of the bridge and direct the forces down the supports to the ground.
• **Ebert Bridge**	Stayed girder bridge	The bridge is built like a trussed beam. The forces are transferred to the supports and carried to the ground.
• **Astoria Bridge**	Girder beam bridge	Trussed piers support the roadway and carry the forces to the ground.
• **Golden Gate Bridge**	Suspension bridge	The cables support the deck of the bridge and direct the forces down the supports to the ground.

The Science of Structure

(Famous Bridges continued)

To extend the exercise, ask students what factors they think determine the type of bridge that is built. (Answers could include the length of the structure, the maximum load that the bridge will carry, and the materials available to construct the bridge. The area over which the bridge is built is also important. For example, over deep water where the construction of supporting piers is difficult, a suspension bridge might be appropriate.)

Direct students' attention to the picture on page 291. Call on a volunteer to read aloud the information about it. Ask students to turn back to page 267 to compare the two bridges. Students should recognize that even though both bridges are suspension bridges, the new Tacoma Narrows Bridge is built with a warren girder to keep it stable.

Answers to
Further Analysis (page 291)

• Bridge of Sighs
(a) Every part of the bridge below the walkway is under compression from the weight of the bridge itself and from the weight of the people walking on it. The two buildings connected to the bridge act as buttresses against the lateral and downward force of the bridge.
(b) None of the bridge is under much tension.
(c) It is unlikely that it would be affected by shear forces.

• Thousand Islands Bridge
(a) The towers are under compression from the weight of the bridge, the cables, and the traffic moving over the bridge.
(b) The cables are under tension from the weight of the roadway suspended from them.
(c) The roadway could experience shear forces caused by severe weather or heavy traffic. The towers could also experience shear forces caused by fast moving water or the shifting of the earth where the towers are anchored.

Bridge of Sighs, Italy (top left); Thousand Islands Bridge linking Ontario, Canada with New York State (top right); Ebert Bridge over the Rhine, West Germany (center left); Astoria Bridge, Oregon (center right); Golden Gate Bridge, San Francisco (bottom)

• Ebert Bridge
(a) The towers are under compression from the forces directed to them by the girders and trusses, from which the bridge is suspended.
(b) The girders and trusses are under tension from the weight of the roadway.
(c) The roadway could experience shear forces caused by severe weather or heavy traffic. The towers could also experience shear forces caused by fast moving water or shifting soil where the towers are anchored.

• Astoria Bridge
(a) The trussed piers are under compression from the roadway they support.
(b) Some of the horizontal beams in the piers may experience tension from the tendency of the vertical beams to pull apart.
(c) The roadway may be subject to shear forces caused by heavy traffic loads, heavy winds, earthquakes, and shifting piers. The piers, anchored to the river bottom, may also experience shear forces from rapidly moving water or the shifting of the river bottom.

Compare this photo of the Tacoma Narrows Bridge with the photo on page 267. The old bridge pitched and rolled in a light wind until it collapsed. What changes were made in the structure of the bridge? How do these changes prevent the bridge from failing again?

Tacoma Narrows Bridge, Washington (rebuilt in 1950)

Further Analysis

1. Choose two of the bridges on page 290 and consider the following questions for each.
   (a) Which parts of the bridge are under compression? What are the compressive forces?
   (b) Which parts are under tension and what are the forces causing this tension?
   (c) Which parts might experience bending or twisting? What would be the possible cause of these shear forces?
2. Visit a bridge and sketch it, showing all the parts that give it strength and support. Label your sketch to show where you think tension, compression, and shear may occur. What type of bridge is it?

Follow Up

Assessment

1. Suggest that students make poster diagrams of beams, trusses, and arches. The diagrams should show how the forces are distributed through each of the structures.
2. Display several pictures of bridges and have students identify them as to type, using the classification system in their texts. Then have students identify some of the structures used in the construction of the bridges, such as arches, I-beams, and trusses. (If there are bridges in your area, photograph them for students to analyze.)

Extension

1. Suggest that students work in pairs or small groups to make a model of one of the bridges illustrated in the text. Have a spokesperson from each group explain how the forces are distributed in their bridge. Display the model bridges around the classroom.
2. Point out to students that foundations are the chief means of support in most structures. Suggest that they do some research on different kinds of foundations—spread, pier, and pile—and how they are constructed.

- **Golden Gate Bridge**
  (a) The towers are under compression from the weight of the bridge, the cables, and the traffic moving over the bridge.
  (b) The cables are under tension caused by the weight of the roadway suspended from the cables.
  (c) The roadway could experience shear forces caused by wind, heavy traffic loads, or earthquakes. The towers could also experience shear forces caused by fast moving water or the shifting of the bottom of the San Francisco Bay where the towers are anchored.

The Science of Structure

Answers to Challenge Your Thinking

The photographs are discussed in clockwise order.
- The Gateway Arch in St. Louis, Missouri, is 192 meters high.
- The ancient stone structure uses simple beams supported by columns.
- The bench uses cantilevers.
- The sailing ship features trussed cantilevers.
- The Roman Colosseum uses arches and simple beams.

Challenge Your Thinking

From the photographs below, identify examples of the following structures:
(a) simple beams
(b) cantilevers
(c) trusses
(d) arches

292 **Structures and Design**

THE ART OF DESIGN

4 Beauty in Structures

Our Human-Made Environment—Yesterday and Today

So far you have studied some of the scientific and technical features of structures. But that is not all you must consider when you plan structures for Riverwood. What will your structures look like? Will they blend in with their environment? How will they relate to each other? Engineers, architects, designers, and urban planners look to the past to help them make these decisions. In Exploration 7 you will take a journey through history. This will help you make your own decisions about the building style you would like to see in Riverwood.

EXPLORATION 7

A Journey Through History

On the following pages are photographs and illustrations of some important buildings that were built at different times and places. You are going to write and illustrate your own story about the people and buildings from one of these times in history.

What to Do

1. Team up with one or two classmates, and look through the next seven pages. Choose one period in history that interests you.
2. Research the period that you have chosen so that you can answer the questions below.
   (a) What was the culture of the society like? What kind of government did the society have? Who owned the land? Who owned the buildings? Who built the buildings?
   (b) What purposes did the buildings serve?
   (c) What structural features are evident in the buildings of this period (columns, beams, trusses, arches, domes, cantilevers)?
   (d) What materials were used?
   (e) What construction methods were used?

LESSON ORGANIZER

Objectives

By the end of the lesson, students should be able to:
1. Describe the architectural characteristics of a particular historical period.
2. Explain how the architecture of the past has influenced the architecture of the present.
3. Demonstrate how culture and technology influence the design of architecture.
4. Describe and compare the influence of Walter Gropius and Frank Lloyd Wright on modern architectural design.

Process Skills

analyzing, researching, writing, illustrating

New Terms

none
(Organizer continues on next page)

LESSON 4

✷ Getting Started

This lesson provides students with a look at the historical development of structure and design. Students examine the architecture of different civilizations and cultures, and learn how it has influenced modern design. The lesson concludes with imaginary conversations with two pioneers of modern design, Walter Gropius and Frank Lloyd Wright.

Main Ideas

1. Each major period of history has been characterized by its own unique architecture.
2. Modern architecture is often based on models from the past.
3. Architectural design is often based on available technology.

✷ Teaching Strategies

Our Human-Made Environment

Ask students to name as many different periods of architecture as they can. Obtain photos of local structures such as museums, churches, concert halls, and schools. Point out the period on which each building is based. Ask students what period of architecture they like best and why. Ask them what periods they would choose for their design of Riverwood and why.

Exploration 7

Have students read silently pages 293 and 294 to find out what they are to do. Suggest that students work in small groups so that the responsibilities for the project can be divided among group members.
(Continued on next page)

The Art of Design 293

(Exploration 7 continued)

For example, one student might be responsible for magazine research, another for book research, and another for obtaining pictures and illustrations.

Encourage students to be creative. Point out that including different kinds of illustrations, such as pictures, diagrams, tables, graphs, and charts, can make the story more interesting. Mention that quotations from primary or secondary sources can also make the story seem more real. The background information on each time period that follows is provided as a discussion aid and as a tool to help you evaluate student research.

Nomadic Peoples

Provide students with time to read the material on nomadic peoples and to look at the illustrations. They should recognize that the teepee was built around a cone-shaped framework. A similar shape is seen in the spires on churches, cathedrals, castles, and government buildings.

Nomadic peoples are not a relic of the past. Many nomadic groups survive today. The African Pygmies and Australian Aborigines are examples of hunting and gathering nomads who move from place to place in search of food. The Bedouin herders in the deserts of Arabia and northern Africa move from place to place in order to feed and water their herds of camels, sheep, and goats. Gypsy traders all over the world are nomadic merchants and craftspeople who travel to serve their customers and to sell their wares.

Since nomads move frequently, their structures must be portable. As a result, they are usually made from easily obtainable and lightweight materials that can be put together and taken apart quickly. Such materials often include animal skins, fabrics, and frames made from wooden poles.

EXPLORATION 7—CONT.

3. Using your knowledge of the period, write your own story. Include at least one illustration. You may choose one of the buildings shown on these pages or choose one from another source. The length of your story should not be more than two to three pages.

Sources of Information

- Encyclopedias—look up information on architecture.
- Libraries—ask the librarian to help you find books on architecture and on the specific period of history that you are studying.
- Your social studies teacher—ask your social studies teacher to help you find textbooks that describe the period of history that you have selected.

Nomadic Peoples

Nomadic people who formed many of the earliest human communities did not build permanent dwellings. Their lives involved constant movement in search of food and other materials.

The teepee shown above is similar in design to those used by the nomadic Plains Indians of North America. These people had a highly developed society but no permanent architecture. What is the shape of the framework that gives support to the teepee? Where else can you see a structure shaped like this? Do you know of any nomadic cultures in modern times?

A nomad family in the South Gobi, Mongolia—outside their *ger*, the felt-covered, collapsible tent of Mongolia.

(Organizer continued)

Materials

none

Science Discovery Videodisc

Disc 2, Image and Activity Bank, 5–4

Time Required

two class periods

Early Civilization (9000–1000 B.C.)

When people learned to raise crops, they were able to settle in one location. Staying in the same location gave them the chance to build permanent buildings. These first settlements represent the earliest civilizations.

These ruins of the Kingdom of Cush are located in the Sudan.

The Great Pyramid of Cheops

Early Civilization

The Cush kingdom traces its roots back to a region along the Nile in the present-day Sudan. The Egyptians greatly influenced this region, which was an important trading center. Nevertheless, the pottery, jewelry, and other ornaments found among these ruins reflect a distinct culture.

Ancient Egypt was one of the world's earliest civilizations, originating along the Nile River about 5000 years ago. Even though the Egyptians built great cities, they are best known for the pyramids they constructed as tombs for their rulers.

Throughout most of its history, ancient Egypt was ruled by kings, who were believed to embody the god Horus in human form. Forty-two local provinces called *nomes* were governed by *nomarchs* who were appointed by the king.

Ancient Egyptian society was made up of three main social classes. The upper class consisted of the royal family, wealthy landowners, doctors, high-ranking government officials, priests, and army officers. The middle class consisted mainly of merchants, craftworkers, and manufacturers. The lower, and largest, class consisted mainly of unskilled laborers.

The Egyptians built their houses with bricks of dried mud. The trunks of palm trees were used as columns to support the flat roofs. Many city houses had three or more floors.

The oldest and largest stone structures in the world can be found among Egypt's pyramids. They were built from huge limestone blocks, weighing more than 1.8 t (metric tons) each. The ancient Egyptians also built most of their temples from limestone. Many of these structures are noted for their huge decorative columns that were constructed to hold up massive beamed ceilings.

The Art of Design

Ancient Greece

Ancient Greece flourished about 2500 years ago. Yet its achievements in art, science, philosophy, and government continue to influence the western world.

Ancient Greece never organized itself into a nation. Instead, it was made up of many city-states, each consisting of a city or town and the surrounding area. Probably the best known of these city-states was Athens.

Greece's mild climate enabled people to live in simple one- or two-room houses with walls made of sun-dried bricks and floors of hard-packed earth. The wealthy lived in larger, more elaborate, and more comfortable homes built around courtyards.

The architecture for which the ancient Greeks are best-known is reflected in their temples, the designs of which centered around the column. The Greeks developed three basic kinds of columns—Doric, Ionic, and Corinthian. Each had its own distinctive decoration and style. The design of a temple consisted of arrangements of columns surrounding a long chamber, which usually housed the sculpted figure of the god or goddess for whom the temple was built. The best known temples were constructed on the Acropolis in Athens around 400 B.C.

An acropolis was the religious and military center of a city-state, usually built on the top of a hill. The original Athenian Acropolis was partially demolished by a Persian invasion in 480 B.C., but was later rebuilt in even greater splendor. Among the new buildings was the Parthenon, dedicated to the patron goddess Athena. The Erechtheion was built to honor the legendary founders of the city. Another temple honored Athena as the goddess of victory. Two theaters and several minor temples were built on the slopes of the hilltop. The entrance to the Acropolis was through a large, roofed gateway called the Propylaea.

The influence of Greek architecture can often be seen in public buildings. For example, the White House, the Capitol, and the Lincoln and Jefferson Memorials in Washington, D. C. were patterned after the magnificent temples of ancient Greece.

Ancient Greece (800–338 B.C.)

The architecture of ancient Greece has had an enormous influence on designers for many centuries. This drawing is from a model of the Acropolis at Athens, Greece, as it looked about 450 B.C. Name some buildings you have seen that appear to be patterned after the buildings of ancient Greece.

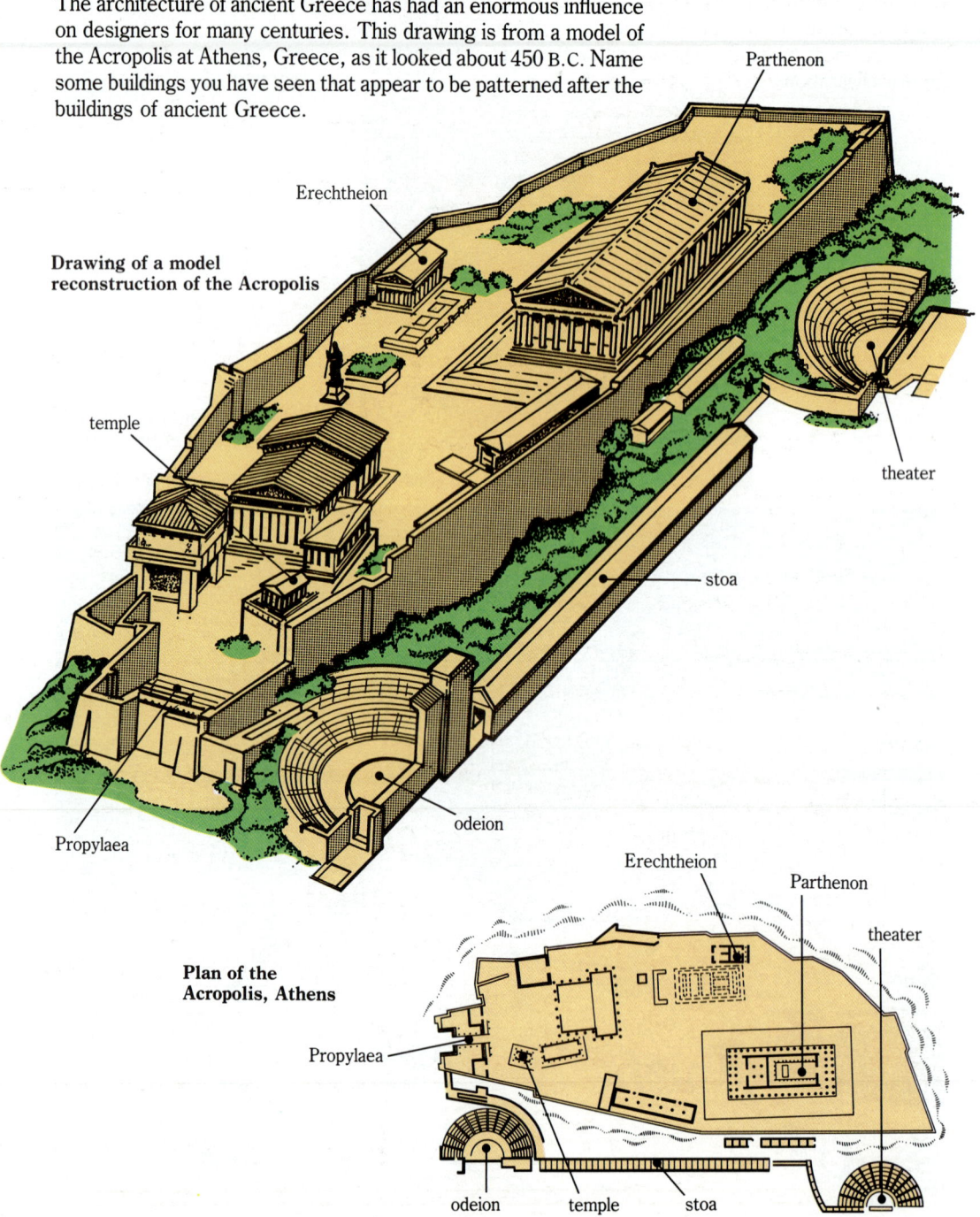

Drawing of a model reconstruction of the Acropolis

Plan of the Acropolis, Athens

296 Structures and Design

Ancient Rome (27 B.C.–A.D. 476)

Ancient Rome is characterized by its classic structures. In the center of the reconstructed model, you can see the Roman Circus Maximus. On the upper right side of the model, you can see the famous Colosseum. At the bottom of the model is the aqueduct that carried water to the citizens of ancient Rome. Many of the classic structures are still standing in modern Rome.

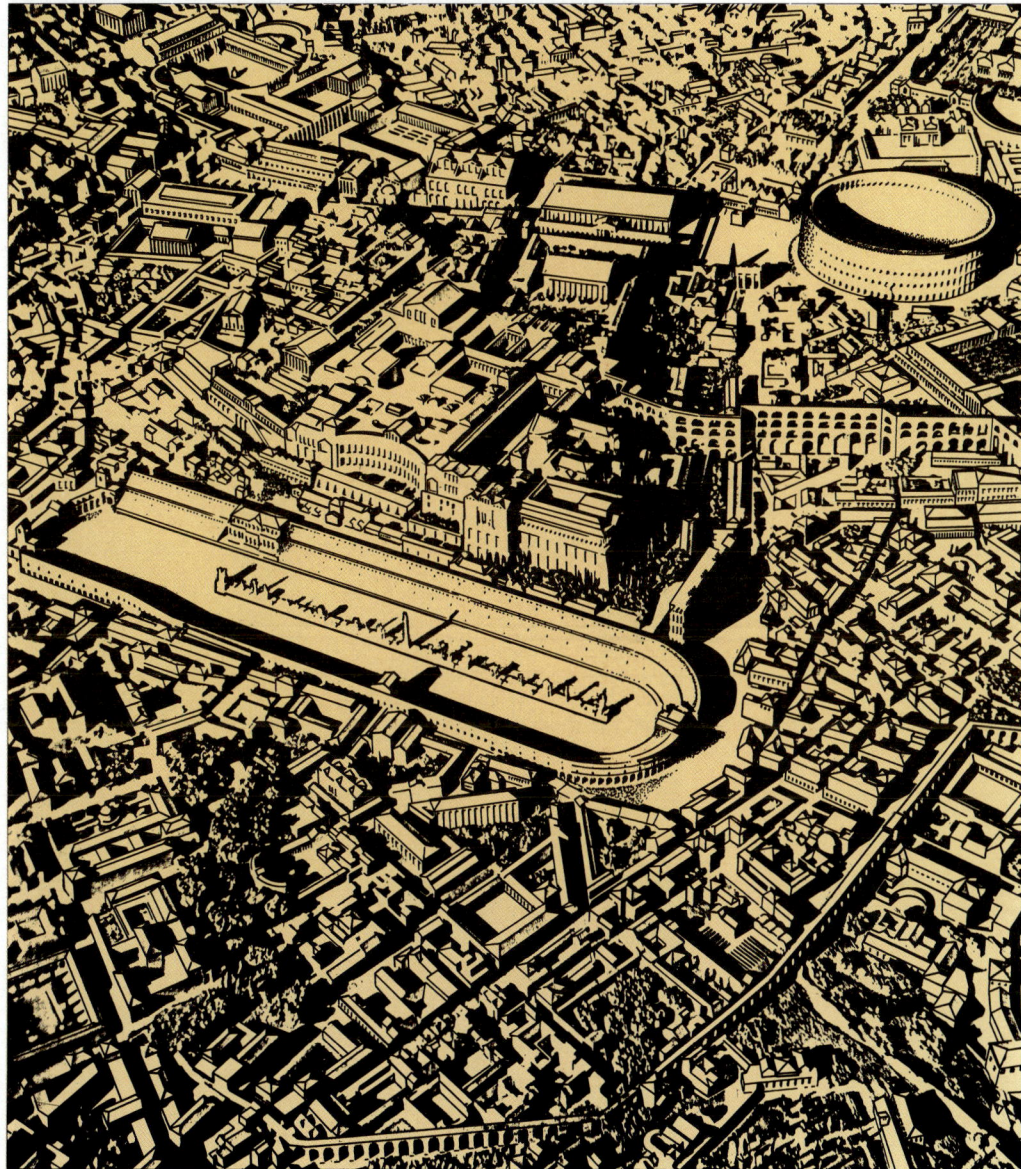

Ancient Rome

Ancient Rome was the capital and largest city of the Roman Empire. At its height in the A.D. 100s, the Roman Empire included all the land around the Mediterranean Sea, the land that stretched north to the British Isles, and all the land east to the Persian Gulf. It encompassed from 50 to 70 million people, and had to accommodate many different customs and languages. The ability of the Romans to hold this diverse empire together is considered an achievement in itself.

At its height, nearly one million people lived in ancient Rome. No other city had ever been as large. Most of the people in Rome lived in crowded apartment buildings that were from three to five stories high. A wealthy family would have lived in a house built around an atrium or courtyard. The atrium was spacious and had an open roof that let in light and air to the surrounding, windowless rooms. Larger houses had a second courtyard called a peristyle, which served as a garden. The poor, in the surrounding farming areas, lived in huts made of sun-dried bricks.

Much of ancient Roman architecture was adapted from ancient Greece. This includes the large temples surrounded by rows of columns. The Romans, however, also created many new kinds of structures, such as the public amphitheaters that were designed to hold large crowds of people.

The construction of large buildings was made possible by the Roman use of the arch and the invention of concrete. Arches were used to support such structures as bridges and aqueducts. Domes and arched roofs, called vaults, allowed for the construction of large, open, interior spaces. Arches, vaults, and domes eliminated the need for columns to hold up roofs. Instead, the weight of the roof rested on the outer walls of the building. The invention of concrete provided the strong building material that was needed for the outer walls.

The Art of Design 297

Medieval Europe

Also called the Middle Ages, this historic period in western Europe spanned ancient and modern times. Life in medieval Europe revolved around the control of land and was based on the system of feudalism. Large areas of land were owned by powerful lords. Small armies of faithful knights defended the land, while peasants farmed it. These feudal estates were bound together by the influence of the Christian church.

Two major types of architecture were developed during the medieval period—Romanesque and Gothic. Romanesque architecture began in the late 800s and reached its peak during the 1000s and 1100s. The most impressive and important Romanesque buildings were churches. A typical Romanesque church had thick walls, heavy curved arches, and rows of columns built closely together. There were few windows or openings to let in light. A pointed tower rose where the transept crossed the nave.

Gothic architecture flourished from the mid-1100s to the 1400s. Gothic architects developed a system of construction that enabled them to design churches with thinner walls than had been possible in Romanesque architecture. The ribbed vault or dome was one of the most distinctive characteristics of Gothic churches. Many churches had flying buttresses, which were brick or stone arched supports built against the outside walls. The buttresses strengthened the walls and permitted the use of large stained-glass windows that allowed light to enter the building.

The cathedral of Notre Dame is considered a masterpiece of Gothic architecture. It was built between 1163 and 1250. Notre Dame was one of the first buildings to have flying buttresses.

Medieval Europe (1100s–1500s)

Medieval Europe is characterized by the famous cathedrals that were built during this period. Is there a church or other building in your state that appears to be patterned after the cathedrals of medieval Europe? In your Journal, make a list of examples you can think of.

Notre Dame Cathedral, Paris, France

Renaissance (1300s–1600s)

After the Middle Ages, a widespread cultural revolution occurred in Europe. It was called the Renaissance, and it affected science, the arts, and education. Architects and artists, like Donato Bramante and Michelangelo, worked together to create beautiful buildings. Do you see any similarities between Roman and Renaissance architecture?

◄ Chateau de Fontainebleau, France

St. Peter's Church, Rome ►

The Cathedral of Florence, Italy ▼

Renaissance

The word *renaissance* comes from a Latin word that refers to the process of rebirth. To scholars and artists of the Renaissance, it meant a rebirth in interest in the classical culture and art of ancient Greece and Rome. The influence of the Renaissance on future generations proved to be immense. Most historians agree that the modern era began with this period of history.

Renaissance architecture began in Italy in the 1400s and gradually spread throughout Europe, eventually reaching the New World. A group of Italian scholars of classical culture is credited with beginning the movement. They considered classical culture superior to their own. Soon, architects began studying the ruins of ancient Rome and Greece, and began modeling their designs on classical buildings. They incorporated the use of classical columns as well as Roman vaults and domes.

The dome of the Cathedral of Florence, designed by Filippo Brunelleschi, is an early example of Italian Renaissance architecture. The cathedral itself was begun in 1296 in the Gothic style.

One of the greatest building projects of the late Renaissance was the construction of St. Peter's Basilica in Rome. It was started in 1506 and completed in the 1600s. Ten architects worked on the church during its construction, including Donato Bramante and Michelangelo.

In the early 1500s, the Renaissance spread from Italy to France. The finest examples of French Renaissance buildings are châteaus, or castles, such as the Château de Fontainebleau built in the early 1500s by King Francis I.

The Art of Design

Modern Civilization

The modern period of architecture is generally considered to have begun in the late 1800s. New materials allowed architects to develop and design the first completely new structural styles in centuries. American architects greatly influenced international architecture for the first time. One example is the skyscraper, which was first developed in the United States.

During the late 1800s and early 1900s, Chicago became the center for modern architecture in the United States. A fire in 1871 had destroyed much of the city. The rebuilding of Chicago provided architects with the opportunity to experiment with new designs utilizing the new materials that were then available. The world's first metal-frame skyscraper, the 10-story Home Insurance Building, was constructed in downtown Chicago during this period.

In recent decades, architects have started many new movements with names such as New-brutalism, Post-modernism, and Deconstructivism. Often these movements have been short-lived.

Post-modernism, which began in the United States in the 1960s, revived the use in modern buildings of historic elements such as classical columns and arches. These were often used in unconventional and playful ways.

Modern Civilization

Advances in technology have changed how we build structures. Structural steel and elevators have made building skyscrapers possible. Pre-stressed concrete is used to make beautifully curved roofs for stadiums, theaters, and other buildings. Even huge, enclosed stadiums can now be built through the application of modern knowledge of materials and aerodynamics.

▲ Sydney Opera House, Australia

◄ Superdome, New Orleans

▼ New York skyline

EXPLORATION 8

The Bauhaus School of Design, Dessau, Germany, headed by Walter Gropius and later, Ludwig Mies van der Rohe. Although this style of building is common today, it was very unusual when it was built. How does it compare to school buildings constructed before 1935 in your community?

Conversations with Walter Gropius and Frank Lloyd Wright

Daniel and Karen are two junior-high students who became very interested in modern architecture. They did research and created an imaginary interview with two pioneers of modern design. As you read their interviews, make a list in your Journal of the most important ideas. Later you can apply these ideas to your plans for Riverwood.

Daniel and Karen met Walter Gropius (1883–1969) first. Imagine it is March 15, 1952. Gropius is the head of the Department of Architecture at Harvard University. Listen in as Karen and Daniel conduct their interview.

Karen: We are pleased that you could take time to meet with us. We have heard so much about you. Many architects in the United States give you the credit for developing the modern ideas of architecture and design.

Walter: Thank you. The new International Style of architecture and design is the result of ideas from many artists, designers, and engineers from around the world. Much of the pioneering work was done at the Bauhaus School of Design in Germany when I was its director.

Daniel: What was it like working there?

Walter: It was very exciting. Many young artists and designers worked there. Several of them became famous for their work.

We came up with many new ideas about the way things should be designed. We considered buildings, furniture, books, and much more. Unfortunately, most of us had to leave the country when the Nazis came to power. I first went to England in 1934 and then came to Harvard University in 1937. Here, I have had the good fortune to work with many talented architects.

Karen: What did you and the other designers achieve?

Exploration 8

Direct students' attention to the picture at the top of the page. Encourage them to suggest words to describe how the building looks and makes them feel—shiny, open, glassy, light, airy, metallic, free, and so on. Some students may recognize that compared to earlier buildings, this structure uses much more glass and metal to create an open and airy feeling.

Encourage students to speculate as to how it would feel to live or work in a building like this. Then have them read silently the imaginary interview with Walter Gropius. Or, you may wish to call on volunteers to role-play the interview for the class. When students have finished reading, involve them in a discussion of the interview. You may wish to use questions similar to the following:

- Did Walter Gropius work alone on his designs? *(No, he worked with other architects, designers, and artists.)*
- With what does Gropius link art and design? *(He links art and design with modern technology.)*
- How did Gropius think of structures? *(He thought of structures as forms of sculpture.)*
- To what did Gropius believe designers should adapt their designs? *(He believed they should adapt their designs to mass production.)*

Students may find it interesting to know that the term *The International Style* came from the title of a book by the same name. It was written by two American architects in 1932. They had reviewed the architecture of the previous 10 years and concluded that a new international style had developed in several countries. This style was characterized by building designs that had geometric shapes, white walls, and flat roofs. There was little or no exterior ornamentation. Most were constructed from reinforced concrete and had large windows to create a light, airy feeling.
(Continues on next page)

The Art of Design

(Exploration 8 continued)

Encourage students to discuss the photographs at the top of the page. Tell them that Vladimir Tatlin, the artist who created the sculpture on the left, was part of a movement called Russian Constructivism in which the architects had grand ideas but lacked the technical knowledge to build what they designed. Few will view the Tatlin sculpture as similar to any building they have ever seen. The interior of the Guggenheim Museum in New York City is somewhat similar to this sculpture. The museum's interior is also a simple spiral, yet it is larger at the top than at the bottom. If possible, obtain a photograph of the interior of this museum and show it to students. Ask: What might be the reason for a spiral shape in a museum? *(People can view paintings in a continuous flow of movement, from top to bottom.)*

The Naum Gabo sculpture may remind some students of modern furniture. Have them compare the sculpture to the photographs of furniture on page 303. Encourage students to express whether or not they feel the designs are similar. Ask them if the designs evoke the same kinds of feelings. Then have students read the imaginary interview with Frank Lloyd Wright. When students have finished reading, involve them in a discussion of the interview. You may wish to use questions similar to the following:

- What was Frank Lloyd Wright's lifelong interest? *(He wanted to tackle the problem of the house as a shelter.)*
- How did Wright's views about architecture differ from those of Gropius? *(Gropius believed that architecture should reflect modern technology and be designed for mass production. Wright believed that buildings should blend in with their natural environment, and that their construction should include materials from that environment.)*
- What did Wright call his kind of architecture? Why? *(He called it "organic" architecture because it linked structures with nature.)*
- Which kind of architecture, that of Wright or Gropius, do you think you like better? Why? *(Accept all reasonable answers. Students' responses should reflect an understanding of both kinds of architecture.)*

EXPLORATION 8—CONTINUED

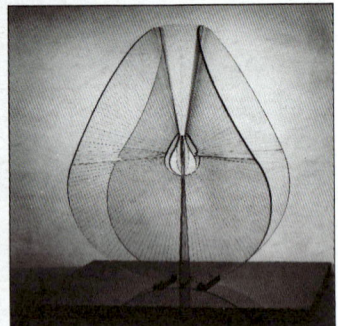

◄ A model for a building that was never constructed. Can you imagine how it might have looked? Have you seen any buildings like this sculpture?

▲ Naum Gabo—Translucent Variation on Spheric Theme. This is a 1951 version of the 1937 original. The sculptor used plastic, glass, and wire to create a sense of space and weightlessness in his modern sculpture.

Walter: Those of us who worked at the Bauhaus School linked art and design with modern technology. The sculptors were particularly important in leading the way. They were using modern materials. From these artists we have learned to think of structures as forms of sculpture.

Daniel: What are your goals as a teacher of design and architecture?

Walter: I want people to recognize that everything built by humans has a design. Some people use a lot of decoration in their designs. However, I prefer simplicity. I believe designers should be able to adapt their designs to mass production.

Daniel and Karen's next stop was in Taliesin, Wisconsin. There they met Frank Lloyd Wright (1869–1959), the most famous American architect. The date is July 15, 1955.

Daniel: When did you begin your career in design, and with whom did you work?

Frank: In 1887 I lived in Chicago and began to work with Louis Henri Sullivan. He was a pioneer in the design of large commercial buildings. In those days, we were learning to build all those skyscrapers. We tried to make them look friendly by including many of the decorative elements of earlier architecture.

After six years with Louis, I set up my own studio where I could devote myself to the designing of homes. I wanted to tackle the problem of the house as a shelter. That has been a life-long interest for me. Of course, I continued to design many public buildings, including an art gallery and a university campus.

Provide students with a few moments to study and compare the pictures of Fallingwater and the Bauhaus School of Design. Then involve them in a discussion of how the two differ, and how they reflect the ideas of their creators. *(The Bauhaus School of Design reflects Gropius's idea of combining mass production and modern technology with design. Wright's Fallingwater reflects his views that architecture should be combined with nature.)*

Armchair designed for Bauhaus by Marcel Breuer, 1924.

Modern furniture design, inspired by the work of the Bauhaus School.

Karen: It is known that you have opposed the International Style of architecture. How do your views differ from those of Walter Gropius and his followers?

Frank: Modern architecture has been influenced by Gropius' views and my own. I believe homes and buildings should blend in with their natural environment. Their construction should include materials from that environment. I call this organic architecture because it links structures with nature.

It seems to me that many architects and designers forget about nature when they design. So there is a definite contrast between my views and those of Gropius. I think a building's relationship to nature is one of its most important characteristics.

Fallingwater, Bear Run, Pennsylvania (1936). Designed by Frank Lloyd Wright. How has Wright used cantilevers to create a home that blends in with its environment?

Answer to caption: The cantilevers are attached to the side of the house near a hill, giving the impression that the house is a part of the land.

Follow Up

Assessment

1. Have students use the information on the historical development of structure and design on pages 294 to 300 to make a timeline. Their timelines should highlight at least one major feature of the architecture from each of the periods.
2. Have students make posters to illustrate the architectural features of a particular period of time. Labels and brief descriptions should be included for each feature. Display the completed posters around the classroom.

Extension

1. Suggest that students work in pairs or small groups to make models of famous buildings. Suggest that they model one of the structures pictured in the lesson or choose a building of their own to model. When students are finished, call on several volunteers to describe their models to the class, pointing out the most important architectural features. Then display the models around the classroom. Each model should be accompanied with a brief description identifying the name of the building and where it is, or was, located.
2. Suggest that students write short biographies of some of the most famous modern architects, such as Frank Lloyd Wright, Henry Hobson Richardson, William Le Baron Jenney, Walter Gropius, Le Corbusier, Ludwig Mies van der Rohe, Paul Rudolph, and Louis Kahn. Have students organize their biographies into a booklet, make a cover, and think of a title. The booklets could be displayed in the classroom for others to enjoy and read at their leisure.

The Art of Design

Lesson 5

✴ Getting Started

In this lesson, students observe and discuss the urban environment. They evaluate the views of an urban architect, and present their own ideas on how urban environments can be improved. The lesson presents students with a series of research activities that focus on the challenges of urban design. Through these activities, students learn about the problems of the inner city, the role of city councils in combating these problems, and the function of zoning laws in protecting the integrity of the urban environment and the safety of the people who live there. The lesson concludes with students constructing a model city using the knowledge they have acquired throughout the unit.

Main Ideas

1. The location and design of structures in an urban setting is a social as well as an environmental concern.
2. Most cities have laws to regulate the construction and design of buildings to ensure a pleasant and safe environment.
3. How buildings and open spaces are designed can have an important impact on the social environment of a community.

5 A Question of the Environment

The photos on these two pages illustrate what architects and planners can do to make the urban environment more pleasant for people. Describe which features of these structures make them pleasant to view or to visit.

LESSON 5 ORGANIZER

Objectives

By the end of the lesson, students should be able to:
1. Describe the relationship between urban design and the environment.
2. Recognize the importance of creating an urban environment that conforms to the needs of the people living there.
3. Evaluate how well an urban setting meets social as well as environmental concerns.

Process Skills

analyzing, evaluating, designing models, constructing models

New Terms

none

304   Structures and Design

The designers of Riverwood have to consider what kind of city they want to live in. They have to know what regulations apply that may restrict the kinds of buildings they can build. An example of such a regulation would be a *zoning law* that does not allow commercial buildings in a residential area.

✳ *Teaching Strategies*

Involve students in a discussion of their community. Encourage them to identify what they like and dislike about it. Invite them to speculate about ways in which their community could be improved. Then call on a volunteer to read the lesson introduction at the top of page 304.

Direct students' attention to the photographs. Ask: What makes the structures in the photographs pleasant to view and to visit? *(Most students will probably identify such things as open spaces, plants, decorations, and building design features.)*

To extend the discussion, have students describe buildings that they have enjoyed visiting. Some students may have photographs of urban buildings and open spaces that they can share and discuss with the class.

Call on another volunteer to read the paragraph at the top of page 305, and have students discuss the picture. Ask students to name the features in the design of Rockefeller Center that they find appealing. Point out to students that the plaza can be used for outdoor dining in the summer and ice skating in the winter. Ask if any students have ever visited Rockefeller Center, and if so, invite them to share their impressions.

Materials

Exploration 9: uncooked spaghetti, string, balsa wood, index cards, glue, cardboard, plastic wrap, metric ruler, tape

Science Discovery Videodisc

Disc 2, Image and Activity Bank, 5–5

Time Required

two class periods

The Art of Design

Consider the Views of Phyllis Lambert

Have students read the page silently or call on a volunteer to read it aloud, pausing before the Lambert quotations. Direct students' attention to the photograph of the Seagram Building and have them suggest words that describe how it looks—tall, glassy, plain, simple, bright, airy, and so on. Encourage students to express whether or not they like this kind of architecture, and have them provide reasons for their points of view. Help students to recognize that the beauty of architecture, like that of art, is often in the eye of the beholder. Then provide students with some time to consider the comments by Phyllis Lambert.

Call on several volunteers to summarize Lambert's comments. Then involve students in a discussion of Lambert's ideas and opinions. You may wish to use questions similar to the following:

- According to Phyllis Lambert, who were the first to cry out about the real problems of the cities? *(The people living in the cities were the first to recognize urban problems and to suggest solutions.)*
- How does Phyllis Lambert describe citizens' groups? *(She describes them as groups made up of people who want to make sure that their neighborhoods are not disrupted.)*
- What does Phyllis Lambert consider to be her major task? *(Making sure that there is an effective means for action and participation by citizens' groups in the decisions that affect their daily lives.)*
- Do you agree or disagree with Phyllis Lambert's opinions? Why or why not? *(Answers may vary, but most students will probably agree with Phyllis Lambert's opinions because they take into account the needs and wishes of the people that are affected by what urban architects do.)*

Consider the Views of Phyllis Lambert

Canadian architect Phyllis Lambert has had an influence on the appearance of several North American cities. One of her greatest achievements was her work as the Director of Planning for the Seagram Building in New York. She worked for the renowned architects Mies van der Rohe and Philip Johnson. The building they created helped to establish the International Style as a major trend in North American architecture.

She is now setting a different trend as a conservationist, working to restore and preserve buildings and entire communities.

Think about and discuss the following comments that she has made:

It was not the architects who first publicly complained of urban disaster... Those who first cried out and pointed to the real problem and created the basic structures of reform have been citizens' groups. These groups are made up of people who are concerned about not having their neighborhoods disrupted... Such citizen groups not only make better urban developers than people in government and business, but also become more competent themselves in directing their own lives... I consider that my major task ... is seeing to it that there is an effective means for action and participation by citizens' groups in the decisions that affect our daily lives.

Phyllis Lambert

The Seagram Building, New York City

Challenging Research Activities

Here are some activities that may interest you and some of your friends.

1. Ask your parents or other adults how the appearance of your city has changed in the past years. What areas of the city have had new buildings constructed? Find out if any buildings or neighborhoods were torn down.

2. From your council representative, find out the laws your city has about the kinds of buildings that can be built in your neighborhood. These are called zoning laws. Look for answers to these questions:
   (a) What kinds of buildings are allowed in your neighborhood (commercial buildings, apartment buildings, single family homes, factories, office buildings)?
   (b) Is there a limit on the height of a building?
   (c) How close together can builders construct homes and other buildings?

3. Find out about the work of historical preservation groups in your city. What buildings and neighborhoods have been preserved because of their work?

4. Consider the design problems of urban centers. What is meant by the term *inner city?* Find out if your city council has plans to combat the problems of the inner city by changing the environment. If you do not live in a large city, research the history of design and structures found in Harlem in Manhattan, New York.

Sources of Information

- The library
- Architectural firms
- The planning department of the city government
- Parents and other adults

What could residents and city officials do to improve the environment here?

A Project for Everyone

In a sketch, redesign a familiar environment (your room, the school gym, the classroom, or the school). Decide how the changes you propose would affect other people who share that environment.

Exploration 9

Planning a town is a major undertaking, and for that reason you may wish to have groups of three or four students work together and submit their plans to the class. Once a plan has been decided upon, the entire class should be involved in making the model of Riverwood.

Establish an area of the classroom where the model can be constructed. Then suggest that different students, or groups of students, take responsibility for modeling a particular part or aspect of the community. For example, a few students might work together on a particular street or area. Or, one group of students could be responsible for constructing buildings, another for constructing streets and sidewalks, and another for constructing parks and recreational areas.

Follow Up

Assessment

1. Have students use pictures from magazines to make posters showing urban buildings and settings designed to make the urban environment more pleasant. Each picture should be accompanied by a caption that describes what it shows.
2. Have students record their ideas about what features are important in the design of an urban environment. (Answers could include safety, appearance, materials used, interaction with the environment, and practicality.)

Extension

1. Students may enjoy designing their ideal home. Remind them to apply the ideas and concepts they have learned from the unit. Invite volunteers to share their finished designs with the class.
2. Suggest that students take a survey to find out what different kinds of architecture are represented by the buildings in their neighborhood or community. Two or three students could work together to identify and assess the architecture in one section of your town or community. Students could then pool their information to use in a class presentation.

A Design for Riverwood

You have now considered many problems in designing and building structures. You know about some of the scientific and engineering ideas involved.

If you were making decisions about a home or other structure, you might be influenced by your knowledge of the history of architecture and design. In making your decisions, you would be more likely to consider the effects on other people. You would want any structure you built to be safe and to meet the standards of zoning laws. In these respects, you now think like a professional designer. More importantly, you can now play an active role in future decisions about your environment.

It is time to apply some of your new knowledge to the design of Riverwood. As soon as you have a plan for the community, proceed to Exploration 9.

EXPLORATION 9

Constructing a Model of Riverwood

1. Submit your own plan for Riverwood to your class. Explain how it meets the following criteria:
   (a) usefulness
   (b) beauty
   (c) safety
   (d) the interests of people who live in Riverwood
   (e) the concerns of those who live in the surrounding neighborhoods
2. As a class, decide which plan best meets the needs of the Riverwood community.
3. Build models of the structures that will appear in Riverwood. As a class, decide:
   (a) what materials are to be used in building the structures (spaghetti, string, balsa wood, index cards, cardboard, or other materials)
   (b) what materials are to be used for structures that need enclosure, such as buildings and homes (paper, cloth, plastic wrap, or other materials)
   (c) the size of the models
   (d) how the models will be tested for strength
   (e) the architectural style or styles that will be used
4. On your own, or with a partner:
   (a) Build a model of the framework for one of the new structures for Riverwood. Test it for strength. Your model must be safe under a load that is one and one-half times greater than the maximum load that might be expected. Remember to consider the effects of natural disturbances as well as the weight placed on the structure.
   (b) If your model is not safe, repeat (a) using more material or a different design.
   (c) Add walls, windows, and roofs as needed.
   (d) Add decorations to your structures in harmony with the architectural style or styles agreed upon by your class.

Challenge Your Thinking

1. Take a look at the buildings around your school or home. Do you see the use of triangles, circles (or parts of circles), rectangles, and squares in the buildings? Do you think that these shapes have anything to do with support or strength? Record your findings in your Journal.
2. Try your hand at architecture in the style of either Frank Lloyd Wright or Walter Gropius. Sketch a home for a family of four. Be sure to include bathrooms, a kitchen, a garage, a place for a washer and dryer, and anything else that the family might need.
3. Below is a photograph of the White House. Analyze the structure, and determine which period of architecture it represents or imitates. Is it attractive to look at? Why or why not? How does it fit into its natural environment?

Answers to Challenge Your Thinking

1. Answers will vary. Encourage students to make sketches of their observations in their Journals, along with an explanation of how each shape contributes to the whole. The contribution could be structural, ornamental, or both.
2. Encourage students to be as creative as they like. Students can begin this project by making a list of the needs of the structure and of its future occupants. Ask students to make sketches from all angles of the structure, including an aerial view, in which the interior can be seen. Tell students that graph paper is sometimes helpful to sketch on when deciding proportions in a structure.
3. The White House represents the Neoclassical style, which takes its influence from both ancient Greece and Rome. Students will most likely notice the effect of the columns on the curved portico, which extends out into the yard. This design element accentuates the landscaping and the fountains.

The Art of Design 309

UNIT 5
Making Connections

Summary for
The Big Ideas

Student responses will vary. The following is an ideal summary.

Factors that contribute to the success or failure of a structure include strength and thickness of materials chosen, arrangement and type of design features, and amount and type of environmental forces that interact with the structure. Environmental forces that can contribute to a structure's failure include wind, floods, changes in temperature, earthquakes, and rusting.

Forces affecting a structure can be described as tensile, compressive, or shear. A tensile force causes stretching, as in a cable supporting a bridge; a compressive force causes compaction, as in a column supporting a roof; and a shear force causes bending or twisting, as in scissors cutting paper. Materials can be strengthened to resist forces by altering their shape (materials formed into tubes or rectangles are stronger), laminating (layered materials are stronger), and adding corrugation or tube girders.

Some simple elements of design include beams and cantilevers. A beam is a simple structure used to span the distance between two points. A cantilever is a beam that is fixed at one end. Cantilevers can be used in decks, signs, and bridges.

Bridges can be given strength and support through the use of beams, trusses, cables, girders, and columns. Arches and domes are stronger than simple beams because the force of gravity is displaced to the side supports. The arch or dome is better able to resist the force of gravity.

Architecture of different eras reflects the technology, geography, climate, and cultural traditions of society through the ages. For example, teepees developed by early cultures represent the need for mobility and indicate the types of natural materials that were available. The pyramids of Egypt reflect the importance of the afterlife to the Egyptian society. The carefully planned cities of the ancient Romans show the importance of society, as well as the influence of earlier Greek architecture. Architecture has always been an extension of early designs. Today, new materials and technology have provided endless possibilities.

Viewing architecture through history shows that structure and design can combine to form a structure that is both beautiful and enduring. In creating a structure, designers must consider not only the function of the building itself, but also the needs of the community and of the environment.

Unit 5 Making Connections

The Big Ideas

In your Journal, write a summary of this unit, using the following questions as a guide:

- What makes a structure strong?
- What causes a structure to collapse?
- How do you distinguish between tensile, compressive, and shear force?
- How do the three different forces affect a structure?
- How can shape, lamination, corrugation, and tube girders increase strength?
- What is the difference between a beam and a cantilever? Where would you find each?
- How are bridges given strength and support?
- Why are arches and domes stronger than simple beams?
- What are some characteristics of the different architectural periods?
- How can structure and design make buildings beautiful?
- What must engineers and builders consider before constructing a building?

Checking Your Understanding

1. Study the skeleton of the horse shown below. Then answer the questions.

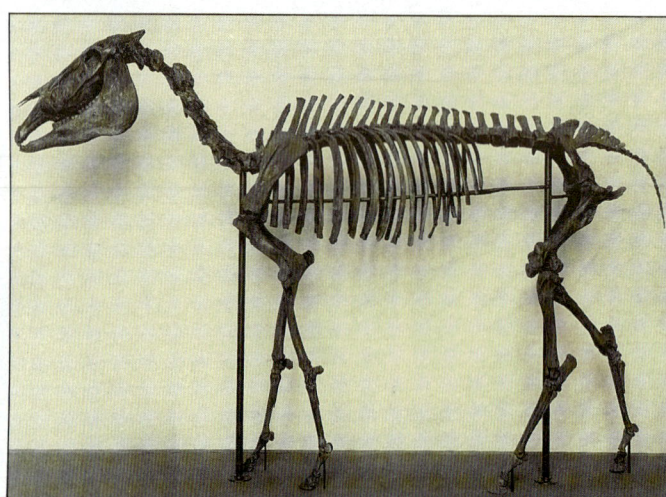

(a) Which parts of this structure function like columns?
(b) Which parts function like beams?
(c) Which part functions like a cantilever?

(d) If the horse had a rider, which parts of the horse would be under compression?

(e) Which parts would be under tension?

2. Design a solution for this structural problem: Support a glass of water by using three stainless steel knives and three soft drink bottles. Each knife can touch only one bottle. Using a paper cup, try out your design. Draw the support arrangement you created.

3. Draw a concept map using the following words: *nomadic, stone, skyscraper, teepees, modern, architecture, Parthenon, animal skin, Ancient Greek,* and *reinforced concrete.*

Reading *Plus*

Knowing about structure and design helps you understand how buildings, bridges, highways, and skyscrapers are constructed. Now read pages S77–S88 of the *SourceBook* to explore the variety of architecture found in other cultures.

Updating Your Journal

Reread the paragraphs you wrote for this unit's "For Your Journal" questions. Then rewrite them to reflect what you have learned in studying this unit.

About Updating Your Journal

The following are sample ideal answers.
1. Structure could refer to the actual building, while design could refer to the way in which a structure's parts are organized.
2. Steel frames, using the design elements of the arch and the truss, interact to support the Eiffel Tower.
3. Buildings should be both structurally sound and pleasing to look at; in addition, a building should complement its environment.
4. The larger a building is, the more important its structural design becomes. For example, the higher a building is, the more wind it must resist and the more weight it must distribute through its supporting structures.

Answers to

Checking Your Understanding

1. (a) The parts of the horse that function as columns are the bones associated with the legs and the cartilage between these bones.
   (b) The backbone functions as a beam.
   (c) The neck and tail function as cantilevers. Some students might also point out that the ribs function as cantilevers.
   (d) The legs and cartilage between the bones would be compressed.
   (e) The back would be under tension.
2. Solutions to this problem may vary. One solution would be to have the knives act as cantilevers, forming a triangle, in which each knife overlaps and supports the knife ahead of it.
3.
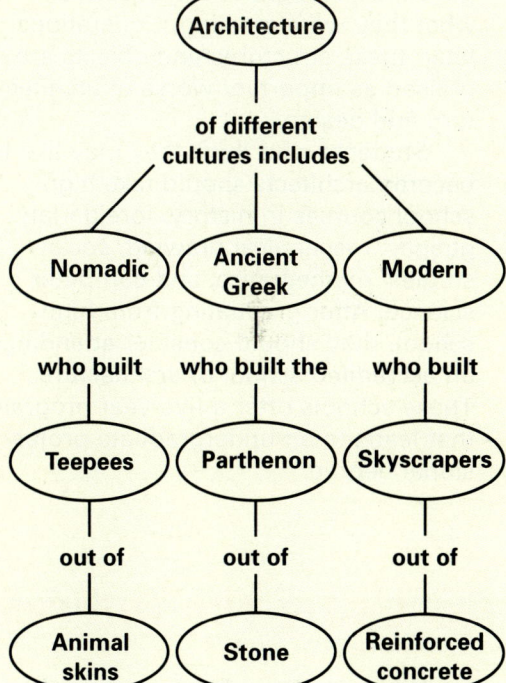

Making Connections 311

Science in Action

♦ Background

Architecture is an art form that dates to prehistoric times. A society's culture is often reflected in its architecture. For example, the importance the ancient Greeks gave to discipline and harmony was reflected in architectural design that was balanced and orderly. The Medieval period was greatly influenced by religion. As a result, architects designed magnificent, towering cathedrals.

Sometimes an architectural style is developed without the help of professional architects. For example, a group of people known as the Shakers designed buildings and furniture to reflect their belief in a simple lifestyle, and gave little thought to the beauty of what they were creating. Generations later, these structures and objects are praised as important works of architecture and design.

Students who think they may like to become architects should take high school courses in history, foreign languages, mechanical drawing, social studies, mathematics, and computer science. After graduating from high school, they should consider attending an accredited school of architecture. These schools offer a five-year program that leads to an undergraduate professional degree.

Science in Action

Spotlight on Architecture

Sydney Dumerasq, like his father, grandfather, and great grandfather before him, is an architect. He and a team of architects he works with design homes, buildings, and other structures. We met with Sydney at his office. He showed us how he uses a computer to help him design buildings.

Q: What do you find most appealing about your job?

Sydney: The creative aspect. You take a site with nothing on it, put your imagination to work, and actually design a building that wasn't there before. That building may stand for generations.

Q: What are some of the considerations you have when designing a building?

Sydney: First of all, I always have to bear in mind that I have two clients. I have the people who are paying the bills, but I also have the general public as a client. If the building is going to stay for generations, people are going to have to look at it for that time. People will live in it and work in it. It must be safe in all respects.

Q: What school subjects turned out to be the most important for your career?

Sydney: Everything. You should study absolutely anything and everything. Never turn down an extra course. Always keep your eyes and ears open.

Q: How do these relate to architecture?

Sydney: History helps a lot. For example, when you study Roman civilization or Greek civilization you will probably become familiar with their architecture. Geography is also very important because buildings are related to their environment. And mathematical concepts and tools are essential.

Q: Which aspects of science are important in architecture?

Sydney: Physics is especially important, particularly the physics of how structures work and under what conditions they stand or fall. Chemistry is also important. How do some materials react with others? How much does a given material expand or contract when the temperature changes? You don't need to know anything in great detail but you need to know the basic things.

Q: What qualities are most essential for architects?

Sydney: Optimism and perseverance.

♦ Using the Interview

When students finish reading the interview, evaluate their understanding by asking questions similar to the following:

- What does Sydney Dumerasq find most appealing about his job? *(The creative aspect. Taking a site with nothing on it and creating a building that was not there before.)*
- Who are the two clients Sydney Dumerasq has when he works on a project? *(He has the people who hired him and are paying the bills. Also, he has the general public who must look at and live with the building for generations.)*
- Why does Sydney Dumerasq feel that it is important for architects to have a basic knowledge of physics and chemistry? *(Architects should understand the physics of how structures work and under what conditions they will stand or fall. Chemistry is important in order to understand how materials react with one another and what the basic properties of different materials are.)*
- What qualities does Sydney Dumerasq feel are most essential for architects to have? *(optimism and perseverance)*

Some Project Ideas

1. Build a scale model of the home you designed. Use simple materials such as cardboard, craft sticks, and balsa wood.
2. Design a solution to a real construction problem. To find such a problem, visit homes and buildings in your area. Perhaps an additional room or a new set of stairs is needed. Design a solution and construct a model. If the solution is a good one, you might even convince someone to apply it. If it is your home, you might show your parents your suggestion and see if they agree.
3. Consider the special needs of the handicapped. Find out what architects include in their designs in order to accommodate handicapped people. Redesign your airport or local grocery store so the needs of the handicapped are met.

4. Investigate the factors that can make a building energy efficient and design improvements to an existing building.
5. Talk to your teachers about desirable characteristics of a classroom, a laboratory, or other school facility. If you have already identified a problem that needs a solution, design an improvement to the existing school or classroom structure. Check with teachers, other students, and the principal. Making an improvement to the school could be a class or school project.

For Assistance

Look under the heading *Architecture* in the yellow pages of your telephone directory. Also, there may be a Department of Architecture at a university in your region. Once you have identified a project and have given it some thought, arrange to visit an architect who is willing to advise and encourage students doing architectural projects.

Useful advice and assistance can also be obtained from builders and from businesses that sell building materials.

Industrial arts teachers in your school can be especially helpful. They can advise you on how to go about designing a solution, including the choice of materials. Later they can give you advice about constructing your solution (either a model or the real thing). You may be able to arrange to do your work in an industrial arts workshop.

◆ Using the Project Ideas

Provide students with some time to look at the project ideas. Be prepared to answer any procedural questions they may have. Then have students decide on a project to complete. To ensure that a variety of projects have been chosen, you may wish to have students present their ideas to you before they begin.

Set aside time for students to share their completed projects with the class. Also, you may wish to consider having students use the results of their projects to create an architectural environment in the classroom. For example, students who have drawn building designs to solve special problems could use their drawings to make bulletin-board displays. Students who have made models could arrange them on display tables. An architectural reading center could be set up for students to use at their leisure.

◆ Going Further

Ask students to investigate the following questions:

- Who are some of the most famous architects in the United States?
- What architectural form is uniquely American? When did it develop? Who was largely responsible for it? *(the skyscraper; late 1800s–early 1900s; architects in Chicago)*
- Who do you think should decide how land is used? The government, individual owners, business people, environmental groups, and citizen groups all have a stake in the way that land is used. Whose rights do you think are most important and why?

Science in Action   313

Science and the Arts

✦ Background

Florence is considered by historians to be the birthplace of the Renaissance. During the period from about 1300 to 1600, it became the home for some of the greatest painters, sculptors, and writers in the world. Among them were Leonardo da Vinci, Fra Angelico, Michelangelo, Giovanni Boccaccio, Dante, and the architect Filippo Brunelleschi.

In the 1420s, Brunelleschi and his circle of friends began to articulate a set of ideas that would define the character of Renaissance art. For example, they believed that the classical models were ideal, especially in sculpture and architecture. Brunelleschi himself was considered to have rediscovered the secrets of Roman architecture. In addition, he and his followers believed that mathematics could unlock the secrets of the world.

The dome on the Cathedral of Florence, which Brunelleschi designed and constructed, dominates the city to this day. The citizens of Florence so cherished the majestic dome that rose high above their city's streets that "dome-sickness" became the word to describe their nostalgic feelings when they were away from home.

Science and the Arts

An Architectural Marvel

In the early 1400s the people of Florence, Italy, had a real problem on their hands. Their plan for enlarging and rebuilding their cathedral had called for a magnificent dome, 42 meters in diameter, to be built at one end of the cathedral. They had begun this enormous project over a hundred years before, and now they didn't know how to finish it.

The new cathedral was to be the largest in the world. It would establish the power and prestige of Florence over all the other cities in central Italy. It was a matter of great civic pride. But if they couldn't figure out how to build the dome, they would be the laughing-stock of the neighboring cities.

The dome of the Florence cathedral presented a major engineering problem.

The Problem

The usual method of building a dome, according to the techniques known at the time, was to build a supporting structure of wood, lay brick over it, and then remove the wooden structure. But for such a huge dome, there was no way to get timber long enough or strong enough to build the supporting structure.

An Artist's Solution

In 1418 the cathedral officials announced they would give a cash prize of 200 florins to anyone who could figure out how to construct the dome. One person who entered the competition was a famous artist and goldsmith who had turned his attention to architecture. His name was Filippo Brunelleschi (broon uh LESS kee). He demonstrated, by means of models, how the dome could be built without a supporting structure. His models were carefully studied, along with his inventions for lifting construction materials into place. In 1420 the decision was made. Brunelleschi was made the chief architect of the project.

Brunelleschi's solution to the dome problem consisted of using a double shell with ribs formed by pointed Gothic arches. The double shell construction saved weight, and the pointed arches kept the side-thrusts to a minimum. The spaces between the ribs were filled in with bricks, one layer at a time, in a series of circles of decreasing diameter.

Completed in 1436, after sixteen years of work, the dome assured Brunelleschi's fame. Scholars and architects still study the Florence cathedral for insights into Brunelleschi's methods.

The dome was built using a double shell.

✦ Promoting Observational Skills

Ask students to carefully observe the photographs in order to answer the following questions.
- What does the size of the people on the lantern suggest to you about the size of the structure? *(The dome is very large—42 m in diameter.)*
- What features are added for ornamental effects? Do these structures have any other function? *(The buttresses and arches are attractive, and they also help to support the structure.)*
- What effect does the use of domes have on the appearance of an interior space? *(Domes allow for a large interior space to exist without supports that obstruct the view. The large space gives the ceiling an appearance of weightlessness.)*

✦ Discussion

When students have finished reading the feature, involve them in a discussion of what they have read. You may wish to use questions similar to the following:
1. What was the method of building a dome in the 1400s? *(The method was to build a supporting structure made of wood, lay brick over it, and then remove the wooden structure.)*
2. What were the problems in building the cathedral's dome? *(The quantity of wood needed could not be lifted to the top of the church; because of the size of the dome, the wood would not be strong enough to support its own weight.)*
3. What was Brunelleschi's solution to the problem? *(He suggested that the dome could be built without using a supporting structure. He would use a double-shell with ribs formed by pointed Gothic arches.)*

Brunelleschi's design for the "lantern" at the top of the cathedral shows his artistic side.

The Crowning Touch

The *lantern* on top of the dome was designed by Brunelleschi, but it was not completed until after his death. A lantern, in architectural terms, is an open structure on the top of a building to let in light and air. The graceful design reflects Brunelleschi's past experience as an artist and goldsmith.

Find Out for Yourself

Brunelleschi was an artist with a scientific mind. His mathematical interests led him to the rediscovery of *linear perspective*, a method artists use to show depth on a flat surface. This method had been known to the Greeks and Romans but was lost during the Middle Ages. Do some research to find out about this method of drawing.

Brunelleschi took an interest in the works of the ancient Romans. He filled his notebooks with sketches and data of their buildings. He was particularly impressed by a beautifully preserved domed building known as the Pantheon. This ancient building influenced him in his thinking about the cathedral project.

Brunelleschi wasn't the only person interested in ancient architecture. Ancient Greek and Roman ideas were beginning to have a major impact on European art and architecture in many ways. What can you find out about this historical development, which was known as the *Renaissance?* What was its influence on architecture?

This ancient Roman building, the Pantheon, convinced Brunelleschi that the cathedral dome could be built.

♦ Critical Thinking

To promote logical-thinking skills, you may wish to ask questions similar to the following.

1. When Brunelleschi decided to use arches in the construction of the dome, what relationship between an arch and a dome do you think he recognized? *(He must have recognized that an arch is like a slice, or cross-section, of a dome.)* How would these arches give added support to the dome? *(They would help carry the weight of the dome to the walls of the building.)*

2. Do you think it was a good idea to make Brunelleschi chief architect of the project? Why or why not? *(Most students will probably agree that it was a good idea to make Brunelleschi the chief architect because he understood the design and the plans the best. No one else would have known more about the procedures and methods than Brunelleschi himself.)*

♦ Going Further

Florence is considered to be the birthplace of the Italian Renaissance. Do some research to discover what the city was like during this time. You may wish to use the following questions to guide your research.
- Who were some of the famous people who lived and worked there during this period?
- What attracted great artists, writers, and sculptors to Florence?
- Who financed the many great artistic and architectural projects in Florence?
- Is there any art or architecture from the Renaissance period still in Florence? If so, what are some examples?

Share what you discover with your classmates by making fact sheets or bulletin-board displays.

Science and the Arts 315

UNIT 6 The Restless Earth

✴ Teaching the Unit

☀ Unit Overview

In the unit *The Restless Earth,* students are introduced to the earth's geological processes of change. In the first section, *Shake, Rattle, and Flow,* students are encouraged to explore the mechanism of plate tectonics as it relates to mountain-building, earthquakes, and volcanism. In the section *The Role of Rocks,* students study rock characteristics, the three main rock-forming processes, and the rock cycle. The nature of fossils, hypotheses about mass extinction, and the interpretation of fossil evidence are discussed in the final section, *Fossils—Records of the Past.* Students also develop their own relative and absolute time scales using familiar events to understand the development of the geologic time scale.

☀ Using the Themes

The unifying themes emphasized in this unit are **Scale and Structure, Patterns of Change,** and **Evolution.** The following information will help you weave these themes into your teaching plan. A focus question is provided with each theme as a discussion tool to help you tie the information in the unit together.

Scale and Structure is evident throughout Unit 6. In Sections 1 and 2, students review the plate structure of the earth's crust and the characteristic structures of the three rock groups. In Sections 2 and 3, the Richter Scale (measuring earthquake intensity) and the geologic time scale invite further examination of the theme.

Focus question: *How does the internal structure of the earth relate to the processes that shape it?*

Students should understand that the internal structure of the earth provides the basis for the process of plate tectonics. The plate structures, floating and moving on the molten mantle, obey the laws of physics.

Patterns of Change can be discussed in conjunction with Sections 1 and 2. In Section 1, students are introduced to the many changes in the earth's crust, such as mountain-building, movement along faults (earthquakes), and volcanism. Section 2 describes patterns of change leading to lithification. Intrusions and the rock cycle are other patterns explored in Section 2.

Focus question: *How does the formation of igneous, sedimentary, and metamorphic rocks form a larger pattern of change?*

Students should understand that the processes forming each of these rock types occur continuously, are interrelated, and form a pattern called the *rock cycle.*

Evolution can be discussed in relation to Sections 1 and 3. The continents have evolved and continue to evolve through the process of plate tectonics. The fossil record indicates that conditions on earth have changed over time.

Focus question: *What does the fossil record show about the history of the earth's biotic and abiotic factors?*

Students should understand that the fossil record shows that environmental conditions on earth have changed throughout time and that living organisms suited to those conditions have evolved over many generations.

☀ Using the *Science Discovery* Videodiscs

Disc 1 *Science Sleuths, The Misplaced Fossil*
An amateur paleontologist has found a dinosaur bone from the Cretaceous Age (65–140 million years ago) in a Tertiary Stratum dating back to only 10 million years ago. The Science Sleuths must analyze the information in order to verify or disprove the paleontologist's finding.

Disc 2 *Image and Activity Bank*
A variety of still images, short videos, and activities are available for you to use as you teach this unit. See the *Videodisc Resources* section of the **Teacher's Resource Binder** for detailed instructions.

☀ Using the *ScientificPlus SourceBook*

Unit 6 focuses on rocks and the minerals that comprise them. The three types of rocks—igneous, sedimentary, and metamorphic—and their place in the rock cycle are examined. Students learn about the composition and properties of rock-forming minerals. Finally, the unit discusses the earth's history, as recorded in rocks, and rock-dating techniques.

PLANNING CHART

SECTION AND LESSON	PG.	TIME*	PROCESS SKILLS	EXPLORATION AND ASSESSMENT	PG.	RESOURCES AND FEATURES
Unit Opener	316		observing, discussing	For Your Journal	317	Science Sleuths: *The Misplaced Fossil* Videodisc Activity Sheets TRB: Home Connection
SHAKE, RATTLE, AND FLOW	318			Challenge Your Thinking	340	
1 Making Mountains	318	2 to 3	observing, inferring, formulating hypotheses, making models	Exploration 1	324	Image and Activity Bank 6–1 Resource Transparency 6–1
2 The Earth Breaks Apart	325	3	observing, inferring, formulating hypotheses, analyzing	Exploration 2 Exploration 3	326 331	Image and Activity Bank 6–2 TRB: Resource Worksheet 6–1 Resource Transparency 6–2 Graphic Organizer Transparency 6–3
3 Volcanoes—Holes in the Earth	332	2	observing, making a word puzzle, researching, drawing conclusions	Exploration 4	338	Image and Activity Bank 6–3 TRB: Resource Worksheets 6–1 and 6–2 Activity Worksheet 6–3
THE ROLE OF ROCKS	341			Challenge Your Thinking	359	Graphic Organizer Transparency 6–7
4 Rocks in the Past	341	2 to 3	observing, inferring, classifying, using a key	Exploration 5 Exploration 6	343 345	Image and Activity Bank 6–4 TRB: Resource Worksheets 6–4 and 6–5 Resource Transparency 6–4 Graphic Organizer Transparency 6–5
5 Hot Rocks	346	3	observing, inferring, analyzing, compiling a dictionary	Exploration 7	346	Image and Activity Bank 6–5 TRB: Resource Worksheet 6–6 Graphic Organizer Transparency 6–6
6 Rocks From Sediments	351	3	observing, inferring, investigating, predicting	Exploration 8	353	Image and Activity Bank 6–6 TRB: Activity Worksheet 6–7
7 Changed Rocks	355	2	observing, inferring, synthesizing, interpreting data	Exploration 9	357	Image and Activity Bank 6–7 TRB: Activity Worksheet 6–8
FOSSILS—RECORDS OF THE PAST	360			Challenge Your Thinking	369	
8 Rocks Reveal a Story	360	2	observing, analyzing, inferring, formulating hypotheses			Image and Activity Bank 6–8 TRB: Activity Worksheet 6–9 Resource Transparency 6–8
9 Telling Time With Rocks	365	2	sequencing, classifying, recognizing relationships, interpreting data			Image and Activity Bank 6–9 TRB: Activity Worksheet 6–10 Graphic Organizer Transparencies 6–9 and 6–10
End of Unit	370		applying, analyzing, evaluating, summarizing	Making Connections TRB: Sample Assessment Items	370	TRB: Activity Worksheet 6–11 Graphic Organizer Transparency 6–11 Science in Action, p. 372 Science and the Arts, p. 374 *SourceBook*, pp. S89–S110

*Time given in number of class periods.

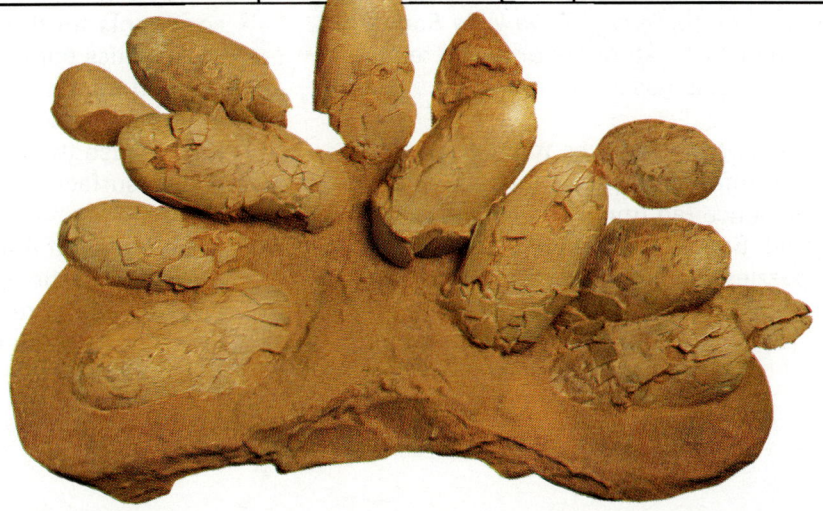

✳ Meeting Individual Needs

☀ Gifted Students

Pose one or more of the following "what if" questions to students:
- What if the substances forming the continental crust were much harder and more resistant to erosion?
- What would happen if the average temperature inside the earth's mantle was much higher?
- What would the earth's surface be like if there were 20 smaller crustal plates instead of six major ones?
- What would the earth be like if it were a solid ball without a molten mantle and core?

Have students write paragraphs that explain their responses. Encourage them to consider the effect the changes would have on living organisms.

☀ LEP Students

1. Individually or in small groups, have students compile a rock collection and write a short paragraph (in English or their dominant language) about the "life story" of each of the rocks collected.
2. Have students tell the story of an earthquake or volcanic eruption using a comic strip format. Students can create or gather a series of drawings or photographs to illustrate the sequence of events. Then encourage them to write a simple dialogue or commentary to accompany the pictures.
3. At the beginning of each unit, give Spanish-speaking students a copy of the *Spanish Glossary* from the *Teacher's Resource Binder*. Also, let Spanish-speaking students listen to the *English/Spanish Audiocassettes*.

☀ At-Risk Students

Unit 6 offers many motivating activities for students to gain first-hand knowledge of the processes taking place inside the earth. In *Explorations 1* and *2*, students build and manipulate models of mountain-building processes and earthquakes. In *Exploration 5*, students examine rock samples and organize their own classification scheme. Lesson 9, *Telling Time With Rocks*, can be personalized by having students organize relative and absolute time scales based on important events in their own lives.

You Be the Scientist reinforces the thought processes used when inferring a past sequence of events from physical evidence (an important task in geology and paleontology). Encourage students to create their own footprint puzzles to exchange with their classmates.

For special recognition, give students an opportunity to create more detailed models. For example, a student could build, paint, and label a three-dimensional relief map of their region.

☀ Cross-Disciplinary Focus

Literature and Creative Writing

A number of popular works of fiction have been written on the subject of "a voyage to the middle of the earth." One example is Jules Verne's *Journey to the Center of the Earth*, published in 1864. You may wish to have students read excerpts from this book as the basis for a class discussion. Then ask students to write their own stories about a journey to the center of the earth. They should incorporate facts that are known today about the interior of the earth.

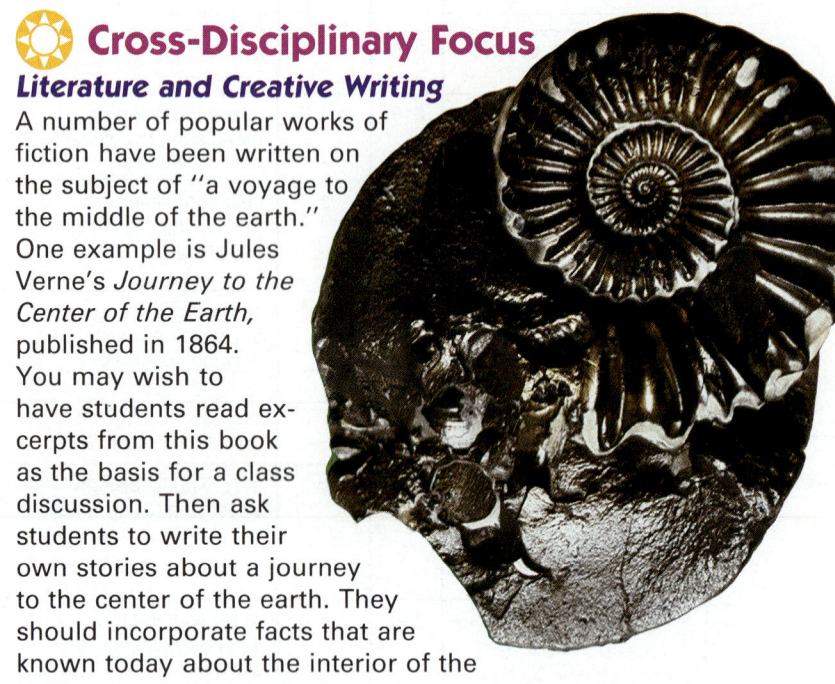

Art

Materials from the earth have been used by artists since ancient times. Prehistoric rock carvings and paintings made with crushed mineral pigments have been found throughout the world. Marble, a metamorphic rock, is used by sculptors. So are other rocks, such as granite and soapstone. Ask students to design a work of art using earth materials. In conjunction with the design, have them write an explanation of how they took advantage of the specific properties of the material.

☀ Cross-Cultural Focus

World Ceramics

Ceramics have been made by people on all of the continents for thousands of years. The materials used in ceramics come from the earth. Clay is a moist, easily molded material made up of fine particles of hydrous aluminium silicates and other minerals. High temperatures cause the mineral particles to fuse, resulting in a glassy surface. This process is known as vitrification. Have students investigate the clay deposits and ceramic traditions of Asia, Africa, North and South America, Europe, and the South Pacific. Students can choose an individual culture to investigate, or a group could work together to study a particular geographic region.

Earthquake Impact

A powerful seismic event, such as an earthquake with a magnitude of 7.0 or greater on the Richter Scale, affects a large area of the earth's surface, across international boundaries. Have students research an event such as the September 19, 1985 earthquake (magnitude 8.1) that was centered in Mexico City. Students should discover what aspects of Mexican culture diminished or increased the effects of the quake. Ask: What were the buildings made of? How was the dense population of Mexico City affected? What kind of economic and social effects occurred? How far away from Mexico City was the earthquake felt? Bilingual students should be encouraged to consult foreign language periodicals. Students could use maps to show the epicenter of the earthquake, and they could consider how different cultures in surrounding nations were affected.

Resources

Bibliography for Teachers

Barrow, Lloyd H. *Adventures with Rocks and Minerals: Geology Experiments for Young People.* Hillsdale, NJ: Enslow Publications, 1991.

Bramwell, Martyn. *Volcanoes and Earthquakes.* New York, NY: Watts, Franklin, Inc., 1986.

Brownlee, Shannon, and Dennis Overbye. A Special Report. "Did Comets Kill the Dinosaurs?" *Discover* (May 1984): 21–32.

Bullard, Fred M. *Volcanoes of the Earth.* Austin, TX: University of Texas Press, 1984.

Eicher, Don L., A. Lee McAlester, and Marcia L. Rottman. *The History of the Earth's Crust.* Englewood Cliffs, NJ: Prentice-Hall, 1984.

Gould, Stephen Jay. "Sex, Drugs, Disasters, and the Extinction of the Dinosaurs." *Discover* (March 1984): 67.

Hildebrand, Alan R. and William V. Boynton. "Cretaceous Ground Zero." *Natural History* (June 1991): 46–53.

Metz, Robert. "How to Make a Great Geological Impression." *The Science Teacher* (November 1981): 41–42.

Bibliography for Students

Bates, Robert L. *Stone, Clay, Glass: How Building Materials Are Found and Used.* Hillside, NJ: Enslow Publications, 1987.

Lauber, Patricia. *Volcano: The Eruption and Healing of Mount St. Helens.* New York, NY: Bradbury Press, 1986.

Stommel, Henry and Elizabeth Stommel. *Volcano Weather: The Story of 1816, The Year Without a Summer.* Newport, RI: Seven Seas Press, 1983.

Symes, R.F. *Rocks and Minerals.* London, England: Alfred A. Knopf, 1988.

Vogt, Gregory. *Predicting Earthquakes.* New York, NY: Watts, Franklin, Inc., 1989.

Films, Videotapes, Software, and Other Media

Earthquakes.
  Videotape.
  Coronet Film and Video
  108 Wilmot Road
  Deerfield, IL 60015

Fossils: Evidence of Life.
  Videotape.
  Focus Media, Inc.
  839 Stewart Avenue
  Garden City, NY 11530

Resources in Its Crust.
  Videotape.
  Coronet Film and Video
  108 Wilmot Road
  Deerfield, IL 60015

Rocks and Minerals: The Hard Facts.
  Videotape.
  Focus Media, Inc.
  839 Stewart Ave.
  Garden City, NY 11530

Volcano.
  Videotape.
  Focus Media, Inc.
  839 Stewart Ave.
  Garden City, NY 11530

Volcanoes and Earthquakes.
  Software.
  For IBM PC.
  IBM Direct
  PC Software
  Department 999
  One Culvert Road
  Dayton, NJ 08810

The Earthquake Simulator.
  Software.
  For Apple II family.
  Focus Media, Inc.
  839 Stewart Avenue
  Garden City, NY 11530

Rock Doctor.
  Software.
  For Apple II family.
  Didatech Software, Ltd.
  3812 William Street
  Burnaby, BC V5C 3H9

UNIT 6

✳ Unit Focus

Pose the following questions to students: What do you think the earth will look like in 2 million years? Will it look as it does today, or will it be different? *(Students will probably respond that it will be different somehow.)* Ask them to predict the changes they think will occur. List the changes on the chalkboard. After a sufficient list has been generated, ask students to explain the predictions they made. *(Students may mention global warming, erosion of the earth's surface, and other occurrences.)*

Explain that processes inside the earth, as well as those on the surface, will affect the earth's future shape.

✳ About the Photograph

This photograph of the Grand Canyon clearly shows horizontal layers of sedimentary rock in the canyon walls. These rocks are mostly limestone, freshwater shale, and sandstone. Although the canyon appears red from a distance, the colors of the rock layers include tan, gray, pink, light green, brown, slate gray, and purple. The oldest rocks in the deepest part of the canyon are granite and schist that are over 1.7 billion years old.

Unit 6 — The Restless Earth

SHAKE, RATTLE, AND FLOW
1. Making Mountains, p. 318
2. The Earth Breaks Apart, p. 325
3. Volcanoes—Holes in the Earth, p. 332

THE ROLE OF ROCKS
4. Rocks in the Past, p. 341
5. Hot Rocks, p. 346
6. Rocks from Sediments, p. 351
7. Changed Rocks, p. 355

FOSSILS—RECORDS OF THE PAST
8. Rocks Reveal a Story, p. 360
9. Telling Time With Rocks, p. 365

316 The Restless Earth

For Your Journal

Write a paragraph summarizing your thoughts about each of the following:
1. How were the mountains formed?
2. What processes cause the face of the earth to change?
3. What information can rocks tell us about the earth?
4. What is the Grand Canyon made of?
5. What forces might have caused the formation of the Grand Canyon?

✵ Using the Photograph

Have the students study the photograph of the Grand Canyon. Ask: Do you see any patterns in the canyon? *(A pattern of horizontal bands is visible in the canyon walls and buttes.)* Ask them to select a highly visible band and follow it as far as they can through the photograph. A prominent bright red band occurs throughout the canyon about midway down the canyon wall. This is known as the Redwall Limestone. Although this limestone is actually gray-blue in color, it has been stained a bright red by iron oxides from layers above it.

Explain that the Redwall Limestone built up in a wide, ancient sea during the Mississippian Period, about 345 million years ago. Tell the class that the horizontal layers were built up over millions of years, forming a sequence with the most recent layers on top.

✵ About Your Journal

Students should answer the Journal questions to the best of their abilities.

These questions are designed to serve two functions: to help students recognize that they do indeed have prior knowledge about the topic, and to help them identify any misconceptions they may have. In the course of studying the unit, these misconceptions should be dispelled.

The Restless Earth 317

Lesson 1

SHAKE, RATTLE, AND FLOW

1 Making Mountains

✷ Getting Started

This lesson focuses on the internal forces created by moving plates that cause land to be uplifted into mountains. Students review the basic concepts of the theory of plate tectonics.

Main Ideas

1. The external forces of weathering and erosion are offset by internal forces that cause the land to be uplifted.
2. The plate tectonic theory states that mountains are formed when plates collide.
3. Rocks under stress either bend to form folds or break to create faults.
4. Mountain-building is an ongoing process.

✷ Teaching Strategies

A knowledge of weathering and erosion is a prerequisite for these lessons on mountain-building. Without these concepts, students will have difficulty understanding the cycle of mountain building. Review these concepts by involving students in a class discussion before beginning the lesson.

Among the earth's most spectacular features are the mountains. At 8848 m, Mount Everest is the tallest of all, but it has many rivals of impressive size scattered over the face of the earth. Throughout the ages, people have marvelled at and wondered about mountains.

Lesson Organizer

Objectives

By the end of the lesson, students should be able to:
1. Identify the forces that counteract weathering and erosion.
2. Describe the most recent theory of mountain-building.
3. Construct models that illustrate how folds and faults are formed.

Process Skills

observing, inferring, formulating hypotheses, making models

318   The Restless Earth

The Rise and Fall of Mountains

We know from early writings that people have been fascinated by mountains for thousands of years. Early civilizations tried to answer the same questions about mountains that we raise today. Examine each of the ideas described on pages 319 and 320, and discuss them with someone.

Sarah's Idea

Just think, someday there will be no mountains to climb.

Why is that?

Last year, when we studied erosion and weathering in science class, the teacher told us that mountains are worn down about 30 cm every 1000 years. Since Mount Everest is the tallest mountain in the world, at 8848 m, by the time it is worn down to sea level, all the other mountains will be gone too.

Do you agree with Sarah's reasoning? If not, how would you convince her that she is wrong? According to her figures, how long would it take Mount Everest to wear down?

Back to the Greeks

The ancient Greeks and Romans believed that the mountains were formed by giants who wanted to reach the heavens. The giants then piled one peak on top of another to form a stairway to the sky. Zeus, the Greek king of the gods, struck down the peaks with his thunderbolts, and scattered the remains into rugged mountain chains. Then Zeus imprisoned the giants beneath the mountains. During their struggle to escape, the giants broke the earth's crust and hurled liquid rock against their enemy, Zeus.

Could mountains have been formed in this way? What explanation does the myth give for earthquakes? for volcanoes?

Zeus, with thunderbolts in his right hand

New Terms

Lithosphere—the outer crust of the earth.
Plate tectonics—the theory that the earth's crust is made up of rigid plates that move; the motion of these plates causes continental drift.
Fault—the boundary between two rock sections that have been displaced relative to each other.

Materials

Exploration 1: 2 colors of modeling clay, knife, metric ruler, rolling pin or wooden dowel, wax paper

---

The Rise and Fall of Mountains

Use the following questions to ascertain students' prior knowledge of the mountain-building process.
- What are mountains made of? *(rock, sometimes covered by soil, snow, and ice)*
- How were mountains formed? *(Rocks were uplifted by tectonic forces or volcanism.)*
- Do mountains last forever? *(No, they wear down by weathering and erosion.)*

Answers to
In-Text Questions

Sarah's Idea
Have students read the conversation on page 319. Then divide them into pairs or small groups to discuss the questions. Reassemble the class and have groups report their ideas.
- Sarah is incorrect because processes of uplift are also continually occurring. As old mountains are worn down, new ones are formed.
- It would take between 31 and 32 million years for Mount Everest to wear down if no uplifting processes occurred.

Back to the Greeks
- Mountains probably were not formed this way; however, to people in ancient times, this may have seemed a plausible explanation.
- The myth explains earthquakes and volcanoes as the actions of giants trapped beneath the mountains.

(Answers continue on next page)

Teacher's Resource Binder
Resource Transparency 6–1

Science Discovery Videodisc
Disc 2, Image and Activity Bank, 6–1

Time Required
two to three class periods

Shake, Rattle, and Flow

Answers continued

Agricola's Proposal

Call on volunteers to read aloud the text and monologue on page 320. Then involve students in a discussion of Agricola's ideas, which are summarized in statements 1 through 5.

1. Students should identify this proposal as one that makes sense in some cases. However, the most obvious weakness in Agricola's idea is that it does not provide an explanation for how new mountains are formed. Specifically, Agricola did not explain "uplift."
2. Students should recognize that this proposal does not make sense. Mountains are found in locations other than along seacoasts and desert lands. If mountains were formed by the piling up of sands, there would be observable evidence, such as huge sand dunes in these areas. Pressure and heat are required to change sand and other sediments into rock. Such forces are not found on the earth's exterior.
3. Students should recognize that this proposal does not make sense. There is no evidence that suggests the existence of underground winds. If they did exist, it seems very unlikely that they would have the force to push rocks into mountains.
4. Students should recognize that this proposal does not make sense. Earthquakes, although powerful, cannot force up tons of rock as high as the elevations that mountain ranges reach.
5. Students should identify this proposal as one that makes sense in some cases. Mountains can be formed by volcanic eruptions, although such an explanation would account for only one kind of mountain. For example, Mount Shasta in northern California and Mount Vesuvius in Pompeii, Italy, were formed from volcanoes.

Direct students' attention to the photograph at the bottom of page 320. The steep-walled canyon was sculpted by the action of moving water.

Agricola's Proposal

In 1546, Agricola, a German doctor, suggested that mountains came into being in five different ways:

1. through the eroding action of water
2. through the piling up of sands along seacoasts and in desert lands
3. through underground winds
4. through earthquakes
5. through volcanic fires

Agricola believed, however, that most mountains were formed by the eroding action of water. Read how he described it. Do you think any of Agricola's ideas make sense? Are there ideas you disagree with? Why?

It is evident that many and indeed most mountains owe their origin to the action of water. The little brooks first wash away the surface soil and then cut into the solid rock. By carrying it away grain by grain, the water can finally cut even a mountain range in two, removing great blocks of rock in the process.

In a few years, a small stream can dig a deep depression or river bed across a level or gently sloping plain. Over the course of years these stream beds reach an astonishing depth, while their banks rise majestically on either side. From those banks small fragments of rock are continually detached by rain or frost. Large rocks that have cracks and fissures fall because of their grand weight and land in the stream below. In this way the steep cliffs gradually recede and convert into gentle slopes, and so the original plain converts into a series of elevations and depressions. The elevations are called mountains and the depressions valleys.

(Adapted from De Ortu et Cavsis Subterraneorum, *Book*, 1546)

Could water do this?

320   The Restless Earth

Plate Tectonics and Mountain-Building

The theory of continental drift was proposed in 1911 by the German scientist Alfred Wegener. It helps explain the gradual motion of the continents as they drift in the ocean. The Canadian geologist J. Tuzo Wilson took this idea one step further.

Wilson suggested that the outer crust of the earth (the *lithosphere*) is broken up into many large rigid plates and several smaller ones. The continents ride on top of these plates. It is the motion of the plates that causes continental drift. Wilson's idea became the theory of **plate tectonics**. The map shows the boundaries of these plates.

J. Tuzo Wilson

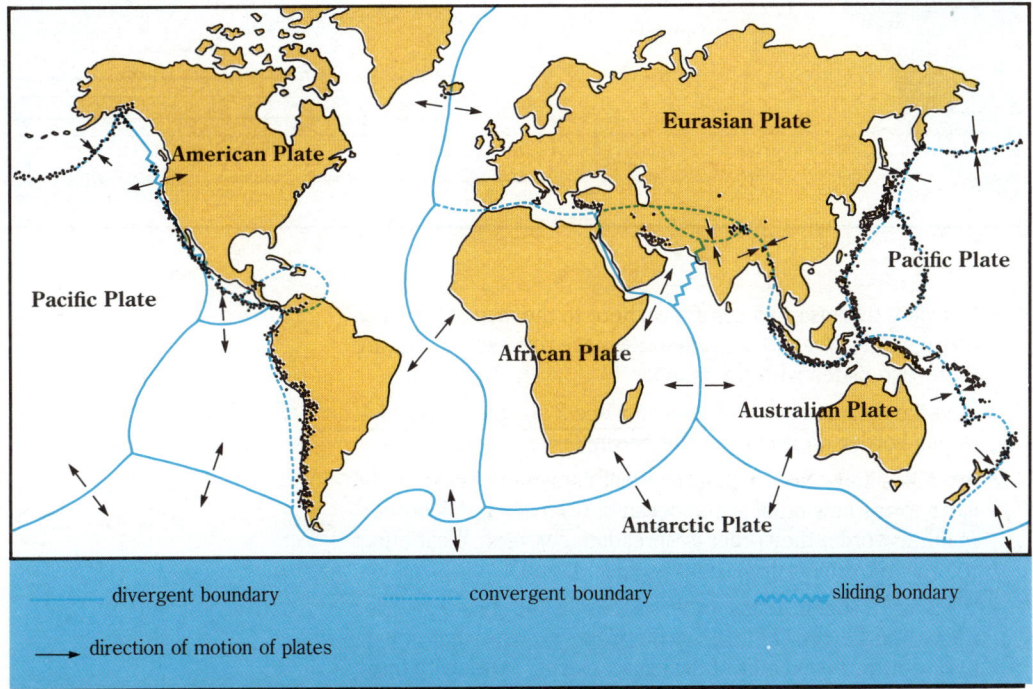

What is happening at the boundaries of the plates, according to the theory of plate tectonics? Check the dictionary for definitions of *divergent* and *convergent*. Then do the Thought Experiment on page 322 to find an answer to this question.

Plate Tectonics and Mountain-Building

Direct students' attention to the map on page 321, and encourage them to identify the continents and oceans. Ask them to count the large and small plates. Then point out the legend at the bottom of the figure. Make sure that students understand the boundary line markings. Be certain that students notice the arrows on the map.

Spend some time discussing how movement occurs at the different kinds of boundaries. The dark areas on the map indicate the location of earthquakes.

Some students may have difficulty understanding that crust is found under the oceans as well as on the continents, and that both continents and oceans ride on plates. You may have to demonstrate this concept using tangible objects, such as pieces of cardboard.

As a quick review of the main features of Wegener's theory of continental drift, write the following paragraphs on the chalkboard. Leave blanks for the underlined words and have students add the missing terms collaboratively. Students will probably be unfamiliar with some of this information. Through discussion, guide students' critical thinking to help them fill in the information.

In 1911, Alfred Wegener proposed that 300 million years ago all of the <u>continents</u> were joined together in one large land mass called <u>Pangea</u>. Wegener suggested that, at some point, the <u>continents</u> split apart and began to <u>drift</u>, or move away, from each other. Wegener put together his evidence with evidence collected by other scientists to support his theory of <u>continental drift</u>. The evidence was convincing since the continents fit together so well: <u>rocks</u> in eastern North America are of the same type and age as those found in Scotland and Scandinavia; and <u>fossils</u> of ancient and similar organisms have been found on both sides of the Atlantic, in <u>Africa</u>, and South America.

Little serious attention was paid to Wegener's ideas until about 19<u>50</u>. After studying the ocean floor, scientists discovered that not only were the <u>continents</u> moving, but the <u>ocean</u> floor was drifting as well. This resolved some of the problems scientists had experienced with Wegener's <u>theory</u> of drifting continents.

Shake, Rattle, and Flow 321

An Earth-Shaking Thought Experiment

Ask students to compare the illustrations on pages 321 and 322. *(The illustration on page 322 shows only the plates, not the overlying continents and oceans.)*

You may wish to take this opportunity to discuss the problem of representing the spherical earth on a flat piece of paper. An atlas may provide examples of different ways to solve this problem. Be sure to point out that the Mercator projection (used on page 321) shows Europe and North America larger than their actual size in relation to the other continents. Use a globe to help illustrate this concept.

Answers to In-Text Questions

- Each puzzle piece represents a large, rigid plate. To help students understand how two pieces can have the same number, trace the puzzle pieces, cut out the rectangle, and form a cylinder by matching the plates.
- The common border between piece 2 and pieces 3 and 4 is in the middle of the Atlantic Ocean.
- Pieces 2, 3, and 4 will be forced apart by the new crust forming at their common boundary. The Atlantic Ocean will get wider.
- As new crust is added at the common boundary of plates 2, 3, and 4, the other plates will collide or slide over, under, or past each other.

Making a model of the diagram using cardboard or stiff paper will help students understand what happens to the pieces when plate movement continues over millions of years. Some trends students may suggest are: plates 3 and 5 will move up (north); plate 4 will move to the right (east); plate 2 will move to the left (west); and plate 1 will decrease in size.

An Earth-Shaking Thought Experiment

Think of the earth as a jigsaw puzzle with six huge pieces and several smaller ones.

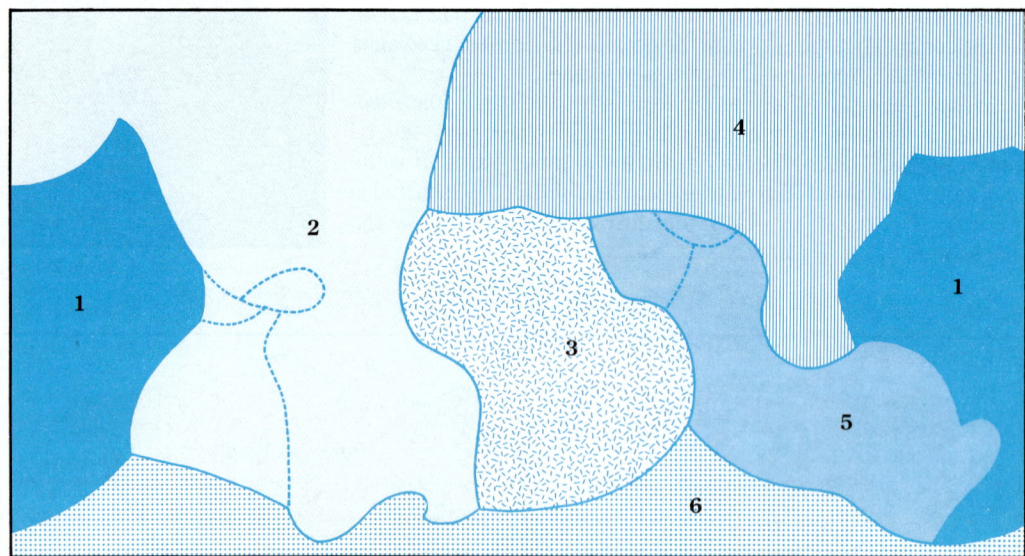

- Compare the jigsaw pieces shown here to the map of the earth on page 321. What does each puzzle piece represent? Why are there two plates with the number 1?
- Find the common border between piece 2 and pieces 3 and 4. Is this border on land or in the ocean?
- Scientists know that molten rock that comes from deep in the earth forms new crust at the ocean bottom along this border. In other words, the border is spreading outward. What effect will this movement have on pieces 2, 3, and 4?
- What will happen to the other pieces in the jigsaw puzzle? Check the key on the map of the plates to see what is happening at the boundaries of the jigsaw pieces. Work with three of your classmates to draw a map predicting the change in the plates that will occur millions of years from now. Where will each plate be? How will the puzzle change?

Applying the Theory of Plate Tectonics

Review these observations and explanations made by geologists. Then apply the theory of plate tectonics in order to answer the questions.

- As the new crust made of younger rock is formed, the ocean floor spreads. This causes the plates to move. They move from 5 cm to 20 cm per year. Estimate how long it has taken the continent of Africa to drift away from South America. Check an atlas or a world map to determine the current distance between these two continents.

- When an oceanic plate drifts up against a continental plate, the oceanic plate slides under the continental plate. Look at the map on the opposite page to see where this may be happening. Which puzzle pieces are involved? This type of motion may cause various disruptions. What might they be?

- When two continental plates collide, the crust of the earth buckles and moves upward. Locate a place where this may be happening. Which puzzle pieces are involved?

Rocks Under Stress

It is difficult to comprehend that there are forces strong enough to push the earth's crust into mountains. Yet when geologists study existing mountains, they find rocks which have been bent, folded, crumpled, and broken, as in the pictures below. The following Exploration illustrates how the theory of plate tectonics explains the folding of the rocks in mountains.

Applying the Theory of Plate Tectonics

Have students read the material on page 323. Involve them in a discussion of the theory of plate tectonics and the accompanying questions.

Answers to
In-Text Questions

- If Africa had drifted away from South America and this had occurred at the rate of 10 cm per year, then it would have taken 60 million years for the continents to get to their present positions.
  (6000 km × 100,000 cm/km × 1 yr/10 cm = 60,000,000 years)

- When an oceanic plate collides with a continental plate, it slopes under the continental crust. Puzzle pieces involved are 1 and 4, 1 and 5, and 1 and 2—near Alaska. The various disruptions that may occur are mountain-building, earthquakes, volcanoes, folds, and faults. Invite students to speculate as to what happens. You need not go into detail about earthquakes or volcanoes at this time.

- When two continental plates collide (puzzle pieces 3 and 4 and pieces 4 and 5), uplifting processes cause mountain-building. A range like the Himalayas may result.

Rocks Under Stress

Evidence of folds and faults in rock suggests that tremendous forces must exist beneath the ground. The models students will make in the following Exploration should help explain how folds and faults occur. It is important to relate each model to the actual situation shown in the photographs. In this way, students will gain insight into the mountain-building processes.

Shake, Rattle, and Flow

Exploration 1

Divide the class into small groups and distribute the materials. The modeling clay can be rolled flat using a rolling pin or wooden dowel. Caution students not to press the strips too close together. If possible, have each group make a fold model and a fault model for comparison.

Part 1

4. This model illustrates folding. The layers of clay represent different layers of rock. The forces might originate from two colliding plates, or from one plate slipping under another.

Part 2

1. The boundary formed when the rock broke and one side moved away from the other.
2. The collision of plates caused the rocks of the continental crust to buckle and break.
3. Students should measure the distance along the fault between the two dark and thick lines. They should measure from the top edge of the band on the left to the top edge of the band on the right, a distance of approximately 2.2 cm. Therefore, the distance of offset is: 2.2 × 75 cm, or 165 cm.

Answer to
Drawing Conclusions

Possibilities for future mountain sites exist where plates are closing in upon each other, or where oceanic crust is sliding under continental crust. These locations are primarily in the western Pacific, the Aleutian Islands, and in Central America.

EXPLORATION 1

Folds and Faults

Models can help explain how folds and faults are formed in the crust.

Part 1
You Will Need
- 2 colors of modeling clay
- a knife

What to Do

1. Cut or flatten the clay into strips approximately 1 cm × 3 cm × 10 cm. Make two strips of each color.
2. Stack the strips, alternating the colors.
3. Place one narrow end of the block against a wall. With your hand, apply pressure to the other narrow end.

4. Compare our model to the picture on the previous page. What process does this model illustrate? What do the layers of clay represent? Where do the forces that deform rocks originate?

Part 2
What to Do

1. Look at the photograph above. Why do you think the boundary in the rock formed as it did? Use another clay model to demonstrate your answer.
2. Compare your model to the photograph. The boundary between two rock sections that move relative to each other is called a **fault**. Explain how the fault in the rock might have formed.
3. If 1 cm in the photograph corresponds to 75 cm in the actual rock, find out how far the rocks have moved along the fault.

Drawing Conclusions

Rocks deep within the earth are under extreme pressure and temperature. As a result, they behave like clay, bending when forces are applied. On or near the surface, however, rocks tend to break rather than bend. This is because the pressure and temperature are not as great on the surface of the earth.

In this Exploration, you have found out that the formation of folds and faults can result in mountains. Mountains can be formed in other ways as well. For example, Mount Shasta in northern California was formed by the action of volcanoes.

It appears that mountain-building is an ongoing process. Where do you think the next mountain will be formed?

Assessment

1. Have students describe in words or with pictures how the formation of folds and faults can result in mountains. *(At a fold, pressure can push rock upward. At a fault, one plate pushes above the other.)*
2. Have students use two index cards to simulate the following situations:
   (a) A boundary opening, causing the plates to separate
   (b) Two plates colliding, causing buckling and folding
   (c) Two plates colliding, with one plate being pushed under another
   (d) Two plates sliding by each other

Extension

Have students locate relief maps, photographs, or other information on nearby mountain areas. Ask them to determine how and approximately when these mountains were formed.

2 The Earth Breaks Apart

A Picture Study
Carefully examine the picture above. What observations can you make about the fence in the foreground? Make several inferences that might explain the appearance of the fence. Do your observations support your inferences? What added information would you need to verify your inferences?

LESSON 2 ORGANIZER

Objectives
By the end of the lesson, students should be able to:
1. Explain current scientific ideas about why earthquakes occur.
2. Describe how seismographs measure earthquake strength and location.

Process Skills
observing, inferring, formulating hypotheses, analyzing

New Terms
Elastic rebound theory—explains how rocks spring back to their original shape after they have been deformed by tectonic forces.
Focus or hypocenter—the point at which stress breaks the friction lock between two plates of the earth's crust.
Shock or seismic waves—the stored energy released in the form of intense vibrations during an earthquake.
(Organizer continues on next page)

LESSON 2

✷ Getting Started

This lesson focuses on how the theory of plate tectonics explains the origin of earthquakes. Students examine location, strength, and prediction of earthquakes.

Main Ideas
1. Earthquakes occur when plates slide past each other, collide, or move apart.
2. The elastic rebound theory explains how rocks spring back into their original shape after they have been deformed by seismic forces.
3. Seismographs provide a method for determining earthquake strength.

✷ Teaching Strategies

A Picture Study
The photograph was taken after the San Francisco earthquake in 1906. The scene is along the San Andreas fault where a slip of several meters occurred.
Observations of the fence:
- It appears to be wooden.
- The gap between the two sections is about twice the height of the fence.
- The height of the fence appears to be the same on both sides of the gap.

Inferences about the fence:
- The gap may have been caused by a storm.
- The gap may have been caused by a mudslide.
- The gap may have been caused by an earthquake.

If students can estimate the age and location of the photograph, they could research historical records to verify their inferences.

Shake, Rattle, and Flow 325

Exploration 2

The procedure for this Exploration is similar to that used in *Exploration 1*. However, students must understand that this model represents a type of fault that contributes to earthquakes.

Answers to Questions

1. The portions of clay represent sections of the earth's crust.
2. The layers of clay represent the layers of rock in the earth's crust.
3. The boundary between the clay portions simulates the fault where the plates are moving against each other in opposite directions.
4. If a fence crossed a fault, it would be severely twisted and torn apart. Now an explanation for the picture study on page 325 should become clear.
5. The model shows that the two portions of clay have moved in opposite directions from each other. This is similar to the photograph on page 325, which shows the San Andreas Fault.
    The San Andreas fault is the boundary where the Pacific plate slides past the North American plate. During the San Francisco earthquake of 1906, the Pacific plate slipped northward by several meters.

EXPLORATION 2

Another Kind of Fault

In Exploration 1, you made models that helped explain how mountains are formed. In this activity you will make a model of a different type of fault that will help explain how earthquakes occur.

You Will Need

- the clay block from Exploration 1
- a knife

What to Do

1. Separate the layers of clay carefully, and reshape each layer into rectangles measuring 1 cm × 3 cm × 10 cm.
2. Restack the layers as before.
3. Cut the block lengthwise into two equal pieces.
4. Holding the left portion of the block in your left hand, push the other block forward with your right hand until the blocks slide apart, as in the photograph below.

Questions

1. What does each block of clay represent?
2. What does each layer represent?
3. What is simulated by the boundary between the two clay blocks?
4. What would happen to a fence if it crossed the fault?
5. Compare your model to the picture of the fault in the photograph on page 324.

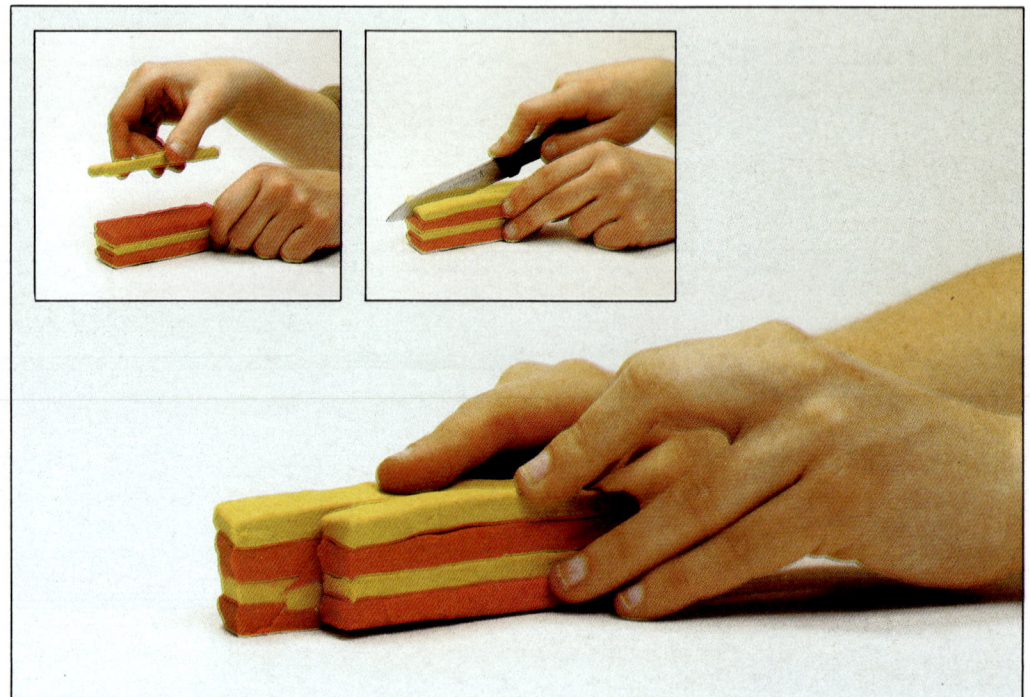

(Organizer continued)

Magnitude—the strength of an earthquake.
Richter Scale of Magnitude—Scale used to measure the strength of an earthquake based on the amplitude of seismic waves.
Seismologists—people who study earthquakes.
Seismographs—instruments that record the vibrations of the earth during an earthquake.
Epicenter—the position on the surface of the earth directly above the focus of an earthquake.

Materials

Exploration 2: the clay model from *Exploration 1*, knife, ruler
What Causes Earthquakes?—An Elastic Theory: the clay model from *Exploration 2*, rubber bands, toothpicks
Exploration 3: Journal, blank-outline world maps

Teacher's Resource Binder

Resource Worksheet 6–1
Resource Transparency 6–2
Graphic Organizer Transparency 6–3

Science Discovery Videodisc

Disc 2, Image and Activity Bank, 6–2

Time Required

three class periods

Another Picture Study

Starting at the bottom of the photo, trace the course of the stream as it crosses the fault. How does this picture support your inference about the fence in the first Picture Study on page 325?

What Causes Earthquakes?—An Elastic Theory

What do earthquakes and elastic bands have in common? H.F. Reid was one of the investigators of the earthquakes that devastated San Francisco in 1906. He devised a theory to explain the cause of earthquakes. That theory is called the **elastic rebound theory.**

Look at each drawing and the related facts on the next page, and you too will discover the connection between earthquakes and rubber bands!

After the San Francisco earthquake

Another Picture Study

Have students study the photograph of the stream crossing the San Andreas fault in central California. The streambed provides evidence that the ground moved along the fault in a direction perpendicular to the streambed. The area at the top of the photograph is located on the Pacific Plate, which is slowly moving northward (to the right in the photograph) relative to the North American Plate (in the foreground). This type of long-term movement along a fault produces repeated earthquakes. Students should conclude that this photograph supports their inference made about the fence on page 325. An earthquake did occur.

What Causes Earthquakes?— An Elastic Theory

Have students read the material on the bottom of page 327 and study the photograph of the San Francisco earthquake.

Students may have some difficulty relating the sketches on page 328 to the factual information. You may wish to perform a demonstration of the changes involved using the clay model from *Exploration 2* and a rubber band. One way to do this is to anchor the rubber band to each clay portion with toothpicks. When the portions are moved in opposite directions to one another, tension will be produced in the rubber band. This is analogous to the energy stored in rocks along a fault, which are held in place by friction. After sufficient tension builds up, the rubber band will break and snap back, much like the rocks after the friction lock is broken. For the purposes of the demonstration, you may wish to release one end of the rubber band from its anchor in the soft modeling clay, rather than stretching it to its breaking point.

Shake, Rattle, and Flow 327

Diagrams A–D

Students may have difficulty interpreting the diagrams. To aid their understanding, ask students to use their clay models of faults from *Exploration 2.*

Place the two portions of clay in their original positions. Place a row of toothpicks directly opposite each other on the inner edge of each piece of clay. Have students gently slide the clay portions apart as in *Exploration 2.* As the stress builds up, the model illustrates diagrams B and C. When the clay portions slip, the model represents diagram D. The toothpicks that were opposite each other have now changed positions in relation to one another.

The meaning of the term *focus,* or *hypocenter,* should be clarified, since students may be familiar with the term *epicenter* from media reports of earthquakes. The *epicenter* is the point on the earth's surface directly above the focus, or hypocenter. The focus can be far below the surface of the earth.

If students are curious about the interior of the earth, this is an appropriate time to teach them the information. Much of what is known about the earth's interior comes from monitoring the two types of seismic waves that travel out from the focus of an earthquake. By studying these waves, scientists have discovered that the earth has a core of very dense material and that the inner part of this core is solid.

You Be the Teacher

This is a writing exercise in which students are asked to explain the elastic rebound theory to a classmate. The task will help students to sort out their ideas about the forces causing earthquakes.

The diagrams represent aerial views of the boundary between two plates, P and Q.

Diagram A: The Facts

- P and Q are blocks of crust slipping in opposite directions along a fault.
- The broken line is a street lying across the fault.
- The rocks forming blocks P and Q are locked by friction created by the pressure of surrounding rock layers. The friction keeps the blocks from moving.
- The rocks store the energy of the frictional force just like a stretched rubber band stores energy.

Diagram B: The Facts

- As stress builds up, the rocks begin to strain.
- The street bends with the rock.
- The rocks continue to build up energy from the pressure.

Diagram C: The Facts

- Finally, the stress breaks the friction lock at the **focus** or **hypocenter.**
- On each side of the focus, the rocks spring back into their original shape. This is similar to how a rubber band returns to its original shape after you've stretched it.
- The stored energy is released in the form of intense vibrations known as **shock waves** or **seismic waves.** This release is what we call an earthquake.

Diagram D: The Facts

- The street is displaced at the fault.
- The blocks have moved in opposite directions.

You Be the Teacher

To determine whether her students understood the elastic rebound theory, Ms. Morris asked them to use the theory to explain the photograph of the stream on page 327. Manuel was absent from school when Ms. Morris explained the theory, and now he needs some help.

Imagine that you are a member of Ms. Morris's class and that Manuel has come to you for help with the assignment. Write down what you might say or do to explain the theory to him.

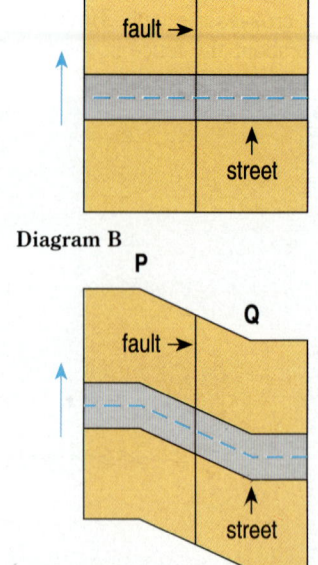

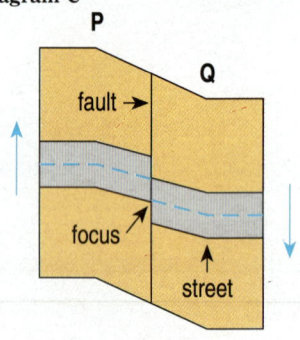

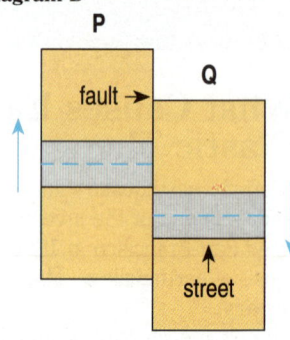

Measuring the Strength of Earthquakes

Have you ever felt the earth vibrate? Perhaps you have felt the ground shake as a train passed by. Earthquakes also cause the earth to vibrate. As you can imagine, huge amounts of energy are released when the earth moves like this.

Every earthquake is given a number describing its **magnitude,** or strength. But what does this number really mean? Measuring the energy released in an earthquake is a long, complicated process. To permit rapid calculation of the strength of an earthquake, **seismologists** (people who study earthquakes) have adopted the **Richter Scale of Magnitude.**

The number expressing the magnitude of an earthquake is calculated by means of a formula. The information contained in the formula is provided by instruments known as **seismographs.** These instruments record the vibrations of the earth during a quake. An increase by just one number in the measurements corresponds to a tenfold increase in earthquake strength. For instance, an earthquake that is 7.4 on the Richter Scale is about ten times stronger than an earthquake with a magnitude of 6.4.

A vertical seismograph and the recording drums

Answers to Analysis, Please!

Have students study the table on page 330 and answer the questions. They may work individually or in groups.

1. 8.0 to 8.6
2. 2.0 to 3.4
3. 7.0 to 7.3
4. "Seism" is a Greek noun meaning earthquake. "Seismo" is the combining form. For example, a seismogram is the record of an earthquake recorded by a seismograph. Other words include:
   - **Seismography**—the branch of seismology dealing with the mapping and description of earthquakes.
   - **Seismology**—the study of earthquakes.
   - **Seismic**—about or from earthquakes.
5. The epicenter is the position on the surface of the earth directly above the focus of an earthquake.
6. The affected area makes up a rectangular cube.
   Volume of affected area:
   1000 km × 100 km × (2 × 50 km) = 10^7 km^3
   10^7 km^3 × 10^9 m^3/km^3 = 10^{16} m^3
   Energy released per cubic meter: 100 J
   Energy released in earthquake:
   10^2 J × 10^{16} = 10^{18} J, or 10^{15} kJ
   The earthquake would be very severe (8.0 or greater on the Richter Scale).

The Modified Richter Scale of Magnitude

Magnitude	Estimated number of earthquakes recorded each year	Estimated damage
10.0	Possible but never recorded	Would be felt all over the earth
9.0	Possible but never recorded	Would be felt in most parts of the globe
8.0 to 8.6	Occur infrequently	Very great damage
7.4 to 7.9	4	Great damage
7.0 to 7.3	15	Serious damage; railway tracks and bridges bent
6.2 to 6.9	100	Widespread damage to most structures
5.5 to 6.1	500	Moderate to slight damage
4.9 to 5.4	1400	Felt by everyone within the affected area
4.3 to 4.8	4800	Felt by most
3.5 to 4.2	30,000	Felt by a few
2.0 to 3.4	over 150,000	Not felt, but recorded

Analysis, Please!

1. What is the magnitude of the strongest earthquake ever recorded by seismographs?
2. Which magnitude of earthquakes occurs most often?
3. An earthquake that measures ___?___ on the Richter Scale will bend bridges and railroad tracks.
4. What does the prefix *seismo*, as in *seismograph*, mean? Find other words that use this prefix and explain them.
5. Look at the diagram, then state what you think the **epicenter** of a quake is.
6. About 100 J of energy are released from each cubic meter of rock at the time of an earthquake—the equivalent of a firecracker per cubic meter. This may not seem like very much energy, but consider the following situation:

   Suppose the fault is 1000 km long, extends 100 km downward, and bends streets as far as 50 km on either side of the fault. How many cubic meters of rock would be under strain? How much energy would be released when the rocks break the friction lock?

Seismologists in the field

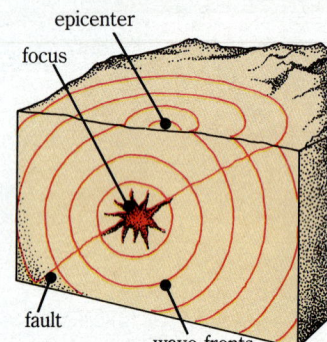

EXPLORATION 3

Where are the Earthquakes?

Here is a research project. Consult the school library or the local public library for information recorded on earthquakes during the past fifteen years. In your Journal, plot their locations on a simple sketch map of the earth.

Next compare the pattern, if one exists, to the map on page 321 that shows the earth's plates. What is happening at the edges of the plates? Review the elastic rebound theory on page 328. Why are the plates moving in opposite directions? What might P and Q represent?

The map below shows the epicenter locations for earthquakes between 1969 and 1979. Did earthquakes after 1979 occur in similar locations? Are you surprised by the number of earthquakes recorded?

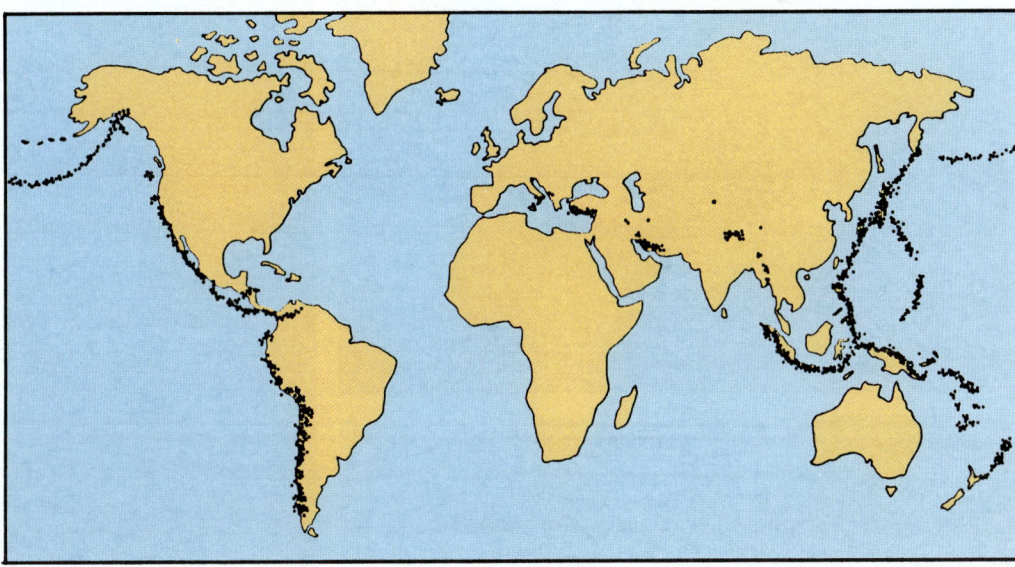

Exploration 3

The map on page 331 shows earthquake patterns during the 1970s. Students are asked to research current data about earthquakes to supplement the data supplied on the map.

Answers to
In-Text Questions

- The plates are moving in opposite directions.
- The forces of plate tectonics cause stress in the rocks—stress that is released by movement at the plate boundary.
- P and Q represent adjacent plates.
- Students should find that quakes continue to occur in the same general location, near plate boundaries.
- Students should discover that earthquakes occur more frequently than they had thought.

✶ Follow Up

Assessment

1. Ask students to use clay and toothpicks to make a model to show what would happen to a fence located on a fault during an earthquake. The toothpick fence in their model should cross the fault line. Then students will need to move the clay portions in opposite directions from each other.
2. Look at the map of the world on page 331. Where would you suggest that someone avoid living if they did not want to experience an earthquake? Explain the reasons for your answer.

Extension

Have students research local earthquakes (or earthquakes of interest to them) by contacting their library, a science museum, a nearby university, or other community agency. Have them collect information on the number of earthquakes, when they occurred, their size, location of epicenter, and effects on people and property.

Shake, Rattle, and Flow

LESSON 3

✵ Getting Started

This lesson focuses on the different kinds and causes of volcanoes. Students also examine the relationship of plate tectonics to volcanoes.

Main Ideas

1. Volcanoes generally occur along plate boundaries and are created when plates collide or move apart.
2. Volcanic eruptions occur in different ways and produce different results.

✵ Teaching Strategies

Volcanoes— Holes in the Earth

Have students study the photographs on page 332. Involve them in a discussion of the sequence of events visible in the photos. Ask students to imagine what other things might be taking place that are not visible (e.g., sounds, temperature changes, and pressure changes).

3 Volcanoes—Holes in the Earth

Mount St. Helens erupting

LESSON 3 ORGANIZER

Objectives

By the end of the lesson, students should be able to:
1. Describe different types of volcanic eruptions.
2. Identify the location of past and present volcanoes.
3. Infer that volcanoes occur at the edges of plates.
4. Describe the origin of volcanoes using the plate tectonic theory.

Process Skills

observing, making a word puzzle, researching, drawing conclusions

New Terms

Basalt—a volcanic rock that, as plates move apart, either moves quietly into the gap or erupts from below to fill the gap.

Magma—molten rock, lighter than the rock around it, that triggers earthquakes and creates volcanoes as it rises.

Lava—magma that emerges on the earth's surface in either a quiet or an explosive outburst.

Pumice—a dark rock filled with tiny holes created when gas separates from the lava.

332   The Restless Earth

A Survivor Speaks—Mount St. Helens, May 18, 1980

At the moment of the eruption, David Crockett was there.

He shouted into his recorder, "The road exploded in front of me . . . it's hard to breathe . . . my eyes are full of ash."

Grabbing his equipment, he left the car and raced up the near ridge which served as a wall between him and the blast. Since it had been clear-cut of timber, he was not threatened with falling timber but a blizzard of ash turned day into night. "Right now I think I'm dead," he recorded, coughing and fighting for his breath. It was then he realized that the deadly blast had gone over him, leaving him uninjured. He dropped to his knees and thanked God for his miraculous escape.

(from *Mount St. Helens: A Sleeping Volcano Awakes,* Marian T. Place, 1981)

A Survivor Speaks—Mount St. Helens, May 18, 1980

Create an atmosphere in the classroom in which students can relate to the words of eruption-survivor David Crockett. For example, ask students to imagine what it would be like to be in the middle of a volcanic eruption as shown in the photographs on pages 332 and 333. Elicit descriptive expressions such as hot, dark, hard to breathe, choking, and so on. Then call on a volunteer to read the survivor's words with feeling.

You may wish to have students research to find other magazine or newspaper reports of the eruption. If students have access to a college or large public library, they may be able to find articles from newspapers in cities located near the eruption site. These will contain more details of the impact of the eruption on the nearby communities. Have students present their research through oral reports, bulletin-board displays, role-playing, a simulated newscast, or other means.

Primary Source
Description of change: excerpted from *Mount St. Helens: A Sleeping Volcano Awakes,* by Marian T. Place, 1981.
Rationale: excerpted to illustrate the experience of an eruption-survivor.

Materials

Exploration 4: atlas, world map, Journal

Teacher's Resource Binder

Resource Worksheets 6–1 and 6–2
Activity Worksheet 6–3

Science Discovery Videodisc

Disc 2, Image and Activity Bank, 6–3

Time Required

two class periods

Shake, Rattle, and Flow

Mount St. Helens

Call on a volunteer to read aloud page 334. Direct students' attention to the photographs. Involve them in a discussion of how people perceive risk. Accept all reasonable answers to the following questions:
- Why did some of the residents ignore the predictions?
- How much time will it take before they will feel safe again living close to a volcano?
- Why might people choose to live in volcanic areas?

Use analogies to make the figures describing the effects of the eruption of Mount St. Helens more meaningful. For example, help students visualize the size of the debris-column. Locate a familiar place about 15 km away from your school. To help students appreciate the size of the crater, identify nearby locations that are approximately 1 and 2 km away.

This lesson would be enhanced by showing a film or videotape that illustrates the power of a volcano.

Answer to
In-Text Question

The scientific predictions about a possible eruption were based on a variety of evidence. Scientists monitor volcanoes such as Mount St. Helens by first studying regional geophysical and geological surveys and then by dating and geochemical investigations that reveal the volcano's history. Once the seismic data is known, scientists record the slope of the volcano for contractions and expansions by using electro-mechanical tilt-meters and field-surveying techniques. Changes in horizontal and vertical distances are measured by laser-beam technology. Swelling of the ground and earthquakes precede, accompany, and follow volcanic eruptions. These changes can be recorded using seismic instruments.

Mount St. Helens after the eruption

Billions of volcanic ash particles, like the one above, covered the area.

Earth tremors and escaping gases had warned of a possible eruption for several weeks. However, Mount St. Helens had been dormant for over 120 years. Few of the people living in the shadow of the volcano expected it to erupt this time. Certainly no one anticipated such an eruption!

When Mount St. Helens did erupt, the explosion was tremendous. It released 500 times the energy of the atomic bomb dropped on Hiroshima in 1945. Much of the mountain was blown away. One cubic kilometer of debris—rock, ash, and steam—shot up in a column 15 km to 18 km high. The eruption left a crater 1 km wide, 2 km long, and 0.3 km deep.

We not only have first-person accounts from survivors like David Crockett, but we also have pictures of the eruption taken before, during, and after its occurrence. Television brought the destructive power of the explosion to people around the world.

But many scientists had been confident that Mount St. Helens would erupt this time. As early as 1975, in fact, some geologists were predicting an eruption before the end of the century. How could they foresee this?

A truck destroyed by the blast

Clouds of volcanic ash

Re-seeding the mountain

In June of 1991, a major volcanic eruption occurred at Mount Pinatubo in the Philippines. Just 90 km northwest of Manila, the eruption left over one hundred thousand people homeless. Ash shot 30 km into the air and lava poured down the mountain at 100 km per hour.

A Volcanic Word Search

No doubt you have studied or read about volcanoes before. You may also have seen films of eruptions or visited a volcano. Make a list of all the words you can remember that are associated with volcanoes. As you read the following pages, add new words to your list. Then make a word search puzzle in your Journal like the one started here. The words can run in straight lines in any direction—up, down, across, and diagonally. Fill in all the unused boxes with other letters, and then give the puzzle to a friend to solve.

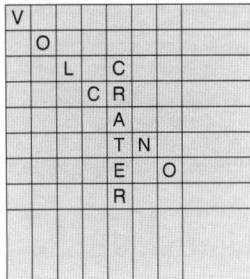

Mount Pinatubo

Call on a volunteer to read the material about the recent volcanic eruption in the Philippines. Involve students in a discussion of the details of the eruption. Then have them compare and contrast this eruption to the one that occurred at Mount St. Helens. Have students research the eruption using these questions as a guide.
- Which one was more severe?
- Which caused more damage to homes and property?
- Which caused more people to become homeless?
- Had predictions been issued about each eruption? If so, why were they ignored?
- What can each area do to protect its inhabitants in the future?

A Volcanic Word Search

Suggest to students that they write all the new terms about volcanoes in their Journals. The descriptions of famous volcanoes on pages 336 and 337 provide some of the vocabulary associated with volcanoes. These pages also provide background information on the different types of eruptions that give rise to the cone. Some insight is provided into the death and destruction that can result from a volcanic eruption. Any terms describing the effect of volcanic eruptions on world climate could also be included. Then, at the end of the lesson, students can create their own word-search puzzles and exchange them with their classmates.

Shake, Rattle, and Flow 335

Famous Volcanic Eruptions

Encourage students to research the names of famous volcanoes throughout the world. Invite them to compete individually or in a team to compose the longest list of names.

Students should use the material on the next few pages to compile a list of "Facts About Volcanoes" (e.g., volcanoes can be quiet or explosive). Memorization of dates, names, and characteristics of individual volcanoes is *not* encouraged.

Volcanoes are commonly classified according to their eruption status, in addition to descriptions of their shape and composition. The following terms are used:

- **Active**—a volcano that has erupted recently, is currently erupting, and is expected to erupt in the future. Mauna Loa and Mount St. Helens are active volcanoes.
- **Dormant**—a volcano that has had no recent eruption but that could erupt in the future. Mount Ranier, in Washington, is a dormant volcano.
- **Extinct**—a volcano that has not erupted in historic time and is not expected to erupt in the future. Mount Mazama, in Oregon, home of Crater Lake, is extinct.

Vesuvius, A.D. 79

Mount Vesuvius, located in southern Italy in the city of Pompeii, is an example of a *composite cone* volcano. Mount St. Helens, featured on pages 332 to 335, is another example of this kind of volcano. These volcanoes form large mountains, 1800 to 2400 m above their bases. They are built of alternating layers of lava flow, volcanic ash, and cinders. They are also called *stratovolcanoes*.

The eruption of Vesuvius was recorded by an eyewitness, the Roman writer Pliny, who was only 17 years old at the time of the eruption. His account was the first of many writings on the subject, which continue to the present day. There are many magazine articles available with color photographs of the remains of the Roman cities of Pompeii and Herculaneum, which were covered by ash and mud in this historic eruption. You may want to locate some of these articles for classroom use, or have students locate them as a research project.

Famous Volcanic Eruptions

Vesuvius, A.D. 79

This is history's most famous volcanic eruption. For two days, dense clouds of black ash and pumice poured from the vent of Vesuvius. The Italian city of Pompeii was buried under 5 m to 10 m of mud, ashes, and pumice. Several thousand people were trapped and killed.

Vesuvius, like Mount St. Helens, is a *composite cone volcano*, or *stratovolcano*. Such a volcano is made up of alternate layers of lava, ash, and cinders, rising about 2 km above the base. Vesuvius is a highly explosive volcano even today.

Krakatoa, 1883

Volcanic eruptions can be described as "quiet" or "explosive." On August 27, 1883, an explosive volcano erupted. A volcanic island in Indonesia literally blew apart. The explosion was so loud it could be heard 5000 km away, in Australia. The eruption left a hole in the ocean floor over 350 m deep and 10 km wide. Ash and steam rose to a height of 80 km and circled the globe, producing red sunsets for several years.

Krakatoa, 1883

An explosive eruption, such as that of Krakatoa, often leaves behind a distinctive landform known as a *caldera*. A caldera is a depression, often circular in form, caused by the complete collapse of a volcano. Crater Lake, in Oregon, is another example of a caldera.

So much ash was released into the atmosphere from the eruption of Krakatoa that the level of solar radiation was reduced by about 10 percent (as measured in Europe). Involve students in a discussion of the possible ramifications of such a change. *(Decreasing solar radiation produces a decrease in temperature.)* The recent eruption of Mount Pinatubo in the Philippines has been credited with producing enough ash to affect global temperatures in a similar way.

Mont Pelée, 1902

On May 8 at 7:50 A.M., Mont Pelée, a volcano on the island of Martinique, exploded. A huge cloud of glowing ash and steam, believed to have had a temperature of at least 1000 °C, rose and hung in the air for a moment; then it roared down the slope to engulf the town of St. Pierre. In less than two minutes the city was in ruins with 30,000 people dead. There were only two survivors.

Paricutín, 1943

On February 20, 1943, an astonished farmer observed a column of white fumes rising from a crack in a cornfield in Mexico. By the next day a cone of cinders and lava had grown 8 m high. At the end of the first week the cinder cone was 128 m in height; immense quantities of lava were jetting from it into the sky. This volcano continued to be active until 1952, when the cinder cone reached a height of 375 m.

Surtsey, 1963

The island of Surtsey, off the coast of Iceland, began as a crack in the ocean floor in November, 1963. Although spectacular, the formation of Surtsey can be described as a "quiet" eruption. In this type of eruption, lava flows through cracks or fissures in the earth's crust. As Surtsey was born, the lava flow contacted the seawater and caused numerous explosions.

Surtsey, 1963

New crust is continually being formed at the mid-Atlantic ridge. Volcanoes occur in Iceland because Iceland is located on the mid-Atlantic ridge. The formation of new crust by volcanic activity in this area works to push the adjacent plates in opposite directions.

Answer to caption: The formation of Surtsey was considered "quiet" because no explosive discharge of dust, ash, and gas took place.

Why is the Surtsey volcano described as "quiet"?

Mont Pelée, 1902

Mont Pelée, located on the island of Martinique in the Caribbean Sea, is an example of a *lava dome* volcano. Thick, pasty lava forms domes in the shape of steep-sided, craggy knobs or spines over a volcanic vent. Other lava domes can be short, steep-sided lava flows called *coulees*.

Another type of volcanic island is formed by a *shield* volcano. The Hawaiian Islands are a cluster of shield volcanoes and are some of the largest volcanoes in the world. Mauna Loa, Hawaii, is 8500 m from ocean bottom to peak. Short and broad shield volcanoes are built almost entirely of lava flows. From a main vent, or group of vents, lava pours out on top of preceding flows. The result is a broad, gently sloping cone, 4500 to 6000 m high and 5 to 6 km wide.

Paricutín, 1943

This volcano, in a farmer's field in Paricutín, Mexico, is an example of a *cinder cone*. A cinder cone is a volcano formed through the eruption of cinders that fall back on the vent, forming a crater and a cone. Lava pours out of the top after excess gas in the magma below the earth's surface has escaped.

Shake, Rattle, and Flow 337

Exploration 4

Have students study the photograph on this page for evidence of volcanic activity.

Answer to caption: The shape and type of material that make up the cone provide evidence that volcanic activity has occurred. Also, the lack of snow in an otherwise snowy landscape is a hint that an underground heat source is present.

Students may work individually or in small groups to complete this Exploration. Have world maps and atlases available. When the activity has been completed, reassemble the class to discuss their findings and the final two questions of the Exploration.

Answers to
In-Text Questions

Active volcanoes occur in the western continental United States. After their studies of earthquakes and plate boundaries, students should not be surprised by the fact that Mount St. Helens is located on the west coast. *Other famous volcanoes:*
- Aconcagua (located in the Andes on the Chilean-Argentine border)
- Colima (southwestern Mexico)
- Haleakala (Island of Maui)
- Mauna Kea (Island of Hawaii)
- Mount Lassen (northern California)
- Popocatepetl (central Mexico)
- Stromboli (Italy)

It should not be difficult for students to decide that volcanoes and plate boundaries are associated, and that volcanoes and earthquakes tend to occur in the same locations.

As an extension to *Exploration 4,* you may wish to work with a geography or social studies teacher. An integrated lesson could be developed concerning the social and economic importance of cities located near major volcanoes.

338 **The Restless Earth**

EXPLORATION 4

Where Are the Volcanoes?

Can you find evidence of volcanic activity in this photograph?

Where are the active volcanoes in the United States? Does it surprise you that Mount St. Helens lies on the west coast?

Is there a reason why volcanoes occur only in certain places? Scientists believe they have an answer to this question. Here is how you can share in their discovery.

In an atlas, locate as many famous volcanoes as you can. Add at least five more names to the list given below. Then sketch another map of the world in your Journal, and mark the locations.

Fujiyama (Japan)
Mayon (Philippines)
Hekla (Iceland)
Etna (Italy)
Katmai (Alaska)

The more volcanoes—active, dormant, or extinct—you can locate, the more easily you will be able to understand the scientists' explanation.

When you have completed your map, describe in general terms where volcanoes occur. Compare this map to the earthquake map you made in Exploration 3, and to the map on page 321 showing the earth's plates. Is there any similarity among the maps? Why do you think volcanoes occur only in certain places?

The Cause of Volcanoes—Those Plates Again!

When the earth's crust shifts restlessly or a volcano erupts somewhere, the explanation usually lies in the movement of the earth's plates.

As plates move apart, volcanic rock called **basalt** either moves quietly into the gap or erupts from below to fill the gap. This is one way in which volcanoes are formed.

Another way that volcanoes are formed is by the collision of plates. When plates collide, one of them is usually driven under the other. When the underlying plate reaches the hotter temperatures of the interior of the earth, it begins to melt. This melted rock, known as **magma,** is lighter than the rock around it. Thus, it slowly rises, triggering earthquakes and creating volcanoes. The magma that emerges on the surface—in either a quiet or a violent outburst—is called **lava.** The violence of the volcano is determined by the gases in the magma. If the gases can escape gradually, as they do in the volcanoes of Hawaii, there is usually no explosion. Instead a constant flow of lava allows the volcano to erupt quietly. In the case of a stratovolcano like Mount St. Helens, however, the magma gases are bottled up and they burst forth in a tremendous explosion.

As the gas bubbles from the magma (now called lava, since it is on the surface), it creates **pumice,** a type of lightweight rock filled with tiny holes. Quite often, the pumice gets broken up into volcanic ash, cinders, and dust. These small fragments make up the clouds that billow high above a volcano. Eventually they rise into the atmosphere and spread around the earth, often causing spectacular sunsets.

A Field Trip

Imagine that you are taking a field trip with your class to a volcano. You climb to the top of the volcano and look over the edge. How might you tell if the volcano is active, dormant, or extinct?

Suddenly you can hear rumblings under your feet, and the ground begins to move. In your Journal, record the sequence of events inside the volcano as you think they would occur.

Lava forms polygonal columns (top). A lava flow (bottom)

The Cause of Volcanoes— Those Plates Again!

As a synthesis exercise, ask students to rewrite this material in their own words as if they were explaining the causes of volcanoes to a friend or family member.

A Field Trip

This is a creative writing activity. Students should explain the differences between active, dormant, and extinct volcanoes.

Follow Up

Assessment

1. Ask students to explain why earthquakes and volcanoes tend to occur in predictable locations around the world. *(Movement at plate boundaries causes both earthquakes and volcanoes.)*
2. Have students decide whether these statements are accurate. Students should correct any of the incorrect statements.
   (a) Volcanoes can occur only in certain places.
   (b) Volcanoes occur when plates move apart or come together.
   (c) Once volcanoes erupt they become dormant. *(False. Volcanoes can erupt several times.)*
   (d) Scientists cannot predict when volcanic eruptions will occur. *(False. The presence of earthquakes, ash, and gas, as well as detection equipment, all help predict volcanic eruptions.)*
   (e) Volcanoes can be either quiet or explosive.
   (f) Volcanic eruptions occur only on land. *(False. Volcanic eruptions also occur in the ocean.)*
   (g) The violence of a volcano is determined by the gases in the magma.
   (h) Volcanic eruptions can affect the world's climate.

Extension

1. Present students with the following scenario: You are a journalist assigned to the site of a volcanic eruption. After days of reporting on death and destruction, you have been told to write an article with an opening sentence that reads, "While volcanoes often destroy both lives and property, they also benefit the earth." Complete the article.
2. Have students research various volcanic rocks and their past and present uses. *(Examples include pumice, basalt, and obsidian.)*

Shake, Rattle, and Flow 339

Answers to
Challenge Your Thinking

1. The mountains in the photo on the right are older; they show signs of years of erosion and weathering. Old mountains are rounded with gradual slopes. Younger mountains can be recognized by their sharp, ragged peaks.
2. (a) The plates are separating. You would expect to find volcanoes in Iceland.
   (b) Plates are colliding in Alaska, the Caribbean, the west coast of South America, the Mediterranean Sea, and the Philippines. You could expect to see earthquakes, mountain-building, and volcanoes.
   (c) Yes. The Himalayas began to form 25 million years ago when the plate carrying India ran into the Eurasian Plate. The Rockies were formed when colliding crustal plates caused oceanic crust to slide under continental crust, causing a folding of the continental crust.
   (d) Sliding boundaries occur along the west coast of North America and across the Indian Ocean between the African and Eurasian plates. Earthquakes can be expected in this area.

Challenge Your Thinking

1. Compare the photographs of the mountains shown. Which mountains are older? What evidence supports your inference? What factors differ for each mountain?

2. Sharpen your logic skills! Use the basic facts presented to deduce the answers to the questions that follow.

Facts	Predictions
(a) The Atlantic Ocean is getting wider. New ocean crust is being formed.	What is happening to the plates at the mid-Atlantic boundary? What evidence of crustal movement would you expect to find in Iceland?
(b) Plates collide.	Find three examples of convergent plate boundaries. What kinds of activities will happen here?
(c) When plates collide, the continental crust thickens and buckles as the plates are compressed.	Locate the Himalayas and the Rocky Mountains. Could they have been formed this way?
(d) Plates slide by each other.	Find two places where plates are sliding by each other. What do you predict will happen there?

340 The Restless Earth

THE ROLE OF ROCKS

 ## Rocks in the Past

The Uses of Rocks

Have you seen pictures of the impressive arrangement of huge rocks at Stonehenge in England? If so, you have probably wondered how people in those early days constructed such a sophisticated observatory. Study the photograph and explanation to discover what those ancient people could do with this vast arrangement of rocks.

Stonehenge

Stonehenge was built in southern England around 1650 B.C. The people who built this stone observatory apparently used it to keep track of the motions of the sun, moon, and stars. By standing at the center of the circle, an observer sees the sun, moon, or stars line up in between specific gaps in the stone. The sighting stones outside of the circle increased the accuracy of the measurement.

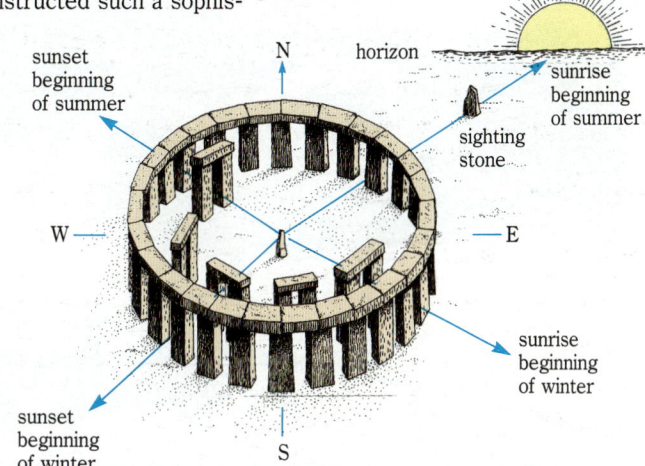

LESSON 4

✸ Getting Started

In this lesson, students are introduced to the study and classification of rocks. The types of rocks formed by sedimentary, igneous, and metamorphic processes are introduced.

Main Ideas

1. Rocks are used for a variety of purposes.
2. Rocks exhibit a variety of textural and compositional characteristics by which they can be classified.
3. Earth scientists classify rocks based on the way they are formed.
4. All rocks are classified into three groups: igneous, sedimentary, and metamorphic.

✸ Teaching Strategies

The Uses of Rocks

Use the picture study to involve students in a discussion about:
- The different ways rocks have been used throughout history
- The significance of rocks in everyday life
- The practical uses of rocks
- Their significance in ancient cultures

Have students study the photographs on pages 341 and 342. Ask them to identify the uses of rocks in each one.

(Continues on next page)

LESSON ORGANIZER

Objectives

By the end of the lesson, students should be able to:

1. Identify various ways rocks have been used.
2. Name some characteristics of rocks that can be used to distinguish one rock from another.
3. Devise a classification scheme for rocks based on observed physical properties.
4. Distinguish between igneous, sedimentary, and metamorphic rocks by using a key.

Process Skills

observing, inferring, classifying, using a key

(Organizer continues on next page)

The Role of Rocks 341

(The Uses of Rocks continued)

- At Stonehenge, a circle of large sandstone blocks was built for use as an observatory. The remains of 900 other stone rings (built between 4000 and 1000 B.C.) have been found in the British Isles.

  Astronomically aligned stone rings with a different design were built by ancient Native Americans. The best known example is the Big Horn Medicine Wheel in Wyoming.

- The Great Wall of China was built as a military fortification. The major part of the Great Wall was built in only seven years under the supervision of Chinese military engineer Meng T'ien (221 B.C.).

- The pyramids of Egypt served as tombs and monuments, and were designed to represent the quadrants of the northern celestial hemisphere as four flat, triangular planes.

- The small photographs on page 342 show examples of the use of rocks as art material and as tombstones.

  Other uses of rocks include: tools, fuel (coal), jewelry, shelter (slate roofs, rock-walled dwellings), road surfaces, breakwaters, retaining walls, and fences.

Study this collage of pictures of rocks. How and why have these rocks been used? Think of more uses for rocks. Create a drawing of at least one other use for rocks.

Great Wall of China

Egyptian pyramid

Sculpture of Ramses at Abu Simbel

Tombstones

(Organizer continued)

New Terms

Igneous rock—a rock formed by the cooling and solidification of magma or lava.

Sedimentary rock—a rock formed from the weathered pieces of other rocks.

Metamorphic rock—any rock changed due to exposure to extreme heat and/or pressure.

Materials

Exploration 5
Part 1: magnifying glass, pencil, paper, 2 different types of rocks (each about 4–6 cm in diameter), metric ruler, Journal
Part 2: rocks from Part 1, index cards
Exploration 6: magnifying glass, rock samples

Teacher's Resource Binder

Resource Worksheets 6–4 and 6–5
Resource Transparency 6–4
Graphic Organizer Transparency 6–5

Science Discovery Videodisc

Disc 2, Image and Activity Bank, 6–4

Time Required

two to three class periods

The Restless Earth

Getting Acquainted With Rocks

Imagine picking up a rock just outside your home. What does it look like? Where did it come from? How did it form? Studying the rock can give you clues about its age, its formation, and the changes in the earth it has experienced.

The nature of rocks can give you information about many of the things you wish to know about the earth. In this activity, you will begin to explore the nature of rocks.

Part 1
Examining Rocks

You Will Need
- a magnifying glass
- 2 different types of rocks, each about 4–6 cm in diameter

What to Do

1. Divide into groups of four. Each group will have eight rocks to study.
2. Select two rocks from those given to your group. Study and record your observations about each on a sheet of paper. Your description should be clear enough so that your classmates can identify the two rocks.
3. When everyone in your group has finished writing the descriptions, mix up the rocks on a table. Try to pick out the two that you wrote about. Can you find them?
4. After finding your own rocks, put them back on the table and mix them up again. This time, exchange descriptions with another member of your group. Find the rocks described on your classmate's sheets. Were you successful? If not, perhaps the descriptions need to be improved. If you were successful, could you suggest anything to add to or delete from the descriptions? Repeat this exchange with each member of your group.
5. As a group, compare the descriptions written by each member. Which properties of the rocks did each describe? Which observations were the most helpful? least helpful?
6. Using all the observation sheets of your group, make up a new sheet listing those properties of the rocks that were most helpful in identifying them. Below is an example.

Part 2
Classifying Rocks

You Will Need
- rocks from Part 1
- index cards

What to Do

1. Join with another group, so you now have sixteen rocks to work with. Choose one property that half of the rocks share, and divide the rocks into two groups according to this property. Using an index card, make a label naming the property for each group.
2. Determine how each of the two large groups can be divided into smaller and smaller groups that have more similar characteristics. Make labels each time you subdivide the groups.
3. Continue until you have separated all the rocks into either groups having one rock or groups whose properties are all the same.
4. Describe your classification scheme in your Journal. Then display your rocks and labels in the classroom.

Property (Characteristic)	Observations Related to Property
Color	Ranged from plain gray to black, white, and red
Composition	The gray rock was smooth. The mixed rock was made up of small rocks stuck together. The red rock had shiny crystals and was rather rough.
etc.	etc.

Exploration 5

Invite students to bring in rocks that differ markedly in appearance, or provide such samples yourself. Divide the class into small groups and distribute two rocks to each student to examine. Answers throughout *Exploration 5* will vary from group to group depending on the rock samples available.

Part 1

Collect observations from each group and write them on the chalkboard. Draw students' attention to the number of materials of which the rock is composed, the way the materials fit together, their size, shape, orientation, and color.

Question 5 can be the subject of class discussion as the students group their observations using different properties. Properties may include color, composition, texture, luster, shape, and so on.

Students may require help when distinguishing between a rock and a mineral. Rocks are usually composed of more than one mineral, and thus appear to be a combination of materials. Explain how a conglomerate rock's composition differs from that of a collection of minerals. Conglomerate rocks are made of boulders, pebbles, and sand held together. Minerals are made of only one kind of material.

Part 2

Spend some time reviewing classification systems. Use a collection of common objects such as shoes, cars in the parking lot, and markers to demonstrate how to devise a hierarchical classification scheme.

The rock classification scheme should be developed according to the students' suggestions. No one scheme is necessarily better than another.

Provide index cards for each group so that they can make appropriate labels for display purposes.

The Role of Rocks

The History of Rock

Call on a volunteer to read the top of page 344. Review the meaning of the terms *igneous, sedimentary,* and *metamorphic.*

Rock Groups

Geologists subdivide rocks on the basis of origin. Igneous rocks form when molten rock cools. Sedimentary rocks are secondary in origin. They are composed of material formed when older igneous, sedimentary, and metamorphic rocks disintegrate due to weathering and erosion. Under appropriate conditions of heat and pressure, igneous and sedimentary rocks can be changed into the third rock group, metamorphic rocks.

Most students will be familiar with the three groups of rocks. The next few lessons deal with the rock types and the differences among them.

The History of Rock

Each rock you examined has a history to tell. The crucial points of the history include:

- how the rock was formed
- where it was formed
- how it ended up in your hands

A sedimentary rock

The observations you made about the rocks you gathered are clues to discovering the history of your rocks. In this section, you will learn to interpret the data and reconstruct the events that led to the rocks' formation. The history of each rock will include one of the following three descriptions.

Deep within the earth, *molten* materials slowly cooled and hardened, forming crystals that were easily visible; an **igneous** rock was born.

Particles of sand and soil were carried by wind and water and eventually deposited on an ocean bottom. Increasing pressure from the accumulated weight of the particles, combined with chemicals from the sea, cemented the particles together to form a **sedimentary** rock.

An igneous rock

Heat and extreme pressure reorganized the crystals of an already existing rock to change it into a new type of rock, a **metamorphic** rock.

Metamorphic rock

Rock Groups

Most of you would have little trouble classifying rock bands into groups according to the type of music they play. But can you classify rocks into groups? There are so many different kinds of rocks that scientists once searched for ways to organize them. Read the three descriptions above again.

On what basis have geologists classified rocks? Perhaps a clue will help: *ignis* is the Latin word for "fire."

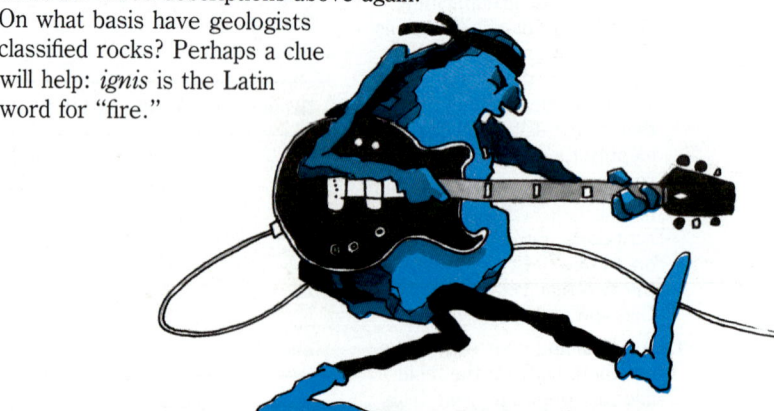

344   The Restless Earth

EXPLORATION 6

Rock Group Identification

It can be difficult to decide whether a rock is sedimentary, igneous, or metamorphic, but the identification key shown will help you to classify the rocks.

You Will Need
- a magnifying glass
- rock samples

What to Do

In your Journal, make a table with the following headings for recording your results: *Sample Number, Observations, Type of Rock.* As you study each rock, record your observations and identification in a table like the one below.

Sample Number	Observations	Type of Rock
1	flat sheets	?

Choose one rock to begin your study. Starting with 1(a) and 1(b), pick the sentence that best describes your rock. Then look at the number at the end of that line and go to the step number indicated. Continue until you reach a line that identifies the rock as either sedimentary (S), metamorphic (M), or igneous (I). Repeat this procedure for eight different rocks.

You may find that some rocks do not seem to fit readily into any category. Expert help may be required. Remember, identifying rock types is not easy!

Identification Key

1. (a) The rock is grainy or is made of more than one material. — 2
1. (b) The rock is made of only one material. — 5
2. (a) The particles or grains are closely fitted together (interlocking). — 3
2. (b) The particles or grains are held in place by natural cement (non-interlocking). — S
3. (a) All the grains are of the same type. — M
3. (b) The grains are of two or more different types. — 4
4. (a) The grains are arranged randomly. — I
4. (b) The grains are arranged in bands. — M
5. (a) The rock is glassy or porous. — I
5. (b) The rock is made of strong, flat sheets. — M

2(b)   3(b)   4(a)

4(b)   5(a)   5(b)

Exploration 6

The basic characteristics used for rock identification are mineral composition and texture, grain or crystal size, size and shape, and orientation of material in a rock. The *Identification Key* works well for igneous and sedimentary rocks, but not as well for metamorphic rocks. You may need to screen a variety of rock samples ahead of time for characteristics described in the key.

If students have not used a key before, help them by working through an example with them. To clarify different parts of the key, display samples illustrating the various characteristics. Students can then compare their samples with the ones on display.

Each group should have 3 to 5 samples, with at least one of each type of rock included. Number each sample, and keep a record of the numbers given to each group and the type of rock represented by each number.

When students have completed the activity, involve them in a discussion of their results. A pattern may emerge that will allow you to identify local bedrock. If your inferences about the rock types do not agree with what you know about the area, it may be that these rocks have been brought in from other areas (for fill or for landscaping purposes) or that they may have been deposited by rivers or glaciers.

✴ Follow Up

Assessment

1. Present students with the following scenario: Imagine that you are in the business of selling rocks. Make a list of all the ways your rocks could be used. Now write an advertisement to promote the sale of your rocks.
2. Ask students to compare how scientists classify rocks to the scheme they developed in *Exploration 5.*

Extension

Have students research rock resources in their state. On a map, have them locate the sites of quarries, gravel mines, and other similar operations.

The Role of Rocks

LESSON 5

★ Getting Started

In this lesson, students study the formation and characterization of igneous rock. They discover the relationship between rate of cooling and size of crystals by simulating the formation of igneous rocks.

Main Ideas

1. Igneous rocks are formed when molten rock cools.
2. The size of crystals in igneous rocks is determined by the rate of cooling.
3. Igneous rocks that form on the surface of the earth are called extrusive or volcanic rocks.
4. When cooling of molten rock occurs deep within the crust, the igneous rock formed is called intrusive or plutonic rock.

★ Teaching Strategies

Exploration 7

Students may hypothesize that the rocks on page 348 look different because they formed under different conditions.

Before beginning the Exploration, discuss safety precautions with students. Safety goggles should be worn, and students must take care to avoid burns.

Remind students to follow the directions in the sequence they are given so that the paper cup in the sand and the sloping surface are ready for the melted stearic acid.

Divide the class into small groups and distribute the materials. Be sure that all questions about the procedure have been answered.

(Continues on next page)

5 Hot Rocks

EXPLORATION 7

Exploring Igneous Rocks

Before doing Exploration 7, look at the rocks on page 348. All of the rocks shown there are igneous. At one time each rock was molten. Many contain interlocking crystals. Why do they look so different? Performing the Exploration that follows will help you answer this question.

A Simulation: Forming Igneous Rocks

This activity will help you understand how igneous rocks are formed. Work in groups of three or four.

Be Careful. *Since you will be using hot materials, you must use caution. Safety goggles must be worn! Use care when handling stearic acid.*

You Will Need

- a metal ring
- a ring stand
- wire gauze
- sand
- 3 beakers (400 mL)
- test-tube tongs
- a small paper cup
- a piece of paper towel
- a piece of glass or aluminum foil
- 10 g stearic acid
- an alcohol lamp or bunsen burner
- ice
- water
- a knife
- a scoop
- 2 test tubes

What to Do

1. Fill a beaker three-quarters full of sand. Scoop out a hole in the sand large enough for the paper cup, and place the cup in the hole.

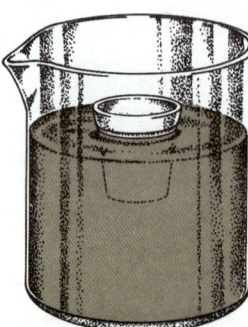

2. Prop up a piece of glass or foil on a folded paper towel. It should form a slight slope.

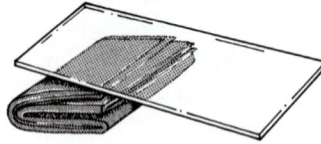

3. Fill a beaker halfway with ice and water.

4. Set up the ring stand. Place the wire gauze on top of the metal ring and the burner below it. Fill the third beaker halfway with water and place it on the wire gauze.

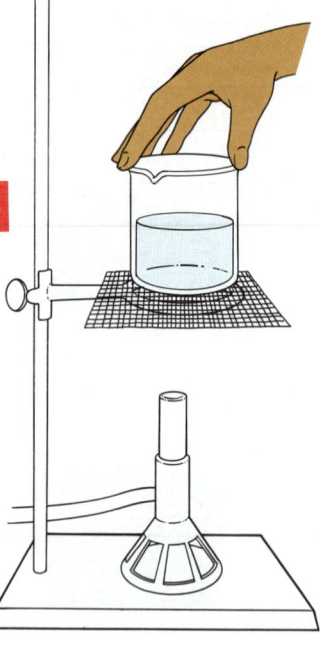

LESSON 5 ORGANIZER

Objectives

By the end of the lesson, students should be able to:

1. Explain how differences in the cooling of igneous rocks affect their structure.
2. Recognize the differences between extrusive and intrusive rocks by examining the size of the grains or crystals.
3. Infer how and where igneous rocks were formed by examining the size of the grains or crystals.

Process Skills

observing, inferring, analyzing, compiling a dictionary

346 The Restless Earth

5. Place 5 g of stearic acid into each test tube and stand the test tubes in the beaker of water. Heat the water until all of the stearic acid melts.

6. Using the test tube holder, carefully pour a small portion of the melted stearic acid onto the sloping glass or foil, allowing it to run down the slope.

7. Quickly pour the rest of the stearic acid from the first test tube into the paper cup sitting in the sand. Observe both samples of stearic acid as they cool. Record your observations in your Journal.

8. Pour the melted stearic acid from the second test tube into the beaker of ice water. Record your observations.

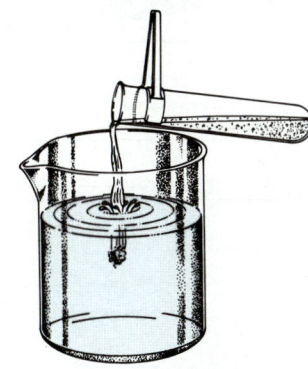

9. When the stearic acid in the ice water is completely cool, cut it in half with a sharp knife.

Interpreting Your Observations

1. Compare the appearance of the stearic acid in the paper cup, the stearic acid poured down the slope, and the stearic acid poured into the ice water. In your Journal record the differences you observe. What do you think caused these differences?

2. Take a close look at the stearic acid that you cut in half. Is there a difference between the appearance of the material in the center and the material on the edge? If so, what do you think caused the difference? Was there a difference in the time it took for the edges and the center to cool? Why?

3. This experiment is meant to simulate the formation of igneous rocks. What is simulated by each of the following?
   (a) the melted stearic acid in the cup
   (b) the sand
   (c) the melted stearic acid that was poured down the sloping glass or foil
   (d) the melted stearic acid that was poured into the ice water

---

(Exploration 7 continued)

Once the stearic acid is melted (melting point 28 °C), students must quickly pour it down the slope and into the paper cup. Assign different group members to observe each situation.

Answers to Interpreting Your Observations

1. The stearic acid in the cup has the largest crystals. The stearic acid on the slope cools rapidly and may not form crystals at all. The stearic acid in the ice water forms a lump with a glossy surface due to rapid cooling. Rate of cooling accounts for the differences in appearance.

2. The acid closest to the ice water cools fastest, thereby protecting the inner material that cools more slowly. The rate of cooling explains the difference.

3. (a) The liquid stearic acid in the cup simulates molten rock formed deep within the earth's crust.
   (b) The sand simulates the surrounding rock that prevents heat from escaping quickly.
   (c) Pouring the melted stearic acid on the slope is similar to what happens when lava pours out onto land surfaces.
   (d) Pouring the melted stearic acid into the ice water is similar to lava flowing into the ocean. It forms pillow-lava as the water rapidly cools the lava and prevents crystals from forming.

New Terms

Extrusive or volcanic rocks—igneous rocks formed from the molten lava that flows onto the surface of the earth.

Intrusive or plutonic rocks—igneous rocks formed from magma that tends to cool slowly, deep within the earth's crust.

Materials

Exploration 7: metal ring, ring stand, wire gauze, 300 mL sand, three 400-mL beakers, test-tube tongs, small paper cup, paper towel, piece of glass or aluminum foil, 10 g stearic acid, alcohol lamp or Bunsen burner, ice, water, knife, scoop, 2 test tubes, safety goggles, matches, Journal

Fiery Language: Journal

Teacher's Resource Binder

Resource Worksheet 6–6
Graphic Organizer Transparency 6–6

Science Discovery Videodisc

Disc 2, Image and Activity Bank, 6–5

Time Required

three class periods

Pedro's Report

Call on a volunteer to read aloud the top of page 348. Involve students in a discussion of what to add to make Pedro's report more complete. Students should suggest that he include what took place when the stearic acid hardened in the paper cup.

Direct students' attention to the photographs of the rocks on the bottom of the page. Invite them to make inferences about the rate at which each rock cooled, based on Pedro's report and their own results from *Exploration 7*. Students may suggest that these rocks could be grouped by color.

Answers to
In-Text Questions

- **Basalt** has small crystals, so it must have cooled quickly.
  Background: Basalt is formed when highly fluid, low-silicate magma escapes to the surface as lava and cools quickly. Oceanic crust, formed from lava flows at midocean ridges, is mostly basalt. Basalt is dark-colored because it contains the minerals iron and magnesium.
- **Obsidian** has very small or no crystals, so it must have cooled very quickly.
  Background: Obsidian forms when lava cools so quickly that no crystals can form, in effect forming a natural glass. Although chunks of obsidian appear dark, a thin slice is nearly colorless. Obsidian, granite, rhyolite, and pumice are similar in mineral content.
- **Rhyolite** has small crystals, so it must have cooled quickly.
  Background: Rhyolite is formed when a viscous, high-silicate lava cools quickly at or near the earth's surface.
- **Granite** has large crystals, so it must have cooled slowly.
  Background: Granite is formed by viscous, high-silicate magma that hardened slowly beneath the earth's surface.
- **Pumice** has very small or no crystals, so it must have cooled very quickly.
  Background: Pumice forms from lava containing hot gases. The network of holes (formed by escaping gases) make the rock so lightweight that it will float on water.

Pedro's Report

Here is what Pedro wrote in his report. What would you add to make the report more complete?

"This was one of the most interesting experiments we've done. When the melted stearic acid cooled quickly by being poured into cold water, a rubbery mass was made. Our teacher said that when lava flows over the earth's surface, it cools down fast and the rock formed has extremely small crystals. When the stearic acid cools more slowly, larger crystals are formed. This must be what happens when magma cools slowly beneath the earth's crust."

Now examine the six igneous rocks pictured here. On the basis of Pedro's report, classify each rock according to the rate at which it cooled. Use the terms *quick* or *slow* to describe the rate that each rock cooled.

On the basis of what other properties could you group these igneous rocks?

basalt

obsidian

rhyolite

granite

pumice

gabbro

- **Gabbro** has large crystals, so it must have cooled slowly.
  Background: Gabbro is formed by highly fluid, low-silicate magma that hardens slowly beneath the earth's surface. Gabbro is similar to basalt in composition.

Fiery Language

There are many terms used in connection with igneous rocks. You have seen some already: lava, magma, molten. Beginning with these words, compile a dictionary of rock terms in your Journal.

The Formation of Igneous Rocks

When igneous rocks form from the molten lava that flows onto the surface of the earth, they are called **extrusive** or **volcanic** rocks. Basalt is the most abundant rock of this type on earth. Other volcanic rocks include rhyolite, obsidian, and pumice. These extrusive rocks are made up of tiny crystals.

When lava flows into the ocean, it forms pillow-like shapes about one meter across. Would this lava have small or large crystals? Why do you think so?

Igneous rocks have different forms.

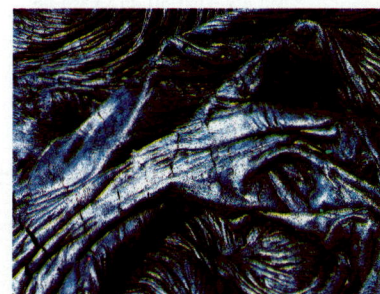

Fiery Language

Encourage students to develop definitions by using their own words rather than a textbook's glossary or dictionary. After the unit is completed, students can brainstorm a number of ways to state the definitions and then refine a final definition for each term.

The Formation of Igneous Rocks

Call on a volunteer to read aloud the material. Then direct students' attention to the photographs on page 349. Involve students in a discussion of the photographs, using them to illustrate points discussed in the text.

The photograph at the top right on page 349 shows a type of solidified lava with a wrinkled surface. It is known as *pahoehoe,* meaning "ropey" in the Hawaiian language. This lava was flowing slowly enough so that the surface hardened without breaking. In many cases, the hot interior lava continued to flow even after the surface had hardened. Eventually, a hollow structure was created. It is called a lava tube.

The photograph at the bottom left on page 349 shows a columnar basalt formation. Similar formations located elsewhere have been given fanciful names, such as the "Giant's Causeway" in Northern Ireland, and the "Devil's Postpile," in California.

The photograph at the bottom right on page 349 shows another basalt formation located in Wyoming. It is known as the "Devil's Tower." This formation is a hardened basalt plug that was the interior of an ancient volcanic cone. The cone has long since eroded away, leaving the plug as a free-standing tower.

Rock formed from lava that flowed into the ocean would have very small or no crystals.
(Continues on next page)

(The Formation of Igneous Rocks continued)

Have students copy the illustration on page 350 into their Journals and write the labels in their correct positions. Assign letters to each label and have students place them on the drawing. Note that a, c, and e refer to extrusive igneous rocks; b, d, and f refer to intrusive igneous rocks.

a. extrusive igneous rock
b. rocks formed from the slow cooling of magma
c. rocks formed from the fast cooling of lava
d. rocks with large crystals
e. rocks with small crystals
f. intrusive igneous rock

Refer to page S162 for a corrected illustration.

Assessment

1. Bring in samples of extrusive and intrusive igneous rocks. Have students observe the samples, study their characteristics, and then write a rock's "life story."
2. Ask students to explain the following observation: "Granite can be observed covering large areas of the earth's surface." *(Students should understand that granite formed beneath the surface of the earth. It is exposed because of the erosion, over many thousands of years, of the layers above it.)*

Extension

Ask students to bring in samples they think may be igneous rocks. Then give them the following key to use in identifying the rocks.

1. Coarse-grained: particles clearly visible to the unaided eye
   a. Mostly light-colored minerals: Granite
   b. Mostly dark-colored minerals: Gabbro
2. Fine grained: size of sugar grain or smaller
   a. Frothy, porous: Pumice
   b. Solid, may have rounded bubbles filled with crystals
      i. Light colored, gray, pinkish, purplish: Felsite
      ii. Dark-colored, black, greenish, brownish: Basalt

Liquid rock, or magma, tends to cool slowly deep within the earth's crust. Recall the stearic acid experiment from Exploration 7. Based on your experimentation, what size do you think the crystals of these rocks would be? Rocks formed in this way are known as **intrusive** or **plutonic** rocks. Granite is the most abundant example. Gabbro is another common plutonic rock. This illustration shows the formation of igneous rock by the cooling of magma. Copy the drawing into your Journal, and place the following labels in the correct positions:

- extrusive (volcanic) igneous rocks
- rocks formed from the slow cooling of magma
- rocks formed from the fast cooling of lava
- rocks with large crystals
- rocks with small crystals
- intrusive (plutonic) igneous rocks

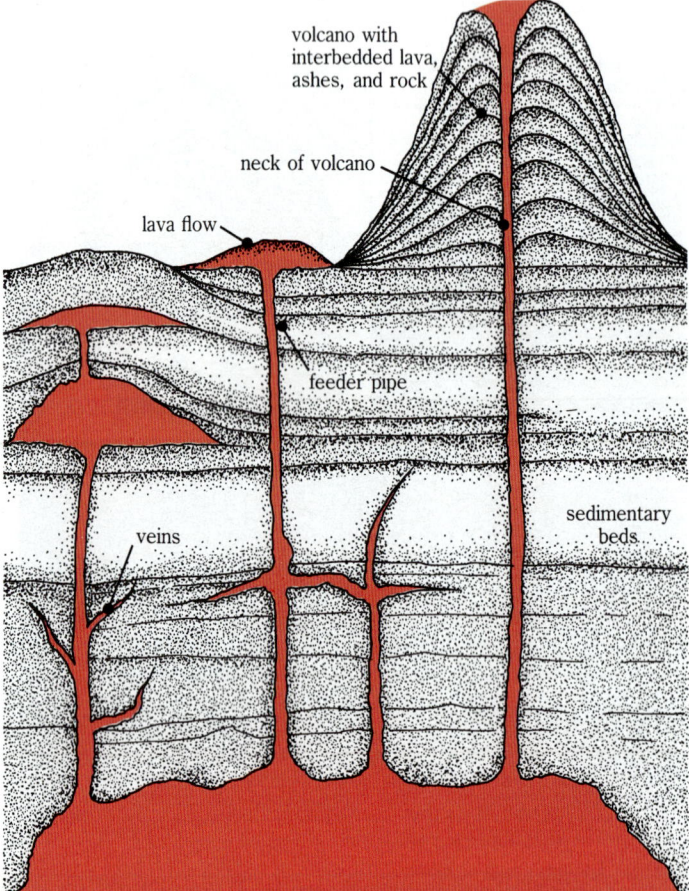

6 Rocks from Sediments

Hanging Out With Granite-Man

"Yeah, yeah, yeah . . . so I was on my way to the concert when I ran into my old friend Sam Sandstone.

'Where ya been?' I asked him.

'Up the river,' he says.

So I say, 'Naw, really? Tell me what's happen' up there lately.'

So he starts tellin' me about these rocks, man, that fall into the river. Now, the river, being one swift dude, carries these rocks, like FAST, down toward the ocean. On the way, the heavy rock dudes drop to the bottom first, the pebbles settle out just a little later, and the sand goes swirling and twirling on top. Then the river dumps these cruisers into the ocean where the heavy salt water presses them together.

'Solid?' I ask him.

'Totally solid,' Sam says. 'These dudes have their own group, man. They are their own kind of rock, if ya know what I mean.'

'Man,' I say, 'That's too cool!'

'My sediments exactly,' says Sam."

Granite-Man certainly tells a good story, doesn't he? Think about the process that Sam described. What do you think sediments really are? Why do they separate out? How does the speed of a river contribute to the formation of sedimentary rocks? What does the ocean have to do with sediment? Reread what Granite-Man has to say with these questions in mind.

Where Are Sediments Deposited?

Examine the illustration to the right. Think about why sediments get deposited in layers. Sketch the drawing in your Journal, and label the locations where sediments might be deposited.

Where would you expect to find larger materials, such as stones, pebbles, and gravel? Where might sand be deposited? Where would you expect to find deposits of fine materials such as silt and mud?

LESSON 6 ORGANIZER

Objectives

By the end of the lesson, students should be able to:

1. Identify possible locations for the deposition of sediments.
2. Observe that different-sized particles in water settle at different rates.
3. Infer that some sedimentary rocks are formed when water evaporates, leaving solid material behind.
4. Infer that the number, size, and shape of the sediments making up sedimentary rocks are clues to their origins.

Process Skills

observing, inferring, investigating, predicting

(Organizer continues on next page)

LESSON 6

✷ Getting Started

In this lesson, students examine the different ways sedimentary rocks are formed.

Main Ideas

1. Sedimentary rocks are formed from sediments transported by wind, water, or ice to the deposition site.
2. Sediments suspended in water settle out at different rates, forming layers.
3. Sedimentary rocks can be formed by evaporation, cementation, compaction, drying, and crystallization.

Answers to
Hanging Out with Granite-Man

- Sediments include solid material that is physically transported by water, ice, or wind, as well as dissolved substances chemically precipitated from water. Types of sediments include rocks, pebbles, sand, salt, and clay.
- Due to their greater mass, the heavier particles settle to the bottom first.
- A swift river can transport larger particles of sediment than a slower river. These larger pieces of sediment will eventually form rock.
- The ocean is the final resting place for the finer sediments.

Answers to
Where Are Sediments Deposited?

Direct students' attention to the diagram and accompanying labels. The shelf is the gently sloping submerged edge of a continent, extending to a depth of about 200 m.

(Answers continue on next page)

The Role of Rocks

Answers continued

The slope is the region of steep slope between the shelf and the ocean bottom.

Have each student draw the diagram in their Journal and add the appropriate labels. Students should recognize that physical transportation and sedimentation generally follow a downhill trend in response to gravity.

- Stones, pebbles, and gravel deposits could be found at river bends.
- Sand could be deposited at river bends, lake and river deltas, beaches, barrier islands, and the ocean areas of shelf and slope.
- Deposits of fine material such as silt and mud could be found in the lake, river delta, and ocean areas of shelf and slope.

✴ Teaching Strategies

Getting to Know the Sedimentary Rock Group

- Shale, sandstone, coquina, and conglomerate are obviously made up of sediments.
- Shale is made up of silt and clay. Sandstone forms from sand. Coquina contains carbonates and fragments of fossils and shells cemented together. Conglomerates contain pebbles and rock fragments cemented together.
- Gypsum and limestone do not appear to be composed of sediments. Gypsum is formed from the precipitation of sulfates. Limestone is formed from the precipitation of carbonates.

Getting to Know the Sedimentary Rock Group

The formation of sedimentary rocks is closely associated with water. One type forms when water carries soil, pebbles, and other particles to the ocean floor where these sediments become rock. The second method involves chemicals dissolved in water. By evaporation and precipitation of substances like calcium carbonate, sedimentary rocks can form.

Several sedimentary rocks are pictured here. However, try to work with actual rocks if samples are available. Then consider the following questions.

- Which of the rocks are obviously made up of sediments?
- What kinds of sediments make up these rocks? (shells of sea animals, salt, clay, sand, pebbles, etc.)
- Which rocks do not appear to be composed of sediments? (These are the kind formed by changes in the water's solutes.)

If you need help, check the library for a geologist's field guide or a geology textbook.

shale

sandstone

coquina

conglomerate

gypsum

limestone

(Organizer continued)

New Terms

Lithification—the process by which soft sediments become hard rock.

Cementation—the process by which the spaces between particles of loose sediments are filled with a cementing agent.

Compaction—the process by which the pressure of overlying sediments or the pressure from the earth compresses the space between the particles of sediments.

Drying—a process by which liquid is evaporated, leaving behind solid sediments.

Crystallization—the process by which solids form from a saturated solution.

352 The Restless Earth

EXPLORATION 8

Learning About Layered Rocks

Earth scientists tell us that certain sedimentary rocks are formed from layers of sediment. In other cases, evaporation plays an important role in the formation of sedimentary rocks. These activities will help you find out more about both processes.

Part 1
You Will Need
- about 6 mL of soil that contains a variety of particle sizes
- 10 mL of water
- a test tube with a stopper or a glass jar with a tight lid

What to Do
1. Place the water into the test tube or jar, and pour the soil into the water.

2. Before shaking the mixture, predict what will happen. Which particles will settle out first—the largest or the smallest? Sketch what you think the results will be.
3. Now shake the mixture vigorously for one minute. Observe the results. Does it resemble your sketch; was your prediction correct? If not, draw a corrected sketch.

Part 2

Marilyn read that if 100 g of seawater evaporates completely, then 3.5 g of dissolved solids remain behind. She was curious to know whether the solid material left behind looked and tasted like table salt. Marilyn collected some seawater and left it in the sunlight until all the water had disappeared. She discovered that the material that remained was not as white as table salt, although it tasted the same. Should she have tasted the salt to identify it?

When Marilyn told her teacher, Mr. Delaney, about her experiment, he told her that what she had collected was mostly ordinary table salt. However, other compounds like magnesium, sulfur, calcium, potassium, bromine, and iodine were also present.

Marilyn asked Mr. Delaney to explain why huge salt beds are found under certain areas of the ocean. What is your theory about how such large salt beds are formed?

Exploration 8

This Exploration is intended to help students understand how sediments form layers. *Part 1* focuses on clastic sediments, while *Part 2* focuses on chemical sediments.

Part 1

Clastic sediments are made up of particles broken and weathered from pre-existing rock. Rocks such as conglomerate, breccia, sandstone, and shale are formed by the deposition of clastic sediments.

Divide the class into small groups and distribute the materials. Have students sketch their prediction of what will happen first. They should discover that the largest particles settle out first. If their predictions were not correct, students should draw a corrected sketch.

Part 2

Chemical sediments are precipitated from solutions, primarily the ocean. They include limestone, dolomite, halite, and gypsum. Students studied salt precipitation in *Unit 3, Solutions*.

Marilyn should not have tasted the salt since she did not know what it contained. In answer to the last question, students may suggest that inland seas evaporated, leaving their salt deposits behind.

Materials

Where Are Sediments Deposited?
Journal
Getting to Know the Sedimentary Rock Group: geology field guide or textbook, sedimentary rocks
Exploration 8: 6 mL of soil (with various particle sizes), 10 mL water, test tube with stopper or glass jar with tight lid, clock or watch, graduated cylinder

Teacher's Resource Binder

Activity Worksheet 6–7

Science Discovery Videodisc

Disc 2, Image and Activity Bank, 6–6

Time Required

three class periods

The Role of Rocks 353

Layers of Sediment Become Rock

Ask a student to read aloud page 354. Involve students in a discussion of the new terms *lithification*, *cementation*, and *compaction*.

Answers to In-Text Questions

Explanations for *drying* and *crystallization* include:
- Drying—a process by which liquid is evaporated, leaving behind solid sediments.
- Crystallization—a process by which solids form from a saturated solution.

Ask students to compare and contrast these various processes of lithification. Then have the class discuss the ideas suggested by Loni, Juan, and Randy. Encourage students to think of ideas of their own, and relate these to the processes of lithification.

Assessment

Bring in samples of sedimentary rocks. Have students study the samples and explain how each was lithified. (*Sandstone: cementation of sand; shale: compaction of mud and silt; gypsum and limestone: crystallization of a solute; conglomerate: cementation of pebbles and rock fragments; coquina: cementation of shells of marine organisms*)

Extension

1. Have students research and write a report on the different forms of limestone.
2. Have students experiment with "making rocks."
   (a) Mix 500 g of coarse sand, a handful of pebbles, and 150 g of cement in a milk carton. Add water slowly and stir until a porridge-like mixture is formed. Store the container until it is hard (1 to 3 days).
   (b) Follow the same procedure as before, but omit the pebbles to make a rock resembling sandstone.

A Rocky Problem

Patrick and Nancy discussed the last Exploration they did in class.

Nancy said, "I'm confused. I can see how the layers formed in the test tube, but trying to understand how pieces of dirt end up as rock on the side of a cliff is beyond me."

"Maybe it would help if we tried to make some rocks," Patrick suggested. "Let's ask Mr. Delaney to help." Their teacher then suggested that they read the information below for some hints about what to do.

Layers of Sediment Become Rock

When soft sediments become hard rock, **lithification** has occurred. One form of lithification is called *cementation*. In this process, the spaces between particles of loose sediments are filled with a cementing agent that is carried in solution by the water. As the cementing agent fills the gaps between the particles, the loose materials are cemented together to form sedimentary rocks.

During *compaction*, the pressure of overlying sediments compresses the space between the particles. As the particles are pressed closer together, they stick to each other forming sedimentary rocks.

There are still other lithification methods. *Drying* and *crystallization* are two of them. How do you think these processes of lithification work?

Patrick and Nancy asked their classmates to help them come up with some ideas to simulate lithification. Loni remarked that popcorn can be made to stick together when mixed with honey. Juan had watched his neighbor make concrete the weekend before, and he thought this was a good simulation. Randy suggested pressing small marbles of modeling clay together.

What was each student thinking about in proposing these ideas? Can you come up with your own lithification model?

7 Changed Rocks

The Meaning of Metamorphic

Metamorphosis means change. You might have used the word to describe the life cycle of frogs or insects. For example, when a caterpillar spins a cocoon and comes out as a butterfly, metamorphosis has occurred. What do you think a **metamorphic** rock is like? Look at the examples of change at the right.

For rocks, metamorphism occurs when the environment changes, such as when there are movements in the earth's crust. How could igneous and sedimentary rocks change into a new kind of rock? Read "Historical Flashback!" to find out.

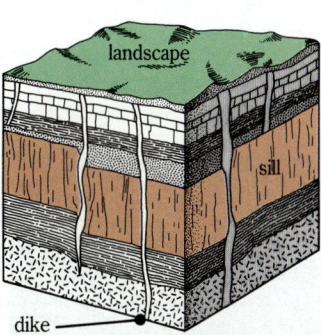

(a) caterpillar   butterfly

(b) ingredients   cake

Historical Flashback!

Charles Lyell became a geologist in a roundabout way. He was born in 1797 in Great Britain. Lyell attended Oxford University to study law, but was very interested in geology as a hobby. This scientific interest led him to the latest geological theories. These theories had been proposed by James Hutton just before Lyell was born. One of Hutton's theories concerned **intrusions.** Hutton thought that magma sometimes pushed upward into the cracks in existing rocks. Then the magma cooled and formed igneous rocks, including granite. Such rocks are called intrusions because the magma *intrudes* (is forced into) the rock.

Two types of intrusions exist. **Sills** are intrusions that form in horizontal cracks in an existing rock structure. **Dikes** are intrusions that form vertically across existing rocks.

In the following two photographs, locate what appear to be sills and dikes. Where did they originate?

LESSON 7 ORGANIZER

Objectives

By the end of the lesson, students should be able to:
1. Explain that metamorphic rocks are changed rocks.
2. Identify the agents of metamorphism as heat, pressure, and fluids.
3. Explain that metamorphic rocks are usually different in structure and texture from their original form.
4. Infer that any one rock type can be changed into another form.

Process Skills

observing, inferring, synthesizing, interpreting data

(Organizer continues on next page)

LESSON 7

✷ Getting Started

This lesson focuses on the formation of metamorphic rock. The concept of the rock cycle is also introduced.

Main Ideas

1. Rocks subjected to heat and pressure beneath the surface of the earth will change their original texture and composition to become metamorphic rocks.
2. Rocks can be changed into another form when conditions in their environment change; thus, rocks can be recycled.

✷ Teaching Strategies

The Meaning of Metamorphic

Involve students in a discussion of *metamorphosis* by directing their attention to the pictures of the caterpillar and butterfly. Review the differences between physical and chemical changes. Students may infer that metamorphic rocks have been changed by environmental conditions.

Historical Flashback!

Call on a volunteer to read this passage that describes the effects of intruded magma on existing rock. This is known as *contact* metamorphism. Metamorphism of large areas is called *regional* metamorphism, and it is usually associated with movements of the earth's crust.

The large photograph on page 355 shows a few *dikes*. A *sill* is shown in the inset photograph. Both originated as magma deep within the earth.

(Continues on next page)

The Role of Rocks 355

(Historical Flashback! continued)

Introduce page 356 with a discussion of "before and after" situations. Then call on a volunteer to read the page aloud. Direct students' attention to the changes in the photographs. Invite them to match the original rocks with their metamorphic counterparts. Ask students to suggest what brought about the changes. *(Some students might remember heat and pressure as agents of change from earlier study.)* Write on the chalkboard, "Rocks change when their environment changes," or "Rocks change because of changes around them." Involve students in a discussion of what these statements mean. Challenge students to identify how a rock's environment might change. *(Exposed rock may weather and erode; rocks underground may be subjected to heat and pressure due to crustal movements.)*

Explain that three primary agents bring about metamorphism: heat, pressure, and fluids.

Answer to

In-Text Question

Students should observe that the texture and organization of grains are different for the metamorphosed rocks. Marble and quartzite (both metamorphic rocks) have larger grains and a more crystalline appearance than their sedimentary parent rocks, limestone and sandstone. Quartzite crystals appear to be more flattened and compressed than the grainy sandstone. Slate appears to have a finer texture, a shinier surface, and a more layered structure than shale.

Hutton observed that the rocks surrounding an intrusion had been through a process of change. The reason for this seemed simple: molten rock is very hot, and the heat affected the sedimentary or igneous rock surrounding the intrusion.

The rocks surrounding the intrusion do not become hot enough to melt. However, certain chemical changes take place as gases and fluids from the cooling magma move into cracks, joints, or small pores within the rocks. These surrounding rocks then change in appearance and characteristics.

Lyell was convinced that Hutton's theory was correct. He placed these "changed" rocks into a third group, which he named metamorphic. Here are some pictures showing original rocks and their metamorphosed forms:

- shale and slate
- limestone and marble
- sandstone and quartzite

What differences do you see between the metamorphic rock and the original rock?

Charles Lyell

shale   limestone   sandstone

slate   marble   quartzite

(Organizer continued)

New Terms

Intrusions—magma that has pushed upward into cracks in existing rocks.
Sills—intrusions that form in horizontal cracks in an existing rock structure.
Dikes—intrusions that cut vertically across existing rocks.

Materials

Exploration 9: 4 pairs of rocks (each pair consisting of an igneous or sedimentary rock and its metamorphosed form), Journal

Teacher's Resource Binder

Activity Worksheet 6–8

Science Discovery Videodisc

Disc 2, Image and Activity Bank, 6–7

Time Required

two class periods

| EXPLORATION 9 |

Rocks: Before and After

You know that metamorphic rocks are changed igneous or sedimentary rocks. In this activity, you will compare original rocks with their metamorphosed forms.

You Will Need

- 4 pairs of rocks, each pair consisting of an igneous or sedimentary rock and its metamorphosed form

What to Do

1. Examine each pair and compare the properties of each. Record your observations in a table in your Journal.
2. When you have finished your examination, check the distinguishing property or properties that you identified against the rock identification key on page 345. Do your observations include all these characteristics?
3. How closely do the metamorphic rocks resemble the original rocks that they formed from?

Can Rocks Be Recycled?

Do you suppose that metamorphic rocks can be changed back into their original forms? Can igneous rocks be changed into sedimentary rocks? In other words, can rocks be recycled? Discuss the following questions with a classmate, and decide whether you think rocks can be recycled.

1. How are igneous rocks formed?
2. What conditions change igneous rocks into metamorphic rocks?
3. How are sedimentary rocks formed?
4. What is the source of sediments?
5. What conditions change sedimentary rocks into igneous rocks?

What pattern, or cycle, have you discovered? Using the words and phrases listed below, write out a description of how recycling occurs.

plate tectonics
molten rock (magma or lava)
metamorphic rock
heat
cooling
uplift
igneous rock

erosion
pressure
sedimentary rock
melting
cementing
sediments
intrusions

Can Rocks Be Recycled?

Divide the class into pairs. Ask them to read page 357 and answer the questions. After a specified amount of time, reassemble the class to discuss their responses to the questions and their descriptions of rock recycling. Emphasize that all rock types can be changed into another form.

Answers to
In-Text Questions

1. Igneous rocks are formed when magma cools and hardens.
2. Exposure of igneous rock to excess heat and pressure will cause metamorphism. Erosion can break igneous rocks into sediments that can become sedimentary rock and subsequently be metamorphosed.
3. Sedimentary rocks are formed by compaction and cementation of sediments or by crystallization of solutes.
4. Sediments are eroded from exposed rocks.
5. Exposure of sedimentary rock to excess heat and pressure can turn it into magma, which when cooled and solidified, becomes igneous rock.

A sample description: *Cooling of molten rock* forms *igneous rocks*. *Weathering* and *erosion* change igneous rocks into *sediments*. *Cementing* of these sediments produces *sedimentary rocks*. When these rocks undergo burial, *heat*, and *pressure* due to *plate tectonics*, they *melt*. The melted rock can rise up through cracks and cool, causing an *intrusion*. If the rocks are not completely melted, they can be *uplifted* by huge crustal movements. Once on the surface, the rocks can be weathered and repeat the cycle again.

Exploration 9

To prepare students for this Exploration, involve them in a discussion of the features they will look for in the rock pairs. *(Possible observations: color, general appearance, composition, crystal size, the way the crystals lock together, presence of banding or layering, distortion of layers, and so on)*

Metamorphism is characterized by new minerals, new textures, new structure, or a combination of these. The metamorphic rock is often so changed that it is difficult to identify the original rock.

Obtain as many pairs of rock samples as possible. When the activity is completed, reassemble the class to discuss the distinguishing properties of each rock pair.

The Role of Rocks

Those Plates Again!

Draw a rough sketch on the chalkboard similar to the diagram on page S162. Place the letters (a to e) on your diagram.
(a)—oceanic plate moving under the continental plate
(b)—part of continental plate under great pressure because of collision
(c)—rocks of oceanic plate being heated and melted
(d)—magma rising
(e)—igneous and metamorphic rocks eroding

Follow Up

Assessment

1. Ask students to describe the characteristics they would look for in order to identify a metamorphic rock. *(obvious banding patterns and alternating layers of dark and light minerals)*
2. Imagine a conversation between James Hutton and Charles Lyell. Write a dialogue in which Lyell tells Hutton exactly why he agrees with the theory of intrusions, and why he decided to name a third group of rocks *metamorphic*.

Extension

1. Have students research other recycling processes that occur in nature. Ask them to compare those processes to the rock cycle. *(Possibilities include: the water cycle, the carbon-oxygen cycle, and the nitrogen cycle.)*
2. Certain metamorphic rocks, such as schist and gneiss, can be produced from many different types of rock by metamorphic processes. Have students research the variety of original rocks that can be changed to schist and gneiss, and the conditions that produce the change in each case. *(Shales, impure sandstones, and basalt can metamorphose into schists; conglomerates, shales, and granites can metamorphose into gneiss.)*

Those Plates Again!!

The theory of plate tectonics supports the idea that rocks can be recycled. The collision of plates creates conditions that cause the formation of both igneous and metamorphic rocks. Look at the diagram below. It shows the collision of a continental plate with an oceanic plate. With a partner, find the locations of each of the following occurrences:

- The oceanic plate is moving underneath the continental plate.
- Parts of the continental plate are under great pressure because of the collision. This leads to the formation of metamorphic rocks.
- The rocks of the oceanic plate are being heated to a high temperature and are melting.
- Magma is rising in cracks in the continental plate, forming new igneous rock.
- Igneous and metamorphic rocks erode and then are carried to the bottom of the ocean. This is the first step in forming sedimentary rock.

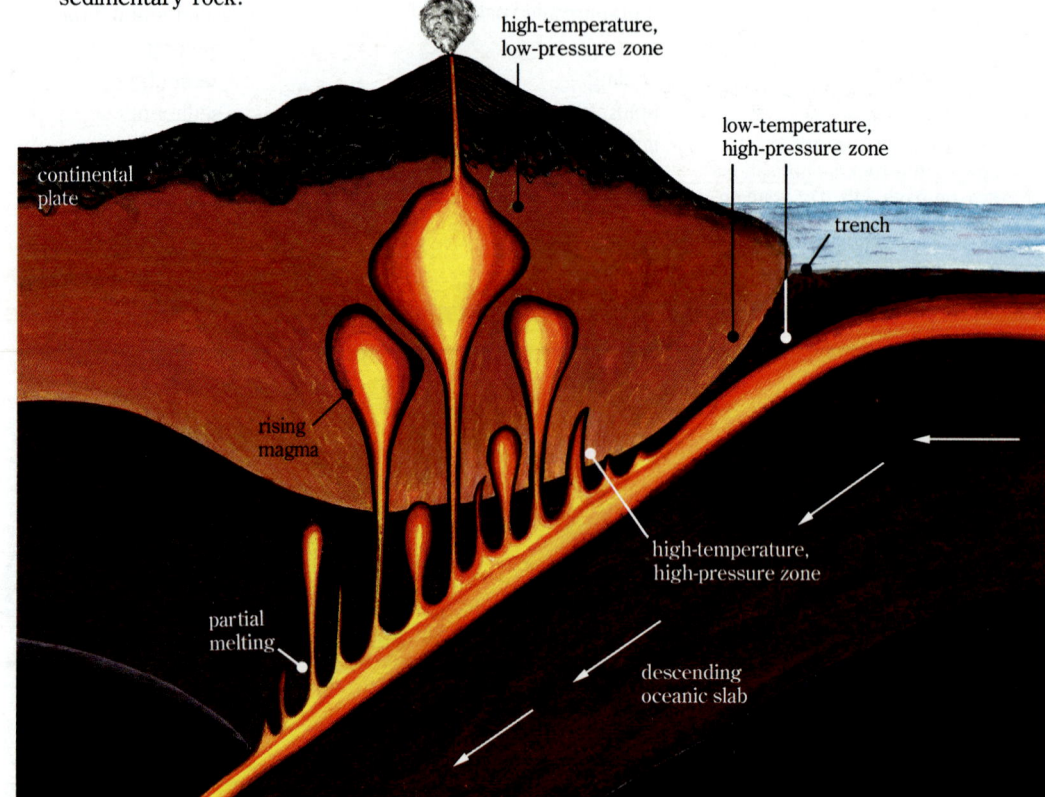

358 The Restless Earth

Challenge Your Thinking

1. Copy this diagram into your Journal. Do not write your answers in the book. For each arrow, fill in the process that occurs as one type of rock is recycled. The first answer has been completed for you.

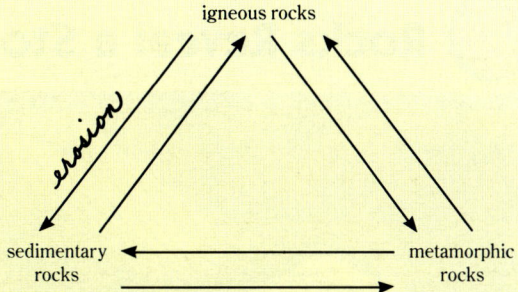

2. Nigel is having trouble classifying some rocks. Using the following information, help Nigel determine whether the rocks are igneous, sedimentary, or metamorphic.

   Rock 1: This rock is grainy and has a lot of different particles. I think I see half a pebble stuck in there!

   Rock 2: This rock is grainy too, but the grains fit together tightly. There are at least four different grains here—two of them are gray, one is white, and one is black. You can see that each color has its own band.

   Rock 3: This rock isn't grainy at all, but it is made of several flat layers.

   Rock 4: I like this rock because it's smooth and glossy. It's only made of one material and it isn't grainy either.

3. Imagine that you are a talk show host. You have invited the following people to appear on your program:

   (a) a geologist who lives in the Rocky Mountains

   (b) Charles Lyell and James Hutton

   (c) a rock collector who is organizing a field trip for junior high school students

   Make a list of questions you would ask each of your guests about geology.

Answers to Challenge Your Thinking

1.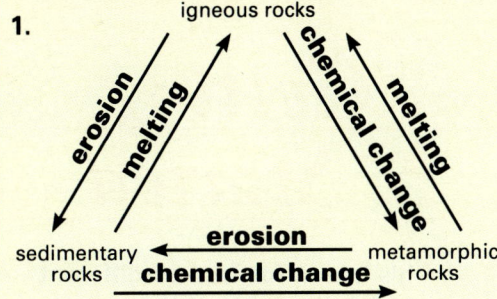

2. Rock 1: sedimentary
   Rock 2: metamorphic
   Rock 3: metamorphic
   Rock 4: igneous

3. Answers will vary. Students' questions should focus on the role of rocks in the lives of the guests on the talk show.

The Role of Rocks 359

Lesson 8

✷ Getting Started

This lesson focuses on different types of fossils and how they are formed. Students assume the role of scientist as they formulate hypotheses to explain dinosaur extinction. The lesson concludes with students interpreting ancient animal tracks.

Main Ideas

1. Fossils can provide information about ancient plants and animals.
2. Fossils can be classified into four different groups.
3. The environment of an organism determines the type of fossilization.
4. Many hypotheses have been proposed to explain the extinction of dinosaurs.

✷ Teaching Strategies

Call on a volunteer to read about fossils on page 360. Then direct students' attention to the illustration. Most students will find the size of the creatures hard to believe. Elicit comments from students about fossils and what they can tell about the history of earth. Encourage them to do some library research on the geological history of their state and to identify any characteristic fossil plants or animals.

FOSSILS—RECORDS OF THE PAST

8 Rocks Reveal a Story

Imagine the scene: dense clumps of trees and huge ferns crowd the many swamps and river banks. Amphibians large and small crawl on the land and swim in the warm waters. The air is alive with the sounds of giant insects, but no birds can be seen. A many-legged arthropod resembling a giant cockroach makes its way across a mudflat.

Can you imagine that such a scene took place in your state? Fossil evidence suggests that plants and animals such as these lived throughout North America. Three hundred million years later, the scene has changed dramatically. You have learned about the forces that brought about some of these tremendous changes. Now you are going to learn how scientists use fossils to reconstruct the events of the past. Fossils are abundant across the United States, and provide a rich source of information about the plants and animals that lived long ago.

What does fossil evidence suggest about your state's history? Can you discover more about it? Consult an encyclopedia or your librarian for information about the prehistoric environment of the United States.

LESSON 8 ORGANIZER

Objectives

By the end of the lesson, students should be able to:
1. Describe how fossils are formed.
2. Identify different fossil types.
3. Hypothesize why dinosaurs became extinct.
4. Observe and interpret fossilized tracks.

Process Skills

observing, analyzing, inferring, formulating hypotheses

New Terms

Fossil—the remains of prehistoric animals and plants preserved in rock.
Paleontology—the study of fossils, ancient life forms, and their evolution.

360 The Restless Earth

A Classroom Collection

Do you live in an area where fossils are found? Perhaps you have seen fossils embedded in the rock. The fossils could be as small as a leaf or as large as a dinosaur. Look carefully at the rocks around your neighborhood. Also, look at the stones used in constructing large buildings—sometimes fossilized leaves or insects can be seen. If possible, make a display of fossils brought in by your class.

How Fossils Are Formed

Fossils are the remains of prehistoric animals and plants preserved in rock. The rocks containing fossils lie undisturbed beneath the surface of the earth for millions of years. Finally, through weathering and erosion, the fossil-bearing rocks are exposed on the surface.

Types of Fossils

All the fossils you have seen were formed in one of four different ways. Therefore, fossils fall into four groups:

(a) the actual bones, shells, or other remains

(b) an imprint or mold of the remains

(c) a cast, made when a mold was filled with sediment that later hardened into stone

(d) some form of track or trail left by an animal

At least one example of each of these groups is pictured here. Examine each picture and identify the group it belongs to. Record your conclusions in your Journal.

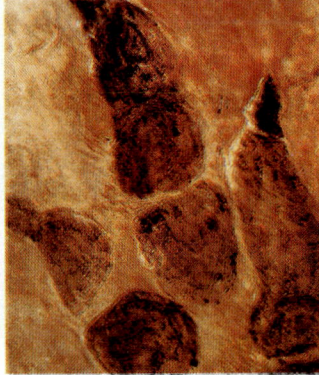  

A Classroom Collection

Encourage students to look for fossils embedded in rocks or stones in their neighborhood. Good places to look for fossils are sedimentary rock outcrops. Local museums or universities may be able to suggest possible nearby fossil-hunting sites.

How Fossils Are Formed

Involve students in a discussion to ascertain their prior knowledge about fossils. Identify and correct any possible misconceptions before proceeding.

Types of Fossils

If possible, provide fossil samples that illustrate the four different types of fossils. Call on a volunteer to read the material about fossil types. Ask other students to describe how each fossil was formed, referring to what was just read. Imprints, molds, and casts can be demonstrated using plaster, mud, silt, clay, or putty.

Students may have difficulty understanding how a mold becomes a cast. Help them to understand that, in nature, filling of the mold by mineral matter deposited by ground water is a very slow process. The identification exercise at the bottom of the page can be used to reinforce student familiarity with the different fossil types.

Answers to

In-Text Questions

(a) The dinosaur skeleton is an example of actual remains.
(b) The remains of the plant are imprinted.
(c) A cast of the remains of the fish was made.
(d) The footprint is a track left by an animal.

Materials

Types of Fossils: Journal
Who Gets Fossilized? Journal

You may wish to collect information on fossils in your area and other materials needed for the Extension activities.

Teacher's Resource Binder

Activity Worksheet 6–9
Resource Transparency 6–8

Science Discovery Videodisc

Disc 2, Image and Activity Bank, 6–8

Time Required

two class periods

Who Gets Fossilized?

Call on a volunteer to read the top half of page 362. Then divide the class into small discussion groups to do the classification exercise. Students need to determine the conditions that must exist before organisms become fossils. Students should conclude that the organism's environment is the key to who gets fossilized.

Show the class several objects such as shells, twigs, bones, and leaves. Ask students to consider good and poor environments for fossil development. Suggest that they think about the effects of natural agents that tend to destroy soft and hard body parts.

After students have grasped the importance of the environment in fossilization, have them do the exercise.

Answers to

In-Text Questions

(a) Poor—The bacteria present in the swamp will break down the wood until it is decomposed.

(b) Poor—The tracks in the sand are easily washed or blown away.

(c) Fair—Shells can be buried and preserved.

(d) Good—The rapid deposition of sediments will bury the horse's remains.

(e) Poor—Its lack of skeleton or hard parts and its habitat in a relatively high-energy near-shore environment make it an unlikely prospect for fossilization.

(f) Poor to Fair—It depends on the speed of the moving water in the stream, the location of the clam in the stream, and the type of sediments present.

(g) Fair—Fine sediments may preserve the tracks, depending on the location of the mud.

(h) Good—Tar acts as a preservative.

(i) Poor—Soil bacteria will decompose the reindeer.

(j) Poor—There is little opportunity for a protective covering to preserve the remains of the tree.

Who Gets Fossilized?

Examples of life are all around us, yet the chance of a single organism being preserved in the geologic record is very unlikely. Why is this so?

There are many ways in which the remains of organisms can be destroyed. If they are left lying on the surface, they can be attacked by scavengers or bacteria. Bones or shells can be crushed by overlying sediments, or they can be eroded along a seashore or in a stream bed. Why, then, do some survive?

If the organism has hard body parts, or if it is covered by some protective material like soft mud or sand shortly after its death, the remains have a better chance of being preserved.

Using the list below, sort the organisms into these three categories: *Good chance for fossilization*, *Fair chance for fossilization*, and *Poor chance for fossilization*. Record your work in your Journal.

(a) a tree growing in a swamp
(b) deer tracks in the sand
(c) a seashell on a beach
(d) a horse drowned while crossing a river's flood plain
(e) a jellyfish in the ocean
(f) a freshwater clam in a stream
(g) dinosaur tracks in mud
(h) a small animal caught in a tar pit
(i) a reindeer killed by disease on an island
(j) an evergreen growing on a rocky ledge

Something to Think About

In which rock group would you expect to find the greatest number of fossils? the fewest fossils? Give an explanation for your answers.

Something to Talk About

Imagine that you have been transported several million years into the future. Discuss with your classmates what you might find in the rocks that exist in your area. What are some human-made objects that might survive? Which ones might not? What could a scientist of the future learn about our civilization from these fossils?

Answer to

Something to Think About

The greatest number of fossils are found in sedimentary rocks. The fewest fossils are found in igneous and metamorphic rocks. This is because of the extreme heat and pressure that cause these rocks to form.

Something to Talk About

Remind students that the geologic processes occurring today are the ones that are shaping the future landscape and will affect the preservation of any human-made objects. Direct groups to think of objects that might survive, and those that might not. They should be prepared to defend their choices.

362 The Restless Earth

Interpreting the Fossil Record

Perhaps the most fascinating fossil remains are those of the dinosaurs. You know that the dinosaurs disappeared from the earth. This occurrence marked the beginning of our own geological era—the Cenozoic Era.

What Killed the Dinosaurs?

There are many hypotheses that try to explain the sudden death of the dinosaurs 65 million years ago. Can you think of any yourself? Science relies on hypotheses that can be tested. A hypothesis will begin to be accepted only after a great deal of data has been collected and examined and the hypothesis still seems correct. Many hypotheses have been suggested to explain the extinction of the dinosaurs. Some of them are listed below.

(a) Radiation from an exploding star burned them.
(b) Flowering plants developed lethal poisons that the dinosaurs ate.
(c) The climate became too hot or too cold for reproduction to be successful.
(d) A huge asteroid collided with the earth, creating a vast dust cloud that blocked the sun. This lessened photosynthesis and drastically lowered the temperature.
(e) An epidemic of some sort killed them.

How many of these hypothesis could be tested? How likely do you think they are?

What Killed the Dinosaurs?

Before students read the hypotheses about dinosaur extinction in the middle of the page, ask them to offer their own extinction hypotheses. Record each suggestion on the chalkboard without evaluation. When student ideas are exhausted, refer them to those in the text.

Ask if all the hypotheses on page 363 are equally acceptable. Emphasize that good hypotheses can be tested and will often lead to further questions. Although students will not have the background to decide conclusively whether the hypotheses are testable, review the process of hypothesis formulation and testing.

The only way to test any of the dinosaur-extinction hypotheses is to look for evidence in the fossil record (or in rock layers) that is consistent with the hypothesis.

(a) Radiation-induced changes in rocks of that age would support this hypothesis. However, since such changes have not been found, there is no way to establish that heavy doses of radiation occurred.

(b) This hypothesis is hard to test since plant materials (and any poisons) from that time have decomposed. One of the strongest arguments against it is that flowering plants existed millions of years before the dinosaurs became extinct.

(c) Evidence of climatic changes could be sought in the fossil record.

(d) Evidence of a large plant "die off" could be sought in the fossil record. Remains of the dust cloud could be found in sedimentary rock layers. Evidence for this theory exists in the form of an iridium layer in the rock record at the Cretaceous-Tertiary boundary. However, critics argue that dinosaurs were dying off before the asteroid struck, and evidence from volcanic eruptions suggests that the dust clouds created by the impact would not have long-term effects.

(e) This hypothesis is difficult to test. Although dinosaur remains could be examined for signs of death by disease, many diseases would not be detectable in the fossil remains.

Fossils—Records of the Past

Answers to
You Be the Scientist

Have students work in small groups to study the footprint illustrations and answer the questions.
1. Two animals were involved.
2. They were moving to the left in the illustration.
3. One animal was larger than the other.
4. These animals could have been dinosaurs that walked on two legs.

The story the footprints tell:
- The paths of a large and small animal intersected.
- There was a struggle.
- The large one walked away.

Follow Up

Assessment

1. Bring in a collection or photographs of fossils. Have students study the items and classify them according to the four types of fossils.
2. Present students with the following scenario: Imagine that you have just returned from a fossil field trip. Describe some of the things that you found and explain why you think they became fossils.

Extension

1. Have students make models of fossils (molds, casts, or impressions) from everyday objects.
2. Have students use one of the following titles to write a report: Coal Age Forests, Dinosaurs of My State, Petrified Wood, La Brea Tar Pits, Frozen Fossils, Insects in Amber, or A Guide to Fossil Fuel.

You Be the Scientist

Paleontology is the study of fossils, ancient life forms, and their evolution. Suppose you are a paleontologist whose task is to interpret the fossilized footprints in the drawings shown. What can you discover about the creatures that made these tracks?

Use these questions to get you started.

1. How many animals were involved?
2. In which direction did each animal move?
3. What can you tell about the size of each animal?
4. How many legs did each animal have?

Now, piece together these drawings so that they follow a logical sequence of events. What story do these footprints tell?

364 The Restless Earth

Telling Time With Rocks

The Rock Record

You know that geologists are interested in events that took place long before humans were around to record them. You also know that the rocks of the earth's crust contain evidence of these events. In this lesson, you will see how this evidence has been organized into a geological history of the earth. The first step is to learn what is meant by *relative time*.

A Relative Time Scale

In your Journal, place the events in this list in the order that they occurred. Put the event that occurred the longest time ago at the top of the list.

(a) You started kindergarten.
(b) Columbus discovered North America.
(c) Elvis hit the top of the charts.
(d) The first humans landed on the moon.
(e) Marie Curie won the Nobel Prize for her discovery of the elements radium and polonium.
(f) The Berlin Wall was torn down.
(g) Magellan sailed around the world.
(h) Rap music became popular.
(i) You were born.
(j) Today.

Did you have trouble deciding the order of events? Is the time span between all the events the same? Did most of these events occur recently or before you were born?

When you list events in the order in which they occurred, you make a relative time scale. You do not need the exact dates; you simply need to know what happened first. This gives you a scale based only on the sequence of events. Such a scale does not tell you how much time is involved or exactly when an event occurred. However, it does show whether one given event happened before or after another.

LESSON  ORGANIZER

Objectives

By the end of the lesson, students should be able to:
1. Explain the concept of relative time.
2. Describe the historical development of the geologic time scale.
3. Name the specific divisions of the geologic time scale.
4. Identify when specific events occurred based on the geologic time scale.
5. Describe how dates in geologic time have been established.

Process Skills

sequencing, classifying, recognizing relationships, interpreting data

(Organizer continues on next page)

LESSON 9

★ Getting Started

This lesson focuses on the concepts of relative and absolute time. The geologic time scale, with its eras, periods, and epochs, is also examined.

Main Ideas

1. A relative time scale is a "before and after" scale for a sequence of events. It does not provide information about the amount of time involved.
2. The events of earth's history have been placed in order on the geologic time scale.
3. The subdivisions of the geologic time scale are based on recognizable changes, such as changes in life forms and episodes of mountain-building.

★ Teaching Strategies

The Rock Record

Call on a volunteer to read aloud the introductory material on page 365. Ask students to explain, in their own words, what the rock record means.

A Relative Time Scale

Working individually or in pairs, have students write the events (a to j) in chronological order. *The correct sequence of events is: b, g, e, c, d, i, h, a, f, j. (Depending on students' age, events h and a could be interchangeable.)*

Help students understand that a relative time scale gives only one kind of information: whether one event occurs before or after another. An exact history requires specific dates.

Fossils—Records of the Past   365

The Geologic Time Scale

Call on a volunteer to read the introductory material on page 366. Involve students in a discussion that emphasizes the analogy between the sequencing activity on page 365 and the development of the geologic time scale by scientists.

Subdividing the Geologic Time Scale

This reading is designed to help students understand how events are classified under certain headings. Call on a volunteer to read the material. Elicit ideas from the class as to how the different events presented on page 365 might be grouped. Have students add more events to each group, and then subdivide them again. Dates are still needed before an accurate picture can be presented. *(Possible eras include: Era of Exploration, Era of Electronics, and Personal Era. Students should group events by similarities or common factors.)*

Explain to students how scientists subdivided the information they had on rock formations and layers into sequence. The specific names of eras, periods, and epochs are not meant to be memorized. The information is presented so that students can use the time scale in a later exercise. Emphasize that the dividing line between subdivisions is not a sharp one, but rather a transition zone.

The Geologic Time Scale

Just as you ordered the historical events on page 365, geologists and paleontologists have interpreted the sequence of events of the story told in sedimentary rock. In doing this, they developed the **geologic time scale.** On this scale, rocks are placed in order from oldest to youngest. The geological and biological events recorded in these rocks are also placed in order.

Subdividing the Geologic Time Scale

Like many ideas in geology, the geologic time scale started out simply. Then, as more data became available, the scale became more complex. First, rock formations and layers were placed in order. Then geologists faced the task of subdividing this time scale. To see how they approached this task, look back at the relative time scale you made earlier. Classify the events into various groups, calling each group an "era." For example, you might consider Columbus' discovery of North America and Magellan's voyage as "the era of exploration." Do some eras contain only one event? If so, think of at least two more events to add to each era. Now, subdivide the eras into smaller groups, giving each its own name.

On what basis did you make your subdivisions? Geologists subdivide their time scales based on the period of formation of certain rocks. They call each subdivision an **era.** There are four eras. Here are their names, starting with the oldest era.

Precambrian: the era of only primitive life forms on earth
Paleozoic: the era of the earliest complex life forms on earth
Mesozoic: the era of the "middle" life forms on earth
Cenozoic: the era of the most recent life forms on earth

The four eras are then subdivided into **periods.** For instance, the Mesozoic Era consists of three periods: the Triassic, the Jurassic, and the Cretaceous. Periods themselves are further divided into **epochs.**

The dividing line between any two time units—between one era and the next, or one period and the next, for example—is based on a recognizable change. The change could be anything from a change in life forms to a period of mountain-building. For instance, the extinction of the dinosaurs separates the Mesozoic Era from the Cenozoic Era. The Paleozoic Era is considered to have ended with the uplift of the Appalachian Mountains. It is important to keep in mind that the dividing line is never sharp or sudden. Rather, it spans the time that it took for a major change to occur.

(Organizer continued)

New Terms

Geologic time scale—a scale on which rocks are placed in order from oldest to youngest.
Era—a subdivision in the geologic time scale. There are four eras: Precambrian, Paleozoic, Mesozoic, and Cenozoic.
Periods—subdivisions of an era.
Epochs—subdivisions of a period.

Materials

A Relative Time Scale: Journal

Teacher's Resource Binder

Activity Worksheet 6–10
Graphic Organizer Transparencies 6–9 and 6–10

Science Discovery Videodisc

Disc 2, Image and Activity Bank, 6–9

Time Required

two class periods

Making Sense of What You Read

1. How is the geologic time scale subdivided?
2. What is the purpose of subdividing the large units of time?
3. How do geologists decide where one division ends and a new one begins?
4. How sharp is the dividing line between subdivisions?
5. What is still missing from the geologic time scale? (Hint: Check the next paragraph.)

Geologic Time Scale

Era	Period	Epoch
Cenozoic	Quaternary	Holocene (recent)
		Pleistocene
	Tertiary	Pliocene
		Miocene
		Oligocene
		Eocene
		Paleocene
Mesozoic	Cretaceous	
	Jurassic	
	Triassic	
Paleozoic	Permian	
	Pennsylvanian	
	Mississippian	
	Devonian	
	Silurian	
	Ordovician	
	Cambrian	
Precambrian		

Clocks in Rocks

The geologic time scale merely puts the events of the earth's history in sequence. It does not reveal anything about the dates the events occurred. The age of the earth itself was an unanswered question for the early geologists. For hundreds of years, scientists debated the age of the earth, but it was not until about 1850 that serious study of these matters began.

Ernest Rutherford's laboratory

Various techniques for dating the earth were proposed. One of the simplest was proposed at the the end of the nineteenth century. It related to the accumulation of salt in the ocean. A scientist named John Joly measured how much salt was being added to the ocean each year. By estimating the total amount of salt in the ocean, he determined that the earth was about 100 million years old. Unfortunately, this number was not accurate.

The discovery of radiation finally helped answer the question. In 1906, the famous physicist Ernest Rutherford first discovered the age of uranium in his laboratory at McGill University in Montreal. Geologists now had a method for putting dates on the geologic time scale, and even for discovering the age of the earth. The result is pictured on the next page.

Answers to Making Sense of What You Read

Divide the class into pairs or small discussion groups to study the geologic time scale and answer the questions.

1. The geologic time scale is subdivided into eras, periods, and epochs.
2. Large units of time are subdivided in order to provide more specific information. Such units are less cumbersome and easier to use when communicating information among scientists.
3. The division is based on a recognizable change. For example, at the end of each period, in Europe at least, marine sedimentation stopped because of an uplift in the land. Also, the earth was inhabited by a different set of organisms in each period, as revealed by the fossil content.
4. The dividing line is not sharp, but rather a zone of transition.
5. Actual dates are missing.

Clocks in Rocks

The concept of radioactive dating is complex and has been presented here only to explain how dates were finally assigned on the geologic time scale. The radioactive clock is read by measuring the daughter elements. The daughter elements are created from the original, broken-down (decayed) parent elements. Uranium is a parent element, and lead is a daughter element. If the amount of the daughter element can be identified and measured, and if the rate of decay is known, then the point at which there was only a parent and not a daughter element can be calculated.

Have students research to investigate how scientists have arrived at the age of the earth. Students may be interested to know that meteorites and moon rocks have allowed modern scientists to calculate the age of the earth as 4.5 billion years old.

Fossils—Records of the Past 367

Answers to
Investigating the Geologic Time Scale

1. The longest era was the Precambrian. The Cenozoic is the shortest.
2. The Precambrian provides the least fossil evidence; the Cenozoic the most. There was little, if any, life in the Precambrian Era. Simple marine plants and animals may have existed.
3. Animals and plants first appeared on land during the Silurian Period about 400 million years ago.
4. (a) Dinosaurs were on earth for about 135 million years.
   (b) They appeared on land during the early Triassic Period and became extinct at the end of the Cretaceous Period, 65 million years ago.
5. The first forests began growing in the late Mississippian Period and continued during the Pennsylvanian Period, 345 million years ago.
6. Due to better dating techniques, more information may become available on the Precambrian Era. This would lead to further subdivisions as well as more refined dating of the other eras. New fossil finds may also correct current dates.

Investigating the Geologic Time Scale

Relative and absolute time and history of the earth

Use this diagram that shows the eras, periods, and epochs, and the times when they occurred, to answer the following questions:

1. Which was the longest era? the shortest era?
2. Which era provides the least fossil evidence? the most fossil evidence?
3. When did plants and animals first appear on land?
4. (a) How long were the dinosaurs on earth?
   (b) When did they become extinct?
5. In which period did the first forests grow? How long ago was it?
6. Suggest some changes your grandchildren might see when they study the geologic time scale.

Follow Up

Assessment

1. Have students make a relative time scale using events from their own lives. Then ask them to convert it to an absolute time scale by adding specific dates.
2. Humans have certain periods in their lives such as infancy, childhood, adolescence, and adulthood. Have students compare these to the periods in the geologic time scale.

Extension

1. Have students create an alternate illustration of the geologic time scale by using adding machine tape or another continuous source of paper. A convenient scale is 1 cm equals 1 million years.
2. Have students plan a field trip to a fossil-collecting area. Local natural history museums may have specific information on nearby areas. Check with the museum personnel about any regulations regarding fossil collecting.

368 The Restless Earth

Challenge Your Thinking

1. Test your skills as a paleontologist! Give one or more possible explanations for each of the following statements.
   (a) Some sedimentary rocks contain a large number of marine fossils.
   (b) Rocks of the Precambrian Era have few, if any, fossils.
   (c) Fossils of plants are less common than animal fossils.
   (d) The fossils of similar species are found in Africa, China, Paraguay, and the United States.
   (e) Fossils are seldom found in igneous rocks.

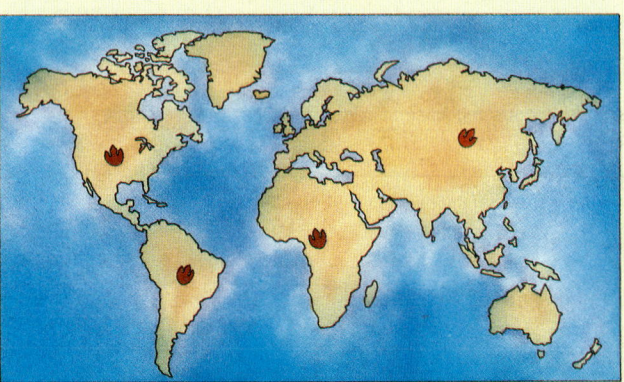

2. Fossil evidence leads scientists to believe that, one hundred million years ago, the western plains were covered by a warm, shallow sea. Check the geologic time table to discover what animals existed then. Imagine that you were one of these animals and could express yourself. Write a description of what life and your surroundings were like.

3. Kristi and Julianne went hiking in the Rocky Mountains. While gathering rocks for their classroom's geology display, they came across a fossil of a small fish. How can you explain that?

Answers to Challenge Your Thinking

1. (a) Marine fossils tend to accumulate on shallow sea bottoms where they are covered by fine marine sediments.
   (b) Probably few living organisms existed at that time. Small and simple organisms fossilize poorly. During the Precambrian Era there were several episodes of igneous intrusion, metamorphism, and mountain-building, so any existing ancient life forms would have been destroyed.
   (c) Plants have fewer hard parts than animals.
   (d) It is theorized that these areas were once joined as Pangea.
   (e) Molten igneous material is very hot and normally destroys organic matter on contact.

2. One hundred million years ago, the earth was in the Cretaceous period. Dinosaurs and other reptiles were prominent. Students should conclude that dinosaurs would thrive along a warm sea coast.

3. Millions of years ago, a stream, lake, or ocean might have been present in the region where the Rocky Mountains are now. A small fish could have been fossilized and preserved at the bottom of the lake or streambed. When mountain building occurred, the fossilized fish was uplifted along with the other rocks as part of the Rocky Mountains.

Fossils—Records of the Past

UNIT 6
Making Connections

Summary for
The Big Ideas

Student responses will vary. The following is an ideal summary.

Mountains are formed when continental plates collide and the earth buckles upward (uplift), or from the action of volcanoes.

Earthquakes are caused by the release of energy from the friction lock between blocks of crust slipping in opposite directions. The energy is released as shock waves that originate at the focus and radiate outward.

The elastic rebound theory explains how plates of rock locked by friction return to their original form after being strained by the pressure of surrounding rock layers. As the crustal plates spring back into their original shape, the stored energy is released as intense vibrations called an earthquake.

Movement of the earth's crust causes volcanoes to occur. When plates separate, volcanic rock fills the gap. When plates collide, some of the rock involved melts into magma that rises to the surface, causing a volcano.

The theory of plate tectonics describes the locations of sliding, convergent, and divergent boundaries between earth's floating crustal plates. At these boundaries, earthquakes and volcanoes occur.

Igneous rocks are made either of one material and are glassy or porous, or they are made of interlocking grains that are randomly arranged. Igneous rocks are formed from molten rock. Sedimentary rocks are made of several kinds of rock grains that are cemented or lithified together. These rocks form by the erosion and weathering of other rocks. Metamorphic rocks may be made of one material and arranged in sheets, or of several different materials with grains that are closely interlocked and/or arranged in bands. Metamorphic rocks form when heat and pressure from the environment chemically change an existing rock.

Unit 6 Making Connections

The Big Ideas

In your Journal, write a summary of this unit, using the following questions as a guide.

- How are mountains formed?
- What causes earthquakes?
- What is the elastic rebound theory, and how does it relate to earthquakes?
- What causes a volcano to erupt?
- What does the plate tectonics theory tell us about the location of earthquakes and volcanoes?
- How do you distinguish among igneous, sedimentary, and metamorphic rock? How is each formed?
- How are rocks recycled?
- How are fossils formed? Where are fossils the easiest to find? Why?
- What is a relative time scale?
- How is the geologic time scale divided?

Checking Your Understanding

1. Fill in the blanks for each of the following sentences.

   (a) When an earthquake occurs, a ___?___ measures the magnitude of force. Strong quakes often happen at ___?___ ___?___ where plates meet.

   (b) The change in location of continents can be explained by the theory of ___?___ ___?___. In this theory ___?___ boundaries are points of collision and ___?___ boundaries are points of separation.

   (c) In a volcano, ___?___ is released either through small cracks on the crust or a large vent. When it emerges, it is called ___?___. A ___?___ volcano no longer erupts with this material.

Rocks are recycled through erosion and weathering into sedimentary rock, through melting into igneous rocks, and through chemical changes into metamorphic rock.

Fossils are formed by preservation of actual bones or shells, imprinting or molding of the remains, forming a cast, or preservation of a trail left by an animal. Fossils are easiest to find in sedimentary rock because the heat and chemical changes that form other rock types often destroy delicate remains.

A relative time scale is a list of events in the order that the events happened. Actual dates are not included.

The geologic time scale is divided into four major eras: Precambrian, Paleozoic, Mesozoic, and Cenozoic. The eras are further divided into periods, and periods are divided into epochs.

370   The Restless Earth

2. How good are you at making connections? How are the words in each pair related to each other?

   (a) plate tectonics and earthquakes
   (b) the Cenozoic era and the extinction of dinosaurs
   (c) the formation of igneous rock and volcanoes
   (d) shale and slate

3. Write a story that traces the formation of one of the following items.
   - a sedimentary rock
   - a fossil of a prehistoric turtle
   - the Himalayan mountains in India
   - a volcanic island

4. Draw a concept map that relates the following words: *basalt, pumice, metamorphic, shale, rocks, igneous, slate, limestone, sedimentary,* and *marble.*

Reading *Plus*

Now that you've been introduced to the movements of the earth's crust and the rocks of which that crust is composed, let's take a closer look at the make up of rocks. On pages S89–S110 of the *SourceBook,* you will find new and interesting information about rocks and the stories they reveal. Read to discover more about our restless earth.

Updating Your Journal

Reread the paragraphs you wrote for this unit's "For Your Journal" questions. Then rewrite them to reflect what you have learned in studying this unit.

About Updating Your Journal

1. Mountains are formed by the buckling of the crust that takes place at the boundaries of colliding tectonic plates. This is called uplift. Volcanoes, which also occur at plate boundaries, can form mountains.
2. Shifting tectonic plates, mountain-building, erosion, and deposition all cause the earth to change. *(These processes reinforce the **Patterns of Change** theme.)*
3. Rocks contain clues about when they were formed, as well as the sort of conditions under which they formed. Sedimentary rocks may contain fossils, which reveal the type of living things that existed at the time the rocks were formed.
4. The Grand Canyon is a deep gorge through a thick section of layered sedimentary rock. At the bottom of the gorge is the Colorado River.
5. The Grand Canyon formed as the Colorado River eroded downward through layer after layer of rock.

Answers to
Checking Your Understanding

1. **(a)** seismograph, sliding boundaries
   (b) plate tectonics, convergent, divergent
   (c) magma, lava, dormant
2. Answers will vary. Sample answers include:
   (a) The movement of crustal plates is explained by the theory of plate tectonics. When two plates meet and slide past each other, an earthquake occurs.
   (b) The extinction of the dinosaurs separates the Mesozoic era from the Cenozoic era, which is the era of the most recent life forms on earth.
   (c) Molten rock that cools forms igneous rock. Molten rock reaches the earth's surface through the eruption of volcanoes; after erupting, the molten rock cools to form igneous rock like basalt or pumice.
   (d) Shale is a sedimentary rock. Heat and pressure from surrounding rock can cause chemical changes in shale that change it into slate, a metamorphic rock.
3. Answers will vary. Discussion of the formation of sedimentary rock should include erosion, weathering, transport by water or wind, and lithification. Discussion of fossilization of a prehistoric turtle should include finding the preserved remains of the shell or the formation of a cast. Students should address the turtle's place on the geologic time scale. Discussion of the Himalayan mountains should include the theory of plate tectonics, uplift, and mountain-building. Discussion of a volcanic island should include the theory of plate tectonics, magma, and a quiet or explosive eruption.
4. See concept map on page S162.

Making Connections 371

Science in Action

✦ Background

Geologists perform a variety of pure and applied research tasks in the laboratory and in the field. Geology laboratory work includes radiometric dating, microscopy of thin rock sections, and geochemical analysis. Field work includes making observations and collecting data at rock outcrops all over the world.

A Bachelor of Science degree in geology is the basic qualification for work as a geologist. For supervisory positions in industry, or for university and college teaching positions, a master's degree and Ph.D. are required.

Science in Action

Spotlight on Geology

Rebecca Jamieson has a Bachelor's degree, a master's degree and a Ph.D. in geology. She enjoys her work as a professor and research scientist at a university.

Q: How did you become interested in geology?

Rebecca: My interest began with rock collecting trips with my parents and family friends. I also became quite interested in the outdoors, particularly camping. But even though I always had an interest in rocks and fossils, I certainly didn't think then that I would become a geologist. That came later after I enrolled in a geology course at the university.

Q: What do you find most exciting about your work?

Rebecca: Aside from teaching, I find that field work is by far the most stimulating work that I do as a geologist. You go out in the countryside, look at rocks that you haven't seen before, and try to understand them. It's a real challenge. You can't ignore anything.

Q: What are field trips like?

Rebecca: Well, for example, I was recently on a field expedition to the Bay of Islands area in Canada. One of the things that geologists always want to do is to find out how old rocks are. But the kinds of rocks found there don't usually have fossils in them, so we had to use dating techniques that can only be done in the laboratory. We collected big buckets of old fragments of oceanic crust to take back with us to the lab for dating.

Next we took a float plane to a nearby lake. Then we had to climb a mountain, where we got more samples and carried them back down with us and out through the woods. Finally we took a boat to look at more rocks

✦ Using the Interview

Have students silently read the interview on pages 372 and 373. Then involve them in a discussion of what they have read. You may wish to evaluate their understanding of the interview by asking questions similar to the following:
- Based on Dr. Jamieson's descriptions, what is it like to be a geologist? *(You enjoy a physically active, outdoor lifestyle, travel to remote locations, and spend periods of concentrated study at a laboratory.)*
- What was the research question Dr. Jamieson was investigating in the Bay of Islands area? *(Students can infer from the text that Dr. Jamieson and her team were investigating the age of the islands and how they were formed.)*
- What data did they collect to help them answer this research question? *(rock samples from oceanic crust, a mountaintop, and the bay)*

- Dr. Jamieson made the statement "You can't ignore anything." What do you think she meant by that? *(Answers may vary, but possible responses might include that a geologist in the field needs to make a wide variety of observations, pay close attention to details, and think of the "big picture" at any given study site.)*
- Dr. Jamieson also discussed the high school courses that a future geologist would find most useful. Why do you think she made these recommendations? *(College coursework and future work in geology requires a strong math background and an understanding of chemistry, physics, and biology. History is useful in relating human factors to aspects of the landscape, and English courses help develop written communications skills.)*

at the Bay of Islands. But we got caught in a terrible thunderstorm. It was so dark you couldn't even see the rocks. In the hour going back by boat we were soaked. It was terrible, but that doesn't happen too often.

You can tell that most geologists travel a lot. The rocks won't come to you. You have to go to them. So I get to visit many different places.

Q: What subjects should high school students take if they're interested in studying geology at a university?

Rebecca: Geology is in many ways a science that brings together the other sciences. For this reason, it's really important to have a good background in math, chemistry, physics, and biology. History and English are also important. I encourage students to take all these courses.

Some Project Ideas

Rock and mineral collections can be the basis of many worthwhile research projects. Here are some of the possibilities:

1. Collect samples from rock outcroppings near your home. Look along the sides of highways, roads, and railroad tracks. Consult a geological map, a reference book on the geology of your area, or a geologist to find out (a) whether the rocks you collect are sedimentary, igneous, or metamorphic; (b) what minerals are present in them, and (c) how old they are. Compare this information to the geological history of your area. Is there more than one major rock type near your home? Which type is oldest? Which is youngest?

2. Collect stones, pebbles, and fragments of boulders near your home. With the help of a reference guide to rock types, identify the different types. How did these rocks get to the location where you found them? Where did they probably originate? You may need the help of a geologist to answer such questions.

3. Geology collections make excellent displays. In a display you can summarize what you learned by making use of maps, drawings, and photos, as well as samples.

For Assistance

Reference books about geology, paleontology, and geophysics are good resources for information. In addition, geology teachers at high schools and universities can provide information and advice.

✦ Using the Project Ideas

1. Geological maps of your local area can be obtained from the United States Geological Survey. Excellent guidebooks to easily accessible rock outcrops ("roadside geology") are available for many areas. If students can locate a rock outcrop with visible layers, the lowest layers are usually the oldest.

2. Students may locate many stones near their homes that were not transported by natural processes. Such stones originate in quarries or gravel mines and are brought in by trucks or trains for road building, construction, and landscaping uses. Students can find local rocks at nearby outcrops. Rocks transported by natural processes can be found near streambeds and bodies of water. For safety reasons, recommend that students explore such areas with a responsible adult and obtain necessary permission in advance.

✦ Going Further

Ask students to find answers to the following questions:
- What kinds of rocks are found in your area? How would a geologist study these rocks?
- What do geophysicists study? geochemists?
- What industries make use of geological information?
- Who should decide whether a newly discovered mineral deposit should be mined? What should the role of government be?

Science and the Arts

◆ Background

Gems are rare or unusual varieties of minerals that are valued for their color, luster, transparency, and hardness. These properties make them desirable for jewelry.

Diamonds are the hardest natural substance on earth. The only thing that will cut a diamond is another diamond. Diamonds are a pure form of the element carbon. They have been crystallized in such a way that the carbon atoms form a cubic structure. This cubic structure gives the diamonds their extreme hardness.

Kimberlite, the source of diamonds, is an intrusive igneous rock that forms when iron-magnesium-silicate magma is injected into overlying rocks. The injected material hardens in the shape of a cylinder or pipe. Several hundred kimberlite pipes are known to exist but only a few contain diamonds.

Rubies and sapphires are forms of the mineral corundum (crystalline aluminum oxide). Pure aluminum oxide is colorless, but the presence of small amounts of trace minerals creates a variety of colors.

Emeralds are a form of crystalline beryl aluminum silicate, with trace amounts of chromium or vanadium. Beryl aluminum silicate forms gem-quality stones of other colors, also. One example is the blue-green gem known as aquamarine.

Science and the Arts

Jewelry

The land masses of the earth are constantly on the move, producing tremendous heat and pressure. Under these conditions certain precious stones may be formed—like diamonds, emeralds, and rubies. Volcanoes, for example, may contain a rock known as *kimberlite,* which often contains diamonds.

Precious stones as they are found in nature bear little resemblance to the beautiful finished products of the gem-cutter's art. This art requires patience, skill, and—at times—nerves of steel!

The Cullinan Diamond

The largest diamond ever discovered was found in Pretoria, South Africa, by a workman in a mine. He was attracted by something reflecting the sun, which he first thought was a large chunk of glass. It turned out to be a diamond weighing 3106 carats—about half a kilogram—and measuring 5 × 6 × 10 cm. The diamond was named the Cullinan Diamond after the founder of the mine. It was bought by the Transvaal government for $750,000 and presented to King Edward VII of England in 1907 for his sixty-sixth birthday.

Cutting the World's Largest Diamond

The Cullinan Diamond was sent to Joseph Asscher, a famous diamond-cutter in Amsterdam. The task of cutting it was truly nerve-wracking. With larger diamonds, the first cut is often made by cleaving the diamond along its grain. The difficulty with this is that the grain is invisible. Asscher studied the diamond for months and, finally, decided on his line cleavage. He drew a V-shaped groove along that line.

The tension was almost unbearable as he lifted his mallet to strike the metal rule he held along the groove. If his line of cleavage was not correct, the world's largest diamond could shatter completely. He brought the hammer down and, strangely, the metal rule broke in two! The diamond was still intact, but Asscher's nerves were shattered. He went to a hospital to recover.

◆ Promoting Observational Skills

Have students study the gemstones grouped together in the photo on page 374. Then ask:
- What are the similarities and differences between cut and uncut stones?
- How does cutting a gemstone enhance its appearance?

◆ Discussion

When students have finished reading the feature, involve them in a discussion of what they have read.

1. What is the story of the Cullinan Diamond? *(Students should retell the story presented in the feature, highlighting the most important details.)*
2. What is involved in the process of cutting a diamond? *(Students should mention the need to study the diamond, the task of cleaving the diamond along its grain, and the eventual cutting of the large diamond into smaller ones.)*
3. Where is the Cullinan Diamond mounted and displayed? *(It is mounted in the imperial scepter of the British government. It is on display in the Tower of London.)*
4. What are the goals of diamond cutting? *(Bringing out the full beauty and value of the diamond, and maximizing the brilliance and "fire" of the diamond.)*

When he finally had the strength and courage to try again, Asscher brought his personal physician with him. This time, his mallet struck cleanly, and the diamond cleaved exactly along the line he had chosen. Asscher himself did not learn of his success until some time afterward, for he had fainted at the moment he struck the blow!

The Cullinan Diamond was eventually cut into nine major gems and ninety-six smaller ones, plus a number of fragments. The largest gem, the Cullinan I or Star of Africa, is mounted in the imperial scepter of the British government. It is on display in the Tower of London.

The world's largest diamond

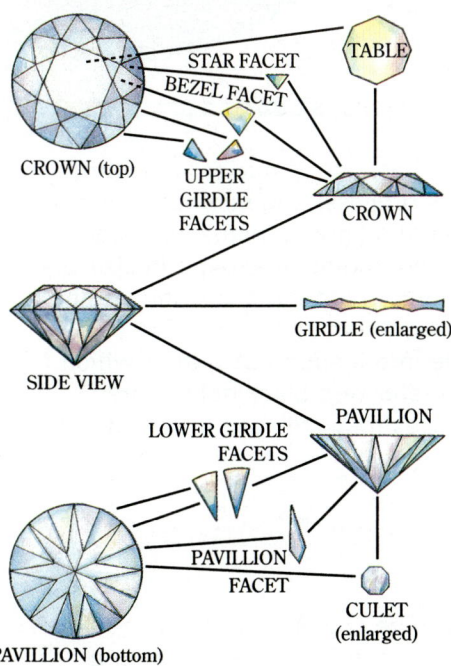

The Diamond Cutter's Art

A diamond is judged by the four C's: color, cut, clarity, and carat weight. Perfect cutting can bring out the full beauty and value of a diamond. The traditional cuts are designed to maximize the brilliance and "fire" of the diamond. One of the most popular cuts is the *brilliant cut* used in many rings. In this cut there are 58 facets, or flat surfaces, including the tiny one at the tip called the *culet*.

Find Out for Yourself

Why do you think people like to wear gems? What can you find out about the history of jewelry? Do some research on different gemstones and precious metals. Where are they found? What are their chemical compositions? How are they formed? What are different methods and styles of cutting and polishing them?

◆ Think About It

Pose the following questions:

1. Why do you think gemstones are rare? *(This is an open-ended question. Students should recognize that very precise conditions are required to produce gemstones, and that these conditions are unusual in the constantly changing crust of the earth.)*

2. Why have people made these colorful minerals a valuable commodity? *(At first, people responded to the beauty of these natural materials, and then social and economic factors increased their desirability.)*

3. Why are gemstones usually cut into particular shapes? *(The way a stone is cut enhances the way it interacts with light. This shows off the stone to its best advantage.)*

4. Do you think the way a gemstone is cut is an artistic or a scientific decision? Explain. *(Gem cutting is an art based on many scientific principles: crystal structure, optics, inertia, and friction.)*

◆ Going Further

- The history of jewelry is a good topic for a project that integrates social studies and science. Ancient societies liked to adorn themselves with jewelry, and unusual stones became items of trade for thousands of years. Have students report on gemstone varieties and trading routes and then display their research findings on a large map of the world.
- Have students research the location of major gemstone mines.
- Ask students to choose a particular gemstone and write its "life story." This should include the formation of the gemstone.

Science and the Arts

UNIT 7 Toward the Stars

✷ Teaching the Unit

☀ Unit Overview

In the unit *Toward the Stars,* students examine the structure of the universe. In the first section, *The Observers,* students seek their own explanations for common astronomical events and explore the theories used to explain these events. In *The Earth Moves,* students investigate how the tilt of Earth's axis and its orbit around the Sun create the seasons. In the third section, *Exploring the Solar System,* students investigate the components of the Solar System by examining the discoveries of three space probes—*Voyagers 1* and *2* and *Magellan.* In *Our Universe,* students learn about the composition and life span of stars and about the big-bang theory. Students also take an imaginary journey through the universe.

☀ Using the Themes

The unifying themes emphasized in this unit are **Scale and Structure, Patterns of Change,** and **Evolution.** The following information will help you weave these themes into your teaching plan. A focus question is provided with each theme as a discussion tool to help you tie the information in the unit together.

Scale and Structure is a dominant theme because of the unit's focus on the composition and size of the universe. Students explore how the Solar System fits into the universe by examining galaxies and the universe itself.

Focus question: *How does the scale and structure of the Solar System relate to the scale and structure of the Milky Way galaxy?*

The Solar System consists of one star about which nine planets and several smaller bodies revolve, while the Milky Way galaxy is composed of billions of stars that revolve about a central point. The Solar System is a very small part of the extremely large Milky Way galaxy.

Patterns of Change is integral because many astronomical patterns result in predictable changes, such as Earth's revolution around the Sun and the life span and eventual death of stars.

Focus question: *What seasonal pattern emerges as Earth orbits the Sun?*

As Earth orbits the Sun, spring, summer, autumn, and winter follow one another in an annual pattern.

Evolution is a pervasive theme because students explore the process by which a celestial body, or the universe itself, evolves over time. By examining the big-bang theory, students learn about one way astronomers think that the universe may have begun, and they learn how it continues to evolve.

Focus question: *What happens during a star's life span?*

Stars produce energy because of nuclear reactions in their cores. During this process, huge amounts of energy are released as heat and light. Eventually, a star runs out of nuclear fuel and begins to expand in size until it becomes a red giant. In time, it cools and begins to collapse under its own gravitational force. The collapsing star may explode into a supernova, after which it may become a neutron star or a black hole. Stars with relatively small masses may collapse into white dwarfs.

☀ Using the *Science Discovery* Videodiscs

Disc 1 *Science Sleuths, The Lost Mining Probe*
It's the future. The World Space Police have gotten a tip that an illegal mining probe may have been launched into a restricted area of space. The Science Sleuths must analyze the evidence for themselves to determine whether the probe was sent and, if so, exactly where it is now.

Disc 2 *Image and Activity Bank*
A variety of still images, short videos, and activities are available for you to use as you teach this unit. See the Videodisc Resources section of the **Teacher's Resource Binder** for detailed instructions.

☀ Using the *SciencePlus SourceBook*

Unit 7 focuses on space and space exploration. Students learn about the composition and energy production of the Sun and other members of the Solar System. The unit highlights types of stars, galaxies, and some of the stranger objects in space. The unit concludes with a brief look at cosmology.

PLANNING CHART

SECTION AND LESSON	PG.	TIME*	PROCESS SKILLS	EXPLORATION AND ASSESSMENT	PG.	RESOURCES AND FEATURES
Unit Opener	376		observing, discussing	For Your Journal	377	Science Sleuths: *The Lost Mining Probe* Videodisc Activity Sheets TRB: Home Connection
THE OBSERVERS	378			Challenge Your Thinking	391	
1 What Is Astronomy?	378	1	analyzing, formulating definitions, evaluating, writing			Image and Activity Bank 7–1
2 An Ancient Science	382	4	observing, evaluating, formulating hypotheses, investigating	Exploration 1 Exploration 2	382 388	Image and Activity Bank 7–2 TRB: Activity Worksheet 7–1 Resource Transparency 7–2 Graphic Organizer Transparency 7–1
THE EARTH MOVES	392			Challenge Your Thinking	405	
3 A Scientific Revolution	392	2 to 3	using models, analyzing, inferring, drawing conclusions	Exploration 3	393	Image and Activity Bank 7–3 Resource Transparencies 7–3 and 7–4
4 Motions and Their Effects	398	4	analyzing, inferring, illustrating, drawing conclusions	Exploration 4 Exploration 5	399 403	Image and Activity Bank 7–4 Resource Transparency 7–5
EXPLORING THE SOLAR SYSTEM	406			Challenge Your Thinking	421	
5 Visitors from Space	406	2	observing, analyzing, inferring, drawing conclusions			Image and Activity Bank 7–5
6 The Space Probes	411	2	observing, analyzing, inferring, drawing conclusions	Exploration 6	415	Image and Activity Bank 7–6 TRB: Activity Worksheet 7–2 Graphic Organizer Transparency 7–6
7 Colonizing the Solar System	416	1	observing, analyzing, formulating hypotheses, researching	Exploration 7	418	Image and Activity Bank 7–7 TRB: Activity Worksheet 7–3 Graphic Organizer Transparency 7–7
OUR UNIVERSE	422			Challenge Your Thinking	439	
8 Messages from the Stars	422	2	observing, analyzing, investigating, using models	Exploration 8 Exploration 9	424 428	Image and Activity Bank 7–8 Resource Transparencies 7–8 and 7–9
9 Earth's Place in the Universe	429	2	observing, analyzing, formulating hypotheses, inferring	Exploration 10	430	Image and Activity Bank 7–9 TRB: Activity Worksheet 7–4 Resource Transparency 7–10
10 A Likely Beginning	436	1	observing, analyzing, inferring, summarizing	Exploration 11	438	TRB: Resource Worksheet 7–5
End of Unit	440		applying, analyzing, evaluating, summarizing	Making Connections TRB: Sample Assessment Items	440	Science in Action, p. 442 Science and Technology, p. 444 *SourceBook*, pp. S111–S132

*Time given in number of class periods.

Meeting Individual Needs

Gifted Students

1. Have students consider the following questions:
   - What keeps planets and their satellites in orbit?
   - Why don't planets fall into the Sun or fly out into space?

   Challenge students to apply the principles of *centripetal force* to answer these questions.

2. Point out to students that if they look carefully at the night sky, they may be able to see faint, hazy patches of light called *nebulae*. Suggest that they do some research to discover:
   - What are nebulae composed of?
   - Where did they come from?
   - How large are they?
   - How far away are the nebulae that we see?
   - How are nebulae classified?

3. Explain to students that Einstein's *general theory of relativity* is used in studies of the universe, especially in studies of events that occur in extremely strong gravitational fields. Invite students to discover what the general theory of relativity says will happen to starlight in the presence of a large object like the Sun.

LEP Students

1. Suggest that students make poster diagrams of the Solar System. Their diagrams should include the following: the inner planets, the outer planets, the Sun, and the asteroid belt. Explain that each part of the diagram should be identified with a bilingual label that presents an interesting fact about it. When students finish, review their work for science content, with minimal emphasis on language proficiency.

2. Have students make a bilingual dictionary of astronomy terms. Terms might include: planet, comet, star, meteor, galaxy, solar system, satellite, orbit, and asteroid. Have students illustrate their entries.

3. At the beginning of each unit, give Spanish-speaking students a copy of the *Spanish Glossary* from the *Teacher's Resource Binder.* Also, let Spanish-speaking students listen to the *English/Spanish Audiocassettes.*

At-Risk Students

Unit 7 provides students with many motivating activities that introduce the laws and concepts that govern the Solar System and the universe. In Lesson 3, *A Scientific Revolution,* students make a model of the Solar System based on the ideas of Copernicus and use it to explain celestial motions, including the retrograde motion of Mars. In Lesson 4, *Motions and Their Effects,* students make a model to discover the position of Earth relative to the Sun during each of the seasons. In Lesson 6, *The Space Probes,* students travel with the *Voyager* spacecrafts as they explore the Solar System. In Lesson 9, *Earth's Place in the Universe,* students take an imaginary trip through the Local Group of galaxies to learn about the size of the universe.

For special recognition, give students the opportunity to explain how Earth's two motions affect everyone who lives on Earth. For example, they might make poster diagrams to show how Earth's rotation accounts for day and night and how Earth's revolution around the Sun accounts for the seasons.

Cross-Disciplinary Focus

Social Studies

Point out to students that many ancient cultures had a considerable understanding of how the Solar System works. Have students research how some of these ancient civilizations used their observations of movement of the Sun, the Moon, and stars to create calendars or astronomical monuments. Examples include: Stonehenge in southern England, the great pyramids of Egypt, the American Indian Medicine Wheel in Wyoming, the stone calendar representing the cosmology of the Aztecs, and the Maya-Toltec Caracol, which was probably used as an observatory.

Language Arts

Challenge students to write newspaper articles about important events in the history of astronomy such as the trial of Galileo; the discovery of Uranus, Neptune, or Pluto; the first spacecraft to land on Mars; the first shuttle flight; or the *Voyager 2* fly-by of Saturn or Jupiter. Remind students that a good newspaper article tells the who, what, where, when, why, and how of the story. When students have finished, have them organize their work into an "Astronomy Newsletter."

Mathematics

Suggest to students that they make a scale diagram of the Milky Way. They should use the following information to prepare their scale:
- The Milky Way galaxy is about 100,000 light-years across.
- It is about 10,000 light-years thick at the center.
- It is about 3,000 light-years thick where Earth is located.
- The center of the galaxy is about 30,000 light-years from our Solar System.

Photography

Some students might enjoy the challenge of trying to take nighttime photographs of the Moon. Suggest that they talk to a friend or relative who is familiar with photography or do some reading to learn about timed exposures. Have students present a "photographic exhibition" of their work for their classmates.

 ### Cross-Cultural Focus

African-Americans in the Space Program

Suggest that students write biographies on African-Americans who are part of the space program. People they may wish to consider include: Dr. Mae Jemison, the first African-American woman selected for space flight; Guion S. Bluford, Jr., the first African-American to fly in space; and Robert McNair, the second African-American to fly in space. To extend the activity, have students write biographies on members of other cultural groups who are involved in the space program, such as Franklin Chang-Diaz (Costa Rican) and Rodolfo Neri (Mexican).

International Space Research

Point out to students that the United States has conducted joint research efforts with other countries through the European Space Agency (ESA). Suggest that students find out what international organizations are involved in space programs and what their plans are for future research projects.

 ### Bibliography for Teachers

Baker, David. *The Henry Holt Guide to Astronomy.* New York, NY: Henry Holt and Company, 1990.
Frazier, Kendrick. *Solar System.* Alexandria, VA: Time-Life Books, 1985.
Kerrod, Robin, ed. *The Heavens.* Chicago, IL: World Book, Inc., 1987.
_____. *The Illustrated History of NASA, Anniversary Edition.* New York, NY: Gallery Books, 1988.
McAleer, Neil. *The Mind-Boggling Universe.* Garden City, NY: Doubleday and Company, 1987.
Morrison, Philip and Phylis. *Powers of Ten: About the Relative Size of Things in the Universe.* New York, NY: Scientific American Library, 1982.
Sagan, Carl, and Ann Drayan. *Comet.* New York, NY: Random House, 1985.
Time-Life Books. *Moons and Rings.* Alexandria, VA: Time-Life Books, 1991.
Time-Life Books. *Stars.* Alexandria, VA: Time-Life Books, 1989.
Time-Life Books. *Voyage Through the Universe.* Alexandria, VA: Time-Life Books, 1989.
Trefil, James S. *Space Time Infinity.* Washington, D.C.: Smithsonian Books, 1985.

 ### Bibliography for Students

Asimov, Isaac. *Ancient Astronomy.* Milwaukee, WI: Gareth Stevens Publishing, 1989.
_____. *Asimov's Guide to Halley's Comet.* New York, NY: Walker and Company, 1985.
_____. *How Was the Universe Born?* Milwaukee, WI: Gareth Stevens Publishing, 1989.
Dunlap, Storm. *Astronomy.* New York, NY: Exeter Books, 1984.
Gallant, Roy A. *101 Questions and Answers About the Universe.* New York, NY: Macmillan Publishing Company, 1984.
_____. *The Macmillan Book of Astronomy.* New York, NY: Macmillan Publishing Company, 1986.
Whyman, Kathryn. *Solar System.* New York, NY: Gloucester Press, 1987.

 ### Films, Videotapes, Software, and Other Media

Astronomical Society of the Pacific.
390 Ashton Avenue
San Francisco, CA 94112
Write for their catalogue of education materials, including videos, software, slides, posters, and laserdiscs.

The National Science Teachers Association.
1742 Connecticut Ave.
Washington, DC 20009
Write for catalog, which includes some excellent astronomy and Earth Science resources.

Project STAR.
Center for Astrophysics
60 Garden Street
Cambridge, MA 02138
NSI-sponsored project to develop excellent new teaching activities and materials for middle school and high school astronomy.

Planetarium on Computer: The Solar System. Software. For Apple II Family.
Focus Media, Inc.
839 Stewart Avenue
Garden City, NY 11530

The Creation of the Universe. Videodisc.
Voyage to the Planets. CD-ROM. Software. For Apple II Family.
MECC Etc.
6160 Summit Dr. N.
Minneapolis, MN 55430

UNIT 7

✷ Unit Focus

Challenge students to name as many different kinds of celestial bodies as they can (moons, planets, stars, galaxies, comets, asteroids, meteors, and so on). Keep track of their responses on the chalkboard. Then ask students how many of these objects they have actually seen. Put a check next to each item on the list that is identified. Call on volunteers to describe each of the objects for the class.

Ask students to respond with a show of hands if they have ever identified a constellation in the night sky. Call on volunteers to name the constellations they have seen. Keep track of their responses by listing them on the chalkboard. Then call on volunteers to describe the night sky and how it makes them feel when they look at it.

✷ About the Photograph

The photograph is a multiple exposure of a total eclipse of the Sun. On February 26, 1979, the Moon totally eclipsed the Sun for two minutes over Winnipeg, Manitoba, in Canada. The photograph shows successive stages of the eclipse as the Sun is covered by the disc of the Moon. Notice that the city has turned on its street lights even though it is daytime.

Unit 7 — Toward the Stars

THE OBSERVERS
1. What Is Astronomy? p. 378
2. An Ancient Science, p. 382

THE EARTH MOVES
3. A Scientific Revolution, p. 392
4. Motions and Their Effects, p. 398

EXPLORING THE SOLAR SYSTEM
5. Visitors from Space, p. 406
6. The Space Probes, p. 411
7. Colonizing the Solar System, p. 416

OUR UNIVERSE
8. Messages from the Stars, p. 422
9. Earth's Place in the Universe, p. 429
10. A Likely Beginning, p. 436

For Your Journal

Write a paragraph summarizing your thoughts about each of the following:
1. If you could travel from Earth into outer space, what might you pass first? next?
2. How do you account for day and night? summer and winter?
3. Why are kilometers often inadequate for measuring distances in space? What do we use in their place?
4. What motions of the Moon make the Sun look like it does in this photograph?

✴ Using the Photograph

Ask students to explain the different motions that are shown in the photograph of the eclipse. *(The rotation of Earth causes the Sun and Moon to appear to move together through the sky. The Moon's monthly revolution around Earth causes it to move into position between Earth and the Sun.)*

When it is totally eclipsed, the Sun reveals a white band (the corona or "crown") that it does not appear to have in the other images. Ask students to explain this. *(The Sun is surrounded by glowing gases that are normally not visible because the Sun itself is so bright. Students familiar with photography may be aware that the images of the partially eclipsed Sun were shot through very dark filters to keep the Sun from overexposing the film.)*

✴ About Your Journal

Students should answer the Journal questions to the best of their abilities.

These questions are designed to serve two functions: to help students recognize that they do indeed have prior knowledge about the topic, and to help them identify any misconceptions they may have. In the course of studying the unit, these misconceptions should be dispelled.

Toward the Stars 377

LESSON 1

Getting Started

This lesson sets the stage for the rest of the unit by encouraging students to examine what they know about astronomy. The lesson begins with a hypothetical situation in which a team of astronomers has made a startling new discovery. A newspaper description of their discovery is used to help define the scope of astronomy. The lesson concludes with students using a series of photographs to generate their own questions about astronomy.

Main Ideas

1. Astronomy is the study of the universe and its components.
2. New discoveries in astronomy are continually being made.
3. Scientists who study the universe and Earth's place in it are called astronomers.

Teaching Strategies

Imagine!

Before students begin reading, direct their attention to the photograph and call on a volunteer to describe what it shows. Ask students if they have ever used a telescope. If so, call on a volunteer to explain what he or she saw when using one.

If any of your students have visited an observatory, have them share their experiences with the class. Then have students read pages 378 and 379.
(Continues on next page)

THE OBSERVERS

1 What Is Astronomy?

Imagine!

You are an astronomer atop the Mount Kea volcano in Hawaii, using a telescope to take photographs of the heavens. The next morning while studying the photographs you took the night before, you discover a most amazing sight.

Quickly, you call other members of the team for consultation. Together, you decide that you have photographed two galaxies colliding in outer space. Even more exciting, it looks as though there's a dark, mysterious "black hole" within the galaxies. A breakthrough? You and your fellow astronomers write an announcement about your discovery for the media.

LESSON 1 ORGANIZER

Objectives

By the end of the lesson, students should be able to:
1. Suggest a definition for astronomy.
2. Recognize the kinds of questions that are relevant to the study of astronomy.
3. Demonstrate an understanding of what astronomy is by identifying some of the objects observed and studied by astronomers.

Process Skills

analyzing, formulating definitions, evaluating, writing

New Terms

Astronomy—the study of the universe and the objects in it.

How important is their discovery? What is a galaxy? a black hole? a light-year? a quasar? A simple answer is that they are all part of the science called **astronomy**.

Materials

What Is Astronomy? Journal

Science Discovery Videodisc

Disc 2, Image and Activity Bank, 7–1

Time Required

one class period

(Imagine! continued)

When students have finished reading, call on a volunteer to summarize the newspaper article. Then direct students' attention to the questions at the bottom of the page. Point out that each of the new terms mentioned in the questions will be discussed later in the unit.

Answers to

In-Text Questions

- A *galaxy* is a huge system of stars, dust, and gases held together by gravity.
- A *black hole* is an invisible object believed to exist in space. The gravitational force of a black hole is so strong that nothing can escape it, not even light.
- A *light-year* is the distance that light travels in one year, about 9.46 trillion kilometers.
- A *quasar* is an extremely bright object thought to be the brilliant origin of a galaxy.

(Continues on next page)

The Observers

(Imagine! continued)

After reading the newspaper article on the discovery of a possible black hole, direct students' attention to the photographs on pages 380 and 381. Involve them in a discussion of what each one shows. Invite students to describe in a sentence or two what is depicted in each photograph. Then ask them to generate phrases that complete the statement "Astronomy is _____."

The phrases generated could reflect what is involved in the study of astronomy (e.g., "Astronomy is calculating the distance to stars.") They could reflect on what astronomers do (e.g., "Astronomy is studying the stars with a telescope.") The phrases could also reflect the kinds of objects that make up the universe (e.g., "Astronomy is the study of galaxies.") Invite volunteers to share and discuss their phrases with the class.

What do you think astronomy is? In your Journal, write as many phrases as you can to complete the statement "Astronomy is ___?___." The photos on this page and the next will help you add to your list.

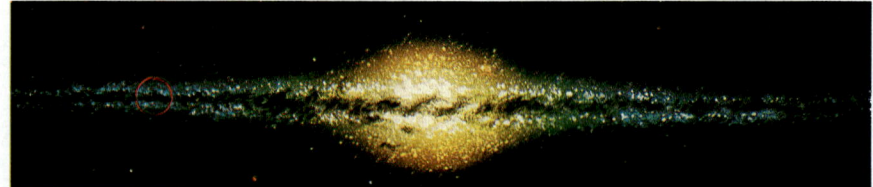

Artist's impression of the Milky Way galaxy. The red circle shows the location of our Solar System.

Astronaut on our nearest neighbor—the Moon

Volcanoes on Mars

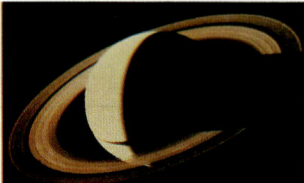

Saturn's rings

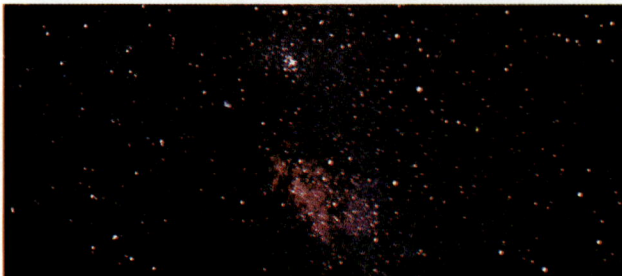

Star cluster

The Discovery shuttle, with the Hubble space telescope aboard

380 Toward the Stars

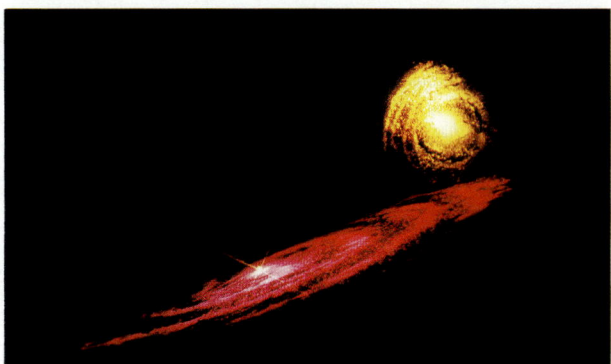
Artist's impression of colliding galaxies

Solar eclipse

Ask an Astronomer

What do astronomers do? If you could ask an astronomer some questions about the universe and Earth's place in it, what would you ask? In small groups, make a list of questions that really interest you. The drawing shows some questions to get you started.

Share your group's questions with the other groups. As you study this unit, you'll discover answers to many of these questions.

Ask an Astronomer

The task of generating questions can be done in small, cooperative-learning groups. You may wish to have students write their questions on slips of paper and display them on a bulletin board. Group them under categories such as those that pertain to the universe as a whole, to the Solar System, or to human technology. As students complete the lessons, have them answer the questions that apply. If some questions remain unanswered after the unit is completed, have volunteers do some research to discover and share the answers with the class.

✹ *Follow Up*

Assessment

1. Suggest to students that they take a survey among family members and friends to find out what they think astronomy is. Suggest that students organize the responses and share their information with the class.
2. Students may enjoy writing a poem to describe the science of astronomy. Have volunteers share their poems with the class. Then have students organize their poems in a booklet, design a cover, and think of a title. The booklet should be displayed where others can read it at their leisure.

Extension

1. Suggest to students that they look for articles on astronomy in magazines and newspapers. They could cut out or photocopy the articles and use them to make a bulletin-board display entitled "Astronomy Today."
2. Students may enjoy doing research to discover more about black holes, galaxies, and quasars. Have them share what they learn by presenting oral reports to the class or by writing fact sheets.

The Observers 381

Lesson 2

✦ Getting Started

This lesson focuses on some common astronomical events that can be observed in the night sky and explores the theories used to explain them. Students complete and discuss *A Celestial Quiz* about astronomical observations they may have made. Then students complete several activities to verify or disprove statements in the quiz. Several picture studies expand on these observations and introduce ideas that are essential for understanding the concepts developed in the unit. The lesson concludes with students examining and evaluating some of the theories proposed by early Greek and Egyptian astronomers.

Main Ideas

1. Many observable astronomical events can be explained by the movements of the planets, stars, and other heavenly objects.
2. Early astronomers developed several theories to explain their observations of the night sky.
3. Two models used to explain the motions of heavenly objects are the heliocentric and geocentric models of the Solar System.

(Teaching Strategies follow on next page)

2 An Ancient Science

Observing the Heavens

Did you know that astronomy is the oldest science? Since the earliest times, humans have looked toward the stars. In fact, ancient observers were excellent viewers of the skies. By careful observations, early Babylonians, Egyptians, Aztecs, and Mayans could accurately predict the locations of heavenly bodies. They relied on the skies for answers to such questions as when to plant crops and when the seasons would change.

You are like the ancient astronomers when you look up at the sky on a clear night. What you see is what they saw. Make a list of all the heavenly bodies these ancient gazers might have seen.

Did you include in your list NJYLCRQ, KMML, QRYPQ, QFMMRGLE QRYPQ, and the QSL (or SUN, if you crack the simple code)? Ancient peoples saw all of these. In the following Exploration, you will think about and investigate some of the motions that occur in these heavenly objects. Part 1 has a list of observations that all sound like they might be true; some may even be observations you have made yourself. The trick is, some are false. Your mission is to discover which statements are not true. In Part 2, you will have the chance to check out some of these statements experimentally.

EXPLORATION 1

Heavenly Motions

Part 1

A Celestial Quiz

Work in groups of three. Decide if each statement is true or false. Is there complete agreement on each statement among members of your group?

1. Each day the Sun rises due east and sets due west.
2. Every 7 days, the length of daylight changes by several minutes.
3. In your area, there are times of the year when the Sun is directly overhead.

4. Your winter shadow at noon is shorter than your summer shadow at noon.
5. Shadows shorten as the morning progresses.

LESSON 2 ORGANIZER

Objectives

By the end of the lesson, students should be able to:

1. Describe some astronomical events they can observe in the night sky.
2. Describe some of the observations made by astronomers of ancient civilizations.
3. Explain the observed motions of the planets, the Moon, and stars in relation to the movement of Earth.
4. Evaluate models of the universe as put forth by Aristotle, Aristarchus, and Ptolemy.
5. Explain the difference between heliocentric and geocentric models of the Solar System.

Process Skills

observing, evaluating, formulating hypotheses, investigating

New Terms

Retrograde motion—a backward motion in which a planet appears to make a loop in the sky, going backward for a while before continuing its forward motion.

382 Toward the Stars

6. Like the Sun, many stars appear to rise and set at various times of the night.
7. If you look at the Big Dipper at two different hours of the same night, its orientation (position) in the sky will change.
8. If you see the Moon in the early evening in the east, it must be a full Moon.
9. On two consecutive evenings at the same hour, the Moon is farther east and bigger on the second evening.
10. You always see the same side of the Moon from your position on Earth.
11. All stars that you see appear to be the same color.
12. The Moon appears larger when it is seen on the horizon than when it is seen higher in the sky.
13. The only time you can see the Moon is at night.
14. The Moon is closer to Earth than the stars are.
15. Human behavior is affected by the phase or shape of the Moon.
16. Not all stars are equally bright.
17. The star patterns, or constellations, change shape from night to night.
18. Planets are indistinguishable from stars when seen with the naked eye.
19. Some constellations are seen only in certain seasons.
20. Planets wander in and out of the constellations.

Do we always see this face of the Moon from Earth?

Did everyone agree on these statements? Which ones created the greatest debate? In the next part of this Exploration, you will check out some of these statements as an observer. From your investigations, you may be able to suggest other observations that could be added to the Celestial Quiz.

Materials

Observing the Heavens: Journal
Exploration 1
Part 2: modeling clay, paper-towel tube, piece of cardboard, binoculars, Journal

Teacher's Resource Binder

Activity Worksheet 7–1
Resource Transparency 7–2
Graphic Organizer Transparency 7–1

Science Discovery Videodisc

Disc 2, Image and Activity Bank, 7–2

Time Required

four class periods

✸ Teaching Strategies

Direct students' attention to the lesson title. Encourage them to speculate as to how old the science of astronomy might be and why people became interested in it. Point out that people everywhere observe the sky at night. It is natural that they are curious about what they see. Then have students suggest ways in which the science of astronomy might be useful. *(Answers may vary, but possible responses include: astronomy allows us to predict the tides, when the Moon will be full, and when the seasons begin and end.)*

Observing the Heavens

Have students work in small, cooperative-learning groups to list the kinds of heavenly bodies ancient observers might have seen in the night sky. Encourage students to describe the motions, whether real or apparent, associated with each one. *(The coded words are: planets, Moon, stars, shooting stars, and Sun.)*

Answers to
Exploration 1

Part 1
1. False. The Sun rises due east and sets due west only on the dates of the vernal and autumnal equinoxes.
2. True. The exact number of minutes by which the length of daylight changes depends on the latitude and on the time of year. The rate of change is greater the farther one is from the equator. Also, the rate of change slows around the time of the solstices.
3. False. Unless you live within 23 1/2 degrees north or south of the equator, the Sun is never directly overhead.

(Answers continue on next page)

The Observers 383

Answers continued

4. False. Due to Earth's tilt on its axis, the summer's noon shadow is shorter than the winter's noon shadow. This is because the angle at which the Sun's rays strike Earth's surface is closer to perpendicular in the summer than in the winter.
5. True. Shadows shorten as the Sun rises and lengthen after noon.
6. True. Stars appear to rise and set at various times of the night because of Earth's rotation.
7. True. The position of the Big Dipper seems to change because of Earth's rotation.
8. True. The Moon appears to be full when Earth is almost exactly between the Sun and the Moon. In this orientation, the Moon rises in the east at about the same time the Sun sets in the west.
9. True. Each evening the Moon will appear east of its previous position and will be more toward the full phase. The apparent size of the Moon varies with the phase of the Moon. During early evening, when most students will view the Moon, the Moon appears larger on the second night since it is approaching the full phase.
10. True. It takes the Moon the same amount of time to turn once on its axis as it takes it to circle Earth.
11. False. Stars vary in color—bluish, white, yellow, or red.
12. True. Although the size of the Moon does not change, it appears to be larger when it is near the horizon. The apparent difference in size is an optical illusion.
13. False. The Moon can often be seen during the day if its path is not too close to the path of the Sun.
14. True. The mean distance from Earth to the Moon is 384,403 km. The mean distance from Earth to the Sun is about 150 million km. The nearest star, excluding the Sun, is Alpha Centauri, which is 4.3 light-years away.
15. Answers will vary. Certain scientific studies purport to show a relationship between the phases of the Moon and human behavior, but there is no agreement on the scientific validity and merit of these studies.
16. True. One way scientists classify stars is by their brightness.

EXPLORATION 1—CONTINUED

Part 2
Checking It Out

Choose at least two of the following observational activities to help you verify statements from the Celestial Quiz. When reporting your conclusions, indicate which of the statements on the Celestial Quiz you have confirmed or disproved. What other observations have you made?

1. The first sundial that was used to determine the passage of time was a simple upright stick. Using modeling clay, mount a paper towel tube on a piece of cardboard, as shown below. For several days during the next week, place it outdoors at the same time each day. Be sure to put the cardboard in exactly the same position each time. Draw the shadow's position and length at that time, and record the date for each day. Does the direction of the shadow change? The length of the shadow? What does the shadow indicate about the Sun's position in the sky?

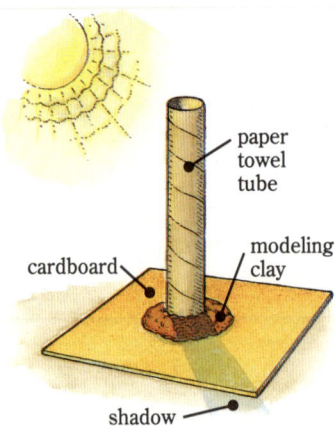

2. Use the shadow stick from #1 to find out how shadow length and position change during the day. What direction does the shadow point in early morning? at noon? a little before the Sun sets? Will there ever be a time when the Sun is directly overhead and therefore not cast a shadow? How does the length of the shadow vary?

3. Observe the rising or setting Sun. **Caution:** *Don't look directly at the Sun.* Several times during the week, record the time when the Sun first appears above or disappears below the horizon (or the top of a building). Make sure you view the Sun from the same position each time. Use some fixed object, such as a tree, to determine the exact point at which the Sun rises or sets. Does this point change? If it changes, in what direction does it change? Would this direction be the same in spring as in fall? Does the time of rising or setting change? By how much?

17. False. No change in star patterns can be detected from night to night. However, over thousands of years, changes can be seen as stars move relative to each other.
18. False. Planets can be detected since they move relative to the more distant stars. Movement of stars relative to each other cannot be detected with the naked eye.
19. True. Some constellations can be seen only during certain seasons due to Earth's tilt on its axis and its revolution around the Sun.
20. True. Against the background of stars, planets appear to wander through the constellations as they revolve around the Sun. Because they are so much closer to us than are the stars, planets appear to move more quickly across the sky as they travel around the Sun.

Part 2

1. The shadow's direction and length will change slightly over several days during most of the year. Little, if any, change would be expected if the activity is performed near the winter or summer solstice. Observing the shadow demonstrates that the Sun's position in the sky changes continually, since its rays must come from a different direction to cast a different shadow. This activity does not verify or disprove any *Celestial Quiz* statements.

4. Over the period of a week, chart the position and shape of the Moon. Do this at the same time each night. As the month progresses, in what direction does the Moon move? How does its shape change from night to night?

5. On a clear night, spend some time observing the stars. Are all stars equally bright? Are they all the same color? Using a star chart, try to identify the brightest stars.

6. Sight a star near the western horizon. Follow its position at various times on the same night. Does it set as the Sun does? Do stars rise in the east?

7. Locate the constellation of the Big Dipper in the night sky. Draw what you see on a piece of paper. An hour or two later, observe the Big Dipper again and draw what you see, using the same piece of paper. Did its shape change? Did its orientation change? Is there a point in the sky about which it and other stars seem to be revolving?

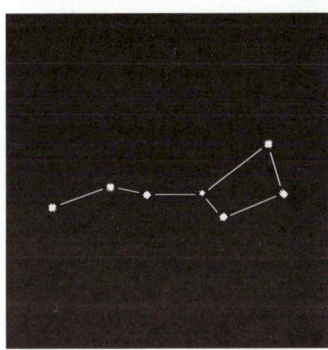

Lining up two fixed points with a star to see if the star moves

8. Line up two fixed points, such as a house and a tree, with a star. Each night make another sighting at the same time from the same spot. What do you observe? Does the star's position appear to be different from night to night? If so, in what direction does it move?

9. With or without binoculars, carefully observe the features of the Moon. Make a diagram of some of the features you see. Do the features change from night to night, or do they remain the same?

Think About It

Look over the statements in the Celestial Quiz again. Are there some statements not yet verified? What observational activities could you plan to check them out? Are there other statements you could add to the Celestial Quiz? How would you verify these new statements? Are there still other observational projects you would like to carry out? What are they?

2. Since the Sun never goes north of the Tropic of Cancer, shadows will always have a northward cast. In early morning, the shadow will point to the northwest. At noon, it will point north. A little before sunset, it will point to the northeast. In the U.S., there is never a time when the Sun is directly overhead. This activity verifies statement 5 and disproves statement 1 on the *Celestial Quiz*. (Statement 3 cannot be verified unless the observations are continued for a year, or at least through two solstices.)

3. The point at which the Sun rises and sets changes from day to day. It seems to move from south to north from December to June and from north to south from June to December. The times of rising and setting may vary up to a few minutes a day, depending on geographical location and the time of the year. This activity verifies statement 2 and disproves statement 1 on the *Celestial Quiz*.

4. The Moon seems to move from west to east when viewed at the same time over several nights. The Moon changes from new Moon to full Moon and back again in 29.5 days. If performed before the Moon is full, this activity proves statement 9 on the *Celestial Quiz*.

5. Stars are not all the same brightness or color. Some may appear bluish, while others may appear white, yellow, or red. The brightest stars are: Sirius, Canopus, Alpha Centauri, Arcturus, Vega Capella, and Rigel. Canopus and Alpha Centauri are not visible from mid-northern latitudes. This activity verifies statement 16 and disproves statement 11 on the *Celestial Quiz*.

6. A star sighted near the western horizon will seem to set in the west as the Sun does. Most stars seem to rise in the east and set in the west. Stars near Polaris (the North Star) remain above the horizon and appear to circle Polaris in a counter-clockwise direction. This activity verifies statement 6 on the quiz.

7. From night to night, the Big Dipper will not change shape, but it will change its orientation. The Big Dipper and other stars seem to revolve about the North Star, or Pole Star. This activity verifies statement 7 and disproves statement 17 on the quiz.

8. The star's position does change; it rises about four minutes earlier each night. The star's position moves to the east each night. This activity does not verify or disprove any of the statements on the *Celestial Quiz*.

9. The features in the visible portion of the Moon do not change from night to night. This activity verifies statement 10 on the quiz.

Answers to Think About It

Statements in *Part 1* not verified or disproved by the activities in *Part 2* include: 3, 4, 8, 12, 13, 14, 15, 18, 19, and 20. Encourage students to develop and share strategies to either verify or disprove these statements.

A Celestial Picture Study

These picture studies introduce concepts and ideas that will be discussed in greater detail later in the unit. Emphasis should be on promoting observational skills instead of on giving explanations or inference-making.

Answers to
Picture Study 1

Students may be interested to know that calendar wheels, such as the one shown here, are some six centuries old. Their exact purpose is not certain, but it is believed that they were used as crude calenders to predict the summer solstice. The people who made them were probably the ancestors of such Native American groups as the Crow, Gros Ventre, and Blackfoot.
- Yes, the Sun is higher in the sky at certain times of the year. The Sun is highest in the summer and lowest in the winter.
- The position on the horizon where the Sun rises and sets changes from day to day. (It moves southward from June to December and northward from December to June.) By observing these changes over a period of time, early "astronomers" were able to predict with some accuracy where the Sun would rise and set.

A Celestial Picture Study

The following illustrations and their accompanying stories will add to the large number of observations you have already made. Team up with another person to study each illustration, read its story, and consider the discovery questions. How many new ideas about the motion of the heavenly bodies can you discover?

Picture Study 1

Native Americans in Wyoming's Bighorn Mountains used a calendar wheel to predict the day when the Sun would reach its highest point in the sky. According to legend, "the growing power of the world" is strongest on the day the Sun is highest. On this day, the rising Sun was in line with two piles of rocks, or cairns.

- Is the Sun higher in the sky at one time of the year than another?
- Does the position on the horizon where the Sun rises and sets change from day to day?

Picture Study 2

In 1610 a famous scientist named Galileo made the diagram of the Moon shown below by using a new invention called a telescope. (You will be hearing more about Galileo later in the unit.) Next to his diagram is a recent photograph of the Moon in a similar phase.

- Does it appear that the face of the Moon you see at night is the same as the one that Galileo saw almost 400 years ago?

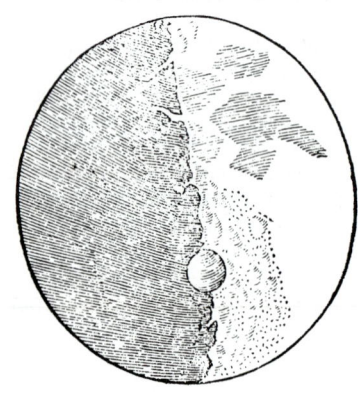

Answers to
Picture Study 2

The strongest of Galileo's telescopes magnified only about 33 times, but that was enough for him to make some startling observations. For example, he was the first to see Saturn's rings, four of Jupiter's moons, and spots on the surface of the Sun.

- Galileo's diagram of the Moon is very similar to the more recent photograph. He was able to identify such Earth-like features as mountains, valleys, and craters.

Picture Study 3

If the lens of a camera is held open for a time while aimed at the night sky, each star becomes a star track.

- Do stars appear to revolve? If the arrow points to the beginning of a star track, in what direction are the stars revolving?
- What other observations regarding stars can you make from this photograph?

Picture Study 4

The early observers noted objects in the skies that they called "wanderers." These are the planets. At that time, only five were known. How many do we know about now? This time-lapse photograph shows the motions of several planets over a long period of time.

- What strange motions do planets seem to make that the Sun and Moon do not?
- Does each planet seem to follow more or less the same path that the other planets do?

Picture Study 5

Star charts show the positions of stars and constellations for a particular time of the year. Study the star charts for both a summer and a winter sky.

- What features are common to both charts? How do they differ?

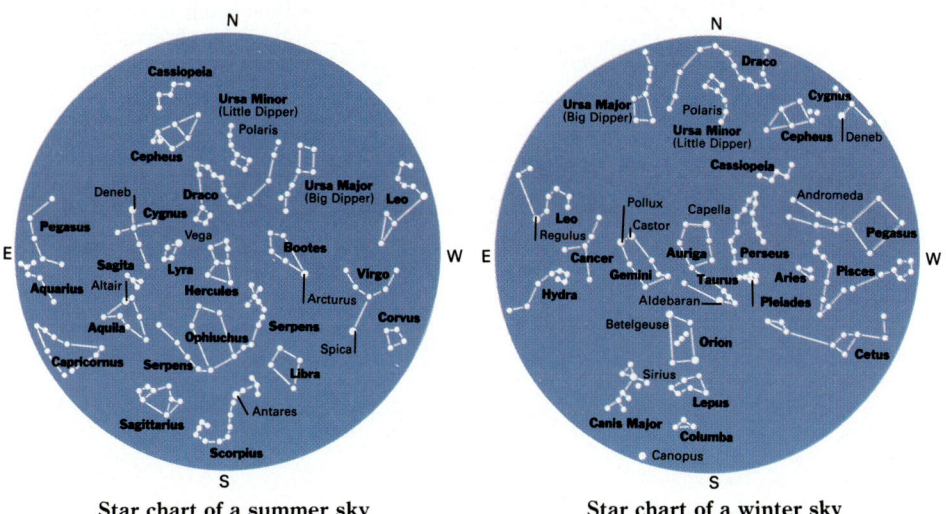

Star chart of a summer sky   Star chart of a winter sky

Answers to
Picture Study 4

Nine planets, including Earth, are now known to orbit the Sun.

- A time-lapse photograph of the planets in our Solar System shows that some of them, particularly Mars, appear to make strange looping motions that neither the Sun nor the Moon make. This kind of motion is called *retrograde motion*. It is an optical illusion caused by the difference in the orbital speeds of the planets. (Retrograde motion will be discussed in greater detail later in the unit.)
- The photograph shows that the orbits of the planets follow more or less the same path. This indicates that the planets orbit the Sun in similar planes. That is, if you looked at the Solar System on edge, the planets, with the exception of Pluto, would appear to be in the same "slice" of space.

Answers to
Picture Study 5

You may wish to point out to students that the cardinal directions are printed on the star chart. Remind students that to use the chart, they must hold it above their heads with the cardinal directions aligned accordingly. The circle enclosing the chart represents the horizon.

- The location of Polaris (the North Star) is the same in both charts. Both charts indicate the path of the Sun, the cardinal directions, and the horizon. Many of the constellations appear on both charts.
- Students should observe that there are a number of constellations that appear on only one chart and that from one chart to the next, all constellations have changed in position (orientation).

Answers to
Picture Study 3

- In a time-lapse photograph of the night sky, stars seem to revolve counterclockwise around a central point, forming circular paths. It is Earth's rotation that causes these patterns in the night sky. You may wish to point out to students that a similar photograph taken in the Southern Hemisphere would show the stars revolving clockwise.
- Careful examination of the photograph shows that the star tracks vary in brightness and color. The varying widths of the star tracks indicate that the stars are of different brightnesses. Stars that make complete circular paths do not set (i.e., they don't go below the horizon).

The Observers

Making Sense of the Skies

To summarize what students have learned so far, you may wish to involve them in a discussion of their observations and explanations. Then provide time for students to work with a partner to develop a theory that explains many of the observations they have made so far. Call on volunteers to share their theories with the class. (Accept all reasonable ideas without comment.) Point out to students that early astronomers arrived at many different theories to explain what they observed in the heavens. In the next Exploration, they will learn about some of these theories.

Exploration 2

Have students read the introduction to the Exploration silently. Then divide the class into pairs. Suggest that they answer all of the questions raised and draw each of the models as they are described. When students have finished, have them share and discuss their answers and diagrams as a class.

1. Aristotle

Aristotle is considered to be one of the greatest and most influential thinkers of all time. He was born in Stagira, a small town in northern Greece. When he was about 18, Aristotle entered the Academy—Plato's school in Athens. He studied there for the next 20 years. Around 334 B.C., Aristotle founded his own school, the Lyceum.

Aristotle's belief that everything was made up of combinations of the four elements (air, earth, fire, and water) lasted for nearly 2000 years. It was not until the 17th century that scientists began to insist that direct observation and experimentation, rather than abstract philosophy, should be the basis of scientific thought.

In a treatise entitled *On the Heavens*, Aristotle explained his ideas on how the heavenly bodies moved. Although it may be difficult to believe today that he could have had such ideas, his views were based on a combination of philosophy and what he observed.

Making Sense of the Skies

If you lived 2000 years ago at a time when there were no telescopes or spaceships, and the only means of observing celestial objects was with the naked eye, how would you explain your observations so far this unit? With a friend, weave your explanations into a theory that explains as many of your observations as possible. Can your theory explain the reasons for night and day? the star tracks? why the calendar wheel works? Are there observations your theory cannot explain? Does this mean that your theory is not useful?

In Exploration 2, you will be able to compare your theory with those of a number of early thinkers and observers.

EXPLORATION 2

The Early Observers

Living in Greece and Egypt over 2000 years ago were many keen observers and thinkers. How did they explain their observations of the heavens? As you read about three of them, think of the observations they may have been basing their ideas on.

1. Aristotle (384–322 B.C.)

Aristotle (pronounced air us TAHT ul) was the most influential thinker of his time. He suggested that everything in the universe is composed of four elements: earth, air, fire, and water. He proposed that each element has a natural resting place, with earth at the center, surrounded by spheres of water, air, and fire.

His model of the universe contained a total of eight spheres revolving around Earth. The first seven spheres carried the Moon, the Sun, and the five known planets. Aristotle proposed that beyond these seven spheres, the stars were tiny lamps fixed to another revolving sphere.

Aristotle

He theorized that the sphere carrying the Sun revolved around Earth each day, causing day and night. The motion was much like following the threads on a corkscrew. With each revolution around Earth, the Sun was carried higher (or lower) in the sky. He proposed that this movement causes the seasons.

- How is this theory based on everyday observations?
- Try drawing a diagram based on Aristotle's ideas.

Answers to

In-Text Questions

- Aristotle's theory seems based on every day observations because, to an observer on Earth, it appears that Earth is motionless and that everything else moves around it. Two observations that might give the impression that celestial objects are attached to spheres revolving around Earth:
1. observing the Sun moving across the sky when Earth does not seem to move
2. seeing the stars near the axis of rotation (near Polaris) move in a circle each night. Also, the fact that the Sun's position in the sky rises and falls and then rises again through the cycle of a year might give the impression that Earth moves up and down within (and along the rotational axis of) this sphere.

2. Aristarchus of Samos (310–230 B.C.)

Aristarchus (pronounced air us TAR kus) combined mathematics and good observation to come up with another theory that is surprisingly close to the picture of the Solar System known today. He suggested that the Sun is the center of the universe—not Earth. He proposed that Earth and the other planets revolve around the Sun, causing the year. Day and night, Aristarchus suggested, are caused by the rotation of Earth on its axis. He was the first to measure the distance to the Sun and Moon. He concluded that the Sun is much farther away and larger than previously thought. The stars, he reasoned, are unimaginably far away. *Although his theory explained observations as well as Aristotle's theory, it was not accepted by the people of that time and was forgotten for nearly 2000 years.*

- What do you think about Aristarchus' theory?
- Why do you think that good ideas and reasoning are sometimes disregarded in favor of other ideas?
- Try drawing a diagram that illustrates Aristarchus' ideas.

3. Ptolemy (about A.D. 127–145)

Ptolemy

Ptolemy (pronounced TAHL uh mee) supported the ideas of Aristotle. He reasoned that if Earth really moved, the tremendous winds caused by Earth's movement would blow birds off their perches and leaves off the trees. He supported Aristotle's idea that the Sun revolves around a stationary Earth by noting that every day, we see the Sun actually moving across the sky.

But Ptolemy recognized that Aristotle's model had two serious flaws. First, it could not explain why the "wanderers," or planets, varied in brightness from year to year. If they orbited Earth at fixed distances, their brightness should not vary greatly. Second, the paths that the planets traveled were more erratic than the paths suggested by Aristotle's theory. A planet such as Mars, for instance, appears to make a loop in the sky going backwards for a while before continuing its forward motion. This curious backward motion is called **retrograde motion.**

As seen from Earth, Mars appears to make a looping motion in the sky.

2. Aristarchus of Samos

Aristarchus was a brilliant but little-known natural philosopher who became a master of geometry and a devout observer of the skies. Based on clever reasoning and observation, Aristarchus was the first to arrive at the radical conclusion that Earth revolved around the Sun and that the Sun, not Earth, was the center of the Solar System.

During his lifetime, Aristarchus was reviled for his ideas and was threatened with prosecution for alleged impiety against the prevailing views of the time.

Answers to
In-Text Questions

- Good ideas and reasoning are often disregarded when they do not blend neatly with prevailing notions of what is true.

3. Ptolemy

Ptolemy, also known as Claudius Ptolemaeus, made most of his astronomical observations in Alexandria, Egypt. He compiled his ideas in a 13-volume work which became known as the *Almagest,* a Greek-Arabic term which means "the greatest."

Much of what Ptolemy taught was pure astrology, a listing of Egyptian and Babylonian beliefs that the positions of the planets and stars controlled everything on Earth. In spite of his mysticism, Ptolemy was a tireless observer. He catalogued the celestial latitude and longitude of over 1000 stars. He also discovered the irregularity of the Moon's orbit. The *Almagest* survived the collapse of Greece and endured the fall of the Roman empire to become the backbone of the Roman Catholic church's view of the universe. The heavens, as defined by Ptolemy, were perfection, with Earth at its center. This view prevailed until the 16th century.

Answers to
In-Text Questions (page 390)

- Direct students' attention to the diagram showing how Ptolemy explained *retrograde motion.* Help them to recognize that Ptolemy suggested that Mars moved in its own orbit, called an *epicycle,* as it orbited Earth. When Mars was on the far side of its epicycle, it would appear fainter, and it would move in one direction. As Mars came around its epicycle, it would appear to be closer, brighter, and to be moving in the opposite direction.
- Another explanation for retrograde motion could be that Earth is moving faster in its orbit than Mars is. Much like a faster car on the inside lane of a racetrack, as Earth passes Mars, Mars seems to go backward.

Answers to
Thinking Further

1. Students may recognize that the Earth-centered theory proposed by Aristotle and Ptolemy accounts best for their observations, even though it is incorrect.
2. Accept all reasonable responses.
3. The Earth-centered view placed Earth and humans in a special position.
4. *Heliocentric:* the Sun is at the center of the universe. *Geocentric:* Earth is at the center of the universe.

 Follow Up

Assessment

1. Encourage students to bring in diagrams and pictures that illustrate observations they and astronomers can make about the motion of heavenly bodies, including Earth. Such contributions may be used to make interesting bulletin-board displays.
2. Present students with the following scenario. "Imagine that an alien has just arrived from a distant planet. The alien knows a lot about the universe but doesn't know what it looks like from Earth."

   Ask students to explain what the alien is likely to see regarding the motion of heavenly bodies. Suggest that students role-play this scenario after they have written their responses as an essay.

Extension

1. Suggest that students make a timeline highlighting people who have influenced the history of astronomy. Each name should be accompanied by a major accomplishment. (Possibilities include: Pythagoras, Eudoxus of Cnidus, Aristotle, Heraclides of Pontus, Aristarchus of Samos, Hipparchus, Ptolemy, Copernicus, Tycho Brahe, Galileo, and Isaac Newton. Encourage students to recognize in their timelines the contribution of women and diverse cultural groups.)
2. Suggest that students make an "astronomical calendar" showing events that can be observed for several months or throughout the year. Suggest that their calendars include information on the phases of the Moon, when and where planets will be visible, when eclipses will occur, and so on. (This information can be found in almanacs and astronomy periodicals.)

EXPLORATION 2—CONTINUED

Ptolemy's solution was to add another motion to a planet. He suggested that a planet not only revolves in a circular orbit around Earth but also makes a smaller revolution about points on its main orbit—like a small wheel turning on the circumference of a larger wheel. To get a good picture of Ptolemy's idea, imagine the rocking seat of a ferris wheel actually making a complete turn as the ferris wheel itself turns. The motion of the seat is like the motion of a planet in Ptolemy's model.

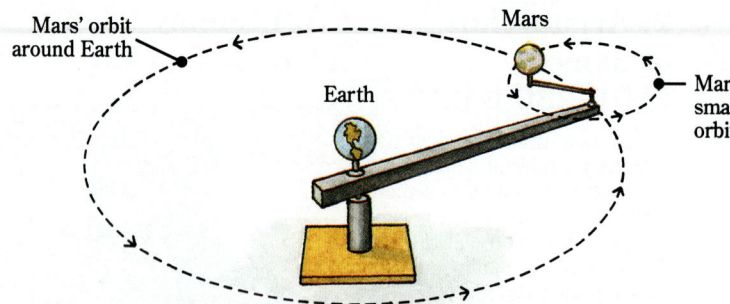

Ptolemy gave Mars two motions. He thought that Mars moves in small circles as it moves around Earth in a larger orbit.

- Study the model above that shows how Ptolemy thought Mars moved. How does Ptolemy's "solution" explain the changing brightness and occasional backward (retrograde) motion of a planet such as Mars, as viewed from Earth?
- Can you think of another explanation for retrograde motion?

Thinking Further

1. Is the theory you proposed on page 388 similar to any of the early Greek and Egyptian observers' theories? Which theory seems to account best for your observations of the heavens? Why?
2. Imagine that you lived at the time of Aristotle, but your ideas matched those of Aristarchus of Samos. What could you say to convince Aristotle that Earth revolves around the Sun? Remember, they had no telescopes and spaceships back then.
3. For 1500 years, people believed that Earth was the center of the universe. Why do you think this was such a popular view?
4. Aristarchus of Samos proposed a *heliocentric* model of the Solar System. Aristotle and Ptolemy both proposed *geocentric* models. Based on their theories, what do you think these terms mean?

Challenge Your Thinking

1. Many early astronomers suggested that Earth is flat. For example, one astronomer pictured Earth as a flat circular disk floating in a sea of water. Another early observer considered Earth to be a flat object freely suspended in space. What may have been some reasons for thinking Earth is flat? What are some reasons for thinking Earth is spherical?

2. Polaris (the North Star) will appear noticeably higher and higher above the horizon the farther north you travel, until at the North Pole, it is directly overhead. Does this observation support a flat-Earth or a spherical-Earth theory? Study the diagrams to the right and explain how they help support your answer.

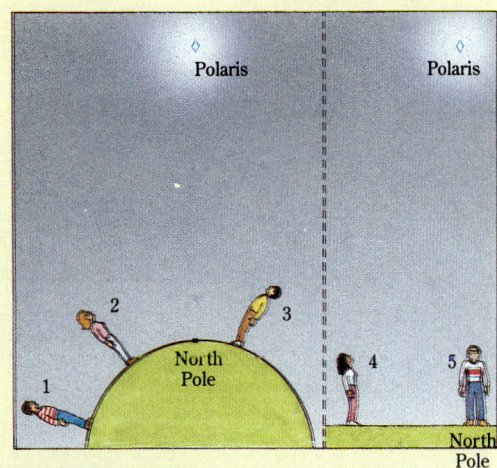

3. The scenes below show various positions of Polaris over a horizon. Copy the scenes into your Journal and label each one with the observer (from the spherical-Earth diagram) who could have seen such a view. Which scene below is closest to what Observer #4 would see? Would the view that Observer #5 sees be much different from that of Observer #4?

Answers to Challenge Your Thinking

1. Reasons for thinking Earth is flat:
   - It looks flat.
   - If the Earth were spherical, we might fall off when our rotation takes us sideways or upside down.

   Reasons for thinking Earth is spherical:
   - We can't see all the way across it, even with a telescope.
   - As ships disappear over the horizon, the top of the mast is the last to disappear.
   - As you approach a point on the horizon, objects behind that point rise into view.
   - Earth's shadow on the Moon during an eclipse is curved.
   - The Sun's angle above the horizon at noon is lower as you go north. At times, the Sun does not rise above the horizon at the North and South Poles.

2. The observation could, by itself, be used to support both theories. However, the amount of change as you move a given distance north on a spherical Earth is greater than the amount of change that would occur if Earth were flat. Therefore, the observation supports the spherical-Earth theory. From the diagrams, you can see that Observer #1 will see Polaris just above the horizon, while Observer #2 will see it at about 45 degrees to 50 degrees above the horizon. Observer #4 will see Polaris as being more directly overhead than Observer #3 will.

3. The scene on the left would be seen by Observer #2. The middle scene would be seen by Observer #3. The scene on the right would be seen by Observer #1. The middle scene is closest to what Observer #4 would see. Observer #4 would see almost the same view that Observer #5 would.

The Observers

Lesson 3

✴ Getting Started

This lesson focuses on the heliocentric model of the Solar System. The lesson begins with Galileo on trial for defending the ideas of Copernicus. Students are asked to prepare a defense for Galileo based on what they have learned. To prepare, they must make a model of the Solar System based on Copernicus' ideas. They will use it to determine whether the Copernican theory explains celestial motions that the theories of Aristotle and Ptolemy do not.

Main Ideas

1. Religious beliefs and scientific ideas are sometimes in conflict.
2. The Copernican model of the Solar System can be used to explain the motions of the planets and the apparent motions of the Sun and stars.

✴ Teaching Strategies

Remind students that they read about Aristarchus in the previous lesson. Call on a volunteer to briefly explain the model Aristarchus developed. *(the Sun-centered theory)* Ask students if they recall what happened to Aristarchus' theory. *(It was forgotten for nearly 2000 years.)* What theory was believed instead? *(Ptolemy's Earth-centered theory)*

The Trial of Galileo

Have students read about Galileo's trial to discover what they are to do in his defense. Then have them work in small, cooperative-learning groups to discover answers to the questions on page 393 and to complete the Exploration.

THE EARTH MOVES

3 A Scientific Revolution

The Trial of Galileo

It is June 1632. The atmosphere in the courtroom is electric. The greatest scientist of the time, Galileo Galilei, is on trial for teaching the views of Copernicus: that the Sun is the center of the Solar System—not Earth. In this model, Earth is just one of a number of planets that revolve around the Sun. This is the second time Galileo has been on trial for his views. After the first trial, he was ordered by the Roman Catholic Church not "to hold or defend" Copernicus' theory. At the second trial, Galileo must answer to charges that he has willfully disobeyed this order.

The verdict—guilty on all counts. The sentence is house arrest for the remainder of his life. Legend says that as Galileo was led sadly from the courtroom, he was heard to mutter ". . . and still the Earth moves."

LESSON ORGANIZER

Objectives

By the end of the lesson, students should be able to:
1. Explain the motions of the planets and the apparent motions of the Sun and stars by using the Copernican model of the Solar System.
2. Explain the apparent retrograde motion of Mars by using the Copernican model of the Solar System.
3. Explain what is meant by the *plane of the ecliptic.*
4. Draw a scale model of the distances of the planets from the Sun.

Process Skills

using models, analyzing, inferring, drawing conclusions

Imagine you are living at that time and have been asked to be part of a team of lawyers preparing for the trial of Galileo. The task is great, and much needs to be done. Here are some thoughts that your team has jotted down. Are there other questions that may be important in working toward an acquittal ("not guilty")?

- Who was Galileo? What was his background?
- Who was this man Copernicus? Where did he come from? What were his ideas? Why does Galileo consider himself a Copernican?
- Why would Church officials care whether the Sun went around Earth or vice versa?

EXPLORATION 3

Copernicus' Model: The Planets and the Ecliptic

To prepare for the trial, you and your team need to make a model of the Solar System based on Copernicus' ideas. You want to see whether the Copernican theory explains celestial motions that the theories of Aristotle and Ptolemy could not explain.

Part 1
The Model and Its Use

At the time of Galileo, only six planets were known. How many are known now? Galileo taught that the six planets orbited the Sun. Your model will show the orbits of the first four planets you would see if you started at the Sun and traveled outward. The distance of each planet from the Sun is expressed in terms of Earth's average distance from the Sun. This distance is referred to as one astronomical unit (AU), and it is equal to 150 million kilometers. (Try writing out this figure.) The time it takes for each planet to revolve once around the Sun is given in Earth days.

Your model will have a scale of 1 AU = 10 cm. Using this scale, Earth would be 10 cm from the Sun on your model. Why?

Observers from the earliest times noticed that, when seen from Earth, the planets moved against a background of 12 different groupings of stars, called **constellations.** These constellations make up what is popularly known as the **zodiac.** The early observers knew that the planets did not wander randomly through the heavens. They all appeared to move counterclockwise through the zodiac, which consists of the constellations Leo, Virgo, Libra, Scorpio, Sagittarius, Capricorn, Aquarius, Pisces, Aries, Taurus, Gemini, and Cancer. Your model would not be complete without the 12 constellations of the zodiac.

Planet	Average Distance (AU)	Time It Takes to Revolve Once Around the Sun (in Days)
Mercury	0.4	88
Venus	0.7	225
Earth	1.0	365
Mars	1.5	687

Answers to
In-Text Questions

- Galileo (1564-1642) was an Italian astronomer and physicist. He is considered to be the first to effectively use the telescope to discover important new facts about astronomy. Galileo became convinced that the theories proposed by Copernicus were correct, and he spent much of his life proving them.
- Copernicus (1473-1543) was a Polish astronomer who revived and expanded on the ideas of Aristarchus. Like Aristarchus, Copernicus believed that Earth and all the other planets revolve around the Sun.
- Roman Catholic Church philosophy rested on the notion that Earth and humans were the center of all things. Removing Earth from the center would make Earth and humans seem less significant.

Answers to
Exploration 3

Part 1
The six known planets at the time of Galileo were Mercury, Venus, Earth, Mars, Jupiter, and Saturn. Today, Uranus, Neptune, and Pluto complete the list of planets circling the Sun. Students should recognize that Earth will be drawn 10 cm from the Sun in their model because it is 1.0 AU away. Be sure students understand that the term *zodiac* identifies the 12 constellations.

(Answers continue on next page)

New Terms

Zodiac—circular backdrop of 12 constellations through which the planets appear to move, located along the ecliptic.
Rotation—the turning of an object about its own axis.
Revolution—the motion of one object around another object in a circle, ellipse, or curve.
Plane of the ecliptic—the plane in which Earth orbits the Sun.

Materials

Exploration 3
Part 1: pencil; compass; protractor; meter stick; metric ruler; glue or tape; star charts; 4 sheets of unlined paper glued or taped together to form one large rectangle; several small Styrofoam balls or marbles; modeling clay

Teacher's Resource Binder

Resource Transparencies 7–3 and 7–4

Science Discovery Videodisc

Disc 2, Image and Activity Bank, 7–3

Time Required

two to three class periods

Answers continued

1. Students should use the information in the table on page 393 to determine the size of the "orbits" in their model. Suggest that students draw their circles on the lower half of the paper. They will need the extra room at the top of the paper for *Part 2* of the Exploration. You may wish to point out that based on the scale they are using, every tenth of an AU is equal to 1 cm. Therefore, the orbit of Mercury will have a 4-cm radius, the orbit of Venus will have a 7-cm radius, the orbit of Earth will have a 10-cm radius, and the orbit of Mars will have a 15-cm radius. The outermost circle in the model represents the orbit of Mars.
2. Since the diagram is a representation of the Copernican, or heliocentric, model of the Solar System, the Sun is located at the center.
3. Remind students that there are 360 degrees in a circle. Since there are 12 constellations, each one will occupy one-twelfth, or 30 degrees of each circle. You may wish to demonstrate how to use a protractor to divide the circumference of a circle into 12 equal sections.
4. Students may recognize that the names of the constellations are also the names of the birth signs used in astrology. They may have seen these names in newspaper columns, in books on horoscopes, on birthday cards, and so on. Be sure students understand that astrology is based in mysticism. Astronomy is a science based on observation.
5. Astronomy magazines, such as *Sky and Telescope* and *Astronomy*, include monthly star charts that students can use to learn the correct location of Earth in their models.

(Answers continue on next page)

EXPLORATION 3—CONTINUED

You Will Need

- a pencil
- a compass
- a protractor
- a meter stick
- star charts
- glue or tape
- four sheets of unlined paper, fastened together to form one large rectangle
- several small plastic-foam balls or marbles to represent the planets
- modeling clay to hold the plastic-foam balls or marbles in place

What to Do

1. Turn your glued paper so that its longer direction is up and down. Using a compass, draw a circle with a radius of 15 cm. The center of the circle should be in the lower half of the paper so that the circle comes within 4 cm of the bottom of the page. Which planet would have the orbit you just drew? Now draw other circles to represent the orbits of the remaining planets from the table on page 393.

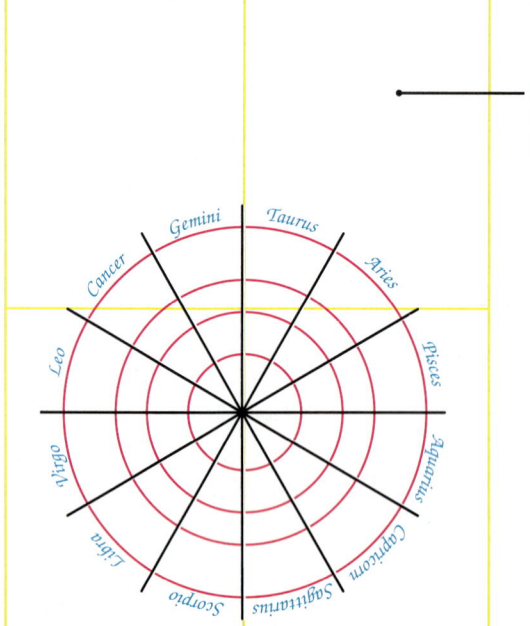

four pieces of paper glued together

2. Where is the Sun in this model?
3. Using a protractor and ruler, divide the circles into 12 equal segments. Each segment will form an angle of 30 degrees at the center of the circles. Extend each line 4 cm beyond the orbit of Mars.
4. In the segments outside the orbit of Mars, place the names of the constellations that form a backdrop to the moving planets. Refer to the diagram above to place the zodiac constellations in their proper order. Place Taurus at the top of your model, as shown in the diagram. Each constellation will occupy one 30-degree segment, although in reality they vary in size. Of course the stars making up these constellations are actually a tremendous distance beyond the orbit of Mars. Have you come across the names of these constellations before? Where?
5. Examine a star chart for this month. Which constellations of the zodiac are visible in your night skies? Where must Earth be placed on your model such that those constellations would be visible this month? Use a marble or plastic-foam ball to represent Earth, and place it in its proper location on the model. Use a small piece of modeling clay to hold it in place.

Interpreting the Model

1. From this model, infer what ideas Galileo taught that eventually caused him problems with the Church authorities.

2. Demonstrate on the model the motion of Earth over one year. To make the model fit with observation, the ball must be moved in a counterclockwise motion around the Sun. What might you observe about the stars' positions because of Earth's motion?

3. Use another plastic-foam ball or marble to represent Mars. Place Earth and Mars in their respective orbits with the constellation of Taurus as their backdrop of stars. Move Earth through one year. In this time, how far has Mars moved? After the planets have moved this distance, will Mars be visible from Earth?

4. Any model must show why you experience night and day. How would you move the ball to demonstrate this? The motion you just showed is called **rotation**. What is the difference between rotation and **revolution**? Refer to the picture of the star tracks on page 387 and the observations you made of the Big Dipper. In what direction must Earth be rotating to cause these effects?

5. As the year progresses, new constellations of the zodiac appear in the eastern sky, while others disappear in the western sky. Use your model to explain this observation.

6. Copernicus (and later Galileo) inferred that the Sun and planets lie in the same plane. In other words, they inferred that the Solar System is essentially flat. This plane is called the **plane of the ecliptic.** Where is the plane of the ecliptic in your model?

Are the people on the ride rotating, or revolving?
Is the ball rotating, or revolving?

7. In your opinion, is this a good model? Could it be used as an exhibit in the retrial of Galileo? Does it explain the retrograde motion of Mars in a satisfactory way? Is it a simpler explanation than the two motions that Ptolemy suggested? Part 2 of this Exploration will help you discover some answers.

Answers continued

Interpreting the Model

1. Galileo taught that the Sun was the center of the Solar System and that the planets revolved around the Sun. This idea caused Galileo problems with the Church because it removed Earth and humans from the center of all things.

2. The stars appear to change their position throughout the year. Different constellations are seen at different times of the year.

3. Since the Martian year is almost twice as long as Earth's, it will have moved halfway around its orbit and be on the opposite side of the Sun from Earth after one Earth year. Mars will not be visible from Earth because the Sun will be in the way.

4. To show night and day on the model, "Earth" must be rotated on its axis in a west-to-east direction. Rotation, or the turning of an object about its own axis, is different from revolution. Revolution occurs when one object circles or moves around another, as when Earth revolves around the Sun. As a result of Earth's west-to-east rotation, the star tracks appear to be going in a counterclockwise direction.

5. As Earth rotates on its axis, constellations appear to rise in the east and set in the west, just as the Sun does. As Earth revolves around the Sun, new constellations will appear in the eastern sky, and old ones will disappear from the western sky.

6. In this model, the paper represents the plane of the ecliptic. The Sun and planets, with the exception of Pluto, lie more or less within this plane.

7. Most students will probably agree that this is a good model because it can be used to explain many of the daily and yearly astronomical events observed from Earth.

Answers to captions: The people on the carnival ride are revolving; the ball is rotating.

The Earth Moves 395

Part 2

Before students begin reading, remind them that they learned about the *retrograde motion* of Mars in Lesson 2. Call on a volunteer to explain retrograde motion. *(the apparent backward movement of an object)* Determine if students can recall Ptolemy's explanation for this phenomenon. If they can't, refer them to Lesson 2. *(Ptolemy proposed that Mars not only revolved around Earth, but also revolved in a smaller orbit, as shown on page 390. He theorized that this created retrograde motion.)* Then involve students in a discussion of how Copernicus might explain why Mars appears to reverse direction and go backward in its orbit for a while. (Accept all reasonable responses.) Have students read the introduction to *Part 2* to discover if their ideas were correct.

Remind students to refer to the diagrams in their texts as they determine the locations of Earth and Mars during a period of 100 days.

EXPLORATION 3—CONTINUED

Part 2
The Retrograde Motion of Mars

What causes Mars to appear to go backward in its path for a while, before continuing in its original direction? A Copernican would say that it is simply the result of the faster Earth catching up to and passing Mars. This causes Mars to appear to loop backward in the sky when viewed against the more distant stars.

You can demonstrate this by following the progress of Earth and Mars over a period of 100 days on your model. Every 10 days on the model, note where these planets lie among the more distant stars.

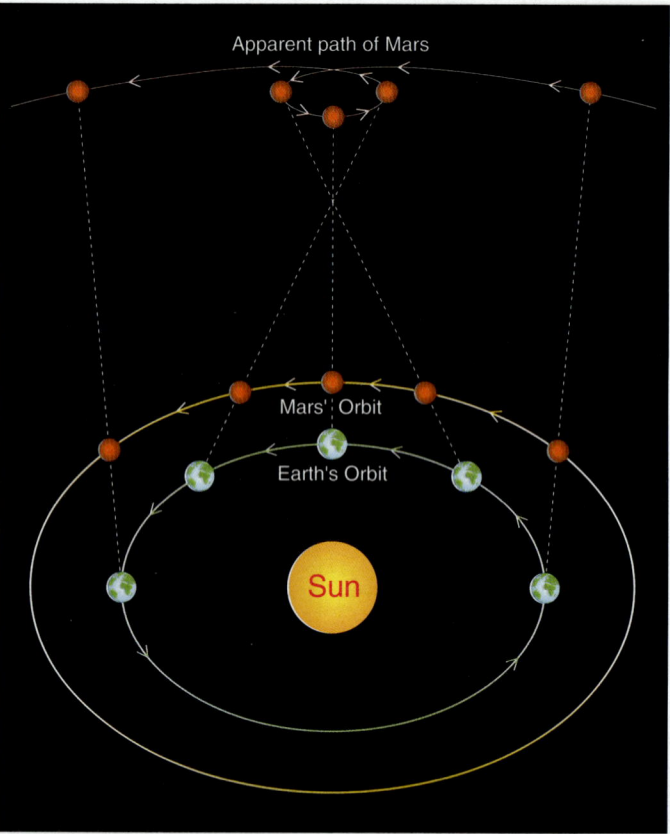

396 Toward the Stars

What to Do

1. The top of your page will represent where the stars appear to be when viewed from Earth. Call this the *star line*. In reality, this is a tremendous distance beyond the orbit of Mars. As Mars and other planets move, they will appear to move among the stars along this line.

2. Using a pencil, draw a dot to represent Earth where its orbit crosses the boundary between Taurus and Aries, as shown in the diagram below. This is the starting position of Earth. Label this position as Earth 1.

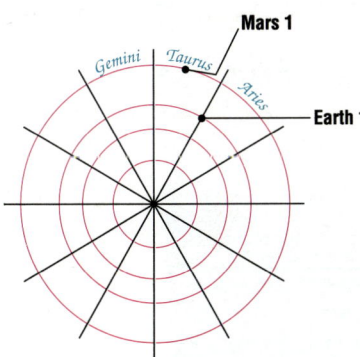

3. Draw Mars on its orbit so that it is halfway between the Taurus/Aries and the Taurus/Gemini boundaries. Label this position as Mars 1, as shown in the diagram above.

4. In 365 days, Earth will travel once around the Sun—a distance of about 63 cm on your model. How far will it travel in one day? Where will Earth be on the model in 10 days? Can you convince yourself that it has moved a distance of 1.7 cm from its original position? Using a compass, measure off this distance and label this position as Earth 2. Repeat this step to locate Earth's position in 20 days, 30 days, 40 days, and so on up to 100 days. Label these positions as Earth 3, Earth 4, and so on up to Earth 10. Be sure that Earth is revolving counterclockwise around the Sun.

5. Where will Mars be after 10 days, 20 days, and so on up to 100 days? Taking into account the larger orbit of Mars and the fact that Mars takes about twice as long as Earth to make the journey around the Sun, you can discover that Mars will move 1.4 cm along its orbit in 10 days. Mark off 1.4-cm segments from the original position of Mars, moving in a counterclockwise direction. Label the ends of these successive segments as Mars 2, Mars 3, and so on up to Mars 10.

6. Draw a line connecting Earth 1 with Mars 1. Extend this line to the star line at the top of the page. This is where Mars will appear to be for an observer on Earth.

7. Repeat this for each new position of Earth and Mars. Where does Mars appear on the star line for each position of Earth and Mars? Does Mars appear to change direction when viewed against the background of stars? In reality, does it go backward?

Defending Galileo

Are you ready to develop your defense of Galileo? How can what you just discovered be used to convince a jury that it is Earth that moves around the Sun—not the other way around? Practice your defense. Try explaining to an imaginary jury why retrograde motion is a natural consequence of Earth's motion. In the pages that follow, you will investigate in greater detail the effect of the motions of Earth as suggested by Copernicus and later by Galileo. This information may provide crucial evidence for the retrial of Galileo.

Defending Galileo

Reassemble the class and involve them in a discussion to summarize what they have learned so far. Completing *Exploration 3* should have helped students realize that the Copernican model provides a plausible explanation for retrograde motion. A geocentric model cannot explain Mars' retrograde motion.

Assessment

1. Suggest that students work in pairs to draw a cartoon or comic strip explaining the Copernican model of the Solar System. Display the finished cartoons around the classroom for others to enjoy.
2. Challenge students to make a model using their fellow students to demonstrate Mars' apparent retrograde motion (i.e., students will play roles of the background stars, Earth, and Mars). Have students demonstrate their "living" model for the class.

Extension

1. Suggest that students do some research to discover the meaning of the terms *celestial equator* and *celestial poles*. Have them make a poster diagram to explain the terms to the class.
2. Point out to students that the names of the constellations are associated with heroes, heroines, and beasts in ancient mythology. Suggest that they choose one of the constellations and do some research to discover the story of the character it represents. Have them share what they learn by presenting oral reports to the class or by making bulletin-board displays.

Answers to

In-Text Questions

4. Students should recognize that the distance Earth moves on their model every 10 days can be calculated in the following manner:
   63 cm ÷ 365 days = 0.17 cm/day
   0.17 cm × 10 days =
   1.7 cm every 10 days

5. The distance Mars moves on the model can be calculated in the same way. If necessary, remind students that the circumference of a circle is determined by the formula:
   pi × diameter.
   3.14 × 30 cm = 94-cm orbit
   94 cm ÷ 687 days = 0.14 cm/day
   0.14 cm × 10 days =
   1.4 cm every 10 days

7. As students connect the positions of Earth and Mars, they should see that as Earth moves past Mars, Mars appears to reverse direction against the backdrop of stars.
   The model shows that Mars continues in its orbit without actually reversing direction. Students should conclude that the retrograde motion of Mars as observed from Earth is an optical illusion created by the different rates at which the two planets are revolving.

The Earth Moves 397

Lesson 4

Getting Started

In this lesson, students investigate how the tilt of Earth's axis and Earth's orbit around the Sun affect the seasons. Students begin by examining what happens as Earth travels around the Sun. Then they discover how Johannes Kepler deduced that the orbits of the planets are ellipses. The lesson concludes with students role-playing the retrial of Galileo using what they have learned in Lessons 3 and 4.

Main Ideas

1. As Earth revolves around the Sun, the tilt of Earth's axis causes the seasons to change.
2. The Sun appears at its highest and lowest points during the summer and winter solstices.
3. During the vernal (spring) and autumnal equinoxes, the Sun is directly above the equator.
4. The planets revolve around the Sun in elliptical orbits.

Teaching Strategies

Display a globe and have volunteers identify and locate the equator, the Tropic of Cancer, and the Tropic of Capricorn. Ask students to identify in which hemisphere each lies. Point out that in this lesson they will learn more about the significance of these imaginary lines.

Many Journeys in a Lifetime

Have students read page 398. Then involve them in a discussion of the observations they would make as tour guides on a trip around the Sun.
(Continues on next page)

 **Motions and Their Effects**

Many Journeys in a Lifetime

How many times have you traveled around the Sun? According to the Copernican theory, you make this journey once a year. The journey covers a distance of 942 million kilometers. If you were a tour guide for this journey, what sights would you point out? What major events would you experience? Besides the changing constellations of the zodiac, what other signposts would tell you where you are on your journey? For example, think about changes you experience throughout the day and year.

Referring to the model of the plane of the ecliptic you made in Exploration 3, write down three observations you would point out as tour guide for this trip. Then try Exploration 4, in which you will model such a journey.

 LESSON 4 ORGANIZER

Objectives

By the end of the lesson, students should be able to:
1. Explain why the length of daylight changes throughout the year.
2. Describe how the tilt of Earth on its axis creates the seasons.
3. Identify the position of Earth during the summer and winter solstices and the vernal and autumnal equinoxes.
4. Illustrate how a planet orbits the Sun by drawing an ellipse.

Process Skills

analyzing, inferring, illustrating, drawing conclusions

New Terms

Axis—a real or imaginary line about which an object rotates. Earth's axis runs from the North Pole to the South Pole, straight through the center of Earth.

398 Toward the Stars

EXPLORATION 4

A Trip Around the Sun

You Will Need

- a large plastic-foam ball (20 cm in diameter or larger)
- an elastic band
- a plastic cup
- 2 toothpicks
- a light source
- your model of the ecliptic from Exploration 3
- a protractor
- a compass

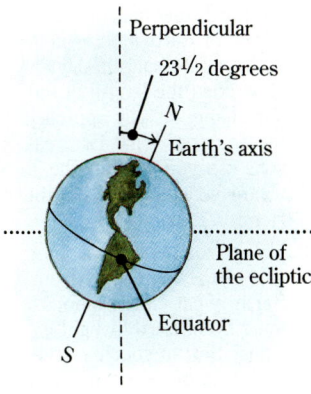

What to Do

1. Form groups of four or five. Set up the model as shown in the diagram below. The elastic band around the large plastic-foam ball represents the *equator* of Earth. The toothpick represents the axis about which Earth rotates. Earth's **axis** is an imaginary line that runs from the North Pole to the South Pole, straight through the center of Earth. The axis is perpendicular to the plane of the equator. The light source represents the Sun.

2. You'll be setting up four situations that represent different positions on your journey at specific times of the year. At each position of Earth, you'll have several missions to accomplish and some thought-provoking questions to consider.

3. In setting up the models, keep two things in mind:
   - The motion of Earth around the Sun is counterclockwise when viewed from over the North Pole of Earth, as you can see from the diagram above.
   - Regardless of Earth's position around the Sun, Earth's axis always points in the same general direction. At any position along Earth's orbit, Earth's axis is parallel to where the axis was (or will be) at any other position in the orbit. Earth's axis is tilted at an angle of 23½ degrees from the perpendicular, toward the plane of the ecliptic. The axis always appears to point toward the star Polaris.

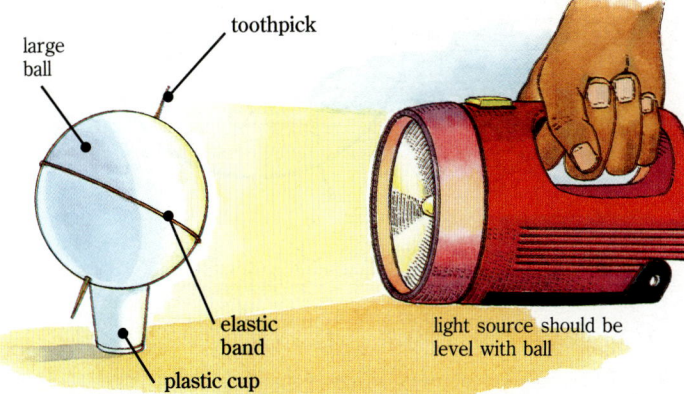

What season is represented in this set-up?

Materials

Exploration 4: large plastic-foam ball (20 cm in diameter or larger), rubber band, plastic cup, toothpicks, flashlight or other light source, models from *Exploration 3*, compass, protractor to measure the 23 1/2-degree angle

Exploration 5: sheet of corrugated cardboard or a stack of several sheets of cardboard, unlined paper, 2 thumbtacks, 40-cm piece of string, pencil or marker, metric ruler

Teacher's Resource Binder

Resource Transparency 7–5

Science Discovery Videodisc

Disc 2, Image and Activity Bank, 7–4

Time Required

four class periods

(Many Journeys in a Lifetime continued)
(Observations might include: sunrises and sunsets, changing phases of the Moon, changing seasons, changing lengths of daylight throughout the year, changing positions of the planets, and the retrograde motion of Mars.)

Exploration 4

Have students read pages 399 to 401. Then review each of the pictures. For the successful completion of the Exploration, it is important that students understand what is illustrated in each picture.

Note that students must position the toothpicks such that, if extended, they would meet in the center of the plastic-foam ball.

You may wish to set up the activity in one of the following ways:
- As described in the text, have groups of four or five complete the activity using their own sets of materials.
- Have groups share one or two light sources set up in the classroom.
- Set up four separate stations for groups of students to use. Each station would represent one of the positions in the Exploration. If this method is used, caution students not to change the orientation of the globes.

Answer to caption: Summer

The Earth Moves 399

Answers to

Position 1: The Winter Solstice

Have student groups complete the activity and discuss and compare their results. Point out that the winter solstice marks the first day of winter and the shortest day of the year.

1.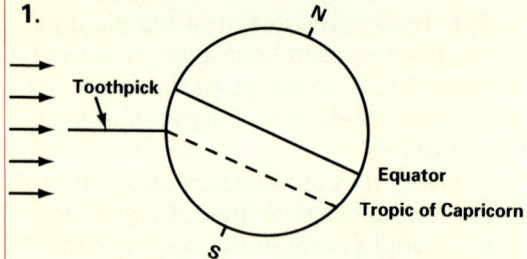

2. The North Pole will remain in darkness for 24 hours because it tilts so far away from the Sun that the Sun's rays do not strike it. In contrast, the South Pole is in continual daylight because it is tilted toward the Sun.

   As you approach the equator from the North Pole, there are more hours of daylight.

   The winter season in the Northern Hemisphere is the result of (1) the fewer hours of sunlight during the days and (2) the tilt of the North Pole away from the Sun, causing the rays of sunlight to be less direct there than they are at any other time of the year.

3. Students should infer that Leo, Cancer, and Gemini will be visible, and they may include Virgo and Taurus as well. According to star charts, Cancer, Gemini, Taurus, Aries, and Pisces are most likely to be seen in the night sky on the winter solstice.

EXPLORATION 4—CONTINUED

Position 1

December 21 or 22—The Winter Solstice

In this position, Earth's axis is tilted at an angle of 23½ degrees from the perpendicular so that the Northern Hemisphere is pointing away from the Sun. Notice where on Earth the Sun is directly overhead. Do this by inserting a toothpick in a position where it will not cast a shadow. The toothpick should be positioned so that if extended, it would pass through the center of the plastic-foam ball. This latitude is 23½ degrees south of the equator—a latitude called the *Tropic of Capricorn*. At this time of the year, the constellation Capricorn cannot be seen from Earth because it is in back of the Sun.

1. In your Journal, use a compass to draw a diagram of Earth similar to the one shown here. Label the equator and the Tropic of Capricorn on your diagram.
2. Demonstrate a day by rotating Earth one full rotation. Which of Earth's poles is in darkness 24 hours a day? How does the length of the day change as you approach the equator? What two reasons can you give for the winter season in the Northern Hemisphere?
3. If Capricorn is on the other side of the Sun, away from Earth, what constellations must be in Earth's night skies? Set up your model of the plane of the ecliptic (Exploration 3) to represent this situation.

Position 2

March 20 or 21—The Vernal Equinox

Move Earth counterclockwise one quarter of the way around the Sun. Be careful to keep the axis pointing in the same direction at all times. Revolving Earth this far means that one-quarter of a year has passed. In this position, people are experiencing equal hours of day and night, all over Earth. Can you see why?

1. At what latitude is the Sun directly overhead at noon? Find out by inserting a toothpick so that it does not make a shadow. Draw a diagram like the one you drew representing Earth at the winter solstice. Label the latitude at which the Sun is directly overhead at noon.
2. Rotate the model Earth in this position so that the opposite side of the model is now experiencing sunlight. Is the Sun still directly overhead at the same latitude?
3. Go to your model from Exploration 3 and set it up to represent the vernal equinox. What constellations are visible in the night sky?

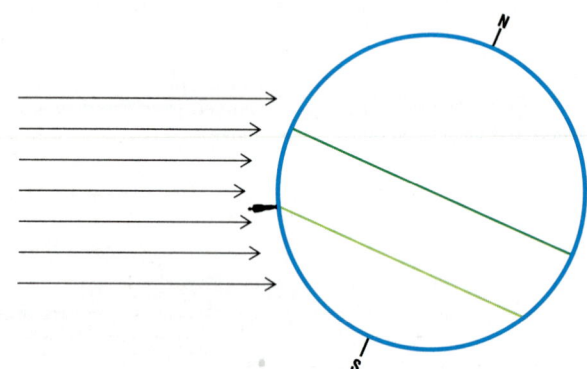

Answers to

Position 2: The Vernal Equinox

The term *equinox* comes from a Latin word meaning *equal night*. The word *vernal* means *of spring*. The vernal equinox marks the beginning of spring.

1. A toothpick inserted into the equator will not cast a shadow because the "Sun" (light source) is directly overhead. Students' diagrams will be similar to the illustration on page 400, only the person will be standing on the equator.
2. Caution students that as they rotate "Earth," they must be careful to keep the axis pointing in the same direction. They should discover that when they rotate Earth in its vernal-equinox position, the Sun is always directly overhead at the equator.
3. Students should infer that Scorpio, Libra, and Virgo will be visible. According to star charts, Leo, Cancer, and Gemini are also visible.

400 Toward the Stars

Position 3

June 20 or 21—The Summer Solstice

Earth has moved so that it is now directly opposite to where it was on December 21 or 22. Earth's axis should be parallel to what it was on that date. This is the first day of summer—the longest day of the year in the Northern Hemisphere.

1. The Sun's hottest rays are hitting Earth 23½ degrees north of the equator at a latitude called the *Tropic of Cancer*. How could you prove that this is so? Draw a diagram like the one you drew representing Earth at the winter solstice, only this time, label the Tropic of Cancer. Will Earth be on the same side of the Sun as it was for the winter solstice?
2. Which pole is in darkness 24 hours a day? in sunlight 24 hours a day? What two reasons can you give for the warmer temperatures in the Northern Hemisphere?
3. Go to your model of the plane of the ecliptic and set it up to represent the summer solstice. What constellation is in back of the Sun? in Earth's night sky?

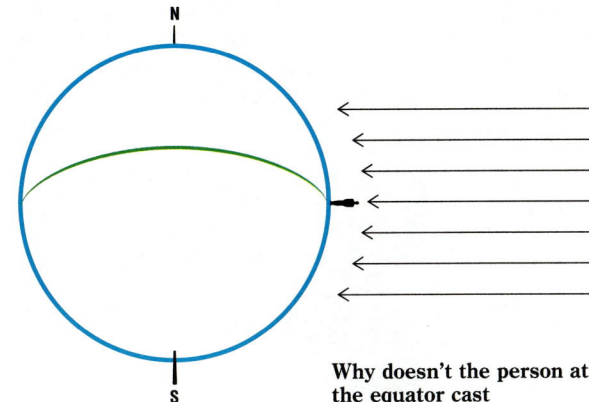

Why doesn't the person at the equator cast a shadow?

Position 4

September 22 or 23—The Autumnal Equinox

In this position on your journey around the Sun, the Sun's hottest rays are again hitting the equator at noon, as they were on March 20 or 21. This means that on this day, the Sun is directly overhead at noon. Again, everyone on Earth is experiencing equal hours of daylight and darkness.

1. During the complete day, is the Sun always directly overhead at the equator at noon? How could you find out?
2. If you were at the North Pole looking toward the Sun, would you see it rise above the horizon during the day, as it does where you live?
3. During a complete year, through what range of latitudes do the Sun's hottest rays move? Where would you most like to live?
4. Set up your model of the plane of the ecliptic to represent the autumnal equinox. What constellations are visible in the night sky at this point in your journey?

Now that you have made this trip, you know enough to be the tour guide on the next trip. What are the major sights and events you would point out on your next journey around the Sun? Record what you plan to highlight.

Answers to

Position 3: The Summer Solstice

During the summer solstice, there are 13 hours and 13 minutes of daylight at 20 degrees latitude, 14 hours and 30 minutes at 40 degrees latitude, and 18 hours and 30 minutes at 60 degrees latitude.

1. When a toothpick is inserted into the Tropic of Cancer, it does not cast a shadow, proving that the "Sun" is directly overhead. Direct rays are the hottest.

Position 3
June 20 or 21—The Summer Solstice

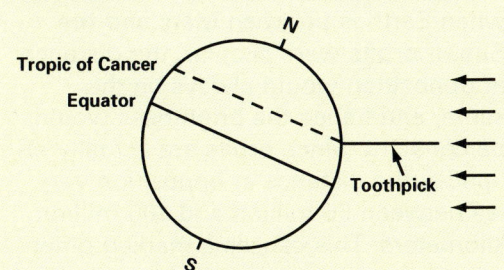

Earth will be on the opposite side of the Sun from where it was during the winter solstice.

2. The South Pole will remain in darkness for 24 hours because it tilts so far away from the Sun that the Sun's rays do not strike it.
   The North Pole will be in continual daylight because it is tilted toward the Sun.
   The warmer temperatures in the Northern Hemisphere are the result of (1) the tilt of the North Pole toward the Sun (so that the daylight hours are longer) and (2) sunlight strikes Earth more directly there than at any other time of the year.
3. From their model, students may suggest that Cancer, Gemini, or Taurus is behind the Sun. The actual answer is Gemini. According to star charts, Scorpio, Libra, and Virgo are also visible.

Answers to

Position 4: The Autumnal Equinox

The autumnal equinox marks the beginning of autumn.

1. During a complete day, the Sun is always directly overhead at the equator at noon. This can be shown to be true by inserting a toothpick into the equator and noting that it does not cast a shadow.
2. The Sun at the North Pole rises above the horizon, but it remains closer to the horizon there than it does where most of the students live.
3. During a complete year, the Sun's hottest rays, or perpendicular rays, move from 23 1/2 degrees south latitude (the Tropic of Capricorn) to 23 1/2 degrees north latitude (the Tropic of Cancer), for a total of 47 degrees.
4. Students should infer that Taurus, Aries, Pisces, and Aquarius will be visible.

Kepler's Puzzle

Have students read about Kepler and his attempts to discover the shape of orbital paths. Talk about how his discovery also provided an explanation for why the brightness of Mars seems to fluctuate. You may wish to remind them that Ptolemy's theory accounted for this phenomenon by suggesting that Mars circled in a small orbit of its own as it revolved around the Earth. Ptolemy's theory suggests that if Mars was on its own orbit, there would be times when it would be farther away from Earth during that orbit. Then as Mars came into view, it would appear faint. Mars would appear brighter if it was near Earth on Mars' own orbit. Students should realize that the Copernican model they have been working with does not support Ptolemy's ideas.

Johannes Kepler (1571-1630) was a brilliant German astronomer. He constructed a model of the Solar System based on a very complex theory involving five different geometric shapes. He sent a copy of his theory of circular orbits to the Danish astronomer Tycho Brahe in Prague. Brahe invited Kepler to come to Prague and work as his assistant.

Within a year, Brahe died, and Kepler was given complete access to Brahe's lifetime accumulation of observations and data. Kepler soon discovered that Brahe's work supported the ellipse as the correct orbital shape of the planets.

Eight years later, Kepler published two laws based on Brahe's observations of the orbit of Mars. The first law stated that every planet moves in an elliptical orbit with the Sun as one of the foci. The second law stated that all of the planets speed up in their orbits as they approach the Sun and slow down as they move away from it.

Mars varies in brightness for two reasons. One, which could be explained by Copernican theory, is that Mars and Earth each orbit the Sun at different speeds. As a result, both planets are on one side of the Sun at certain times, and at others, they are on opposite sides of the Sun. Mars is much brighter when it is close than when it is far away.

Mars also varies in brightness when both planets are in *opposition*, that is, when Earth is between Mars and the Sun. If orbits were circular, the distance at opposition would always be the same, and hence the brightness would be constant. Since orbits are actually elliptical, the distance at opposition varies between 56 million and 100 million kilometers. This causes a marked difference in brightness that could not be explained by the Copernican model of circular orbits.

Kepler's Puzzle

The models you made in Explorations 3 and 4 lend support to Copernicus' theory. But during Galileo's time, those who thought that Copernicus' theory was valid still faced one serious problem: If the orbits of the planets were perfect circles, why does the observed brightness of Mars differ so much each time Earth comes between it and the Sun? The variation is much greater than it would be if the orbits were perfect circles.

Kepler musing over possible shapes for orbital paths

A German astronomer named Johannes Kepler happened on an answer to the problem in 1605 as he was trying to discover the shape of planets' orbits. For three years, Kepler attempted to fit the information known about Mars' orbit to circles of all sizes. At that time, everyone assumed that the planets traveled in perfectly circular paths because to them, the circle represented perfection. But no matter how hard he tried to make it work, no circle would fit the data. Kepler had fallen into the trap of anticipating what he believed to be the correct answer, thus blinding himself to possible alternatives. Has this ever happened to you?

In frustration Kepler finally resorted to an idea that he had previously disregarded—a shape called an *ellipse*, which looks like a stretched-out circle. When the data known about Mars' orbit was applied to this shape, it fit perfectly. Mars' orbit did not describe a circle but an ellipse, instead!

With this discovery, Kepler was able to describe the paths of all revolving objects bound by gravitational forces. His discovery applies to comets, planets, moons, human-made satellites, and even distant stars that have been caught up in each other's gravitational forces.

In the next Exploration, explore the shape for which Kepler at first had such low regard. Then answer the question that had long baffled astronomers: why does the brightness of Mars vary?

EXPLORATION 5

Investigating an Ellipse

You Will Need

- a sheet of corrugated cardboard or a stack of several sheets of cardboard
- unlined paper
- two thumbtacks
- a 40-cm piece of string
- a centimeter ruler

What to Do

1. Press the thumbtacks into a sheet of unlined paper that has been placed on top of the cardboard. Loosely tie a piece of string around the thumbtacks, as shown in the photograph below. Place a pencil through the string and trace out an ellipse. What you'll see is the shape that planets travel as they orbit the Sun. Using the same piece of string, can you make ellipses with different shapes? How?

2. Measure the long axis of several ellipses. What is the relationship between the positions of the thumbtacks and the shape of the ellipse? What must be the position of the thumbtacks to make a circle?

3. At the position of one of the thumbtacks, draw a diagram representing the Sun. This location is a focus of the ellipse. (The plural of focus is *foci*. An ellipse has two foci.) When Earth is closest to the Sun, it is 147 million km from the Sun. At this distance, it is winter in the Northern Hemisphere. At its farthest distance, it is 152 million km from the Sun. Earth reaches this position during the summer in the Northern Hemisphere.
   (a) Place this information on your diagram. Label the summer and winter positions of Earth.
   (b) How do you explain that when Earth is closest to the Sun, it is winter in the Northern Hemisphere? (Think of what you found out at Position 1 in Exploration 4.)

4. How does Kepler's discovery explain why Mars changes in brightness when seen from Earth?

5. Through his observations, Kepler made another discovery about planets moving in their elliptical orbits: The planets do not move with constant speeds. A planet increases in speed as it gets closer to the Sun. Compare the speed of Earth in its orbit during spring, summer, fall, and winter. If Earth's orbit were circular, what might you infer about its speed?

Drawing an ellipse

Exploration 5

To do this activity, each student needs to have his or her own set of materials. Monitor how students are making their ellipses by moving among them and observing their progress. If necessary, point out that the string must be longer than the distance between the two thumbtacks. As they draw, students should hold the pencil upright against the string so that the string is always taut.

Encourage students to describe the shape of their ellipses. Help them to recognize that an ellipse is shaped like an oval whose ends are the same size.

Answers to In-Text Questions

1. Ellipses with different shapes can be drawn by changing the distance between the two thumbtacks.
2. The closer together the thumbtacks, the more circular the ellipse becomes. The farther apart the thumbtacks, the longer the ellipse becomes. A circle will result if the thumbtacks are placed on top of each other (in effect, if only one thumbtack is used).
3. (a)

   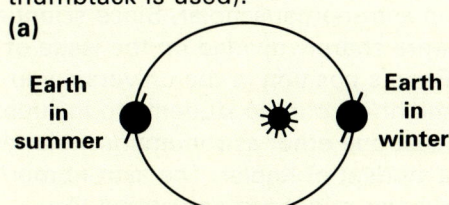

   (b) It is winter in the Northern Hemisphere (even though Earth is closest to the Sun at this time) because Earth's axis is tilted away from the Sun. As a result, the Sun's rays do not strike the Northern Hemisphere directly, and there are fewer hours of daylight.
4. Kepler's discovery shows that differences in Mars' brightness as seen from Earth are due to Mars' elliptical orbit. Mars appears brightest when Mars and Earth are closest together. At their greatest distance apart, Mars appears dimmest.
5. In the Northern Hemisphere, Earth's orbital speed increases during the summer and fall (from the summer to winter solstice) since Earth is getting closer to the Sun at that time. It decreases during the winter and spring (from the winter to summer solstice) as Earth gets farther from the Sun. Because of Earth's elliptical orbit, the time interval between the vernal equinox and the autumnal equinox is longer than that between the autumnal equinox and the next vernal equinox. The distance between Earth and the Sun is closest in January—about 147,100,000 km (as opposed to 152,100,000 km at its greatest distance). Therefore, Earth completes the semi-ellipse from the autumnal equinox to the vernal equinox faster than it does the opposite semi-ellipse. If Earth's orbit were circular, Earth's orbital speed would be constant.

A Mock Trial

The scenario the students are to enact will not be historically correct. Galileo would not have had a lawyer of his own. Questions would have been asked by an Inquisitor, a person chosen by the Roman Catholic Church. The Inquisitor may have tried to act impartially, but he would certainly have thought that Galileo was wrong.

Call on volunteers to make up the six teams and the jury. One role should be assigned to each team. The roles include: Galileo, his lawyer, the prosecutor, the Church official, a merchant, and an astronomer/scholar. Since scientists were sharply divided on the issue of Earth's position in the universe, you might encourage students to include at least one other astronomer/scholar and a student of Kepler. The astronomer/scholar might say something like: "There is no doubt that Galileo is a good experimenter. But his conclusions contradict the unsurpassed work of Aristotle, Ptolemy, and other reputable scholars. Most scientists support their theories."

Each team is responsible for determining the viewpoint to be argued by its character. Students not on teams can be jury members. Having students participate in this mock trial should allow them the opportunity to summarize and analyze the main concepts presented in Lessons 3 and 4.

Follow Up

Assessment

Point out to students that the equinoxes and solstices can be predicted to the minute. Suggest that they find the exact time of the vernal and autumnal equinoxes and summer and winter solstices for the current year.

Ask them to make a poster diagram to show Earth's position at each of these times. (Have almanacs available so that students can locate this information.)

A Mock Trial

You have all the physical evidence needed for the retrial of Galileo. Which were the more persuasive arguments in support of the Copernican theory? What questions would you ask of the following witnesses who have been asked to testify at the retrial? Are there other witnesses you would call? What other witnesses could the prosecutor call?

Set up a mock retrial of Galileo with different class members playing the different roles. Form at least six small teams. Each team will be responsible for filling out the details of one or more roles and for choosing which team member will play that role. Those not playing a role will be the jury. Besides Galileo, his lawyer, and the prosecutor, include at least the following three roles. You may, of course, add other roles to the list.

- Church official: "I believe absolutely in the teaching of Aristotle. Earth, and therefore humans, are at the center of all things."
- Merchant: "What do I know of moving planets or suns? I believe what I *see*. I see the Sun moving, so I believe it moves."
- Astronomer/Scholar: "Galileo is very persuasive in his arguments. But I would like to examine both theories more carefully."

Extension

1. Explain to students that Earth is not perfectly round and that it wobbles on its axis. Suggest that they do some research on each of these phenomena and share what they learn by making fact sheets.

2. Challenge students to do some research on the orbits of Mars, Earth, Venus, and Mercury. Then have them use this information to make a poster diagram showing the elliptical orbits of the four planets. Their diagrams should indicate how far and how close each of the planets is to the Sun.

Challenge Your Thinking

1. You've made many observations of what happens in the Northern Hemisphere as a result of the motions of heavenly bodies. What happens in the Southern Hemisphere? Which way will the stars appear to revolve there—clockwise or counterclockwise? On days when you wear a winter coat to school, what would Australian students be most comfortable wearing?

2. The observations below are all true. Using plastic-foam balls, toothpicks, and any other materials you think you'll need, show why each observation is true. Work with others in groups so that you can share ideas as scientists do when they work on a problem. Then, with the aid of a diagram, explain in writing why each is true.

    The Observations:
    (a) The Sun rises in the East and sets in the West.
    (b) The Sun is higher in the sky in the Northern Hemisphere in the summer than in the winter.
    (c) In the summer in the Northern Hemisphere, the North Pole has 24 hours of daylight.
    (d) The Moon goes through phases from new Moon (when it is not visible in the nighttime sky) to full Moon (when the full face of the Moon can be seen).

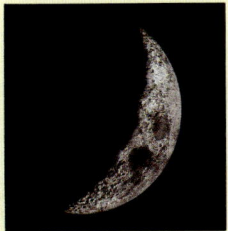   

Crescent moon **Quarter moon** **Gibbous moon** **Full moon**

   (e) Each night the Moon rises later (by approximately 50 minutes) than it did on the previous night.
   (f) The same face of the Moon is always visible from all parts of Earth.

3. In July 1991, a complete eclipse of the Sun was visible in Hawaii and Mexico. With a diagram or your model, show how a total eclipse of the Sun occurs and why one is not seen every year.

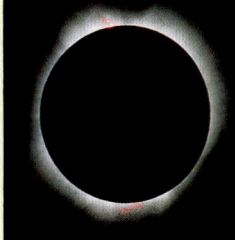

Answers to Challenge Your Thinking

1. In the Southern Hemisphere, the stars appear to revolve clockwise about the South Polar axis. Australian students would be wearing summer clothing when it is winter in the Northern Hemisphere.

2. (a) Because Earth's rotation is from west to east, the rays of sunlight cross Earth's surface from east to west, making the Sun appear to rise in the east and set in the west.
   (b) The Sun appears higher in the sky when the N. Hemisphere is tilted toward the Sun (summer) because the Sun is closer to being directly overhead than when the N. Hemisphere is tilted away (winter).
   (c) Due to Earth's tilt on its axis, the Sun's rays shine 24 hours a day for all points north of 66½°N latitude (inside the Arctic Circle). This happens continuously on and around the summer solstice. The N. Pole has 24 hours of daylight from the vernal to the autumnal equinox.
   (d) When the Moon is between Earth and the Sun, we "see" only its darkened half. As the Moon's orbit takes it around Earth, we see a larger and larger portion of the lighted surface, until the Moon is full (when Earth is between the Moon and the Sun) and we see its entire lighted face.
   (e) Since the Moon orbits Earth once every 27⅓ days, it is approximately 1/27 of its way farther around Earth each time Earth completes one rotation (24 hours). To reach the point where the Moon rises, Earth must rotate a little farther each day. This takes about 1/27 of a day (about 50 minutes).
   (f) The Moon rotates once on its axis in exactly the same time it takes to make one revolution around Earth. Both motions occur in a west-to-east direction.

3. A total solar eclipse occurs only when the Moon's orbit takes it exactly between the Sun and Earth. A total eclipse does not occur every year because the Moon's orbit is oriented at a slight angle to the plane of the ecliptic.

Lesson 5

Getting Started

In this lesson, students discover the difference between meteors, meteorites, and comets. By comparing meteorite-impact craters on Earth to those on the Moon, students conclude that similar forces created both. By learning that natural forces, such as weathering and erosion, have destroyed most of the craters on Earth, students conclude that the absence of these forces on the Moon explains why so many craters are found there. The lesson continues with discussions about comets and concludes by examining the history and movements of Halley's comet.

Main Ideas

1. Meteors are lumps of rock and/or metals that sometimes enter Earth's atmosphere.
2. A meteorite is a meteor that strikes Earth's surface.
3. Meteorites have formed craters on Earth that are similar to those on the Moon.
4. Much evidence of meteorite impacts on Earth has been destroyed by weathering and erosion.
5. Comets are frozen masses of water, dust, gases, and other materials.

Teaching Strategies

Have students respond by a show of hands if they have ever seen a "shooting star." Call on a volunteer to share his or her experience with the class. Invite class members to discuss what shooting stars are like, where they come from, and what they are made of. Point out that they will discover if their ideas are correct as they read Lesson 5.
(A teaching strategy for Meteors and Meteorites follows on next page)

EXPLORING THE SOLAR SYSTEM

5 Visitors from Space

Meteors and Meteorites

The photo above shows evidence of visitors from space to Earth. Do you know what these visitors are?

Try viewing the Moon through binoculars. What is its most distinguishing characteristic? Is there any evidence of water? of wind and dust storms? What do you suppose caused the craters on the Moon? Much can be learned about your nearest neighbor by simple observation.

The Moon's features are similar to those on Earth. Compare the picture of the crater in Arizona, shown above, with a Moon crater, shown on the right. How are they similar? Do you see any differences? Could the craters found on both surfaces have been formed in the same way? You can simulate one possible cause for the craters in the following way. From a height of 1 m, drop several milliliters of water onto a bed of flour. What do you observe? What could have been "dropped" on Earth and on the Moon to create the craters there?

Lumps of rock and/or metals that sometimes enter Earth's atmosphere are called **meteors.** If large enough, they crash to Earth instead of burning up completely in the atmosphere. Meteors that reach the ground are called **meteorites.** The craters are visible evidence of their impact, even though the meteorite itself may not be found. That's because the heat generated on impact can vaporize a meteorite. How does your flour-and-water activity simulate this part of the process of crater formation?

Moon crater

LESSON ORGANIZER

Objectives

By the end of the lesson, students should be able to:
1. Identify the differences between a meteor, a meteorite, and a comet.
2. Explain why many meteorite-impact craters are visible on the Moon but not on Earth.
3. Describe the composition of meteors and comets.
4. Explain the origin of comets.
5. Describe the relationship between comets and meteor showers.

Process Skills

observing, analyzing, inferring, drawing conclusions

406 **Toward the Stars**

Much of the evidence of meteorite impacts on Earth has been destroyed through weathering and erosion by air and water. How could weathering and erosion be simulated in the flour-and-water activity? Why do there seem to be so many more craters on the Moon?

Have you ever seen a "shooting star"? If you have, what you saw was actually not a star at all, but something within Earth's atmosphere. (Recall that stars are located great distances away from Earth.) What observation might support the statement that "shooting stars" are a nearby phenomenon? What you saw were rock particles or "dust" that had entered the atmosphere. The heat created as the particles streaked through the atmosphere caused the meteor to glow white-hot, burning itself up before it could reach Earth's surface.

Meteor shower

During certain times of the year, meteor showers are quite common. The photo above shows a meteor shower. Many "shooting stars" can be observed in a short time-period. When this happens, Earth is passing through debris left behind by another visitor from the far outreaches of the Solar System. What might this visitor be? Read on.

Meteors and Meteorites

Have students read page 406 and discuss the questions. Most students will agree that the Moon's most distinguishing characteristic is its many craters. Most will agree that there is no real evidence of wind or water on the Moon. They should conclude from the photographs that the Arizona crater and the Moon's craters are similar in that they are both deep and round. The Moon's craters, however, appear rougher.

Divide the class into small groups and distribute the materials so that they can complete the simulation activity suggested on page 406. Students should conclude that craters on Earth and the Moon may have been created when objects from space struck their surfaces.

Have students finish reading about meteors and meteorites. Students can simulate weathering and erosion by gently blowing on the flour and spraying it with water. They should conclude that the absence of weathering and erosion on the Moon has left its craters intact, while Earth's erosional forces have destroyed most of the craters on Earth.

New Terms

Meteors—lumps of rock and/or metals that sometimes enter Earth's atmosphere.
Meteorites—meteors that have made impact with Earth's surface.
Comets—ice masses that consist of frozen water, other frozen material, dust, and frozen gases that slowly orbit the Sun.
Asteroids—rocks and boulders that have been observed in the Solar System. They are different from comets in that they have no tails.

Materials

Meteors and Meteorites: cookie sheet or wax paper, flour, water, eyedropper, meter stick

Science Discovery Videodisc

Disc 2, Image and Activity Bank, 7–5

Time Required

two class periods

Exploring the Solar System 407

The Hairy Star

Have students read page 408 silently. Then direct their attention to the photograph of the Bayeaux tapestry. Help them to locate the comet and involve them in a discussion of the people's reactions to it. Encourage students to speculate as to why people were fearful of comets. (Answers may vary, but possible responses include: they did not understand what comets were; they were afraid that a comet might hit Earth; they believed that comets were bad omens.)

As students continue reading the lesson, they should discover the answers to some of the questions at the bottom of page 408. At this time, use the responses provided below to guide students toward the correct answers.

Answers to
In-Text Questions

- Comets travel around the Sun in an elliptical path. Some make the journey in less than seven years. Others may travel in such a large orbit that it takes thousands of years to complete the orbit. Astronomers consider comets to be part of the Solar System.
- The head, or nucleus, of a comet probably consists of frozen gases, ice, and dust—something like a dirty snowball. As a comet approaches the Sun, heat causes the outer layers of the nucleus to vaporize. This vaporization releases the dust and gases that form the tail. A comet's tail may extend for 160 million km across space. Solar wind pushes the comet's tail so that it is always pointing away from the Sun.
- Asteroids, also called minor planets or planetoids, differ from comets in that they are made mostly of rocky material. As a result, they do not form comet-like tails. Most asteroids are found between the orbits of Mars and Jupiter, forming a band called the *asteroid belt*.
- Eventually, a comet will vaporize or burn away, but this may take millions of years.

408 **Toward the Stars**

The Hairy Star

The ancient observers were familiar with "hairy stars"—stars with tails. From the photo at the right, can you see why it might be called a hairy star? Of course, what they saw were not stars at all, but another visitor from the outer reaches of the Solar System. If you knew Greek, you would know that the ancient observers were talking about **comets**. *Comet* comes from the Greek word for "hair." The material in the tail of a comet gets left in space and can be responsible for meteor showers.

Throughout history, comets were thought to be a symbol of bad luck and disaster. To look at one was to invite something terrible to happen. For instance, the comet of 1066 appeared right before King Harold of England was overthrown by William the Conqueror. The comet was seen as a bad omen for the losing side. The arrival of this comet is recorded in the Bayeaux tapestry, shown below. Notice the comet at the top of the tapestry and the reactions of the people.

A comet

Today, comets are regarded as objects of interest. Both amateur and professional astronomers spend a lot of time trying to be the first to locate a new one. As a result of this interest, scientists now know the answers to questions such as the ones below. As you read further, see if you can discover answers to these questions too.

- Do comets revolve around the Sun, as do the other members of the Solar System?
- Why do comets have "tails"? Rocks and boulders called **asteroids** have been observed in the Solar System, but they have no tails.
- Are comets and asteroids made of the same material?
- What eventually happens to a comet?

Comets—What Are They?

The tail of the comet is a clue to its makeup. Think for a moment about what substances, when vaporized by the Sun's heat, can be made to glow. Did you name ice? If so, you've discovered one part of a comet's makeup.

In the far outreaches of the Solar System, trillions of frozen ice masses are held in a slow orbit around the Sun by their gravitational forces. Besides frozen water and other frozen material, these ice masses consist of dust and frozen gases. The orbits of some of the ice masses send them speeding toward the Sun, at which time they may be seen by observers on Earth. As the comet nears the Sun, the rise in temperature causes moisture and dust particles to be swept from the head of the comet. These particles form the comet's tail. The tail becomes visible when the particles reflect and scatter light.

The Comet of a Lifetime

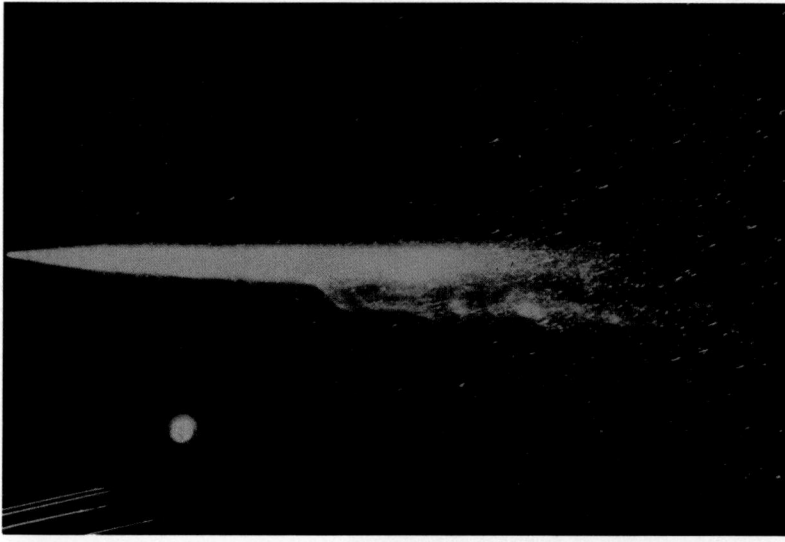

Halley's comet

In 1682 Edmund Halley (HAL ee) observed the comet that now carries his name. After calculating its orbit around the Sun, he predicted that the comet would return in about 76 years. Few believed him, and he did not live to see if his prediction would come true, but it did. The return of the comet in 1759 proved that comets are members of the Solar System and that they revolve around the Sun.

Comets—What Are They?

Comets, like asteroids, are remnants of the orbiting cloud of gases and dust that gave rise to the Solar System 4.6 billion years ago. Today, astronomers believe that comets exist in two forms. In the more familiar form, they are visible to the naked eye and have long tails that may stretch for millions of kilometers across space. In their second form, the volatile gases have been burned off, and the comet is reduced to a rocky nucleus that is almost indistinguishable from an asteroid. In this form, comets can be seen only with the aid of a telescope.

Research into the origin of comets began in 1577 when the Danish astronomer Tycho Brahe declared that comets traveled in paths far beyond the Moon and not just within Earth's atmosphere. Toward the end of the 18th century, Isaac Newton concluded that comets traveled in elongated elliptical orbits. During the same period, Edmund Halley tracked the orbits of 24 comets. In 1950, the Dutch astronomer Jan H. Oort suggested that a cloud of comets exists at the far outreaches of the Solar System. There, the vast majority of comets, estimated at about 100 billion, are located thousands of times farther away than the planet Pluto is from Earth. Held by the Sun's weak gravitational pull, comets in the Oort cloud follow huge orbits that take thousands of years to complete.

The Comet of a Lifetime

By calculating the orbit of a comet he observed in 1682, Halley proved that it was the same one astronomers had seen in 1531 and 1607. He correctly predicted its return in 1758. The comet, named after Halley, has returned at approximately 76-year intervals ever since. On May 21, 1910, Earth passed through the tail of Halley's comet.

The first reported sightings of this comet were made by Chinese astronomers around 240 B.C. Its appearance in A.D. 1066 is recorded in the Bayeaux tapestry.

(Continues on next page)

Exploring the Solar System

(The Comet of a Lifetime continued)

Direct students' attention to the diagram of the orbit of Halley's comet. Help them to locate the Sun and Earth's orbit. Earth's orbit appears circular because from this perspective the distinction between an ellipse and a circle cannot be detected.

Answers to
In-Text Questions

1. By traveling in elliptical orbits around the Sun, comets appear to obey Kepler's discovery regarding the paths of revolving bodies.
2. The comet travels at its greatest speed as it approaches the Sun and at its slowest speed when it is farthest from the Sun.
3. Halley's comet should return in 2061. Students may answer 2062 (because 76 plus 1986 yields 2062), but explain that the comet has an orbit of *about* 76 years.
4. At present, the comet is making its way to the outer limits of the Solar System, where its material is not being vaporized and illuminated by the Sun.

✦ Follow Up

Assessment

1. Ask students to take a survey to determine how many people they know have seen a shooting star. They should also ask each person to describe a shooting star. Have students bring the results of their surveys to class to share and discuss.
2. Ask students to make a diagram of the orbit of a comet, showing its relationship to the planets, the asteroid belt, the Sun, the Oort cloud, and the position of the comet's tail as it circles the Sun. Call on volunteers to explain their diagrams to the class. (Include the asteroid belt, the Oort cloud, and the position of the comet's tail as it circles the Sun—if you mentioned these points in the discussions. They were presented only in the teacher commentary, not in the pupil's text.)

Halley's comet made its last return in 1986. Examine the diagram below, which shows Halley's orbit. You'll learn a great deal about comets by using the diagram to answer the following questions.

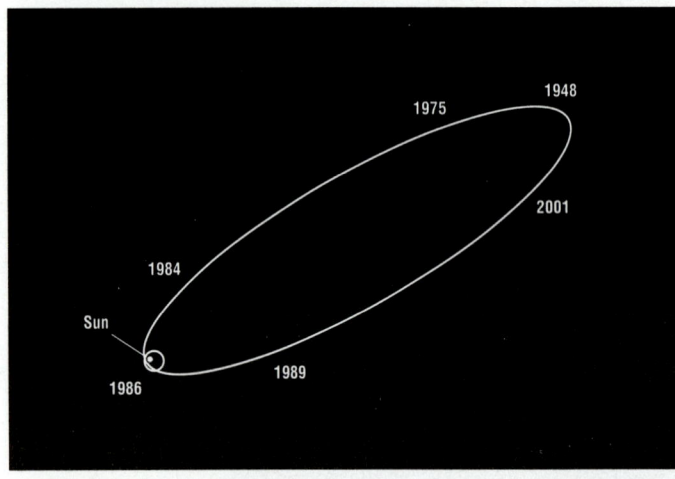

The orbit of Halley's comet. Earth's orbit is the small ellipse at the left.

1. Does a comet appear to obey Kepler's discovery regarding the paths of revolving bodies?
2. Where is the comet traveling at its greatest speed? at its slowest speed?
3. When will Halley's comet return to be seen again by observers on Earth?
4. Why can't Halley's comet be seen with the naked eye at the present time?

Halley's comet

Extension

1. Have students do some research to discover when they can expect to view a meteor shower in their area. Students should keep track of the number of meteors they observe. Have them discuss the experience on the following school day. (Annual meteor showers include: Orionid—Oct. 22; Leonid—Nov. 17; Geminid—Dec. 12; Quadrantid—Jan 3; Lyrid—Apr. 21; Eta Aquarid—May 4.)

2. Point out to students that on March 13, 1986, the European Space Agency's *Giotto* spacecraft passed within 600 km of Halley's comet and took 2000 photographs of it. Suggest that they do some research to find out what the space agency discovered about the comet. They can use what they learn to make a bulletin-board display.

6 The Space Probes

The *Voyager* Discoveries

Early voyagers sailed the seas and discovered new lands. They returned to report their findings. The *Voyager* spacecrafts are also discovering new lands, but they report by sending back photographs and information that will keep scientists busy for years to come.

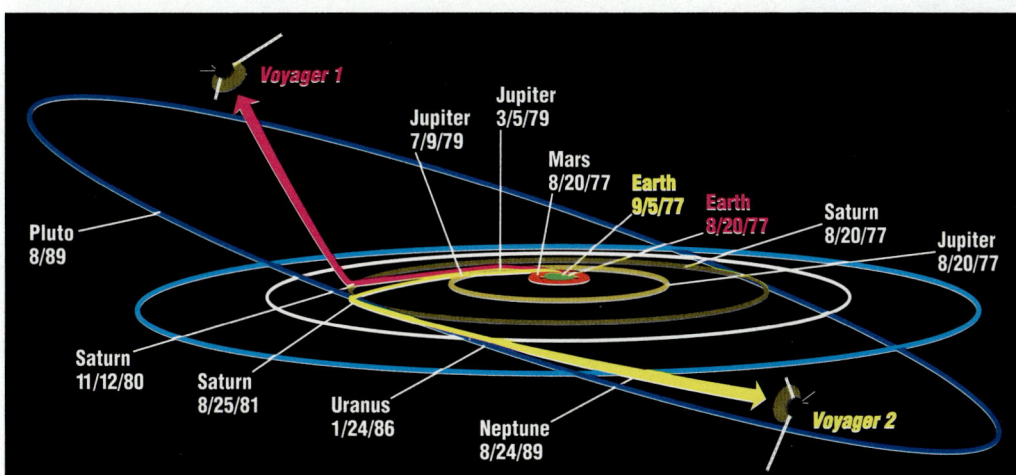

Your flight plans—the paths of *Voyagers 1* and *2*. The white labels show the positions of the planets on the dates indicated.

Imagine being on a journey that takes you from Kennedy Space Center in Florida, past Mars, and finally near the giant planets of Jupiter, Saturn, Uranus, and Neptune. You would see sights that human eyes have never seen before.

If either *Voyager* spacecraft had a crew and you were a member, how would you respond to the sights and experiences as they unfold? Use the picture study on pages 412 and 413 to write a personal account as you and *Voyager* encounter different members of the Solar System. Your log could center on the times and events shown on these pages.

Where you're heading: the giant planets. Compare their sizes to Earth's size.

LESSON 6 ORGANIZER

Objectives

By the end of the lesson, students should be able to:
1. Describe in general terms the distances that separate the planets.
2. Identify some of the principal features that distinguish one planet from another.
3. Explain how technology has been used to gather data about the planets.
4. Identify the names and achievements of several space probes.

Process Skills

observing, analyzing, inferring, drawing conclusions

New Terms

none

(Organizer continues on next page)

LESSON 6

✷ Getting Started

In this lesson, students examine the discoveries of three space probes—*Magellan* and *Voyagers 1* and *2*. The lesson begins by having students imagine that they are crew members on one of the *Voyager* spacecrafts. Through a picture study, they examine the planets and satellites that *Voyagers 1* and *2* encountered. The lesson concludes with an Exploration in which students examine a photograph of the surface of Venus taken by the space probe *Magellan*.

Main Ideas

1. The planets are separated by vast distances.
2. Each planet has its own unique features and characteristics.
3. Technology has provided up-close views of the planets and their satellites.

✷ Teaching Strategies

Direct students to the lesson title and read it aloud. Involve them in a discussion of what a space probe is and help them to formulate a definition. *(a spacecraft designed to explore outer space)* Ask students if they know the names of any space probes, and list their responses on the chalkboard. Point out that they will learn about the discoveries of three space probes as they complete Lesson 6.

The *Voyager* Discoveries

Have students read page 411. When they have finished reading, discuss the diagram of the flight paths of *Voyagers 1* and *2*. Note that the green area at the center of the diagram represents the orbits of Earth, Mercury, and Venus around the Sun.
(Continues on next page)

Exploring the Solar System   411

(The Voyager Discoveries continued)

In the discussion, you may wish to use questions similar to the following:
- When were the spacecraft launched? *(Voyager 1 was launched on August 20, 1977. Voyager 2 was launched on September 5, 1977.)*
- Did the spacecraft travel in the same direction? *(They left Earth in the same direction but later traveled in opposite directions.)*
- Which planets did each spacecraft fly by? *(Voyager 1 flew by Jupiter and Saturn. Voyager 2 flew by Jupiter, Saturn, Uranus, and Neptune.)*
- About how long did it take *Voyager 2* to reach Uranus? *(about eight and a half years)*
- According to the diagram, what happened to the spacecraft? *(They left the Solar System and continued on into space.)*

Be sure that students understand how they are to use the picture study on pages 412 and 413. This activity can be done individually, in pairs, or in small groups. As students create their logs, encourage them to include any additional questions they may have for later research and discussion. You may wish to gather materials for a resource center on space exploration for students to use as they work. Provide class time for students to share and discuss their ideas.

Answers to
In-Text Questions

Day 1: Such words as excited, anxious, nervous, exhilarated, and thrilled would probably be appropriate.

Day 215: Using the data from page 393 about Mars' and Earth's rates of revolution around the Sun, students will see that in 215 days, Mars will have traveled almost one-third of the way around its orbit, situating it too far from *Voyager 1* for Mars to be visible.

Day 295: The asteroid belt lies between the orbits of Mars and Jupiter. It is the region where most of the asteroids in the Solar System can be found—perhaps as many as 50,000. The possibility of an asteroid striking the spacecraft makes passing through the asteroid belt potentially dangerous.

(Answers continue on next page)

Day 1

Blast-off: How would you feel to be leaving on such a journey?

Day 215
Crossing the orbit of Mars: Examine the flight path of *Voyagers 1* and *2* on page 411. Note where Mars was on August 20, 1977— the day *Voyager 1* left Cape Kennedy. Will you be able to see Mars as you cross its orbit? Will *Voyager 2*?

Day 295
Entering the asteroid belt: What is the asteroid belt? Will this be dangerous?

Day 475
Leaving the asteroid belt: Will you feel relieved?

Jupiter from a distance of 30 million km

Day 570
Looking at Jupiter from a distance: What do you see? Describe Jupiter to someone who has not seen it before. How many different features can you describe?

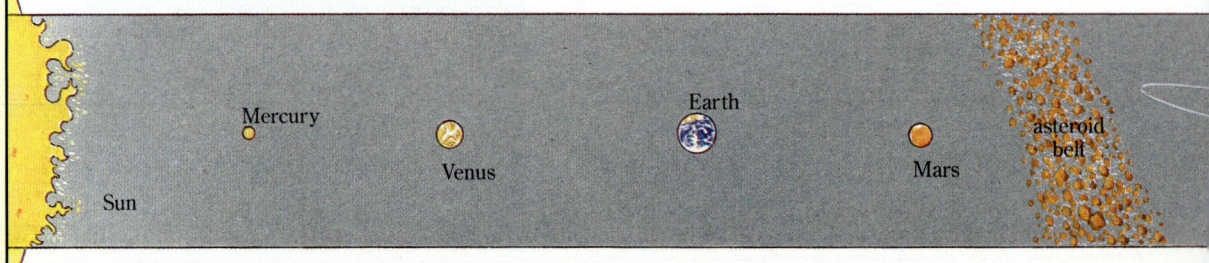

(Organizer continued)

Materials
The Voyager Discoveries: Journal

Teacher's Resource Binder
Activity Worksheet 7–2
Graphic Organizer Transparency 7–6

Science Discovery Videodisc
Disc 2, Image and Activity Bank, 7–6

Time Required
two class periods

Day 630

Observing the Great Red Spot: What words could describe your view? What is the Great Red Spot? How large is it?

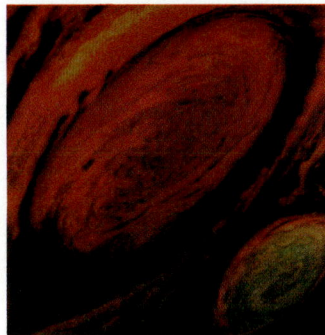

The Great Red Spot

Day 650

Examining some satellites of Jupiter: Jupiter has 16 satellites, one of which is about the size of our moon. What is a satellite? How are the satellites shown here different? Which one appears to be volcanically active? Which one could be covered with a thick layer of fractured ice?

Io

Europa

Ganymede

Day 662

On the way to Saturn: What will you do for the next 905 days?

Day 1567

Encountering Saturn: How is Saturn similar to Jupiter? different from Jupiter?

Portion of Saturn's rings, in false color

Answers continued

Day 475: A sense of relief should be felt upon leaving the asteroid belt.

Day 570: From a distance, Jupiter may be described as having an atmosphere of swirling orange and cream-colored bands. Embedded in the bands is a giant red spot that looks like a huge storm. Jupiter has 16 satellites.

Day 630: The Great Red Spot looks very much like what it is—a giant storm. It covers an area of about 26,000 to 40,000 km long by 15,000 km wide.

Day 650: A *satellite* is a heavenly body that revolves around a planet or other larger heavenly body. The Moon is a satellite of Earth. Io, Ganymede, and Europa are three of the 16 satellites of Jupiter. Io is seen with plumes of sulfur-rich vapor from a volcanic eruption. Europa's surface looks like a cracked egg shell and is covered with a layer of ice. Ganymede is covered with impact craters, rock, and ice.

Day 662: Accept all reasonable answers, which might include data analysis, preparatory work for the next encounter, and spacecraft maintenance. Point out to students that the trip from Jupiter to Saturn will take over two and a half years.

Day 1567: Saturn and Jupiter are both very large planets. Each appears to be covered with a banded atmosphere. However, Jupiter's atmosphere appears to be made up of swirling clouds, while that of Saturn appears to be smooth and calm. The most prominent feature of Saturn is its ring system. The most prominent feature of Jupiter is the Great Red Spot.

Exploring the Solar System

Answers to
Research Activities

1. The account written by Carl Sagan begins on page 151 of *Cosmos*. This book is an excellent resource about the Solar System and outer space.
2. Uranus is about 2,866,900,000 km from the Sun. It has a diameter of about 50,800 km. Fifteen satellites circle the planet. It also has a ring system. Uranus is the only planet that is tilted on its side. It makes one complete orbit around the Sun every 30,685 Earth days. *Voyager 2* discovered that Uranus has a relatively quiet atmosphere with few clouds and no evidence of storm systems.

   Neptune is about 4,495,000,000 km from the Sun. It has a diameter of about 49,500 km. Eight satellites circle the planet. Neptune is an extremely cold planet with an upper atmosphere temperature of −220 °C. Its atmosphere is extremely turbulent, with storm systems comparable to those on Jupiter.
3. Students may enjoy organizing their "space messages" into a booklet for others to enjoy.

Answer to caption: In the picture of Uranus, the Sun must be on the left because the left side of Uranus is illuminated.

Research Activities

1. In his book, *Cosmos*, Carl Sagan wrote a similar account of what a captain of the *Voyagers* might have said if the *Voyagers* had a crew. His account makes interesting reading.
2. Before leaving the Solar System forever, *Voyager 2* made two more encounters: Uranus on January 24, 1986, and Neptune on August 24, 1989. Find out what you can about these distant planets.

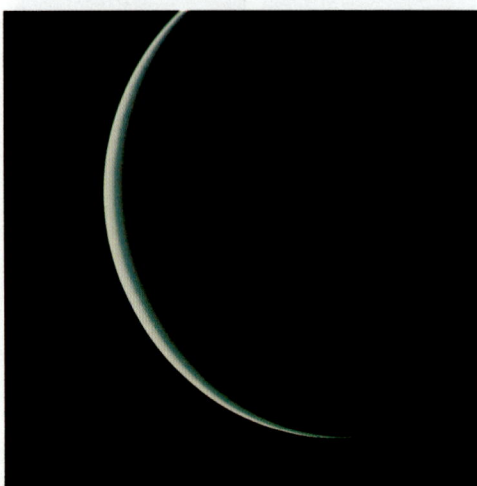

Uranus seen as a crescent. Where must the Sun be in this picture?

The Great Dark Spot on Neptune. This storm system is as large as Earth's diameter.

3. *Voyagers 1* and *2* will drift forever among the stars unless they are captured by a star's gravitational field. Perhaps, in the future, an alien from a distant star system may chance on the spacecraft and wonder about the world that sent the craft on its epic voyage. That alien will be able to learn something of Earth by listening to a gold-plated record that carries "The Sounds of Earth." The record begins with greetings in over 60 languages. The alien will be able to hear the roll of thunder, the roar of a volcano, crashing surf, and gurgling mud. Living sounds include that of whales, chimpanzees, and Chuck Berry singing his rock song, "Johnny B. Goode."

   Design a wordless message for another spacecraft that will eventually leave the Solar System. What would you place in the message that would give aliens some idea of what life was like on Earth?

Magellan—Mapping a Distant World

Ferdinand Magellan, a Spanish explorer, mapped the coastlines of the Americas in the early 1500s. The spacecraft *Magellan* is mapping a more distant land—a planet called Venus. As a result of the pictures sent back to Earth, more will eventually be known about the surface of Venus than that of Earth. That's because much of Earth is covered with water, forest, and ice, which hide its surface features.

The surface of Venus is hidden from observers on Earth by a thick layer of clouds. Some early astronomers suggested that rivers, streams, and perhaps even living things may lie beneath the clouds. Could this be true? Before giving your opinion, examine the information regarding Venus on page 416.

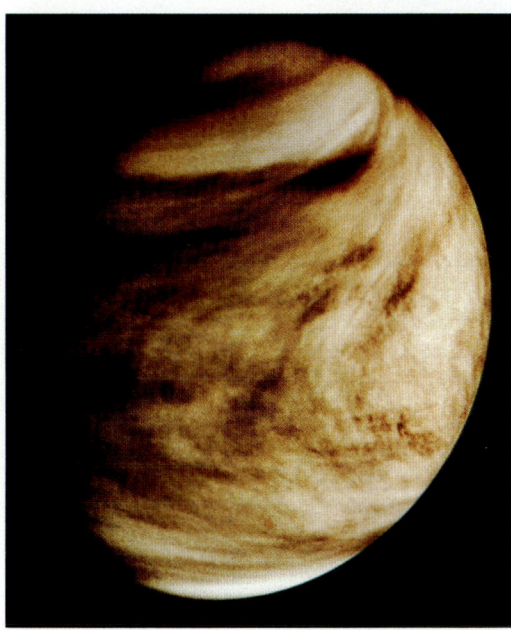
Clouds cover Venus to a depth of 60 km.

How can *Magellan* penetrate the cloud-cover to see the surface below? It uses microwave radiation, which can penetrate clouds. The reflected signals are formed into pictures by computers and then sent back to Earth.

EXPLORATION 6

Analyzing the Data

Scientists are busy interpreting the pictures sent back by *Magellan*. Using the photos on this page, be a scientist and interpret the data yourself. Perhaps you can answer these questions:

1. What surface features do you recognize on Venus that are similar to those on Earth?
2. How is the surface of Venus similar to or different from that of its near neighbor, Mercury? Can you think of reasons that would explain your observations?

The surface of Venus, photographed by *Magellan*

View of Mercury

Magellan—Mapping a Distant World

Have students read about the *Magellan* space probe. If possible, display some pictures of Venus taken by *Magellan* and call on students to describe the planet's appearance.

The *Magellan* probe was launched in May, 1989, and reached Venus 15 months later. During 1991, the probe mapped more than 90 percent of the Venusian surface. Venus is like Earth in size, density, and probably composition, but its surface is covered by a dense atmosphere of carbon dioxide. The planet's temperature averages about 482 °C. Surface water, if there ever was any, has long since boiled away, leaving behind a landscape that is completely arid and incapable of supporting life. Venus has few craters, but seems to be covered with fractures, ranging from elaborate networks of fine cracks to giant canyons thousands of kilometers long.

Involve students in a discussion of the possibility of there being life on Venus. There is no water or plant life on Venus, and its extremely high temperature precludes the possibility of life as we know it.

Answers to

Exploration 6

1. The surface of Venus has canyons, valleys, and plateaus that look similar to those on Earth.
2. The surface of Mercury is covered with craters and almost free of cracks. The cracked surface of Venus is relatively free of craters. The dense atmosphere of Venus (which exerts a surface pressure 90 times greater than that of Earth's atmosphere) causes most meteors to burn up or break apart before hitting the Venusian surface. The almost total absence of an atmosphere on Mercury allows meteors to constantly bombard its surface. The cracks in the Venusian surface may be due to past and present volcanic activity, which is absent on Mercury.

✹ Follow Up

Assessment

1. Have students make planetary data tables. The tables should include such information as the distance of each planet from the Sun, each planet's diameter, the number of Earth years it takes each planet to orbit the Sun, and the number of satellites revolving around each planet.
2. Ask students to take a survey of adult family members and friends to discover how they feel about the space program. The survey should include questions about missions with and without crews.

Extension

1. Suggest that students do some research on space probes that were not mentioned in the lesson (such as *Surveyor 5*; *Pioneer X*; *Mariners IV, IX,* and *X*; and *Vikings I* and *II*).
2. Have students work in groups to make models that demonstrate the relative sizes of the planets or the relative positions of the planets from the Sun. Point out that students will have to do some research, determine a scale, and plan a model that works. (For size: a possible scale is 1 cm = 10,000 km. For distance: 1 cm = 10 million km.)

Exploring the Solar System

Lesson 7

✷ Getting Started

In this lesson, students investigate the possibility of establishing a colony on Mars. In order to determine if Mars is the best choice for such a project, students must examine the characteristics of the other planets as well. Students then choose one of three projects to research and complete—planning a mission to Mars, creating a travel poster for some place in the Solar System, or designing an imaginary creature that might exist on another planet.

Main Ideas

1. Permanent space stations and a colony on Mars will probably be a reality in the next century.
2. Mars is more like Earth than any of the other planets.

✷ Teaching Strategies

List the names of the planets on the chalkboard. Then ask students which planet they feel is the best candidate for a space colony. As you point to each of the planets, ask students to raise their hands if they think it is the best choice. Tally the votes to decide which planet is preferred. Point out to students that in this lesson they will investigate the possibility of building a space colony.

Project Mars

Call on a volunteer to read aloud pages 416 and 417. Provide students with time to study the data table at the top of page 416. Then involve them in a discussion of why Mars is the best candidate for colonization. Help students to recognize that, based on the information in the table, the Martian environment is most similar to that of Earth.

(Continues on next page)

---

7 Colonizing the Solar System

Project Mars

Of all the planets, Mars holds the greatest hope of sustaining a human colony. Why is this so? Why not a colony on Venus, Jupiter, or Pluto? Examine the following table of information to find some answers.

Mars	Venus	Jupiter, Saturn, Uranus, and Neptune	Pluto
• Very cold—Average temperature: $-23\,°C$ • Extremely dry • Evidence of flowing water in the past • Icecaps • Thin atmosphere of carbon dioxide • Former atmosphere may have reacted with soil and rocks	• Surface temperature: $500\,°C$ • Very dense atmosphere of carbon dioxide • Atmospheric pressure 90 times that on Earth • Clouds of sulfuric acid	• Gaseous • Consists mainly of hydrogen and helium • Violent storms	• $1/100$ the volume of Earth • Extremely cold: $-230\,°C$ • Receives little sunlight • Surface may be covered with layer of frozen methane

By the year 2020, NASA hopes to land astronauts on Mars—the "red planet." By 2061—the next return of Halley's comet—many scientists foresee a permanent colony on Mars living in greenhouse-like structures called *biospheres*. Studies at Biosphere 2 in Arizona, shown below, may prepare scientists for a permanent space station on Mars.

LESSON 7 ORGANIZER

Objectives

By the end of the lesson, students should be able to:
1. Compare and contrast Mars and Earth.
2. Describe some of the major features of each of the planets.
3. Discuss the possibility of establishing a colony on Mars.
4. Identify a plan for establishing a colony on Mars.

Process Skills

observing, analyzing, formulating hypotheses, researching

New Terms

none

There is even speculation that it may be possible to "green" Mars by releasing the gases that are locked up in the rocks and soil, and by melting some of the water in the permafrost and ice-caps. If this is to happen, many more space developments must take place, such as those shown below. What do you think of these plans? Of what value is each? How would you add to or modify them?

Artist's conception of a permanently occupied space station

1992	Mars orbiter launched. Its intended mission was to circle Mars, sending back measurements and images. However, in August 1993, NASA lost contact with the probe. NASA hopes to launch another probe in 1996.
1998	A permanent space station built to orbit Earth, using the space shuttle as a transporter of people and materials.
2010	Permanent space station on the Moon.
1998–2014	Numerous robot landings on Mars.
2014–2020	First astronauts land on Mars.
2020–2030	First research colony established on Mars.

Perhaps the events may proceed as described in this hypothetical news forecast.

The first Earthlings will arrive on Mars almost exactly four centuries after the Pilgrims landed near Plymouth Rock. After a three-month flight from Earth, the Mayflower, *a nuclear-powered rocket not yet built, will deposit them on the Martian surface in a prefabricated colony designed to shelter 12 to 14 astronauts for a year.*

Would you like to be one of these Earthlings? What training would you have to undergo? This may be something that you'd like to investigate further.

Materials

Exploration 7: markers, poster board

Teacher's Resource Binder

Activity Worksheet 7–3
Graphic Organizer Transparency 7–7

Science Discovery Videodisc

Disc 2, Image and Activity Bank, 7–7

Time Required

one class period

Exploring the Solar System 417

Exploration 7

Call on a volunteer to read aloud the instructions for the Exploration and answer any questions students may have. Then have students decide which of the three projects they wish to do. You may decide to allow them to work in pairs or small groups. Encourage students to determine a way to share the results of their projects with the class.

Activity 1

Suggest that each member of a group assume the responsibility for a part of the project. Tasks might include gathering resource materials for the project, drawing a diagram of the living quarters on Mars, planning the trip, and deciding what to take on the mission.

Answers to

In-Text Questions

1. Some students may feel that large sums of money should be spent because of the many benefits that will be gained from the mission. Others may feel that the money should be spent on solving the problems on Earth before establishing colonies on Mars.
2. Students should make this determination based on what an individual has to contribute to the overall mission.
3. The first tasks to be done should provide oxygen, heat, shelter, food, and water for the people living in the colony.
4. Water and oxygen will probably have to be brought from Earth initially. Eventually, the biosphere should become self-sustaining with an established water cycle and carbon dioxide–oxygen cycle. Plants would be necessary for establishing these cycles.
5. At first, food will have to be brought from Earth. Later, food will be grown in the biosphere.
6. Oxygen, water, and all organic material should be recycled. Ideally, everything would eventually be recycled.
7. Oxygen, water, food, growth medium for plants, seeds, plants, clothing, fuel or means to produce energy, building materials, and medicine are some of the essential materials that would have to be brought from Earth.

EXPLORATION 7

Researching the Solar System

In this Exploration, you will choose from three different kinds of projects. Each requires you to research a part of the Solar System. In Activity 1, you can be part of a team that's planning a mission to Mars. In Activity 2, you can play the role of an advertising agency and develop a travel poster for one part of the Solar System. Or in Activity 3, you can create an imaginary alien that could survive in conditions that exist somewhere else in the Solar System.

Activity 1
Planning a Mission

Form groups of four and be part of a planning team for a mission to Mars. Your task is to create some guidelines and procedures that will ensure the success of the mission. Here are some questions you might want to consider:

1. Should vast sums of money be spent on this venture? Explain your reasoning.
2. Who will be part of the first 12-member team to spend a year on Mars?
3. What will be some of the first tasks to be done?
4. Where will their water and oxygen come from?
5. Where will they obtain food?
6. What materials will have to be recycled?
7. What essential materials must be brought from Earth?

418 Toward the Stars

Activity 2

Developing a Travel Poster

The year is 2065. The colonizing of Mars has been a success, and an increasing number of people are visiting this outpost as tourists. The usual length of stay is 1 year. Travel agencies are advertising this and other excursions throughout the Solar System.

You could spend a lifetime on Mars and never run out of things to do or see. Jump as high as a small tree in Mars' low gravitational field.

A thin coating of frozen carbon dioxide on Mars

Hit a golf ball more than 1.5 km. Climb the highest mountain in the Solar System—the volcano, Olympus Mons.

Olympus Mons—a volcanic mountain on Mars

With an elevation of 27 km, Olympus Mons is three times higher than Mt. Everest. Tour the grandest canyon in the Solar System—5208 km long (longer than the U.S. is wide) and in places, three times deeper than the Grand Canyon on Earth. Enjoy the long Martian summer, which is almost twice as long as that on Earth. Take a side excursion to Phobos and Deimos, Mars' satellites, for spectacular views of the red planet. But before venturing forth, be sure to listen to the weather report. Dust storms are frequent on Mars, with winds as high as 500 km/h!

Research a component of the Solar System. You could choose a planet or one or more of its many satellites. Comets, asteroids, and the Sun would make interesting studies as well. Develop a travel poster that would encourage others to visit the site (or perhaps to keep them away).

Activity 2

Encourage students to use the sample poster as a model for their own posters. Challenge students to be creative. For example, some students may wish to advertise trips through the Solar System instead of to just one place. Others may wish to include pictures from magazines or drawings of their own. Some students may want to include poems or make collages.

To extend the activity, suggest that students make travel brochures. Display several brochures for them to use as models.

Point out to students that they will have to do some research in order to determine what to include in their posters. Explain that the information in their posters must be based on facts but that this requirement doesn't preclude creativity. Their posters should be informative and interesting. Display the completed posters around the classroom.

Exploring the Solar System

Activity 3

This activity provides students with a chance to let their imaginations soar. Even so, they must make the alien's characteristics suitable to the environment in which it will live. Therefore, students must research what that environment is like.

Encourage students to read the Martian waterseeker description. Ask them to identify how it describes the ways the animal has adapted to its environment.

When students have finished their work, have them display their aliens around the classroom. As an alternative, suggest that students organize their work into a booklet or portfolio, design a cover, and think of a title. Then display the booklet or portfolio where class members may enjoy it at their leisure.

Primary Source
Description of change: excerpted from *National Geographic Picture Atlas of Our Universe,* by Roy A. Gallant, p. 45.
Rationale: excerpted to provide an example of a creature adapted to conditions that differ from those on Earth.

Assessment

1. Have students write stories about a day in the life of a Martian pioneer. Explain that the stories may be entertaining but should be based on facts about the Martian environment and the problems of living on Mars. Call on one or two volunteers to share their stories with the class.
2. Suggest that students work in groups to create a model of a biosphere that could be used for a colony on Mars. Supply students with a variety of materials such as clay, papier maché, various containers, pipe cleaners, tape, and construction paper to make their colonies look as realistic as possible. When models are complete, have a member from each group present it to the class.

EXPLORATION 7—CONTINUED

Activity 3

Create an Alien—Plant or Animal

Whisper-thin winds hiss along a dry, dusty canyon. Deadly ultraviolet radiation pours from an unshielded Sun. Nighttime cold reaches -80 °C. Perfect weather for a fellow like the Martian waterseeker. *Its parasol tail can lift three meters in Mars' low gravity, shading it from ultraviolet sunburn. The long snout can probe for pockets of ice under dried-up channels. And the giant ears, needed to hear well in the thin air, also serve as blankets: In Mars' frigid nights the waterseeker stays snug by clamping its ears tightly around its whole body.*

Research a part of the Solar System and create your own imaginary alien plant or animal that is adapted to the conditions there. Describe why it is well adapted to its environment. A good place to start: think of a living plant or animal that you are familiar with. Consider what adaptations would be necessary for it to survive in a more hostile world.

Extension

1. Suggest that students do some research to discover what scientists have learned about Mars from space probes that have already been there. The *Mariner* and *Viking* missions collected data about the Martian environment.
2. Some students may enjoy doing research on what the expected benefits might be from colonizing Mars. *(Possibilities include: manufacturing special products, mining, and deep-space research.)*

Challenge Your Thinking

Here are some unusual facts about the members of our Solar System. For each fact, there is a question to consider or an activity to do.

1. Venus has a most unusual day. It is longer than its year. Its year is 225 Earth days, while its day is 243 Earth days.

   Using plastic-foam balls to represent Venus and the Sun, demonstrate the year and day of Venus.

2. All planets except one rotate about an axis that is somewhat close to being perpendicular to the plane of the ecliptic. For example, you know that Earth's axis is tilted only 23½ degrees from perpendicular. But Uranus' axis is tilted 98 degrees! Perhaps Uranus was hit by a huge body in the past, and the blow changed its tilt.

   Speculate on Uranus' rather unusual seasons that result from this tilt.

   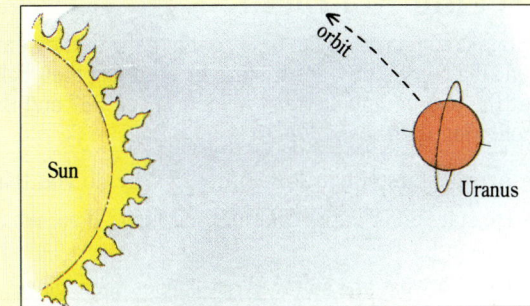
   Will the same pole of Uranus always point toward the Sun?

3. Pluto is the outermost planet, but not all the time. For 20 of the 248 years it takes to orbit the Sun, Pluto's orbit lies inside that of Neptune.

   Illustrate this observation by drawing a diagram of the possible orbits of Neptune and Pluto around the Sun.

4. Earth can be considered a miracle planet. It is just the right size and the right distance from the Sun to support life as we know it.

   What would be the consequences of Earth's being much larger? much smaller? closer to the Sun? farther from the Sun?

   View of Earth from *Apollo 17*

Answers to Challenge Your Thinking

1. Student models should show that Venus takes longer to rotate on its axis than it does to revolve around the Sun.
2. Given Uranus' distance from the Sun, the difference between summer and winter temperatures that would result from heating by the Sun's rays would be very small. However, due to its tilt on its axis, each of Uranus' poles has constant daylight for nearly 42 years and thus very long summers. (Uranus' year is 84 Earth years.) Each pole also experiences nearly 42 years of constant darkness, so Uranus has very long winters as well. The equatorial regions should experience four definite seasons.

   Answer to caption: No, each pole points to the Sun for 42 Earth years.

3. The illustration on the bottom of page 430 of the pupil's text shows the orbits of Neptune and Pluto.
4. If Earth were much larger, it would have a much denser atmosphere due to its greater gravity, and it would have a different mixture of atmospheric gases (far more hydrogen, for example). These characteristics would make life far different from that on Earth today, if not impossible. If Earth were much smaller, it would have no atmosphere, due to its lesser gravity. This would leave it dry and lifeless, like Mars. If Earth were much closer to the Sun, it would be much hotter, and water would be vaporized. This would result in a dry, lifeless surface similar to Mercury or Venus. If Earth were much farther from the Sun, it would be much colder, and its water would be frozen. This, too, would leave it dry and lifeless.

Lesson 8

Getting Started

The focus of this lesson is the life span and composition of stars. The lesson begins with a discussion of the size of stars, using the Sun and Earth for comparison. The lesson continues with students exploring the relationship between the composition of a star and the nuclear reactions that occur within it to produce light and heat. Then students examine the relationship between the temperature of a star and its color. The lesson continues with an overview of the "life" of a star, including discussions of supernovas, red giants, neutron stars, and black holes. Finally, students perform a simulation of the life span of stars.

Main Ideas

1. Stars vary in size, color, and temperature.
2. The color of a star provides information about its size, temperature, and evolution.
3. Stars pass through stages that depend in part on their mass.
4. The life history of a star takes place over billions of years.

(Teaching Strategies follow on next page)

OUR UNIVERSE

8 Messages from the Stars

Seeing the Light

Stars communicate with us through the light they emit, so you can find out a great deal about stars by studying their light. Look again at the photograph of star tracks on page 387 and think about the following questions:

- Are all stars of equal brightness? Why do you think this is so?
- Do stars differ in the color of their light? If so, what could be the reason for this?
- Exactly what are stars? What causes their glow?

You'll discover answers to these questions in the following reading and Exploration.

The Closest Star

You are very familiar with the Sun as an emitter of light and heat. Did you also know that it's a star? For those living on Earth, the Sun is a special star because it supplies the energy necessary for life. However, among the stars that you observe on a clear night, the Sun is just a typical star. In fact, it is smaller and less bright than many stars you see! On a clear night, identify several bright stars. If you were watching these stars from the same distance that separates you and the Sun, you would see that many stars are larger and brighter than the Sun. The largest stars are larger than the orbit of Jupiter.

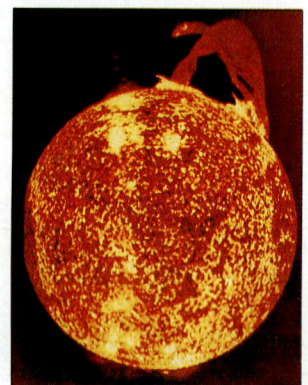
The Sun under magnification

Hot gases rising from the Sun's surface

Lesson Organizer

Objectives

By the end of the lesson, students should be able to:

1. Describe the size of stars, using comparisons to Earth and the Sun.
2. Explain the relationship between the color of a star and its temperature.
3. Describe the relationship between the composition of stars and the nuclear reactions that occur within them to produce light and heat.
4. Describe the life span of stars and what eventually happens to them.

Process Skills

observing, analyzing, investigating, using models

New Terms

Supernova—a very large star that explodes, producing a bright light.

Red giant—a very large and relatively cool star in its final stages of life.

Neutron star—a very small star consisting of the remnants of an exploded star. It contains the mass of several Suns, compressed to the size of a small asteroid.

Black hole—a small and dense object formed from a massive collapsing star. It has a gravitational pull so strong that even light cannot escape.

422   Toward the Stars

To get an idea of the size of the Sun, compare it with the size of Earth. How much larger than Earth is it? Is it 100 times as large? 1000 times? Actually, it is capable of holding more than a million Earths. Now think about how large that makes the brightest stars in the sky!

All stars, including the Sun, are made up chiefly of hydrogen gas, along with some helium gas. Although we haven't sampled the core of any star, scientists think that heat and light are produced in stars in the following way. When the gases in a star reach a tremendously high pressure and temperature, *nuclear reactions* occur within the star's core. In these reactions, hydrogen atoms fuse together to form another substance—helium. In doing so, a tremendous amount of energy is released. Some of that energy becomes the heat we feel and the light we see from stars. Humans have duplicated this process in a hydrogen bomb. If each star releases this much energy, imagine the amount of energy released from a constellation!

One easily recognizable constellation is Orion the Hunter, shown below. Orion's most distinguishing characteristic is a group of three stars that makes up the belt of the Hunter. Find Orion on the star chart for winter skies on page 387. Better still, locate it in your night sky. In the Northern Hemisphere, Orion is visible all winter.

Can you locate the constellation Orion? the star Betelgeuse? Use the diagram on the right to help you.

Orion

Materials

Exploration 8: large nail; Bunsen burner; light source with a clear bulb; 3 D-cells (2 fresh, 1 weak); flashlight bulb; small-gauge electric wire; tongs, matches, star chart, safety goggles
Exploration 9: 2 white balloons; 2 small, white, plastic-foam balls; red, yellow, and black felt-tip markers; a straight pin

Teacher's Resource Binder

Resource Transparencies 7–8 and 7–9

Science Discovery Videodisc

Disc 2, Image and Activity Bank, 7–8

Time Required

two class periods

✴ Teaching Strategies

List the following names on the chalkboard: Alpha Centauri, Altair, Betelgeuse, Aldebaran, and Antares. Ask students if they know what these words identify. If no one responds, point out that they are the names of stars. Then involve students in a discussion of what a star is and help them to formulate a definition.

Seeing the Light

Encourage students to write their answers to the questions in their Journals. These can be referred to and revised after they complete *Exploration 8*.

Answers to

In-Text Questions

- Stars differ in brightness because of their distance from Earth, their size, and their energy output.
- Stars differ in the color of their light because their surface temperatures are different.
- Stars are heavenly bodies consisting of huge balls of glowing gas. Nuclear reactions taking place within the core of a star cause it to glow.

The Closest Star

Monitor students' understanding of how a star produces heat and light by involving them in a discussion of the process. You may wish to point out that the nuclear reaction that occurs in a star is called *fusion* because it is a coming together, or fusing, of atoms. This is the opposite of nuclear *fission*, which takes place in nuclear reactors on Earth. Nuclear fission causes atoms to break apart.

Exploration 8

Divide the class into small groups and distribute the materials. Caution students that they will be working with very hot materials and should use extreme care to prevent accidental burns. If you are concerned about your students' abilities to carry out this Exploration in a responsible manner, you may want to perform it as a demonstration.

Answers to
In-Text Questions

1. The nail should turn from red to orange as it gets hotter. If enough heat is applied, it will eventually turn white.
2. The filament in the clear light bulb is white as it glows. The filament is white and therefore has a higher temperature than the red nail. By adding more heat to the nail, it will eventually glow white.
3. Students should discover that with the weak dry cell, the filament burns with a dull red color; with one fresh dry cell, the filament burns with a bright red or orange color; with two fresh dry cells, the filament becomes almost white. Cooler objects emit red light. As an object becomes hotter, its color gradually changes from red, to orange, to white.

Can you find several very bright stars in Orion? Each one is brighter and much more massive than the Sun. The reddish-colored star in Orion is called Betelgeuse (BET ul jooz). What does this red color tell you about Betelgeuse? Exploration 8 will help you answer this question.

EXPLORATION 8

The Temperature of a Star

You Will Need

- a large nail
- a Bunsen burner
- a light source with a clear bulb
- 3 D-cells (2 fresh, 1 weak)
- a flashlight bulb
- small-gauge electric wire
- a pair of tongs

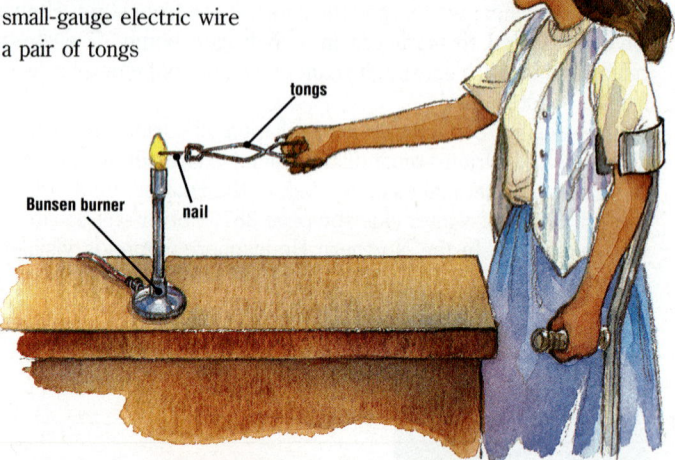

What to Do

1. Using a pair of tongs, heat a nail in a Bunsen burner until it begins to glow. What color is it when it first begins to glow? Can you make it glow at any other color? How?
   Be Careful: *To avoid accidental burns, use extreme care when handling these hot materials.*

2. Now turn on the clear light bulb. What color is the filament as it glows? Which has a higher temperature, the nail or the filament in the light bulb? What can you do to make the nail glow as bright as the filament in the light bulb?

3. What is the relationship between the temperature of the glowing object and its color? If you're not sure, try this. Attach a weak dry cell to a clear flashlight bulb using conducting wires. Observe the color of the filament. Feel the temperature of the bulb. Now attach a fresh dry cell to the bulb and repeat your observations. Finally, attach two fresh dry cells and make similar observations. Note the color of the filament. Feel the temperature of the bulb. What conclusion can you draw about the relationship between the color of light and the temperature of objects emitting the light?

Questions

1. Now do you see what information we can get from the color of a star? What color are the stars with relatively high surface temperatures? with relatively low surface temperatures?
2. Polaris, the North Star, appears white in color. Rigel, a star in the constellation of Orion, appears blue. The Sun is yellow. Betelgeuse is red. Arrange these stars in order from highest to lowest surface temperature.
3. Spend some time looking at the stars. How many can you see that are bluish? red? yellow? What color are most stars?
4. Using a star chart, identify the names of some of the more distinctive stars.

History of a Star

If you've ever spent time around a campfire or warmed yourself in front of a fireplace, you know that when the fire dies down, the embers will glow for a while and the ashes will eventually cool. Is this the fate of stars as well? Do they have a life span in which they produce light and heat and then die? If so, what remains? Read on for answers to these and the questions below. Use the numbers in the right margin of "The Life of a Star" to help you in your search.

1. What is a red giant?
2. How do neutron stars and black holes begin?
3. What is a supernova?
4. How do stars produce the heat and light energy they give off?
5. What is the eventual fate of the Sun?
6. What is the relationship between the mass of a star and the way it will end?

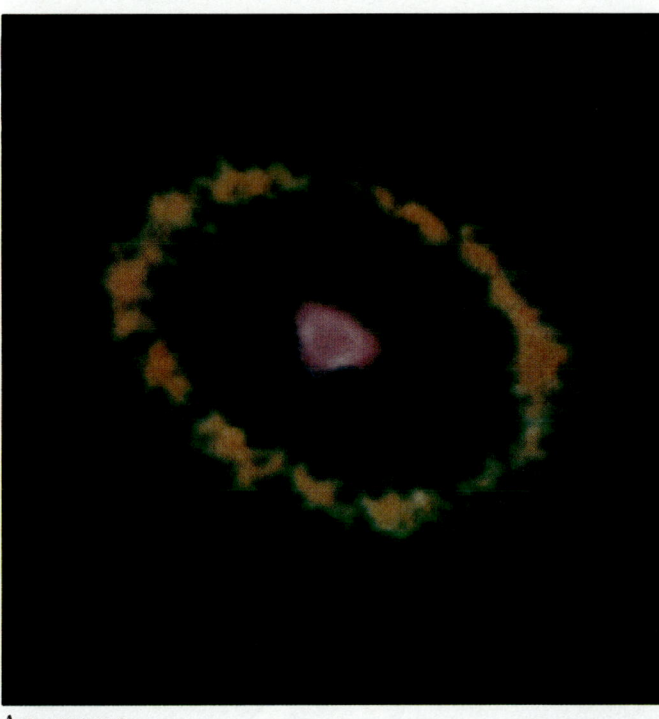

A supernova

Answers to
Questions

1. The color of a star is an indication of its surface temperature. Stars with relatively high surface temperatures are blue-white or white; stars with relatively low surface temperatures are red or orange.
2. You may wish to point out to students that the color of the hottest stars is blue-white. The order of the stars from highest to lowest surface temperatures is: Rigel, Polaris, Sun, Betelgeuse.
3. Most stars appear white or bluish. Have students compare and discuss their observations.
4. Students' responses may vary, but possibilities include: Aldebaran, Altair, Arcturus, Betelgeuse, Capella, Castor, Deneb, Polaris, Pollux, Regulus, and Vega.

History of a Star

Have students read silently pages 425 to 427. Be sure they understand that the answers to the questions can be found by identifying the corresponding numbers in the margins on pages 426 and 427.

Answers to
In-Text Questions

1. A red giant is a very large and relatively cool-temperature star that is nearing its final stages of life. Stars become red giants as they use up their hydrogen and begin burning other fuels.
2. Neutron stars and black holes begin as very massive stars that explode into a supernovas after their red giant stage.
3. A supernova is an exploding star that occurs when a very massive star uses up its nuclear fuel and collapses in on itself. When it explodes, it will be extremely bright for several weeks. Then it leaves behind a neutron star or black hole.
4. Stars produce heat and light energy during nuclear reactions that occur in their cores. First, hydrogen atoms are fused to create helium. When hydrogen is used up, atoms of other elements will be fused. During the process, huge amounts of energy are released as heat and light.
5. Eventually, the Sun will run out of nuclear fuel and begin to expand in size until it reaches the red giant stage. In time, it will cool and contract under its own gravitational forces until it becomes a white dwarf.
6. The greater the mass of a star, the smaller and denser it will become when it finally collapses. The most massive stars will explode into supernovas and then become black holes. Other, very massive, stars will become neutron stars after they are supernovas. Stars with much less mass, like the Sun, may collapse and become white dwarfs.

Our Universe 425

The Life of a Star

Determine student understanding of the material by involving them in a discussion of what they have read. If necessary, review the major points. Divide the class into small cooperative-learning groups to discuss the answers to the questions on page 425. Then reassemble the class and call on volunteers to share and discuss their ideas.

Students may be interested in learning how stars are formed. Explain that scientists do not yet fully understand the process, but they believe that stars begin as a cloud of interstellar gas made up mostly of hydrogen and dust. These materials have been added to and mixed with the remains of stars that have exploded and with the gases thrown out from the surfaces of giant stars.

As a result of gravity, a part of the giant interstellar cloud begins to contract into a ball. It continues to contract for millions of years as gravity pulls it together. Eventually, the pressure of the gas at the center of the ball is so great that it becomes extremely hot. When the temperature reaches about 1,100,000 °C, nuclear fusion begins to occur, and huge amounts of energy are produced. As the gas that surrounds the core of this new star gets hotter, it begins to glow.

The amount of mass in the contracting interstellar cloud determines the kind of star that will evolve. A mass of about 1/10 of the Sun produces a red, faint star. A mass of about 50 times that of the Sun produces a blue, bright star.

(Continues on next page)

The Life of a Star

Can stars be seen in the daytime sky? Sometime during the next 10,000 years, a star will explode in the constellation of Orion. The exploding star, or **supernova,** will be so bright that it should be visible during the day. That star is Betelgeuse. (3)

Betelgeuse started out as a bluish and very massive star with a mass 20 times that of the Sun. It was a **supergiant.** But now, after burning for 10 million years and having exhausted its hydrogen nuclear fuel, it is nearing the end of its life span. Betelgeuse has expanded to a diameter 700 times that of the Sun. It is now a **red supergiant.** To observers on Earth, Betelgeuse appears red—the color emitted by the cooler surface gases, far removed from the core. (4) (1)

As its nuclear fuel is used up and its core cools, Betelgeuse will collapse in on itself. This is due to the extreme gravitational forces of its large mass. The resulting shock wave will send a large amount of Betelgeuse into space. This is the supernova that will be observed on Earth. (3)

What remains of Betelgeuse will be a very dense body called a **neutron star.** A neutron star contains the mass of several Suns, compressed to the size of a small (30 km) asteroid. (Ceres, the largest of the asteroids, is 1000 km across.) A thimbleful of a neutron star would have a mass of a hundred million tons! And what happens to the material thrown out into space? It mixes with the stuff of which new stars and planets are formed. (2)

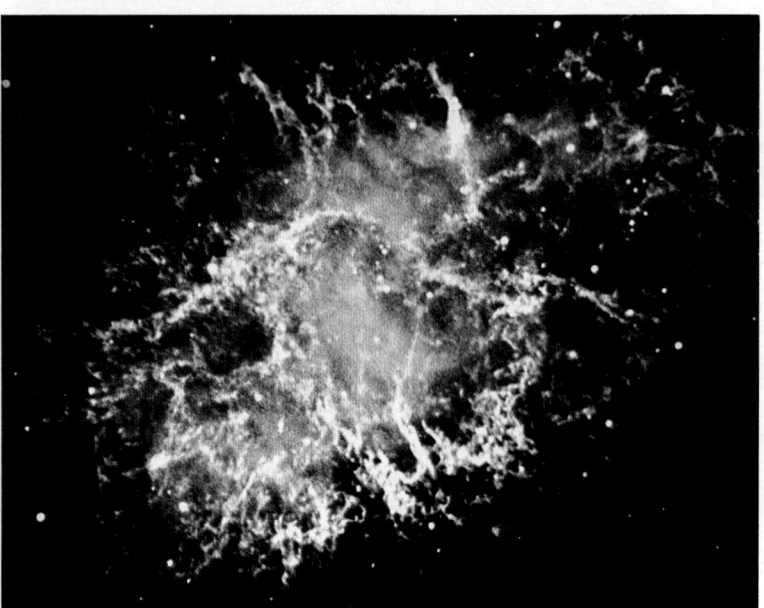

The Crab nebula is the remnant of a supernova recorded on July 4, 1054.

If the mass (and therefore the gravitational forces) of the collapsing star is great enough, it may collapse even more to form a smaller and denser object called a **black hole.** The gravitational forces are so extreme that no object—not even light—can escape the black hole. Since no light can escape from this super-dense object, it cannot be seen in the usual ways—hence its name, a black hole. (6) (2)

Our star, the Sun, will not end its life in a giant explosion or supernova. This is a fate that is reserved for the more massive stars such as Betelgeuse. But as the Sun's hydrogen nuclear fuel runs out and the Sun starts to burn other fuels, the Sun will expand in size. Eventually, it will fill the orbits of the inner planets and perhaps even Earth's orbit. When this happens, the Sun will reach the red giant stage. (5) (6)

Once the Sun has become a red giant, the oceans and atmosphere of Earth will have boiled away. In time, as the Sun goes through the remainder of its nuclear fuel, it will cool and contract under its own gravitational forces until it becomes a **white dwarf**—a dim but extremely dense star. This scenario will not happen for another 5 billion years. Perhaps by then, humans will have migrated to another planet, to the moons of Jupiter, or to a star system far beyond this one—one with a quieter star. (1) (4)

Solar explosion

You can be thankful that the Sun is a yellow star and not a white one such as Sirius, the Dog Star. Sirius is the brightest star in the night skies. If viewed from a distance equal to that from Earth to the Sun, Sirius would be 23 times brighter than the Sun, even though it is only twice the Sun's mass. Sirius is using its energy source at such a rapid rate that its life span is estimated to be less than a billion years. The Sun, on the other hand, is already 4.5 billion years old and is expected to be around for another 5 billion years. (6)

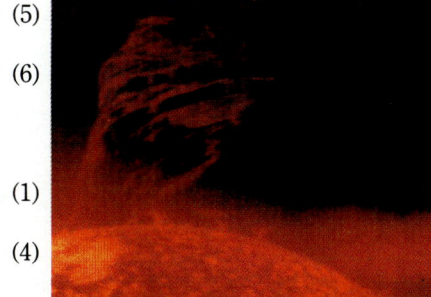
Sirius, the Dog Star

Checking It Out

With a partner, discuss the answers you found to the six numbered questions on page 425. In your own words, describe what you learned.

Now you're ready to do one of the following:

(a) Write a short autobiography of a star.
(b) Depict the life story of a star in the form of drawings only.

(The Life of a Star continued)

The greater a star's mass, the brighter it appears, the higher its temperature, and the faster it changes. For example, a star with a mass of about 10 times that of the Sun will change in a few million years. A star with a mass of about 1/10 that of the Sun will take hundreds of billions of years to change.

One of those changes may be the eventual collapse of a star into a black hole. When a star becomes a black hole, its size decreases considerably. For example, the Sun has a diameter of 1,392,000 km. To become a black hole, its mass would have to be compressed to a diameter of less than 6 km. Such an increase in density would create a tremendous increase in gravitational pull, so that even light could not escape it. The Sun will not become a black hole.

Students may wonder how scientists know there are black holes if they cannot see them. Explain that one way to determine where there is a black hole is to observe its gravitational effect on other objects. For example, in the early 1970s, astronomers discovered a star that orbited what seemed to be an invisible object in the constellation Cygnus. The best explanation was that the object was a black hole. A black hole may also attract and retain nearby comets, planets, and other heavenly bodies.

Exploration 9

Distribute the materials and have students discuss the Exploration in small groups. Have the groups plan the sequence they will follow for each simulation. Once they are satisfied with their procedures, have students conduct their simulations within their own groups. The questions are intended to guide student thinking as they plan their simulations. Definitive answers are not called for. The following information provides background information to help you discuss the questions with your students.

Stars begin as collapsing clouds of gas in interstellar space and so are relatively small early in their life span. As the amount of hydrogen in the core decreases, the center of a star contracts, and the temperature and pressure at the center rise. Meanwhile, the temperature of the outer part of the star gradually drops and the star expands greatly, becoming a red giant. A star with about the same mass as the Sun will collapse inward upon itself to form a white dwarf. A star with more than about eight times the mass of the Sun becomes a supergiant, which may then explode into a supernova. Depending on the mass that remains after the supernova explosion, the star will become either a neutron star or a black hole. The ball in the simulations represent the fate of the star: white dwarf, neutron star, or black hole.

The following briefly describes one way to conduct each of the simulations.

A smaller star
- Color one balloon yellow.
- Place a small, white ball inside the balloon.
- Inflate the balloon slowly, so it starts out small and remains that way for some time.
- Expand the balloon completely to represent a red giant; color the balloon red.
- To represent a white dwarf, release the air and take out the ball inside.

A larger star
- Color one white ball black and place it in the balloon.
- Blow up the balloon to a large size.
- After a short time, blow up the balloon some more. Color it red.
- Use a pin to explode the balloon. Throw away the balloon, but keep the black ball to represent a black hole.

428 **Toward the Stars**

EXPLORATION 9

Simulating the Life Span of Stars

You Will Need
- 2 white balloons
- 2 small white plastic-foam balls
- red, yellow, and black felt-tip markers

What to Do
Form teams of three or four to demonstrate the history of two stars—one the size of the Sun, the other a much larger star. In your simulation, you will use balloons to represent each star at various stages of its life. Decide how to use the remaining materials listed at the left. Then figure out how to sequence the steps of the simulation. The following questions provide some hints as to how you might plan your simulation.

Questions
1. How large will each star be early in its life? How large will each be at various stages in its life? How will you present this in your simulation? The illustration below may help with this part of the simulation.

The relative sizes of stars and their remnants

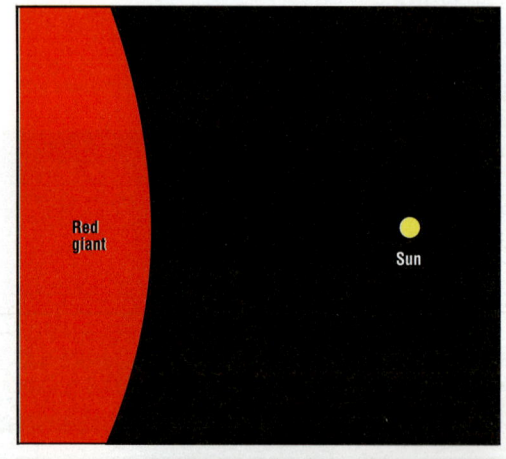

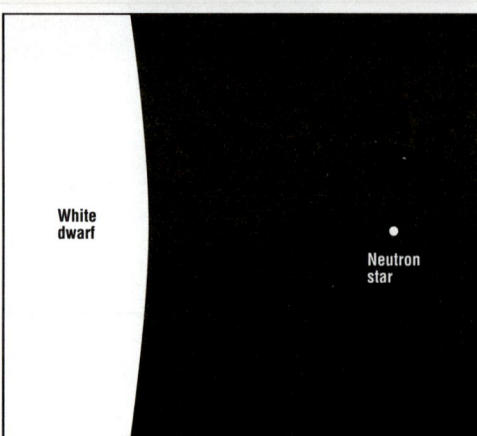
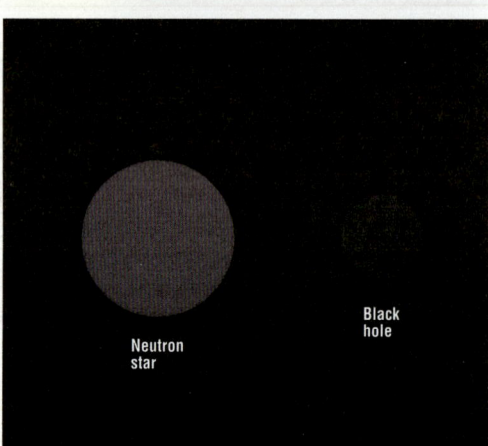

✵ Follow Up

Assessment
1. Have student groups create murals showing the Sun's evolution, from origin to white dwarf.
2. Remind students that a star's energy comes from nuclear fusion. Have them make diagrams to illustrate the process.

Extension
1. Explain to students that our Solar System is located in the Milky Way galaxy. Suggest that they do some research to discover what the Milky Way galaxy is, where it is, how big it is, how many stars it includes, and where the Sun and the Solar System are located within it. Invite students to share what they learn by giving a report to the class.
2. Remind students that the Sun is the closest star to Earth. Suggest that they do some research to discover the composition of the Sun's interior and what is happening on its surface. Their research should include information about the convection zone, radiation zone, core, photosphere, sun spots, and solar flares.

2. How and when will you use the markers in your simulation?
3. What does each plastic-foam ball represent in the simulation?
4. Will the star explode in a supernova? How would you represent this in your simulation?
5. How long will each star exist before its life is over? How will you represent this in your simulation?

When your team has perfected the simulation, demonstrate it to the rest of the class. Ask for feedback on your simulation. What improvements could you make? Also, give feedback on the other teams' simulations.

Earth's Place in the Universe

At the Speed of Light

The stars remind you how much larger the universe is than the Solar System. But just how large is the universe? To answer that question, you'll be taking another imaginary trip in the next Exploration. Only on this trip, you'll be traveling at the speed of light—that's 300,000 km/s (kilometers per second)! At that speed, time takes on a new dimension. Traveling at this speed, you will age much less than if you stayed on Earth.

The ticket has been bought for the trip, and you're ready to go. The brochure tells you that each time you increase your distance from Earth by 100 times, you will stop. At these stops, you'll have a chance to look back in the direction you just came from—back, in other words, toward the Solar System and Earth. You'll be looking back at Earth's place in an immense universe.

Earth

LESSON 9 ORGANIZER

Objectives
By the end of the lesson, students should be able to:
1. Describe Earth's place in the Solar System, Milky Way galaxy, and the universe.
2. Explain the differences between the Solar System, the Milky Way galaxy, and the universe.
3. Explain what astronomical units and light-years are and how they are used.

Process Skills
observing, analyzing, formulating hypotheses, inferring

New Terms
Light-year—the distance light travels in one year—about 63,240 AU, or 9.5 trillion kilometers.

(Organizer continues on next page)

LESSON 9

✷ Getting Started

In this lesson, students explore a small part of the universe. Traveling at the speed of light, they embark on an imaginary journey through space. At each stop along the way, students look back at the vast distances they have traveled. They discover that the Sun is part of a much larger accumulation of stars called a *galaxy,* and that our galaxy is but one in a Local Group of galaxies moving through the universe. The lesson concludes with students taking a few minutes to think about the wonders of the universe they have just seen.

Main Ideas
1. The Sun, Earth, and Solar System occupy only a tiny part of the universe.
2. Despite the huge number of objects of which it is composed, the universe is mainly empty space.
3. The vast distances of space are measured in astronomical units and light-years.
4. The light from stars in our galaxy was generated thousands of years ago. The light from other galaxies was generated millions and even billions of years ago.

✷ Teaching Strategies

Ask students what the term *universe* means to them. List their ideas on the chalkboard. Then ask them what they think their place is in the universe. Where do Earth, the Sun, and the Solar System fit? Help them conclude that everyone and everything is part of one vast universe. Point out that in this lesson they will learn about the size and composition of the universe.
(A teaching strategy for At the Speed of Light follows on next page)

Our Universe 429

At the Speed of Light

The information on page 429 introduces *Exploration 10*, which comprises most of the lesson. Have students read the page silently or call on a volunteer to read it aloud.

Remind students that the last imaginary trip they took was "A Trip Around the Sun" in Lesson 4. This time, they will be traveling much greater distances. Point out that no one can really travel at the speed of light, but in order to travel the great distances in this imaginary trip, it is necessary to pretend that traveling at the speed of light is possible.

Exploration 10

Stop 1

Divide the class into small groups to read, analyze pictures, and respond to questions. They should recognize that the inner planets are Mars, Earth, Venus, and Mercury.

Answers to
In-Text Questions

1. Earth, Venus, and Mercury are 1 AU or less from the Sun. By definition, Earth is 1 AU from the Sun. Venus and Mercury are less than 1 AU from the Sun because they are between Earth and the Sun. Actually, Mercury is 0.39 AU from the Sun and Venus is 0.72 AU away. Mars is 1.52 AU from the Sun.
2. Only rarely do the planets line up on the same side of the Sun. It is unlikely that they would do so on this occasion.

Stop 2

Suggest that students look at the diagram and identify the outer planets as Jupiter, Saturn, Uranus, Neptune, and Pluto. You may wish to point out that the five outer planets are much farther from the Sun than are the four inner planets. Except for Pluto, the outer planets are also much larger than the inner planets. The planet that is usually farthest from the Sun, Pluto, cannot be seen at this stop. Note that until the year 2000, Pluto is closer to the Sun than Neptune.

EXPLORATION 10

Heading for the Stars

Stop 1: You have traveled about 8 minutes since leaving Earth

In this time, you have traveled a distance equal to that from Earth to the Sun. This is how long it takes light to reach Earth from the Sun, when traveling at a speed of 300,000 km/s. The distance from Earth to the Sun, you'll remember, is 150 million km—one astronomical unit (AU). Looking back toward Earth, can you see the four inner planets in their orbit around the Sun? What are they?

1. How many planets are 1 AU or less from the Sun?
2. Would all planets be on the same side of the Sun as Earth?

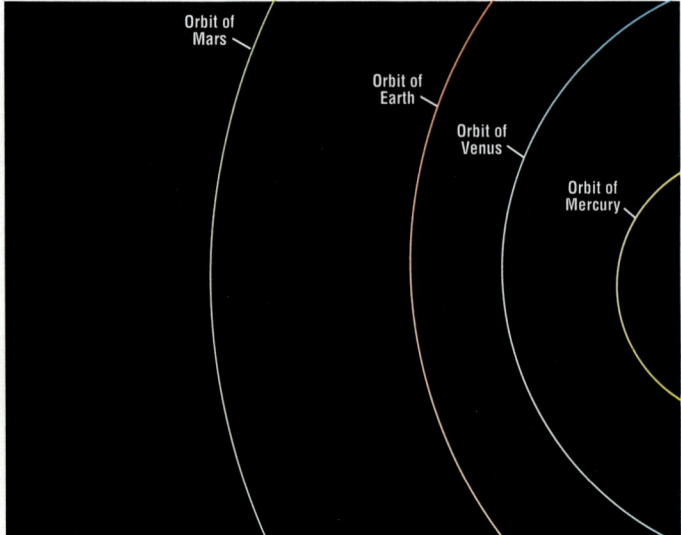

Stop 2: You have traveled 14 hours since leaving Earth

You are 100 AU from Earth. The inner planets are hidden in the brightness of the Sun. All of the outer planets are visible except for the outermost planet, which is so small that it cannot be seen. Which planet is that?

1. What does 100 AU mean?
2. If you are looking down over the north pole of the planets, in what direction would you expect them to be moving around the Sun?

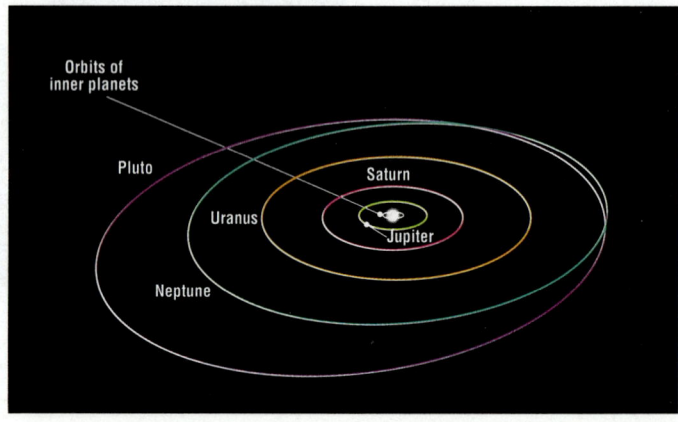

(Organizer continued)

Milky Way galaxy—our home galaxy, the galaxy to which the Sun and our Solar System belong.

Local Group—a family of galaxies of which the Milky Way is a member.

Materials
A Conversation with Huck: Journal

Teacher's Resource Binder
Activity Worksheet 7–4
Resource Transparency 7–10

Time Required
two class periods

Stop 3: You have traveled 0.16 years (58 days) since leaving Earth

Looking back at the Sun from a distance of 10,000 AU, you see just one very bright star in a sea of blackness. Just as astronomers cannot see any planets around other stars (because any planets that may be there do not reflect enough light), so you cannot see the planets of the star called the Sun.

The distances you have traveled are now so large that kilometers and even astronomical units are becoming cumbersome. A new unit is needed—the **light-year**. A light-year is the distance light travels in one year. That distance is equal to 63,240 AU.

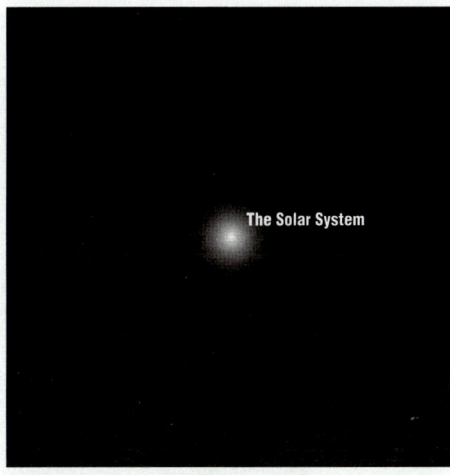

Since you have traveled 0.16 years at the speed of light, you have been traveling for 0.16 light-years.

At this stop, how far are you from Earth in kilometers?

Stop 4: You have traveled 16 years since leaving Earth

Sixteen years have passed since you left Earth. You have traveled a distance of 16 light-years. Now the view looking back toward the Sun includes other stars. You can see Sirius—the brightest star seen from Earth's Northern Hemisphere. From this distance, you notice that what you're seeing is actually two stars: Sirius and its companion white dwarf. The closest star to Earth, Alpha Centauri, can also be seen. It is only 4.3 light-years from Earth. On Earth, it can be seen anywhere south of Miami, Florida.

1. In AU's, how far are you from Earth?
2. Sirius is 8 light-years from Earth. How long would it take for light from Sirius to reach Earth?

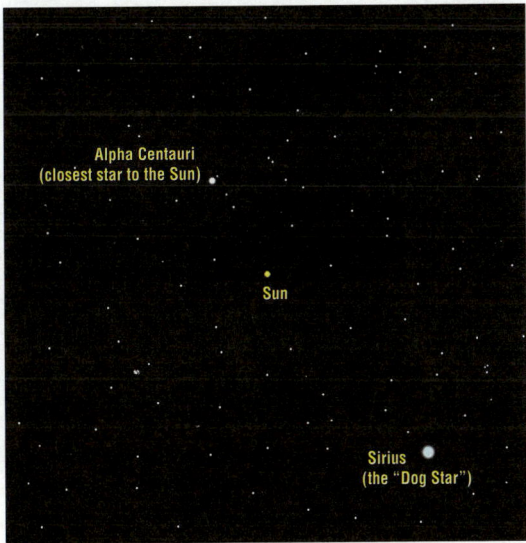

Answers to

In-Text Questions

1. 100 AU is 100 times the average distance from Earth to the Sun, or 15 billion kilometers.
2. The planets are revolving counterclockwise around the Sun.

Stop 3

Explain to students that astronomers usually use astronomical units to measure distances within the Solar System and light-years to measure distances outside the Solar System. You may wish to point out that a light-year is equal to about 9.5 trillion kilometers. If students are curious as to how to arrive at this figure, invite one or more volunteers to come to the chalkboard and calculate it. Or have the class use their calculators to see who can arrive at the answer first.
365 days × 24 h/day × 60 min/h × 60 s/min × 300,000 km/s = 9.5 trillion km

Answer to

In-Text Questions

63,240 AU × 0.16 = 10,118.4 AU
10,118.4 AU × 150,000,000 km = 1,517,760,000,000 km from Earth

Answers to

Stop 4

1. You are 1,011,840 AU from Earth.
2. It would take 8 years.

Our Universe 431

Stop 5

Have students locate the area of space on the winter star chart (on page 387) that is shown in the illustration on page 432. Helpful points of reference include the Orion constellation and the Pleiades.

As students compare the star chart to the illustration, they should recognize that because they are looking back at the stars, right and left are reversed. That is, a star that appeared off to their right on Earth will now appear off to their left.

Students may be interested to know that the Pleiades are one of many star clusters in the Milky Way galaxy. They belong to a category called *open clusters*, or *galactic clusters*, that have from ten to a few hundred stars and include some of the youngest stars in the galaxy.

Without a telescope, the six brightest stars in the Pleiades can easily be seen as a closely spaced group. With the aid of a telescope, as seen here, they can clearly be seen, surrounded by clouds of dust and gas. The earliest recorded reference to this star cluster is in ancient Chinese records dating from 2357 B.C.

The ancient Greeks named these stars for the seven half-sisters featured in an ancient myth. According to the myth, they were saved by Zeus from the pursuit of the giant Orion by being transformed into a group of celestial doves.

EXPLORATION 10—CONTINUED

Stop 5: You have traveled 1600 years since leaving Earth

You are now at a distance of 1600 light-years from Earth. Some of the stars that are so familiar to observers from Earth are now behind you. There is the constellation of Orion with the red star, Betelgeuse. You can also see the Pleiades, a cluster of stars in the constellation Taurus. Had you lived at the time of the dinosaurs, 150 million years ago, you would not have been able to see this group of stars. They are a cluster of young stars that formed relatively recently from clouds of dust and gas.

Examine the star chart on page 387. Compare the stars shown in the star chart with those you can see 1600 light-years from Earth.

The Pleiades—still surrounded by the gases and dust from which they were formed

Stop 6: You have traveled 160,000 years since leaving Earth

At a distance of 160,000 light-years from Earth, a surprising thing has happened: As you have entered emptier space, all of the stars you have been observing have merged into a spiral-shaped structure. This is the **Milky Way** galaxy, our home galaxy. A galaxy is a huge system of stars, dust, and gases held together by gravity.

From a distance of 160,000 light-years from Earth, you see the Sun as just one of 1000 billion stars. The Sun is located near a spiral arm called the Orion Arm (named for the Orion constellation), about two-thirds from the center of the Milky Way. You now realize that on a clear night on Earth, all of the brighter stars you saw were relatively close to Earth, in the same spiral arm as the Sun.

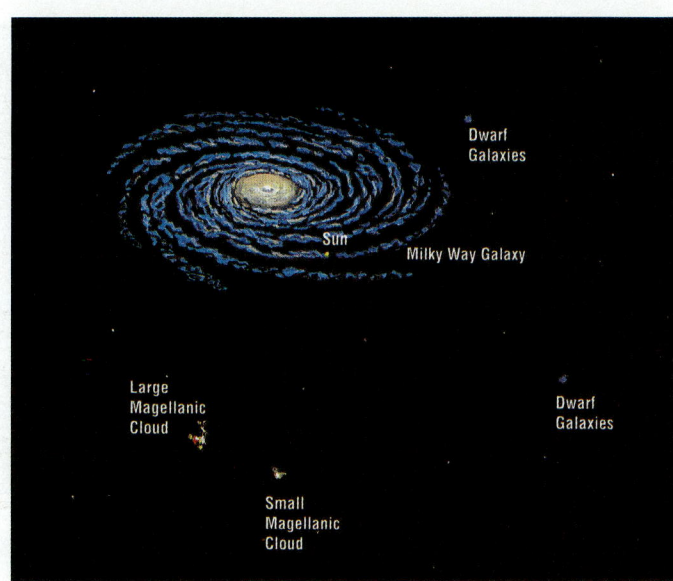

1. A galaxy has been described as an island universe. What does this mean?
2. The Milky Way galaxy is 100,000 light-years across. Approximately how far from the center is the Sun?
3. Where do most of the stars seem to be concentrated in the Milky Way galaxy?

An "island universe"

Stop 6

Point out to students that a portion of the Milky Way galaxy can be seen without a telescope. Ask if any of them have seen the milky-looking band of stars stretching across the night sky. Call on volunteers to describe what it looks like. Point out that in addition to stars, the Milky Way galaxy contains huge dust and gas clouds that block out the light from the stars behind them.

Students may be interested to know that the Milky Way galaxy is about 100,000 light-years across and about 10,000 light-years thick at the center. It becomes much flatter toward the edges. Our Solar System is located about 33,000 light-years away from the center of the galaxy. The distance between the stars in our section of the galaxy averages about 5 light-years. Stars at the center of the galaxy are about 100 times closer.

The stars, gas, and dust in the Milky Way orbit the center of the galaxy, very much like the planets orbit the Sun. For example, the Sun orbits the center of the galaxy once every 250 million years. Almost all of the bright stars in the Milky Way orbit in the same direction. For this reason, the entire galactic system appears to rotate about its own axis.

Answers to
In-Text Questions

1. Just as an island is cut off from other land areas, a galaxy is cut off from other galaxies. A galaxy is surrounded by mostly empty space in much the same way that an island is surrounded by water.
2. If the diameter of the galaxy is about 100,000 light-years across, its radius is about 50,000 light-years. The Sun is about 2/3 from the center of the galaxy. Two-thirds of 50,000 is about 33,000 light-years.
3. Most of the stars appear to be concentrated in the center of the galaxy.

Stop 7

Students may be interested to learn that the Local Group consists of 3 spiral galaxies: the Milky Way, the Triangulum Spiral (M33), and the Andromeda Spiral (M31); four irregular galaxies, including the Large Magellanic Cloud and the Small Magellanic Cloud; and about 25 elliptical galaxies, most of which are relatively small. The Local Group, in turn, is part of a larger grouping called the Virgo Cluster, which contains thousands of galaxies of all types.

Answers to
In-Text Questions

1. Since a light-year is the distance light travels in one year, the light reaching Earth from the Andromeda galaxy left about 2.2 million years ago.
2. Since galaxies are so far away, astronomers are observing what they looked like when the light that is being seen on Earth left them. For example, when astronomers look at the Andromeda galaxy, they are seeing what it looked like 2.2 million years ago.
3. The letter will reach Earth 16 million years from now.
4. The cosmic address may include the following elements:
   Local Group of Galaxies
   The Milky Way Galaxy
   The Orion Arm
   The Solar System
   Earth
   U.S.A.
   State
   Town
   Street
5. Have students share and discuss their ideas. Their responses should indicate an understanding that objects in the universe are separated by vast distances and that our Solar System is but one very small part of that universe.

Stop and Think

After students have had time to read and discuss the questions with their partners, reassemble the class and share and discuss their ideas.

EXPLORATION 10—CONTINUED

Stop 7: You have traveled 16,000,000 years since leaving Earth

You are now 16 million light-years from Earth. The Milky Way galaxy is now seen as part of a family of galaxies called the **Local Group.** All are held together by mutual gravitational forces, just as gravitational forces in our Solar System hold the planets in orbits around the Sun. You may be familiar with one of the galaxies—the Andromeda galaxy, or M31— since it can be seen from Earth with the naked eye. It is the farthest object that can be seen from Earth without the aid of a telescope.

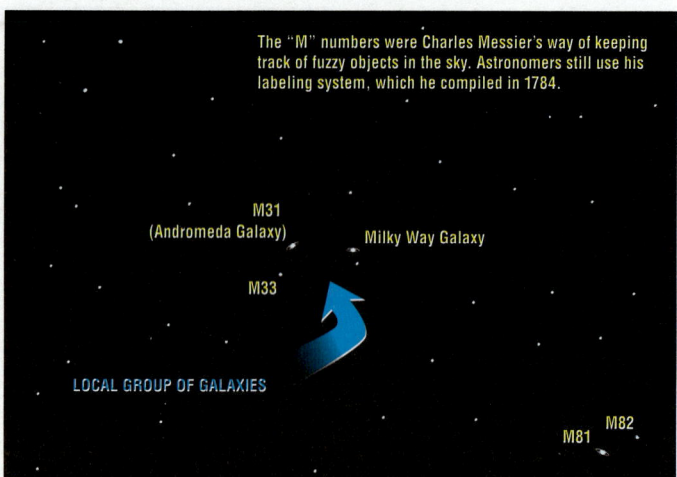

The "M" numbers were Charles Messier's way of keeping track of fuzzy objects in the sky. Astronomers still use his labeling system, which he compiled in 1784.

1. Andromeda is 2.2 million light-years from Earth. Astronomers study it through the light it emits. How long ago did that light leave Andromeda?
2. Studying distant galaxies is described as looking back into the past. What does this mean?
3. You decide to pause on your trip to write a letter back home. When will your letter arrive if it travels at the speed of light?
4. How would you address your letter? In other words, what is your cosmic address back home?
5. What are the two most important ideas this trip has conveyed to you?

The Andromeda galaxy, a typical spiral galaxy

Stop and Think

This is where your journey into the universe stops. As you look around you from the depths of space, your view will take in 100 billion galaxies, consisting of possibly 1 billion trillion stars. In his book, *Cosmos,* Carl Sagan described it this way:

> *A handful of sand contains about 10,000 grains, more than the number of stars we can see with the naked eye on a clear night. But the number of stars we can see is only the tiniest fraction of the number of stars that are. Meanwhile the Cosmos is rich beyond measure: the total number of stars in the universe is greater than all the grains of sand on all the beaches of the planet Earth.*

Answer to
In-Text Question

Life forms on other stars and solar systems would probably be similar to those on Earth if we assume that similar laws of nature hold true.

Primary Source
Description of change: excerpted from *Cosmos,* by Carl Sagan.
Rationale: excerpted to highlight the vast quantity of stars in the universe.

A Conversation with Huck

"We had the sky, up there, all speckled with stars, and we used to lay on our backs and look up at them, and discuss about whether they was made, or only just happened."
—Huck Finn in Mark Twain's book, *Huckleberry Finn*

Around one of the stars, the Sun, you know that life exists on a planet called Earth. Only one other star has been found to have a planet orbiting around it.

Find a partner and discuss whether you think life exists elsewhere in the universe. What are your reasons? Would you like to know for sure? If we assume that the laws of nature hold true on other stars and possible planetary systems as they do here, should life forms be similar? Why?

When you look up in the sky at night, you also see it speckled with stars. But in studying this unit, you have had a "closer look" at them than Huck did. What do you suppose Huck was wondering about as he looked up at the sky? Join in a discussion with Huck. Write down what you and he might say.

There are more stars in the universe than there are grains of sand on all the beaches on Earth.

A Conversation with Huck

You may wish to introduce this material by having students describe how the night sky appears to them. Encourage students to be as scientific or as poetic as they wish. Then have them complete the exercise. When students have finished, call on volunteers to share what they wrote with the class.

To extend the activity, invite students to role-play a conversation with Huck, working with a partner to use what they wrote as a script.

Primary Source
Description of change: excerpted from *Huckleberry Finn*, by Mark Twain.
Rationale: excerpted to raise the question of how the universe began.

 *Follow Up*

Assessment

1. Suggest that students make a travel itinerary or timeline for the trip described in the lesson. The itinerary should provide a brief description of each stop and indicate how much time it has taken to get to each point along the journey. Encourage students to be creative in how they present their information. Display the finished itineraries around the classroom.
2. Have students make diagrams to illustrate where Earth fits into the universe. Explain that their diagrams should begin with Earth and the Moon and end with the Local Group. Call on volunteers to explain their diagrams to the class.

Extension

1. Explain to students that astronomers believe that the most distant objects in the universe detectable from Earth are *quasars*. Suggest that they do some research to discover what quasars are, how they are detected, and how far away they are. Invite volunteers to share what they learn with the class.
2. Have students do some research to learn how the view of the universe has changed from ancient times to the present. Suggest that they use what they learn to make murals or bulletin-board displays for the classroom.

Our Universe 435

Lesson 10

✷ Getting Started

In this lesson, students are introduced to the theory that most astronomers think best explains how the universe began—the big-bang theory. As students read about the theory and learn of its consequences, they create several headings that summarize the main ideas. The lesson concludes by having students simulate one of the consequences of the big-bang theory—the expansion of the universe.

Main Ideas

1. The universe may have begun in an event called the big bang—a cosmic explosion that created the universe and all of the matter in it.
2. The big-bang theory provides an explanation for the expanding universe.
3. According to this theory, we are all products of elements first created within the stars.

✷ Teaching Strategies

Point out to students that as they have progressed through the unit, they have learned about the Solar System, the Milky Way galaxy to which it belongs, and the universe beyond that. Ask them to consider for a moment how the universe may have begun. Then call on volunteers to share their ideas with the class. Accept all reasonable suggestions without comment. Then explain that in this unit, they will learn about a theory that most astronomers think best explains the origin of the universe and how it appears today. **(A teaching strategy for The Big-Bang Theory follows on next page)**

10 A Likely Beginning

The Big-Bang Theory

Scientists have a theory that answers part of Huckleberry Finn's question, the one you just discussed. But like all theories, it will be modified as predictions are tested by experiments and observations. Read the paragraphs below about the big-bang theory. As you read, pretend that you're a newspaper editor who has to think up headings for each paragraph. In your Journal, write down your three- or four-word headings that catch the message of each paragraph.

_____?_____

According to this theory, the universe began sometime between 15 and 20 billion years ago. The universe was formed in an event called the **big bang**. You can think of this as a cosmic explosion that formed the universe. The matter that was formed was hurled outward in all directions by the force of the explosion.

_____?_____

About 14 billion years ago, the universe cooled, and the hydrogen and helium that were formed in the big bang clumped together as a result of gravitational forces. New stars then condensed within these clumps to create what we now see as galaxies. Using a large telescope, astronomers have detected very bright energy sources that are billions of light-years from Earth. One theory is that these energy sources, called **quasars,** are galaxies in their early stages of formation.

_____?_____

Oxygen, carbon, and iron—all of which are found on Earth—were formed within the stars. Some of the more massive stars eventually exploded in an event called a *supernova*. These explosions sent into space material that mixed with the matter from which new stars and planets would form.

Interacting quasars

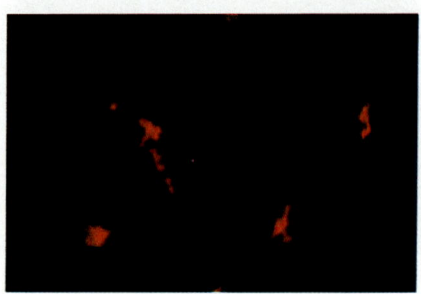
Supernova remnants

Lesson 10 Organizer

Objectives

By the end of the lesson, students should be able to:
1. Describe the big-bang theory as one way of explaining how the universe began.
2. Explain how the "big bang" led to the formation of stars, galaxies, and the Solar System.
3. Identify cosmic microwaves and the expansion of the universe as two observable consequences of the big bang.

Process Skills

observing, analyzing, inferring, summarizing

New Terms

Big bang—a cosmic explosion that took place 15 to 20 billion years ago and created the universe.
Quasars—bright energy sources billions of light-years from Earth; possibly galaxies in their early stages of formation.

436 Toward the Stars

___?___

Approximately five billion years ago a new star, the Sun, and a family of planets formed from gas and dust particles in space, which mixed with heavy elements that were created inside stars. One of the planets that formed—Earth—is the home of humans. Another planet—Jupiter—never reached high enough temperatures to start nuclear reactions in its core, even though it is massive and made of the same stuff as the Sun. The Solar System almost had two suns!

Jupiter nearly became the second sun in our Solar System

___?___

There are a number of predictions suggested by this theory. If the universe began from an explosion of a super-hot fireball, some of that heat should still be detectable. By now, this radiation should have the same wavelengths as *microwaves*. In the 1960s two American physicists, Arno Penzias and Robert Wilson, were working on a way to get rid of static on their microwave antenna system. The sound seemed to be coming from the sky. After trying everything they could think of to eliminate the static (they even tried cleaning bird droppings from the antenna of their radio telescope), they realized that they may have found the radiation left over from the birth of the universe. They may have been listening to the faint echo of the big bang that occurred about 15 billion years ago! Experiments and observations have so far supported the big-bang theory. As so often happens in science, this discovery illustrates how progress is sometimes made by chance instead of by carefully planned experiments.

___?___

Since the galaxies are separating from one another, astronomers reasoned that a big bang may have occurred. In addition, the more distant galaxies are receding at a faster speed than the closer ones. In 1925 Edwin Hubble discovered evidence of an expanding universe. The universe was expanding just as it would be if a cosmic explosion had occurred.

An artist's conception of how the initial explosion may have looked

The Big-Bang Theory

Have students read the introduction to the lesson silently, and be sure they understand what they are to do. Then have them complete the exercise on their own. When they have finished, call on volunteers to share their headings with the class, and involve the class in a discussion of each of the paragraphs. As an alternative, you may wish to have a volunteer read each paragraph aloud, have students think of and write a heading, and then discuss their headings before going on to the next paragraph.

Answer to

In-Text Question

Although headings will most certainly vary, responses should reflect the ideas in the following examples:
- A Cosmic Explosion
- Stars and Galaxies Form
- Planets Form from Exploding Stars
- Formation of the Sun
- An Echo from the Big Bang
- The Expanding Universe

Materials

The Big-Bang Theory: Journal

Exploration 11: large, round balloon; black marker

Teacher's Resource Binder

Resource Worksheet 7–5

Time Required

one class period

Exploration 11

Divide the class into pairs and distribute the materials. Suggest that one student have the job of blowing up the balloon while the other observes what happens.

Have students read the Exploration and then answer any questions they may have before starting the activity. Then have students complete the activity and answer the questions. When they have finished, invite students to share and discuss their ideas.

Answers to Questions

1. The dots did not all separate by the same amount. The farther apart two dots were from one another, the farther they separated.
2. Dots that are farther from one another are separating faster than dots that are side-by-side.
3. This model shows two things: that all matter in the universe is moving outward from a point and that every part of the universe must be moving away from every other part. It predicts that the universe is expanding and that more distant galaxies are receding at a faster rate than nearby galaxies.
4. The model suggests that the universe is hollow, with galaxies only on its outermost edge.

Assessment

1. Have students work in groups to create murals or other classroom displays to illustrate the big-bang theory and its consequences.
2. Suggest that students write an explanation for the following statement: "We are all part of the stars." Have volunteers share their explanations.

Extension

1. Explain to students that there are scientific theories other than the big-bang theory used to explain how the universe began. Suggest that they do some research to discover what some of these other theories are. Have them share what they learn by making oral presentations to the class.
2. Suggest to students that the universe is contracting instead of expanding. Have them write a brief explanation of what this might mean. *(All of the matter in the universe may contract to such a density that it explodes in a big bang. This process of expanding and contracting may happen over and over again. This could be illustrated by the balloon model.)*

EXPLORATION 11

Model of an Expanding Universe

All models are imperfect because they cannot exactly duplicate the real thing. After you construct the following model, think about its uses. Does it help you visualize a difficult idea? How is this model a good one? What are its limitations?

You Will Need

- a large round balloon
- a marker

What to Do

1. Blow up a balloon only partway. This will represent the universe.
2. Now, with a marker, place dots on the balloon, some close to each other, some far apart. The dots will represent the galaxies that are part of the universe.
3. Measure or estimate the distances between the different galaxies.
4. Now blow up the balloon some more and observe what happens to the dots.

Questions

1. Did the dots all separate by the same amount?
2. Which ones separated faster than others? slower than others?
3. Discuss this model in terms of the big-bang theory. What does the model illustrate about the theory? What predictions of the theory are illustrated by the model?
4. How is this model an imperfect picture of the universe?

Challenge Your Thinking

1. Some of stars that we associate with a particular constellation are not part of the same star group at all. Examine the Big Dipper as it was 100,000 years ago and as it is now. Draw a diagram of what you think it will look like in another 100,000 years. Which stars are obviously not part of the same star group? Why do they all look equally far away from Earth?

2. Examine the orbit of the white dwarf around the star Sirius.
   (a) What is a white dwarf?
   (b) What is the shape of its orbit?
   (c) Where is it moving the fastest? the slowest?

3. The following newspaper article is similar to the reading that started the unit. This time, certain terms are underlined. In your own words, describe your understanding of the underlined words.

   Astronomers reported today the discovery of an almost inconceivably massive object within two colliding <u>galaxies</u> 300 million <u>light-years</u> from Earth. If the object is a <u>black hole</u>, it is by far the largest one ever detected. Its mass is estimated to be equivalent to all of the <u>stars</u> in the <u>Milky Way</u>. If the object isn't a <u>black hole</u>, it might be a dormant <u>quasar</u>. More observations are needed to make sense of this new <u>observation</u> in terms of current <u>theory</u>.

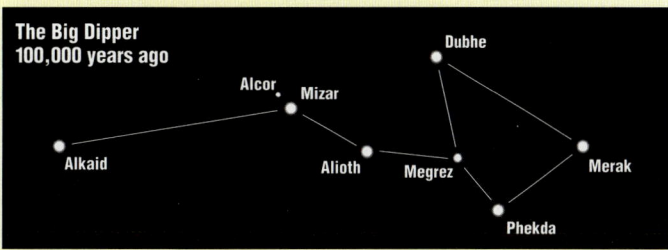

The Big Dipper 100,000 years ago

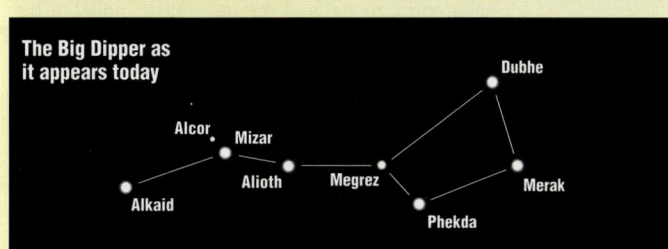

The Big Dipper as it appears today

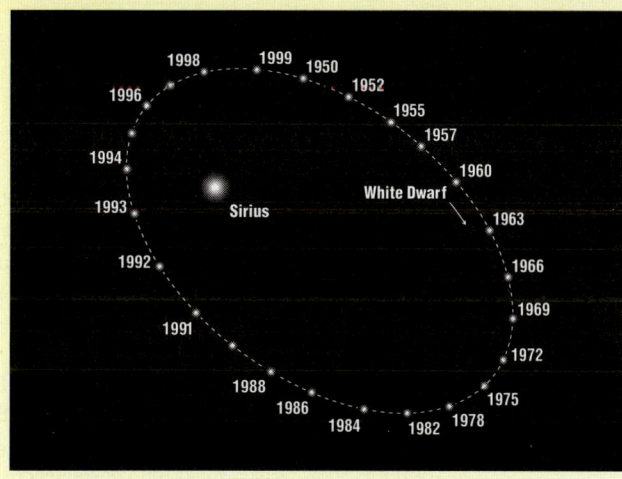

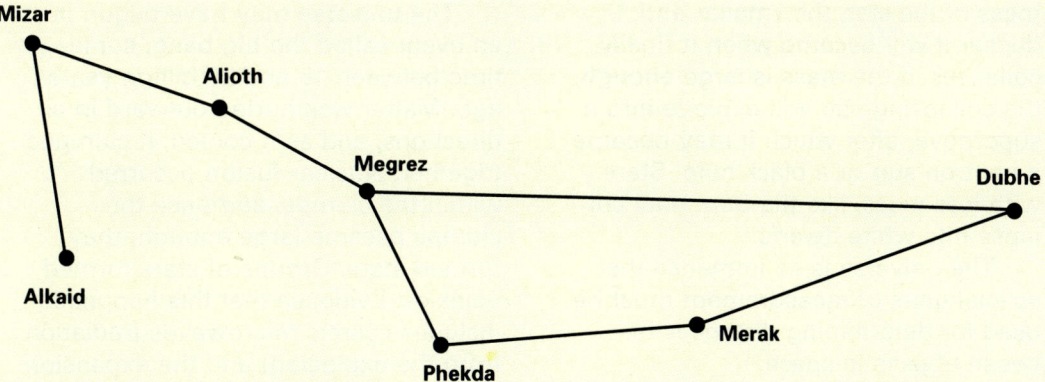

The Big Dipper, 100,000 years from now

Answers to Challenge Your Thinking

1. Alkaid and Dubhe are not part of the same star group as the others, since their positions have changed greatly in relation to the others. The pairs Mizar and Alcor and Alioth and Megrez appear to be the most closely related to each other. See diagram below. They all look equally far from Earth because they are so far away that we cannot visually distinguish their distances and because they all appear about the same size (brightness).

2. (a) A white dwarf is the small and very massive (about the size of Earth) remnant of a star that was once up to eight times the size of our Sun. Once the star used up its hydrogen fuel for nuclear fusion, it began to collapse, appearing first as a red giant. When the gases that expanded to produce the red giant drifted away, a small, massive core called a white dwarf remained.
   (b) elliptical
   (c) The white dwarf moves the fastest as it makes its closest approach to Sirius. It moves the slowest when it is in the portion of its orbit that is the farthest from Sirius.

3. • *galaxies*—a huge grouping of stars that is held together by mutual gravitational forces.
   • *light-year*—the distance that light travels in one year.
   • *black hole*—something with an extremely dense mass that exerts such a large gravitational force that even light cannot escape it.
   • *stars*—gaseous balls made mostly of hydrogen and helium that produce light and heat by nuclear fusion.
   • *Milky Way*—the galaxy in which the Sun and its solar system are located.
   • *quasar*—a very distant body that emits energy at an extremely high rate.
   • *observation*—an event or object that is detected by human senses.
   • *theory*—explanation that makes sense of a wide range of observations. Theories are useful if they suggest predictions that can be tested through observation and experimentation.

Our Universe

Unit 7 Making Connections

Summary for

The Big Ideas

Student responses will vary. The following is an ideal summary.

Astronomy is the study of the Earth's place in space, of the observable make-up of the universe, and of the theories that make sense of our observations of the universe. We learn about the Solar System and the universe by direct observation, using the naked eye, telescopes, space probes, and other technology.

In the heliocentric model, the Sun is at the center of our Solar System, with Earth and other members of the Solar System revolving around it. In the geocentric model, Earth is the stationary center of the Solar System, with the Sun and other members of the Solar System revolving around it. The heliocentric model is the better model because it explains observable phenomena better.

We see different stars in the sky at different times of the year because Earth revolves around the Sun, which makes the Sun appear to pass through the 12 constellations of the zodiac. As Earth orbits the Sun, stars located on the side of the Sun away from Earth are obscured.

The seasons are caused by changes in the angle at which the Sun's rays strike Earth's surface (due to the tilt of Earth's axis) as Earth revolves around the Sun as well as by changes in the amount of daylight.

The Solar System is made up of the Sun and the bodies revolving around it, including the nine planets. The universe consists of solar systems, galaxies, comets, asteroids, meteors, and all of the space between these bodies.

Stars are bodies of gases that give off tremendous amounts of energy in the form of light and heat. Stars produce energy because of nuclear reactions in their cores. Eventually, a star will run out of nuclear fuel and begin to expand in size until it becomes a red giant. In time, it will cool and begin to collapse under its own gravitational forces. The greater the mass of the star, the smaller and denser it will become when it finally collapses. If the mass is large enough, the collapsing star will explode into a supernova, after which it may become a neutron star or a black hole. Stars with less mass, like the Sun, may collapse into white dwarfs.

The universe is so immense that special units of measurement must be used for determining distances between objects in space.

Our cosmic address may include the following elements:
Local Group, Milky Way Galaxy, Orion Arm, Solar System, Earth, Northern Hemisphere, U.S.A., State, Town, and Street

The universe may have begun in an event called the big bang, sometime between 15 and 20 billion years ago. Matter was hurled outward in all directions, and as it cooled, it clumped together. Nuclear fusion occurred within the clumps, and once the clumps became large enough, they formed stars. Groups of stars formed galaxies. Evidence that this happened includes cosmic microwaves (radiation from the explosion) and the expansion of the universe.

Unit 7 Making Connections

The Big Ideas

In your Journal, write a summary of this unit, using the following questions as a guide.

- What is astronomy?
- How do we learn about the Solar System and the universe?
- What are the essential differences between the heliocentric model and the geocentric model of the Solar System? Which is the better model, and why?
- Why do you see different stars in the sky at different times of the year?
- What causes the seasons?
- What objects make up the Solar System? the universe?
- What are stars, and what is their fate?
- How large is the universe?
- What is your cosmic address?
- How might the universe have begun? What evidence do we have?

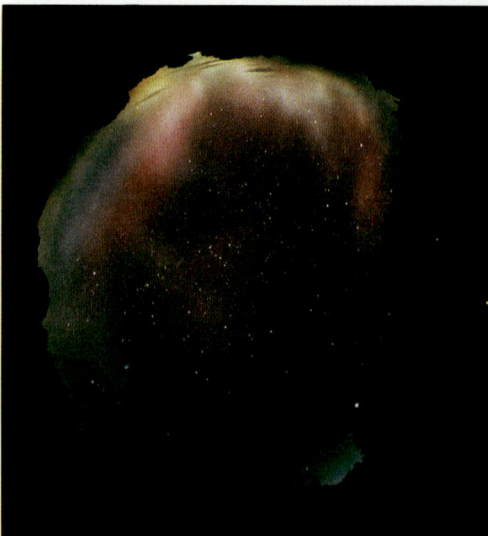

Aurora Borealis (the northern lights)

Checking Your Understanding

1. Where on Earth would you be if:
   (a) The star Polaris is directly overhead?
   (b) The star Polaris never rises above the horizon?
   (c) The Sun is directly overhead once a year?
   (d) The Sun is directly overhead twice a year?
   (e) The stars at night appear to revolve clockwise over your head?
   (f) The stars at night appear to revolve counterclockwise over your head?
   (g) You have equal hours of daylight and darkness?
   (h) You have 24 hours of sunlight?

2. Earth's axis is tilted at an angle of 23½ degrees toward the plane of the ecliptic. What would the consequences be if the tilt were:
   (a) 0 degrees—no tilt at all?
   (b) 45 degrees?
   (c) 90 degrees?

3. Although the other planets lie near the plane of Earth and Sun (the plane of the ecliptic), they are not exactly on this plane.

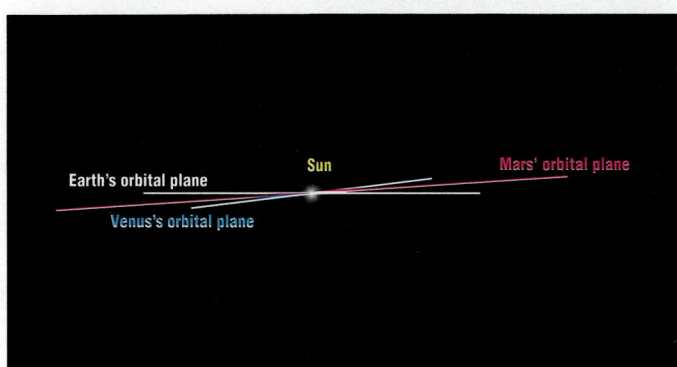

A side view of part of the Solar System, showing three of the planets' orbits

(a) What observations would you make if all of the planets did lie exactly on the same plane?
(b) Earth, the Moon, and the Sun do not all lie exactly on the plane of the ecliptic. If they did, what observations would you make on a regular basis?

4. Draw a concept map that shows the relationships between the following terms: *universe, Betelgeuse, Milky Way, Orion, constellations, stars, galaxies,* and *Polaris*.

Reading *Plus*

You have been introduced to astronomy and to human exploration of the universe. In reading pages S111–S132 in the *SourceBook*, you will take a closer look at members of the Solar System and at how humans are furthering their investigations of space.

Updating Your Journal

Reread the paragraphs you wrote for this unit's *For Your Journal* questions. Then rewrite them to reflect what you have learned in studying this unit.

About Updating Your Journal

The following are sample ideal answers.
1. You would first pass the Moon and then Mars' orbit, the asteroid belt, and the outer planets' orbits.
2. Day and night result from the rotation of Earth about its axis. The seasons are caused by changes in the angle at which the Sun's rays strike Earth's surface (due to the tilt of Earth's axis) as Earth revolves around the Sun and by changes in the amount of daylight. *(The predictability of these changes illustrates the theme of **Patterns of Change**.)*
3. The kilometer is too small a unit to efficiently measure the vast distances involved in studying space. Astronomical units (AU) and light-years replace kilometers.
4. The rotation of Earth causes the Sun and Moon to appear to move together through the sky. The Moon's monthly revolution around Earth causes it to move into position between Earth and the Sun.

Answers to Checking Your Understanding

1. (a) the North Pole
   (b) the Southern Hemisphere
   (c) on the Tropic of Cancer or on the Tropic of Capricorn
   (d) anywhere between the Tropics
   (e) the Southern Hemisphere
   (f) the Northern Hemisphere
   (g) almost anywhere on Earth at the time of equinox passage or at the equator any time
   (h) anywhere north of the Arctic Circle or south of the Antarctic Circle

2. (a) There would be equal hours of daylight and darkness every day; that is, there would be a perpetual equinox. There would be only one season. The Sun's hottest rays would be hitting the equator all the time.
   (b) The Sun's hottest rays would hit Earth as far north as 45°N in the summer and 45°S in the winter. (The Tropics would be at 45°N and 45°S latitudes.) Summers would be much warmer and winters would be much colder. More people on Earth would see the Sun directly overhead. Everyone north of 45°N or south of 45°S would experience 24 hours of daylight at some point during the summer and 24 hours of darkness at some point during the winter.
   (c) Each hemisphere would be in sunlight for six months of the year and in darkness for six months of the year. The seasons would be extreme.

3. (a) The planets would pass directly in front of one another and pass directly behind the Sun. Our view of one or more planets would be blocked by another planet at times. If we looked at the Solar System from the side (in profile), the planets would appear to move back and forth along a single line. The planets would travel along the same arc in the sky.
   (b) There would be a total eclipse of both the Sun and the Moon every month.

4. See concept map on page S162.

Making Connections 441

Science in Action

✦ Background

Planetary geophysics is one of the newest branches of geophysics. Planetary geophysicists compare Earth to other planets in order to better understand Earth and its evolutionary history. To date, most of the studies have focused on Jupiter, Mars, and Venus.

In order to understand what a planetary geophysicist does, it is helpful to take a look at the field of geophysics. Geophysics uses the sciences of geology and physics to study Earth and its atmosphere. It includes such fields as seismology, meteorology, and hydrology. Like their geophysicist counterparts, planetary geophysicists measure the shape, temperature, gravity, electricity, and magnetism of other planets. They use what they learn to answer questions about the origin and history of the planets, including Earth.

Science in Action

Spotlight on Astronomy

Bob Grimm is a planetary geophysicist at a university, where he is involved in research and teaching. His educational background includes a B.A. in Physics and Geology, and a Ph.D. in Geophysics.

Q: What does your daily work involve?

Bob: Right now I spend most of my time looking at pictures of Venus sent back from the *Magellan* spacecraft. With these pictures, I trace geologic structures and piece together what the surface of the planet looks like. Using computer models and mathematical equations, I try to predict surface structures and understand the forces that create them.

Q: What sort of equipment has been developed for your use in observations?

Bob: I am the end-user of robot spacecraft that have gone out on missions and sent back photographs of the planets. These are very complicated, automatic spacecraft that can beam data down to us for years. I also program and use sophisticated computers that can store images of the planets. These computers enable us to zoom in on and roam around the planets' surface features.

Q: Do you ever have to deal with crisis situations?

Bob: When the *Magellan* spacecraft was placed into orbit, it malfunctioned just as it reached Venus. It looked like the whole project might be destroyed! The engineers in spacecraft operations who built the machine were the ones who had to deal with the crisis, but my work was directly affected by the outcome. Fortunately for all of us, the engineers were able to fix the *Magellan*.

Q: Does your job involve traveling?

Bob: Oh yes, I travel a great deal to NASA's Jet Propulsion Laboratory where all robot missions originate. This is also where data is sent after a mission has begun. Eventually, all the data coming back from space gets distributed to the universities, but I go to the center so I can see the data immediately. I also spend a lot of time traveling to the East and West coasts and to Europe, where I meet with other scientists to discuss the information we have gathered and to formulate ideas and theories.

Q: Do environmental issues ever come into play in your research or activities?

Bob: At first it may not seem so, but what we try to do is look at the "whole picture." We try to see the evolution of Earth in a bigger context—where it came from and where it's going. There are many complex issues that can be analyzed, such as how the oceans work with the atmospheres on other planets. The information we obtain from studying these types of questions can give us valuable clues to the way our *own* planet works. We can examine problems like the rising level of carbon dioxide in Earth's atmosphere by looking at Venus, whose atmosphere is mostly carbon dioxide.

✦ Using the Interview

Have students read silently the interview on pages 442 and 443. You may wish to evaluate their understanding of the interview by asking questions similar to the following:

- What was Bob Grimm spending most of his time doing when the interview was conducted? *(studying pictures of Venus sent back from the Magellan spacecraft)*
- What were the pictures helping him to do? *(trace geologic structures and piece together what the surface of Venus looks like)*
- What kind of equipment assists Bob Grimm in making his observations? *(robot spacecraft and sophisticated computers)*
- Why does Bob Grimm's job require him to travel a great deal? *(He travels to NASA's Jet Propulsion Lab to see the data as it is received. He also travels to meet with other scientists to discuss the information they have gathered and to formulate ideas and theories.)*
- How can studying Venus contribute to an understanding of environmental issues on Earth? *(Problems like the rising level of carbon dioxide in Earth's atmosphere can be examined by looking at Venus, whose atmosphere is mostly carbon dioxide.)*
- What does Bob Grimm like about his job? *(intellectual freedom to follow his own interests, the sense of exploration, and doing what no one has ever done before)*

After students have finished discussing the interview, ask them to evaluate whether or not they would like to be a planetary geophysicist.

Q: What activities as a young person led you to your current field?

Bob: Early space exploration in the '60s and '70s—I was greatly inspired in 1969 when *Apollo* landed on the Moon. Space exploration is all so recent, and the United States has always been a leader. That's very exciting to me!

Q: What school subjects turned out to be most important for your career?

Bob: The single most important subject for me was physics because that provided me with the tools that I use most often. Geology was also very important, and so was English. A scientist has to be able to communicate on paper and through presentations. Science is all about *ideas,* and they can only be relayed if the scientist is a good communicator.

Q: What most appeals to you about your job?

Bob: The intellectual freedom to follow my interests. It's a good feeling to know that I'm here because I'm interested in what I'm doing. I come to work each day and say to myself, "What will I think about today?"

The sense of exploration also really appeals to me. Every day, looking at new pictures of planets, I do what no one has ever done before. It's like a hunt—I try to figure something out, bring some relationships together, and soon I have a story to tell!

Q: What personal qualities do you feel are most essential for a person in your field?

Bob: Perseverance. Graduate school is hard. It took me five years, but it was worth it! Projects don't always go right. Sometimes they take so long it's easy to lose track and get "lost." But you just have to stick with it. The results are always worth it.

Creativity is also a very important quality. Sometimes you just have to break the mold and come up with new ideas. Creative talent is invaluable for science!

Q: How would you describe the importance of your work to students?

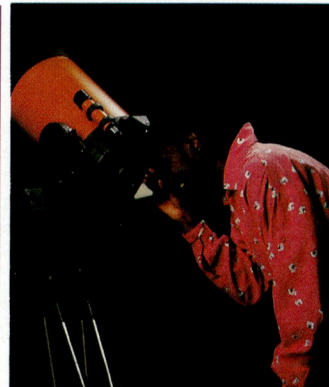

Bob: We *need* to explore. It is difficult today, when there are so many limits on budgets and resources, to advocate tremendous spending on space exploration. However, without continuously questioning our world and wondering about others, humankind has a tendency to become complacent. It is only by questioning that we will learn, and I feel that by asking questions about outer space, we will learn more about our own world.

A Project Idea

Try a planetary mapping experiment. When there is a full Moon, go out and sketch it. Be sure to distinguish the light and dark areas. Then, get a map or globe of the Moon from a map store or a library and look at structures such as lava flows and craters. How does the map compare to your drawing? Determine what the light and dark areas are made of. Does Earth have as many craters as the Moon? Why or why not? What can the number of craters on a planet tell us about its geology?

✦ Using the Project Idea

This mapping experiment gives interested students a chance to look at the Moon and see what structures they can recognize. A good place for them to begin may be to look at satellite photographs of Earth, especially those of *Landsat.* Here they should try to recognize rivers, mountains, and cities. The different perspective will prepare them for the perspective they will need when looking at satellite photographs of the Moon.

Point out to students that maps of the Moon's surface can be found in some encyclopedias and atlases. Suggest that they label as many of the features as they can identify on their sketches.

The students should realize that the bright parts of the Moon that they have sketched are areas that are heavily cratered. The more craters in an area, the older it is. The dark parts are younger structures because they were once craters but have since been filled in with lava. The theory of *geologic superposition* is the principle at work here. It states that the younger structures are on top and the older structures are on the bottom of a surface.

The number of craters on a planet is one indication of how active its geology is. Earth does not have many craters because it is geologically changing faster and more vigorously than the other planets.

✦ Going Further

Financial concerns are often the focus of arguments about space exploration. Continued research and development of space missions is expensive. Many people feel that this money would be better spent on solving immediate domestic problems.

Do research to find out how you feel about this issue, using the following questions as a guide:
- What are the benefits of continued space exploration?
- What is NASA's current budget for space exploration?
- What are the benefits of using that same amount of money to solve domestic issues?

Science and Technology

◆ Background

A radio telescope collects radio waves just as an optical telescope collects light. Optical telescopes use mirrors to collect the light given off by objects in space. A radio telescope uses a huge bowl-shaped reflector to collect radio waves given off by celestial objects. The advantages of a radio telescope over an optical telescope include: its ability to focus on stars, galaxies, and other celestial objects that give off no light at all; its ability to be used in any kind of weather (because radio waves can penetrate clouds and pollution in Earth's atmosphere); and its ability to penetrate the cosmic dust and gas clouds that occupy vast areas of space.

To bring radio waves into sharp focus, a radio telescope must be much larger than an optical telescope because radio waves are much longer than light waves. Because a radio telescope can detect weaker electromagnetic waves than an optical telescope can, a radio telescope can explore farther into the universe.

The information gathered by radio telescopes is often presented as "false-color" images. In the radio image of NGC 1275, for example, blue represents radio waves of the lowest intensity. Red represents the highest intensity.

Science and Technology

The Most Powerful Objects in the Universe

In the 1930s astronomers first began to experiment with **radio telescopes.** These new telescopes picked up radio waves instead of visible light. By the 1950's astronomers had discovered a number of objects in the sky that give off very powerful radio signals. They called these objects *radio stars.*

The Discovery of Quasars

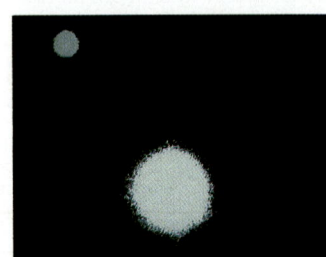

The nearest quasar, 2100 million light-years away

In 1964, however, two astronomers announced that the "radio stars" they had been studying are not stars at all. Actually, they are extremely distant galaxies, known as **quasars,** which are farther away than anything else that has ever been seen. Some object, which the astronomers called an "object X," is at the heart of each of these galaxies. Such objects would be sources of incredibly powerful radio waves, even more powerful than nuclear reactions can produce. These objects would be the most powerful things in the known universe.

Black Holes

Many astronomers now believe that the radio sources inside quasars are objects known as **black holes.** The existence of black holes is more or less taken for granted by many astronomers, although no one has ever seen one. Black holes, if they exist, are in fact invisible!

A black hole, according to the theory, is the result of matter that has been super-compressed. For example, if the sun were compressed from its present diameter of 1,390,000 km down to a diameter of just 6 km, it would become a black hole. The gravitational attraction of such a heavy object would be so great that nothing, *not even light,* could escape from it.

NGC 1275

Astronomers are interested in a particular galaxy known as *NGC 1275,* which is like a quasar in certain ways. It is much closer to the earth than any of the quasars, but like the quasars, it is a strong radio source. Some astronomers believe that there is

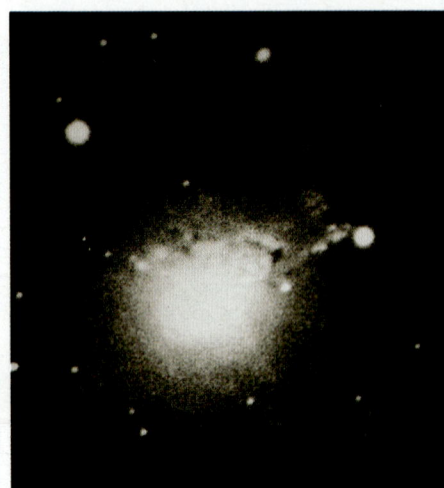

NGC 1275 is a disordered galaxy that emits powerful radio waves.

an enormous black hole, perhaps as large as the sun, at its center. If there is, all the stars and other matter in its vicinity would be attracted to it and crushed. This extreme compression of tremendous amounts of matter would give off large amounts of radio energy.

◆ Discussion

When students have finished reading the feature, involve them in a discussion of what they have read. You may wish to use questions similar to the following:
1. What were the "radio stars" that scientists discovered in 1964? *(extremely distant galaxies known as quasars)*
2. What do many astronomers think is the source of radio waves inside a quasar? *(black holes)*
3. How does a black hole form? *(A black hole is the result of matter that has been super-compressed.)*
4. How can radio telescopes be made extremely large? *(A series of separate radio telescopes can be linked together by computers so that they act like one huge telescope. Such radio telescopes are known as very large array [VLA] telescopes and very long baseline interferometry [VLBI] telescopes.)*
5. Using a VLBI telescope, what did scientists discover about the NGC 1275 galaxy? *(Astronomers found that there may actually be two black holes at the center of NGC 1275.)*

Telescope Teamwork

The largest permanent telescope installation in the world is actually a series of 27 separate radio telescopes linked by computers. When they operate together, they act like a single telescope with a dish diameter of 27 km. Such an arrangement is known as a *very large array* (VLA).

There is a way of getting an even larger system of telescopes together. This is a technique known as *very long baseline interferometry* (VLBI). Using this technique, radio telescopes in different parts of the country, perhaps even on different continents, all work together. Each telescope records data from somewhere in the sky over the same period of time. Then the data tapes are played back together. The result is a set of images from a "telescope" with an effective dish diameter that may be nearly as large as the diameter of the earth!

Colliding Galaxies?

The image on the right was taken by a VLBI team using six radio telescopes in Europe and America. The telescopes all recorded radio wave data from the NGC 1275 galaxy. The astronomers hoped to find evidence of a black hole.

What the astronomers found suggests that there may actually be two black holes at the center of NGC 1275. This information, if true, would support the hypothesis that NGC 1275 is actually the result of the collision of two different galaxies.

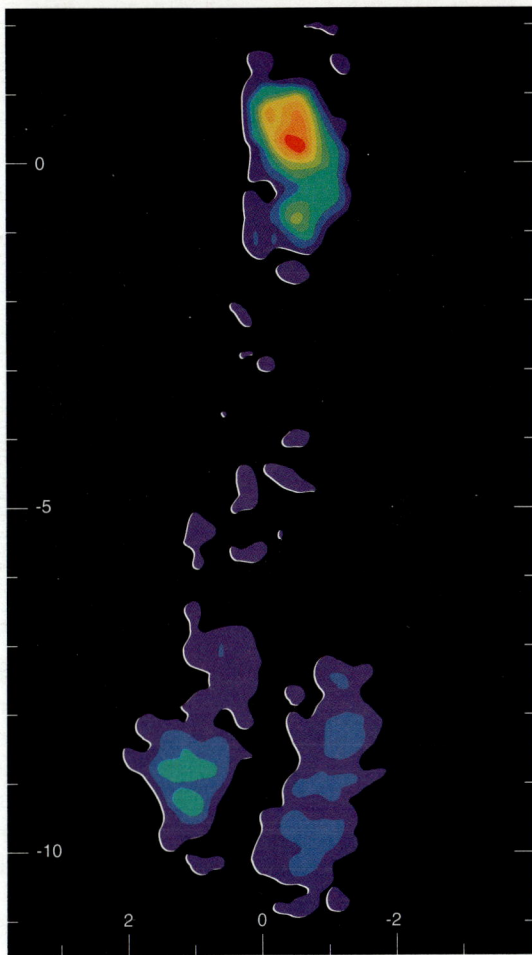

Radio image of core of galaxy NGC 1275. There may be two black holes in the region at the top. The clouds of gas below are being ejected from the galaxy's core at one-fourth the speed of light!

Find Out for Yourself

Do some research to find out about black holes. How do scientists think they might be formed? Some astronomers think there is a black hole in our own galaxy in the constellation *Cygnus*. What is the evidence for this?

There has been some science fiction written about black holes, including even taking trips into a black hole. Does such an idea seem reasonable to you? Write your own science-fiction story about entering a black hole.

✦ Critical Thinking

To promote logical-thinking skills, you may wish to ask questions similar to the following:

1. Why do you think scientists needed to use a VLBI telescope in order to make their discoveries about the NGC 1275 galaxy? *(to collect enough radio waves over a large-enough area of space so that the signals coming from the NGC 1275 galaxy could be analyzed)*
2. Why do you think radio telescopes are used to study some quasars? *(because some quasars give off radio waves, which can be detected and studied only by using radio telescopes)*
3. Why do you think the construction and use of VLBI telescopes became an international effort? *(In order to be effective, the components of VLBI telescopes must be built in different countries. The information gathered by these components must be shared.)*
4. The divisions of the scales in the radio image are milliarcseconds. How large are these units in terms of a degree of arc? *(A second is 1/60 of a minute, which is 1/60 of a degree. A millisecond is 1/1000 of a second. So a milliarcsecond is $1/60 \times 1/60 \times 1/1000$—or 1/3,600,000 of a degree.)*

✦ Going Further

- Explain to students that in addition to quasars and black holes, astronomers have used radio telescopes to discover *pulsars.* Suggest that students do some research to discover what pulsars are and how they relate to quasars. Suggest they share what they learn by making fact sheets.
- Students may enjoy making a poster diagram or model that shows how a radio telescope works. Have students use their models or diagrams to make an oral presentation to the class. Display the models and diagrams in the classroom for others to examine at their leisure.

Science and Technology

UNIT 8 Growing Plants

✹ Teaching the Unit

☀ Unit Overview

In this unit, students examine the relationship between the environment and plant structure and function. In the first section, students consider proposals for the construction of a biosphere on Mars, design their own procedures for testing the germination rate of seeds, and explore the importance, characteristics, and classification of soil. In the second section, students examine the interaction of plants and water, transpiration, and plant nutrition. The section continues with an examination of the life cycle of flowering plants, the process of pollination, and root growth. In the final section, students evaluate the use of pesticides on plants and consider safe alternatives. Then students work in teams to complete a project for the growth and use of plants in interior decoration, outdoor landscaping, or vegetable gardening.

☀ Using the Themes

The unifying themes emphasized in this unit are **Systems and Interactions, Scale and Structure,** and **Patterns of Change.** The following information will help you weave these themes into your teaching plan. A focus question is provided with each theme as a discussion tool to help you tie the information in the unit together.

Systems and Interactions is the unit's dominant theme because of the focus on plant physiology and the interaction of plants and the environment. For example, plants depend on the environment for nutrients, while biological systems within the plant transport those nutrients to all parts of the plant.

Focus question: *How do plants interact with the environment?*

Students should recognize that plants depend upon the environment for water, nutrients, sunlight for making food, and soil for a medium in which to grow. At the same time, plants provide food for animals, give off oxygen through photosynthesis, release water vapor during respiration, and are a source of nutrients for the soil when plants die and decay.

Scale and Structure should be discussed in relation to Lessons 2, 3, 4, 6, and 7, in which students learn how the different structures of plants are specialized to perform different functions.

Focus question: *How are the structures of plants related to their function?*

Students should develop an understanding that structure and function are closely related. For example, roots are ideally structured to push their way through soil in order to absorb water and nutrients. In addition, the flat surfaces of leaves are suited to capture the sun's energy for use during the process of photosynthesis.

Patterns of Change is an integral part of Lessons 2, 6, 7, and 9. In each of these lessons, students are introduced to the many changes that occur during the life cycle of a plant. Also, students explore how changes in the environment affect plants.

Focus question: *What changes occur during the life cycle of a flowering green plant?*

A seed germinates and develops into a sprout. The sprout grows into a mature plant, which then produces flowers. The flowers are pollinated and produce seeds. The seeds are scattered, and the cycle begins again.

☀ Using the Science Discovery Videodiscs

Disc 1 *Science Sleuths,* Green Thumb Plant Rentals #2
The plants from Green Thumb, Inc. (GTI), the plant rental company, are dying at Varisystems (VSI). The owner of GTI suspects a rival company of sabotage. The Science Sleuths must evaluate the evidence to determine why the plants are dying.

Disc 2 *Image and Activity Bank*
A variety of still images, short videos, and activities are available for you to use as you teach this unit. See the *Videodisc Resources* of the **Teacher's Resource Binder** for detailed instructions.

☀ Using the SciencePlus SourceBook

Unit 8 emphasizes the internal structure of plants. It also gives students a closer look at plant growth. Finally, the unit explores the various ways that humans use plants—for food, clothing, and medicine.

445A

PLANNING CHART

SECTION AND LESSON	PG.	TIME*	PROCESS SKILLS	EXPLORATION AND ASSESSMENT	PG.	RESOURCES AND FEATURES
Unit Opener	446		observing, discussing	For Your Journal	447	Science Sleuths: *Green Thumb Plant Rentals #2* Videodisc Activity Sheets TRB: Home Connection
GARDEN INGREDIENTS	448			Challenge Your Thinking	462	TRB: Resource Worksheet 8–3
1 A Garden in Space	448	1	analyzing, problem solving, decision making, writing a report			
2 The Beginning of the Garden	450	2	testing, comparing, investigating, evaluating	Exploration 1 Exploration 2	451 452	Image and Activity Bank 8–2
3 Breaking Ground	454	4	formulating a definition, making data tables, designing an experiment, drawing conclusions	Exploration 3 Exploration 4 Exploration 5 Exploration 6	454 456 457 459	Image and Activity Bank 8–3 TRB: Resource Worksheets 8–1 and 8–2 Graphic Organizer Transparency 8–1
INNER ACTIONS OF PLANTS	463			Challenge Your Thinking	483	
4 Water, Water, Everywhere!	463	3	observing, investigating, manipulating equipment, drawing conclusions	Exploration 7 Exploration 8	463 467	Image and Activity Bank 8–4 Resource Transparencies 8–2 and 8–3
5 Giving Plants a Hand	469	2	investigating, making a data table, making a graph, evaluating	Exploration 9 Exploration 10	470 471	Image and Activity Bank 8–5
6 Flowers and Pollination	473	2 to 3	dissecting, observing, classifying, evaluating	Exploration 11 Exploration 12	474 477	Image and Activity Bank 8–6 TRB: Activity Worksheets 8–4 and 8–5 Graphic Organizer Transparencies 8–4, 8–5, and 8–6
7 Plants from Plant Parts	479	2	decision making, investigating, illustrating, analyzing	Exploration 13 Exploration 14	480 481	Image and Activity Bank 8–7
PLANTS IN YOUR ENVIRONMENT	484			Challenge Your Thinking	493	
8 Medicine or Poison?	484	1	researching, analyzing, evaluating, decision making	Exploration 15	484	Image and Activity Bank 8–8 TRB: Activity Worksheet 8–6 Graphic Organizer Transparency 8–7
9 Make a Green World!	486	6	problem solving, researching, interviewing, decision making	Exploration 16 Exploration 17 Exploration 18 Exploration 19	487 488 489 490	TRB: Activity Worksheet 8–7
End of Unit	494		applying, analyzing, evaluating, summarizing	Making Connections TRB: Sample Assessment Items	494	Science in Action, p. 496 Science and Technology, p. 498 *SourceBook,* pp. S133–S144

*Time given in number of class periods.

445B

✳ Meeting Individual Needs

☀ Gifted Students

1. Have interested students conduct research on the effects of air pollution on plant growth. Suggest that they find answers to the following questions: What kinds of air pollution affect plant growth? How does acid rain affect the health of plants? Why can't plants successfully adapt to air pollution? Have students share what they discover by preparing and presenting oral reports to the class. Encourage class members to ask questions and to discuss the information.

2. Explain to students that plants often respond to their environment with movements called *tropisms*. Two examples of tropism are geotropism and phototropism. (A phototropism is a response to sunlight; a geotropism is a response to gravity.) Have students discover the meanings of these tropisms. Then have them design and perform two activities to demonstrate how geotropism and phototropism function in plants.

☀ LEP Students

1. Provide students with a complete flower, such as a lily or a tulip, and ask them to make a bilingual poster-diagram to show what the different parts of the flower look like. Each part should be identified with a bilingual label. Allow students to refer to a bilingual dictionary, if necessary. When students have finished, review their work for science content, with minimal emphasis on language proficiency.

2. Suggest that students take pictures to illustrate how plants are used by people. Suggest that they organize the pictures into a booklet and write a brief bilingual description of how each picture illustrates they way people use plants.

3. At the beginning of each unit, give Spanish-speaking students a copy of the *Spanish Glossary* from the *Teacher's Resource Binder.* Also, let Spanish-speaking students listen to the *English/Spanish Audiocassettes.*

☀ At-Risk Students

Unit 8 provides students with many motivating activities to help them gain first-hand knowledge of the interaction of green plants and the environment. Students germinate seeds, replant seedlings, grow their own plants, and use plants in interior decoration, landscaping, and gardening. For special recognition, give students the opportunity to participate in a plant-propagation project by making a "Grocery Store Garden." Suggest that they use fruits or vegetables they can buy at the grocery store such as sweet potatoes, the leafy tops of carrots, the tops of pineapples, and so on. Either provide students with space in the classroom for their "garden" or have them grow it at home, making regular reports to the class on their progress. When the plants have grown substantially, have students display them in the classroom.

☀ Cross-Disciplinary Focus

Social Studies

Suggest that students make a career handbook of professions that involve working with plants. The book might include information on horticulture, botany, farming, agricultural engineering, plant breeding, and so on. Suggest that students include a description of each career and provide information on the kind of education a person must have to enter each field.

Language Arts

Have students create a booklet about strange and unusual plants. If possible, they should include a picture of each plant—either a photocopy or a drawing of their own—along with a description of the plant that includes information on where it grows. Unusual plants might include carnivorous plants, plants that grow in strange shapes, plants that are unusual for their size, or plants that have developed unusual adaptations or structures. Display the booklet for other students to read and enjoy.

Health

Students interested in health and fitness may enjoy researching the healthful benefits of specific fruits and vegetables. Suggest that they organize their information in a chart to display in the classroom. The information should include which vitamins and minerals are found in each fruit and vegetable listed in the chart and what the benefits of those vitamins and minerals are.

Mathematics

Have students make graphs comparing the largest crop-producing states in the United States. For example, bar graphs could be used to compare the five largest producers of corn, wheat, or soybeans. To extend the activity, have students make similar graphs comparing the largest crop-producing nations in the world.

Art

Students might enjoy using plant products, such as beans, seeds, or dried flowers and leaves, to make works of art. Encourage students to be creative. Invite volunteers to share their work with their classmates, or set up an area of the classroom as an "art gallery" to display students' work.

Cross-Cultural Focus
Plants from Around the World

Have students work in groups to make murals, bulletin-board displays, or posters that show plants that grow in different parts of the world. For example, divide the class into seven groups, one group for each continent. Each group could focus on the plants that are common in their assigned continent.

To extend the activity, suggest that students discover whether any common houseplants come from the part of the world they have chosen to research.

Gardening in Miniature

Explain to students that bonsai is the ancient oriental art of dwarfing trees and shrubs. Have students research how trees can be made to grow in miniature and present an oral report to the class. Suggest that they bring some pictures to show the results of bonsai techniques.

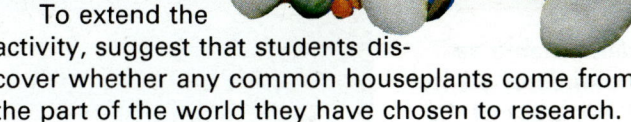

Resources

Bibliography for Teachers

Davis, Brian. *Complete Guide to Garden Plants.* New York, NY: Smithmark Publishers, 1989.

DeJonge, Joanna E. *The Rustling Grass.* Grand Rapids, MI: CRC Publications, 1985.

DeWolf, Gordon P., Jr. *Taylor's Guide to Houseplants.* Boston, MA: Houghton Mifflin Company, 1987.

Eiser, Charles B., Jr. *Of Plants & People.* Norman, OK: University of Oklahoma Press, 1985.

Mabberley, D.J. *The Plant Book: A Portable Dictionary of the Higher Plants.* New York, NY: Cambridge University Press, 1987.

Tompkins, Peter, and Christopher Bird. *The Secret Life of Plants.* New York, NY: Harper and Row Publishers, 1989.

Bibliography for Students

Coldrey, Jennifer. *Discovering Flowering Plants.* New York, NY: The Bookwright Press, 1987.

Cork, Barbara. *Mysteries & Marvels of Plant Life.* Tulsa, OK: Educational Development Corporation, 1984.

Dowden, Anne Ophelia. *The Clover & the Bee.* New York, NY: Harper Collins Publishers, 1990.

Lambert, David. *Vegetation.* New York, NY: The Bookwright Press, 1984.

Lauber, Patricia. *From Flower to Flower: Animals and Pollination.* New York, NY: Crown Publishers, 1986.

Films, Videotapes, Software, and Other Media

Computer Investigations: Plant Growth.
The Plant Growth Simulator, Secondary Version.
Software.
For Apple II Family.
Focus Media, Inc.
839 Stewart Avenue,
Garden City, NY 11530

Plants That Grow From Leaves, Stems, and Roots.
Film and Videotape.
Coronet Film and Video
108 Wilmot Road
Deerfield, IL 60015

Diffusion and Osmosis.
Film and Videotape.
Flowers at Work.
Film and Videotape.
Growth of Seeds.
Videotape.
Roots of Plants.
Videotape.
Encyclopedia Britannica
425 North Michigan Ave.
Chicago, IL 60611

Pollination Biology by Videodiscovery.
Videodisc.
MECC
6160 Summit Drive N.
Minneapolis, MN 55430

445D

UNIT 8

Unit Focus

Have students name as many different kinds of plants as they can. Keep track of the suggestions on the chalkboard. Then ask students to suggest some ways in which the plants might be classified. *(Groupings might include: trees, shrubs, flowering plants, fruits, and vegetables.)* Encourage students to discuss how these different kinds of plants contribute to their life. *(Responses might include: food, shade, beauty, flowers, building materials, medicines.)*

Ask students if they have ever grown plants. Call on volunteers to share their plant-growing experiences with the class.

About the Photograph

This photograph is of an informal perennial garden. It demonstrates one reason for growing plants—to brighten our surroundings. Once a perennial garden is established, the plants return and flower each year.

The designer of this garden has employed several techniques used by landscape architects to construct appealing gardens. These include: massing colors and textures; repeating colors and textures; using pleasing color combinations; varying the height of the plants; and using taller plants behind lower-growing ones.

Unit 8 Growing Plants

GARDEN INGREDIENTS
1. A Garden in Space, p. 448
2. The Beginning of the Garden, p. 450
3. Breaking Ground, p. 454

INNER ACTIONS OF PLANTS
4. Water, Water, Everywhere! p. 463
5. Giving Plants a Hand, p. 469
6. Flowers and Pollination, p. 473
7. Plants from Plant Parts, p. 479

PLANTS IN YOUR ENVIRONMENT
8. Medicine or Poison? p. 484
9. Make a Green World! p. 486

For Your Journal

Write a paragraph summarizing your thoughts about each of the following:
1. How do plants interact with their nonliving environments?
2. What things do the plants in this photograph need in order to grow?
3. A flower and a seed look entirely different. How do you think they relate to each other?
4. How can you grow a garden in space?

✸ Using the Photograph

Ask students to comment on the garden in the photograph. Ask them how they might have designed the garden differently. *(Answers will vary depending on individual tastes.)*

Point out that gardens such as this one are often part of a botanical garden, but they can be established by anyone who knows how to grow plants. Ask students to name things that a gardener should know about in order to grow a garden such as this. *(soil types, nutrients, climate, and how these factors relate to the types of plants grown)*

Point out that perennials are started from seeds, but that they return every year without replanting. Ask students to describe the steps they think they would take to begin and then maintain a perennial garden. Accept all reasonable responses.

✸ About Your Journal

Students should answer the Journal questions to the best of their abilities.

These questions are designed to serve two functions: to help students recognize that they do indeed have prior knowledge about the topic, and to help them identify any misconceptions they may have. In the course of studying the unit, these misconceptions should be dispelled.

Growing Plants 447

LESSON 1

✷ Getting Started

This lesson begins with a scenario set in the year 2152. The United Nations has proposed to build a colony of fully enclosed biospheres on Mars. Interested groups have been invited to submit proposals for the colony. Students are asked to consider responding to the United Nation's request by submitting proposals of their own.

Main Ideas

1. All living things have basic needs that are necessary for their survival.
2. In order for people to colonize other planets, an environment similar to the earth's must be established.

✷ Teaching Strategies

Direct students' attention to the illustrations on pages 448 and 449. Encourage them to discuss what is shown. Accept all reasonable responses without comment. Then have students read the scenario to discover if their ideas were correct. In response to the question at the bottom of page 448, students should recognize that all living things need food, air, and water in order to survive.

Students may be interested to learn that Biosphere 2, a structure similar to the one pictured, was built outside Oracle, Arizona. In September, 1991, eight people began living in the biosphere. Their assignment was to live in it for two years to determine if a similar biosphere can be used as a model for a space colony on Mars.

GARDEN INGREDIENTS

1 A Garden in Space

By A.D. 2140, fusion-powered rockets had made it technically possible to colonize space. The expansion of Earth's population to nearly 10 billion people made emigration to a space colony an attractive possibility.

For several decades, experiments were conducted to determine how to sustain life in space. Some people favored the establishment of orbiting space colonies. The colonies would be close enough to Earth to permit regular visits between space-dwellers and their friends and relatives on Earth.

Other people favored the establishment of colonies on Mars. Travel time to Mars had been reduced, at the closest approach of Mars to Earth, to less than a month.

On January 23, 2152, the President of the United Nations, Elena Kwame Tao, announced the plans of her administration to establish a colony on the planet Mars. The announcement read:

Proposals Invited

During the next 15 years, a vast network of fully enclosed biospheres will be built on Mars. Groups interested in establishing a colony on Mars are invited to apply to the Working Committee for the Colonization of Mars, General Assembly of the United Nations.

Proposals to the Working Committee must incorporate detailed plans for the establishment of the colony. This includes the design and contents of the biosphere that will house the colony.

Could this story really happen someday, or is it only science fiction? The illustration on page 449 shows what such a **biosphere** might look like. The roof has been lifted in the drawing so that you can see the areas inside the structure. A biosphere is an area or region where conditions are suitable for living things to survive. What must living things have for survival, wherever they are? In your Journal, make a list of what you consider to be essential.

LESSON  ORGANIZER

Objectives

By the end of the lesson, students should be able to:
1. Explain what a biosphere is.
2. Describe what a biosphere on Mars might be like.
3. Identify the basic needs necessary for the survival of living things.

Process Skills

analyzing, problem solving, decision making, writing a report

New Terms

Biosphere—an area or region where conditions are suitable for living things to survive.

448 Growing Plants

Pretend that you and a group of your friends have decided to make a biosphere proposal to the United Nations. You should consider the following questions and discuss them with others.

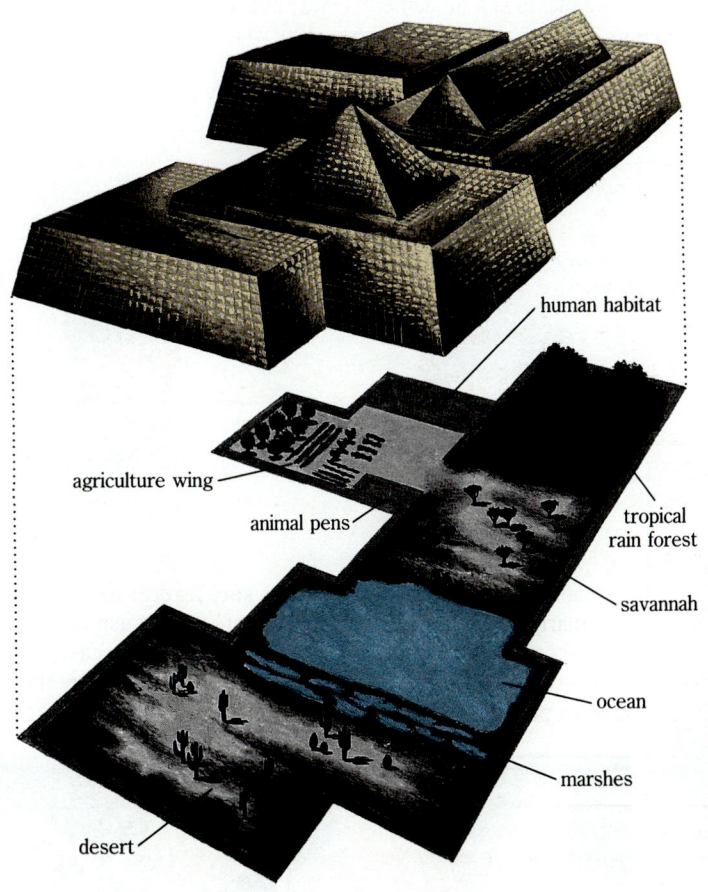

human habitat
agriculture wing
animal pens
tropical rain forest
savannah
ocean
marshes
desert

Think About It!

- How will you control temperatures?
- What will happen to wastes?
- Where will fresh water and fresh air come from?
- How will the plants survive?
- What foods will you need? (Make a list of ten nutritious items.)
- What is the natural source for each of the foods on your list?
- What will your Mars Biosphere have to include so that you can have a continuous supply of food?
- List everything you would need to support the food-producing plants and animals.

Write a report based on your answers. Then present your report and compare it with those of other groups. You might want to change some of your answers after listening to the other groups.

A side view

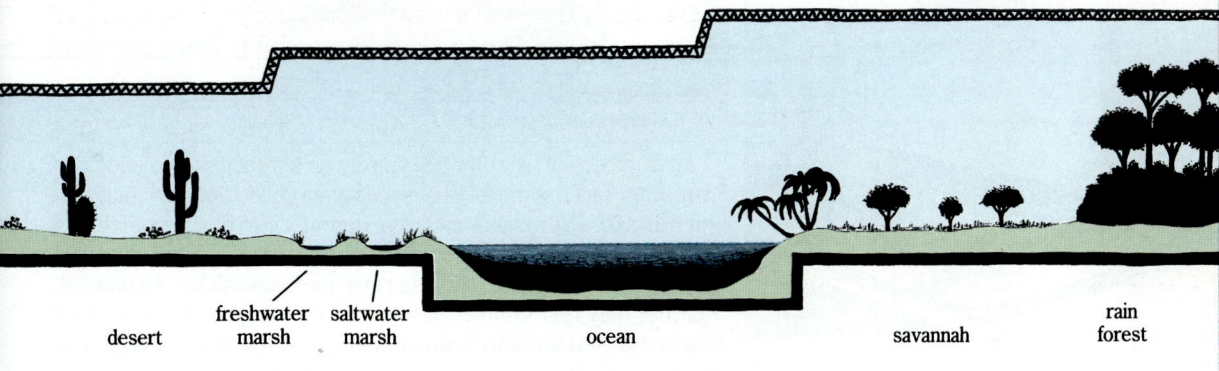

desert — freshwater marsh — saltwater marsh — ocean — savannah — rain forest

Think About It!
Have student groups brainstorm answers to the questions and work together to compose reports to share with the class. Student responses will vary, but they should reflect an understanding of the basic needs of living things and the necessity of devising a means for meeting those needs in their Mars biosphere.

 *Follow Up*

Assessment

1. Suggest to students that they make a poster that illustrates the needs of all living things.
2. Have students write poems about what it might be like to live in a colony on Mars.

Extension

1. Suggest to students that they research NASA's plans to build space colonies.
2. Suggest to students that they find out all they can about Biosphere 2.

Materials
A Garden in Space: Journal

Time Required
one class period

Garden Ingredients 449

LESSON 2

✷ Getting Started

In this lesson, students observe the germination rates of seeds and the growth of plants. As the lesson concludes, students evaluate their results and suggest ways of improving experimental designs.

Main Ideas

1. Provided with the right conditions, seeds will develop into plants.
2. In a given sample, it is unlikely that all seeds will germinate.
3. Careful observations and records must be kept in successful experiments.

✷ Teaching Strategies

Ask students if they have ever had a garden or planted seeds in containers. Call on a volunteer to explain how he or she prepared the soil and planted the seeds. Then involve students in a brief discussion of what factors are necessary in order for seeds to grow. *(warmth and moisture)*

Germination of Seeds

Have students read the paragraph at the top of page 450. Let them discuss whether or not the needs of plants on Earth will be the same as in a Mars biosphere. Help students to conclude that the needs of plants will be the same wherever they grow. Then have them read the rest of the page.

2 The Beginning of the Garden

Germination of Seeds

Growing plants is a fundamental start in planning and preparing to live in an experimental biosphere. But what are the needs of one living thing—a plant—here on Earth? Will these needs be the same as in the Mars biosphere? The best way to start your investigation is to grow a plant of your own.

"I'll grow plants from seeds I collected at home."

"I'm going to bring a package of seeds from the store."

"I'd take seeds and some full-grown plants into space."

Julie collected seeds from flowers such as nasturtiums, zinnias, and marigolds. You may do the same or you can use seeds from a store-bought package, as Nu suggested. In either case, you and your classmates should pick one kind of seed and grow the same kind of plant.

1. Pick the first seeds that form.

2. Dry the seeds.

3. Clean and keep the heaviest, plumpest seeds.

4. Keep seeds in a cool, dry place.

Ali is uncertain whether he can rely on his seeds to develop into plants. In fact, some of the seeds you collect or buy may not **germinate**. (Germinate means to sprout, or to begin to develop.) Imagine how important it is for a farmer to know what percentage of seeds will germinate, and what size the crop will be. To find out, a farmer may test samples of seeds before planting them. You will test seeds for their germination rates, so you won't be disappointed when you begin to grow your own plant.

LESSON 2 ORGANIZER

Objectives

By the end of the lesson, students should be able to:
1. Explain why growing plants is fundamental in preparing to live in a biosphere.
2. Describe how to conduct a test to compare seed germination rates.
3. Describe the growth of a plant from a seed to maturity.

Process Skills

testing, comparing, investigating, evaluating

New Terms

Germinate—to sprout, or to begin to develop.

Growing Plants

EXPLORATION 1

Testing for Germination Rate

You Will Need
- a package or collection of seeds
- materials of your own choice

What to Do

Plan your own procedure for testing what proportion of the seeds will germinate.

Remember, seeds need moisture and warmth to begin to grow. To get you started, here are some factors to consider as you plan your germination test:

- The kind of seeds to use (fruit, vegetables, flowers, etc.)
- The number of seeds to use in the test
- The substance to place them in (on a moist towel, in a jar, in potting soil)

Before carrying out your test, discuss with your classmates how you plan to do it. Exchange advice and ideas. Keep a record of your test in your Journal.

1. Begin by stating the kinds of seeds tested and where you got the seeds. Be sure to note how long they have been stored. Will the number of seeds you use affect your results?
2. Describe the procedure used to test the seeds and include a sketch of your seed samples.
3. Report the results.

Seed Analyst

The germination rates of the seeds that you buy in the store are tested by *seed analysts*. In the table below, the percentage of seeds that germinated in one test are listed. Seed analysts are responsible for finding out the information used in tables like this one.

Seed Germination Levels	
Percentage	**Kind of Seed**
90%	sweet corn
80%	beans, broccoli, cabbage, turnip, mustard, cantaloupe, cucumber, pumpkin, popcorn, watermelon, radish
75%	peas, beets, tomato, onion, asparagus, swiss chard
70%	lettuce
65%	spinach, leek, pepper, eggplant, chives
60%	carrot, parsnip
55%	celery, parsley
50%	dill, sage, summer savory, thyme
35%	watercress

Questions

1. How do the percentages shown here compare with the results from your test in Exploration 1?
2. What do you think may cause variations in the germination rate of seeds?
3. Suggest ways you could find out if a specific factor affects germination rate.
4. You must select a seed analyst to join the Mars Biosphere. Create a want-ad for a newspaper that begins, "A successful applicant . . . ."

Exploration 1

Encourage students to devise a plan to test the germination rate of seeds. Suggest that they work in small groups or in pairs to complete the Exploration. Remember to have students reserve enough germinated seeds to satisfy the requirements for *Exploration 2*.

Seed Analyst

Ask students why a seed analyst's job is important. *(Farmers, foresters, and others rely on information about germination rates to prepare their farms, forests, and gardens.)* Have one student explain how to read the table. For example, students should understand that for sweet corn, 90% of the seeds planted will grow into plants.

Answers to Questions

1. Responses will vary with the type of seeds used in *Exploration 1*.
2. Temperature, light, water, nutrients, space, and breed of plant all affect the germination rate.
3. Answers may vary, but should include the concept of holding all variables constant except for the one being tested.
4. Answers will vary.

Materials

Exploration 1: package or collection of seeds, materials to test the germination of seeds, Journal
Exploration 2: germinated seeds from *Exploration 1*, magnifying glass, potting soil, pots, plant food, gram scale, water, spoon, Journal

Science Discovery Videodisc

Disc 2, Image and Activity Bank, 8–2

Time Required

two class periods

Garden Ingredients

Growing Your Own Plants

Have students share and discuss gardening stories and their lists of plants. Invite them to prepare a master chart on the chalkboard, classifying the plants they have grown as either vegetables, flowers, indoor plants, or others. Encourage each student to contribute at least one plant to the chart. To add interest to the project, bring a couple of seed catalogs to class for students to browse through. Then have students form small groups or choose partners to complete *Exploration 2*.

Exploration 2

Have students use the seeds they germinated in *Exploration 1*. If possible, have each student plant two or more germinated seeds in case one of them fails to grow.

Show students the type of container they should use, or supply all students with the same kind of container. An inexpensive container can be made from cardboard milk cartons by cutting off the top third and poking a few holes in the bottom. You may wish to have students bring in their own soil, or you may provide potting soil for them to use. It would be best if all students used potting soil from the same source.

Have students examine one of the germinated seeds with a magnifying glass. Help them to determine which part of the sprout is the leaf and which part is the root. Then you may wish to demonstrate for the class how to plant one of the germinated seeds. Caution students to be very careful not to break the delicate root or leaf.

When students have finished planting their germinated seeds, involve them in a discussion of the things they should observe and record. Students could make charts indicating the amount of water and fertilizer used. At least once a week, encourage students to make short progress reports. Perhaps a special session could be arranged for unexpected problems that might occur. Invite volunteers to display their plants for the class at appropriate intervals. Remind students to make their observations and recordings on a periodic basis.

Growing Your Own Plants

Have you ever grown plants before? In a group of three or four, make a list of plants that you have grown or have seen others grow. Group the plants under the following headings: *vegetables, flowers, indoor plants,* and *others*. Were the plants healthy? How and where did you plant them? What did you do to help the plants as they grew?

Gardeners learn from the experience of others. Talking to other gardeners is one way to become a better gardener. Another way is reading articles and books written by experienced gardeners. In the following Exploration, you will start growing your own plant from one of the germinated seeds. Keep a daily account in your Journal. Carefully record everything you do and everything you observe about the development of the plant. Do not hesitate to ask questions or to look for help in gardening books.

When the plants are fully developed, have students arrange them in a display. Involve students in a discussion of how they took care of their plants and what they learned about the needs of plants. Encourage them to compare and discuss why variations in growth rate, size, and color may have occurred among the plants they grew.

EXPLORATION 2

Your Plant

You Will Need

- germinated seeds from Exploration 1
- potting soil
- pots
- plant food

What to Do

Choose one of the germinated seeds. Look closely at the seed and be very gentle, since the newly formed root and the leaves are easily damaged. Plant the seed in soil that has been thoroughly broken up. Soil mixtures from a garden store would be useful. Think about the position that the root should be in. Then make a small hole in the soil. Put in the seed, and gently fill in the space around it with soil. How deep should the seed be? Leaves should be showing, as if they had just broken through the soil into the air. Lightly pack the soil so it supports the seedling in an upright position. Add enough water to make the soil moist, but not wet. Some plant food should be added as well. Remember to keep a record of what you do and what you observe as your plant grows.

Things to Observe and Record

- Record the planting conditions. Include the medium (garden soil) that the seed was planted in, the kind and size of the container, and where the container was located.
- Record when and how much you water and fertilize the seed.

- Record the date when you plant the germinated seed and sketch what it looks like.
- Record changes in the plant as it grows, including the
  —height and other dimensions, such as the circumference of the main stem;
  —number, color, and shape of the leaves;
  —number of buds and flowers;
  —characteristics of the flower;
  —signs of pests and diseases.

Special Instructions

1. If there are enough germinated seeds available, plant two or three of them. This gives you an alternative if something should happen to one of your seeds.
2. When the seeds are planted and the soil is moist, measure the mass of the container and its contents. Later you can see how growth affected the plant's mass. When you weigh the container a second time, be sure that the soil has the same amount of moisture in it.

3. Record any changes in growing conditions, such as the amount of light or heat at the location where your plant is growing.
4. When your plant is fully developed, compare your results with those of your classmates.

Other Things to Do

1. Use a series of prints or slides to show your plant's growing stages.
2. Draw a series of sketches of your plant as it grows.

Problems? You can get help from a conservatory or arboretum like this one!

✶ Follow Up

Assessment

1. Have students make bulletin-board displays or murals to show how a plant develops from a seed to a mature plant. Suggest that they use pictures of their own plants as well as pictures from magazines.
2. Have students show you their Journal entries from *Exploration 2*. Evaluate them with regard to: number of observations made, original ideas introduced, methods used to record, and general appearance.

Extension

1. Students may enjoy setting up a reading center that focuses on the variety, care, and growth of plants. Among the types of materials included in the center could be gardening magazines, seed catalogs, library books about plants, newspaper articles about plants and gardening, stories involving plants, and plant poetry.
2. Have interested students start growing some perennial plants. These can be started indoors and then transplanted outdoors. Some of the easiest perennials to grow include: black-eyed Susan, Shasta daisy, coral bells, baby's breath, lupine, poppy, and foxglove. Encourage students to find out which plants grow best in their area.

Garden Ingredients

Lesson 3

✴ Getting Started

This lesson emphasizes the importance of soil for the growth of plants. Students examine the composition, color, and texture of soil. They evaluate soil samples in order to classify them according to soil type and then measure their percolation rates.

Main Ideas

1. Soil is made up of rock and mineral particles and the decaying remains of living things.
2. Soil provides support for the roots of plants and holds the water, air, and nutrients needed for plant growth.
3. Soils with fine particles have a greater water-holding capacity than soils with coarse particles.
4. Factors affecting plant growth include: soil type, temperature, nutrients, water, and the amount of light the plant receives.

✴ Teaching Strategies

Soil—An Important Factor

Have students suggest ways in which people use soil. Then ask students how important they think soil is for plants. Point out that animals depend on plants for food, so soil is extremely important to all living organisms.

Have students read the material and discuss possible answers to the questions. Accept all reasonable responses. Students will have numerous opportunities to refine their ideas about soil as they progress through the lesson.
(A teaching strategy for Exploration 3 follows on next page.)

---

3 Breaking Ground

Soil—An Important Factor

"If you have good soil and enough rain, anything can grow."

This statement may be a slight exaggeration. However, it emphasizes the importance of two factors, soil and water, for a plant's growth.

Soil is familiar; you can see it in a garden, the park, the football field, abandoned lots, flower pots, and other places. But have you ever thought about questions such as these?

- Exactly what is soil?
- Where does it come from?
- Are there different kinds of soil?
- If there are, how do they differ from one another?
- What is the best kind of soil for growing plants?
- Is sand considered to be a soil?
- Is there any danger that our soil might disappear or be damaged? If so, how?

Discuss some of these questions with your classmates. Then consider whether soil will be important in the Mars biosphere. If so, what will the soil be like, how will it be obtained, and how will it be conserved so it will continue to be useful?

EXPLORATION 3

A Close Look at Soil

You Will Need
- a half liter of soil obtained from your backyard, a vacant lot, or a garden
- magnifying glass
- water

Keep the soil securely in a plastic bag or sealed jar. Label it with your name. You will use this soil in many of the activities that follow.

What to Do

1. Along with the particles of various minerals (the substances that rocks are made of), soil contains decayed remains of once-living things. Spread a small amount of your soil sample on a piece of paper. Use a magnifying glass to examine it closely. Do you see mineral particles—pebbles, sand, or even finer particles? What evidence can you find of things that once were living? Try to identify what these things were. Material that was once living is called **humus**. It is an important part of fertile soil.

LESSON 3 ORGANIZER

Objectives

By the end of the lesson, students should be able to:
1. Formulate a definition of soil.
2. Identify different types of soils and the characteristics of each.
3. Investigate the water-holding capacity and percolation rate of soils.
4. Identify the important factors needed for plant growth.
5. Design and perform a controlled experiment to investigate plant growth.

Process Skills

formulating a definition, making data tables, designing an experiment, drawing conclusions

New Terms

Humus—material that was once living; an important part of fertile soil.
Texture—the feel of the soil.
Nutrients—chemicals needed for the functioning and growth of living things.

454 **Growing Plants**

2. Add a small amount of water to your soil—just enough to make it slightly damp. Now rub some of the soil between your fingers. Does it feel smooth? gritty? What other words can you use to describe the feel of the soil? This characteristic of soil is called **texture**.

3. What color is your soil—black, gray, reddish, yellowish, or some other color? Black soils are rich in humus. Gray soils come from areas where there is poor water drainage. Gray soils do not get much exposure to the air. Red or yellow soils indicate good drainage and, consequently, good exposure to air.

4. After making close observations of your soil sample, write a complete description of it. Form a group with three others. Mix up your soil descriptions and distribute them among the members of your group. Try to match the description you receive with the appropriate sample.

What Is Soil?

You have been chosen to write a definition of soil for a new science dictionary. Drawing on your recent experience with soil, write a definition consisting of one or two sentences. Compare it with definitions others have written and with various dictionary definitions of soil. Of all these definitions, which one do you think is best? Why?

Soil Particles

Even soils that look the same may differ in the size of the particles that they are made from. Particle size determines the texture of a soil. In the next Exploration, you will investigate particle size.

Exploration 3

Divide the class into small groups and distribute the materials. Groups should read and complete the Exploration. Discuss with students the three identifying characteristics of soil—composition, texture, and color. After students have written their soil descriptions, challenge them to identify a soil sample using someone else's description.

What Is Soil?

Encourage students to share their definitions of soil, and help them to arrive at a consensus of what soil is.

Soil Particles

This is an introduction to *Exploration 4.* It presents the idea that particle size determines the texture of soil.

Loam—a mixture of sand, silt, and clay.
Percolation rate—the time it takes for a certain volume of water to pass through a sample of soil.

Materials

Exploration 3: 0.5 L of soil, plastic bag or sealed jar, magnifying glass, marking pen, water, labeling tape
Exploration 4: jar (5 cm in diameter and 10 cm high), water, soil sample, watch or clock with second hand, Journal

Exploration 5: 2 large plastic-foam cups, paper towels, glass beaker or bottle, cardboard, dry soil sample, sand, watch or stopwatch, Journal, water, graduated cylinder
Exploration 6: large plastic-foam cups or containers; several kinds of soil; humus; seedlings from Exploration 1; fertilizer; thermometer; spoon; clock or watch; measuring cup or graduated cylinder; water; Journal

Teacher's Resource Binder

Resource Worksheets 8–1 and 8–2
Graphic Organizer Transparency 8–1

Science Discovery Videodisc

Disc 2, Image and Activity Bank, 8–3

Time Required

four class periods, plus a four- to six-week growing period

Garden Ingredients

Exploration 4

Explain that in the previous Exploration, students discovered that soil was made up of several different kinds of substances. In this Exploration, they will take a closer look at some of these substances and determine how much of their soil is made up of each substance. Divide the class into small groups and distribute the materials. Monitor what groups are doing by circulating among them and making yourself available for assistance.

Students should observe that the soil samples, once they are well shaken, settle into distinct layers according to particle size. Different soil samples will have different results. Suggest that students examine and compare the layers formed by each other's soil samples. Encourage them to identify differences in the samples. For example, some samples may settle out into more than three layers, or similar layers of different samples may have different thicknesses and colors. Help students establish an accurate record of the characteristics of their own soil sample by using magnifying glasses and sketches.

Soil and Plants

Before students read this material, ask them how soil interacts with plants. Accept all reasonable responses. Then have students read the material to see if their ideas were correct. Students may find it interesting to know that the nutrients in soil that are used by plants include: nitrogen, potassium, calcium, phosphorus, sulfur, magnesium, iron, manganese, zinc, copper, chlorine, and boron.

Think About It!

Have students read the material silently or call on a volunteer to read it aloud. Involve students in a brief discussion of the question. Accept all reasonable responses, encouraging students to provide reasons for their ideas. Point out to students that they will have an opportunity to test their ideas in the following Exploration.

EXPLORATION 4

Sorting the Soil

You Will Need

- a jar 5 cm in diameter and 10 cm high
- water
- soil from your sample

What to Do

1. Fill the jar halfway with soil. Add water until the jar is nearly full. Put the cover on securely. Shake vigorously for one minute.
2. Place the jar on a table top or desk, and let the soil settle for 15 minutes. Do separate layers form as the soil settles? How many layers do you see in your jar?

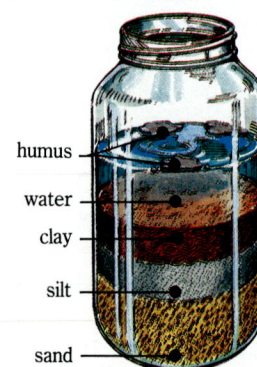

Identify the components of your sample with this information:

- Humus: floats on top; once was living matter
- Clay: very fine particles
- Silt: slightly larger particles
- Sand: still larger particles
- Even larger particles

Compare your jar with those of your classmates. What differences do you observe?

3. Let your water-soil mixture settle for 24 hours, and then observe the layers. In your Journal, make a sketch or a graph showing the proportion of each of the three bottom layers in the jar. It is this proportion that determines the soil type.
4. Look at the Soil Classification Table. What are the sizes of the particles in your jar? You may need to magnify particles of silt and clay to see them individually.
5. Slowly pour off the water and humus from your layered mixture. Spoon off the clay and touch it to feel its texture. Observe the textures of the silt and sand particles, as well. Compare your observations with those given in the table.

Soil Classification Table

Particle	Texture
sand (0.02 mm–2 mm)	Feels gritty when rubbed between fingers. Not sticky when moist.
silt (0.002 mm–0.02 mm)	Feels smooth and powdery when rubbed between fingers. Not sticky when moist.
clay (less than 0.002 mm)	Feels smooth and sticky when moist. Forms hard clods when dry. May remain suspended in water for a long period of time.

Soil and Plants

How does soil interact with plants? Why is this interaction important to plants? In what other ways do plants and soils interact?

Have you ever pulled up a plant by its roots? In this case, what was the soil doing for the plant? The roots of the plant form a complex network. Water and other **nutrients** (chemicals needed for plant functioning and growth) enter the plant from the soil through the smallest roots. Thus, soil performs two important functions for plants:

- It provides support.
- It holds essential nutrients that can be taken in by the network of tiny roots.

Think About It!

A mixture of sand, silt, and clay is called **loam**. A typical mixture is 40% sand, 40% silt, and 20% clay. Mixtures with larger portions of one ingredient are called silty loam, clay loam, etc. How do you think different soil mixtures hold water and other nutrients?

Soil and Water

When you started your study of soils, you wrote a preliminary definition of soil. Your definition was probably more about what soil is than what it does. Now, enlarge your first definition so that it includes the uses of soil.

Do you agree with the statement, "Providing water to the roots of plants is the most important function of soil"? What are some arguments that might support this statement? Consider these questions about how soil and water interact. Use what you already know about soil to come up with the best answers.

- How much water can soil hold?
- Can some soils hold more water than others?
- Which holds more water, a soil mixture or pure sand?
- Which of these allows water to pass through at a faster rate?

Now you will have the chance to test your answers to the questions above.

EXPLORATION 5

The Amount of Water that Soil Holds

You Will Need

- 2 large plastic-foam coffee cups
- paper towels
- a glass, beaker, or wide-mouthed bottle
- your dried soil sample
- sand
- a stopwatch or clock
- a graduated cylinder

What to Do

Before starting this Exploration, let your soil sample air dry by spreading it out on a paper towel for 24 hours.

1. Using a pencil, punch three holes in the bottoms of the two coffee cups.

- coffee cup with soil or sand
- cardboard to fill space so that cup is properly suspended
- beaker

Remember to record all your measurements in your Journal. A table like this one might be helpful.

Quantities	Soil	Sand
Volume of water added	mL	mL
Volume of water that passed through	mL	mL
Volume of water held	mL	mL
Time for water to reach the bottom of the cup	seconds	seconds

Soil and Water

Call on a volunteer to read the material in the left-hand column of page 457. Involve students in a discussion that will further expand their definition of soil. Next discuss the questions about how soil and water interact. Have students record their answers so that they can review, evaluate, and revise them after completing *Exploration 5*.

Exploration 5

Divide the class into pairs or small groups and distribute the materials. Have groups review the steps of the Exploration before they begin. Be sure that they use dry soil samples. Point out that in step 3, they use the weight of a cup of water to pack down the sand or soil in the cup. They should not pour the water into the soil or sand. Students should use the same glass of water to pack each of their samples. Instruct students to measure the level of sand and soil in each cup to make sure that the amounts used are equal. Remind students that it is important to perform the experiment carefully and accurately if the results are to be meaningful.

After the Exploration, assemble the class and involve them in a discussion of their results. Although their results will vary because different soil samples were used, students should discover that more water was held by the soil than by the sand.

Garden Ingredients 457

Answers to

Making Sense of Your Data

1. Students should discover that sand holds much less water than soil does. From this observation, they should be able to predict that silty clay will hold more water than will sandy clay. The water-holding capacity of soil is important because plants must be able to obtain water from soil even during periods of little rain.
2. Students should discover that different types of soil hold different amounts of water. However, all of them will hold more water than will sand. Help students to conclude from their charts that soils with fine particles hold more water than soils with coarser particles.
3. Sand has a greater percolation rate than most soils because it has fewer fine particles than soil mixtures. Students should discover that soils with more coarse particles have a higher percolation rate than soils with more fine particles. Help students conclude that percolation rate is important because plants require air to remain healthy; water retained in the soil can force out all of the air. Some students may also point out that if water cannot drain through the soil, flooding will result during rainstorms. The opposite is also true. There is a problem if too little water is retained.
4. Clay soil can be improved for gardening by adding sand, silt, humus, or a mixture of all three. Sandy soil can be improved by adding clay, silt, humus, or a mixture of all three. In both cases, the resulting mixture provides a soil that retains just enough water to maintain the healthy growth of plants, but not so much as to force out all the air.
5. Answers may vary, but should reflect an understanding that the goal is to have a soil mixture similar to loam—about 40% sand, 40% silt, and 20% clay. Additional amounts of humus, in the form of peat moss or some other organic material, should also be added.

458   Growing Plants

EXPLORATION 5—CONT.

2. Place two circles of paper towel in the bottom of each cup. The holes should be covered.
3. Fill one cup three-quarters full with dry sand and the other cup three-quarters full with dried soil from your sample. Place a glass of water on top of the sand or soil in order to pack it together. Use the same glass of water for each cup. This gets rid of air spaces.
4. Place each cup in the mouth of a beaker, bottle, or glass so that the cup is supported above the bottom of the container.
5. Add 250 mL of water to the sand. Time how long it takes for the first drop of water to drip into the beaker.
6. Repeat step 5 for the soil sample.
7. When water has stopped dripping into both containers, measure the volume of water that escaped from each cup by pouring the water into a graduated cylinder. How much water was held by the sand? by the soil?

Making Sense of Your Data

1. Compare the amount of water held by the soil and by the sand. Was the prediction you made at the start of this lesson accurate? Which do you predict would hold more water: sandy clay or silty clay? Why is the capacity to hold water an important property of soils?
2. Did your soil hold more water or less water than your classmates? Compare your results with those recorded by your class. At the same time, compare the soil types. Create and complete a table like this one in your Journal.

Name	Soil Type	Water-holding Capacity
Mike Annunzio	silty loam	mL
Susan Bates	clay loam	mL
etc.		

3. The time it takes for a certain volume of water to pass through a sample of soil is called its **percolation rate**. Which had a higher percolation rate, sand or soil? Why do you think this is so? Compare the percolation rates of the soil samples of your classmates. Which types of soils have slow percolation rates? fast percolation rates? Why is percolation rate an important characteristic of soil?
4. What could be added to a clay soil to improve it for gardening? to a sandy soil? How could the soil become better for growing plants?
5. What would you use to make a soil mixture suitable for the space biosphere? How much clay or sand would you use?

EXPLORATION 6

Growing the Biggest and the Best

This is a long-term project. Your objective is to find out what it takes to grow the biggest and the best plant in the classroom. You will want to find out how to get the best results in each of these areas:

- Type of plant to use (must grow quickly)
- Type of soil
- Amount of water
- Amount of sunlight
- Temperature
- Distance between plants
- Size of container
- Fertilizer

As you start this project you will need to organize your activities. The class should divide into fifteen different groups. Each group will choose one investigation from those listed in the table shown. Later the groups can pool their results. Doing this will help you find out how to obtain the best results possible. Then you will all be on the way to growing the biggest and the best!

Your class is going to investigate the effect of five variables on the growth of three types of plants. How will you judge the amount of plant growth? Make a decision now. Here are some factors to consider:

- height of plant above ground
- number of leaves
- size of leaves
- mass of plant

The best idea may be to keep a record of as many observations as possible. Make these observations every second or third day, depending on how fast your plant grows. Record them in your Journal. Use several different ways to record your observations: a written description, measurements in a table, graphs, life-size or scale drawings of the plant, or some other means.

Experiment	Plant 1	Plant 2	Plant 3
Investigating different kinds of soil	Group 1	Group 2	Group 3
Investigating different amounts of humus in the soil	Group 4	Group 5	Group 6
Investigating different amounts of water	Group 7	Group 8	Group 9
Investigating different amounts of sunlight	Group 10	Group 11	Group 12
Investigating use of fertilizers	Group 13	Group 14	Group 15

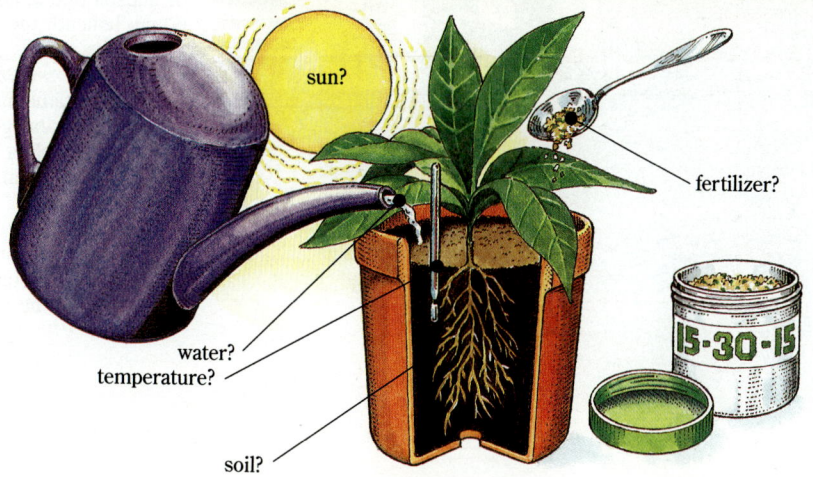

Exploration 6

To monitor each group's ideas and to encourage students to plan their investigations carefully, ask students to submit summaries of their investigations for your approval before the investigations begin. Explain that each summary should include a statement of the hypothesis that is being tested, a detailed description of the procedure, and a list of the materials that will be used. Point out that the summary should make it clear how the chosen variable is to be tested.

Each group of students should use the same method to evaluate plant growth. The best measurement is likely to be the one that includes more than one characteristic.

Allow sufficient time for adequate plant growth. Four to six weeks may be needed. After all of the results are recorded, allow class time for each group to briefly present their conclusions and respond to any questions about procedures and results.

Garden Ingredients 459

Wait a Minute!

Provide students with time to read the material and study the illustration. Then involve them in a brief discussion of Alex's experiment. Help students conclude that Alex's results are unreliable because he did not control all the variables except the one being tested. The variables he should have controlled, but did not, include: the location of the plants, the number of germinated seeds in each container, the container size, the amount of sunlight, the temperature, and the type of soil. Students should be able to suggest ways in which each of these variables could have been controlled. Students should recognize that the only factor that Alex should have varied was the amount of water because that was the variable he was testing.

General Hints

Have students read the section and be prepared to answer any questions they may have. You may wish to provide students with the potting soil and containers they are to use. Milk cartons can be cut in half to make containers. Be sure students poke holes in the bottoms of their containers to provide drainage. Either commercial potting soil or good garden soil can be used. Students who are doing the soil investigations will need several different kinds of soils.

Hints for Each Investigation

Be sure students read the material that applies to them. You may wish to monitor the soils used by those groups that are investigating different kinds of soils. You may also wish to monitor the amount and kind of humus used by those groups investigating the amount of humus in soil. If bagged manure is used, be sure that it is composted manure. Composted manure will have little or no odor.

EXPLORATION 6—CONTINUED

Wait a Minute!

Were you ready to start? Just a word of caution—you have to be careful that you don't do what Alex did! He was investigating the effect of the amount of water on plant growth. After looking at the illustration, would you say his results are reliable? How would you alter his experiment?

Alex varied the amount of water in the three containers. But he didn't keep the other variables constant. Remember to keep everything the same in each of your trials, except the variable whose effect you wish to measure.

General Hints

1. All groups should use the same kind and size of container. Large plastic-foam coffee cups are good.
2. All groups should use the same kind of garden soil, except those who are investigating soils. (Soil will be provided for you.)
3. All groups will use seedlings that have grown since the germination tests in Exploration 1. By now they may have several leaves. This is a good time to transplant them so they will have room to grow.
4. Water the plants when the soil appears to be dry. Keep a record of the amount of water added. Remember to add the same amount of water to each container, unless you are part of the watering experiment. The amount of water required may be different for different species of plants.

Hints for Each Investigation

Investigating different kinds of soil: Use sand, clay, silt, loam, or subsoil (found about 50 cm to 60 cm beneath the topsoil).

Investigating amount of humus: Add varying amounts of compost, peat moss, or bagged manure to soil that has only a small amount of natural humus in it.

Investigating amount of water: Remember that both the amount of water and the time between watering must be considered.

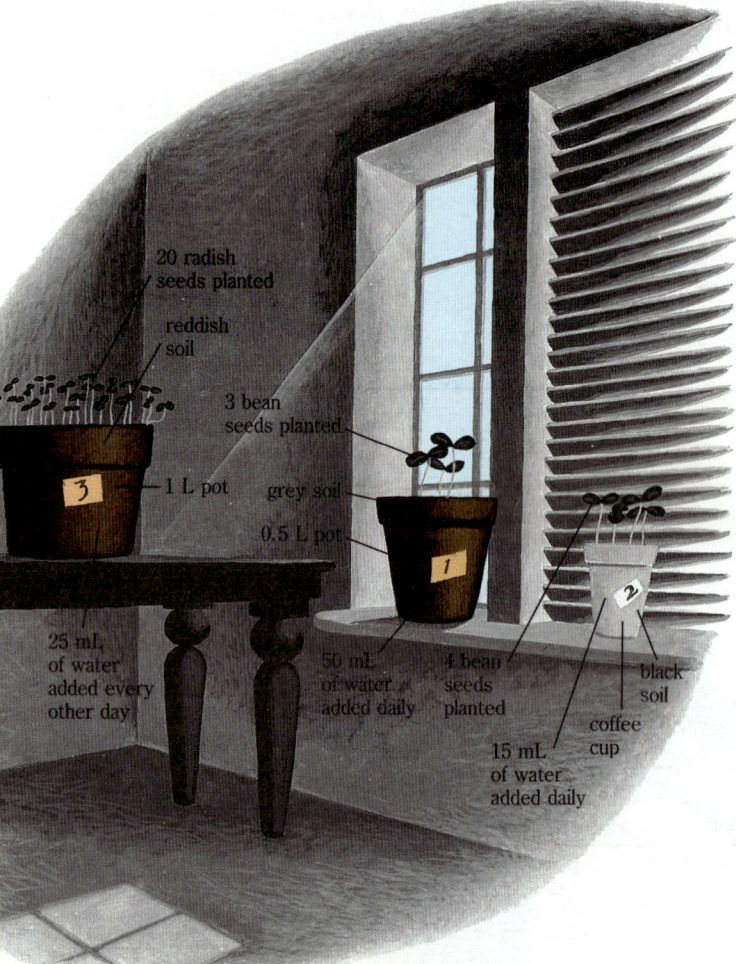

Investigating amount of sunlight: Look for locations with plenty of sunlight and very little sunlight; or vary the amount of time that plants are placed in direct sunlight.

Investigating fertilizers: Use liquid fertilizer, varying either the amount used or the time between applications. Read the directions on the package before using any. The package will also tell you what types of nutrients you are adding to the soil.

Hints for Transplanting Seedlings

1. Put some of the soil that you want to use in the bottom of a container. Lift one plant from beneath the roots, taking care not to hurt the roots. (A spoon or craft stick may help.) Some soil will be removed.
2. Put the plant on the soil in the container and hold the plant upright. Fill the container with more of the same kind of soil. Make the soil firm enough to support the plant, but not so firm that there is no space for air.
3. Add enough water (room temperature) to moisten the soil. Do not pour water directly on the plant. Keep the plant at a constant temperature in a fairly dark place for a few hours. This helps it to recover from the shock of being transplanted. It can then be given the amount of light that you have decided to use in your experiment.

Judging the Results

1. Carefully listen to or read each group's report. In what ways can you learn from their experiments?
2. How might you determine exactly what is the *best* plant growth?
3. How might you improve the design of your experiment?
4. What factors were uncontrollable or difficult to control in your experiment?
5. What would you recommend to a future bean grower? radish grower? mustard plant grower?
6. Which recommendations would be useful for growing any plant?
7. "The best conditions for the growth of one kind of plant are not the best conditions for the growth of other kinds of plants." Do you agree with this statement? Support your answer with evidence from the experiments of all groups.

 **Follow Up**

Assessment

1. Suggest that students make poster diagrams to show the characteristics of the very best soil for gardening. Their diagrams should show the importance of each characteristic and indicate whether they would be supplied by sand, silt, clay, or humus.
2. Provide each student with the name of a houseplant. Explain that they should find out the factors that will promote the best health and growth for the plant name they received. Suggest that they look in books on houseplants to find the information.

Extension

1. Suggest that students research to discover how soil is formed. They should consider the effects of climate, land surface features, plants and animals, and time.
2. Explain to students that plants have developed many ways of adapting to different climate areas, or biomes, around the world. Suggest that they do some research to find out how plants have adapted to conditions in different biomes. Biomes to research include: deserts, tropical forests, Arctic tundra, deciduous forests, and temperate forests. To extend the activity, have students work together to make vegetation maps of the United States or the world for display in the classroom or school library.

Hints for Transplanting Seedlings

Have students read the hints and encourage them to study the illustration. You may wish to demonstrate the procedure for the class. As students actually begin transplanting, monitor how they are progressing and offer assistance as needed.

Judging the Results

Involve students in a discussion of each of the questions. Their responses will vary depending upon their experiences with the investigations. Students should conclude that all plants require the right amount of sunlight and water in order to grow properly, but that different kinds of plants may have different requirements.

Garden Ingredients

Answers to
Challenge Your Thinking

1. Answers will vary. The following are some examples:
   - Soil is an anchor for plants and a provider of water, air, and minerals needed for plant growth.
   - Soil is good for plants if it contains the right amount of humus, a variety of sizes of particles, and water.
   - Soil is made up of rock and mineral particles and the decayed remains of once-living organisms.
2. Answers will vary. Students should include selection of seedlings and soil, careful transplanting of the seedling, addition of nutrients (fertilizers), and provision of sunlight, water, and room to grow.
3. **Answers Across:**
   1. experiment
   2. sand
   3. percolation rate
   4. humus
   5. nutrient

   Answers Down:
   1. biosphere
   2. seed
   3. root
   4. germinate
   5. loam
   6. texture
   7. silt

Challenge Your Thinking

1. How would you complete the sentence, "Soil is . . . "? Use as many words or additional sentences as you need.
2. Prepare an article for a gardener's manual. You need to explain the instructions for raising a seedling from a seed, so a beginning gardener will easily know what to do.
3. Try your hand at the crossword puzzle. Don't write in your book; copy the puzzle into your Journal.

Clues
Across
1. A test of a hypothesis
2. A material in soil that lets the most water pass through soil
3. The time it takes for water to pass through soil (two words)
4. Once-living matter that enriches the soil
5. Chemical needed for growth and functioning

Down
1. A system that supports life
2. This is what the plant begins as
3. The chief water entrance of a plant
4. To sprout
5. Soil containing clay, silt, and sand
6. The way soil feels in your fingers
7. Fine soil particles

462 Growing Plants

INNER ACTIONS OF PLANTS

4 Water, Water, Everywhere!

You know that soil and water are important for the natural growth of plants on Earth. Consider the Mars biosphere. Will plants on Mars need soil and water just like plants on Earth? While soil may not be absolutely necessary, water must always be present for successful plant growth. How does a plant take in water? How does water travel through the plant? How is the water released again into the Earth's environment or into a closed environment such as a biosphere?

This Exploration will give you an opportunity to collect evidence about the interaction of plants and water.

EXPLORATION 7

Water and Plants Interact

Part 1

Entry

Water enters plants through their roots.

You Will Need

- germinated seeds
- a magnifying glass

What to Do

Look at the newly formed root of a seedling. Notice the root hairs, which are extensions of the root cells. Use a magnifying glass to look at them closely.

Each root hair adds to the total surface area of the root. In what way would this be useful to the plant? Imagine that the water in the soil forms a thin layer around each root hair.

Think About It!

Simone wants to plant a garden in an area of hard-packed clay soil. Ray has sandy soil in his garden. Tell these gardeners why their soils need improvement. Advise them about what to add to their soils so that the roots of their plants will have better access to the water in the soil.

LESSON 4 ORGANIZER

Objectives

By the end of the lesson, students should be able to:
1. Describe some interactions between water and plants.
2. Identify the relationship between the structure and function of leaves and roots.
3. Explain the process of osmosis.
4. Explain the importance of leaves in plant transpiration.

Process Skills

observing, investigating, manipulating equipment, drawing conclusions

New Terms

Stomata (singular, stoma)—narrow openings in the leaves of a plant through which gases (oxygen, carbon dioxide, water vapor) pass.

(Organizer continues on next page)

LESSON 4

★ Getting Started

This lesson focuses on the interaction of plants and water. Students examine roots, the process of osmosis, and the function of leaves during the process of transpiration.

Main Ideas

1. Water enters plants through their roots and exits through the leaves.
2. Water moves from one plant cell to another by osmosis.
3. A plant's structure permits the distribution of water throughout the plant.

★ Teaching Strategies

Exploration 7

Have volunteers identify the parts of a plant—the leaves, stems, and roots. Encourage students to speculate as to how water and food get from one part of the plant to another.

Part 1

Suggest that students place a drop of water on the roots and then observe them. Students should recognize that the tiny root hairs greatly increase the surface area through which water can be absorbed.

Answers to

Think About It!

Simone's clay soil is composed of very fine particles that can become packed into a hard, heavy mass. In that condition, air spaces are lacking and water cannot be absorbed into or drained through the soil. Adding some sand can improve soil drainage.
(Answers continue on next page)

Inner Actions of Plants   463

Answers continued

Ray has sandy soil through which water passes quickly, preventing plants from obtaining sufficient moisture. Adding some clay soil and some humus will help retain the necessary moisture. Both gardeners should try to turn their soils into loam by adding substances their soils lack.

Answers to
Part 2

3. Answers will depend on the plant pieces used.
4. Students should observe that the plant material in container A remained crisp or became firmer. The plant material in container B became spongy or limp.
5. In container A, water moved from the container into the plant cells. In container B, water moved out of the plant cells and into the salt solution.

- Students should conclude that in the diagram at the bottom of the page, water would move from section 1 to section 2 to section 3.

(Answers continue on next page)

EXPLORATION 7—CONTINUED

Part 2
Which Way?

The surface of a root hair is a membrane that water can pass through. Water from outside the root hair crosses over the membrane. It enters one of the many cells that make up the root hairs. By crossing several membranes inside the root, the water moves to the other cells in the plant. But water must be able to move out of cells as well as into them. What makes water move in any particular direction through cell membranes? Try this and see.

You Will Need

- two containers
- water
- salt
- plants that pieces may be removed from

What to Do

1. Put some water into container A and a concentrated salt solution into container B.
2. Into each container, place several plant pieces, like a 5-mm thick carrot slice, a fern frond, a lettuce leaf, or a geranium leaf. Be sure to place the same kinds of things into each container.
3. After 20 minutes, observe and compare the plant pieces in the containers. Look again after 1 or 2 hours. Do they resemble the plants that they were taken from?
4. Which plant pieces remained crisp or became firmer? Which ones became spongy or limp?
5. Which direction did the water move in container A? in container B?

Water   Concentrated salt solution

In container A, water moved from the container into the plant cells. In container B, water moved out of the plant cells and into the salt solution. Such water movement is called **osmosis**. What determines the direction that the water will move? It is the concentration of solutes that determines water's movement.

Water moves through cell membranes from areas where water is more concentrated to areas where water is less concentrated. What does this mean? You know that the amount of solute in a solution determines how dilute or concentrated the solution is. Now think about a solution in this way:

Water (the solvent) and any solute dissolved in it are made of tiny particles.

Look at the diagram below. In the first section, all of the particles are water particles. In section 2, there are some particles of water and some particles of solute. That means that the concentration of water is less in the second section. In section 3, there are many particles of solute, so the concentration of water is even lower than it is in section 2. Predict what would happen if the walls between these liquids were cell membranes. Would water move through the cell membranes? In what direction?

Look at the observations that you made when you put the pieces of plants into containers A and B.

In container A, the water inside the plant cells had dissolved substances in it, such as minerals that plants need. The water in

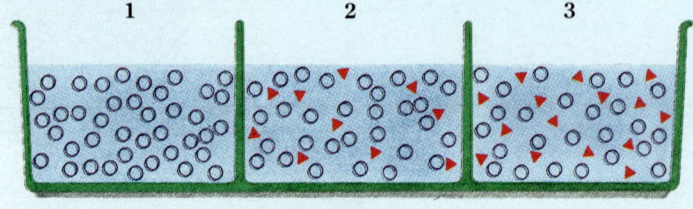

○ water particle   ▲ solute particle

(Organizer continued)

Osmosis—movement of water across a semipermeable membrane from areas where water particles are more concentrated to areas where they are less concentrated.

Photosynthesis—the process in which green plants use energy from sunlight to convert water and carbon dioxide into food for the plant.

Transpiration—the movement of water out of the leaves of a plant.

Materials

Exploration 7
Part 1: germinated seeds, magnifying glass
Part 2: two containers, water, salt, plant pieces, labeling tape, marker, watch or clock
Part 4: stalk of celery, blue food coloring, clear drinking glass, water, Journal
Part 5: geranium or other plant with large leaves, small plastic bag, tape, lamp or sun-lit window
Exploration 8
Part 1: variety of leaves, magnifying glass, Journal

Part 2: microscope, microscope slide and cover slip, leaf, water

Teacher's Resource Binder

Resource Transparencies 8–2 and 8–3

Science Discovery Videodisc

Disc 2, Image and Activity Bank, 8–4

Time Required

three class periods

464   **Growing Plants**

container A did not have these dissolved substances in it. Water moved from container A into the plant cells just as it would move from section 1 to section 2 in the diagram.

In container B, there were fewer dissolved substances in the water inside the plant cells than in the water in the container that had a lot of solute (salt) dissolved in it. Water moved from the plant cells into the solution, just as it would move from section 2 into section 3 in the diagram. How did the movement of water out of the plant cells affect the plant parts? Water movement is always from areas where water particles are more concentrated to areas where they are less concentrated.

Knowledge about osmosis can be useful to gardeners, while lack of knowledge about it can cause problems. Think of some situations in which gardeners need to understand osmosis.

Beverly's Blunder

Beverly put some powdered fertilizer on top of the soil of her potted plant. Then she poured a little water over it. A few days later, the leaves of her plant wilted and turned yellow. Why did this happen? What should Beverly have done differently?

Part 3
A Quick Recovery

A plant that has been deprived of water appears limp and wilted. A wilted impatiens plant usually recovers quickly once water is provided. Explain why water moves into the root cells, and from there to all of the plant's cells. Where is the concentration of water particles greater? What direction is the water moving?

The appearance of the recovered impatiens plant shows that water has moved throughout the plant. Water has made the shape of the plant parts normal again.

Part 4
Special Routes
You Will Need
- a stalk of celery (cut at both ends)
- blue food coloring
- a clear glass
- water

What to Do

Put a few drops of blue food coloring and a piece of celery into a glass of water. Observe the celery over several hours. Describe the top end of the celery. What evidence is there that water and food coloring have moved up the celery stalk? Sketch the stalk of celery, showing any features that you think are important.

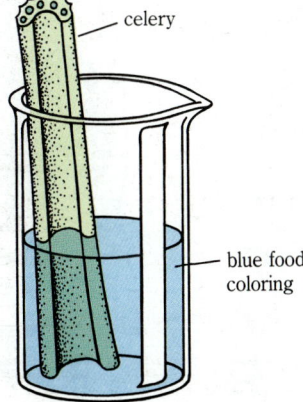

The celery plant has special cells that serve as tubes for the movement of liquids.

Answers continued

Students may recognize that gardeners can use a knowledge of osmosis to determine when plants need water (when they begin to wilt), and how to keep plants fresh after they have been harvested.

Answers to
Beverly's Blunder

The leaves turned yellow because the concentration of fertilizer in the soil caused water to move out of the plant. Beverly should have diluted the fertilizer.

Part 3

Call on a volunteer to read aloud the first paragraph of *Part 3*. Direct students' attention to the illustrations of the impatiens plant. Ask students to respond with a show of hands if they have ever had a plant that wilted like the one in the illustration. Call on a volunteer to explain what he or she did and what happened as a result. Involve students in a discussion of the questions. Students should recognize that once water is supplied to the soil, the concentration of water particles becomes greater outside the root cells. Therefore, the movement of water will be from the soil into the cells.

Part 4

Students should observe that water and food coloring traveled up tiny tubes in the celery stalk. Have students first cut the celery stalk in half and then longitudinally to examine the path of the colored water. Suggest that they use a magnifying glass to examine the celery stalk in detail.

Inner Actions of Plants 465

Part 5

After students have set up *Part 5*, they should make observations after 30 minutes' time and again the following day. Students should see that tiny droplets of water form on the inside of the plastic bag. They should conclude that water taken in by the roots of the plant has traveled through the plant and through the leaves.

At this point, you may wish to briefly review what students know about photosynthesis. Photosynthesis is the process by which green plants use the sun's energy to make food. During photosynthesis, carbon dioxide, water, and light energy from the sun combine in the presence of chlorophyll and certain enzymes to produce glucose and oxygen.

Students might be interested to learn that only about one percent of the water that reaches the leaves is used by the plant in chemical reactions. The rest evaporates, or transpires, from the leaves as water vapor. The process of transpiration cools the leaves. Leaves produce heat as a result of the chemical processes that occur there. The evaporation of water from the leaves also helps pull water from the roots through the stems.

Answers to
Think About It!

1. **(a)** Dry, warm, windy conditions favor the drying of a towel on a clothesline by increasing the rate of evaporation.
   (b) The conditions that help a wet towel to dry also increase the transpiration rate of a plant.
   (c) Answers may vary, but possible responses include: spraying the plant with water, placing the plant in a cooler or shadier environment, keeping the plant in a humid or moist location, and keeping the plant out of the wind and away from drafts.
   (d) Transpiration helps cool the plant.
2. Student responses may vary, but should indicate that the transpired water vapor would have to be condensed into liquid water and then collected in some way. For example, the transpired water vapor might condense on the roof of the biosphere and flow down the sides into pipes that would carry it to collection vessels.

3. **(a)** 50,000 × 0.90 = 45,000 germinated kernels.
   45,000 × 0.66 = 30,000 mature corn plants.
   If 1 corn plant gives off 200 L of water, 30,000 corn plants would give off: 200 L × 30,000 = 6,000,000 L of water.
   (b) A loam that holds water well would be good for corn. Such a soil would contain a small amount of sand and larger amounts of clay, silt, and humus.
4. Desert areas in the Southwest and dry prairies in the Midwest often suffer from water loss. Irrigation and soil treatment help prevent problems.

EXPLORATION 7—CONTINUED

Part 5
Exit

Water reaches the leaves of a plant where it is used in the food-making process called **photosynthesis**. Extra water is released from the leaves into the air. It is easy to collect some of this water.

You Will Need

- a geranium or other plant with large leaves
- a clear plastic bag
- tape

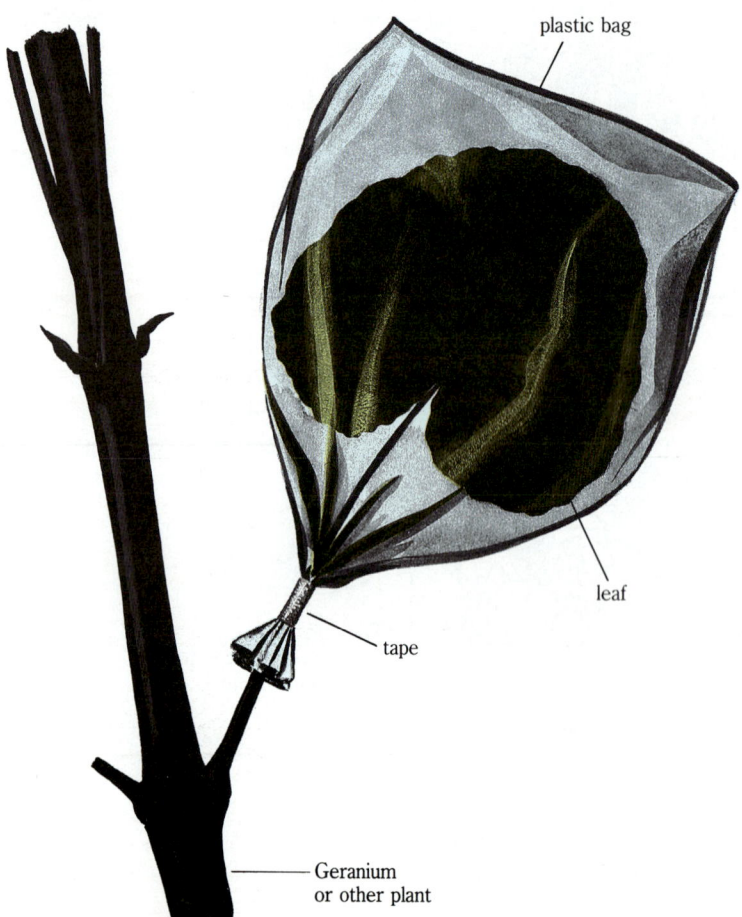

What to Do

Cover one of the leaves of the plant with a plastic bag. Tape it so that no air can enter or leave. Leave the plant in the sunlight or under a bright lamp for half an hour. Look closely at the inner surface of the bag. What do you see?

Transpiration is the term used for the movement of water out of the plant through the leaves.

Think About It!

1. A leaf has been compared to a wet towel on a clothesline.
   (a) What conditions help a towel to dry?
   (b) Would these conditions also increase the transpiration rate of a plant?
   (c) How could you help prevent a plant from drying out while it is transpiring?
   (d) What beneficial effect does transpiration have for plants on a hot day?
2. In the space biosphere, the plants' water supply would have to be maintained. How would you retrieve the water lost by plants through transpiration?
3. Eli and Tanya are farmers. They plant a field of corn with 50,000 kernels. The minimum germination rate for corn kernels is 90%. Suppose that two thirds of the germinated seeds grow into mature corn plants. One corn plant gives off about 200 L of water by transpiration while it is growing.
   (a) How much water could all of the corn plants together lose through transpiration during one growing season?
   (b) What soil type would you consider to be the best for growing corn? Why?
4. Where are the areas in the United States in which water lost from transpiration causes problems for farmers? What can they do about it?

Water and Plants

Consider the evidence that you have gathered about water and plants. You know that water enters a plant's roots, that it moves upward through the plant, and that some of the water escapes through the leaves. Water and plants interact continuously.

In what ways is water useful to a plant?

- Water helps the plant hold its shape.
- Water carries essential dissolved substances throughout the plant.
- Water has a cooling effect as it evaporates during transpiration.
- The chief use of water by plants is for photosynthesis (making food), which takes place in the leaves. For photosynthesis to occur, plants also need carbon dioxide and sunlight. Oxygen is then released by the plant.

Investigate

You might want to investigate the process of photosynthesis by using the resources of a library. If you do, include a look at the process called *respiration* as well. Many people are surprised when they find out that plants respire. Plant respiration resembles our own respiration process; it involves taking in oxygen, which then is used to release energy from stored food. Carbon dioxide is released during respiration.

Your recent observations have shown that leaves are important to plants. The next Exploration gives you a chance to look at leaves of a variety of plants and to consider how leaf structure suits a plant's needs.

EXPLORATION 8

A Look at Leaves

Part 1
Outer Looks

You Will Need
- a variety of leaves
- a magnifying glass

What to Do

1. Examine the upper and lower surfaces of each kind of leaf.
2. Describe each leaf. You may either draw the features of each leaf surface, or write a list of words to describe its features.
3. Exchange your lists or drawings with others to see if they recognize the leaves that you described.

Coleus (top), Nasturtium (bottom)

Answers to
Think About It!

1. Answers will vary depending on the kinds of leaves students examine. Students should recognize that all of the leaves have veins and contain green coloring.
2. The basic shape of leaves provides a large surface area on which to collect the sun's energy, which is essential for photosynthesis.
3. Answers will vary, but may include differences in color, texture, lack of luster, and the way the veins stand out. The upper surface of a leaf is usually smoother than the lower surface. The veins on the lower surface tend to stand out more. The upper surface is usually a darker green. This is because more chlorophyll is concentrated on the upper surface where sunlight strikes the leaves.

Part 2

You may wish to demonstrate how to prepare a slide for the class. Staining the tissue with a little iodine will make it easier for students to distinguish the cells. Hopefully, most students will be able to locate stomata. Students should conclude that gases, such as oxygen, carbon dioxide, and water vapor pass through these openings.

Answers to
Something to Think About

1. Findings will vary depending on plants used. Students should illustrate their observations.
2. Students should recognize that leaves offer a good source of food for both immature and adult insects.
3. Students should conclude that a plant's ability to survive will be limited if too many of its leaves suffer damage. If enough leaves are lost, the plant will be unable to carry out photosynthesis and, thus, unable to make food for itself.

468   Growing Plants

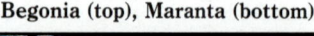

Think About It!
1. In what ways are all the leaves alike?
2. How do you think the basic shape of leaves equips them for photosynthesis?
3. Can you tell the upper and lower surfaces of the leaves apart? How? Why do you think the surfaces are different?

Now, look for clues to answer this question: How do most substances enter and exit a leaf?

Begonia (top), Maranta (bottom)

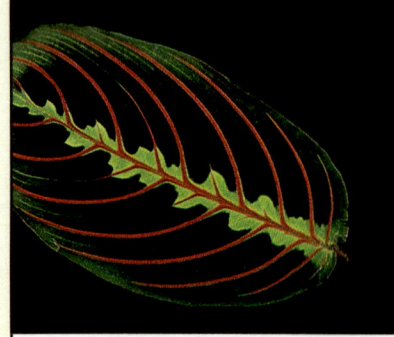

Part 2
A Closer Look

You Will Need
- a glass slide
- a cover slip
- a microscope
- a leaf (a large geranium leaf or a piece of romaine lettuce will be suitable)

What to Do
1. Carefully peel the thin layer from the lower surface of a leaf. (Hint: If the leaf is crisp, bend a piece of it until it "snaps," and then separate a piece of skin from the leaf's surface.)
2. Place the sample on a slide, with the lower surface of the leaf facing upward. Put one drop of water on top of the sample. Gently set the coverslip over your sample, and examine it under the microscope.

Compare what you observe on your slide with the picture shown here. Do you see what look like narrow openings? These are **stomata** (singular, stoma). What substances are likely to pass through these openings?

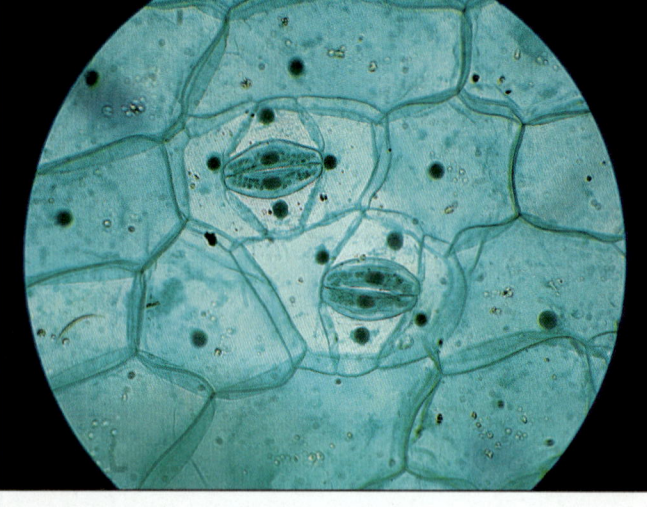

Something to Think About

Look at the leaves of the plants you are growing. Are they all in perfect condition? Look at the leaves of other available plants.

1. Make a record of any damage you find on the leaves. This could include holes, change of color, or a wilted appearance.
2. Caterpillars and other living things find plant leaves very attractive. Explain why you think this is so.
3. What kinds of problems would you predict that your plants will have if their leaves suffer damage from other living things?

Assessment

Suggest that students make a model of a plant from clay, papier mâché, or some other material that shows how water flows from roots to leaves. Display the models around the classroom.

Extension

Using a microscope, suggest that students examine a cross-section of a leaf. If possible, provide them with a prepared slide. Suggest that they use a biology text or some other source to identify the different kinds of cells and tissues they observe.

5 Giving Plants a Hand

What's in a Number?

Are the people on the other side of the fence speaking in code? It certainly looks that way! Actually, they are describing the kinds of fertilizers they use. Each set of numbers refers to the percentages of the three chemical substances most essential for plants. What does 10-52-10 mean? Such a fertilizer contains 10% nitrogen (N), 52% phosphorus (P), and 10% potassium (K). The other ingredients are inactive substances like clay or chemical substances like sulfur that make the fertilizer easy to use. What does 20-20-20 contain? 20-20-20 contains: 20% N, 20% P, and 20% K. What does 15-30-15 contain?

These number groupings are called the N-P-K ratings of fertilizers. Why did one person on the other side of the fence recommend using 10-52-10 first? This fertilizer is especially good for seedlings and for plants that have just been transplanted or repotted. Which chemical is present in the largest amount? This chemical helps plants develop strong root and stem structures. What would happen if a plant lacked this chemical?

Each mixture has its special use. For fertilizing leafy trees and shrubs, 28-14-14 is ideal. The nitrogen helps branches and leaves develop. Nitrogen must also be present for plants to have their usual green color. What symptoms would you expect to indicate a nitrogen deficiency?

LESSON 5 ORGANIZER

Objectives

By the end of the lesson, students should be able to:
1. Identify the three main nutrients found in most fertilizers and explain the purpose of each.
2. Evaluate a fertilizer based on its N-P-K rating.
3. Explain what soil pH is and why it is important to plant growth.
4. Explain how plants can be grown in nutrient solutions.

Process Skills

investigating, making a data table, making a graph, evaluating

New Terms

Hydroponics—growing plants in nutrient solutions instead of soil.

(Organizer continues on next page)

LESSON 5

✴ Getting Started

In this lesson, students are introduced to the three nutrients found in most plant fertilizers—nitrogen, phosphorus, and potassium. The pH of soil is discussed, and the lesson concludes with students applying what they have learned to hydroponics.

Main Ideas

1. Nitrogen, phosphorus, and potassium are the substances needed in the greatest quantities by plants.
2. Plants may develop deficiencies if a nutrient is lacking or if the soil pH is incorrect.
3. Liquid solutions of nutrients can replace soil for plant growth.

✴ Teaching Strategies

What's in a Number?

Ask students what the picture at the top of page 469 illustrates. Then have them read the first paragraph to discover if they were correct. Be sure that students realize that the order in which the nutrients in fertilizers are listed is always nitrogen, phosphorus, and potassium. Students should recognize that 15-30-15 represents 15% nitrogen, 30% phosphorus, and 15% potassium.

Have students read the second paragraph. Students should recognize that phosphorus is the chemical in the fertilizer that is present in the largest amount. Without it, plants would not develop healthy roots and stems. Water and nutrients could not be absorbed properly and the plants would die. After reading the third paragraph, students should conclude that discolored leaves may indicate a nitrogen deficiency.

Inner Actions of Plants 469

Answers to

Increase Your Knowledge

1. Answers will vary. Students should evaluate at least three different fertilizers.
2. Encourage students to find out where organic fertilizers come from and how they compare in nutrient value to chemical fertilizers.
3. Accept all reasonable responses. However, most students will probably agree that they would take chemical fertilizers because they weigh less and would take up less space on the trip and in the biosphere. Some students may point out that there will be human waste, and possibly animal waste, in the biosphere that could be recycled as natural fertilizer.

Have students read the paragraph on soil pH. You may wish to draw a pH scale on the chalkboard (see Unit 1, page 53) to help them visualize what it means to be acidic or alkaline. Students should recognize that a pH of 6.5 is slightly acidic.

Exploration 9

Divide the class into small groups and distribute the materials. Soil samples should be from different areas of the community. Have students complete the Exploration and compare their results to discover how the pH of soils from different areas varies.

To extend the activity, have students draw a map of their area, identify where each soil sample came from, and indicate each sample's pH.

Potassium strengthens a plant, especially when fruits or vegetables are forming. It also increases a plant's resistance to pests or diseases. Potassium may play a part in storing or releasing nitrogen, according to the plant's needs.

Nitrogen, phosphorous, and potassium are not the only substances necessary for plants. There are many others, but these three nutrients are needed in the greatest quantities.

Increase Your Knowledge

1. Find several different kinds of fertilizers by using N-P-K ratings. Look at containers in gardening centers and read catalogues from garden supply stores. Match each fertilizer with its recommended use.
2. Investigate the composition and use of what are called "natural" fertilizers such as manure and compost.
3. Earlier, you chose a soil mixture for the space biosphere. Now consider what fertilizers you would take. Would you take:

    (a) chemical fertilizers only?

    (b) natural fertilizers only?

    (c) both types of fertilizer?

    Explain the reasons for your choice.

The composition of fertilizers is not the only chemistry important to gardeners. Plants have their own special pH needs. Some grow best in "sour" soil, that is, a soil that is acidic (pH value less than 7). Others require an alkaline soil (a pH value greater than 7). For example, cabbage grows best in slightly alkaline soil while rhododendrons thrive in acidic soil. Many plants grow best in soil with a pH of 6.5. Would this soil be acidic, neutral, or alkaline? The pH affects whether nutrients will dissolve in the water around the plant. That's why gardeners must be aware of the nutrients in a fertilizer and the pH of their soil.

EXPLORATION 9

The Litmus Test

You Will Need

- distilled water
- soil samples from different areas
- blue and red litmus paper

What to Do

Perform a simple test on your soil sample. Add a little distilled water to a small amount of soil until it makes a thick paste. Place pieces of red and blue litmus paper on the moist soil and record your observations. If the soil is acidic, blue litmus paper turns red. If the soil is alkaline, red litmus paper turns blue.

Think About It!

So far, you have been thinking about growing plants in soil. Could plants be grown without soil? Read on and see.

(Organizer continued)

Materials

Exploration 9: distilled water, soil samples from different areas, blue and red litmus paper, Journal

Exploration 10: rainwater or distilled water, tap water, rainwater with hydroponic nutrient solution dissolved in it, rainwater with quarter-strength hydroponic solution dissolved in it, acid rainwater (10 mL vinegar per 1 L of water), basic rainwater saturated with gardener's lime, baby food jars or clear plastic cups, vermiculite or clean sand, mustard seeds, eyedropper, ruler, stirring rod, graduated cylinder, graph paper

Science Discovery Videodisc

Disc 2, Image and Activity Bank, 8–5

Time Required

two class periods, plus a one-week growing period

---

470 Growing Plants

Plants Without Soil—The Way of the Future?

Growing a plant without soil sounds impossible! But if you replace soil with something else that can perform each function of soil, then the plant should grow. Therefore, you need to find a way to anchor the plant in an upright position, and to provide water, air, and other essential nutrients for it.

Today some plants are being grown without soil. This process is known as **hydroponics**. Instead of soil, a solution supplies the important nutrients. Perhaps hydroponics would be a useful way to grow food in your biosphere. Consider using this method for at least some of the crops when you write your proposal.

Here is an Exploration that investigates the growth of mustard plants in different nutrient solutions.

EXPLORATION 10

Which Nutrient Solutions Are Best for Growing Mustard Plants?

Before starting, form teams of three or four and choose one of the nutrient solutions listed below to investigate.

You Will Need

- rainwater or distilled water
- tap water
- rainwater or distilled water with hydroponic nutrient solution dissolved in it (See the package for directions.)
- rainwater or distilled water with quarter-strength hydroponic solution dissolved in it
- acid water: rainwater or distilled water with vinegar added (at least 10 mL of vinegar per liter of pure water)
- alkaline water: rainwater or distilled water saturated with gardener's lime
- a solution of your own choice
- baby food jars or clear plastic cups
- vermiculite or clean sand
- mustard seeds
- a dropper

What to Do

1. Fill a baby food jar or a clear plastic cup halfway with an *aggregate* such as vermiculite or well-washed sand. (An aggregate is a mixture of particles like sand or small pebbles, used here to support the plant.)
2. Slowly add your nutrient solution until the top of the aggregate is damp.
3. Place ten mustard seeds in the thin film of water on the surface of the aggregate. Put your jar in moderate sunlight. Each day, use a dropper to bring the water level in the container up to the point where the aggregate is wet on top.

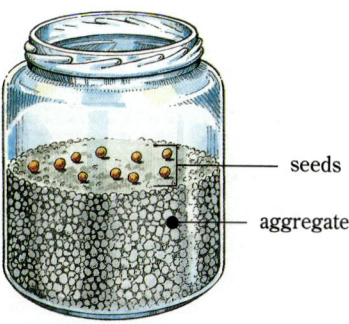

4. At the same time each day, count the number of seeds that have germinated. Record your findings in a table like this one:

Day	0	1	2	3
Number Sprouted	0	0	3	8

Plants Without Soil—The Way of the Future?

Have students read and discuss the material on hydroponics. (They may be aware of this topic if they did the extension activity on hydroponics in Unit 3.) Encourage them to speculate as to whether or not hydroponics would be a better way to grow plants in a biosphere. Then have them complete the Exploration that follows.

Exploration 10

This Exploration helps students understand that plants can get the nutrients and water they require from either soil or solutions of nutrients. The Exploration is easy to set up and results can be seen within a week's time.

Divide the class into small groups and distribute the materials. Each group should investigate one solution. There are six solutions: rainwater and hydroponic nutrient solution dissolved according to package directions, one-quarter strength hydroponic solution and rainwater, the vinegar and rainwater solution, the rainwater and saturated gardener's lime solution, plain tap water, and plain rainwater. (Distilled water can be used instead of rainwater.) Variables, such as the nature of the nutrient solution and the temperature, can be easily controlled.

Inner Actions of Plants

Answers to
Analyze Your Findings

1. Seed germination rate will probably be about the same for all of the nutrient solutions, since germination is not dependent on the nutrient content of the medium.
2. Seedling growth will probably be the best in the nutrient solution made from the label directions. It will probably be the worst in the saturated lime solution.
3. Students will probably observe that seedling growth is less when less than the recommended amount of hydroponic solution is used. If more than the recommended powder is used, plants may grow too fast and become spindly.
4. Student tables should illustrate the most important information from each group.
5. The data will vary depending on the results of student investigations and the types of nutrient solutions used.
6. Questions will vary.

✷ Follow Up

Assessment

1. Suggest that students visit a garden center and make a list of the fertilizers that are for sale, noting the N-P-K rating for each one. Then have students determine the specific uses of each of the fertilizers.
2. Suggest that students take some soil samples from potted plants around their home or from the home of a friend and test the pH of each one. Suggest they do some research on the plants to determine if the pH levels are suitable.

Extension

Point out to students that there are two methods of hydroponics—water culture and aggregate culture. Suggest that they do some research to find out how these methods differ. They should share what they discover by making poster diagrams to display around the classroom.

EXPLORATION 10—CONT.

5. Once your seeds have sprouted, measure the height of the five largest seedlings at the same time each day. Then calculate their average height.

6. Plot a graph of the average height of a seedling against the time in days.

Analyze Your Findings

Share your results from the Exploration with the members of other teams.

1. Did the average time for germination of the seeds vary with the kind of nutrient solution used? If so, list the nutrient solutions in order according to the time it took for the seeds to germinate in each.
2. How was seedling growth affected by different types of nutrient solutions? Which solutions produced the most plant growth? Which produced the least plant growth?
3. How did the amount of hydroponic powder dissolved in the water affect plant growth?
4. Suppose your class wished to present a report of your findings to another class that is also doing this Exploration. Design a table and record the most important data found by each of the teams in your class. Will another group of students clearly understand the way you have set up and labelled parts of your table?
5. Draw some conclusions from class data that you can present to the other class.
6. Have you thought of other questions that you might investigate about hydroponics? What are some of them? Choose one of these questions and design an experiment that will help you answer it.

Tomatoes growing in a hydroponic research laboratory

6 Flowers and Pollination

New Plants From Seeds

You have been growing plants from seeds. You know that seeds are special parts of plants produced specifically to make new plants. You have worked with seeds in germination tests and have seen how plants can grow from seeds. A moment's thought will probably remind you of things you've known for years about flowers and seeds. But where did the seeds come from? A flower? How are flowers and seeds related to each other?

Look at the illustration of a flower on page 475. See if you can answer these questions.

1. What parts of a flower are often brightly colored and may attract bees or birds?
2. What parts at the base of the blossom enclose the flower bud before it opens and are often green?
3. Which parts are known as the female parts of the flower? What do they do for the plant?
4. Which parts are known as the male parts of the flower? What do they do for the plant?
5. Where do the seeds form?
6. What has to happen in order for seeds to form?

Pollination and Seed Production

Pollination is the transfer of pollen from the tip of the *stamen* of a flower to the tip of the *pistil* of the same or another flower. After this, fertilization, and later fruit and seed production, can occur. This completes the life cycle of a plant.

In nature, pollination is often aided by the action of wind and insects. Sometimes people can help make pollination successful. This is especially useful for plants grown indoors. How could you make certain that plants would be pollinated in the biosphere?

Most flowers have stamens and pistils in the same blossom. Only a slight movement is required to pollinate such plants as the tomato, pepper, eggplant, pea, or bean. Some plants, however, have separate male and female blossoms. These include cucumbers and squash. You can recognize the female blossoms by the miniature fruit *(ovary)* that appear beneath the flower.

LESSON 6 ORGANIZER

Objectives

By the end of the lesson, students should be able to:
1. Identify the parts of a flower.
2. Describe the process of pollination.
3. Describe the life cycle of a flowering plant.
4. Explain the difference between self- and cross-pollination.
5. Explain how cross-pollination is used as a plant-breeding technique.

Process Skills

dissecting, observing, classifying, evaluating

New Terms

Pollination—the transfer of pollen from the tip of the anther to the tip of the stigma of the same or another flower.

(Organizer continues on next page)

LESSON 6

✷ Getting Started

This lesson focuses on pollination and the life cycle of flowering plants. Students examine the structure of a variety of flowers and identify examples of self- and cross-pollination.

Main Ideas

1. Pollination is the beginning of the process by which flowering plants produce fruits and seeds.
2. The flowers of some kinds of plants are self-pollinating; those of other kinds must be cross-pollinated.
3. Carefully controlled cross-pollination can result in new varieties of plants.

✷ Teaching Strategies

New Plants From Seeds

Have a volunteer read the lesson introduction on page 473. Then, using the illustration on page 475, have students discuss the questions.

Answers to In-Text Questions

1. the petals (corolla)
2. the sepals (calyx)
3. The pistil—stigma, style, and ovary—make up the female parts of a flower. It is here that the seeds and fruit develop.
4. The stamen—anthers and filaments—make up the male parts of a flower. This is where pollen is produced.
5. Seeds form in the ovary.
6. For seeds to form, pollen must enter the ovary and fertilization must take place.

(A teaching strategy for Pollination and Seed Production follows on next page.)

Inner Actions of Plants 473

Pollination and Seed Production

Have students read the material silently. When they have finished, involve them in a discussion of pollination. Be sure students understand that pollination and fertilization are not the same thing. In fertilization, a sperm cell in a grain of pollen travels through the pistil of a flower down a pollen tube to unite with an egg cell in the ovary.

If possible, bring a few flowers to class and demonstrate how easily pollen can be shaken from the anthers. Students should be able to conclude that in a biosphere, pollen could be transferred by having people shake the plants, by providing an artificial wind, or by using a colony of bees.

Exploration 11

Direct students' attention to the pictures on page 474, and involve them in a brief discussion of how the parts of these flowers are alike and how they are different. For example, all of the flowers have petals, but the petals are different colors and shapes.

At this point, you may wish to introduce the concept of perfect and imperfect flowers. Explain that a flower that contains either male or female parts but not both is called an *imperfect flower*. If the flower contains a stamen but no pistil it is called a *staminate flower*. If it contains a pistil but no stamen it is called a *pistillate flower*. A flower that contains both male and female parts is called a *perfect flower*.

(Continues on next page)

EXPLORATION 11

Parts of Flowers

You Will Need
- a variety of cut flowers
- a razor blade or sharp knife

Clockwise from above, Buttercup; Carolina Rose; Oriental Lily; Zinnia

---

(Organizer continued)

Materials
Exploration 11: variety of cut flowers, single-edged razor blade in a holder or sharp knife, Journal

Teacher's Resource Binder
Activity Worksheets 8–4 and 8–5
Graphic Organizer Transparencies 8–4, 8–5, and 8–6

Science Discovery Videodisc
Disc 2, Image and Activity Bank, 8–6

Time Required
two to three class periods

What to Do

1. Study each flower.
   (a) Determine whether the flower is male, female, or both.
   (b) Identify the stamen, pistil, and ovary. If necessary, carefully dissect the flower with a razor blade or sharp knife.
2. Sketch and label the flower.
3. What features of each flower increase the chance that it will be pollinated in nature?

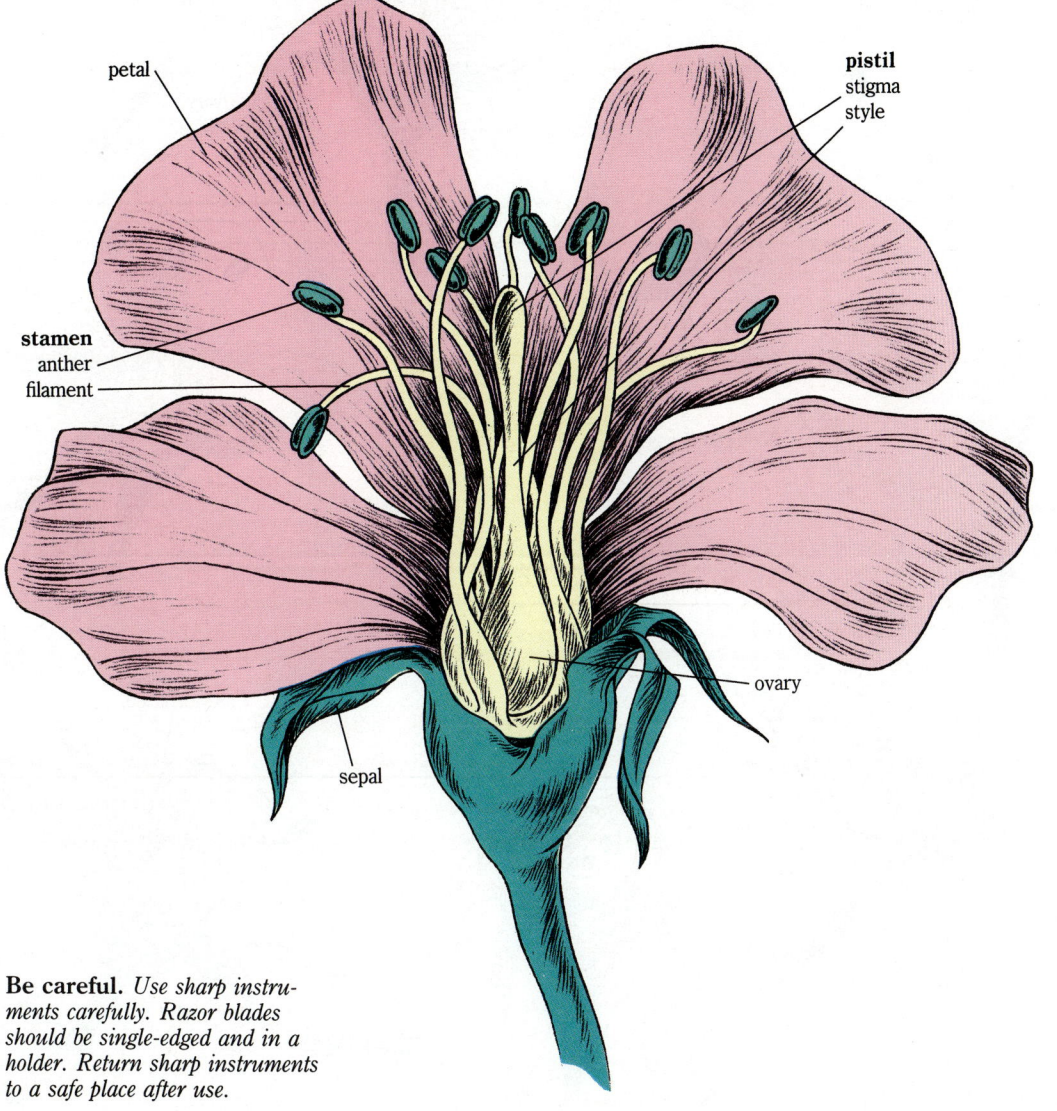

Be careful. *Use sharp instruments carefully. Razor blades should be single-edged and in a holder. Return sharp instruments to a safe place after use.*

(Exploration 11 continued)

If possible, provide students with a wide variety of flowers to examine. Before they begin the activity, caution students to be careful using razor blades and knives. Set up stations around the room so that students can travel from one station to the next, investigating a different flower at each one. If there are examples of both perfect and imperfect flowers, be sure that all students have a chance to compare at least one of each type. Monitor the activity by walking among students and confirming their observations. Encourage students to share their results with one another.

When students have completed the activity, involve them in a discussion of how the specific features of the flowers increase their chances of becoming pollinated. For example, pollen is easily knocked from anthers attached to long filaments. Insects easily collect pollen from short stamens that lay flat against the petals. A tall pistil is likely to catch pollen from the wind. A short pistil among tall stamens will easily become covered with pollen dropping from the anthers.

You may wish to display student sketches around the classroom or use them to make a bulletin-board display.

Inner Actions of Plants 475

A Picture Story

Before students begin writing, you may wish to have them review the "plant-cycle diagram" as a class. Begin the discussion with the germination of a seed, and call on students to continue the explanation until the entire cycle has been discussed. If necessary, elaborate on parts of the process. For example, you may wish to explain to students that the pollen grain produces the pollen tube and gives rise to the sperm cell. The eggs develop within the ovary. An egg and a sperm unite to form a new seed. Point out that the fleshy part of a fruit is the enlarged wall of the ovary that was once a part of the flower.

Answers to
In-Text Questions

Students' descriptions of how seeds form and develop into new plants should include the following steps:
- A seed germinates and produces a seedling.
- The seedling grows into an adult plant that produces flowers.
- Pollen grains from the anthers of the flower are transferred to the stigma during pollination.
- A pollen grain grows a pollen tube down the pistil. The sperm passes through the tube to reach the ovary.
- During fertilization, the sperm unites with the egg.
- The fertilized eggs develop into seeds as the ovary develops into a fruit.
- The fruit continues to grow until it ripens and falls from the plant.
- The fruit eventually releases the seeds, and the process begins again.

EXPLORATION 11—CONTINUED

A Picture Story

In your own words, describe how seeds form and develop into new plants. Use this "plant-cycle diagram" as your source of information.

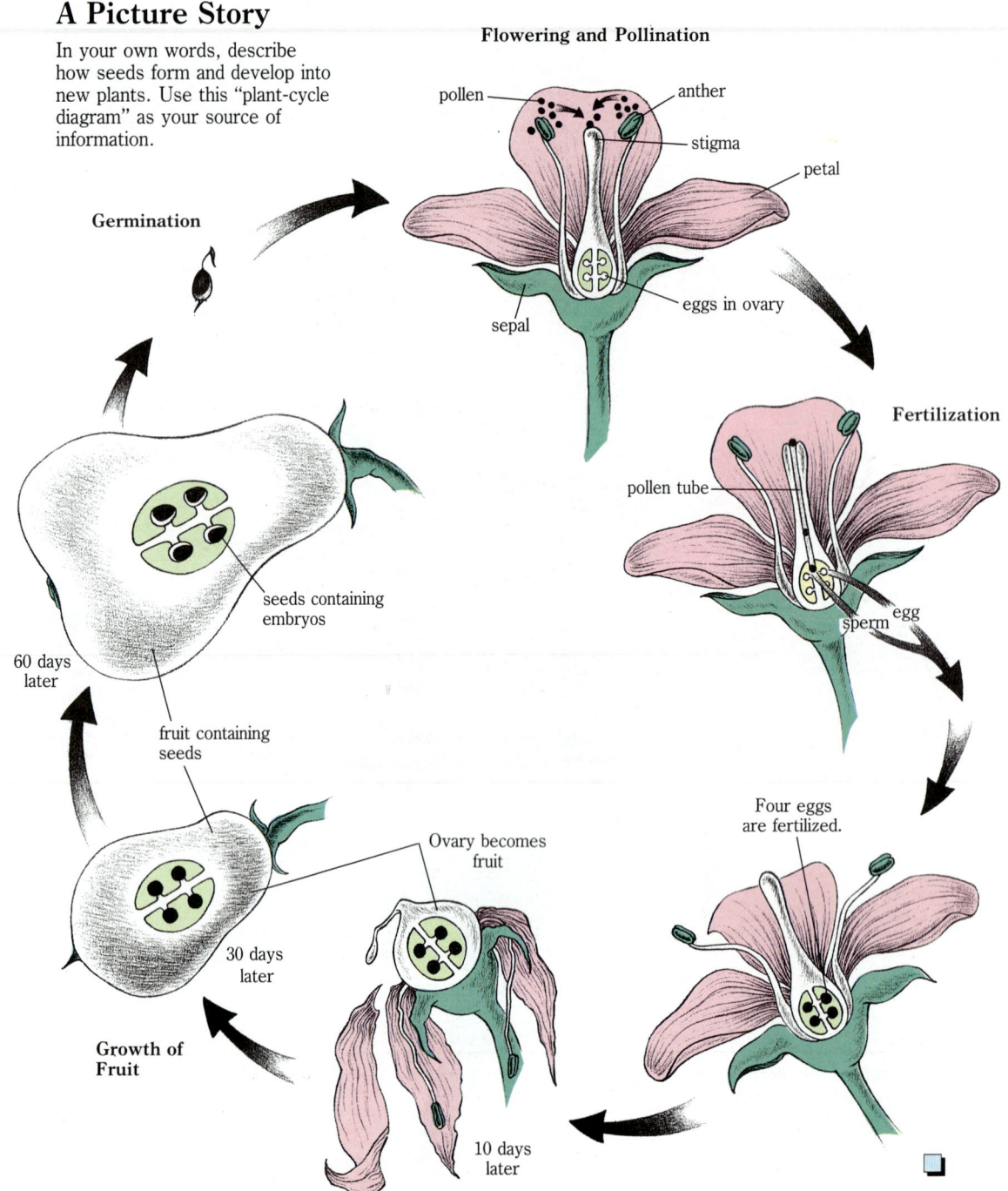

476 **Growing Plants**

EXPLORATION 12

What Kind of Pollination?

Look at the examples of some of the ways that pollination takes place. Decide which show *self-pollination* (pollen and egg come from the same plant) and which show *cross-pollination* (pollen is transferred to a different plant).

Part 1
Peach Tree

Josh performed this experiment, which was recommended in his gardener's guidebook.

1. Pinch off petals and anthers.

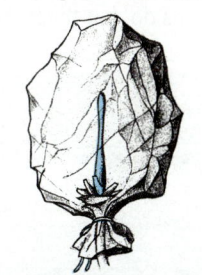

2. Cover the pistil with a bag.

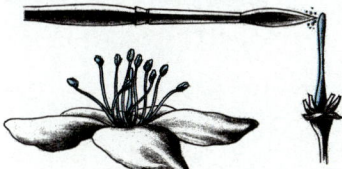

3. Get anthers from another tree.
4. Using a brush, transfer pollen to the stigma and cover it with the bag for a short time.

Part 2
Nasturtium Plant

Nasturtiums have perfect flowers. (They have both male and female parts.) A bee or hummingbird carries pollen from the anthers to the stigma of the same flower as it probes for nectar.

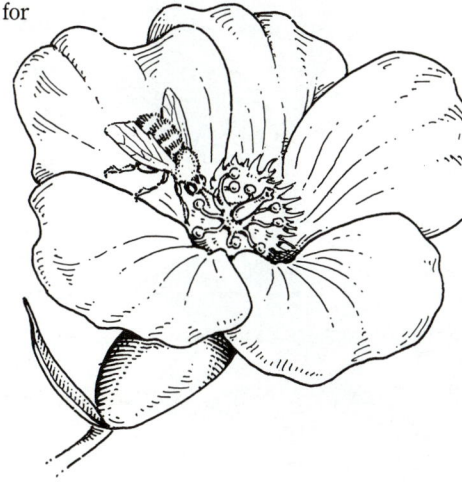

Part 3
Holly Plant

Each plant has only male flowers or only female flowers. Pollen must be transferred by the wind or an insect.

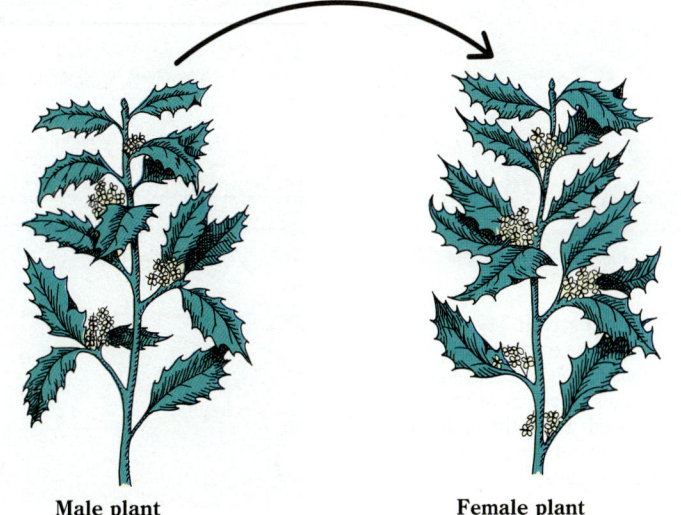

Male plant → Female plant

Answers to
Exploration 12

Part 1
This illustration shows an example of cross-pollination because the pollen came from one tree and the egg came from another. Be sure that students recognize that two different trees (or plants), not just two different flowers, must be involved for cross-pollination to occur.

Part 2
This illustration is an example of self-pollination because pollen from the anthers of the flower is transferred to the stigma of the same flower. Examples of perfect flowers in which self-pollination occurs include: buttercup, wild rose, petunia, morning glory, and lily.

Part 3
This is an example of cross-pollination because pollen from the anthers of one plant is transferred to the stigma of another plant. You may wish to point out that this example differs from the example in *Part 1* because the flowers of one plant are all male and the flowers of the other plant are all female.

(Answers continue on next page)

Inner Actions of Plants

Answers continued

Part 4

This is an example of self-pollination because pollen from the anthers of one flower is transferred to the stigma of another flower on the same plant. Point out that this example differs from the example in *Part 2* in that a single cucumber plant has both male and female flowers. Examples of imperfect flowers that self-pollinate include: pumpkin, squash, holly, willow, and watermelon.

Cross-Pollination as a Plant Breeding Technique

Point out to students that cross-pollination can be used to develop varieties of plants with special characteristics. These plants are called *hybrids*. Have students begin reading, pausing before the cross-breeding method is discussed. Students should recognize that farmers might like to change the length of the growing season of wheat from 110 days to 90 days. Have students suggest a method for getting a wheat with the desirable qualities of both S and W. Accept all reasonable responses. Students should continue reading to see if their ideas were like the one suggested on page 478. Ideally, the new wheat that is produced would resist disease, survive low temperatures, produce high yields, and mature in 90 days.

EXPLORATION 12—CONT.

Part 4
Cucumber Plant

This plant has separate male and female flowers, called *imperfect* flowers, on the same plant.

female

male

Suggestion: Try your hand at making certain that either self-pollination or cross-pollination takes place. Save any seeds produced. Plant them, and enjoy the results of your breeding technique.

Cross-Pollination as a Plant Breeding Technique

Cross-pollination often requires a delicate touch. A scientist transfers pollen from the anthers of a flower on one plant to the stigma of a flower on another plant. Here is what might be done with two varieties of wheat, Wheat S and Wheat W.

Wheat S	Wheat W
• Resists disease	• Subject to certain diseases
• Survives low temperatures	• Dies in low temperatures
• Produces well (high yield)	• Has a low yield
• Matures in 110 days	• Matures in 90 days

Wheat S has one quality a farmer might like to change. What is it? Before you read further, propose a way to get a wheat with the desirable qualities of both S and W.

Here is one method. It is called *cross-breeding*.

1. Grow Wheat S and Wheat W.
2. When flowers appear, remove the anthers from the flowers of Wheat S. Cover the flowers with small bags.
3. Several days later, transfer pollen from the anthers of the Wheat W flowers to the stigmas of the Wheat S flowers.
4. At harvest time, collect the seeds from Wheat S.

Plants grown from these seeds will have various qualities of Wheat S and Wheat W. What combinations do you think might be produced?

✳ Follow Up

Assessment

1. Students may enjoy making a model of a flower that includes all the parts labeled in the diagram on page 475. Use clay, papier mâché, or a variety of other materials. Encourage students to be creative.
2. Suggest that students find out the names of some plants that are self-pollinating and some that are cross-pollinating. Have them use their information to make poster diagrams of the two kinds of plants.

Extension

Explain to students that inside every seed is a tiny embryo. Suggest that they do some research on seed embryos and then complete the following activity. Soak some dry lima beans for a few hours or overnight until they double in size. Separate the two halves of one of the seeds, and locate the embryonic plant with its tiny leaves and root. Draw and label the parts in a diagram of the seed embryo.

Plants From Plant Parts

Can you grow a new plant without a seed? You can if you use roots, stems, leaves, buds, or twigs from an old plant. This way of growing new plants is called *vegetative reproduction*. Read the brief descriptions of the methods used in each example. Look at the illustrations and identify the method shown. Tell which part of the plant is being used. Remember, for each kind of plant, only some of the plant parts are suitable for vegetative reproduction.

The methods used to grow new plants from plant parts include:

1. Division—the roots of the plant are divided and planted separately.
2. Layering—plants send out new roots from points on their stems that have been left in contact with the earth. These parts are separated from the old plant and used for new plants.
3. Cuttings—a piece of a plant is cut off and replanted. This is only successful if the stem will readily grow new roots.
4. Grafting—a bud or twig from an existing plant is transferred to another plant that already has a root system.

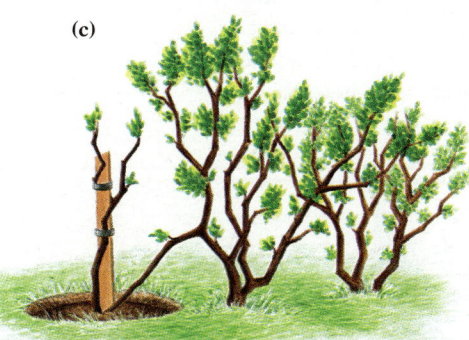
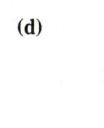

(a) (b) (c) (d) (e)

Biosphere Decision

The members of the Biosphere Proposal Committee must decide what kinds of plants to choose and whether seeds, seedlings, or mature plants should be taken into the biosphere. Some members think that they will need plants in all three stages. What do you think?

LESSON  ORGANIZER

Objectives

By the end of the lesson, students should be able to:
1. Identify several different methods of vegetative reproduction.
2. Describe and use vegetative techniques to grow new plants.
3. Explain how roots increase in length and penetrate the soil.

Process Skills

decision making, investigating, illustrating, analyzing

New Terms

none

(Organizer continues on next page)

LESSON  7

✳ Getting Started

This lesson focuses on the vegetative reproduction of plants. Students identify different methods of vegetative reproduction and then grow new plants from cuttings of their own.

Main Ideas

1. Many plants are capable of reproducing vegetatively.
2. Roots are the first structures to form in most methods of vegetative reproduction.

✳ Teaching Strategies

Ask students what would happen if you placed the end of a leaf or stem in some moist sand or a glass of water for a week or two. *(It would probably begin to grow roots.)* Explain that this is one way to grow new plants without using seeds. Have students read the introduction and discuss the drawings.

Answers to

In-Text Questions

The method of vegetative reproduction each illustration shows:
(a) cuttings   (d) division
(b) grafting   (e) cuttings
(c) layering

Biosphere Decision

Accept all reasonable suggestions. In reality, seeds would most likely be taken to a biosphere because a large number could be stored in a small space, and they would not be susceptible to damage during the journey.

Exploration 13

One day before doing the Exploration, ask students to water the plants that will be used.

Divide the class into pairs and distribute the materials. Have students read and discuss what they are to do. A volunteer should demonstrate how to plant a cutting. When students finish planting their own cuttings, have them cover the containers with clear plastic wrap and place the cuttings where they will receive indirect sunlight. Allow the cuttings to remain in the potting mix for at least a week before examining for roots. Suggest that students use a magnifying glass to examine several of the newly rooted plants. Ask them to observe what the roots look like and to determine where growth occurs. Encourage students to transplant some of their cuttings into flowerpots and potting soil.

EXPLORATION 13

Vegetative Reproduction

You Will Need

- mature plants
- scissors
- potting mix (equal volumes of perlite and vermiculite)
- a large, shallow container in which to plant the cuttings
- clear plastic wrap
- a large, deep container

What to Do

1. Water the plants the day before taking cuttings from them. This will save the plants from added stress.
2. You will root several cuttings in one container to save space and material. The potting mix should be moist, but not wet, or the cuttings might rot.
3. Fill the container with the mix. Then dig small holes about 2 cm deep with your finger or a pencil. Put the holes about 4 to 5 cm apart.
4. Take a stem or leaf cutting as described. Always hold the plants by the leaf, not by the stem, to avoid damaging the plant.

For stem cuttings, cut the stem with scissors just below the third or fourth pair of leaves. Strip off the lower pair of leaves so two or three sets of healthy leaves are left. The new roots will probably grow from the place where the leaves were removed.

For leaf cuttings, cut off a leaf with its stem attached. For a snake plant, cut off a 5-cm section from the middle of the leaf.

5. Carefully place the cutting into a hole in the mix. Fill in the hole so the cutting is held securely.
6. Put a piece of clear plastic over the whole container until the new plants have rooted.
7. When the new plants begin to grow, carefully remove them from the rooting container and transplant each into a deeper container. Use a general purpose soil mixture. Remember the precautions to take when transplanting.

Try some of these plants.

Stem Cuttings
 Impatiens
 Coleus
 Geranium
 Swedish Ivy

Leaf Cuttings
 Snake plant
 African violet
 Begonia

(Organizer continued)

Materials

Exploration 13: mature plants (for stem cuttings: impatiens, coleus, geranium, Swedish ivy; for leaf cuttings: snake plant, African violet, begonia); scissors; potting mix (equal volumes of perlite and vermiculite); large, shallow container; clear plastic wrap; water; magnifying glass; large, deep container, metric ruler

Exploration 14: sprouted mung beans, radish seeds, or mustard seeds; fine-tipped waterproof marker; metric ruler; magnifying glass; paper towels; water; Journal

Science Discovery Videodisc

Disc 2, Image and Activity Bank, 8–7

Time Required

two class periods, plus a one-week growing period

Root Growth

For every method of vegetative reproduction described, what plant part must develop first (if it is not already there)? The root is the first part of the plant to emerge from the seed as it germinates. When a new plant forms from part of an old one, the roots must develop if the new plant is to live. The next Exploration will help you understand how roots grow.

Germinating barley seed

EXPLORATION 14

The Growth of a Root

You Will Need

- sprouted mung beans, radish seeds, or mustard seeds
- a fine-tipped waterproof marker
- a ruler
- paper towels
- water

What to Do

Examine mung beans, radish seeds, or mustard seeds that have already sprouted on moist paper towels. They should have roots. Gently put marks 2 mm apart on the entire length of one root. Use the waterproof marker. Watch for several days as the root lengthens. Then measure the distance between the ink marks on the root.

1. In your Journal, draw the root, showing the distances you recorded between marks.
2. Where is growth taking place in the root? What evidence leads you to think so?
3. Look for root hairs on the new root. Water enters the plant through the root hairs. What other use could root hairs have for a growing root as it moves through the soil? It may help to look at this photograph of a seedling.

Root Growth

Have students read silently the introductory paragraph. Then direct their attention to the photograph, and have them identify what it shows. *(a germinating barley seed with leaf and roots)* Point out to students that the primary roots are covered with tiny root hairs. Call on a volunteer to remind students about the function of the root hairs for a young plant. *(to increase surface area for water absorption and provide tiny pockets of air around the root)* Have students discuss how the roots in the photograph are similar or different from the roots that grew on their cuttings in the previous Exploration.

Exploration 14

Of the three choices, mung beans will provide the largest roots and will probably be the easiest to work with. Remind students to keep the seedlings moist. You may wish to have students perform the Exploration with all three seeds and then compare the results to determine which roots grew the fastest. Once the seeds have germinated and the roots have been marked, two or three days should be enough growing time for noticeable growth to occur.

Answers to

In-Text Questions

2. Students should discover that the marks closest to the root tip have moved the farthest apart, indicating that the root grows near the tip.
3. Students should conclude that the root hairs are able to enter almost any crevice in the soil and extract water and minerals.

Inner Actions of Plants

Growing Through the Soil

Encourage students to respond to the questions before they read the answers at the bottom of the page. Students might find it interesting to use a magnifying glass to look for the root cap at the end of a root on a germinated seed. If possible, have students observe a root hair under a microscope.

Assessment

1. Suggest that students survey adult family members and friends to find out if they have ever grown plants vegetatively. Have students make a list of the plants, the methods used, and the success rate of the methods surveyed.
2. Have students write a description of one of the methods of vegetative reproduction. They should not identify which method they are describing. Then have them exchange papers, and identify the method described on the paper they receive.

Extension

1. Suggest that students make poster diagrams to compare the way roots and stems grow. Point out that they will have to do some research on stems. Suggestions for places to look for information include biology books, botany books, and encyclopedias. Display the finished poster diagrams around the classroom.
2. Some students may enjoy trying their hand at propagating plants using one of the methods (other than cuttings) described in the lesson. Suggest that they do this at home. Have students present an initial outline and then make weekly reports on their progress. Students should display the rooted plants in the classroom for others to examine.

Growing Through the Soil

How does a root grow in length? Why aren't roots damaged when they push through hard soil? Examine the diagram below for answers.

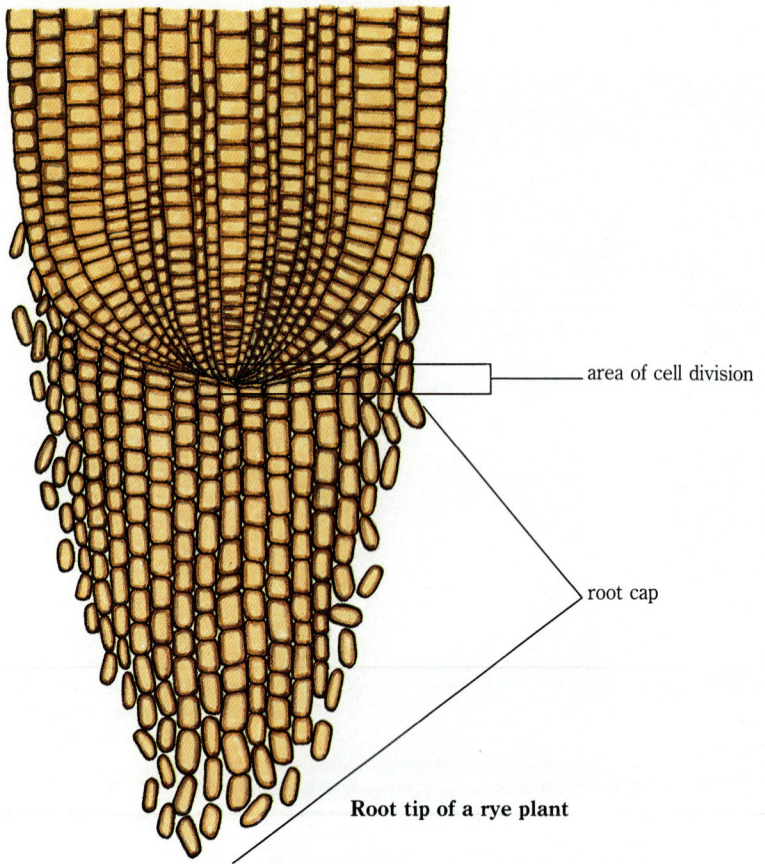

Root tip of a rye plant

The growth of the root takes place as new cells are formed in a process called *cell division*. Most of the new cells cause the root to grow longer. At the same time, however, there is a loose collection of cells forming a *root cap*. Look at the diagram of the tip of a root. These cells protect the area of cell growth. They get scraped away by soil particles as the root forces its way through the soil.

482 Growing Plants

Challenge Your Thinking

1. Light affects plant development. Choose one of the following ideas and outline an experiment that would support or disprove the idea.

    (a) Does the shape of a tree have much effect on the amount of light it gets? Think about a pine tree and a maple tree.

    (b) Jeremy stated, "I think plants will grow faster in red light than in green light, and I am going to try to prove it."

    (c) Poinsettias bloom naturally in December. Dionne said, "If I can give a poinsettia plant the light conditions it would normally get in November and December, then I could make it bloom in summer or any other time I want."

2. Out of Order!

    (a) In your Journal, redraw this illustration, putting things in order. Be careful! Most, but not all, of the parts are out of place.

    (b) Label all of the stages shown. Two are labelled for you.

    (c) Why is this called a cycle?

The life cycle of a bean plant

Seeds are removed and dried.

The seed contains a young plant.

Answers to
Challenge Your Thinking

1. Answers will vary. Students should make sure that only the variable being tested is allowed to change. They should also test a large sample size or repeat their experiments several times.

2. (a) & (b) See diagram below.
   (c) This is called a cycle because it has no beginning and no end. Plant life continues over and over again through the production of seeds.

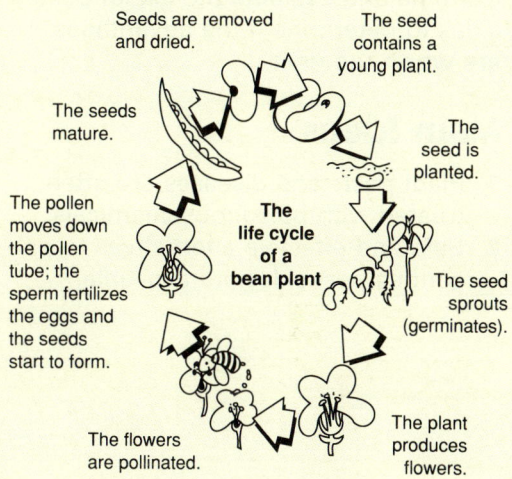

Inner Actions of Plants 483

LESSON 8

✦ Getting Started

In this lesson, students evaluate the positive and negative aspects of using chemical pesticides on plants. Through a series of activities, students examine and evaluate the use of pesticides to determine if the advantages are worth the risks.

Main Ideas

1. Plant pests and diseases are often treated with poisonous chemicals.
2. Safe and effective alternatives to poisonous pesticides are available.

✦ Teaching Strategies

Ask students what they would do if plants in their homes became infected with insects or some other pest. *(Many students will probably respond that they would use a pesticide or poison to kill the pests.)* Next take an opinion poll to determine how many students believe that pesticides can be harmful and how many think that they are perfectly safe. Then have students read the scenario at the beginning of the lesson.

Exploration 15

Students could work in small groups to read the Exploration, and then do the library research needed to answer the questions. Involve the class in a discussion to share what they have discovered. This is an excellent opportunity for students to do some problem solving and use decision-making skills.

If the class becomes divided on the use of pesticides versus other alternatives, involve students in a debate of the issues.

PLANTS IN YOUR ENVIRONMENT

8 Medicine or Poison?

Plant Medicines: Use With Caution!

As Mom came in the back door, three voices delivered some bad news.

"The cat's sick," said Maria.

"Poisoned!" Angela called out from the living room.

"...and the vet said it could be from something we used to kill pests in the garden!" This was from Carlos who was getting dinner ready.

Later, Mom said, "The vet says the cat will be all right this time, fortunately. But we must not let this kind of thing happen again. We'll have to be more careful about using poisons, or we'll be hurting ourselves next."

The Herrera family did something about their concerns. They began by learning about the situation. In this Exploration you will help the Herrera family identify some potential problems with pesticides and some possible alternatives that they might use.

EXPLORATION 15

Danger in the Garden

Part 1
Identifying the Issue

Sometimes plants in the Herrera garden were attacked by pests or diseases. Maria made a chart showing which chemicals had been used in the garden.

Area Number	Plants	Pest/Disease	Treatment
1	marigolds	slugs	metaldehyde bait
2	roses	aphids	dimethoate
		Japanese beetles	diazinon
3	daisies	spittlebugs	malathion
4	lilacs	leaf miners	diazinon
5	phlox	powdery mildew	benomyl

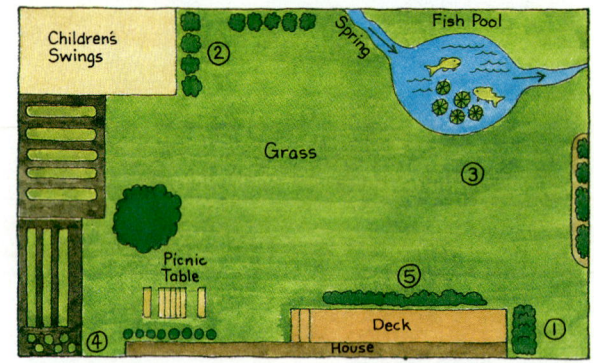

1. Find out if the pesticides (substances used to kill pests) named in the table can have effects on other living things. You might get help from people at the following locations:
   (a) Environmental Protection Agency
   (b) A garden supply store
   (c) An agricultural institute or university

2. Find out how chemicals used in agriculture can affect living things, especially if they get into drinking water or are breathed in with the air.

3. Look at the map of the Herrera's garden. Consider questions like these:
   (a) Would spray be carried by air currents? Where might it land?

LESSON ORGANIZER

Objectives

By the end of the lesson, students should be able to:

1. Identify several poisonous pesticides and their effects on people and pets.
2. Suggest safe alternatives for some commonly used pesticides.
3. Evaluate the positive and negative aspects of using poisonous pesticides or their safe alternatives.

Process Skills

researching, analyzing, evaluating, decision making

New Terms

none

---

484   Growing Plants

(b) How long will the chemical remain on the plants? in the ground?
(c) How will I rinse the sprayer I use? Where will I put the rinse water? Where will I put the empty poison bottle?

Add five more questions that might be asked.

The Herrera family felt they were ready to attack the problem. Mom said, "We must not spray poisons thoughtlessly, but we do want to get rid of pests and diseases from our plants. I want to find out if there are other ways of dealing with them."

Carlos said, "Janice's father has ladybugs. He said they eat pests on the roses. He doesn't use chemicals in his yard."

Mom said, "Ask him for any other ideas he has. We could find some safe alternatives."

Part 2
Identifying Alternatives

1. How can the Herrera family eliminate plant pests and diseases without using poisons? Two alternatives are illustrated here.

Find at least five other ways to help plants stay healthy without using poisons. Ask gardeners you know for alternate methods. Read a magazine devoted to organic gardening.

2. Make a collage of all the methods you discover.

Part 3
Researching

Look at gardening magazines and books. You need to answer questions like these:

- Can the use of all garden poisons be avoided?
- Are there safe ways to use some poisons to kill pests?
- Do alternatives have some disadvantages?

Part 4
Decision Making

1. Work in groups to discuss alternative treatments for the Herrera's plant problems.

2. Decide what action you would recommend for the future treatment of pests and diseases.

3. If you still find it necessary to use some poisons in your garden, name them and list the precautions that you would take to avoid possible dangers from them.

Part 5
Taking Action

The Herreras decided that they could eliminate some pesticides from their back yard. They found a collection depot where they could take these poisons, and they gladly got rid of them.

Next season the Herreras will be prepared. The cat and all of the family will be safer.

Part 6
Evaluating

Make a list of questions to ask the Herreras next year about their new ways of caring for plants.

Follow Up

Assessment

1. Suggest that students visit a local gardening center. Have them make a list of the names of the pesticides sold. Ask them to read the labels to determine which pesticides are poisonous and which are safe to use. Students should prepare fact sheets to show what they have learned.

2. Students may enjoy talking with local gardeners to discover which pests are most common in their area. Invite students to report their findings.

Extension

Point out to students that one of the major drawbacks of pesticides is that they can contaminate the food chain. Suggest they research to find out how this happens.

Materials

Exploration 15: books, magazines, and other resource material on gardening; poster board; scissors; glue or tape; Journal

Teacher's Resource Binder

Activity Worksheet 8–6
Graphic Organizer Transparency 8–7

Science Discovery Videodisc

Disc 2, Image and Activity Bank, 8–8

Time Required

one class period

Plants in Your Environment 485

LESSON 9

★ Getting Started

In this lesson, students work in teams to complete a project in which they devise a plan for interior decoration, outdoor landscaping, or vegetable gardening. The lesson concludes by having students return to the Mars biosphere reports they wrote at the beginning of the unit.

Main Ideas

1. A plant's environment provides the conditions in which it must live and grow.
2. The growth and use of plants to create, improve, or change an environment requires careful planning.

★ Teaching Strategies

Ask students to suggest ways that people use plants to improve the indoor and outdoor environments around them. *(Plants are used indoors to make rooms, offices, and lobbies more attractive. Plants are used outdoors to landscape yards, parks, and roadways, to attract birds and butterflies, to prevent erosion, and to grow food.)*

A Design Project

Have students read the material silently. Explain that after they have completed *Explorations 16, 17,* and *18,* students who are interested in the same project will form teams to work on it.

Small Scale Designs

Have students describe the environment in an imaginary Mars biosphere. They should recognize that light, temperature, moisture, and nutrients will still be needed for plants to grow no matter where the plants are.

486 **Growing Plants**

9 Make a Green World!

A Design Project

Now you have a chance to do some designing on a larger scale. You will choose a project to do with other members of a team. The project requires you to plan for the growth and use of plants in one of three situations:

(a) indoors, at home or in school;

(b) outdoors, around a home or other building;

(c) in a vegetable garden, as a source of food.

Explorations 16, 17, and 18 will introduce you to these projects. Choose the one that suits you best after you have completed the Explorations.

Small Scale Designs

You have cared for plants and observed them closely. A plant's environment provides its living conditions. Plants have special needs for light, temperature, and moisture. An African violet grows well in soft, filtered light, while a cactus thrives in full sunlight. People who have learned to provide the right environment may grow both of these plants in the same house, but probably not on the same windowsill. There can be a variety of environments in one room.

What kind of environment will there be in the imaginary Mars biosphere? Consider the surroundings your biosphere plants will need.

LESSON 9 ORGANIZER

Objectives

By the end of the lesson, students should be able to:

1. Describe the best conditions in which to grow specific kinds of plants.
2. Describe the steps necessary in planning how to grow and use plants in indoor or outdoor environments.
3. Analyze an environment in order to determine the kinds of plants that grow best in it.
4. Identify the factors needed in order to grow plants in an artificial environment.

Process Skills

problem solving, researching, interviewing, decision making

New Terms

none

EXPLORATION 16

A Greener Room

Han-Ling and Benjamin worked together on a plan for their classroom. They studied the conditions the room provided for green plants so that they could select plants that would grow well there.

What They Used

- graph paper and a pencil

What They Did

1. They noted where windows, doors, and furniture were located in the room.
2. They looked outside and found where trees, shrubs, and buildings partially blocked the sunlight.

Complete their study. You will begin with step 3.

3. In your Journal, draw a detailed floor plan showing what their classroom might look like. A sample sketch is shown at the right. You may add your own details.
4. Determine low, medium, high, and very high light conditions. Indicate these on the floor plan.

Low light—the light level that normally exists in a shaded window.
Medium light—the light level of an area within three meters of a sunny window that is partially blocked by buildings or trees.
High light—the light level that exists in front of an unobstructed window.
Very high light—the light level for an area that gets 3-5 hours of direct sunlight everyday.

5. Han-Ling used the following chart to choose these six plants for the room: heart leaf philodendron; red ivy; aluminum plant; prayer plant; coleus; and snake plant.

   (a) Indicate in your drawing where you would advise her to put these plants.
   (b) The classroom has space for four more plants. Help Benjamin select the next four plants. Then indicate in your Journal where he should put them.

Genus Name	Common Name	Light Level
Adiantum	maidenhair fern	medium
Aloe	aloe	very high
Asparagus	asparagus fern*	medium
Coleus	coleus*	very high
Dracaena	dragon tree	medium
Hedera	ivy*	high
Hemigraphis	red ivy	low
Maranta	prayer plant	medium
Mimosa	sensitive plant*	high
Philodendron	heart leaf philodendron*	low
Pilea	aluminum plant*	medium
Sansevieria	snake plant	low
Saxifraga	strawberry-geranium*	very high
Yucca	false agave	high

*can be used in hanging containers

Exploration 16

Have students read and complete the Exploration. Be sure they understand that they are to complete the drawing of Benjamin's and Han-Ling's classroom, not their own. Their own classroom will be used for one of the team projects later in the lesson. Have students share their finished drawings and choices of plants.

Materials

Exploration 16: graph paper, metric ruler, pencil, Journal
Exploration 17: tracing paper, pencil, Journal
Exploration 18: metric ruler
Exploration 19: graph paper, metric ruler, tape measure or meter stick, colored pencils or markers, tracing paper, Journal

Teacher's Resource Binder

Activity Worksheet 8–7

Time Required

six class periods

Plants in Your Environment 487

Exploration 17

Provide students with some time to analyze and compare the two landscape drawings. Involve them in a discussion of how the drawings are alike and different. Ask students to consider the significance of each feature, such as the drainage of water on the property, the areas that are sunny and those that are shady, and the areas that are windy and those that are protected from the wind.

Answers to
In-Text Questions

Have students complete and discuss the *What to Do* items. Their responses may include the following:
1. Shrubs appear along the front of the house in both drawings. The driveway and sidewalk are in the same place.
2. Flowering shrubs, evergreen shrubs, a walkway, and flowering trees or dwarf fruit trees have been added along the north side of the property. A fence and hedge have been added along most of the southern edge of the property. Shade trees have been planted in front of the house. A deck has been added to the back of the house. A pool has been placed in the low, wet area of the backyard. A vegetable garden, rose garden, and play area have been created along the back of the property. A large shade tree has been planted in the lower left corner of the lot. Flower beds have been planted along the deck and rear of the house.
3. After students have traced the final plan, have them discuss their ideas for the items that follow. Responses will vary, but may be similar to the following.
   (a) If you want to sit in the sun, you might choose the rose garden, the middle of the back lawn, or the deck.
   (b) Standing on the front walkway near the house or standing on the deck near the back of the house might provide ideal places in which to talk to a friend out of the winter wind.
   (c) Sitting under the shade tree at the southwest corner of the house or walking along the dwarf fruit trees along the north edge of the property might provide good places for a summer breeze.
4. (a) The designer planned a shady garden along the north side of the house.
   (b) The designer planned a sunny garden on the northeast corner of the property.
   (c) The designer planned a rose garden and flower beds on the east side of the property.
5. The gardens and pool are designed to provide beauty. The trees provide shade for comfort. The play area was provided for entertainment.

EXPLORATION 17

Planning a Landscape

Think about the area around the outside of a house. How could you design a garden that would take advantage of the natural features of the landscape? The following Exploration will help you make your plans.

You Will Need
• tracing paper and a pencil

What to Do
Compare the professionally landscaped lot with the original plans.
1. What features are the same?
2. What changes were made?
3. Trace the final plan in your Journal. Use lower case letters to show where you could go to:
   (a) sit in the sun
   (b) stand talking to a friend, out of the winter wind
   (c) find a breeze in summer
4. Use capital letters to show where the designer planned:
   (a) a shady garden
   (b) a sunny garden
   (c) a decorative garden
5. What human factors, such as beauty and comfort, did the designer consider when planning the landscape?

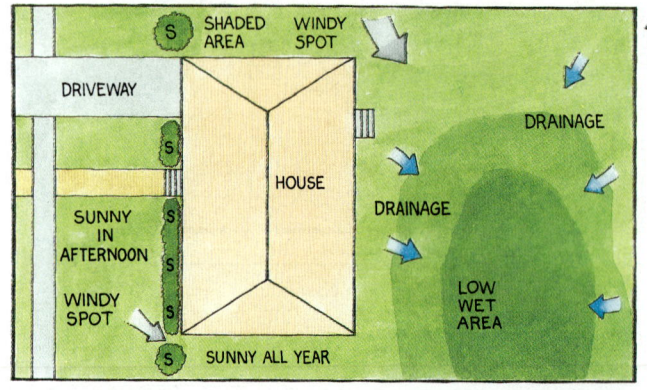

A Base Map

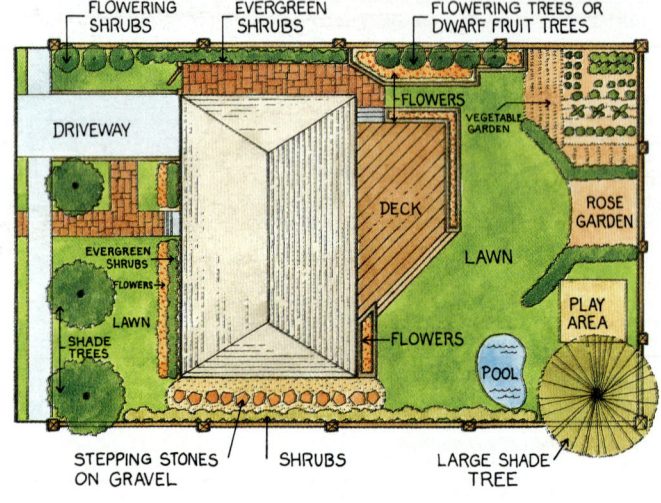

The Final Plan

488 **Growing Plants**

EXPLORATION 18

Theo's Garden

The landscape shown in Exploration 17 includes a vegetable garden. Look again at that area of the plan. Describe the light and wind conditions that you would expect to find there. Then read how Theo planned his vegetable garden.

Theo and several of the neighbors in his apartment building rented spaces for gardens in a field outside of town. Theo planned to grow the things he likes to eat.

1. He drew an accurate diagram of his garden by letting 1 cm on the paper represent 50 cm of the actual garden plot. The diagram he drew shows the plants he intends to grow. Measure the length and width of the diagram. How large will his garden plot be? A garden this big may be large enough to provide food for four people.

2. Theo showed the diagram to his neighbor, Mr. Chandra, who always grows his own vegetables. Mr. Chandra suggested that Theo:

   (a) plant the garden well away from any large trees
   (b) plant 1 m or farther from a road or sidewalk
   (c) avoid an area at the bottom of the hill
   (d) buy young tomato, pepper, and melon plants instead of starting with seeds

   Discuss this advice with several others. Speculate on the reasons Mr. Chandra had for each suggestion.

3. Consider the advantages of having a garden that supplies you with fruit and vegetables. But remember, benefits always have costs. What are the costs of growing your own food?

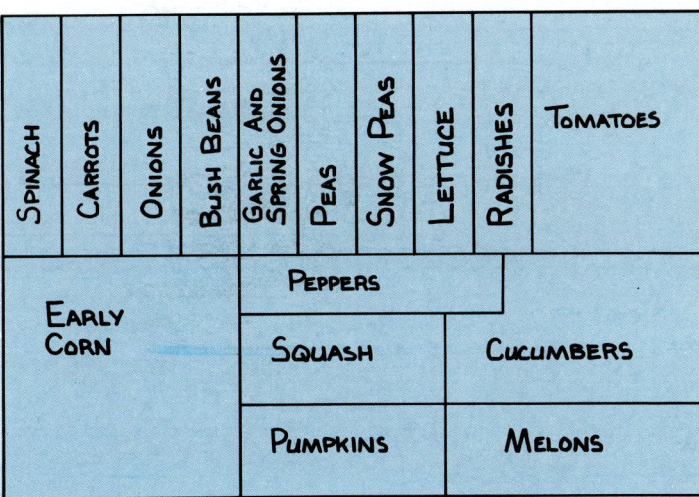

Exploration 18

Divide the class into small groups. Ask them to read the Exploration, discuss the questions, and formulate answers. After a specified amount of time, reassemble the class to share their findings.

Answers to

In-Text Questions

Students should realize that the garden in the final plan in *Exploration 17* on page 488 is in a "high" light area because of its location on the northeast corner of the property and the absence of any trees or structures around it. The base map indicates that there is little wind and that the drainage is good.

1. The garden measures about 4 m × 6 m.
2. (a) The garden should be away from trees so that it will not be shaded by them.
   (b) The garden should be planted away from roads and sidewalks to avoid damage from people and vehicles.
   (c) Avoid low-lying areas so that the garden will not become flooded during rainfalls. Higher elevations provide better drainage.
   (d) These plants take a long time to grow from seeds. Therefore, it is better to begin with plants that have been commercially grown. Also, the growing season in many parts of the country is not long enough to grow these plants outside from seeds.
3. Answers may vary, but possible "costs" include: hard work, having to stay home during the summer to take care of the garden, paying for water and other supplies (such as fertilizer and garden tools), and having to harvest the vegetables by hand.

Plants in Your Environment 489

Exploration 19

Have students read the project rules on page 490. In order to ensure that they understand what they are to do, provide time for them to ask questions or to ask for clarification.

Point out to students that at this time they should be prepared to use all the experience and knowledge they have gained during the course of the unit. Suggest that they will add to their knowledge by doing some of the suggested research and conducting the interviews.

You may wish to ask the school librarian to make resource material available to students, either in the classroom or at special stations in the school library. Organizing the project materials for display will also require your attention. One possibility is to have three-sided display units similar to the one shown in the illustration on page 490.

Allow three full class periods for teams to work together on their projects. Monitor what students are doing by occasionally visiting each group and asking questions about the progress of their project.

Choice 1

If students seem to be having trouble deciding how to divide the work among members of the team, you may wish to suggest the following. Two students can measure the classroom or school interior and be responsible for making a scale drawing. One or two students can visit local flower shops and nurseries to obtain helpful information. A couple of the students can do the library research on common indoor plants, their needs, and their origins. Everyone can request cuttings from various sources and begin growing their own plants. All members should meet to share their information and work together to design a final plan.

Choice 2

Suggest to students that they decide what their goals are in redesigning the landscape of the school grounds. They should record these goals so that they can refer to them as they work on the project. Point out that as the project progresses, they may discover that changes in their original goals have to be made. Explain that this is a normal part of any problem-solving process. Remaining flexible and open to new ideas is important if the best results are to be obtained.

Point out that steps 2 and 3 can each be accomplished by a few students. Step 4 provides another good place for members of the team to divide the work. One or two students can be responsible for each of the items and then report back to the group. Also, mention that they are not confined to using only the items listed in step 5. They will probably want to include additional items in their final plan.

EXPLORATION 19

Your Team Project

Now it is time to choose one of the three design projects described in Explorations 16, 17, and 18.

The following rules apply to each project.

1. Each project will be a team effort. Each member of the team will make their own contributions.
2. When the project is completed, it will be presented in an exhibit. The exhibit can include posters, photographs, illustrations, charts, diagrams, models, and other things you think will help explain what you have learned.
3. Each project must include interviews with people who have helped you. Interviews should be written up and included with the exhibit.
4. Each member of the team will obtain information for the project from at least two written sources (books, pamphlets, magazine articles) in addition to encyclopedias. This information will be included in the project report along with a list of the sources of information used.

Let's find out more about the choices you have.

Choice 1
A Greener Room or a Greener School

1. Develop a plan for your classroom or the school interior. Follow the procedure you used in Exploration 16. Be sure to share the work among the members of the group.

2. Find out more about indoor plants by visiting flower shops and nurseries where you can see the plants and ask questions about them. Talk to your parents and neighbors about the plants they have.
3. Read books, brochures, and pamphlets on indoor plants. Public libraries, flower shops, and nurseries have information to lend you.
4. In "Plants from Plant Parts," you learned how to grow plants from cuttings. Request free cuttings from parents and neighbors, or from flower shops and nurseries, and grow your own indoor plants.
5. Most indoor plants grow outdoors in tropical or semitropical parts of the world. Find out as much as you can about the origins of your favorite indoor plants.

Choice 2
Landscape Gardening

Here is your chance to redesign the school grounds. Each member of your group should participate in creating the overall plan. Then individual gardens and spaces should be assigned to members of the team who can plan in more detail.

1. Decide how you would like to make use of the school grounds. What are your goals in redesigning the landscape?
2. Measure the dimensions of the site. On graph paper, sketch any existing features.

490 Growing Plants

3. Lay tracing paper over the sketch and begin to redesign the area. Make a rough outline of your own design. This will be a preliminary plan. It should be redone after you have researched plants and landscape gardening.
4. Your research should include:
    (a) visits to landscaped lots
    (b) interviews with landscape architects and gardeners (Large nurseries usually employ them.)
    (c) visits to nurseries
    (d) reading about landscape design (Borrow books from a public library, or pick up pamphlets from your local nursery.)
5. Your landscape design should include plans for each of the following:
    (a) a variety of trees and shrubs
    (b) each of these kinds of flower gardens:
        (i) a shady garden
        (ii) a sunny garden
        (iii) a cool garden (near a fountain, pond, or stream)
        (iv) a hot, dry garden

Choice 3
Growing Your Own Food

If you prefer growing edible plants to decorative ones, this project is for you. You will design a garden. Later, if you wish, you can carry out the plan and grow your own fruits and vegetables.

Each member of the group should learn about three different kinds of fruits and vegetables. The garden you plan should include the three kinds you have selected and any other vegetables and fruits that you would like to grow.

The following procedure is suggested for this project.

1. Select a location for the garden. It should get at least three hours of full sun during the growing season.
    You may select one large area to be divided among the members of your group or separate areas for each member. Each member should have an area of at least ten square meters. If your group decides to divide one large area among the members, leave a 1-meter path around each garden area.
    Draw a sketch of the garden site that you have selected. Measure its dimensions and include these on your sketch. Include nearby buildings and trees in your sketch, and note the amount of light that each part of the garden will get during the growing season.
2. Study the soil at your garden site. What kind of soil do you have? How much of your soil consists of clay? of sand? What difference does this make?
    Use a soil testing kit or ask an experienced gardener to determine if your soil is rich in nutrients.
3. Research for this project should include:
    (a) reading books on fruit and vegetable gardening
    (b) interviewing experienced gardeners
    (c) visiting nurseries and gardens

Have each member of your group research and report on one or more of the following topics:
    (a) needs of different plants for light, water, and nutrients
    (b) planting instructions for different kinds of plants

Choice 3

If students are planning to actually have a garden, this activity will have to be completed by the time planting season begins in their area. If students are having difficulty deciding where to select a location for their garden, suggest that they consider part of the school yard, a vacant lot, or an area on the property where one of the members lives. Point out that unless the team actually plans to grow a garden, the location can be anywhere that is accessible.

Again, help students to divide work responsibilities, if necessary.

Design for Outer Space: Biosphere Proposal

Remind students that their biospheres will be located on Mars. Students should find out something about the Martian environment in order to design a successful and practical plan.

In order to complete their proposals, students will have to draw on all that they have learned in the unit. Suggest that it may be helpful if they review particular lessons as they write their proposals.

You may wish to display several of the best proposals around the classroom for students to read and enjoy. Or, have a contest in which the class determines which group's proposal is the most realistic.

✷ Follow Up

Assessment

1. Present students with this scenario: The library in your school needs some new plants. The librarian has asked your class for some suggestions. What factors will you advise the librarian to take into consideration before choosing any plants? Refer to the chart on page 487 to make some specific plant recommendations for the library.

2. Present the following scenario to students: Your community is planning to help landscape the area around the old city hall building in your town. Make suggestions about which plants to keep and which to remove. What new features would you like to see added to the landscape around the building? Explain the factors you have taken into consideration.

Extension

1. Suggest that students investigate what is meant by the following terms: *long-day plants, short-day plants, perennials,* and *annuals.* They should share what they learn by making fact sheets available to the class.

2. Point out to students that many kinds of outdoor plants must be able to survive very cold temperatures. Suggest that they do some research to discover what mechanisms these plants use to stay alive. Students should also find out how gardeners protect outdoor plants during cold weather.

EXPLORATION 19—CONT.

(c) the type of soil and level of acidity (pH) needed

(d) the kinds of fruits and vegetables that grow best in your area

(e) chemical versus organic fertilizers

(f) common pests and plant diseases and how to prevent or control them

(g) the nutritional value of the various fruits and vegetables

Share your research with your group. Use the information that the group has gathered to learn more about the three plants you have selected.

4. Develop a design for your garden and sketch it. Use graph paper. In your written report, include your decisions about items (a) to (g) in step 3. At the end of the report, list all the sources of information you used including books, pamphlets, interviews, and visits.

5. The exhibit should include your plan and report, together with photographs, illustrations, samples, and other materials that will help you communicate what you have learned to other prospective gardeners.

Testing Your Design

You have completed a design. Why not find out how it works? The summer ahead will give you a perfect chance to test your design. In the fall, you can enjoy showing the results!

Design for Outer Space: Biosphere Proposal

Now it is time to submit your proposal for a space biosphere.

1. Review the biosphere report that you wrote earlier. Make any changes needed so that the report reflects what you now know about growing plants. Share your ideas with the other members of your group.

2. Together, list the problems that you would expect to face when building a biosphere.

3. Discuss how to prepare for the problems that you have listed.

4. Write your own biosphere proposal showing your plan for growing plants and providing for the people who will live in the biosphere.

492 **Growing Plants**

Challenge Your Thinking

1. Design a newsletter for the Environmental Gardener's Group that points out specific ways to treat plant diseases and pests without poisons. Illustrate your letter with drawings of sick and healthy plants. Be sure to explain the benefits of environmentally safe pest control.

2. What's wrong with this picture? Randolph and Amy want all of their plants to grow strong and healthy. But someone planted the seeds in the wrong places! For each plant shown, describe the environment that it needs to live in. Then explain how Amy and Randolph should rearrange the plants.

Answers to Challenge Your Thinking

1. Answers will vary. Students should demonstrate a strong understanding of alternatives to pesticides.
2.
   - African violets—require moderate light levels. They should be placed on the table.
   - Geraniums—tolerate high light levels. Could move to the window sill or remain on the table.
   - Red ivy—requires low light levels. Should be placed away from the windows, perhaps on the table on the right of the room.
   - Cactus—requires high light levels. Should be placed near a window.
   - Orange tree—requires space and high light levels. Should be placed outdoors.
   - Coleus—requires high light levels. Should be placed near the bright window sill.

Plants in Your Environment

UNIT 8
Making Connections

Summary for
The Big Ideas

Student responses will vary. The following is an ideal summary.

A biosphere must provide light, air, water, space, warmth, and food for living things to survive. It must also allow for waste disposal and for recycling of resources.

Soil is a mixture of rock and mineral particles, clay, silt, sand, and humus. Soil provides support for the plant, and it supplies nutrients and water to the roots.

Water and nutrients are absorbed across the cell membranes of root hair cells. By osmosis, water concentrated outside the root moves into and through the plant.

Leaves provide the main surface where food is made from light energy during photosynthesis. Excess water is also released through the stomata of leaves in a process called transpiration. Transpiration helps cool the plant from the heat released during chemical processes.

The N-P-K rating identifies the percentage of nitrogen, phosphorus, and potassium in fertilizer. Nitrogen helps give plants their green color, and it helps branches and leaves develop. Phosphorus helps plants develop strong root and stem structures. Potassium strengthens a plant and increases resistance to pests and disease.

Plants can be grown in a nutrient solution in a process called hydroponics. The plant must be anchored in an upright position, and the solution must provide the proper amount of nutrients, water, and air.

Pollination is the transfer of pollen from the stamens of a flower to the pistil of the same or another flower. Pollen is produced in an anther, which is the structure at the tip of a stamen. Bright petals and the production of nectar attract animals and insects to the flower. The structure and placement of the stamens and anthers cause the animals and insects to brush up against the anthers in order to reach the nectar. Pollen that adheres to them may be carried to another flower (cross-pollination) and be deposited on the stigma of a pistil.

Vegetative reproduction is a method of growing a new plant from a part of an existing plant. Division, layering, cuttings, and grafting are means of vegetative reproduction. Roots are the first plant part to develop during vegetative reproduction; they must form for the new plant to live.

Instead of using pesticides, you can use safe alternatives. A cardboard collar or wood ashes around the stem can prevent damage from pests. You can also prevent damage from pesticides by properly disposing of poisonous wastes and contaminated water.

Light level, moisture, and temperature of plants should be considered when growing indoor plants. As well, the design should be pleasing to view. In landscaping, the conditions of the ground, the weather, the position of structures, and beauty should be considered. For a vegetable garden, the richness of soil, types of food grown, season, and climate of the garden should be considered before growing the plants.

Unit 8 Making Connections

The Big Ideas

In your Journal, write a summary of this unit, using the following questions as a guide.

- What must a biosphere provide for living things to survive?
- What is soil and what does it do for plants?
- How are water and other nutrients taken in and transported within plants?
- What functions do leaves perform?
- What does the N-P-K rating of fertilizer tell you?
- How can you grow plants without soil?
- What parts of flowers help pollination occur?
- What is vegetative reproduction? What role do roots play in this process?
- How can you prevent damage from pesticides?
- What factors should you consider in growing indoor plants? in outdoor landscaping? in vegetable gardening?

Checking Your Understanding

1. Megan measured the growth of a leaf over a period of time. Use her data to answer the following questions.

Day	1	3	5	7	9	11	13	15	17
Size of Leaf in Millimeters	2	3	4	6	10	16	24	35	50

(a) Plot this data on a graph.
(b) On day 19, the length of the leaf was 52 mm, and on day 21 it was 53 mm. What conclusion would you draw from these facts?
(c) What value would you expect for day 23?

2. Describe how the following changes would affect life inside the Mars biosphere.
   (a) Eliminate the plants that are edible by humans.
   (b) Eliminate all animal life, except humans.
   (c) Eliminate the ocean.
   (d) Use chemical sprays to control pests.
3. Answer these questions about Matthew's experiment.
   (a) Matthew had two plants of the same kind. He placed one on a windowsill and the other in a closed cupboard. What hypothesis was he testing?
   (b) After two weeks, Matthew noticed that the leaves of the plant in the cupboard were not as green as the leaves of the plant on the windowsill. The plant in the cupboard was also taller and more spindly. What would be a good conclusion for this experiment?
4. Draw a concept map that connects the following words:
   flower, stomata, stamen, leaf, plant, root hair, pistil, and *root*

Reading *Plus*

You have learned about how plants grow and how to help them grow healthy and strong. Next you will explore the functions of plant parts and the many different uses humans have for plants. Read pages S133 to S144 in the *SourceBook* to increase your understanding of plants.

Updating Your Journal

Reread the paragraphs you wrote for this unit's "For Your Journal" questions. Then rewrite them to reflect what you have learned in studying this unit.

About Updating Your Journal

The following are sample ideal answers.
1. Plants use water to cool themselves and in chemical processes like photosynthesis. Plants also need light for photosynthesis. Plants use soil as a support, a source of nutrients, and a source of water.
2. Plants need water, light, nutrients, air, and space to grow.
3. In the plant cycle, a seed germinates and will eventually become a plant with a flower. The flower is responsible for reproduction and for pollination. After pollination, fertilization occurs, and the result is an embryo in the form of a seed. *(This illustrates the theme of **Patterns of Change**.)*
4. Through careful control of the environment, a plant can grow anywhere as long as its basic needs are met.

Answers to
Checking Your Understanding

1. (a)
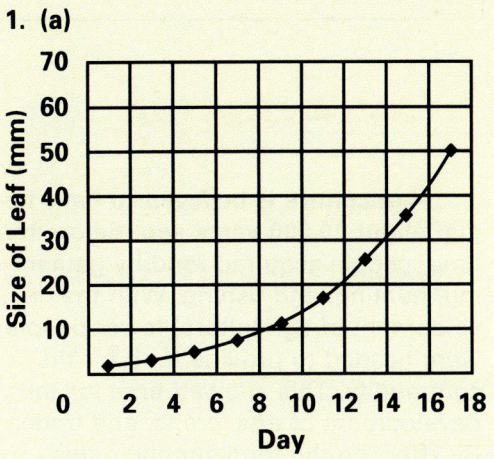

   (b) The growth of the leaf is beginning to slow.
   (c) 53 or 53.5 mm
2. Answers will vary. Students should recognize that all biotic and abiotic features in a biosphere are dependent on each other. Loss of one food source or a major environment disrupts every other environment within the biosphere. Students should also note that everything put into the food supply, such as pesticides on plants, will filter through every system in the biosphere.
3. (a) Matthew was trying to test the hypothesis that plants need light for healthy growth.
   (b) Lack of light causes leaves to lose some of their color and causes stems to grow longer and thinner.
4.
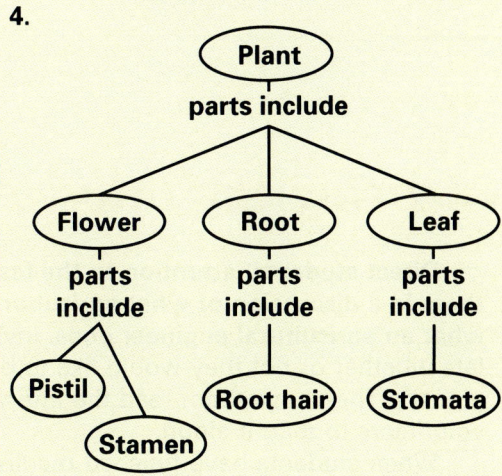

Making Connections 495

Science in Action

✦ Background

Agriculture is believed to have begun about 10,000 years ago. Before that time, people acquired food by gathering, hunting, and fishing. With the development of agriculture, fewer people were needed to provide food for the community. This allowed time for the development of arts, crafts, and trades. By affecting the food supply in this way, agriculture made the rise of civilization possible.

Prior to the seventeenth century, most farming was done by human and animal labor. In the late 1600s, machines began to assist in the jobs of planting, cultivating, and harvesting. In modern times, the design and development of farming equipment has become one branch of agricultural engineering. This field also involves the design and construction of farm buildings, erosion control, irrigation, and land conservation.

Science in Action

Spotlight on Agricultural Engineering

Christine Gorman is an agricultural engineer. We met in her office, surrounded by photographs of barns and other farm buildings.

Q: How did you become interested in agricultural engineering?

Christine: My interest in engineering probably stems from my talent for and enjoyment of mathematics. As to agriculture, I grew up on a farm and always had an interest in biology.

Q: What does your job involve?

Christine: Specifically, I am involved in the engineering of farm buildings. I give advice to farmers about designs and building materials. For example, yesterday I had to decide whether an old barn was still strong enough to withstand heavy winds and other conditions. I advised the owner that it was unsafe and should be replaced immediately. Tomorrow I will survey the site and tell the farmer what kind of barn I think he should build in that location.

Q: What are some of the benefits and drawbacks of your work?

Christine: One of the best parts is meeting a lot of people. To keep up with their questions and problems, I have to update my knowledge constantly. I am out in the field about half the time, at construction sites and in barns and places like that. One of the challenges is understanding all of the new technology. Sometimes it is hard to keep up with it.

✦ Using the Interview

Direct students' attention to the feature title and involve them in a discussion of what agricultural engineering is and what an agricultural engineer does. Invite students to speculate whether or not they would like to become agricultural engineers. Then have them read the interview silently or call on volunteers to read it aloud.

When students have finished reading, you may wish to evaluate their understanding of the interview by asking questions similar to the following:

- What contributed to Christine Gorman's interest in agricultural engineering? *(She had a talent for and enjoyment of mathematics. She grew up on a farm and always had an interest in biology.)*
- In what specific aspect of agricultural engineering is Ms. Gorman involved? *(She is involved in the engineering of farm buildings. She gives advice to farmers about building designs and materials.)*
- How does Ms. Gorman feel she benefits from her work? *(She meets a lot of people. She is out in the field a good deal of the time.)*
- What does she feel is one of the challenges of her work? *(One of the challenges is understanding all of the new technology.)*
- What does Ms. Gorman feel is a drawback of her work? *(It is sometimes difficult to convince people to try new ideas.)*

One of the drawbacks is that it is sometimes hard to change people's minds. They hesitate to try new designs. You must try to convince them that what you have to offer may be better than what they have now.

Q: Do you ever have to deal with crisis situations?

Christine: Yes. Some of our crises involve farm fires where there is a loss of animal life. Often a fire will occur in the fall, and it is critical to rebuild immediately. Our whole staff has to go out and survey a new barn so that a building can be started the next morning. That is a real challenge to our service.

Some Project Ideas

Nowhere have inventions been more important than in agriculture. Those who live on a farm are surrounded by examples of technology. These may include a variety of machines, an irrigation system, buildings and other structures carefully designed for special purposes. There are many possibilities for interesting projects.

Your classmates might be amazed by the many inventions farmers take for granted. Pick out one example of technology on a farm. Find out what it does, how it does it, when it was invented, and whether there are any better ways of doing the same thing. Propose your own idea for improvements. The results can be presented as an exhibit. You might include diagrams to show how this technology works and how it might be improved.

✦ Using the Project Idea

You may wish to have students work individually or in pairs. Point out that some areas of agricultural technology include:
- tillage, planting, and harvesting equipment
- soil and water management
- processing technology and storage facilities
- livestock care
- livestock and plant breeding
- pest and disease control
- building design, materials, and construction

Suggest that students who choose projects in the same general area, such as farming equipment or building design, work together to create exhibits for the classroom.

As an alternative, you may wish to supervise the whole class in creating an "agricultural engineering" environment by having students make murals, bulletin-board displays, poster diagrams, reading centers, and so on.

✦ Going Further

Ask students to investigate the following questions:
- What are some of the chemical fertilizers used by commercial farmers? How are they applied to the soil?
- How would eliminating chemical fertilizers affect production?
- What has research shown to be the long-term effects of using chemical fertilizers?
- What is the state of the nation's soil as a result of years of using chemical fertilizers?

Science in Action

Science and Technology

✦ Background

Genetic engineering involves artificially changing an organism's genetic material. Genes are located in the cells of all organisms, and provide the information that determines the cell's physical characteristics and growth patterns. By changing an organism's genes, scientists can change the physical traits of an organism and its descendants.

Genetic engineering is a relatively new but rapidly developing science. It was not until the early 1970s that American scientists first developed techniques for isolating and altering genes. By the end of that decade, genetic engineers had discovered how to use recombinant-DNA (artificially altered DNA) to produce small amounts of insulin and interferon. By the early 1980s, large-scale production of both substances was taking place.

During this same period, genetic engineers were making progress in determining how to add and alter genes in higher-level organisms. In one of the first successful attempts, new genes inserted into mice caused them to grow to twice their normal size. In 1986, the world's first license to sell a genetically engineered substance was granted by the United States government. The substance was a genetically-altered virus used to combat a disease in pigs.

Science and Technology

Genetic Engineering

Giant mice. Plants that glow in the dark. It may sound like science fiction, but such things have actually been created by **genetic engineering**. What are the limits of this new technology? At this point, no one can say.

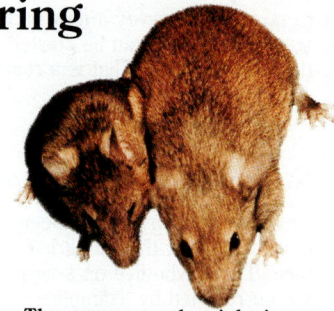

The mouse on the right is a product of genetic engineering.

The World's Tiniest Code

Every living cell of a plant or animal contains enormous amounts of genetic information. This information tells the cells how to develop and what to do. When every cell follows its instructions, the result is a complete functioning organism with all its individual characteristics.

Genetic information is stored in the form of an extremely compact code. In fact, this code is "written" as small as it possibly could be—in the form of arrangements of groups of atoms! These arrangements are embedded in certain very long molecules known as DNA molecules.

A model of a DNA molecule

Strange, But True

The individual units or "words" of the code are called **genes**. Each gene contains a specific coded instruction. For instance, a certain gene in the firefly causes the production of a substance that makes the firefly glow in the dark. If this gene could be transferred to the seed of a plant, what would the result be? A plant that glows in the dark? The answer is yes, as experiments have actually shown!

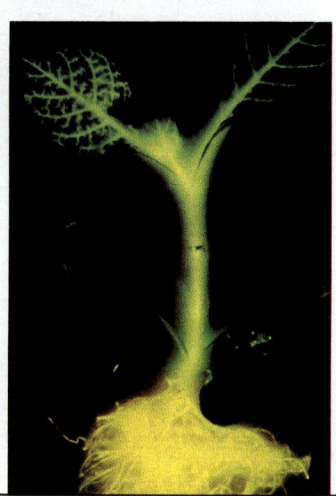

Genetically engineered tobacco plant glows in the dark.

✦ Discussion

When students have finished reading the feature, involve them in a discussion of what they have read. You may wish to use questions similar to the following:

1. How is genetic information stored in living organisms? *(It is stored in the form of an extremely compact code made up of atoms embedded in DNA molecules. The individual units or "words" of the code are called genes.)*
2. How has genetic engineering been used to produce large amounts of insulin? *(By transferring the human gene responsible for the production of insulin to certain bacteria, scientists have turned the bacteria into "insulin factories.")*
3. How do genetic engineers artificially alter the DNA molecules of an organism? *(They locate individual genes in a DNA molecule, cut them out, and then insert them at an appropriate place in a DNA molecule of a different organism.)*
4. Given the scale at which genetic engineering takes place, how do genetic engineers take apart and recombine DNA molecules? *(They use restriction enzymes which have the ability to cut DNA molecules at specific points, depending on the exact nature of the enzyme. The DNA molecules paste themselves back together more or less automatically.)*

Early Results

Genetic engineering isn't just for creating oddities. It has already paid off in important ways. For example, it has made it possible to produce large amounts of *insulin*, which is needed by many diabetics. By transferring the human gene responsible for the production of insulin to certain bacteria, scientists have turned the bacteria into "insulin factories." These bacteria are grown in a laboratory, and they create large amounts of insulin. Today there is enough insulin for everyone who needs it, at lower prices than would otherwise be possible.

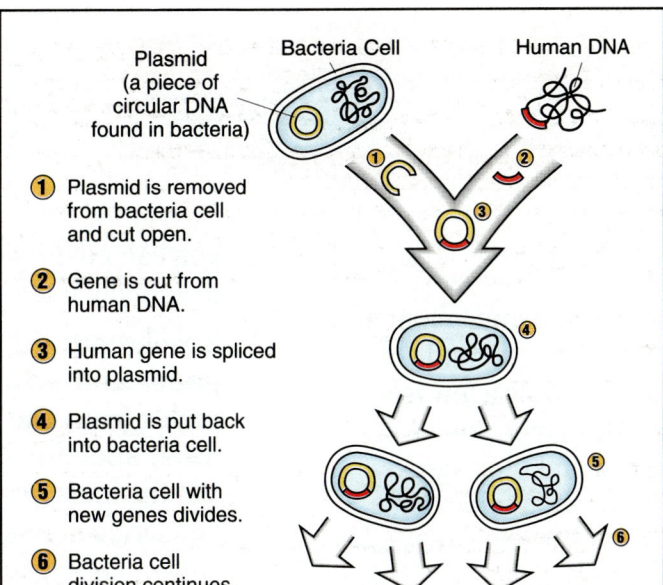

This illustration shows how an insulin-producing gene is transferred to bacteria.

Cutting and Pasting at the Molecular Level

Genetic engineers locate individual genes in a DNA molecule, cut them out, and then insert them at an appropriate place in a DNA molecule of a different organism. The procedure may sound simple, but imagine doing it on the incredibly small scale of a molecule!

The actual work, it turns out, is done by certain enzymes called *restriction enzymes*. Several hundred kinds of restriction enzymes have been identified. These enzymes have the function of cutting DNA molecules at specific points, depending on the exact nature of the enzyme.

The DNA molecules paste themselves back together, more or less automatically, because their cut ends are "sticky." When a number of fragments of cut DNA are mixed together, a certain percentage of them will splice back together in the desired way.

Find Out for Yourself

As an organism grows, many of its cells divide to form new cells. Remarkably, each of the new cells contains exactly the same complicated information as the original cell. This is because the information-bearing DNA molecules have a way of splitting into two halves and then reforming whole new DNA molecules from each half. Do some research on DNA *replication* to find out how this is done.

Write a report on some other accomplishments in the field of genetic engineering. What are future possibilities? What are some ethical considerations that might cause problems for the future of genetic engineering?

◆ Critical Thinking

To promote logical-thinking skills, you may wish to ask questions similar to the following:

1. Why do you think bacteria were used to produce genetically engineered insulin? *(Bacteria are very simple organisms that have a small amount of DNA that is relatively easy to manipulate. In addition, bacteria reproduce rapidly, which makes them ideal for the production of large amounts of substances such as insulin.)*

2. What do you think may be some of the potential dangers of genetic engineering? *(Answers may vary; possible responses include: harmful, uncontrollable bacteria might be produced accidentally; possible environmental damage might result from the deliberate or accidental release of genetically-altered organisms.)*

3. How would you ensure that research into genetic engineering is carried out safely? *(Answers may vary, but students should suggest that all research be monitored and that laws should be passed to carefully regulate the research.)*

◆ Going Further

1. Suggest that students do some research on the history of genetic engineering. They could share what they learn by making fact sheets or timeline displays.
2. Suggest that students take a survey of adult family members and friends to discover what their feelings are about genetic engineering.
3. Have students speculate on the possible benefits genetic engineering could provide for growing healthier plants.

CONCEPT MAPPING

A Way to Bring Ideas Together

What Is a Concept Map?

Have you ever tried to tell someone about a book or a chapter you've just read, and you find that you can remember only a few isolated words and ideas? Or maybe you've memorized facts for a test and then weeks later, you're not even sure what topic those facts are related to.

In both cases, you may have understood the ideas or concepts *by themselves,* but not in relation to one another. If you could somehow link the ideas together, you would probably understand them better and remember them longer. This is something a concept map can help you do. A concept map is a visual way of showing how ideas or concepts fit together. It can help you see the "big picture."

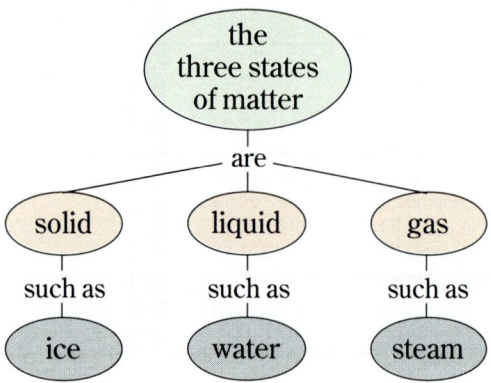

How to Make a Concept Map

1. **Make a list of the main ideas or concepts.**
   It might help to write each concept on its own slip of paper. This will make it easier to rearrange the concepts as many times as you need to before you've made sense of how the concepts are connected. After you've made a few concept maps this way, you can go directly from writing your list to actually making the map.

2. **Spread out the slips on a sheet of paper and arrange the concepts in order from the most general to the most specific.**
   Put the most general concept at the top and circle it. Ask yourself, "How does this concept relate to the remaining concepts?" As you see the relationships, arrange the concepts in order from general to specific.

3. **Connect the related concepts with lines.**

4. **On each line, write an action word or short phrase that shows how the concepts are related.**

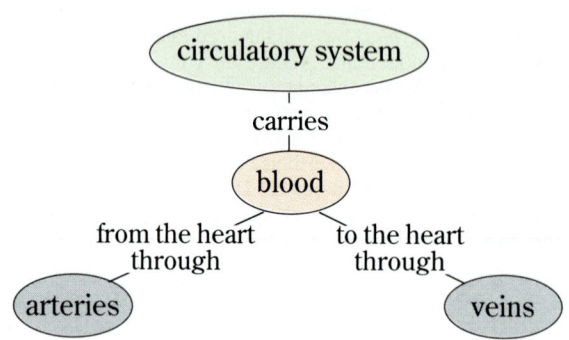

Look at the concept maps on this page and then see if you can make one for the following terms: *plants, water, photosynthesis, carbon dioxide,* and *sun's energy.* The answer is provided below, but don't look at it until you try the concept map yourself.

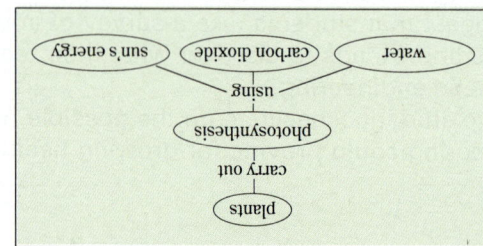

500

SourceBook

This *SourceBook* is designed as a handy reference so you may learn more about many of the concepts you have discovered in your reading of Science*Plus*. For each unit of Science*Plus*, you will find a corresponding unit in the *SourceBook*. Questions on the first page of each unit will help direct your thinking as you read the material.

Contents

UNIT 1 — **S1**
1.1 The Biosphere — S2
1.2 Succession and the Biomes — S8
1.3 Humans and the Environment — S17

UNIT 2 — **S23**
2.1 The Diversity of Modern Life — S24
2.2 The Evolution of Diversity — S35

UNIT 3 — **S45**
3.1 Solutions, Suspensions, and Colloids — S46
3.2 Acids, Bases, and Salts — S51

UNIT 4 — **S61**
4.1 Motion and Force — S62
4.2 Laws of Motion — S68
4.3 Gravitation — S74

UNIT 5 — **S77**
5.1 Architecture in Other Cultures — S78

UNIT 6 — **S89**
6.1 Rocks and the Rock Cycle — S90
6.2 Rock-Forming Minerals — S99
6.3 Stories in Rocks — S105

UNIT 7 — **S111**
7.1 Upward and Outward — S112
7.2 Stars: A Universe of Suns — S118
7.3 One Small Step — S126

UNIT 8 — **S133**
8.1 Plant Structure — S134
8.2 Uses of Plants — S141

Unit 1

In This Unit

1.1 The Biosphere, p. S2

1.2 Succession and the Biomes, p. S8

1.3 Humans and the Environment, p. S17

Reading *Plus*

Now that you understand some of the interactions among organisms and their environment, let's take a closer look at the kinds of interactions that occur in ecosystems. Read pages S2 to S22 and write a report about your role as a human being in your ecosystem. Your report should include answers to the following questions.

1. What biome do you inhabit, and what are some of its characteristics?

2. How do some of the plant and wildlife populations interact with the human population in your community?

3. In what ways could a major environmental problem in your ecosystem be solved?

1-1 THE BIOSPHERE

Biosphere All the life-supporting environments on earth and the organisms in them.

All life exists within the **biosphere**, a thin layer that surrounds the earth. In this layer, all living organisms are born, grow, reproduce, and die. Figure 1-1 shows the vertical dimensions of the biosphere. Yet a narrow range only 200 m thick, extending 100 m above and below the earth's surface, houses almost all of the earth's millions of species of organisms. As you can imagine, the biosphere supports a wide variety of living organisms. Think about what a jellyfish needs to survive. The same biosphere that includes a warm marine environment for the jellyfish includes an icy environment for the polar bear.

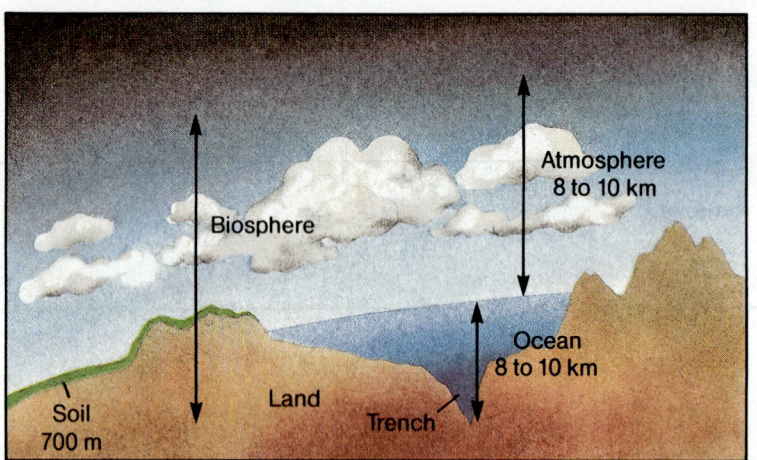

Figure 1-1 The biosphere includes all the places where living things exist.

Ecosystem A group of organisms and their physical environment.

In order to better understand organisms and their environments, it is useful to divide the biosphere into **ecosystems**. An ecosystem consists of all the *biotic* and *abiotic* factors in an area. See Figure 1-2. The kinds of organisms and their interactions, the environment, the population of each species, the size of the area, and many other factors

Figure 1-2 What are the biotic and abiotic parts of this ecosystem?

contribute to an ecosystem. An ecosystem may be as small as a single decaying log in a forest—or it may be as large as the Amazon river!

Think of the different ecosystems that you have visited and interacted with. Did you include places like the park, the beach, and your own backyard? Each of these places contains biotic and abiotic factors that interact in a unique way. The interactions of an ecosystem involve many cycles and pathways that link the earth's organisms to each other and to their abiotic environment. In this section, we will examine some of these interactions.

Populations and Communities

When a coyote preys on a jackrabbit or a caterpillar eats an oak leaf, we say that these organisms are interacting. However, few ecosystems support, for example, only one coyote or one oak leaf. Instead, living organisms are members of interacting groups. Within an ecosystem, the interactions of these groups of organisms affect individual animals, the environment, and other species. Populations and communities are two of these groups.

Populations What is the population of your classroom? You would probably answer this question with a number. But how would you arrive at that number? Would you include the teacher? the potted plants? the fish in the aquarium? In an ecosystem, a **population** represents the number of organisms that belongs to the same *biological species*. So, if your classroom contains plants and fish, it may contain several populations.

Population A group of organisms of the same species.

In addition to the species of the organism, a population is also described by its location and a particular point in time. Was the population of your classroom the same two years ago? ten years ago? Consider the Alaskan brown bears in Figure 1-3. They belong to a different population than do the brown bears that are currently found in Alaska, although the species in each population is the same. Interactions among the individuals of a population contribute to the characteristics of the population and how it changes over time.

Together, the individuals in a population form a *reproductive group*. For a population to survive in an ecosystem, it must have enough individuals for reproduction to occur. If a population becomes too small, the species may disappear altogether. Organisms in small populations may not be able to find each other, or they may require a larger population for the successful growth of their young. For example, the

Figure 1-3 A family of brown bears fishing in the McNeil River of Alaska.

Limiting Factors Environmental elements that stabilize population size and keep species from producing too many offspring.

Figure 1-4 Births and deaths alone do not explain the annual changes in the size of a caribou herd. New animals periodically join the herd, while others leave.

Population Density The number of individuals of the same species in a given area.

wild populations of whooping cranes and California condors became so small that the animals could no longer reproduce in nature. You can see these birds in zoos and wildlife preserves only because people have established breeding programs that bring them together and protect their eggs and hatchlings.

Members of a population also *compete* with one another for certain resources, such as food, light, nutrients, and living space. You have seen competition among members of a population if you have ever watched alley cats fight over territorial rights or a flock of pigeons peck viciously at bread crumbs. The resources that a population needs, plus the climate of a region, are called **limiting factors**. They determine where a particular species can live and how large its population can grow.

Characteristics of Populations One important characteristic of a population is that the number of individuals in it constantly changes. A population grows as new members are born and as adults migrate in from different populations. A population may become smaller as individuals die or migrate to other populations.

Suppose a small herd of caribou, such as the one shown in Figure 1-4, grows in size from 310 to 400 members in one year. During that year, 150 animals are born and 75 die. The difference between 150 births and 75 deaths represents a *net gain* of 75 caribou. However, upon counting the caribou, you discover that the population actually grew by a total of 90 individuals. Where could the other 15 caribou have come from?

Another important characteristic to consider is the number of individuals that occupy a living area. This is called the **population density** of a species. For example, the population density of humans on Manhattan Island in New York City is about 26,000 people per square kilometer. Notice that the population density is not just the *number* of organisms in the population; it also includes the *size* of the area in which the individuals live.

Look at Figure 1-5, which represents two populations of trees that have different population densities. Can you see that the members of Population I have more living space than the members of Population II? What are the advantages of having more living space? What advantages are there to living in a dense population? If a population becomes too dense, it begins to decline because its environment cannot provide enough of its basic needs. Some

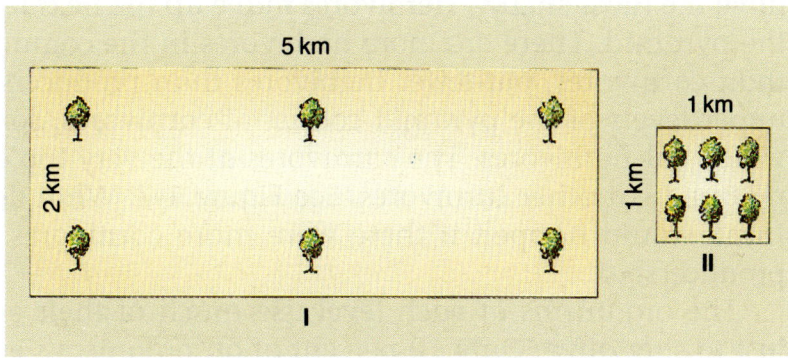

Figure 1-5 Population density is determined by calculating the number of organisms that live in a defined area.

members of the population may move elsewhere, while others may die due to lack of food and shelter.

Communities A *community* consists of all the different populations living and interacting in the same area. For example, all of the plant and animal populations in a forest belong to the same community. See Figure 1-6. Similarly, the populations of fish and marine organisms found on a coral reef form their own community. Remember that the populations of a community depend on one another for food or other nutrients and, in many cases, shelter. Studying communities allows you to learn about the interactions among many different populations that exist together in an ecosystem.

Figure 1-6 The community shown in this photograph is composed of several different populations including grizzly bears, spruce trees, aspen trees, several kinds of grasses, ducks, and many unseen organisms.

Energy Pathways

You are already familiar with two types of energy pathways—food chains and food webs. No matter how complex the food web, a community always needs more producers than consumers. Producers provide the energy necessary for themselves and the consumers. A **food pyramid** shows the energy levels of a food web. Each successive layer of a food pyramid is smaller that the one below it. The bottom of the pyramid represents the producers, which as a group

Food Pyramid A diagram that shows how some factor decreases at each feeding level in a food chain.

have the most energy. Herbivores make up the next level of the pyramid. There are more herbivores in the community than carnivores, but fewer herbivores than producers. The upper levels of the pyramid consist of carnivores, some of which eat herbivores. The carnivores at the very top of the pyramid eat other carnivores. See Figure 1-7. What do you think would happen if there were more carnivores than producers?

The organisms at each level use much of their energy just to stay alive. Only 10 percent of an organism's energy can be passed up to the next level. Therefore, only a few animals are found at the top of the food pyramid. These animals—the *top carnivores*, such as lions, wolves, and tigers—exist only in small populations. They depend on large populations of other animals to survive.

Figure 1-7 Food pyramids show producers at the bottom, then layers of different types of consumers, with a top carnivore on top.

Recycled Materials

Food and energy are not the only important resources an organism must have to survive. Substances from the abiotic environment, including water, carbon, oxygen, nitrogen, and mineral nutrients, are essential for the survival of the biotic factors (organisms) in an ecosystem. These materials are used and then returned to the environment by the activities of living things. In this *recycling* process, there is a constant exchange of nutrients between the biotic and abiotic factors of an ecosystem.

The Water Cycle The sun heats water on the earth's surface and causes it to evaporate into the atmosphere as a gas. This gas (water vapor) cools and forms droplets of water that fall back to the ground as rain, sleet, or snow. See Figure 1-8. How might water interact with the organisms in

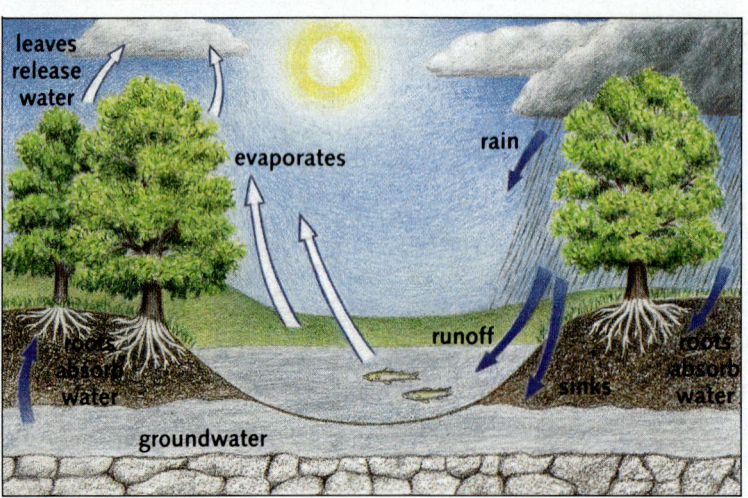

Figure 1-8 The water cycle.

an ecosystem? Plants use water to make food and then release that water back into the atmosphere. Animals drink water, or extract water from their food, and return it to the environment through waste products. In extremely dry areas, organisms such as cactus plants and camels have become adapted for storing water. How do you think this affects the water cycle?

The Carbon Dioxide–Oxygen Cycle The gases carbon dioxide and oxygen provide the basis of chemical interactions for all living organisms. Animals and plants need oxygen in order to release the energy stored in food. During respiration, oxygen is taken in and carbon dioxide is given off. When you breathe, you take in the oxygen that you need to get energy from your food. The waste product of this chemical reaction is the carbon dioxide that you exhale into the environment. Plants complete the cycle—they use carbon dioxide in the environment to make food for themselves and for consumers. See Figure 1-9.

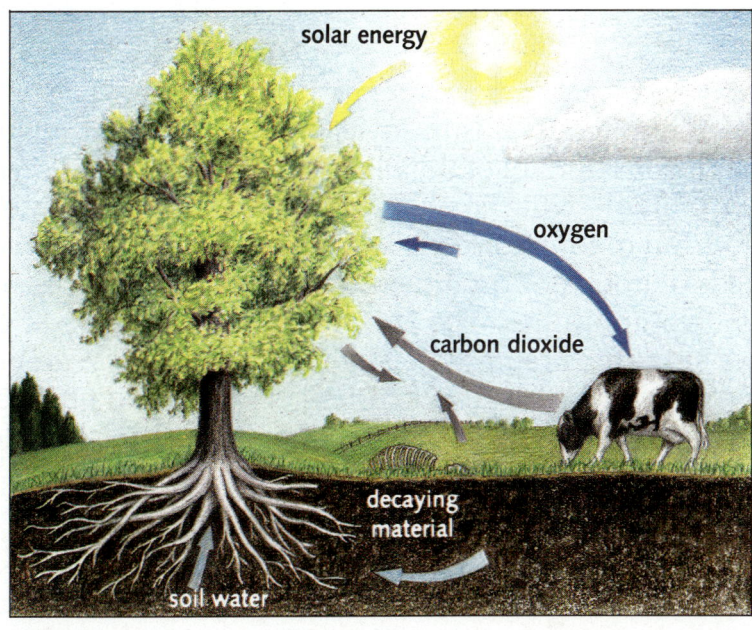

Figure 1-9 The carbon dioxide–oxygen cycle.

Figure 1-10 The lumps on these white clover roots contain nitrogen-fixing bacteria.

The Nitrogen Cycle Nitrogen gas makes up about 80 percent of the earth's atmosphere, and living organisms use it in the production of proteins. However, most organisms cannot use nitrogen directly from the atmosphere. Fortunately, *nitrogen-fixing bacteria* convert the gas into solid compounds that can be absorbed by plant roots. See Figure 1-10. Animals obtain nitrogen by eating these plants or other animals. The nitrogen is returned to the environment through waste products like ammonia and urea. Bacteria that feed on animal wastes produce nitrogen gas that returns to the atmosphere, or nitrogen compounds that remain in the soil. See Figure 1-11.

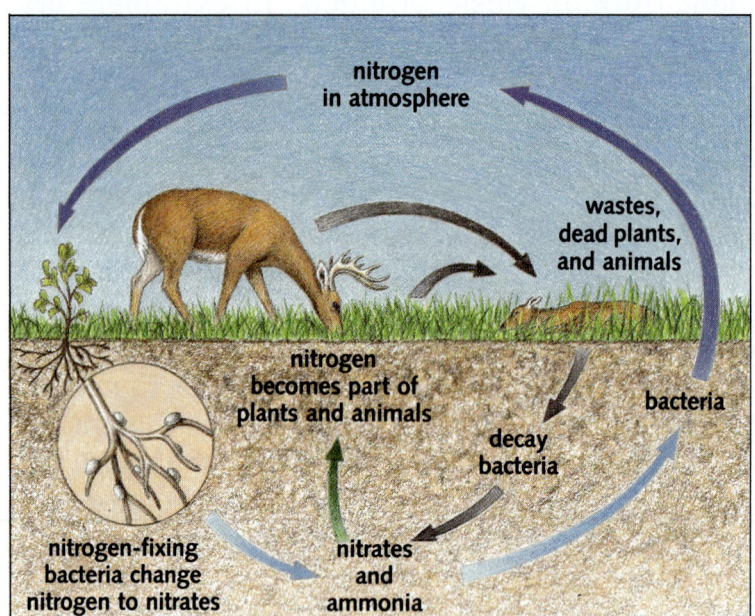

Figure 1-11 *The nitrogen cycle.*

Summary

The biosphere is a thin layer of the earth that includes all organisms and all of the places where organisms can live. It contains all of the earth's ecosystems, including both the biotic and abiotic elements and their interactions. Two interacting groups of organisms in ecosystems are populations and communities. The transfer of energy through an ecosystem can be represented by food pyramids. Water and essential nutrients, such as carbon, oxygen, and nitrogen, also circulate in ecosystems through several complex cycles.

1-2 SUCCESSION AND THE BIOMES

Have you ever wondered how a scene such as the one in Figure 1-12 came to be? Mountain-building, weathering, and erosion gradually molded the land into its present shape. The plant and animal communities that you see here developed slowly over time by the processes of evolution and succession. The makeup of these communities was determined by a variety of limiting factors such as climate, soil composition, disease, and natural disasters.

In this section, you will take a close look at succession as a process by which living communities change over time.

Figure 1-12 *The Sonoran desert of Organ Pipe National Monument in Arizona.*

You will also look at some of the geographic areas that cover major parts of the earth and share similar climates. These areas, which can be identified by their plant life, are called *biomes*.

Ecological Succession

Compare the photographs in Figure 1-13. The view on the left shows part of Glacier Bay National Monument in Alaska. Notice the land closest to the glaciers. The glacier has only recently receded, and the land is lifeless. In the photograph on the right, the land near the mouth of the bay has been exposed for hundreds of years. You can see the forest of spruce trees and can imagine the kind of animal life that inhabits the forest.

Figure 1-13 *The pioneer stage of succession* (left) *as well as a climax community* (right) *can be seen in Glacier Bay, Alaska.*

The glaciers of Alaska's Glacier Bay formed during the last Ice Age. Since that time, a slow, steady melting has taken place, exposing land that was covered by ice for thousands of years. During the next few centuries, a gradual process of change will occur in Glacier Bay as living communities become established on the bare rock. Over time, one community will gradually replace another in an ecological process called **succession**.

Primary Succession Imagine an underwater volcano erupting violently—huge rivers of lava pour out of the ocean floor until a new island appears in the water. This brand new island is barren of plant and animal life. But not for long! Living organisms will invade this new environment as the process of succession begins. Succession that starts in an area where the land has never had organisms living on it is called *primary succession*. In addition to land formed by lava flows and exposed by the retreat of glaciers, sand dunes are also places where primary succession occurs. The types of communities that come and go during primary succession are different in each of these areas, but they all follow similar trends.

Let's take a closer look at primary succession as it proceeds around Glacier Bay, Alaska. Traveling up Glacier Bay

Succession A series of changes that occur in the plant and animal communities of an ecosystem.

Figure 1-14 Lichens completely encrust these rocks to begin the process of primary succession.

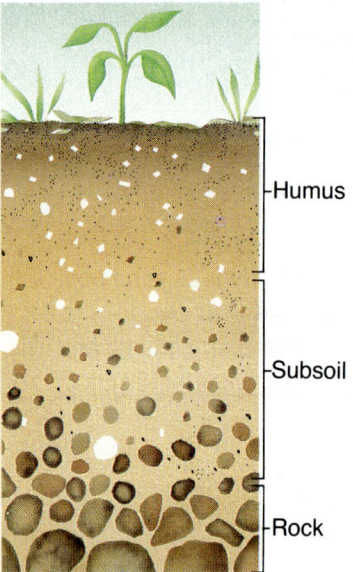

Figure 1-15 Soil is formed during the pioneer stage of succession. Soil is necessary for the next stage of succession to take place.

is like traveling backward in time. By retracing your steps back toward the mouth of the bay, you can investigate the changes in the land as if they were occurring by succession. Communities representing each stage of succession thrive in different parts of the bay. Standing at the edge of the glacier, you will notice that primary succession begins soon after bare, lifeless rock is uncovered by the melting of glacial ice.

Right below your feet, you can see succession beginning at the base of the glacier. *Pioneer species* are the first living things to appear in a previously lifeless area. The harsh habitat of the pioneers is characterized by high winds, extreme cold, bright sun, and heavy rain. There is no soil at this stage. *Lichens*, which consist of a fungus and an alga growing together, are one of the first pioneer species you find. They often completely cover the surface of rocks. See Figure 1-14. In order to get necessary nutrients, lichens produce an acid that breaks down the hard rock. As dead lichens collect and decay, they mix with small pieces of the rock to form a thin crust—the beginning of soil.

A short distance away, *mosses* (another pioneer species) grow in the new soil formed by lichens. Mosses are simple plants that form thick mats. Their decaying parts add to the mixture of dead lichens and rock particles. Mosses also trap bits of dust and dead matter brought in by the wind and waves. More organic matter is added by a variety of animal species, such as insects and birds, that are a part of the moss community. This organic matter enriches the developing soil. See Figure 1-15.

If you climb down from the glacier and cross over the areas of lichen and moss, you will see that the soil becomes richer. The land no longer looks as barren as it did while you were standing on top of the ice. Meadows of grasses and wild flowers have replaced the pioneer species. Wind, birds, and small mammals bring seeds for new plants into the area. The roots of these plants break up rock and trap decaying organic matter that helps form richer layers of soil. Eventually, the soil can support the growth of shrubs such as willow and alder, which thrive in the bright sunlight of open meadows.

If you leave behind the grassy meadows and shrubs, you find that these communities are replaced by forests of white spruce. Spruce seedlings grow well in the shade that the shrubs provide, so the seedlings eventually replace the mature willows and alders. See Figure 1-16. The spruce trees represent the final stage of succession. The community that exists at this stage is called the **climax community**. It includes the population of white spruce, some small shrubs, and the animals that find shelter and food in such a forest.

Climax Community The final stage in the succession of a community.

Secondary Succession
Natural disasters such as fires, floods, diseases, and swarms of insects can disturb a climax community. Agriculture and forestry can destroy these living communities as well. Once the disturbance is over, and if no other disturbances occur, the climax community will slowly re-establish itself. This type of succession, which occurs after an original community has been disturbed or destroyed, is called *secondary succession*. See Figure 1-17.

The pioneers for secondary succession are usually grasses and other weed-like species of flowering plants that are brought in from the surrounding area by wind, mammals, or birds. Grasses and weeds grow quickly, and most have root systems that protect the soil of a disturbed area from erosion. Often, seeds from the plants that originally lived in the area are lying undisturbed beneath the ground. When the environment is ready to support the seedlings, these seeds will germinate and grow. As the plant communities are re-established, the animals that once lived there return as well.

Figure 1-16 In this part of Glacier Bay, spruce trees are beginning to take over the community of sun-loving willow and alder trees.

Figure 1-17 Succession in an old New England field.

Since soil is already present, the process of secondary succession takes place more quickly than primary succession. In many cases, the original climax community can be regained by secondary succession. In some cases, however, it cannot. For example, the eastern forests of the United States were dominated by chestnut and oak trees until the early 1900s. At that time, the chestnut trees were struck by a fungal disease that destroyed all of them by the 1930s. The climax community now consists of an oak and hickory forest.

Terrestrial Biomes

Large areas of the earth's continents share similar climates and soil types. These geographic regions are called **biomes**. Different plants and animals are found in the various biomes. Scientists usually identify these regions by their climax plant species. The six major terrestrial (land) biomes are indicated on the map in Figure 1-18.

Biomes Large geographic areas of similar climate and life forms.

Figure 1-18 *The major biomes of the world.*

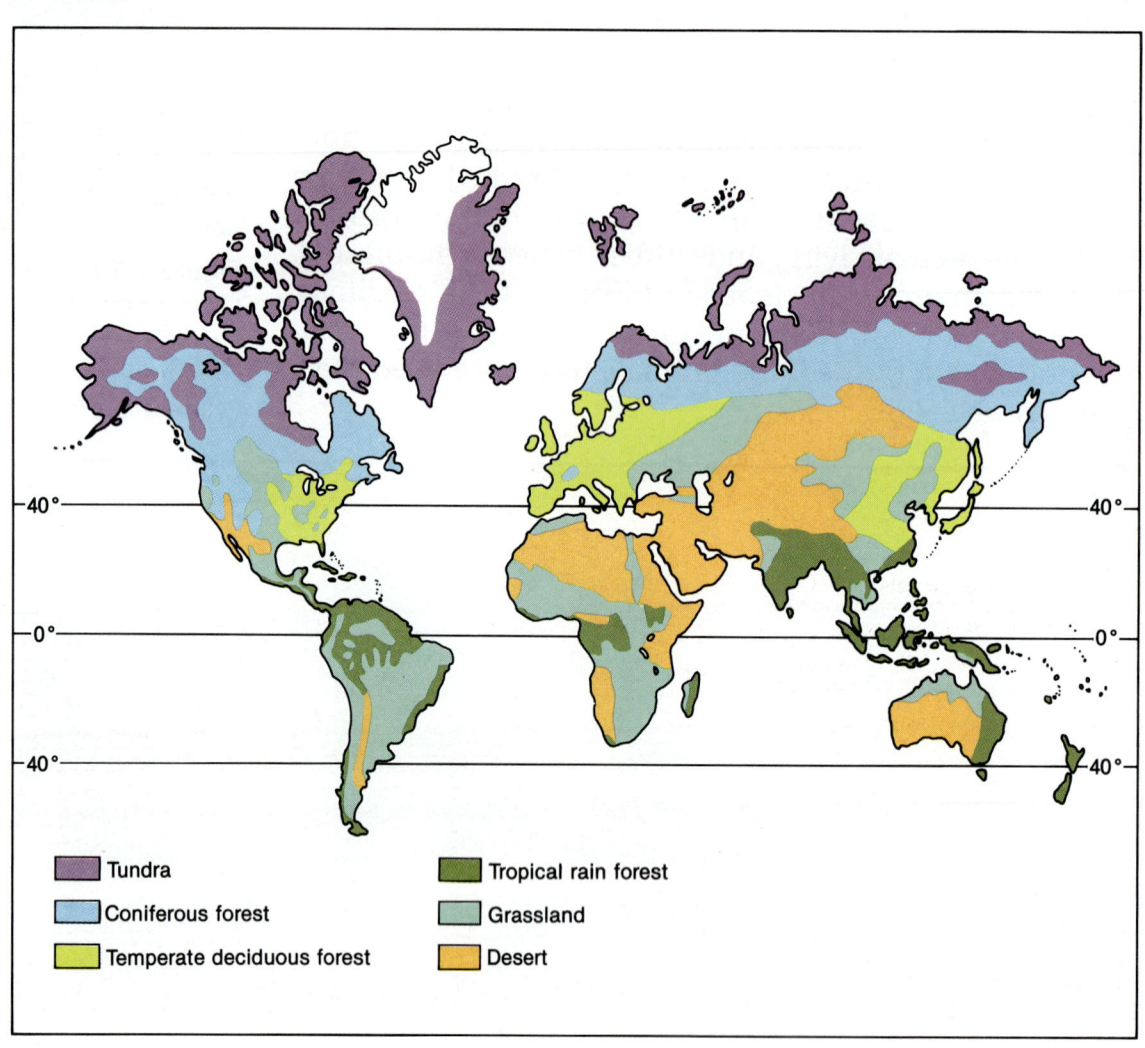

The Tundra
Just south of the arctic icecap is the freezing, harsh biome known as the *tundra*. A land of extreme cold, high winds, and very little (25 cm or less) rain or snow, the tundra has winters where temperatures may drop as low as −40 °C! The tundra receives the least amount of sunlight of

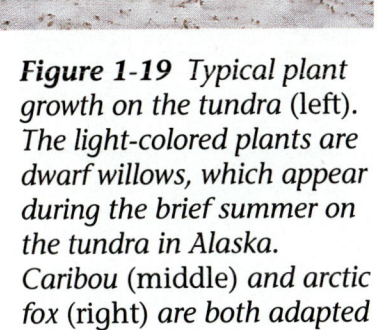

Figure 1-19 Typical plant growth on the tundra (left). The light-colored plants are dwarf willows, which appear during the brief summer on the tundra in Alaska. Caribou (middle) and arctic fox (right) are both adapted to tundra life.

any of the earth's terrestrial biomes. Life on the tundra is limited to hardy species that are adapted to the cold, harsh climate or are able to migrate to warmer regions when necessary. See Figure 1-19. Plant life includes low-growing species of grasses, lichens, mosses, and flowering herbs that can reproduce during the short growing season. Small, woody shrubs of willow and birch grow in some areas. Animal inhabitants include migratory birds, musk-oxen, wolves, grizzly and polar bears, lemmings, and arctic hares and foxes.

The Forests Three kinds of forest biomes cover a large portion of the earth. They differ in climate, amount of rain, temperature range, and growing season. See Table 1-1. These and other limiting factors, such as soil type and humidity, determine the kinds of communities that inhabit each forest biome.

Forest Type	Average Rainfall	Average Temperature	Length of Growing Season
Rain forest	212 cm	High: 35 °C Low: 25 °C	9–12 months
Deciduous forest	75–125 cm	High: 27 °C Low: –10 °C	6 months
Coniferous forest	25–125 cm	High: 20 °C Low: –30 °C	2–5 months

Table 1-1

Figure 1-20 Elk in a coniferous forest in Yellowstone National Park. Note the absence of snow on the branches of the conifers.

The *coniferous forest* biome lies just south of the tundra. The dominant plants of this biome are evergreen *conifers*, or cone-bearing trees, such as spruce, fir, pine, and hemlock. The smooth, slender needles and bendable branches of the trees allow the conifers to shed heavy snows before the trees can be damaged. See Figure 1-20. Animals adapted to this cold biome include snowshoe hares, moose, elk, bears, beavers, and many species of birds.

Moving from the coniferous forest toward the equator, the *deciduous forest* biome is found. This biome has a more humid climate marked by distinct seasonal changes. *Deciduous trees*, like oak, hickory, beech, maple, and elm, are the dominant plant species of this biome. Animal life includes deer, bears, snakes, rabbits, squirrels, and many birds and insects.

The third type of forest biome is the *tropical rain forest* which is found along and near the equator. The hot, wet climate of the tropical rain forest supports the greatest variety of life found anywhere on earth. New species of plants and animals are discovered each year. Some may never be found by humans because they live up to 60 meters above the ground, supported by the ecosystem found in the *canopy* of the tree branches. See Figure 1-21. The animal species include: monkeys, squirrels, bats, snakes, deer, many species of rodents, numerous species of birds, and a vast number of insects.

Figure 1-21 *Typical layers of vegetation in a tropical rain forest.*

Grasslands The grassland biome is usually found in the same general areas as the deciduous forest biome, except that the grasses require less rainfall than the amount necessary for trees. In North America the grasslands are called *prairies*. Great herds of bison once roamed these prairies. Because grassland soils are rich and well developed, much of the world's best farmland is found in this biome. Today, much of the original vegetation has been replaced by fields of grain such as wheat and corn, while great herds of cattle and sheep graze in place of the herds of bison. Still, a variety of small animals such as jack rabbits, ground squirrels, prairie dogs, and mice inhabit the grasslands. They are prey for grassland predators such as coyotes, foxes, snakes, and hawks.

The Desert Deserts occur in geographic areas that receive less than 25 cm of rainfall per year. Rainfall in most deserts is rare; some deserts do not receive any rain for several years at a time. Although not all deserts are hot, the hottest places on earth are found in deserts. For example, Death Valley, California, a part of the Mojave Desert of the southwestern United States, has a record high temperature of 57 °C—in the shade! Desert plant communities contain cacti, small trees, woody shrubs, and many wildflowers. See Figure 1-22. Since moisture is the primary limiting factor for survival in

Figure 1-22 *This desert scene is typical of the southwestern United States and northern Mexico. Shown here is the saguaro cactus, which may reach 15 m in height.*

the desert, the plants that grow there have many adaptations for getting and conserving water. Many desert animals are *nocturnal*, meaning they are only active at night. Kangaroo rats and other rodents, bats, snakes, toads, birds, ants and other insects populate the desert.

Aquatic Biomes

Oceans and seas contain saltwater ecosystems. Lakes, ponds, rivers, and streams contain freshwater ecosystems. Along coastlines, saltwater and freshwater come together to form a special ecosystem called an *estuary*. All of these ecosystems make up the aquatic (water) biomes of the earth. Together, the aquatic biomes cover more than three quarters of the earth's surface. Temperature, sunlight, salinity, water flow, and the amount of available nutrients are important variables that determine how organisms are distributed in the aquatic biomes.

Figure 1-23 *Sea stars have adaptations for life in the intertidal zone.*

The Marine Biome

Covering about 70 percent of the earth's surface, the marine biome is the largest biome on earth. Since it is so large, factors such as temperature, light, and depth of water vary from region to region. The *intertidal zone,* which is closest to the shore, is rich in nutrients that have washed down from shoreline communities. During low tide, the region is exposed to air. Therefore, organisms living in the intertidal zone must be adapted for life underwater as well as in the air. See Figure 1-23. One of the most productive intertidal ecosystems, the estuary, occurs in bays, salt marshes, or swamps where the salinity of the water is between seawater and freshwater. Many marine organisms begin their lives in an estuary, which supports large and diverse communities of organisms such as clams, snails, oysters, and crabs, as well as many plant species.

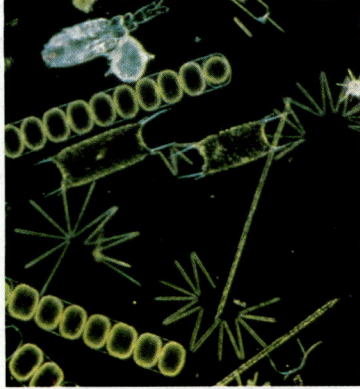

Beyond the intertidal zone is the *neritic zone*, which extends from the low-tide line out to a depth of 200 m. Floating microorganisms, plants, and animals called **plankton** are the most abundant life forms. See Figure 1-24. Other organisms found in this zone are dolphins, sea turtles, fish, and bottom-dwelling crabs and oysters.

The *oceanic zone*, which extends outward from the neritic zone and downward to the bottom of the ocean, is relatively barren of life because there is no light and few nutrients. In this zone, at depths of 2000 m, are some of the strangest-looking organisms on earth. They feed on each

Figure 1-24 *Plant and animal plankton are found floating in the neritic zone.*

Plankton Small organisms that float near the surface of the ocean.

Figure 1-25 A predator of the deep ocean.

other or on dead organisms that fall from the water above. See Figure 1-25.

The Freshwater Biome In freshwater biomes, limiting factors include rate of water flow, oxygen, available nutrients, light, and temperature. These factors vary greatly between streams, ponds, and lakes. Lakes and ponds have nutrient-rich water that supports the growth of surface algae and floating plants. Many species of fish, insects, reptiles, birds, and mammals live on or near the surface of lakes and ponds. See Figure 1-26.

Figure 1-26 A lake is a freshwater ecosystem that contains many different populations.

Life is sparse, however, in swiftly flowing streams, such as the one in Figure 1-27. Cold water and the currents that carry away both the nutrients and organisms are the main

S16

limiting factors. Green plants that attach to rocks live in swift streams, but floating plants and algae merely pass through with the current. The animals of this ecosystem are either excellent swimmers, like trout and bass, or are able attach to rocks and filter nutrients and oxygen from the passing water.

Summary

Succession is the process by which one community is replaced by another over time. Primary succession occurs in areas that have not previously supported life. Secondary succession occurs in areas where an existing community was disturbed or destroyed. The biosphere includes several large geographic areas that are characterized by their climates and the types of plants and animals that live there. The six major terrestrial biomes are tundra, grassland, desert, and coniferous, deciduous, and tropical rain forests. The major aquatic regions include the marine and freshwater biomes.

Figure 1-27 The mountain stream shown here is an example of a moving freshwater ecosystem.

1-3 HUMANS AND THE ENVIRONMENT

Until it died in 1914, the last known passenger pigeon lived in the Cincinnati Zoological Gardens—its calls unanswered by others of its kind. Prior to the mid 1800s, this beautiful bird, shown in Figure 1-28, inhabited the forests of eastern North America in numbers of 3 to 5 billion. How could so large a population become **extinct** so quickly?

In this case, the cause was a migration of an animal species into the birds' ecosystem. This species consumed the oak and beech forests in which the birds nested. The invading species also preyed upon the birds. As the population of passenger pigeons declined, they became unable to reproduce as quickly as they once had. Soon, all of the passenger pigeons were gone—their species extinct. Who were the invaders that caused this extinction? They were human settlers that used the wood of the forests to build their homes and sold the birds to city dwellers to eat as delicacies. Though the extinction of a species is a natural process, human activities in recent years have greatly increased the rate at which species become extinct.

Extinct Refers to a species that no longer exists.

Figure 1-28 Male and female passenger pigeons, as painted by John James Audubon.

The Use and Misuse of Natural Resources

Human societies need a steady supply of food and materials from the earth and its biosphere. These organisms and

Natural Resources All natural substances that humans remove from the environment for their own use.

Pollutants Unwanted materials, such as gases, particles, and chemicals, that are released into the environment.

Figure 1-29 Smoke billowing from factory smokestacks was seen as a sign of progress during the Industrial Revolution. Today, however, people are aware that many materials in this smoke are harmful to plants, animals, and people.

Conservation The protection and wise use of natural resources.

substances are called **natural resources**. Some of our natural resources can be replaced, but others cannot. *Renewable resources* can be replaced at about the same speed we use them. These include trees, crops, livestock, fish, wildlife, water, oxygen, nitrogen, and carbon. *Nonrenewable resources* cannot be replaced at the speed at which we use them. They include soil, minerals, metal ores, and fossil fuels.

Pollution is the process of making the environment unclean with waste products, such as poisonous gases, liquid chemicals, radiation, heat, garbage, and even noise. All waste materials are referred to as **pollutants**. You find them in the air you breathe, the water you drink, and the soil where your food grows. A major factor in the pollution of earth's environment is the amount of waste produced by the people that live on the planet. Americans alone produce about 4 trillion kilograms of wastes each year. In 1989, over a billion kilograms of toxic materials were released into the air by American industries. This amount is equal to about 4.5 kilograms of poison for every person in the United States.

As our human population grows, we require more and more living space, fuel, food, and clothing. Manufacturing, agriculture, mining, and transportation help to meet these needs; they also improve the quality of our lives. However, these processes use up certain resources while their waste products spoil the environment. See Figure 1-29. By practicing **conservation,** the rapid depletion of the earth's natural resources and the pollution of its environment could be slowed down. As members of a world-wide biological community, we must make informed decisions about pollution and the conservation of our resources.

Air Pollution Burning fuels and burning trash are the major causes of *air pollution*. Automobiles, factories, power plants, and other industries are the largest contributors to this pollution. Most coal and petroleum fuels contain sulfur, which when burned, combines with oxygen to put deadly sulfur dioxide gas into the air. Breathing air containing sulfur dioxide gas can cause health problems such as lung diseases. It also causes acid rain.

Water Pollution One of the most dangerous kinds of *water pollution* occurs when *industrial wastes* and untreated *sewage* are dumped directly into streams, lakes, and oceans. See Figure 1-30. Sewage can add bacteria and viruses to water supplies. It also carries household chemicals, such as phosphates from detergents, that cause algae and water plants to grow out of control. Another source of water pollution is the poisonous chemicals used to control pests on crops and lawns. These chemicals frequently wash off the land and into water supplies where they harm both aquatic organisms and the humans who eat these species.

Soil Pollution A third form of pollution is the buildup of harmful substances in the soil. Some *soil pollution* comes from pollutants that fall out of the air. Most, however, is caused by chemicals used to control pests and plant diseases. Poisonous chemicals are often buried in the ground in containers that are supposed to remain sealed. However, many of these chemicals, including nuclear wastes, have leaked into groundwater. Groundwater provides drinking water for more than half of the population of the United States.

Figure 1-30 Wastes from industry may be one of the major sources of water pollution.

The Consequences of Pollution

From a distance, the picture in Figure 1-31 looks perfect. The dark green of tall trees surrounds the deep blue of a crystal-clear lake. But everything is not perfect. There are few birds or frogs around, and the fish are almost gone. This lake, like thousands of others in eastern North America, is dying because its waters have become an unfit habitat for many of the organisms that once lived there. What caused this destruction? Acid rain, a consequence of industrial pollutants, has contaminated the lake.

Figure 1-31 Many lakes in the Adirondack Mountains of New York are suffering from acid rain.

Since the 1700s, over 60 animal species have disappeared from the United States alone. In those days, the extinction rate was about one species every fifty years—it is now one species every year! In addition, there are presently over 400 species in the United States that are **endangered**; the number worldwide is nearly 900 species, according to the U.S. Fish and Wildlife Service.

Why should you be concerned if organisms become extinct? After all, species become extinct naturally. However,

Endangered In danger of becoming extinct.

Figure 1-32 *There will never be another Tasmanian wolf (left). Once it was an important predator in its native Tasmania, an island south of Australia. The giant panda (right) is endangered and may soon follow the Tasmanian wolf into extinction.*

as the rate of extinction increases, the variety of life on earth decreases. See Figure 1-32. New species will develop, but the rate of that development is relatively slow. Every organism occupies a niche in a food chain. If we eliminate a species, could an important food chain collapse and threaten the food web of an entire ecosystem? Remember, humans are at the top of many food chains, so we may be endangering ourselves. It has been said that each time we eliminate a species, we lose a little of ourselves. After all, we are all a part of the same biosphere.

What Can We Do?

As humans, we have the knowledge and the ability to change the environment to meet our needs. Along with this ability comes a responsibility to all other life on the planet. What can humans do to make the future better for all forms of life on earth?

Figure 1-33 *Many states require the exhaust of motor vehicles to be tested for pollution levels. Those that give off too many pollutants must be repaired.*

Citizenship and Participation Individual and corporate efforts over the last 20 years have cleaned up many polluted ponds, lakes, and streams and improved the quality of the air. In addition, *environmentalists* helped pass many laws—at local, state, and national levels—that control the emission of all types of pollutants. See Figure 1-33. Environmentalists are not just scientists and government officials; they are people, just like you and your parents, who are concerned about issues such as pollution and its effects. They teach others, volunteer for clean-up projects, and participate in the law-making process.

One thing you can do is to learn about the causes and consequences of pollution in your own community. Tell your friends and family about what you find out and participate in whatever way you can. For example, picking up litter, such as plastic bags, may help to save the lives of local wildlife, which often die when they become entangled in or accidently eat the plastic. Practicing conservation by recycling and using less water saves community resources. What other ways can you help the environment in your area?

Solid Waste Management One of the most difficult problems facing humans today is the disposal of solid wastes. Some communities still dispose of these wastes, also called refuse or garbage, in *open dumps* where they are left on the ground. Open dumps are often homes to disease-carrying rats, insects, bacteria, and fungi. In addition, rain carries pollutants from these dumps directly into water supplies. Other communities get garbage out of their sight by loading it onto barges that carry it out to sea where it is dumped into the ocean. This practice harms ocean life and causes problems for marine communities. Trash has even been found on the most remote, uninhabited islands of the Pacific Ocean, carried there by waves and currents. These practices must be stopped.

Figure 1-34 The sanitary landfill shown here provides a less harmful way to dispose of solid wastes than an open dump.

Figure 1-35 Community recycling centers provide one way for citizens to recycle refuse such as aluminum cans and newspapers.

Many communities now dispose of solid waste in *sanitary landfills*, like the one shown in Figure 1-34. Others burn wastes in high-temperature furnaces that heat water for use in generating electricity. In landfills, solid wastes are covered with a layer of clay on all sides. The clay helps prevent rainwater from washing through the garbage and carrying pollutants into water supplies. It also keeps disease-carrying rodents and insects away. Many communities recycle materials such as metals, glass, and paper, then sell them to companies that use them to make new items. See Figure 1-35. This practice reduces the amount of garbage that a community must dispose of, and slows down the depletion of natural resources.

Figure 1-36 Using water foolishly is a waste of an important natural resource.

Urban Development The building of cities, towns, subdivisions, and factories.

Figure 1-37 Peregrine falcons are endangered North American predators. These chicks have been incubated and hatched in a breeding program.

Water Conservation The goal of water conservation is to preserve our limited supply of clean, fresh water. Humans use half of the earth's freshwater supply to irrigate agricultural plants. We use water to manufacture products and to generate electricity. In addition, an average person in the United States uses over 700 liters of water daily! What does a person do with so much water each day? Only about two liters is used for drinking. The rest is used for such things as flushing toilets, bathing, food preparation, watering lawns, washing cars, and filling swimming pools.

There are many ways that people can conserve water. One way is to use less water by reducing waste. See Figure 1-36. Leaking faucets and long showers are common ways that people waste water. Another way to conserve water is to protect watersheds where water either enters the ground or is captured by streams. Clearing forests and grasslands for farming and for **urban development** allows water to runoff instead of soaking into the ground. This not only reduces the supply of groundwater, but it causes flooding and degrades the quality of the water with pollutants and fine soil particles.

Wildlife Conservation The goal of wildlife conservation is to protect the wild animals and plants that live on the earth. See Figure 1-37. Setting aside natural habitats is one way to protect wildlife. Wildlife refuges, national forests and parks, and state forests and wilderness areas all provide natural habitats for many kinds of wildlife. Human access is limited in many of these areas in order to reduce the potential for pollution. Laws that permit hunting and fishing only at certain times of the year also protect wildlife. Such laws limit the number, and often the size and age, of animals that can be killed in order to keep those populations healthy. Good conservation practices reduce the effect of humans on wildlife. This allows ecosystems to continue to exist in a more natural way.

Summary

Humans require a variety of materials from the environment for their activities. Many of these activities cause pollution of the air, water, and soil that poisons both wildlife and humans. The result has been loss of habitat for wildlife and the extinction of many species. Some solutions to these problems include community involvement, proper waste management, and water and wildlife conservation.

Unit 2

Reading *Plus*

Now that you have been introduced to some of the many different kinds of organisms and how they are classified, let's take a closer look at this diversity and some ways that it developed. Read pages S24 to S44, and then pretend you are a modern-day Darwin and write an essay about evolution as it relates to humans. Your essay should include answers to the following questions.

1. Why are you classified as a mammal? Be specific and complete. (Remember, to be a mammal, you must first be an animal.)
2. How might humans change over the next few millennia by the process of natural selection?
3. In what other ways, besides selective breeding and resistance to pesticides and antibiotics, are humans affecting the evolution of living things?

In This Unit

2.1 The Diversity of Modern Life, p. S24

2.2 The Evolution of Diversity, p. S35

2-1 THE DIVERSITY OF MODERN LIFE

Did you know that the inside and outside of your body is occupied by many living things that are so small they can be seen only with powerful microscopes? Have you ever imagined that a forest of giants over 30 m tall could consist of algae instead of trees—or that you may someday live on the moon? Even though our own species has the size, shape, and intelligence necessary to send rockets into space, most living things go unnoticed. Already close to 3 million species have been identified and named, though some scientists estimate there may be as many as 30 million different kinds of organisms living on earth today! Let's look at some of the variety of earth's organisms. They are grouped into large categories, called *kingdoms*, which are based on some of the fundamental similarities they share. See Figure 2-1.

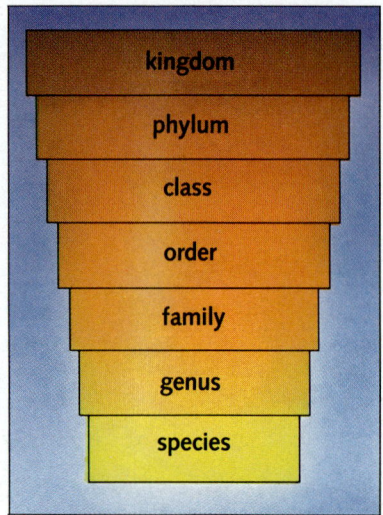

Figure 2-1 *Each of the five kingdoms contains subcategories by which all living organisms can be classified.*

The Monerans

The smallest and simplest life forms are placed in the Kingdom Monera. Monerans are also some of the most important organisms in your everyday life. They are different from all other living things because of their simple cell structure, which has no nucleus or cell organelles. All monerans are single-celled organisms that often live in chains or clusters of cells. Bacteria and blue-green bacteria are the two major groups (phyla) in this kingdom. See Figure 2-2.

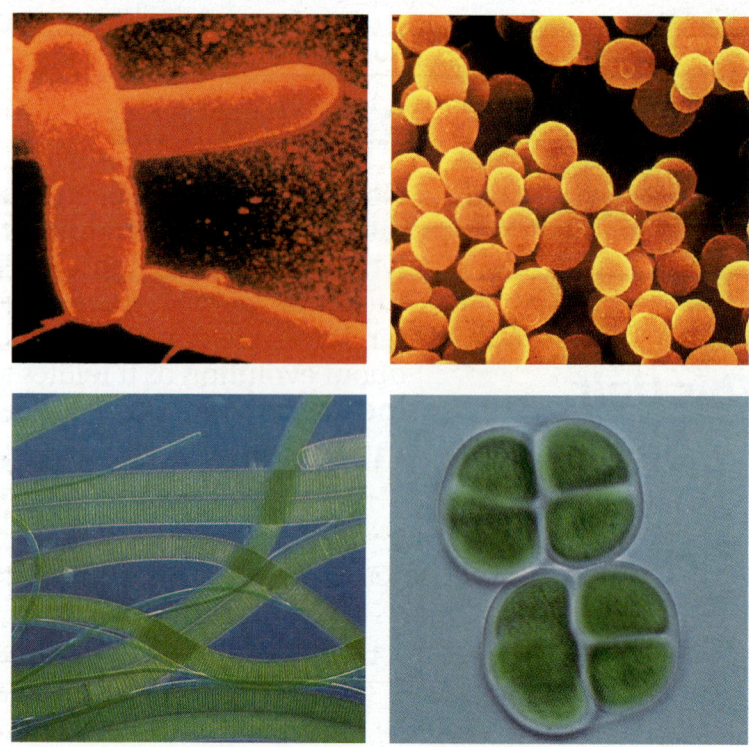

Figure 2-2 *Two examples of bacteria (top) and two examples of blue-green bacteria (bottom).*

Bacteria These monerans are single-celled organisms that absorb food molecules from their surroundings, such as the surface of your skin or the inside of your intestines. Their cells take three basic shapes—round (cocci), rod-shaped (bacilli), and spiral-shaped (spirilli). Many live in pairs, chains, or clusters. Bacteria live in almost every habitat on earth. You may already be familiar with the bacteria that live on or in other organisms and cause disease. However, most bacteria are beneficial by helping animals with digestion, by decaying dead organisms, or by putting nitrogen back into the soil.

Blue-Green Bacteria Blue-green bacteria contain chlorophyll and produce their own food by photosynthesis. They also have different shapes and usually attach to one another in long chains or clusters. Most blue-green bacteria live in fresh water and form the smelly, green scum that you may see on polluted ponds and streams. Some blue-green bacteria form the food-producing member of a lichen.

The Protists

The more advanced single-celled organisms are placed in the Kingdom Protista along with some simple, multicellular life forms. Unlike the monerans, however, protists have much larger and more complex cells that contain nuclei and cell organelles. Although some multicellular forms are quite large, their cells do not specialize into tissues or organs. The protists are divided into several phyla that share similar characteristics. Some of these phyla contain the animal-like protists we call *protozoans*. The others contain the plant-like protists we call *algae*.

Protozoans These protists are all single-celled organisms with no cell walls or chlorophyll. Protozoans *ingest* their food (take it in whole), as you do, and *digest* it inside their cells. Since they must find their food, most protozoans move about in one of three ways—by a flowing motion, such as that used by an amoeba, or by using special structures such as whip-like *flagella* or hair-like *cilia*. Some protozoans, however, have no means of self-locomotion. These protists, many of which cause diseases, simply attach to their host's cells. Protozoans are divided into four phyla based on the way they move. See Figure 2-3.

Algae The algae are protists that have cell walls and chlorophyll, and they make their own food by photosynthesis. Algae may be single-celled, or they may form long chains,

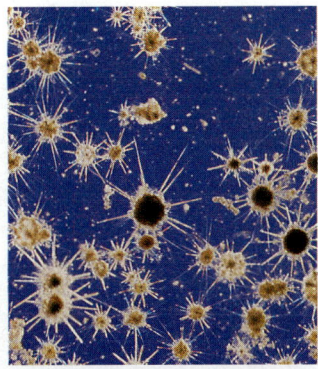
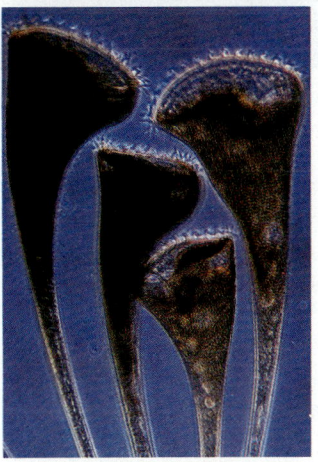
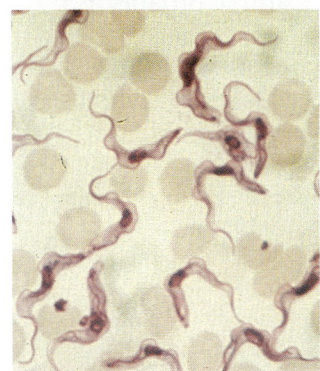

Figure 2-3 *Some examples of protozoans: radiolarians (top), Stentor (middle), and the protozoans responsible for African sleeping sickness (bottom).*

complex colonies, or large masses of unspecialized cells. In some of the largest algae, such as the 30-m giant kelp, the body shape is similar to a plant's. Most algae live in either salt water or fresh water; but, some may be living near you on moist, shady tree bark, on wet cement and rocks, or with a fungus as part of a lichen. Algae are divided into five major phyla based on their color and structure. A sixth phylum contains the *euglena,* which resembles both protozoans and green algae. See Figure 2-4.

Figure 2-4 Volvox (left) *is a colonial green alga. Kelp* (middle) *is an example of a marine brown alga. This example of a red alga* (right) *lives in very deep water.*

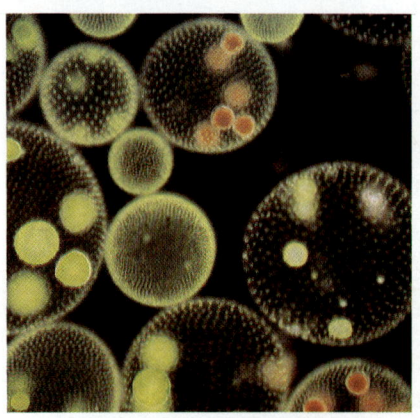  

The Fungi

Some make us sneeze, some live in our bathtubs, and many kinds spoil our food. This is one answer to the question: "What are fungi?". Most members of the Kingdom Fungi are multicellular organisms with a complex cell structure, cell walls, and no chlorophyll. You can find them growing both inside and on top of their food. Fungi eat by releasing digestive chemicals into their food and absorbing the digested food back into their bodies. The fungi are divided into phyla based on their body structure and how they produce **spores**.

Spore A special reproductive cell formed by fungi and other simple organisms.

Thread-like Fungi
Have you ever noticed a black or gray powdery substance growing on bread? These are the spore

Figure 2-5 *The common black bread mold.*

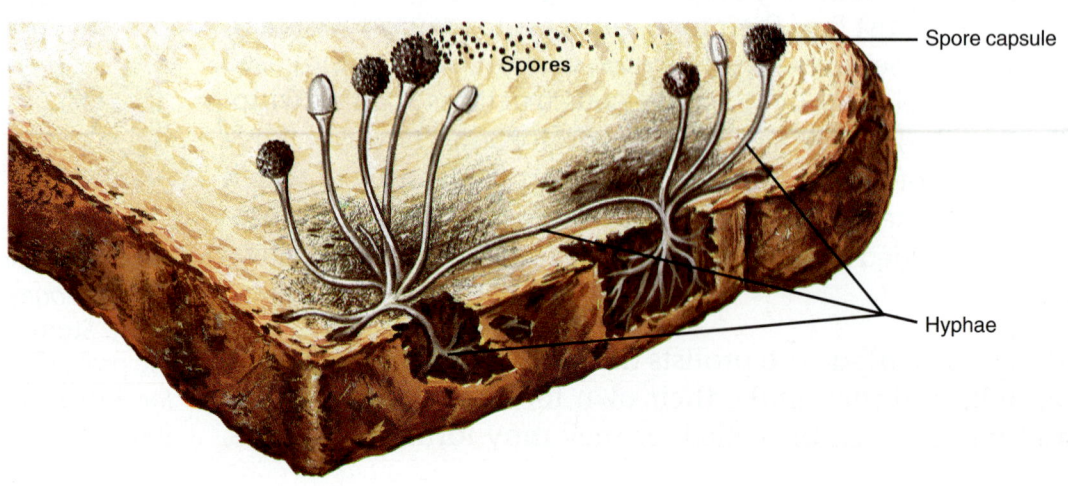

capsules of bread mold, an example of the thread-like fungi. If you look closely at the surface of the bread, you may be able to see the tiny, thread-like hairs or filaments called *hyphae* that make up the bodies of all fungi. Many other hyphae grow inside the bread as well. See Figure 2-5.

Club Fungi Mushrooms, puffballs, and bracket fungi are members of this group. These familiar structures are made up of closely packed hyphae that grow on the surface of a food source and come in a variety of shapes and colors. See Figure 2-6 *(left)*. In mushrooms, spores are found along the *gills* on the underside of the mushroom's cap. You may be able to see these tiny spores, which are often quite colorful, by first cutting off the cap of an opened mushroom. Then place it lower-side down on a white piece of paper and check it after a day or so. The pattern that may appear will be caused by the spores.

Sac Fungi Yeasts, mildews, and morels (a type of edible fungus that resembles a wrinkled mushroom) are members of this group. Many of these fungi are parasites that attack and cause diseases in several kinds of trees, such as elm and chestnut. Some yeasts even cause diseases in humans. The spores of these fungi are found inside tiny sacs or cups that form on the outside of the host. See Figure 2-6 *(right)*.

Figure 2-6 A bracket fungus (left) on a rotting log. The fruiting bodies of a sac fungus (right) resemble small cups.

The Plants
One of the two most advanced groups of organisms, the Plant Kingdom, includes multicellular organisms that have chlorophyll, complex cell structure, and cell walls. Most plants live on land, anchored in soil or attached to something else. Unlike the multicellular algae and the fungi, plant cells specialize to form many different types of tissues, organ structures, and systems. Both the largest and longest-living things on earth are plants. Plants are divided into two main groups based on whether or not they have water-conducting (vascular) tissue.

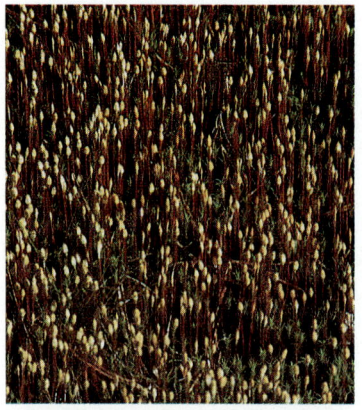

Figure 2-7 *Hair cap mosses (top) and a liverwort (bottom) called* Marcantia.

Figure 2-8 *The club moss* Lycopodium *(left), horsetails (middle), and ferns (right).*

Nonvascular Plants Have you ever noticed a bright green carpet covering a forest floor or growing on a rotten log? This carpet is made of plants that do not have water-conducting tissue. The nonvascular plants must live in moist environments, such as a shaded forest floor, and must also stay very small. The mosses and the liverworts, shown in Figure 2-7, are the two types of plants in this group. Notice that individual mosses and liverworts grow together in tightly packed mats. This helps them to conserve water and also to reproduce. The tiny capsules that appear at the top of a moss plant contain spores.

Vascular Plants Most plants have *vascular tissue* that conducts water, minerals, and food molecules throughout their bodies, which are composed of roots, stems, and leaves. Vascular tissue not only allows these plants to live in drier environments, but also serves as a support system for holding up their stems and leaves. Without vascular tissue, plants like the giant redwoods could not grow to be over 100 meters tall. Five phyla (or *divisions*, as botanists call them) of vascular plants are listed below. See Figure 2-8 for examples of each of the first three groups.

1. The *club mosses* are vascular plants that resemble the mosses. However, their stems are much sturdier due to their vascular tissue, and they grow larger than the mosses. Club mosses can also be found carpeting a forest floor.

2. The *horsetails* are another kind of vascular plant. They have hollow stems with many vertical ridges and joints. A whorl of needle-like leaves appears at each joint. Their stems also feel gritty due to the presence of silica, the material that makes sand. Because they are gritty and coarse, early settlers used them to scour pots and pans.

3. The *ferns* are probably a more familiar kind of vascular plant, since you may have ferns in your house or your garden. Ferns have distinctive leaves, called *fronds*, which consist of a strong central vein and many pairs of leaflets. Their stems remain underground. Contrary to what many people think, the horizontal rows of bumps that appear on the under side of fern fronds are not a disease or bug—they are structures that contain spores.

4. The *gymnosperms*, whose name means "naked seed," are one of the two types of *seed plants*. Seed plants are by far the most common and most successful kind of plants. Gymnosperms produce seeds on the dry scales of cones. Most gymnosperms are *conifers* (cone-bearing plants), such as pine, redwood, juniper, fir, spruce, hemlock, and

Figure 2-9 Cycads (left), ginkos (middle), *and the* larch (right), *are all examples of gymnosperms.*

cypress trees. All have small needle-like leaves and most stay green all year. Some conifers are believed to be over 5000 years old. Other examples of gymnosperms include cycads and ginkgo trees. See Figure 2-9.

5. The *angiosperms*, whose name means "covered seed," are also seed plants. Usually called *flowering plants*, angiosperms are the largest, most diverse group of modern plants. Their seeds are produced inside of a special structure, called an *ovary*, which is part of a flower. See Figure 2-10. An ovary ripens into a **fruit** that surrounds and protects maturing seeds. Fruits also help to *disperse* seeds (spread them around) once they are mature. You have eaten the fruit of many angiosperms including apples, oranges, peaches, tomatoes, eggplant, squash, cucumbers, grains, peanuts, walnuts, almonds, and coconuts. Although their flowers are not as obvious or showy, plants such as grasses, palms, all crop plants, and fruit, nut, and shade trees are flowering plants, too.

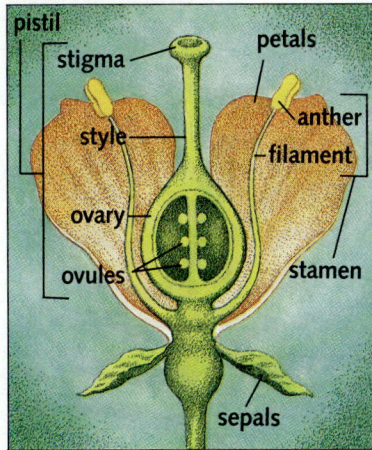

Figure 2-10 The structure of a typical flower showing the ovary and ovules, which will become seeds.

Fruit A ripened ovary that contains the seeds of a flowering plant.

The Animals

You are a member of the Animal Kingdom. How can you tell? You are not a moneran, because you are made of many cells and most of your cells have a nucleus. You are not a protist, because you are multicellular and have complex tissues and organ systems. You are not a plant, because you do not have cell walls and you cannot make your own food. You are not a fungus, because you digest your food inside of your body with a complex digestive system. Therefore, since there is only one kingdom left, you must be an animal.

Animals are divided into nine major phyla and several minor phyla. Eight of these groups contain soft-bodied organisms that are referred to as *invertebrates*. Humans belong to the ninth group of animals, which includes fish, amphibians, reptiles, birds, and mammals—the *vertebrates*.

Invertebrates Invertebrates are soft-bodied animals that have no backbones for internal support. Instead, their bodies stay relatively small, have an outside covering for support, or are supported by the water in which they live. The members of each phylum share characteristic body plans. As you read through the list of invertebrates below, notice how each group is more complex than the previous one. Figures 2-11 to 2-14 show examples of these animal groups.

Figure 2-11 Two simple invertebrates: sponges, (top) and a coelenterate (hydra) (bottom).

1. *Sponges* are the simplest of the animal groups. Adult sponges live in water, attached to some object, and are composed of two layers of body cells. These body cells cling to a network of tiny spikes or fibers that surrounds a hollow central cavity. The walls of a sponge have many tiny pores through which water and tiny bits of food are drawn into the central cavity. Water and wastes leave through a large opening on top.

2. *Coelenterates* are simple animals composed of two specialized cell layers (tissues) that are separated by a jelly-like substance. Hydras, jellyfish, sea anemones, and corals are all types of coelenterates. All live in water and have a hollow, sac-like body with a single opening. This opening, through which food enters and wastes are expelled, is surrounded by tentacles that are lined with stinging cells for capturing and paralyzing live food.

3. *Flatworms* have a flattened body shape, with one body opening, a body cavity with many lobes, and a simple nervous system. Most flatworms are parasites that live

inside of other animals. Blood flukes, liver flukes, and tapeworms are some flatworm parasites of humans. Turbellarians are free-living flatworms, the majority of which live in marine environments. Freshwater turbellarians—the planarians—are distinctive because they have two *eyespots* that are used to detect light.

Figure 2-12 *The tapered body shape of this turbellarian is an adaptation that allows it to move freely and to seek its own food.*

4. *Roundworms* have a rounded body shape, a tube in which food is digested, and two body openings—a mouth for taking in food and another for expelling wastes. Most roundworms are free-living, aquatic organisms that are easily visible in pond water. Many others live in moist soil. Some roundworms are parasites of animals or plants, for example ascaris worms, which live in the intestines of humans and other mammals, and root-knot nematodes, which damage the roots of plants.

5. *Segmented worms* have rounded bodies with two body openings, but their bodies are divided into a series of segments. They have several well-developed body systems that include organs such as a simple "brain" that connects to a nerve cord and a set of five primitive "hearts." The earthworm is a common example of the segmented worms, which also include leeches and marine tube worms.

Figure 2-13 *A roundworm (left) and a segmented worm, the familiar earthworm (right).*

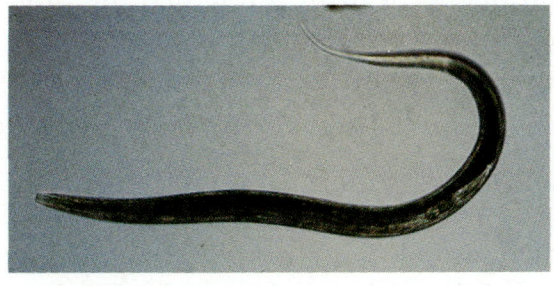 

6. *Mollusks* are soft-bodied animals with well-developed organ systems. Most mollusks make hard, calcium-containing shells that completely cover the body for

Figure 2-14 *Higher-order invertebrates* (left to right): *a mollusk (octopus), an arthropod (crab), and an echinoderm (starfish).*

protection. Clams and oysters are covered by a two-part, external shell, while snails secrete a single, spiralling shell. Octopuses and squids, which are mollusks that have tentacles, do not make a shell for protection. Instead, they swim very fast and shoot out an inky substance to confuse their predators.

7. *Arthropods* are by far the largest group of animals. These organisms are characterized by two, three, or many body segments and numerous jointed *appendages*. They also have a hardened **exoskeleton** for protecting their many, well developed organ systems. Arthropods live successfully on dry land and in water, and many have a complex life cycle. Among the arthropods are lobsters, crabs, shrimp, insects, spiders, centipedes, and millipedes.

Exoskeleton A hard outer covering for protection that is typical of arthropods.

8. *Echinoderms*, also known as the "spiny-skinned" animals, include starfish, sea urchins, sand dollars, and sea cucumbers. Echinoderms live in the ocean and have bodies protected by a spiny skeleton that lies under the skin. All echinoderms move using special structures that resemble suction cups called *tube feet*. Many echinoderms have several flexible body branches, or arms, on which the tube feet are found in rows on the underside. They are famous for their ability to replace missing arms by a process called **regeneration**.

Regeneration The ability of some organisms to grow new body parts.

Endoskeleton An internal system that functions for support of the body in vertebrates.

Vertebrates The vertebrates are animals that have a backbone. This backbone is part of an internal skeleton, called an **endoskeleton**, which provides support for the body and aids in body movements. All vertebrates have a body structure that includes a head, a neck, and a body with appendages. They also have the most highly-specialized organ systems in the animal kingdom. Seven kinds of vertebrates are included in the last major phylum of the Kingdom Animalia. They are:

1. *Jawless Fishes.* Represented by the sea lamprey in Figure 2-15, these vertebrates have a jawless mouth adapted for sucking body fluids from other fishes, an elongated, snake-like body with no appendages, gills for obtaining oxygen, and flexible skeletons of *cartilage*. Like all fishes, they are **cold-blooded** and live only in water.

2. *Cartilaginous Fishes.* The shark in Figure 2-15 represents this group of vertebrates, which have skeletons made of cartilage, two pairs of fleshy fins, gills, and strong jaws with many rows of teeth for tearing and eating flesh. Stingrays and skates are other members of this group, which, like jawless fishes, lives only in the ocean.

Cold-blooded Having a body temperature that changes according to the temperature of the environment.

Figure 2-15 The sea lamprey (left) is a jawless fish, a shark (middle) is a cartilaginous fish, and the angel fish (right) is a bony fish.

3. *Bony Fishes.* Catfish, trout, goldfish, flounder, and eels are examples of bony fishes. These fishes have gills, and they have skeletons made of bone. Most have a streamlined body that is tapered at both ends, and two pairs of fan-like fins. The fins and the body shape of the bony fishes allow them to move easily through the water. They are the most numerous, varied, and successful kind of fishes.

4. *Amphibians.* Frogs, toads, and salamanders are examples of amphibians. The name *amphibian* means "double life." Most amphibians live part of their lives in water and part on land. They have smooth skin that must be kept moist, gills when they are young and lungs when they mature, and two pairs of legs as adults so they can move on land. Like the fishes, their eggs are laid in water, and all forms are cold-blooded.

5. *Reptiles.* Turtles, snakes, lizards, crocodiles, and alligators are the living members of this group, which was once dominated by dinosaurs. The reptiles are truly adapted to life on land. For this purpose, they are covered by hard

plates or scales that prevent water loss by evaporation. Reptiles also have lungs; two pairs of strong legs with clawed toes for digging, climbing, and moving on land; and they lay eggs that are covered with a tough, leathery shell that prevents them from drying out. However, they are still cold-blooded.

Figure 2-16 *Amphibians (left) and a reptile (right).*

6. *Birds.* Birds are recognized by a body covering of feathers and a pair of wings, which gives most of them the ability to fly. In addition, birds lay eggs covered by a hard shell, have a beak for a mouth, scaly legs, and feet with clawed toes. Birds also have light, hollow bones, and enlarged lungs as an adaption for flying. Unlike the fish, amphibians, and reptiles, birds are **warm-blooded**. This allows birds to occupy many different habitats on earth.

7. *Mammals.* Mammals are the most complex kind of vertebrates. They feed their young with milk from mammary glands, have hair somewhere on their bodies, breathe with lungs, and are warm-blooded. While one category of mammals lays eggs, almost all give birth to live young. Mammals are also known for their very advanced nervous system, which includes keen senses and a highly developed brain. Like the birds, mammals also live on land, in the air, and in water.

Warm-blooded Having a body temperature that remains constant despite temperature changes in the environment.

Figure 2-17 *Warm-blooded vertebrates include birds (left) and mammals (right).*

Summary

According to the current biological classification system, the diversity of life on earth today can be separated into five major groups called kingdoms. The members of each kingdom have a few common characteristics. The organisms of each kingdom are further divided into smaller groups that have even more characteristics in common.

2-2 THE EVOLUTION OF DIVERSITY

Imagine going on a long trip and arriving in a strange new place where even the animals and plants look strange. You find giant tortoises grazing like cattle, seafaring lizards, and unusual, flightless birds. Yet, other life forms look much like organisms with which you are familiar. This was the experience of Charles Darwin when he visited the Galapagos Islands near South America. See Figure 2-18. Darwin's voyage inspired him to develop the theory that life has evolved by a process called *natural selection*.

Since Darwin's time, scientists have gathered much information from **fossils** about the life forms that once existed. From this information, they have developed explanations as to how life changes. They have even performed experiments that show how life could have originated from substances that occur naturally on earth. Today the concept that all life has evolved from a few simple forms is a fundamental theme in biology.

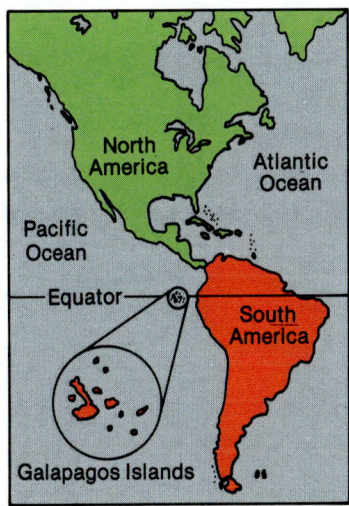

Figure 2-18 The location of the Galapagos Islands.

Fossil The remains or imprint of an organism that once lived.

A History of Life on Earth

If you could travel back to the time before history began, what kinds of organisms would you see? Depending on the time of your arrival, you might see horses the size of dogs, flying reptiles, forests of giant ferns, and even giant insects. Biologists tell us that all the species alive today make up only one percent of all the life forms that ever lived—99 percent are extinct! Some are preserved in the rocks of the crust as fossils. To get the most out of your trip you would need to move quickly, since the earth's history covers an unimaginably long period of time.

Earth's Calendar
During our lives, we mark the passing of each year on a calendar. The study of human history, however, is broken into longer units of time, such as decades, centuries, and *millennia*. To us, a millennium, or 1000 years, is a very long period of time, since very few people live more

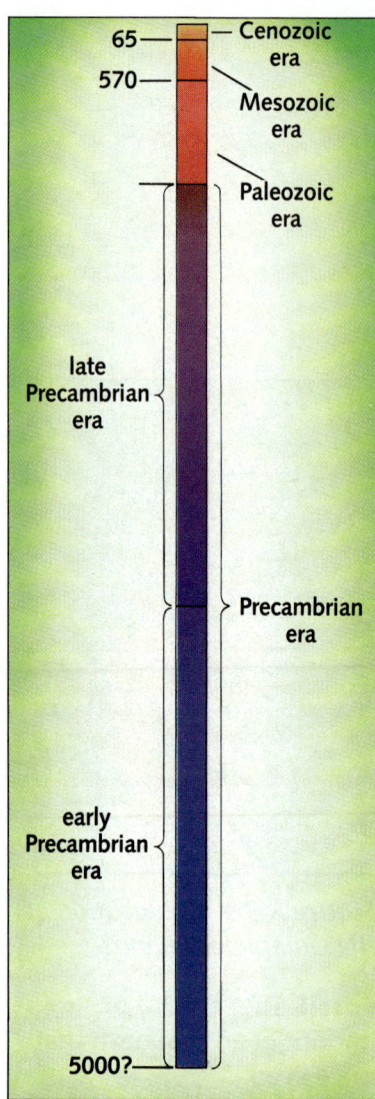

Figure 2-19 *This diagram shows the relative lengths of the geologic eras. The Precambrian era is by far the longest.*

Figure 2-20 *Stromatolites, fossil structures formed by mats of algae, have been found in rocks that are over 3.5 billion years old.*

than a century. The earth, however, is estimated to be over 4.5 billion years old, which is more than forty-five million (45,000,000) centuries of time! How much time is this? If we could squeeze the entire history of the earth into one calendar year, we would be living in the last split-second of December 31 of that year!

Earth's history has been divided into several time periods called *eras*, as shown in Figure 2-19. Each era is further divided into *periods* and then *epochs*. Eras are named for the kinds of life forms that were abundant during that time. Each one ends with a major change in the life forms that existed.

We live in the *Cenozoic Era* (recent life), which began about 65 million years ago. The era before that is called the *Mesozoic Era* (middle life), which lasted about 160 million years. The *Paleozoic Era* (ancient life) came before the Mesozoic and lasted about 345 million years. An abundance of fossils is found in the rocks that formed during these three eras. However, they cover only 13 percent of the earth's history.

The remaining 87 percent of earth's history is often called the *Precambrian Era*. This name comes from the Cambrian Period, which is the oldest period of the Paleozoic Era. The life forms of this era were the first to inhabit the earth. Unfortunately, there are not many fossils of these organisms because most were single-celled or had soft bodies. These kinds of organisms do not make good fossils. However, microscopic examination of fossils, such as the one in Figure 2-20, finally revealed the presence of tiny organisms that lived over 3.5 billion years ago.

How Life Has Changed The earth's history, as recorded in its rocks, reveals that the diversity of life today has evolved from simple to complex forms. This gradual progression is illustrated in Table 2-1. As you study the table, think back to the overview of the five kingdoms that you have just read.

Evolution by Natural Selection

Charles Darwin became convinced that life had changed over time while serving as ship's naturalist on the HMS *Beagle*, a British survey and mapping ship. His job on the ship was to study and collect the different kinds of plants and animals he found in the many places where the ship stopped. See Figure 2-21. In addition, he collected many fossils and made notes of everything he saw. During the long, five-year voyage, Darwin carefully studied these notes and collections.

Figure 2-21 *The marine iguana is one species that Darwin found living on the Galapagos Islands.*

Developing a Theory When he returned home, Darwin began to classify his collections and wondered how organisms could change over millions of years. He knew that farmers selected and bred animals and plants with certain traits, but he did not know what could cause such selection in nature. Then, Darwin happened to read an essay by the noted clergyman Thomas Malthus. The essay discussed Malthus' observations that the number of people in a population gets larger when there is enough food and other necessities, and gets

Era	Period	Million Years Ago	Organisms
Cenozoic	Quaternary	0.025	Complex human societies develop.
		2	Many extinctions occur.
	Tertiary	5	Modern mammals, birds, and sea life appear; early hominids appear.
		24	Grasslands spread.
		37	Primitive apes, elephants, horses, and camels develop. Forests of gymnosperms and angiosperms spread.
		58	Mammals spread rapidly. Fruit-bearing trees become common.
		65	Primates appear.
Mesozoic	Cretaceous	144	Flowering plants and trees appear. Dinosaurs are extinct by the end.
	Jurassic	208	First feathered birds and large dinosaurs appear.
	Triassic	225	Dinosaurs and first mammals appear.
Paleozoic	Permian	286	Conifers and modern insects appear. Many invertebrates become extinct.
	Carboniferous	360	First reptiles appear. Amphibians are common. Coral reefs form. Forests of large ferns develop.
	Devonian	408	Amphibians and insects appear. Gymnosperms appear.
	Silurian	438	Primitive vascular plants, fish, and shell-forming sea animals appear.
	Ordovician	505	First vertebrates appear in the sea. First plants appear on land.
	Cambrian	570	Clams and snails appear. Algae are common.
Precambrian		4,500	Marine invertebrates appear. Bacteria, blue-green bacteria, and algae predominate.

Table 2-1

Figure 2-22 What might the earth be like if human population were not regulated?

Natural Selection The process by which only the organisms that are best suited to their environment survive, passing their traits on to their offspring.

smaller when there is not. Malthus predicted that people would overrun the earth if starvation, disease, crime, and war did not keep their numbers down. See Figure 2-22.

Darwin realized that *all* organisms must compete for food, water, and other necessities in order to survive in what the essay had called a *struggle for existence*. This idea became the key to Darwin's theory of evolution by **natural selection**. He reasoned that the organisms that are better able to compete are the ones most likely to survive and leave more offspring. Those that are not, are less likely to live long enough to produce offspring.

Darwin worked on his theory for twenty years. But, before he finished, he received an essay from a man named Alfred Russel Wallace that stated the main points of the theory that Darwin had worked on for so long. Wallace was also a naturalist and had worked in the jungles of Indonesia. He and Darwin had been writing to each other about their research and their ideas. Darwin quickly finished a paper on his theory and, in 1858, both papers were presented jointly at a scientific meeting in London. The next year, Darwin published his famous book, *On the Origin of Species by Means of Natural Selection*.

Parts of the Theory There are five main parts to the theory of evolution by natural selection. The parts presented here are a bit different from the way Darwin stated them. Our knowledge of genes explains how variations can be transmitted—something Darwin did not know.

1. *Variation.* All members of the same species are somewhat different from one another, or in other words, individuals are not exactly alike in all of their traits. See Figure 2-23. Some differences, like color, may be very noticeable; but many less noticeable traits, such as size, weight, and running speed, can be different as well. These differences

Figure 2-23 Two of these Bengal tigers have a mutation that causes albinism. Would this mutation be helpful or harmful to these animals in the wild?

are called *variations*. Darwin believed that variations were caused by some agent that could be passed from parent to offspring, but he did not know what that agent was. Today, we know that variations result from sections of DNA called *genes*. New variations appear when genes change slightly by the process of **mutation**.

Mutation A change in the DNA of a gene.

2. *Overproduction of Offspring.* Each species produces many more offspring than can survive and reproduce. For example, female fishes lay enormous numbers of eggs. If all of these eggs hatched and the young survived, the waters of the earth would quickly be overrun with fishes. Plants also produce large numbers of seeds, but not all of them grow into adult plants.

3. *Struggle for Existence.* The overproduction of offspring leads to a struggle for survival, since all organisms must compete for a limited amount of food, water, and living space. During this struggle, some individuals will get enough of these necessities to survive and reproduce, and some will not.

4. *Natural Selection.* Individuals with certain variations compete more successfully than individuals that lack those variations. For example, in a jungle, which of the tigers in Figure 2-23 would be more likely to stalk its prey undetected? In other words, which tiger would be the most successful hunter? Would the same be true if the tigers lived on the tundra? As you can see, the environment determines which variations are successful. Individuals with traits that make them successful in their environments have the best chance to survive and reproduce. Offspring that inherit successful variations also have a survival advantage, as long as the environment remains the same. Thus in a population, the percentage of individuals with successful variations increases over time.

Adaptation Due to variation and natural selection, populations and species may *adapt* (change to become better suited) to new conditions in their environments. For example, recall what happened to the peppered moths in England during the Industrial Revolution. Where tree trunks were darkened by pollution, populations of mostly light-colored moths were gradually replaced with ones dominated by dark-colored moths. Because of the variation for dark color, this species was able to adapt to and survive a change in its environment. As many slight, yet beneficial changes accumulate over millions of years, significant changes in structure

Figure 2-24 How did color affect the survival of the peppered moth?

and function can occur. See Figure 2-25. In this way, species that are adapted to particular environments develop.

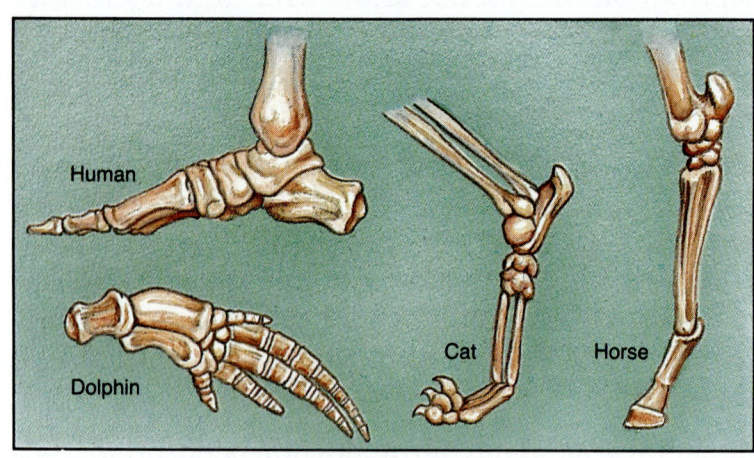

Figure 2-25 *The shape of a mammal's foot shows it has adapted to the way it moves.*

Diversity Directed by Humans

If you could put together the perfect pet, what would it be like? If you could design the perfect vegetable how would it taste? Believe it or not, people have been able to do just that by a process of *artificial selection*.

Selective Breeding The process of making hereditary improvements in animals and plants.

Selective Breeding Humans first began to domesticate plants and animals around 9000 B.C., or about 11,000 years ago. At that time, people began to collect seeds of wild grasses to grow them for food, and tamed certain wild animals, such as cattle, for doing work. Gradually, they started to alter these wild varieties by selecting and raising the offspring of individuals with the most desired traits. Improving plants and animals by choosing and reproducing individuals with certain desirable traits is called **selective breeding**.

For example, by selecting and growing wild wheat plants with the largest seeds, domesticated varieties that always produced larger seeds were eventually developed and thus provided greater amounts of food. See Figure 2-26. These days you must look very closely to see the similarities between wild wheat and its domesticated cousins.

Cattle are another example of selective breeding at work. Individual bulls and cows are selected for qualities, such as strength, size, and meat and milk production, and bred in a controlled manner. Calves that have the most desirable combinations of traits are selected to eventually parent the next generation. There are more than forty different cattle breeds today. In fact, most domestic animals—from cats and dogs to horses, sheep, and pigs, as well as most cultivated varieties of ornamental and food plants—are the result of thousands of years of selective breeding by humans.

Figure 2-26 *This wheat (top), as well as the variety of fruits and vegetables (bottom), was developed from wild strains through the process of selective breeding.*

Resistance Humans have also inadvertently altered many species of insects and disease-causing bacteria by trying to control them with chemicals. Now, many insects and bacteria are *resistant* to these chemicals. When we spray or dust insect pests with chemicals called *pesticides*, most of the insects are killed. However, due to genetic diversity, a few individuals survive the treatment because they have genes that make them resistant to the particular chemicals that were used. These individuals survive the treatment and go on to reproduce, passing this resistance on to their offspring.

In the case of disease-causing bacteria, some have become resistant to certain antibiotics. Bacteria that are resistant to antibiotics survive to produce a resistant strain of bacteria. One way to solve this problem is to alternate treatment with different drugs instead of using the same one over and over again. Resistance keeps scientists busy trying to produce new pesticides and antibiotics to control insects and bacterial diseases.

Patterns of Diversity

Have you ever noticed similarities between members of the same family? What did you notice? How would you describe the similarities? You might describe similar facial features, such as eye color and nose shape. You probably did not describe similar features, such as backbones, two arms, two legs, and the ability to walk upright. Of course these features are common among family members, but they also indicate how all people on earth are related.

Being related means sharing an *ancestor*. For example, your parents are your ancestors. You may know or have pictures of some of your other recent ancestors, such as your grandparents, great grandparents, or great great grandparents. Do you notice any similarities between these ancestors and yourself? See Figure 2-27. You also have other ancestors—ones through which you are related to other

Figure 2-27 Different generations of a family may have some similar features, but new gene combinations make every family member unique.

mammals, vertebrates, animals, and even protists. When two organisms share the same ancestor, we call this ancestor a *common ancestor*.

Determining common ancestry is the key to building modern classification systems. It also helps biologists work out the pathways of evolution. Common ancestry is determined by studying and comparing such things as homologous structures, embryo development, and biochemical molecules. Let's take a look at what these things mean.

Homologous Structures These structures are body parts that have the same basic construction. Let's consider the limbs of vertebrates as an example. Although they vary in size, shape, and purpose, they are all made in the same way. See Figure 2-28. Now, think about how the structure and shape of each of these limbs is helpful to its function. Can

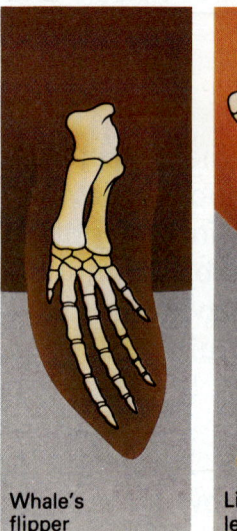
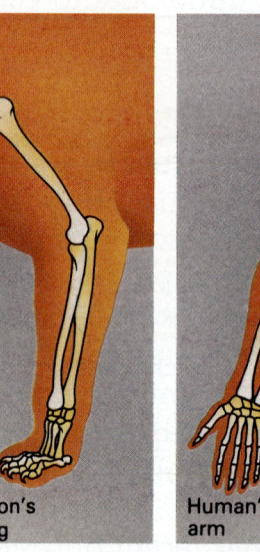
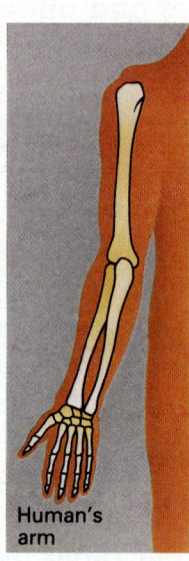
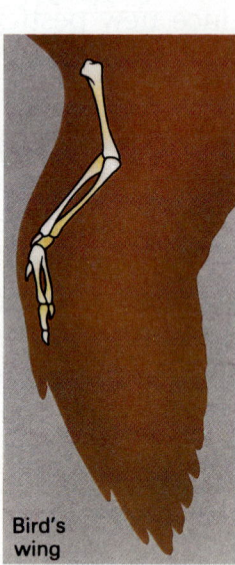

Whale's flipper | Lion's leg | Human's arm | Bird's wing

Figure 2-28 Vertebrate limbs are homologous structures. The flipper of a whale, leg of a lion, arm of a human, and wing of a bird differ in size and shape, but are alike in the arrangement of the bones.

you see adaptation at work? Biologists recognize that these similar structures indicate that these animals have a common ancestor. The variations that led to these adaptations were inherited by each generation from the one before.

Embryo Development Relationships among animals can also be seen by comparing their development before birth. Most multicellular organisms, such as ourselves, begin as a single cell called an *egg*. When the egg is fertilized, it begins to divide rapidly, forming a mass of cells and then an *embryo*. Embryos change shape gradually as they develop. Eventually, they take the basic shape of their parents. Figure 2-29 compares the development of the embryos of three vertebrates. Notice that birds and mammals appear to be more similar (closely related) to one another than to the fish.

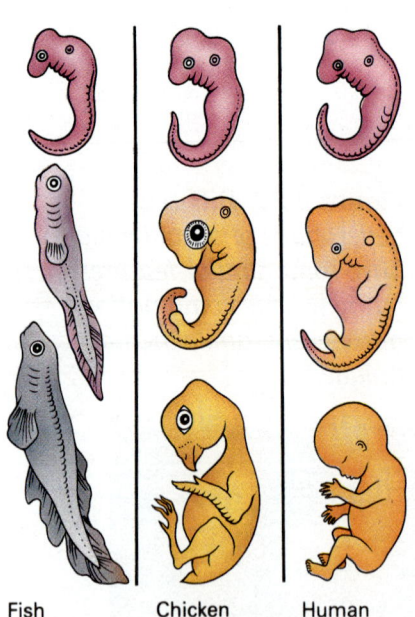

Fish Chicken Human

Figure 2-29 Similarities in the development of fish, chicken, and human embryos suggest an evolutionary relationship.

Evolutionarily speaking, birds and mammals have a more recent common ancestor than do fish and birds, or fish and mammals.

Biochemical Similarities Homologous structures and embryo development were used by Darwin and other early biologists to trace the evolution of life and to classify it. However, these methods are too general to establish the closest relationships. Today, these relationships are determined by looking at the biological molecules that make up organisms. The more closely related the organisms are, the more similar these molecules will be. See Figure 2-30. The most important molecule to look at is DNA, since this molecule has the instructions for making all of the others. These instructions are written in a code that scientists have learned to read. Having DNA at all is the most important similarity among the organisms on earth. The fact that every living thing on earth has DNA is the most convincing evidence that all life on earth may have a common ancestor.

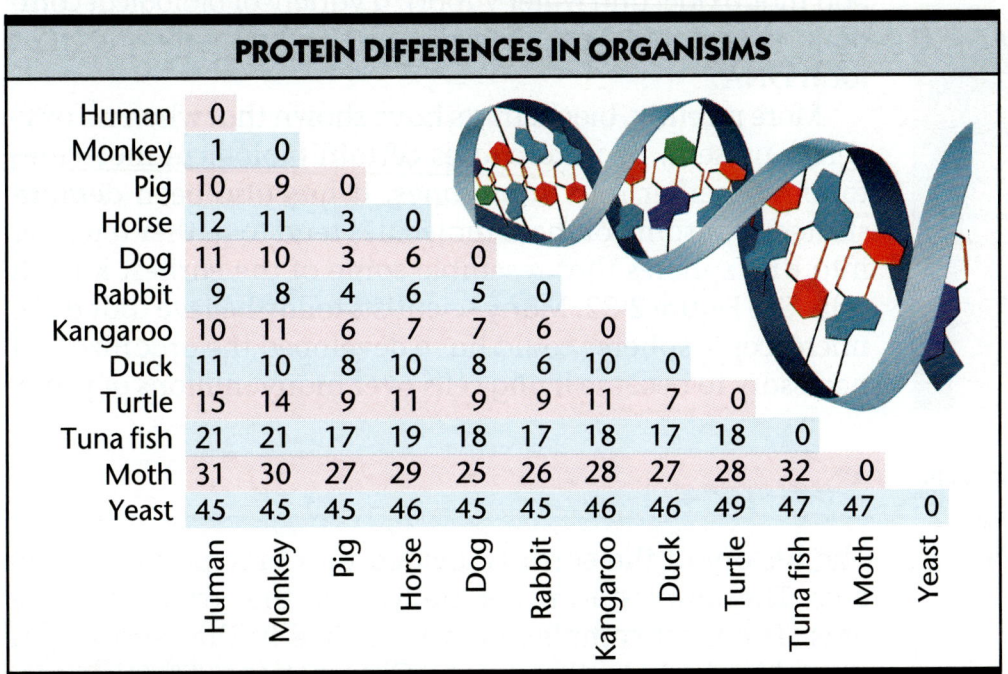

Figure 2-30 This diagram compares differences between a certain DNA-produced protein obtained from different species of organisms.

PROTEIN DIFFERENCES IN ORGANISIMS

	Human	Monkey	Pig	Horse	Dog	Rabbit	Kangaroo	Duck	Turtle	Tuna fish	Moth	Yeast
Human	0											
Monkey	1	0										
Pig	10	9	0									
Horse	12	11	3	0								
Dog	11	10	3	6	0							
Rabbit	9	8	4	6	5	0						
Kangaroo	10	11	6	7	7	6	0					
Duck	11	10	8	10	8	6	10	0				
Turtle	15	14	9	11	9	9	11	7	0			
Tuna fish	21	21	17	19	18	17	18	17	18	0		
Moth	31	30	27	29	25	26	28	27	28	32	0	
Yeast	45	45	45	46	45	45	46	46	49	47	47	0

The Origin of Life

In studying how life could have evolved from the first single-celled forms, it occurred to some scientists that life itself may have developed naturally on the earth. This could not have happened if the earth was the same as it is now. But about 4 billion years ago, conditions on earth were very different. At that time, the earth's surface was covered by erupting volcanoes and shallow, hot seas. There was no oxygen in the

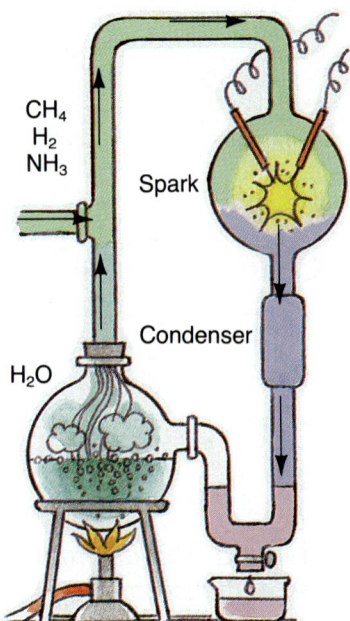

Figure 2-31 An apparatus like the one shown here was used to test Oparin's theory.

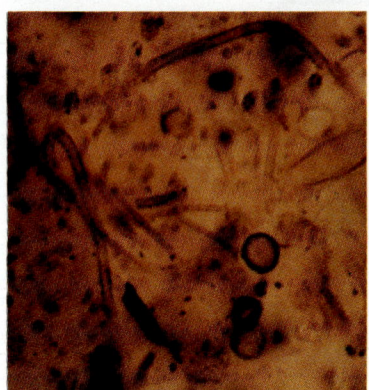

Figure 2-32 These microfossils of spheroid cells and filaments are among the oldest known fossils.

atmosphere, and the daily weather forecast would have included torrential rain and violent electrical storms.

In 1923 a Russian biochemist named Alexander Oparin announced his theory that life could have begun when organic molecules formed in an early atmosphere of extremely hot gases, such as hydrogen, ammonia, methane, and water vapor. These molecules, he said, collected in hot, shallow seas and gradually came together to form the major kinds of biological molecules. After millions of years, the first cells appeared. No one knows exactly how these molecules might have come together to form cells, but the possibility has been demonstrated by several experiments.

In 1953 two American scientists used an apparatus like that shown in Figure 2-31, to heat a mixture of the gases Oparin had suggested with an electrical spark. The experiment produced only two of the molecules used to make proteins. However, in other experiments where the gas mixture was changed to a more likely combination for the earth's early atmosphere (hydrogen, nitrogen, carbon dioxide, carbon monoxide, and water vapor), a variety of biological compounds were formed, including some of the compounds that form DNA.

More recently, biochemists have shown that when no oxygen is present, heating causes certain biological molecules to join together into larger ones. It has also been demonstrated that some of these molecules tend to gather together into tiny spheres that resemble some of the earliest known cells. See Figure 2-32. Many scientists today believe that these microscopic spheres could have developed the organization necessary to become living cells over many millions of years.

Summary

The history of the earth is divided into time periods called eras. The simplest forms of life lived in the earliest of these eras. The most complex organisms live in the most recent era, which includes the present. Observations of this increasing diversity led to the development of the theory of evolution by natural selection. Humans have also been able to change living things by processes such as selective breeding. Evidence of common ancestry such as homologous structures, similar stages of embryo development, and biochemical similarities are used to classify organisms by their evolutionary relationships. Experiments have shown that it is possible that life could have developed naturally on the earth about 4 billion years ago.

Unit 3

In This Unit

3.1 Solutions, Suspensions, and Colloids p. S46

3.2 Acids, Bases, and Salts p. S51

Reading *Plus*

Now that you have been introduced to solutions and non-solutions, let's take a closer look at these mixtures and some of the special properties they display. Read pages S46 to S60 and prepare a narrative of how you encounter various mixtures in your daily life. Your story should include examples related to the following questions.

1. On what are the differences in mixtures based? And how can different types of mixtures be identified?
2. How can the different properties of solutions, suspensions, and colloids be applied to the manufacture of useful products?
3. In what way are acids, bases, and salts chemically related? And what electric properties do they share?

3.1 SOLUTIONS, SUSPENSIONS, AND COLLOIDS

As you know, mixtures are combinations of several substances, each of which keeps its own characteristics. The substances in a mixture can be separated by physical means. You have just learned that a solution is a mixture of two substances, one of which is dissolved in the other. In a solution, one substance "disappears" into the other, and it looks like a single substance. In other mixtures, the non-solutions, particles do not dissolve but float throughout the mixture without dissolving.

Solutions

Solutions can actually contain more than one dissolved substance. A soft drink, for example, is a mixture of water, flavorings, and carbon dioxide gas. However, before a bottle of soda is opened, the soft drink looks like a single substance. Such a mixture is called a **solution**. Every solution has two parts: the *solvent* and the *solute*. As you know, a solvent is the material in which the solute is dissolved. There is always more solvent than any other component in the solution. In a soft drink, water is the solvent. The gas and flavorings in the soft drink are the solutes. It is the gas bubbling out that gives soda its fizz. See Figure 3-1.

Solution A mixture that is formed when the particles of one substance fill the spaces between the particles of another substance, and the particles of both are evenly distributed.

Figure 3-1 The gas in a soft drink bubbles out when the bottle is opened and the pressure is released.

Although many common solutions are made from solutes added to liquids, other solutions are combinations of liquids, solids, and gases. Table 3-1 lists some examples of these types of mixtures.

Types of Solutions In order to describe a solution, you must know the solvent and solute used. For example, a salt solution describes a mixture made by dissolving salt in water.

TYPES OF SOLUTIONS		
Example	Solute	Solvent
air	gas	gas
soda water	gas	liquid
hydrogen in platinum	gas	solid
water vapor in air	liquid	gas
alcohol in water	liquid	liquid
silver amalgam	liquid	solid
sulfur vapor in air	solid	gas
sugar in water	solid	liquid
brass	solid	solid

Table 3-1

You must also know the relative amounts of the materials in the solution. If a small amount of solute is dissolved in a large amount of solvent, the solution is said to be **dilute**. A spoonful of salt in a liter of water makes a dilute salt solution. On the other hand, a large amount of solute in a liter of water makes a **concentrated** solution. A handful of salt in a glass of water makes a very concentrated solution.

You may have observed that there is a limit to how concentrated a solution can be at a certain temperature. For example, if you continue to dissolve sugar in iced tea, the solution will become so concentrated that no more sugar will dissolve. When this happens, the solution is **saturated**. A saturated solution has all the dissolved solute that it can hold at a given temperature. However, if you raise the temperature, the solution can hold more solute. For example, the amount of sugar needed to make a saturated solution of iced tea is less than the amount needed when the tea is hot.

The amount of sugar or salt that will make saturated solutions in 100 g of water at 60°C is shown in Figure 3-2. As you can see, a larger amount of sugar than salt is needed to make a saturated solution at the same temperature. Sugar

Dilute Describes a solution made by dissolving a small amount of solute in a large amount of solvent.

Concentrated Describes a solution that is made by dissolving a large amount of solute in a solvent.

Saturated Describes a solution that has all the solute that it can hold without changing the conditions.

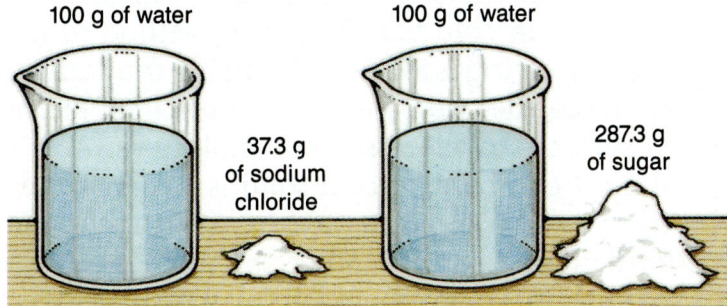

Figure 3-2 Sugar is more soluble in water than is salt. A greater amount of sugar will dissolve in the same amount of water.

Solubility Describes the amount of solute that can be dissolved in a given solvent under given conditions.

is said to have greater **solubility** in water than salt. Solubility describes the amount of solute that can be dissolved in a given solvent under given conditions. The solubility of most solids in water increases as the temperature of the water increases. See Figure 3-3. However, the opposite is true for the solubility of gases in water. Gases are more soluble in *cold* water than in *hot* water. This is why more gas escapes when you open a warm can of soda than when you open a cold can.

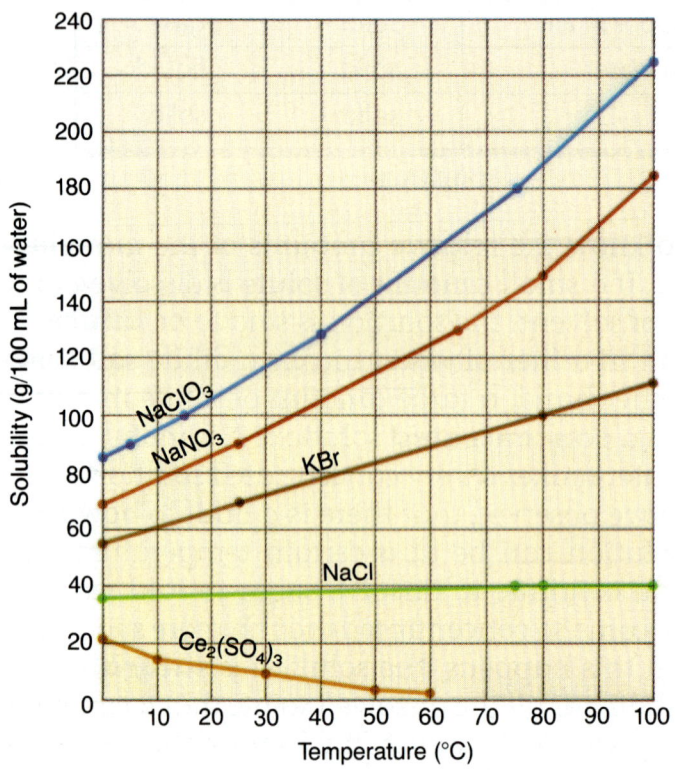

Figure 3-3 *This graph shows the solubility of several substances at different temperatures. Notice that most of the substances have higher solubilities at higher temperatures.*

Rate of Dissolving If you want to dissolve a solid in a liquid, there are three ways you could speed up the process. First, you could raise the temperature. The heat will cause the molecules of the solute and the solvent to move faster. As a result, the two kinds of molecules mix with each other at a faster rate. Sugar, for example, will dissolve faster in hot tea than in cold tea. A second method to increase the speed of dissolving is stirring. A stirring motion helps to mix solvent molecules with the solute molecules, exposing more solute surfaces to more solvent molecules. A third way to speed up the dissolving of a solid is to crush the solid into small pieces. This exposes more surface area of the solid solute to the liquid solvent. Since the molecules of solute can escape into the solvent only from the surface of the solid, this speeds up the rate of dissolving. This is why breaking up a sugar cube helps it dissolve faster. See Figure 3-4.

Figure 3-4 *Using the kitchen utensils shown, how would you increase the rate of dissolving for sugar in ice tea?*

Suspensions

Do you ever use Italian dressing on your salad? If so, you know that the herbs settle to the bottom while the bottle sits on the shelf. You must therefore shake the bottle before using the dressing. After you shake the bottle, the herbs settle again. Particles of clay in water act the same way. After you stir up the clay, the particles eventually sink to the bottom. These mixtures are examples of suspensions. A **suspension** is a mixture in which one of the parts is a liquid and from which visible particles settle out. See Figure 3-5.

Suspension A mixture of a solvent-like liquid and particles that slowly settle out.

Figure 3-5 *A winter-scene knickknack is a suspension of plastic snow in water. No matter how hard you shake it, the snow will settle out.*

Suspensions are used frequently in the pharmaceutical industry. Liquid medicines that require shaking before being taken are suspensions. Some paints contain solid pigments in a liquid. See Figure 3-6. These pigments must be finely ground so the paint does not separate readily. You may have noticed that paint stores have special machines that shake the cans of paint in order to mix the components.

Figure 3-6 *In order to make different colors of paint, pigments are suspended in a liquid.*

Colloids

What do cherry-flavored gelatin, cold cream, lipstick, and shaving cream have in common? They are not really solids or liquids; they are something in-between. All of these substances are colloids. A **colloid** is a type of mixture in which the particles are small enough that they do not settle out or sink to the bottom of their containers, and they are usually too small to be visible.

You do not have to look far to find a colloid. In fact, the cytoplasm in your body's cells is a colloid. Many of the foods you eat are also colloids. Jelly, for example, is one colloid that tastes great on toast!

Not all colloids are mixtures of a solid and a liquid. There are three basic types of colloids, depending on their makeup.

Colloid A mixture in which the particles are small enough that they do not settle out.

1. *Gels* are liquid particles spread out in a solid. Gels flow slowly since they are nearly solid. Some examples of gels are gelatin, jelly, and stick deodorant.

2. *Emulsions* are colloids made of two liquids. You are familiar with many oil–water emulsions, such as mayonnaise, hand cream, and milk. Many creamy salad dressings are also emulsions.

3. *Aerosols* are colloids formed when either solid or liquid particles are suspended in a gas. Fog and smoke are examples of aerosols. A paint-spray can shoots out a colloid of paint particles finely dispersed in a gas, as shown in Figure 3-7.

Figure 3-7 *There are many examples of aerosols in everyday life. The spray paint shown here is an aerosol.*

Instruments called ultracentrifuges can be used to separate the parts of a colloid. You may know of centrifuges as the machines used by hospitals to separate the components of blood for analysis. These centrifuges can spin colloids at very high speeds to separate solids from liquids. This method can be used to study the sizes of the particles in colloids. As the colloid spins, different-sized particles settle out in different layers.

The particles of a colloid scatter light. As you know, this phenomenon is known as the *Tyndall effect* and is one important way to distinguish colloids from solutions. Light scattering can be used to determine the size, shape, and concentration of the particles because all of these qualities affect the intensity of the light passing through. See Figure 3-8.

Figure 3-8 *The jar at the left contains a mixture of water and sodium chloride. The jar at the right contains a mixture of gelatin in water. Which mixture is a colloid and which is a solution?*

Summary

Mixtures contain two or more substances that can be separated by physical means. Solutions are mixtures in which one substance (the solute) dissolves into another (the solvent). Solubility can be expressed by the amount of solute in a given amount of solvent. There are procedures that affect the rate of solution. A mixture with particles large enough to settle out is called a suspension. Colloids are mixtures with characteristics falling between solutions and suspensions.

3.2 ACIDS, BASES, AND SALTS

Many compounds are soluble in water. When an ionic compound dissolves in water, the bonds that hold the ions together are broken. As a result, the positive and negative ions separate from one another and are free to move around the solution. Some molecular compounds can partially break into ions as well. Acids and bases form solutions that are characterized by the presence of certain ions when they are dissolved in water. Salts are chemical compounds that form when solutions of acids and bases are mixed.

Characteristics of Acids

When you think of an acid, you may think of a substance that "eats" through metal. However, not all acids are that strong. In fact, you might have had some acid for breakfast. Orange juice is an example of a weak acid. See Figure 3-9. Carbonated drinks also contain a weak acid.

Figure 3-9 *Citric acid was so named because it was first isolated from citrus fruits.*

One characteristic of acids is that they taste sour. You may have tasted certain mild acids. Most fruits have some acid content. Oranges and lemons contain citric acid; apples contain malic acid. Green, or unripe, fruits taste sour because they have a greater amount of acid than ripe fruits have. Some of the acid in fruit is converted into sugar as the fruit ripens. Another common acid is found in vinegar. Vinegar is a 5 percent solution of acetic acid in water. All of these acids are weak acids.

Many of the weak acids come from living sources and are therefore called *organic acids*. Most organic acids contain a –COOH group as part of the molecule. This group breaks apart into –COO⁻ and H⁺ ions. The presence of hydrogen ions is what makes a compound an acid. Table 3-2 lists some organic acids with which you may be familiar.

ORGANIC ACIDS		
Name	Found in or Produced by	Chemical Formula
Acetic acid	Vinegar	CH_3COOH
Citric acid	Citrus Fruits	$C_5H_7O_5COOH$
Formic acid	Ants	$HCOOH$
Lactic acid	Sour milk	C_2H_5OCOOH
Malic acid	Apples	$C_3H_5O_3COOH$
Salicylic acid	Aspirin	$C_6H_4(OH)COOH$
Tartaric acid	Soft drinks	$C_3H_5O_4COOH$

Table 3-2

SOME COMMON ACIDS	
Name	Formula
Hydrochloric acid	HCl
Sulfuric acid	H_2SO_4
Nitric acid	HNO_3
Phosphoric acid	H_3PO_4

Table 3-3

Acid A compound that produces hydrogen ions in solution.

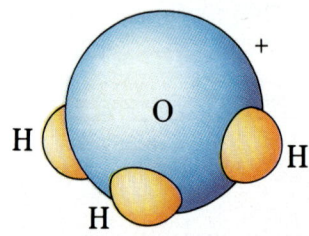

Figure 3-10 Hydronium ions contain one oxygen atom and three hydrogen atoms. The number of hydronium ions in solution determines the strength of an acid.

The Formation of Acids How can you tell if a substance will form an acidic solution when it dissolves in water? The best way is to know its chemical formula. Look at the formulas for some common acids in Table 3-3. Notice that each acid has a hydrogen at the beginning of its formula. When an acid, such as hydrochloric acid (HCl), is dissolved in water, it separates into H⁺ and Cl⁻ ions. In scientific terms, an **acid** is a compound that produces hydrogen ions (H⁺) in solution.

Since a hydrogen atom has one proton and one electron, the positive ion of hydrogen (H⁺) is just a proton. However, protons are not normally found uncombined in solution. What actually happens is that H⁺ combines with a water molecule to form H_3O^+, or a *hydronium ion*. See Figure 3-10. Any solution that contains H_3O^+ is acidic.

The Strength of Acids Why are some acids weak and others strong? The strength of an acid depends on the number of hydronium ions in solution. The more hydronium ions that the acid produces in solution, the stronger the acid. Molecules of strong acids completely break down into ions when dissolved in water, producing many hydronium ions. Therefore, their water solutions contain only ions from the acid; no molecules of the acid are left. The molecules of weak acids, however, do not all break down into ions, so there are not as many hydronium ions in solution. See Table 3-4 for some examples of strong and weak acids.

STRENGTH OF SOME ACIDS	
Strong acids	**Weak acids**
Hydrochloric acid (HCl)	Acetic acid (CH_3COOH)
Nitric acid (HNO_3)	Citric acid ($C_5H_7O_5COOH$)
Sulfuric acid (H_2SO_4)	Carbonic acid (H_2CO_3)
Hydrobromic acid (HBr)	Boric acid (H_3BO_3)

Table 3-4

Concentrated solutions of strong acids are very corrosive. They can react with (eat through) metals and must be handled with great care because they can produce severe burns. The acids listed on the left in Table 3-4 are strong acids. **Concentrated solutions of these acids should never be tasted or touched!**

In addition to reacting with metals, acids also react with substances such as limestone and marble, releasing carbon dioxide. This reaction causes the deterioration of marble buildings and statues. See Figure 3-11. It also occurs in nature, causing the formation of caves and some sinkholes. Where do you think the acid comes from to do this?

Figure 3-11 Acid has eroded the features of this marble statue.

Uses of Acids

Acids are very important to many industrial processes. The production of sulfuric acid, for example, is so important that it serves as an indicator of how well the United States' economy is doing. Sulfuric acid, a strong acid that is a colorless, oily, and dense liquid, is used to make fertilizers and automobile batteries. It is used in tanning leather for your shoes, in treating the paper on which the words of this book are printed, and even in making the vanilla flavoring for cookies, cakes, and ice cream. Sulfuric acid is also an

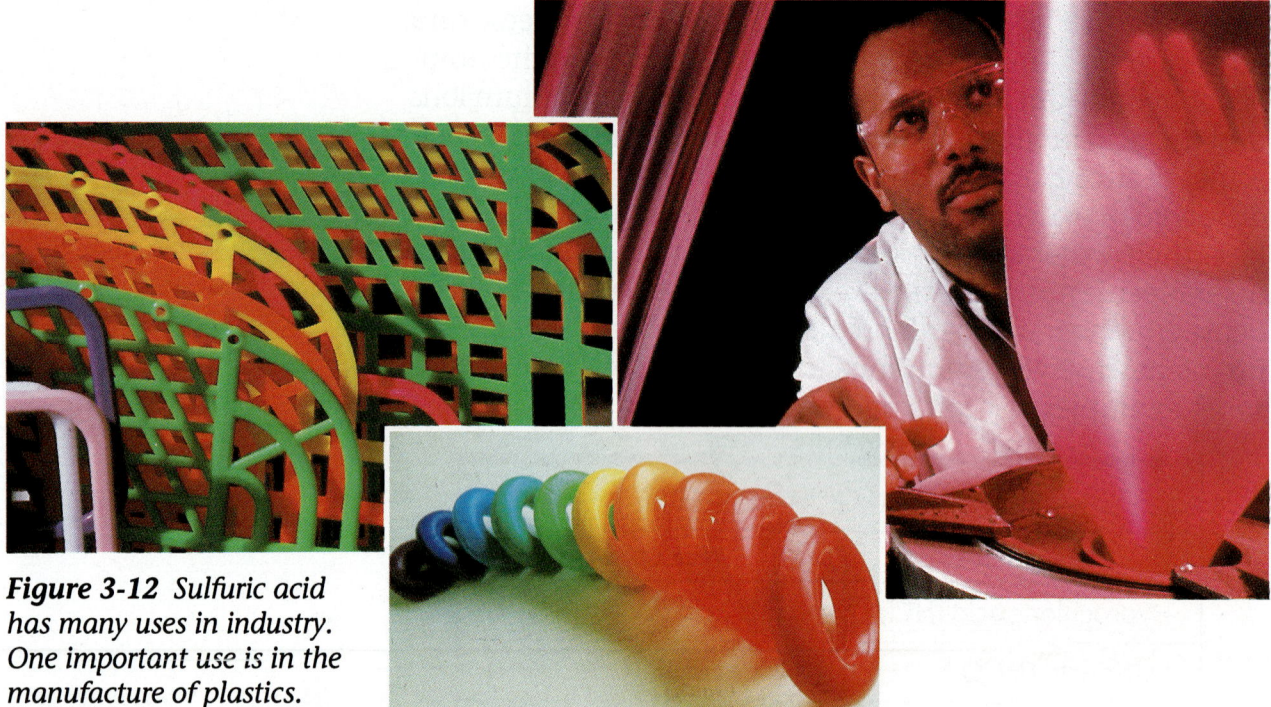

Figure 3-12 Sulfuric acid has many uses in industry. One important use is in the manufacture of plastics.

Figure 3-13 One use of nitric acid is to make dyes such as those shown here.

Base A compound that produces hydroxide ions in solution.

important component in steel production and many other manufacturing processes. See Figure 3-12.

Nitric acid, another strong acid, was first used by the alchemists of the Middle Ages to dissolve metals. They named it *aqua fortis*, which is Latin for "strong water." Today, the uses of nitric acid range from making explosives, such as nitroglycerin and dynamite, to the manufacture of ammonium nitrate, a major solid fertilizer. See Figure 3-13.

Hydrochloric acid, also called *muriatic acid*, is used to clean concrete and to balance the acid content of swimming pools. Like other strong acids, concentrated hydrochloric acid needs to be handled with extreme care.

Characteristics of Bases

In the same way that an acid produces hydronium ions, an inorganic **base** produces *hydroxide* ions (OH^-) in solution. The formulas for many common bases, along with their more common names, are given in Table 3-5. When these bases are added to water, they release negative hydroxide ions. The stronger the base, the greater the number of hydroxide ions that are released into solution. Strong bases are just as dangerous as strong acids. They can cause serious burns and must be handled with care. **Never touch or taste a strong base.** Solutions that contain bases are called "basic solutions." You may have also heard the word *alkaline* used instead of "basic" in reference to such solutions.

SOME COMMON BASES		
Chemical Name	Formula	Common Name
Sodium hydroxide	NaOH	lye, caustic soda
Calcium hydroxide	Ca(OH)$_2$	slaked lime
Magnesium hydroxide	Mg(OH)$_2$	milk of magnesia
Ammonium hydroxide	NH$_4$OH	ammonia water

Table 3-5

Although most bases, such as NaOH, already contain hydroxide ions, there are some bases that do not. Ammonia (NH$_3$), for example, produces a hydroxide ion only when it is dissolved in water.

$$NH_3 + H_2O \rightarrow NH_4^+ + OH^-$$

Since ammonia forms OH$^-$ ions in solution, it is considered a base.

One characteristic shared by all bases is that they have a soapy, slippery feel. Soap solutions, which are mildly alkaline, demonstrate this characteristic. If you have ever had soap in your mouth, you are familiar with another characteristic of bases—their bitter taste.

Bases react with acids in a process called *neutralization*. Neutralization occurs when equal amounts of hydronium ions and hydroxide ions are in solution. These ions produce water, as is shown in the following equation.

$$H_3O^+ + OH^- \rightarrow 2H_2O$$

Once the water is formed, the solution is no longer acidic or basic; it is neutral. This characteristic reaction between acids and bases occurs under many circumstances—even in the human body. When a person has an acid stomach, or heartburn, he or she may take an antacid tablet. The antacid tablet is a weak base that dissolves in the stomach and combines with the excess acid to neutralize it.

Uses of Bases

Although bases are less commonly discussed, they are just as important as acids in many processes. Bases are used for a variety of things, from keeping you clean to unclogging sinks and even making cement. One example of a commonly used base is household ammonia, which is really an

Figure 3-14 *Sodium hydroxide is a strong base that is used in many industrial processes. The paper-making process, shown here, requires great amounts of this chemical.*

ammonium hydroxide solution. This compound is present in many cleaners for its ability to cut grease.

Sodium hydroxide, also known as caustic soda or lye, is an economic indicator, just like sulfuric acid. That is, the amount of sodium hydroxide produced indicates how the economy of the country is doing. Sodium hydroxide is used by many industries, including the paper industry. When paper is made, sodium hydroxide is used to remove pulp fibers from wood. See Figure 3-14.

Many other bases are also commonly used. More than 35 billion pounds of calcium hydroxide are produced in the United States annually. This base is used to make cement, mortar, and plaster. Another base, magnesium hydroxide, is sold as an antacid. These are only a few examples of the many bases used every day.

The pH Scale

pH A measure of the hydronium ion concentration of a solution.

Since you cannot see H_3O^+ or OH^- ions in solution, how can you identify a solution as acidic or basic? Scientists use a pH scale to show how acidic or basic a dilute solution is. The **pH** of a solution is a measure of the hydronium-ion concentration of the solution. The pH of most dilute solutions ranges from 0 to 14. This scale separates acids, bases, and neutral solutions into regions. The middle point, pH = 7, is neutral. A solution with a pH of 7 is neither an acid nor a base. Acids have a pH of less than 7. The more acidic the solution, the lower the number on the pH scale. Figure 3-15 shows the pH values for some common substances. For example, the pH of lemon juice is about 2.3, while tomato juice, which is less acidic, has a pH of 4. Bases have a pH higher

Figure 3-15 *This chart shows the pH of several common substances.*

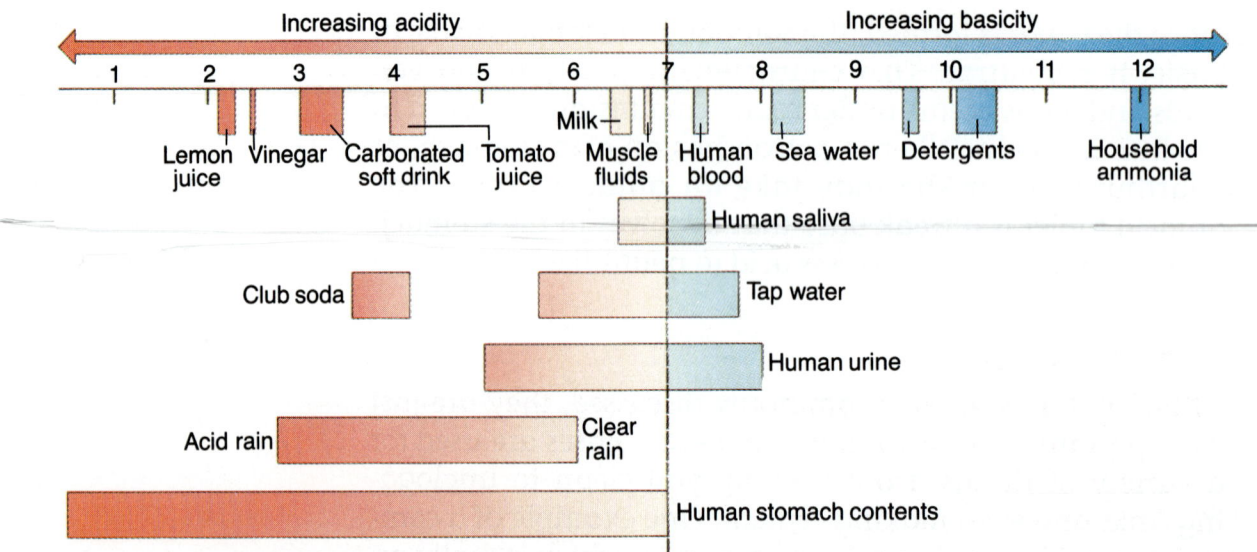

than 7. The higher the number above 7 on the pH scale, the more basic the solution.

Indicators of pH The pH scale is used to determine if a solution is acidic, basic, or neutral. Scientists can determine the pH of a solution with an instrument called a pH meter. A pH meter has electrodes that are placed in the solution. An electric current passes through the meter, and the pH is read from a dial or display. See Figure 3-16.

There are other, less expensive ways in which to determine the pH of common substances. For example, garden stores sell kits to help people determine the pH of their soil. The pH of swimming pools, which is important if the water is to be kept clean and free of bacteria, can also be determined by using a pH kit. Both of these kits work on the basis of color changes. They contain special solutions called *indicators*. An indicator is a substance that changes color in an acid or a base. Some indicators show only two color changes—one for acids and another for bases. Other indicators, however, show various color changes depending on the pH of the substance. See Figure 3-17.

Figure 3-16 *This scientist is using a pH meter to test lake water for traces of acid-rain contamination.*

Figure 3-17 *A sample's pH can be determined by adding indicator and comparing its color to those of standard solutions containing indicator.*

For an indicator to be useful, it must show a very distinct color change over a narrow range of pH. Litmus paper, which is widely used, is a strip of paper dyed with an indicator. It comes in two colors: red and blue. Red litmus paper turns blue in a base. Blue litmus paper turns red in an acid. Another common indicator is phenolphthalein, which is colorless in an acid but turns red in a base.

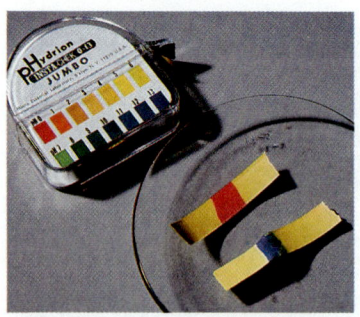

Figure 3-18 *When pH paper touches an acidic or basic solution, it changes color. The pH of the solution may be estimated by comparing the color produced to the scale on the dispenser.*

Some indicator paper is dyed with what is called a *universal indicator*. Universal indicator is a mixture of several different indicators combined so that the paper gives a different color for each pH. The advantage of this type of indicator is that it shows the exact pH of a solution rather than just showing whether the solution is an acid or a base. Paper dyed with universal indicator is called pH paper. See Figure 3-18.

Nature and pH Many plants and animals are very sensitive to changes in pH. Some fish and other aquatic life have died in lakes that have become more acidic than normal due to pollution. Lake water usually has a pH between 6 and 7. However, some lakes in northern states have a pH as low as 3. Many scientists believe that this change in pH has been caused by acid rain, which is produced by the mixing of rain and the pollution from factory smokestacks and automobile exhaust.

Scientists are experimenting with methods to bring the pH of these acidic lakes back to normal levels. In order to do this, large quantities of weak bases, such as ground limestone or sodium bicarbonate, are dumped from planes into the lakes. See Figure 3-19. Scientists hope that the addition of these bases will neutralize the excess acid. The results of these efforts, however, will only be temporary if the sources of pollution are not stopped.

Figure 3-19 *When lakes become too acidic as a result of acid rain, a weak base, such as baking soda, is often added to the water.*

The human body is also very sensitive to changes in pH. The normal pH of human blood is between 7.38 and 7.42. This range is very narrow; a change in the concentration of H_3O^+ ions has a marked effect on the body. If the pH is above 7.8 or below 7, the body cannot function normally. If the pH change is sudden or not corrected rapidly, it can be fatal.

Characteristics of Salts

You are used to calling sodium chloride "salt." Scientists use the term *salt* in a more general way. A **salt** is any ionic compound formed when the negative ion from an acid combines with the positive ion from a base. You may recall that when acids and bases are mixed together, the H_3O^+ and OH^- ions form water. The other two ions that were part of the acid and base form a salt. Look at the following neutralization reaction.

Salt Any ionic compound formed when the negative ion from an acid combines with the positive ion from a base.

$$HBr + KOH \rightarrow H_2O + KBr$$

The potassium and bromide ions combine to form an ionic compound called potassium bromide. All neutralization reactions produce water and a salt.

Not all salts, however, are formed by neutralization reactions. Some are formed by the reaction of metals with acids. For instance, when pieces of magnesium metal are dropped into hydrochloric acid, hydrogen gas is released and a white solid can be collected. This white solid is magnesium chloride, a salt.

$$Mg + 2HCl \rightarrow MgCl_2 + H_2$$

Salts also come in other colors, depending on the particular ions that are combined. See Figure 3-20.

Figure 3-20 The color of a salt depends on the elements that compose it. Pictured here are four different salts, each a different color.

Uses of Salts

Salts are very common compounds. You probably use them every day without even knowing it. There are salts in most canned or processed foods. The salts either improve the flavor of the food, or they help to preserve the food.

Table salt, sodium chloride, is a common chemical compound in the earth's crust. Solid sodium chloride is found in large underground deposits in the form of gray crystals called rock salt. These deposits of rock salt can be mined. Salt can also be obtained by the evaporation of sea water. The oceans contain a huge amount of dissolved salts. Among the various salts taken from the sea, sodium chloride is the most common, although there are many others. The process of removing dissolved salts from sea water is called *desalination*.

Figure 3-21 *Salt is an important part of our diet. It is also used in many industrial processes. Lye, chlorine, hydrochloric acid, and baking soda are some of the chemicals that are made from sodium chloride.*

Approximately 40 million metric tons of sodium chloride are collected each year by desalination. See Figure 3-21.

Salts have the unusual characteristic of lowering the temperature at which water freezes. People often take advantage of this property of salts. For example, sodium chloride is spread on wet streets in the winter to prevent ice from forming at 0 °C. The salted water will need a much lower temperature to freeze. If you have ever made ice cream at home, you have seen the effect that sodium chloride has when added to the ice around the ice-cream freezer. As the ice melts, the salt dissolves in the water, lowering its freezing point. The remaining ice then cools this water to sub-zero temperatures. Since milk, sugar, and the other ingredients in ice cream freeze at a lower temperature than water does, the lower temperature of the salted water allows the ice cream to freeze. Other salts have a similar effect on water, but since sodium chloride is so abundant, it is the cheapest salt to use.

Summary

A solution containing a large number of H^+ ions is an acidic solution. A solution containing more OH^- ions than plain water is a basic solution. When an acid and a base are combined, a neutralization reaction takes place in which the H^+ and OH^- ions combine to form water. The strength of acidic and basic solutions is measured on the pH scale. Another product of a neutralization reaction is a salt. Many different kinds of salts are produced when different acids and bases react. Salts can also be made by other reactions. Salts can flavor and preserve food and also lower the temperature at which water freezes.

Unit 4

IN THIS UNIT

4.1 Motion and Force, p. 62

4.2 Laws of Motion, p. 68

4.3 Gravitation, p. 74

Reading *Plus*

Now that you have been introduced to forces and motion, let's take a closer look at how we describe motion and how motion relates to force. Read pages S62 to S76 and then prepare a demonstration that uses the principles of force and motion you learned about. Your demonstration should illustrate the concepts expressed by the following questions.

1. How do we observe and describe the motion of any object?
2. In what ways can motion be described by physical laws?
3. How are motion, force, acceleration, and gravity related?

4.1 MOTION AND FORCE

The world around us is full of motion. Leaves fall from trees, cars and trucks move along city streets, people walk from place to place. Maybe you walked to school today; maybe you rode the bus. You see motion everywhere. But what exactly is motion?

Observing Motion

Reference Point Any stationary place from which motion can be observed.

Motion A change in position of an object when compared to a reference point.

What do we mean when we say something is in motion? First, we always compare a moving object with an object that appears to stay in place. The object that stays in place is called a **reference point**. On earth, we usually compare a moving object with an object on the ground that appears to be stationary. When an object changes its position with respect to a reference point, it is in **motion**. See Figure 4-1.

Figure 4-1 Relative to the gymnast, what are the reference points in this photograph?

For example, if you look at clouds in the sky, you can tell if they are moving by comparing them to a stationary tree within your view. If the clouds change position with respect to the tree, they are in motion. If there is no change, you know they are still.

Speed Look at Figure 4-2. What reference points do you see in these pictures? In comparing the two pictures, you can see that the balloon has moved past the trees. Now, consider that the time between the two photographs was 20 seconds. During that time, the balloon moved a distance of 100 meters. Every moving object covers a certain distance in a certain period of time. Dividing this distance by the time of travel gives the **speed** of the object. In the case of the balloon, the speed was 5 m/s.

Speed The distance covered by a moving object per unit of time.

S62

Figure 4-2 The balloon has moved past the edge of the trees, as shown by the two photographs. The time between the two photographs was 20 seconds.

Velocity A reference point also allows us to specify direction. Objects can move either *toward* or *away* from a reference point. When we give speed a direction, we call it a **velocity**. People often use the words *speed* and *velocity* as if they have the same meaning. They do not. Velocity is a speed in a particular direction. See Figure 4-3. To completely describe a motion, you must specify two things—the speed *and* the direction of the motion. For example, 60 km/h is simply a speed, while 60 km/h west is a velocity.

Acceleration Most moving objects do not move at a constant speed. When you ride a bicycle, for example, your speed going uphill is usually slower than when you ride on level ground. Your speed may increase as you go downhill. This kind of motion is called **acceleration**. If the speed of a moving object changes, its motion is an accelerated motion. See Figure 4-4.

The word *acceleration* is commonly used to mean an increase in speed. However, to a scientist, acceleration is *any* change in the velocity of a moving object. Therefore, a scientist would say that when a bicycle slows down, it accelerates. We sometimes use the word *deceleration* to indicate this type of motion. However, remember that velocity includes both speed *and* direction. If the bicycle turns a corner

Velocity The speed and direction of a moving object.

Figure 4-3 These jets are traveling at approximately the same speed. However, because they are traveling in different directions, their velocities are different.

Acceleration A change in the speed or direction (or both) of motion.

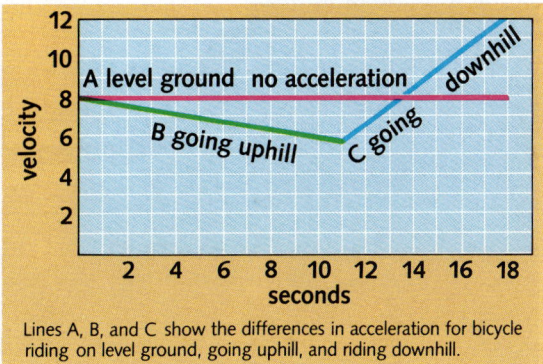

Lines A, B, and C show the differences in acceleration for bicycle riding on level ground, going uphill, and riding downhill.

Figure 4-4 The acceleration or deceleration of a bicycle can be graphed as the relationship between velocity and time.

without changing speed, it is also accelerating. Acceleration, therefore, involves a change in speed, direction, or both. See Figure 4-5.

Frames of Reference

If you look around you, you will notice many objects that seem to be stationary because they do not move with respect to each other. These stationary surroundings are called a **frame of reference**. A frame of reference is used to specify the position and relative motions of objects. Usually, the surface of the earth is the frame of reference for the motions we observe on earth. Most of our everyday activities take place within a limited area bounded by our field of view.

Nevertheless, the earth itself is in motion. While you are reading these words, you are hurtling through space at an incredible speed. See Figure 4-6. We can only describe the motion of an object by comparing it to a reference point within a given frame of reference. For most things, stationary objects on the surface of the earth serve this purpose. We can then ignore other (cosmic) motions.

There are also other, smaller frames of reference that we can use. Imagine that you are riding in a speeding train. The walls and floor of the car you are in are stationary in relation to yourself and can be considered a frame of reference. You can toss a coin into the air and catch it, as you would while standing on the ground, without worrying that the coin is actually going forward at 135 km/h, the speed of the train. As long as the train maintains a constant velocity, all the laws of physics are the same as if you were stopped. In fact, if there were no windows in the train, you might not even notice that you were moving. See Figure 4-7.

Figure 4-5 The cars on this stretch of highway are constantly changing velocity, or accelerating, even when their speed remains constant.

Frame of Reference Any system for specifying the precise location of objects in space.

Figure 4-6 Considering the motion of the earth—as it rotates, revolves, and moves with our galaxy—just how fast are you going when you are "standing still"?

Figure 4-7 What about this photograph tells you that the train is moving?

Forces and Motion

The ancient Greek philosopher Aristotle once said that a moving object would stop moving unless it was pushed or pulled. Do you think this is true? Think about a wagon. As long as you pull it, it will follow you. If you let go of the handle, it will slow down and stop. This little experiment seems to indicate that Aristotle was right. This type of push or pull is called a **force**. Aristotle thought that a force was needed to keep an object moving. However, Aristotle was not aware of a very important aspect of nature—friction.

Great discoveries in science are often made when someone asks the right question. Consider the question, "Why do objects move?". This question was asked by early scientists when they investigated motion. Experiments seemed to indicate that a push or pull was needed to cause motion and that an object with no force acting on it did not move. However, in the 17th century, Isaac Newton saw the problem differently. He asked the question, "Why do objects *stop* moving?".

Newton thought that moving objects are usually stopped by a force, which we call **friction**. For example, there is a force of friction between your shoes and the floor that allows you to walk. If there were no friction, you would have trouble moving forward. See Figure 4-8. Have you ever slipped on a polished floor? As you may have found out, smooth surfaces have less friction.

Force Any push or pull.

Friction The force that opposes motion between two surfaces that are touching.

Figure 4-8 The force of friction between the ground and your shoes helps you walk. Low friction makes it hard to walk on a slippery surface such as ice.

Wheels can also reduce the amount of friction. Have you ever fallen at a roller rink? When one object rolls over another, there is very little friction. But there is always some friction present. How do you think moving objects would behave without friction?

Newton discovered that it is the force of friction that causes a moving object to slow down and, if the force acts long enough, stop. Whenever a single force is applied to an object, the object's motion is changed. If the object is at rest, it moves; if it is moving, it changes velocity. However, if an applied force does not change the object's motion, it must have been *balanced* by another force. Only when forces are unbalanced does an object accelerate. See Figure 4-9.

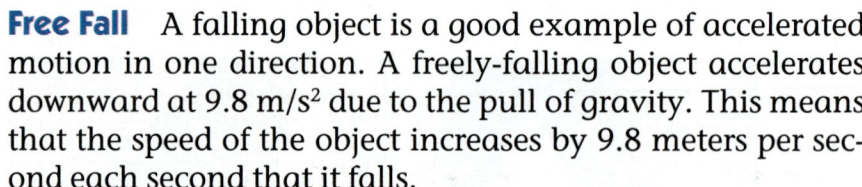

Figure 4-9 Objects remain stationary or in motion at a constant velocity unless acted upon by an unbalanced force.

Types of Motion

Motion of an object can be described by the combination of its speed and direction. Some motion involves constant velocity (constant speed *and* direction), others involve acceleration (changes in either speed, direction, or both), and still others are the combination of two separate motions.

Free Fall A falling object is a good example of accelerated motion in one direction. A freely-falling object accelerates downward at 9.8 m/s^2 due to the pull of gravity. This means that the speed of the object increases by 9.8 meters per second each second that it falls.

Actually, a falling object will accelerate at 9.8 m/s^2 only if it falls in a vacuum. Most objects falling through air will accelerate only for a short time. The acceleration of the object is soon stopped by friction with the air. The greatest speed an object can reach while falling through air is called its *terminal speed*. For example, if you jump from an airplane, you will accelerate until you reach a terminal speed of about 190 km/h. Once you open your parachute, your terminal speed will be reduced by the added wind resistance to a safer rate. All objects falling through the air reach a certain terminal speed once the force of air resistance balances the force of gravity. See Figure 4-10.

Figure 4-10 The upward force of air resistance on the skydivers works against the downward force of gravity.

Circular Motion A race car moving around a circular track, a child on a merry-go-round, and clothes being spun in a washing machine are all examples of objects undergoing *uniform circular motion*. Uniform circular motion refers to motion at constant speed in a circular path. See Figure 4-11.

When an object goes around in a circle, its direction is constantly changing, and it is therefore accelerating. Consider a race car speeding around a curve on the track. The car must constantly change directions to follow the track. Therefore, an unbalanced force is causing the car to accelerate toward the center of the curve. The force that causes objects to move in a circular path is called **centripetal force**. *Centripetal* means "toward the center."

If you were riding in the race car mentioned above, you would feel as if you were being pushed away from the center of the turn. This is a result of your body's resistance to changing velocity. When your body is moving, it tends to continue moving at a constant velocity. See Figure 4-12. As you are moving around a curve in a car, it feels as if you are applying an outward force to the seat belt or the door of the car. However, the seat belt or door is actually applying a centripetal force to you to change your direction of travel.

As a car travels around a corner, the centripetal force is provided by friction between the road and the car's tires. If, for some reason, the tires were to lose traction as the car was moving around the curve, the car would immediately start sliding in a straight line. Without friction between the tires and the road, the car would be impossible to turn.

Motion in Two Directions When a ball is thrown in a horizontal direction, it moves not only outward due to the force of the throw, but also downward due to the pull of gravity. Such an object is called a *projectile*. Projectiles, once they have been thrown, are affected only by the pull of gravity. Their motion can be thought of as a combination of both a constant horizontal motion and an accelerated vertical (downward) motion. The vertical and horizontal motions are independent—that is, they have no effect on each other. The curved path a projectile takes is called its *trajectory*.

One interesting thing about trajectories is that an object thrown sideways will reach the ground at the same time as an identical object that falls straight down. The horizontal speed of the trajectory does not affect the time it takes to hit the ground. Gravity acts equally on objects regardless of their horizontal velocity. See Figure 4-13.

Figure 4-11 When this amusement-park ride spins, its frame provides the centripetal force necessary to keep the riders moving in a circular path.

Centripetal Force A force that causes objects to move in a circular path.

Figure 4-12 What is providing the centripetal force in this situation?

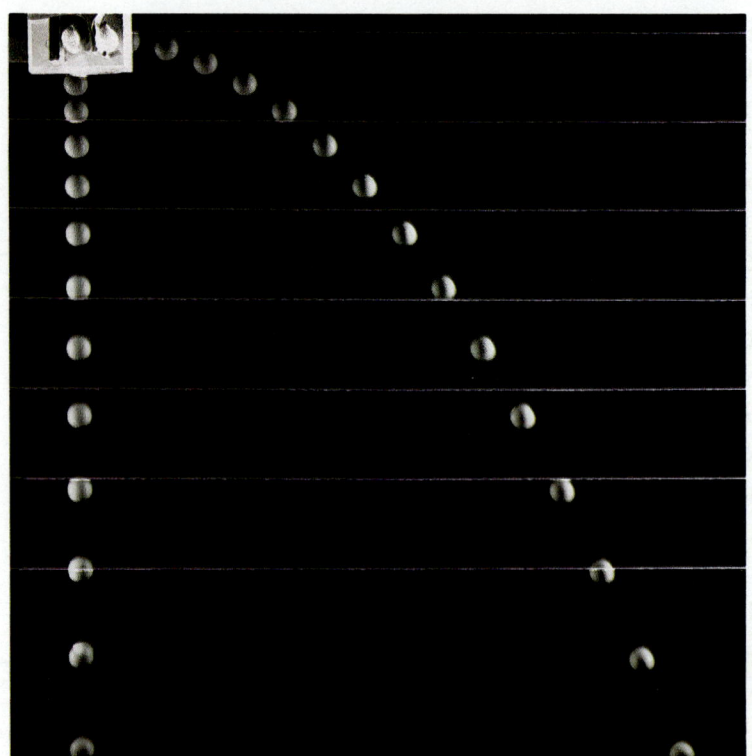

Figure 4-13 *Both balls have the same vertical acceleration due to gravity, but the ball on the right also has a constant horizontal motion.*

Summary

The universe is full of motion. The motion of an object can be described by comparing it to a reference point, which is an object that is considered stationary. The speed of an object is measured by the distance it travels from the reference point in a certain amount of time. Velocity is a measure of both the speed and direction of an object. Acceleration is the change of velocity of an object in either speed, direction, or both. Frames of reference are used to specify the position and relative motions of objects. The motion of any object can be described by the combinations of its speed and direction. Some motion involves constant velocity, while others involve acceleration. Still other motions are a combination of two separate motions.

4.2 LAWS OF MOTION

What is known today about the science of motion is due largely to a book published by Sir Isaac Newton in 1687. Newton's book, *Principia*, describes his findings about moving objects. The basic principles he wrote about are now known as Newton's laws of motion. These and other laws describe the motion we observe and explain the forces behind this motion.

Newton's First Law of Motion

Newton thought that a moving object with no friction or other force acting on it would continue to move in the same direction and at the same speed forever. He decided that a force had to be present whenever there was a change in speed or direction. Therefore, if an object moves at constant speed, no force is needed. To go faster, a force is applied in the direction the object is going; to slow down, a force is applied in the opposite direction. To change direction, a force is applied from the side. Every change in motion thus requires an outside force.

The property of matter that resists any change in motion is called **inertia**. A moving object, in the absence of an outside force, thus continues to move because of its inertia. Similarly, an object that is not moving needs a force to overcome its inertia and start moving. See Figure 4-14. The amount of inertia of any object depends on the **mass** of the object.

The effects of inertia were summed up in what is now called **Newton's First Law of Motion**. It states: *Every object remains at rest or moves with a constant speed in a straight line unless acted upon by some outside force.* This law is also called the law of inertia.

Consider your experience with the wagon. You should now understand that it is the force of friction that slowed down the wagon. If friction was absent, the wagon would have continued to move forever. If the wagon was at rest, it would have remained that way unless a force—provided by your push or pull—changed the situation.

In outer space there is no friction against moving objects. For example, the Pioneer-10 spacecraft, launched from earth in the early 1970s, has now left our solar system. Once it left earth's orbit, its rocket motors were no longer needed. It has since travelled billions of kilometers at a constant speed with no additional push from its engines. See Figure 4-15. Unless some outside force affects it, it will continue out into our galaxy indefinitely. This spacecraft illustrates Newton's first law of motion.

The first law of motion should remind you always to wear a seat belt when riding in a car. If the car stops suddenly, you will tend to keep moving forward. If you are not wearing a seat belt, you might hit the windshield or dashboard. A seat belt makes it less likely that you will be hurt, since it

Figure 4-14 Because of the inertia of the objects on this table, the table setting remains intact momentarily as the table is quickly slid from underneath it.

Inertia The resistance to any change in motion.

Mass The amount of matter in an object.

Figure 4-15 Why does a spacecraft remain in motion even without continuous rocket thrusts?

supplies the force needed to hold you firmly in your seat. See Figure 4-16.

Figure 4-16 *In a car crash, inertia keeps the passengers moving forward until they are stopped by a force. If seat belts are not worn, this stopping force may be provided by the car's windshield.*

Newton's Second Law of Motion

How much force would you have to apply for ten seconds to cause a bicycle to accelerate to a speed of 8 km/h? How much force would be needed to accelerate a car to this speed? You already know that a force is needed to accelerate an object. But the greater the mass of the object, the greater the amount of force needed to cause the acceleration. Therefore, the massive car would need a much larger force to accelerate than the bicycle. See Figure 4-17. This relationship between masses and the forces acting on them was given in **Newton's Second Law of Motion**: *The acceleration of an object of a certain mass is determined by the size of the force acting and the direction in which it acts.*

Figure 4-17 *The smaller mass of the bicycle means that a smaller force is needed to produce the same acceleration.*

All forces act in some direction and, of course, exist in different sizes. Scientists measure the size of a force with a unit called the *newton* (N). You would need a force of about

9.8 N to lift an object with a mass of 1 kg. You can write the second law of motion with one of the most important equations in physics:

$$\text{Force} = \text{Mass} \times \text{Acceleration}$$

or

$$F = ma$$

Suppose you have to push a heavy box across the room. By yourself, you can supply only a small force. You might not even be able to overcome friction to get it started. If ten people pushed, however, the force would be larger and the box would move. Remember that the direction of the force is also important. If five people pushed the box one way and five pushed the other way, what would happen?

The mass of the box is also a factor. If the same ten people supplied a force to a piano case, and then to a shoe box, the acceleration of the shoe box, of course, would be greater. You can see this by rearranging the equation to:

$$F/m = a$$

If the force remains the same and the mass decreases, the acceleration will increase. What will happen to the acceleration if the mass increases?

Newton's Third Law of Motion

How was the Pioneer 10 initially propelled upward against the force of gravity? Hot expanding gases were produced in the rocket engines and forced out the rear of the engine. See Figure 4-18. Rocket engines exert a force on the gases to push them out. The gases also exert an equal and opposite force on the engine, and this force moves the rocket upward. This action is an example of **Newton's Third Law of Motion**, which states: *For every action there is an equal and opposite reaction.*

There are many everyday examples of Newton's third law all around you. A book lying on a table pushes down on the table with a certain force. The table, in turn, pushes up on the book with an equal but opposite force. One force acts on the table, while the second force acts on the book. See Figure 4-19. If you push on your desk with your hand, the desk pushes back. If you push harder, the desk also pushes harder. If you jump off the floor, your feet push against the floor with a certain force. The floor also pushes against your feet with a force that is equal in size, but opposite in direction.

Figure 4-18 How is Newton's third law demonstrated by a rocket engine?

Figure 4-19 The force of gravity is acting on this book, however, the book does not accelerate because it also receives an equal upward force from the table.

Sometimes it is hard to picture these opposing forces. However, there is at least one activity you can try to actually observe them. Imagine that you and a friend are standing on roller-skates on a level rink floor. If you were to push your friend away from you with both hands, you would roll backwards. The force you exerted on your friend would be equal but opposite to the force of your friend pushing back on you. See Figure 4-20. The outward force you feel in a car going around a curve is your reaction force to centripetal force. Can you think of other examples of Newton's third law of motion?

Figure 4-20 *For every action there is an equal and opposite reaction.*

Momentum

If you have ever watched American football on television, you have probably heard an announcer use the word *momentum*. The announcer may have said something like this: "Even though number 34 was hit at the 1-yard line, his momentum carried him into the end zone for a touchdown." In the world of physics, **momentum** is the inertia of an object in motion. More specifically, momentum is an object's mass multiplied by its velocity. This relationship can be shown by the following equation.

$$\text{Momentum} = \text{Mass} \times \text{Velocity}$$

or

$$p = mv$$

Since velocity is one factor in this equation, momentum (like velocity and acceleration) has definite direction.

Since the momentum of an object depends on its mass and velocity, an object with a small mass and a high velocity can have as much momentum as an object with a large mass and a low velocity. In football, for example, a small,

Momentum The mass of an object multiplied by its velocity.

fast runner can be as hard to tackle as a large, slow runner. Runners who are both large and fast are the hardest to tackle due to their momentum. See Figure 4-21.

One of the most interesting things about momentum is that it stays constant unless an unbalanced force is applied. When the momentum of an object changes, it is because the object's velocity, mass, or both have been changed. Since objects usually do not lose or gain mass readily, the momentum of most objects changes as a result of changes in their velocity.

Imagine, for example, that you are back in the skating rink. Suddenly, another skater accidentally runs into you. The push from the other skater is an unbalanced force that causes you to move at a higher velocity. As a result, your momentum increases. When the skater bumps into you, he or she also receives an unbalanced force that causes him or her to move at a slower velocity, decreasing momentum. The momentum that you gained is equal to the momentum lost by the other skater. This example illustrates the **law of conservation of momentum**. This law states that *momentum can be transferred from one object to another but cannot change in total amount*. In other words, when two or more objects collide, the total momentum of the objects is the same after the collision as it was before the collision.

Figure 4-22 demonstrates the law of conservation of momentum. If the ball at one end is pulled out and allowed to swing, it falls back and hits the next one in the line. Almost immediately, the ball at the opposite end of the line flies out, as if by magic. The momentum of the first ball has been passed from one ball to the next until it reaches the end ball,

Figure 4-21 *The momentum of this football player moves him forward even though he is being tackled around his legs.*

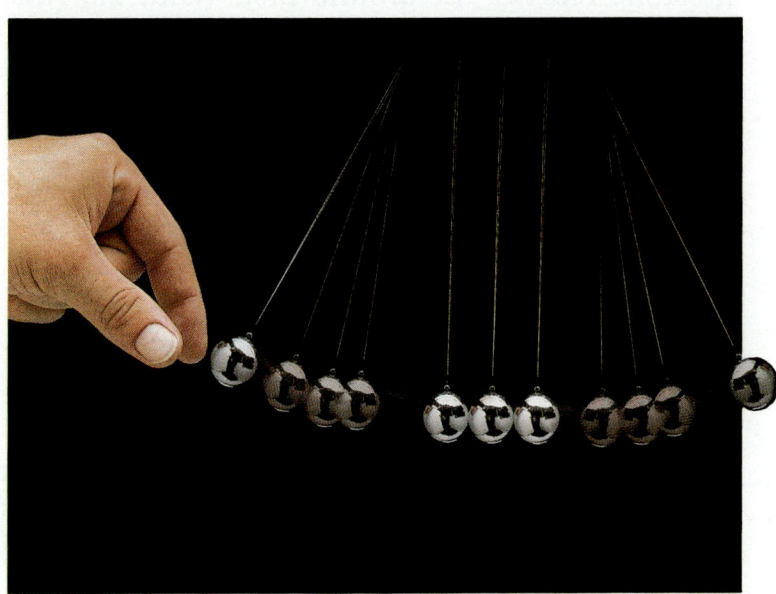

Figure 4-22 *This "Newton's cradle" demonstrates the law of conservation of momentum.*

which is free to move. What do you think would happen if two balls at one end were pulled out and allowed to swing?

4.3 GRAVITATION

What do you think would happen if two stones of different masses were dropped from the same height? Would the larger and heavier one reach the ground first? Or would they both reach the ground at the same time? In 1590 this was the topic of an intense debate among the scientists of Europe. Most scientists believed that the heavier stone would reach the ground first.

Development of Gravitational Theory

Some 2000 years before that time, Aristotle proposed that heavier objects would fall faster. He believed that all matter was composed of four elements. From lightest to heaviest, these elements were fire, air, water, and earth. See Figure 4-23. Each element was thought to have its own natural place in the world. For example, if fire were placed below its natural position, it would tend to rise above air. Similarly, if a stone were dropped, it would pass through fire, air, and water until it came to rest on the ground, its natural place. Aristotle thought that the movement of an object depended on its mixture of these four elements. Therefore, a heavier stone was expected to drop faster than a lighter stone because it contained more of the earth element. Aristotle's physics seemed to fit everyday observations. For example, we know that if we drop a feather and a stone at the same time, the stone reaches the ground first. What did Aristotle not realize?

Figure 4-23 *The four basic elements of the ancient Greeks: fire, water, air, and earth. Some Greeks thought that all matter was made of combinations of these elements.*

By 1589 a young scientist named Galileo had confronted this view of the world. At the age of 25, he had just been appointed professor of mathematics at the university in Pisa, Italy where he began to challenge the opinions of the older professors. As it turned out, Galileo's ideas signaled the beginning of the end for Aristotle's worldview.

According to Aristotle, a 100-kg mass dropped from a height of 150 m would reach the ground by the time a 10-kg mass would fall only 15 m from the same height. Galileo, however, predicted that they would both reach the ground at the same time. We now know that Galileo was right. See Figure 4-24. Although he may never have actually dropped stones from the leaning tower of Pisa as shown in the illustration, he did do experiments that showed that the acceleration of two different masses is equal. See Figure 4-25.

Newton and Gravity

You know from studying Newton's laws of motion that for an object to accelerate, an unbalanced force must act on it. Gravity is the force that causes all things near the earth's surface to fall toward the center of the earth. Newton believed that gravitational force was an attraction between any mass, such as the earth, and the matter near it.

Through his observations and calculations, Newton showed that not only the earth, but all objects in the universe exert pulling forces on each other. For example, the earth and the moon both exert a gravitational pull on each other. Ocean tides are the most obvious effect of the moon's gravity on the earth. The earth's gravity also keeps the moon in orbit about the earth. It is the sun's gravity that keeps all of the planets orbiting in the solar system. Though it would be too small to measure, all the objects around you exert a slight gravitational pull on everything else around them.

Another aspect of gravitational force is that it decreases as the distance between two objects increases. For example, as a spacecraft moves farther from the earth, the earth's pull on it becomes less.

The relationship between gravitational force, mass, and distance is stated in **Newton's Law of Gravitation**: *The gravitational force between two objects depends on their masses and the distance between them.* See Figure 4-26. To be more

Figure 4-24 *Galileo's famous experiment.*

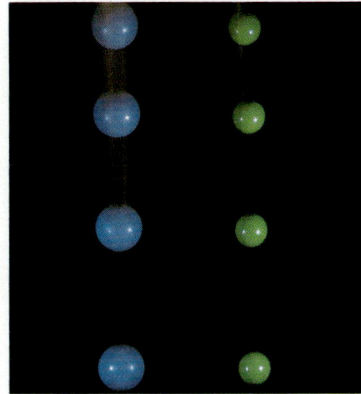

Figure 4-25 *These two objects have different masses, yet they fall at the same acceleration.*

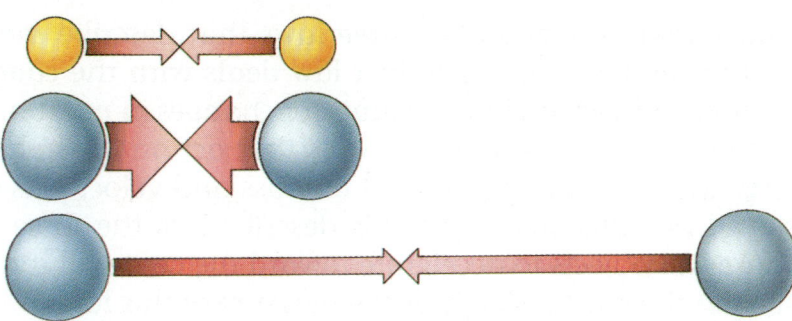

Figure 4-26 *Gravitational force, represented by the arrows, varies with the mass of the objects and their distance from one another.*

specific, a large mass exerts a stronger pull than does a smaller mass at the same distance. So, why do two objects with different masses accelerate and hit the ground at the same time? The object with the greater mass requires more force to accelerate, and because of its greater mass, the force of gravity acting on it is greater. Therefore, everything on earth would fall at the same rate if the effects of air resistance could be removed.

Like many scientific laws, the law of gravitation predicts how some parts of the natural world work. It does not explain what *causes* gravity, but it is a scientific tool that helps *explain* the effects of gravitational forces. For example, Newton's law predicts that the gravitational force on the surface of the moon is about one-sixth of that on the earth. This is because the moon's mass and size are both much smaller than that of the earth. How do you think this would affect you if you lived on the moon? See Figure 4-27.

Figure 4-27 Astronauts on the moon can jump six times higher than they can on earth because their weight is only one-sixth what it is on earth.

Weight A measurement of the gravitational force acting on an object.

Mass and Weight

You know that *mass* refers to the amount of matter in an object. **Weight**, on the other hand, is determined by the force of gravity pulling on an object. Since gravity is a force, it is measured in newtons. On earth, a person with a mass of 50 kg weighs 490 N. On the surface of the moon, the same person would weigh 81.7 N. This does not mean, however, that the person's mass has changed. The *weight* of an object changes as the size of the gravitational force on it changes. But the *mass*, or amount of matter in the object, does not change. As long as you stay in one place, an object with more mass will have more weight. Nevertheless, at any one place on earth, you can determine the mass of an object by measuring the gravitational force pulling on it. For this reason, mass and weight are often used as if they mean the same thing. However, scientists are always careful to distinguish between the terms *mass* and *weight*.

Summary

Isaac Newton formulated three laws that describe motion and its related forces. The first law deals with the cause of motion. The second law accounts for changes in motion. The third law describes pairs of forces that act on different objects. Momentum, the product of mass and velocity, is conserved in collisions. Gravity is described as the force that causes all objects to accelerate toward the center of a mass, such as the earth. Weight is the measure of this force.

Unit 5

Reading *Plus*

Now that you have learned something about the structure, design, and construction of buildings and bridges from the standpoint of western civilization, let's take a closer look at some different kinds of architecture from around the world, past and present. Read pages S78 to S88, and design a collage showing the interaction between culture and design. Your collage should include answers to at least one of the following questions.

1. How might geography and climate affect the architectural styles of a region?
2. In what ways do religion and culture affect architecture?
3. Why might different cultures solve design problems in different ways?

IN THIS UNIT

5.1 Architecture in Other Cultures, p. 78

5.1 ARCHITECTURE IN OTHER CULTURES

The development of design, as seen in the structures of Ancient Greece and Rome, Medieval and Renaissance Europe, and those of the Modern era, represents a history of Western architecture. In other areas of the world, however, different architectural styles were developed by various cultures. While the scientific principles that govern the strength and durability of structures are the same for every culture, there is a tremendous variety of designs that have been used by different peoples through the ages.

The Byzantine Empire

In A.D. 330, Constantine the Great moved the capital of the Roman Empire from Rome to the city of Byzantium and renamed it Constantinople. This new capital (now Istanbul, Turkey) became the center of the great Byzantine Empire. Although the western Roman Empire finally collapsed in A.D. 476, the empire in the East lasted until the Turkish conquest of 1453. At the height of its power, parts of Turkey, Italy, the Middle East, and Northern Africa formed the Byzantine Empire.

Architecture was the greatest form of Byzantine art. One of the architectural masterpieces of the world is the church of Hagia Sophia, built during the 530s. It is a huge building in the form of a cross, measuring 72 meters wide by 76 meters long. See Figure 5-1. Though the outside was composed of plain brick and mortar, intricate murals, mosaics, stone carvings, and metalwork once covered every interior surface. Insets of ivory, silver, and jewels were used to adorn

Figure 5-1 The Church of Hagia Sophia in Istanbul, Turkey. It was converted into a mosque by the addition of four minarets.

the pulpit as well. A huge dome, 56 meters high and 33 meters across, dominates the cathedral. See Figure 5-2. Resting on massive columns instead of walls, the dome illustrates the talent of Byzantine architects who first solved the difficult problem of placing a round dome over a rectangular building.

The Influence of Islam

Religion has influenced the development of architecture in the Islamic world, much like it has influenced western civilization. During the middle ages, the Islamic world stretched from Spain in the west to Samarkand and India in the east. The most important buildings to be designed were meeting places for prayer. These buildings are called *mosques*. Domes and *minarets* helped give the mosques a distinctive appearance. The minaret, a tall and slender tower, is used as a place from which to call people to prayer. See Figure 5-3.

Islam forbids the use of statues and paintings of humans or animals as decorations in the mosques. Instead, mosques are decorated with designs based on flowers, plants, leaves, geometric figures, and Arabic inscriptions. The result was an unparalleled richness of abstract decoration. The influence of Islamic architecture can be seen in the structures of many other cultures. See Figure 5-4.

Figure 5-2 Interior of Hagia Sophia. The Turkish government now maintains the church as a museum.

Figure 5-3 Recently completed, this mosque exhibits the typical dome and minarets that are characteristic of Islamic meeting places the world over.

Figure 5-4 Nestled among modern skyscrapers, this government building in Kuala Lumpur demonstrates the influence of Islam on the architecture of Malaysia.

India: A Contrast of Styles

The subcontinent of India was the place of origin for two of the world's major religions—Buddhism and Hinduism. Much of Indian architecture and art reflects this religious influence. For example, *stupas*, or dome-shaped shrines, were built to hold artifacts and objects associated with Buddhism. The Indian stupa is a large mound-shaped structure covered by brick or stone and surrounded by a fence with elaborate stone gates. See Figure 5-5.

Figure 5-5 A stupa in central India. What do you think the carvings represent?

Figure 5-6 Exterior of a South Indian Hindu temple showing figures of various Hindu deities.

A Hindu temple, on the other hand, is typically composed of a square building with heavy walls. The outer walls of the temple are covered with a multitude of colorful statues and carvings, and the ornamentation is more elaborate than Buddhist ornament. See Figure 5-6.

In 1506, the Mogul Empire, an Islamic empire, rose to power after years of war and conflict in India. The most famous creation of the Mogul period was the work of the emperor Shah Jahan. In the middle 1600s, he built the Taj Mahal, a magnificent tomb for himself and his wife. See Figure 5-7. It is built out of white marble and is studded with gems. Although it is an Islamic building, the Taj Mahal, as well as other Islamic structures in India, incorporated many architectural elements that were borrowed from Hindu temples. These elements include pinnacles, pavilions, and elaborate ornamentation.

Figure 5-7 The Taj Mahal's exterior design typifies the blending of Muslim and Hindu cultures.

Traditional Chinese Design

Traditional Chinese architects seldom distinguished between religious and nonreligious structures. Most buildings—whether temples, tombs, public buildings, or private homes—followed the same basic plan and were characterized by *symmetrical* design. Huge pillars supported curved tile roofs. Therefore, walls were used primarily for privacy, not support. The roof was a major feature of Chinese architecture and generally was made of glazed tiles. Brightly-colored porcelain is also a characteristic of traditional Chinese architecture.

When Buddhism entered China from India, it brought the Indian stupa with it. The Chinese modified the stupa to develop the structure known as the *pagoda*. See Figure 5-8. Each pagoda has the same basic design. They usually have eight sides and many levels, but each higher level is smaller than the one below it. Each level of the pagoda also has its own roof with long upward-curving corners. Pagodas in China are usually constructed of wood, masonry, and highly glazed tile or porcelain.

Chinese gateways, known as Pai-lous, were built as memorials to great men. They usually have one or three openings and were constructed of the same materials as were pagodas. See Figure 5-9. Pai-lous (or gateways) and pagodas are most typical of traditional Chinese architecture.

The last ruler of the Ch'in Dynasty (221–206 B.C.) built many public buildings and extensive fortifications, including part of the Great Wall of China. For several hundred years, roughly one third of all able-bodied Chinese men were forced to help build the wall. It is a massive structure that is 7.6 meters high and 4.5 meters wide, with towers of over 12 meters in height. It stretches east and west for 2240

Figure 5-8 How does this pagoda differ from Western architecture?

Figure 5-9 A typical Chinese pai-lou.

Figure 5-10 The Great Wall of China.

kilometers across northern China and is one of the few structures made by humans that can be seen from space. See Figure 5-10.

Japan: A Natural Approach

In Japan, the architecture is strikingly different from that of Western cultures. Much of Japanese architecture has its roots in Chinese designs. Temples and shrines were constructed in a style similar to Chinese structures, with curving roofs and supporting columns. See Figure 5-11. Traditional Japanese houses are made of wood and have roofs of thatch. However, the use of tile roofs is also common. The gently-curved corners of Japanese roofs add a sense of graceful proportion

Figure 5-11 The Heian Shrine was built in 1895 to celebrate the city of Kyoto's eleven-hundredth anniversary.

that is often imitated in modern Japanese buildings. See Figure 5-12.

The entrances to shrines are often marked by a structure called a *torii*. It is a ceremonial gate that stands on columns with a curved beam across its top. Though similar to the Chinese pai-lou, the torii is typically simpler in its design. The torii in Figure 5-13 marks the entrance to the Meiji Shrine in Tokyo. This shrine is dedicated to the emperor who, in 1868, opened Japan to the West and its new ideas.

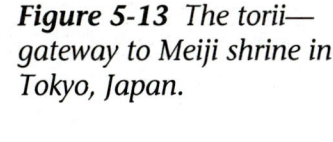

Figure 5-12 *Yoyogi stadium in Tokyo retains the curving roof-lines of traditional Japanese architecture.*

Figure 5-13 *The torii—gateway to Meiji shrine in Tokyo, Japan.*

Besides their religious shrines, Japanese designers also created beautiful gardens designed to imitate nature. These gardens are often miniature representations of the world in which rocks stand for mountains, ponds represent oceans, and sand and gravel represent lakes and rivers. See Figure 5-14.

Temples, shrines, and houses are designed to blend in with their natural surroundings and thereby emphasize harmony between the buildings and nature. See Figure 5-15. The Japanese broke from traditional Chinese architecture by using more asymmetrical designs and a less formal approach that more closely represents the unusual formations found in nature.

Figure 5-14 *A Japanese garden.*

Figure 5-15 *Traditional Japanese homes complement their natural surroundings.*

The Americas Before Columbus

The ancient civilizations of Central and South America developed some unique architectural styles. Although there were several distinctive cultures in these areas, three stand out as important examples of *pre-Columbian* architecture.

Aztecs The Aztecs, who dominated the area of central Mexico at the time of European exploration, built great cities, pyramid temples, marketplaces, and palaces. Altars and plazas accompanied the pyramids as the center of Aztec religious ceremony. See Figure 5-16.

Figure 5-16 A model of the Aztec temple area at Tenochtitlán.

When the Spanish conquistador Hernando Cortés discovered Tenochtitlán in 1519, it was one of the largest cities in the world. Tenochtitlán was a regularly planned city with straight streets. Built on an island in the middle of a lake, it was tied to the mainland by stone-lined causeways and bridges. Its buildings were constructed of rough stone that were plastered with polished and painted *stucco*. The original location of this grand city is present-day Mexico City.

Near Mexico City, in Teotihuacán, an earlier culture had built one of the greatest temple complexes of central Mexico. The people of Teotihuacán built the Pyramid of the Sun around A.D. 100. The base of this pyramid was larger than the Great Pyramid at Giza, Egypt. See Figure 5-17. The

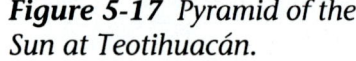

Figure 5-17 Pyramid of the Sun at Teotihuacán.

Figure 5-18 The temple of Quetzalcuatl at Teotihuacán.

Pyramid of the Sun was built with a core of clay, rock, and mud, which was faced with stone slabs that were plastered and brightly colored. Other buildings in this complex were covered with ornate carvings. See Figure 5-18.

Mayans In southern Mexico and northern Guatemala, the Mayan culture developed intricate cities that included temples and other large public buildings. At the peak of development, about A.D. 500, Mayans built great cities centered on tall pyramids of limestone that supported small temples at the top. See Figure 5-19. Besides their architectural achievements, the Mayans also devised a calendar, a form of mathematics, and an intricate writing system.

Incas In South America, the Inca Empire covered the vast area now occupied by the present-day nations of Peru, Ecuador, Bolivia, and Chile. The Incas built fortresses, irrigation systems, and paved roads. They used huge, cut stones to build the fortresses and palaces, many of which were located high in the mountains. These large stones were cut so accurately that they fit together without the need for concrete or mortar. Homes were made of stone, as well, and had thatched roofs.

One of the earliest examples of architecture in the area is the Gateway of the Sun at Tihuanaco. This archway is made of two immense upright slabs of stone that hold up a 3-meter carved stone *lintel*. The lintel by itself weighs approximately 10 tons. See Figure 5-20. Perhaps the best known example of Inca architecture is the mountain-top city of Machu Picchu. See Figure 5-21.

Figure 5-19 *The great stone pyramid built by the Mayans at Tikal, Guatemala.*

Figure 5-20 *Gateway of the Sun, at Tiahuanaco, Bolivia.*

Figure 5-21 *Machu Picchu, Peru. The ruins of an Inca civilization.*

North American Tribes The influence of Aztec and other Mexican cultures also spread to the North American continent. Tribes living along the Ohio and Mississippi rivers built large cities and huge ceremonial mound structures. See Figure 5-22. The largest of these mounds was 30 meters high and covered an area of about 16 acres (an area larger than 12 football fields). These mounds were used for burial as well as to support temples. Some mounds were built in the shape of birds, bears, mountain lions, and other animals.

Figure 5-22 Grass-covered mounds (left) are all that remain of an advanced North American culture whose landscape is illustrated in the drawing (right).

In other parts of North America, various cultures with different kinds of architecture flourished. Many of these differences related to the climate and the availability of materials encountered in these areas. In the hot, dry southwest, Pueblo tribes used *adobe*, a sun-dried brick, to build

Figure 5-23 How are these Pueblo buildings suited to their environment?

flat-roofed communal houses. Many of these houses were several stories high, and some were located under cliffs for protection. See Figure 5-23.

Since many North American tribes were nomadic, their architecture was developed around a need for mobility. Plains tribes, such as the Sioux, used buffalo skin to build cone-shaped *tepees*. These portable tents allowed the tribes to migrate with the buffalo herds. Tribes in the eastern end of the continent used forest materials, such as leaves, reed mats, and bark, to cover a frame of poles. The *wigwam* of the Chippewa tribe was such a structure. See Figure 5-24. The Iroquois used similar methods to build rectangular *long houses*, some of which were 30 meters in length.

Figure 5-24 *Black ash bark and birch bark were used to cover this Chippewa wigwam.*

The Cities of Africa

The architecture of African civilizations is reflected in the development of urban centers. Dominated by walls and passageways, early cities—like Kilwa in Kenya and Kano in Nigeria—had marketplaces, palaces, wide streets, and other structures. See Figure 5-25. The Hausa people, who lived where modern-day Nigeria is now, built cities surrounded by high walls, a common feature in the architecture of Africa. The early inhabitants probably began building walled towns when they first developed iron technology. As a town grew, its residents built a wall around it. The first outer wall

Figure 5-25 *The remains of the city of Kilwa, Africa.*

likely served as a defense against attackers. The second wall often enclosed an area for food storage.

One of the greatest fortress walls was built in Zimbabwe in the 1300s. It has walls 10 meters high made of cut stone laid in a variety of patterns. See Figure 5-26. It is over 240 meters long and consists of some 900,000 large granite blocks.

Figure 5-26 *Ancient walled cities are found in what is now Zimbabwe.*

The development of cities initiated the smaller-scale development of homesteads, such as those of the Kuria along the Kenya–Tanzania border. Here, circular houses with cone-shaped thatched roofs were clustered together and connected with a tall fence that formed a secure yard. The number of houses in the homestead depended on the size of the family.

The use of urban space was as important to Africans as their use of walls. They tried to preserve something of a rural setting, even in the largest cities. Animals were allowed to wander about, and trees shaded streets and plazas. In some towns, people even raised crops along walkways or on small plots of land.

Summary

Many kinds of architecture have developed in different countries around the world. The buildings and cities from Africa, the Orient, India, and the native cultures of North and South America represent the skill and creativity of their designers. The architecture of these civilizations differs from traditional Western architecture.

Unit 6

In This Unit

6.1 Rocks and the Rock Cycle, p. S90

6.2 Rock-Forming Minerals, p. S99

6.3 Stories in Rocks, p. S105

Reading *Plus*

Now that you have been introduced to the movements of the earth's crust and the rocks of which it is made, let's take a closer look at the make up of rock, and the kinds of information it can give us. Read pages S90 to S110 and construct a model representing the rock cycle using as many examples and rock samples as possible. Your model should include concepts represented by the following questions.

1. In what ways do the forces at work on the earth drive the pathways of the rock cycle?

2. How are minerals related to rocks? What characteristics can be used to identify different minerals?

3. How do rocks record change over time? In what ways can we determine the history of the earth using rocks?

6.1 ROCKS AND THE ROCK CYCLE

Have you ever picked up a rock because of its interesting shape, sparkle, or color? Perhaps you own a rock collection. If so, you already know something about the variety of rocks that can be found on the earth. You have learned by now that rocks are the materials that compose the earth's solid crust. Sometimes these rocks form spectacular natural structures, such as the one pictured in Figure 6-1. When small and polished, they can be turned into gems used in making jewelry. Besides being interesting or spectacular, rocks are important to us both scientifically and economically. They not only contain the history of the earth, but are also the source of our planet's riches.

Figure 6-1 Devil's Tower in eastern Wyoming is made of igneous rock.

The Formation of Rocks

Studies of the earth indicate that the earth was probably once made mostly of molten materials. As the earth cooled, the materials at the earth's surface solidified into the first rocks. These rocks were then exposed to forces that caused them to change. Other rocks developed as a result of these changes. By studying the rocks and the forces that change them, geologists have identified three basic processes by which rocks form.

You can simulate rock-forming processes by using candle wax. Suppose you take some old candles and melt them in a pan. If you then pour the melted wax onto paper or into a mold and allow it to cool, the liquid wax will gradually harden into a solid mass. In a similar manner, rocks solidify from molten rock, or *magma*. Rocks that form when the magma cools are called *igneous rocks*. See Figure 6-2. Magma is composed of a mixture of substances called **minerals**. As the magma cools, molecules of the same mineral substances link together to form solid structures called **crystals**. Igneous rocks are solid masses of these mineral crystals.

Now, shave off some thin pieces of the solid wax. If you place a pile of these shavings under a heavy object, such as a book, the shavings will eventually stick together into a single piece. In the same way, most *sedimentary rocks* form from rock pieces that were broken up by weathering processes. See Figure 6-3. Deposits of these pieces then stick together, or *lithify*, by compaction or cementation.

Mineral A naturally occurring substance consisting of a single element or compound.

Crystal A three-dimensional structure in which each face, or surface, has a definite shape and orientation.

Figure 6-2 The dark rocks of the Pacific islands were formed of solidified magma.

Figure 6-3 The mud of this dry lake bed may eventually become sedimentary rock. The mud cracks here may also be preserved in the rock that forms.

Finally, if you warm the remaining solid wax slightly, you should be able to bend it and squeeze it by applying pressure. In a similar fashion, the heat and pressure inside the earth changes both igneous and sedimentary rocks into *metamorphic rocks*. Sometimes the minerals in the rocks are rearranged into a different pattern. In other cases, entirely new minerals are formed.

The Rock Cycle

The formation of igneous, sedimentary, and metamorphic rocks are the individual steps of an endless process that has been going on for billions of years. Old rocks are constantly being made into new rocks by a continuous process called the **rock cycle**. A simplified version of the rock cycle is shown in Figure 6-4.

Rock Cycle The continuous process of change in which new rocks are formed from old rock material.

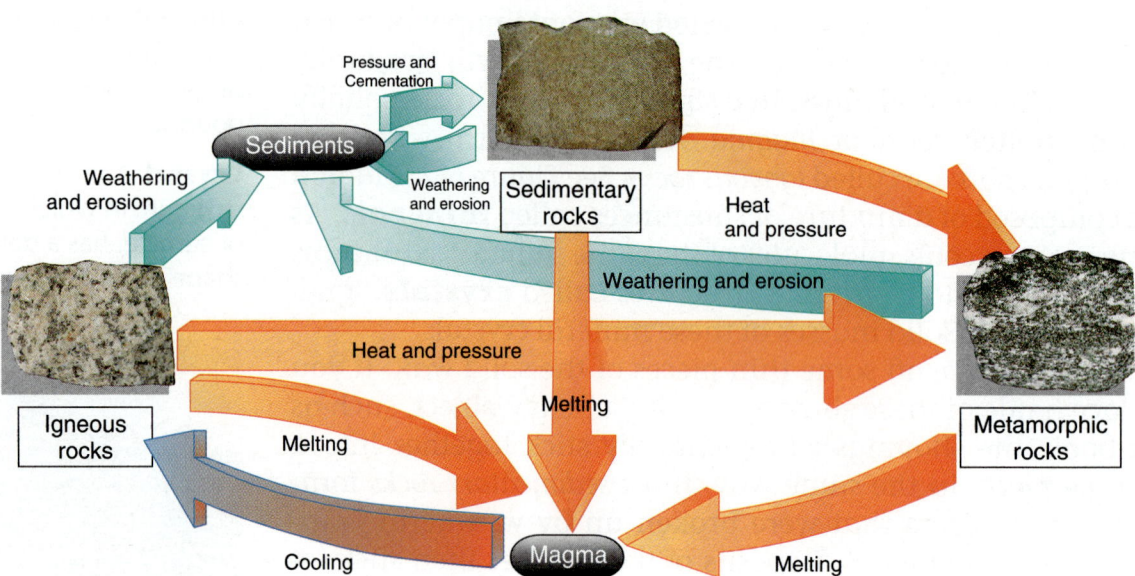

Figure 6-4 The rock cycle is not a one-way process. Any rock may be changed into any other type of rock by going through the cycle.

Figure 6-5 Gneiss is a metamorphic rock formed when granite is put under high temperature and pressure.

As you examine the diagram of the rock cycle, consider that any rock you see has been through at least one pathway in this cycle. When magma cools, for example, it may form the igneous rock, granite. Granite can then be broken by physical and chemical weathering into particles such as sand. Deposits of this sand can be cemented into sandstone, a sedimentary rock. If the sandstone is exposed to heat and pressure, it can be changed into the metamorphic rock called quartzite. Finally, quartzite may later be forced deep into the earth where it can be remelted to become magma once more. This magma, in turn, could then become more igneous rock.

But, what other pathways are there? Suppose granite is not broken down into sand, but exposed to high temperature and extreme pressure. In this case, the granite could become a metamorphic rock called gneiss. See Figure 6-5. The granite itself may also be remelted to become magma.

Scientists believe that much of the rock cycle is taking place at the edges of large pieces of the crust called *tectonic plates*. For example, magma comes to the surface both along the mid-oceanic ridges and near ocean trenches. See Figure 6-6. New igneous rocks form at ridges from dark heavy magma. In the trenches, old rock is being forced down into the mantle. There, the lighter rocks of the crust are melted and mixed with other rock material. This material may once again make its way back to the surface at another underwater ridge.

The rocks of the earth's crust have gone through the rock cycle many times. The rocks near the earth's surface are rarely older than 3 billion years. Since the earth is estimated to be 4.5 to 5 billion years old, most of the original rocks of

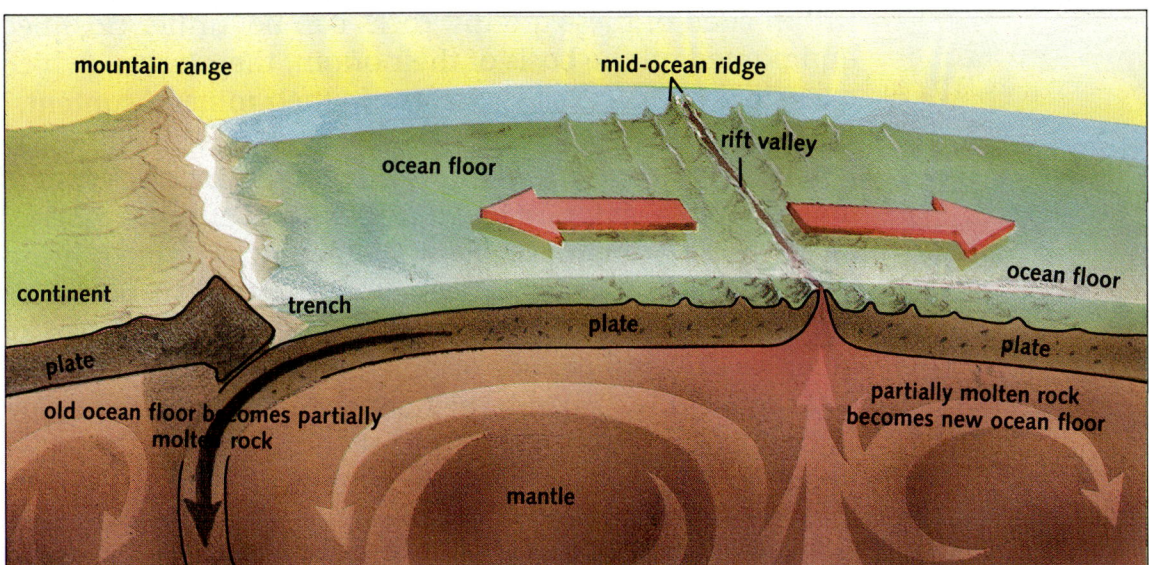

the earth's crust have been changed by the rock cycle since they were first formed.

Classifying Rocks

What about you makes you a *person*? Is it your hair, your face, or perhaps the fact that you have two arms and legs? Is it your personality or IQ? Whatever it is, you have definite characteristics that make you recognizable as a person. Still, you are different from every other person on earth. You are unique!

Rocks also have definite characteristics. For example, look at the photographs in Figure 6-7. You can tell that the objects in these photographs are rocks. But, what is it about these objects that makes them rocks? Just as you share certain characteristics with all other humans, rocks share certain characteristics as well. Now, look closer. Are the two rocks the same? Again, just as you are different from other people, rocks have many differences too.

Like living organisms, rocks can be classified by their similarities and differences. One way rocks are classified is by the way they are formed. On this basis, rocks can be classified as igneous, sedimentary, and metamorphic. In addition,

Figure 6-6 The mid-ocean ridge circles parts of the earth much like a seam circles a baseball. New material is added to the earth's surface at the mid-ocean ridge. In other parts of the world, the plates are forced back into the mantle.

Figure 6-7 Although these rocks look different, they belong to the same rock family. As with brothers and sisters of human families, members of the same rock family share many characteristics.

Figure 6-8 Molten magma from within the earth often comes to the surface as lava. As the lava cools, igneous rock is formed.

within each of these groups, rocks can be further classified and named on the basis of their origin, their *texture* (which refers to crystal or particle size), and their mineral content.

Igneous Rocks As you know, there are two basic types of igneous rocks—*intrusive* igneous rocks and *extrusive* igneous rocks. These names indicate *where* the rocks formed. Intrusive igneous rocks form deep underground in bodies of magma that squeeze in between, or *intrude,* into layers of other rocks. Long periods of time are required for intrusive rocks to solidify, because the heat of the magma dissipates slowly underground. Extrusive igneous rock, on the other hand, come from *lava.* Lava is molten rock that exits, or *extrudes,* from inside the earth through openings such as volcanoes. Above ground, lava cools quickly and soon hardens into extrusive rock. See Figure 6-8.

Although intrusive and extrusive rocks are made of the same types of minerals, they are easily distinguished by their texture. These two types of rock have very different appearances due to their differing textures. For example, look at the two rocks in Figure 6-9. If you look closely, you can see

Figure 6-9 Granite (left) displays coarse texture, while rhyolite (right) displays fine texture.

Figure 6-10 Notice the coarse-grained texture of this sample of gabbro, an intrusive igneous rock.

that granite, an intrusive igneous rock, is composed of large, visible crystals. The slow cooling of intrusive rocks allows large crystals to grow. On the other hand, if an igneous rock cools quickly, it feels fairly smooth and may not have any visible crystals. Geologists say that such rocks, like the rhyolite shown in Figure 6-9, have *fine texture.*

Mineral content affects the color and density of igneous rocks. Basalt, for example, is dark and heavy because it contains large amounts of the elements iron and magnesium. Gabbro, which is also dark gray, is made of the same minerals as basalt. See Figure 6-10. Basalt and gabbro are both made from magma that comes from deep inside the earth. Basalt and other dark extrusive rocks form the crust on the ocean bottom. The lighter-colored rocks, such as granite and rhyolite, contain elements, such as calcium and sodium, that also make them less dense. These minerals make up most of the rocks of the continental crust.

Sedimentary Rocks Most sedimentary rocks are made from sediments, which are pieces of other rocks. For this reason, sedimentary rocks are classified on the basis of the type of sediments they contain. There are three different kinds of sediments that can become sedimentary rocks—clastic, chemical, and organic sediments. Therefore, there are three major types of sedimentary rock.

Clastic sedimentary rocks are made of broken pieces of weathered rocks. These pieces can be large, rounded pebbles, sharp-edged chunks, or soil particles such as sand, silt, or clay. Clastic rocks are found in layers because of the way they are formed. When agents of erosion, such as water and wind, carry weathered rocks away from their source, the particles become *sorted* according to their size. For example, a swift-flowing mountain stream may carry large pebbles and even boulders as well as sand, silt, and clay. As the water slows down, it drops the largest particles first, then gradually smaller pieces follow. By the time the stream becomes a slow-moving river, only the smallest particles are left. This sorting produces layers of particles that are about the same size. Sorting also happens in the ocean when sediments are deposited there.

There are several types of clastic sedimentary rocks based on their particle size. Conglomerates and breccias, for example, are made of large pebbles. Although their particles may be the same size, conglomerates consist of rounded pebbles, while in breccias, the pieces are angular and have sharp edges. See Figure 6-11. Sandstones and siltstones are made of sand-sized and silt-sized soil particles respectively. Shale is composed of clay—the finest of rock particles. See Figure 6-12. In forming sedimentary rocks, larger particles, such as pebbles, sand, and silt, must be "glued" together by

Figure 6-11 Conglomerate (top), *breccia* (middle), *and sandstone* (bottom) *are all clastic sedimentary rocks.*

Figure 6-12 Shale (left) *and siltstone* (right) *are composed of nearly microscopic particles.*

a cementing agent. Shale, however, can become a rock when the excess water is squeezed out of it by compaction.

Chemical sedimentary rocks form when minerals that were dissolved by weathering processes separate from the water and crystallize. In some cases, the minerals separate by the process of evaporation. As water evaporates, the dissolved minerals form crystals and eventually a rock. Rock salt and gypsum are two examples of chemical sedimentary rocks that form by evaporation. See Figure 6-13. When conditions are right, dissolved minerals simply fall out of solution by a process called *precipitation*. Limestone is a common rock that forms in this way. Another familiar rock that precipitates from sea water is chert, which is a dense rock made of quartz. See Figure 6-14. Precipitates such as calcium carbonate and quartz are two of the materials that cement the particles of clastic rocks together. These minerals fill in the spaces around rock particles much like mortar fills the spaces between brick or stone in a wall.

Figure 6-13 *Rock salt* (top) *and rock gypsum* (bottom) *are chemical sedimentary rocks.*

Figure 6-14 *Compact limestone* (left) *and chert* (right) *are both precipitated from sea water.*

Figure 6-15 *This limestone cliff was formed from the shells of millions of dead sea animals, and is therefore a type of organic sedimentary rock.*

Organic sedimentary rocks are a type of sedimentary rock that comes from the action or remains of once-living organisms. For example, limestone (which can also be classified as a chemical sedimentary rock) forms primarily by the action of marine organisms that remove calcium carbonate from the water to build their shells. Layers of limestone can build up in oceans when these organisms die and their shells settle to the bottom. See Figure 6-15. Chalk is a type of limestone that is formed in this way. Coal is also considered to be a type of organic sedimentary rock. As you know, coal is thought to be formed from masses of dead plants that were compressed for millions of years. As this organic material was gradually compressed, it became denser and harder. Two types of coal are considered to be organic sedimentary rocks. They are lignite, a very soft, dark brown coal, and the slightly harder, black bituminous coal. See Figure 6-16.

Figure 6-16 Stages in coal formation. In the last stage, coal becomes a metamorphic rock.

Metamorphic Rocks Just as living things go through physical changes during their lives, many rocks change physically as well. Metamorphic rocks form from other rocks when they are exposed to heat and pressure. Some of these changes are structural changes that result in the rearrangement of the crystals in a rock. In other cases, chemical changes occur, recombining the original minerals of the rock into new ones. See Figure 6-17.

Figure 6-17 Shown here are three examples of metamorphic rocks. In each case, the sample on the right is the metamorphic rock, and the sample on the left is the parent rock from which it formed.

Figure 6-18 *Marble is a nonfoliated metamorphic rock. The striations in this photograph were made when the marble was cut.*

Geologists divide metamorphic rocks into two groups based on how they were formed. *Foliated rocks* have obvious "bands" or "zones" of like materials that indicate they were formed by pressures coming from one direction. These bands often appear to be shiny, due to the melting and reforming of mineral crystals. *Nonfoliated rocks* have no visible alignment, like the marble in Figure 6-18. In this case, limestone was exposed to pressures from several directions.

The amount of foliation, or banding, is used to classify metamorphic rocks. For example, slate, schist, and gneiss are types of foliated rocks. Slate, with the least amount of foliation and no visible crystals, shows the least amount of alteration by the processes of metamorphism. As you can see in Figure 6-19, schist and gneiss have larger crystals and more definite bands and therefore were exposed to greater amounts of temperature and pressure.

When bituminous coal is put under intense heat and pressure, it becomes a harder, more compact form of coal called anthracite. Anthracite is very hard and has few of the impurities found in bituminous coal.

Figure 6-19 *The crystal structure of gneiss (right) is much more defined than that of schist (left).*

Summary

As the earth cooled, some of its molten materials solidified into the first rocks. Rocks that solidify from magma are called igneous rocks. Sedimentary rocks form from pieces of other rocks that were broken up by weathering. Heat and pressure inside the earth changes both igneous and sedimentary rocks into metamorphic rocks. The rock cycle is a continuous process in which old rocks are constantly being made into new rocks. Rocks are classified on the basis of their origin, texture, and mineral content. Igneous rocks are classified as either intrusive or extrusive. There are three different kinds of sedimentary rocks—clastic, chemical, and organic. Metamorphic rocks are divided into foliated and nonfoliated rocks.

6-2 ROCK-FORMING MINERALS

As magma and lava cool, the chemical compounds that make up the molten rock link together to form crystals of certain minerals. Most minerals are made up of at least two kinds of atoms. That is, minerals are usually made up of two or more chemical elements.

Kinds of Minerals

The earth's crust contains over 2000 different minerals, only about 20 of which are very common. These 20 minerals are the basic rock-forming minerals of the earth's crust. Whether or not they are common, all minerals can be classified into two major categories based on their chemical composition.

Silicates Silicon and oxygen are two of the most common elements in the earth's crust. See Figure 6-20. Therefore, it is not surprising that many minerals contain these two elements. Minerals that contain silicon and oxygen are called *silicates*. Silicate minerals make up about 96 percent of the earth's crust. The silicates are found in igneous, sedimentary, and metamorphic rock.

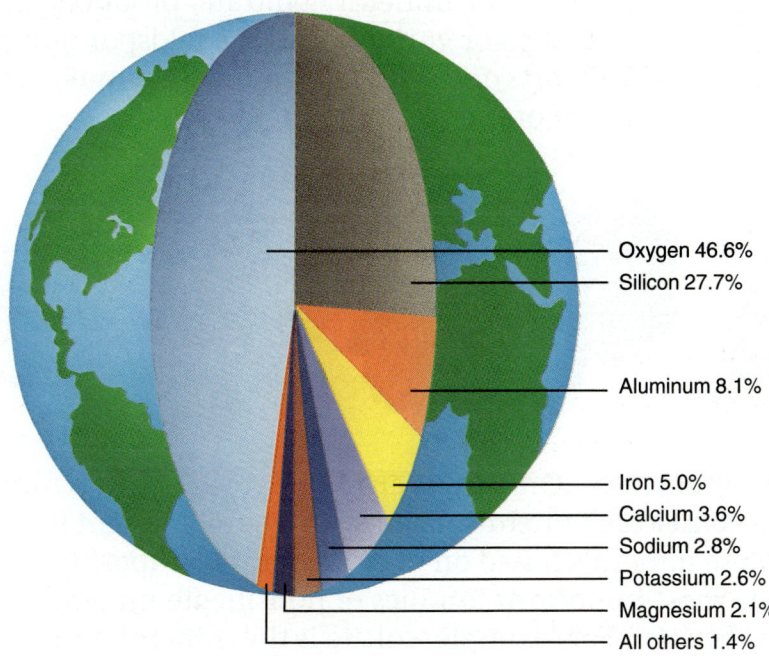

Figure 6-20 The earth's crust is made up primarily of these elements. The graph shows the percentage by mass of each element.

One common silicate mineral that you might be familiar with is quartz. See Figure 6-21. A crystal of quartz is composed of single silicon atoms, each bonded to four oxygen atoms. These atoms form a network in the shape of a *tetrahedron*, which is a three-sided pyramid. When the

Figure 6-21 Pure quartz is colorless.

Figure 6-22 *A silicon atom and oxygen atoms are joined in this pattern. How does the number of oxygen atoms and the number of silicon atoms compare?*

silicon–oxygen tetrahedrons bond together to form chains or sheets, a very hard mineral is formed. See Figure 6-22.

Feldspars are the most abundant silicate minerals. In addition to silicon and oxygen they contain aluminum and calcium, potassium, or sodium atoms in their crystals. The chemical composition and the presence of impurities, such as iron or flecks of the mineral hematite, produces the variety of forms and color variations of the feldspar group. See Figure 6-23. Quartz and the feldspars together make up more than 50 percent of the earth's crust.

Figure 6-23 *Three types of feldspar.*

Nonsilicates Only 4 percent of the earth's crust is composed of *nonsilicate* minerals. There are several different families of these minerals based on their chemical composition. One of the most important families of nonsilicate minerals are the *carbonates*. For example, calcite is the mineral name for calcium carbonate, the substance that makes up limestone. See Figure 6-24. Dolomite is another example of a carbonate mineral. Carbonates are common in sedimentary rocks.

Other mineral families are characterized by different chemical groups. *Oxides* contain oxygen bonded to an element other than silicon, such as iron or aluminum. These

Figure 6-24 *Calcite—a carbonate mineral.*

minerals, for example magnetite (iron oxide), are important sources of metals. See Figure 6-25. *Sulfates*, such as gypsum, contain sulfur and oxygen. You may recall that gypsum is also the name for a chemical sedimentary rock that forms by evaporation. Sometimes the name of a rock and the mineral it is made of are the same. Other types of nonsilicate minerals are the *halides*, such as halite and fluorite, which contain a halogen, and the *sulfides*, such as galena and pyrite, which contain sulfur.

Native Elements Some nonsilicate minerals contain only one kind of element. These rare minerals are called *native elements*. Many of the minerals in this group are valuable metals. Some examples are gold, silver, and copper. See Figure 6-26. Nickel and iron may also appear as native elements, but they are very rare in the earth's crust. However, much of the interior of the earth is believed to consist of uncombined nickel and iron. Diamond and graphite are also native elements because they are made of pure carbon. They differ only in the way the carbon atoms are arranged.

Figure 6-25 *Oxides, such as those shown above, are commercial sources of iron, aluminum, and other metals.*

Figure 6-26 *Native elements (clockwise from the top) copper, sulfur, gold, silver, and lead.*

Identifying Properties of Minerals

Detailed analysis of minerals can be done only in a laboratory with special equipment. However, there are some simple physical properties that you can use to quickly identify a mineral sample.

Color Many minerals have bright colors. For example, cinnabar is red, azurite is a deep blue, serpentine is green, and sulfur is a bright yellow. Unfortunately, color may be one of the least dependable characteristics you can use to identify a mineral. The color of many minerals is due to the presence of small amounts of other elements called impurities.

Figure 6-27 Quartz is found in many colors due to mineral impurities.

Figure 6-28 Metals, such as gold and silver, have a luster that can be easily recognized. Other types of mineral luster are named for common substances, such as pearls and earth (soil).

Figure 6-29 Streak is a good characteristic for identifying minerals since it is not influenced by such things as tarnishing and impurities.

For example, pure quartz is colorless, but with impurities it can be pink, tan, red, purple, or black. See Figure 6-27.

Luster Have you ever admired the shine of a newly waxed car? What you are admiring is the *luster* of the car's surface, or the way it reflects light. The luster of a mineral also refers to the appearance of its surface in reflected light. Each of the minerals has a characteristic luster. Some of the terms used to describe luster—metallic, glassy, waxy, pearly, greasy, or earthy—are related to familiar objects. How would you describe the luster of each of the minerals shown in Figure 6-28?

Streak If you rub a piece of chalk on a sidewalk, it leaves a white mark on the concrete. Some minerals also leave marks when rubbed on a rough surface. The color of the powder in these marks is called the mineral's *streak*. See Figure 6-29. Sometimes the color of the streak is different from the color of the mineral sample. For instance, the streak of golden-colored pyrite is greenish black, and the streak of silver-colored hematite is brick red.

Hardness Another characteristic of minerals is their hardness, or resistance to being scratched. Mohs' Scale of Mineral Hardness is a set of ten standard minerals that is used to compare the hardness of all minerals. See Table 6-1. You probably do not have a sample of each of the minerals on Mohs' Scale, but you can still determine the hardness of an unknown mineral by using some common objects. For example, if a certain mineral can scratch a penny, its

MOHS' SCALE OF MINERAL HARDNESS		
Mineral	Hardness	Common Test
Talc	1	Easily scratched by fingernail
Gypsum	2	Can be scratched by fingernail
Calcite	3	Barely scratched by copper penny
Fluorite	4	Easily scratched with steel knife blade
Apatite	5	Can be scratched by steel knife blade
Feldspar	6	Easily scratches glass
Quartz	7	Scratches both glass and steel
Topaz	8	Scratches quartz
Corundum	9	No simple test
Diamond	10	No simple test

Table 6-1

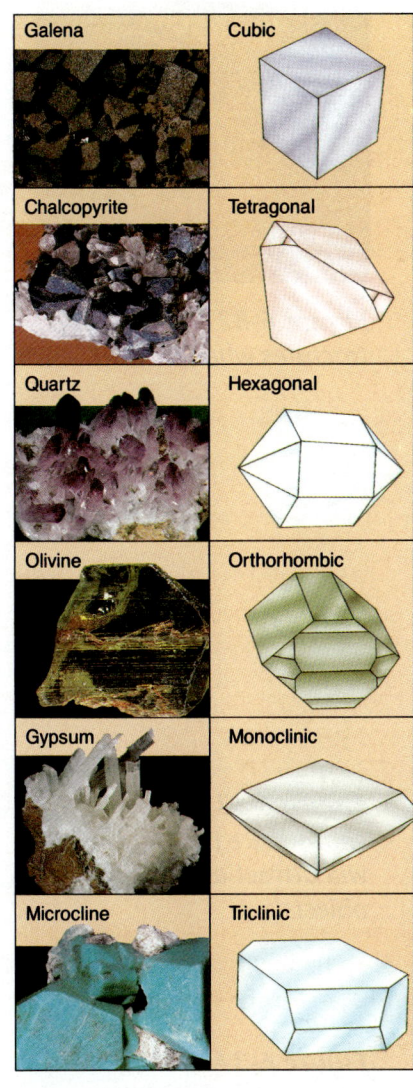

Figure 6-30 *Representative minerals and their crystal shapes.*

hardness is greater than 3. If a steel nail can scratch it, you know its hardness is less than 5.

Crystal Shape Each mineral has a characteristic crystal shape that can be useful in identifying the mineral. This shape is due to the orderly arrangement of the atoms that make up the mineral. See Figure 6-30.

Cleavage and Fracture These terms describe how minerals break. Minerals that always break along flat surfaces have a property called *cleavage*. The cleavage of a particular mineral always occurs in the same way and produces crystal faces that have characteristic angles. Halite, for example, breaks into cubes, while calcite pieces tilt diagonally, and mica peels off in thin sheets. See Figure 6-31.

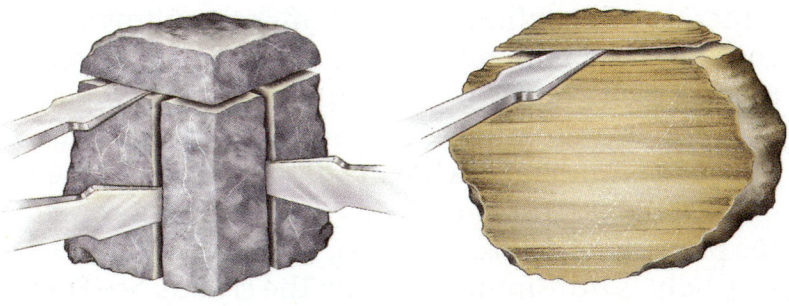

Figure 6-31 *The diagram on the left demonstrates cleavage in three directions, as would be found in galena. The diagram on the right demonstrates cleavage in one direction, which is characteristic of mica.*

Some minerals do not have cleavage but instead break into irregular pieces. This pattern of breaking is called *fracture*. Minerals can show fracture in several different ways. Some fractures have a curved surface that resembles the inside of a clam shell. For example, quartz and obsidian break

Figure 6-32 *Obsidian fractures when it breaks.*

in this way. See Figure 6-32. Fracturing may also produce jagged and uneven surfaces.

Special Properties Minerals show a variety of other interesting properties, some of which are unique to a certain mineral. Magnetite, for example, is a type of iron oxide that is *magnetic*. Rocks that contain magnetite attract iron objects just as a magnet does. See Figure 6-33. Other minerals are *radioactive*, for example, pitchblende, which contains the element uranium. Still other minerals, such as calcite, give off light when exposed to ultraviolet light or X rays. This property is called *fluorescence*. See Figure 6-34. Some

Figure 6-33 *The loadstone shown here is magnetic and will attract a variety of iron objects.*

Figure 6-34 *Notice the change in color of the various fluorescent minerals in this rock as they go from ordinary light (left) to ultraviolet light (right).*

minerals will continue to glow after the ultraviolet light is cut off. These minerals display the property of *phosphorescence*. Calcite has two other interesting properties as well. When a weak solution of hydrochloric acid is placed on a sample of limestone (which is made of calcite) the acid begins to bubble. Furthermore, a very clear calcite crystal shows the property of *double refraction*, which occurs because of the way light is transmitted through the crystal. See Figure 6-35.

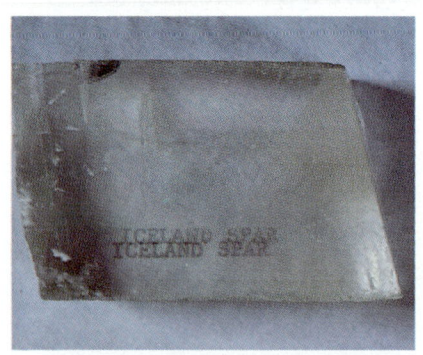

Figure 6-35 *The mineral calcite exhibits double refraction.*

Summary

Minerals can be classified into two major categories based on their chemical composition. Minerals that contain silicon and oxygen are called silicates; those that do not are called nonsilicates. Some nonsilicate minerals are native elements, many of which are valuable metals. The physical properties of color, luster, streak, hardness, crystal shape, and cleavage or fracture can be used to identify a mineral. Other interesting mineral properties include magnetism, radioactivity, fluorescence, phosphorescence, and double refraction.

6-3 STORIES IN ROCKS

Imagine that you find a family diary from 100 years ago. As you lift it, the binding breaks and the pages fall out in bunches. Several bunches fall together, landing in a pile on the floor. Other bunches scatter away from the pile. How are you going to put the diary together again? How will you decide the order in which the pages belong? Is it possible that some pages may be missing? See Figure 6-36.

In a similar way, the earth's history is written in layers of rock. Trying to piece this history together is like putting the diary mentioned above back together in order. However, piecing together earth's geologic history is more difficult because many of the "pages" have been widely scattered and even destroyed. For example, individual rock layers have been separated by processes such as continental drift and mountain-building. Others are lost due to erosion or remelting. Still others are out of order because they were overturned by movements of the crust. Although the story is incomplete, certain clues in the rocks can be used to piece together a fairly complete picture of the earth's past.

Figure 6-36 What clues could you use to put this diary back together?

Uniformity

Have you ever noticed how uniform or consistent some things are? For instance, next time you open a loaf of bread, notice how uniform the texture of the loaf is. A slice of bread from a different loaf of the same brand of bread still shows the same uniformity of texture. The bread-making process insures this uniformity. When the same process is used, even at different times, the results are similar. See Figure 6-37.

Uniformity is an underlying principle of the basic geological processes. In the same way that a bread-making process makes uniform loaves of bread, the geological processes that form rocks and mountains today formed similar rocks and mountains in the past. This idea was first presented as the *principle of uniformitarianism* in 1785 by James Hutton, a Scottish physician and geologist. Simply put, this principle

Figure 6-37 These loaves of bread are uniform in color and texture because they were made by the same process.

states that similar processes, in different times, produce similar results. Observations support Hutton's theory. For example, sediments may be carried away by water or wind and then deposited in new places, but they are always deposited in similar ways regardless of where or when. Over long periods of time, these sediments are compressed and cemented to form sedimentary rocks.

Records of Environmental Change

Did you ever think to look for seashells on top of a mountain? Probably not. But nevertheless, fossil seashells are found in the rocks of mountains throughout the world. See Figure 6-38. For this to be true, either the rocks have moved or the environments in which they formed have changed.

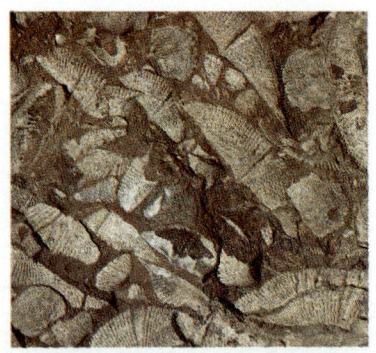

Figure 6-38 *Fossils, such as brachiopods, indicate that the origin of this rock was in an underwater environment.*

Like a mystery story, certain rocks contain clues about what happened in the past. Geologists can "read" these clues to figure out the mystery of their history! With these clues, they can figure out what the environment was like when a rock was formed and how those conditions have changed. For example, what does finding fossil seashells in the rocks of a mountain tell you? A geologist would say that the rocks first formed in an ocean, and then were uplifted to become a mountain.

The kinds of fossils contained in rocks can tell us where the rocks were formed—in an ocean, by a stream, or on land. This is possible since the organisms that live in each place are different from one another. Sedimentary structures, such as ripple marks, mud cracks, and sand dunes, also indicate what the environment was like when and where the rocks were formed.

The shape of mineral grains may also help geologists determine what sort of environment existed when rocks were formed. For instance, the rounded quartz grains of sandstone probably formed on a beach. Geologists know this because the action of waves rounds the sand grains, and sand-sized sediments are often deposited near shore.

The size of the fragments in sedimentary rocks is another clue to the conditions under which the rocks formed. Large fragments are deposited first as a stream slows. In a similar way, the movement of waves also separates fragments by size. Large fragments are deposited close to shore to form sandstone, while smaller particles are carried out into deeper water. See Figure 6-39. Ocean currents may carry clay particles far from land. In quiet water, away from the beach, clay settles to the bottom to form shale. Eventually, all the particles settle to the bottom. In the deepest, quietest parts

Figure 6-39 The lighter sediments are carried much further out into the sea than the larger, heavier sediments.

of the ocean, limestone forms as dissolved calcite precipitates from the water.

Relative Age One clue to the order of the pages in the earth's diary is the *law of superposition*, which simply states: younger rock layers lie on top of older rock layers. In other words, the earth's diary reads from back to front. As the layers of sedimentary and extrusive igneous rocks formed, the oldest layers were deposited first. Successive layers formed one at a time, one on top of the other. Thus, when you look at the layers of rock in a cliff, the law of superposition tells you that the top layer was deposited last; and as you go down the cliff, the layers get older. See Figure 6-40.

Another geologic principle states that layers of sedimentary and extrusive igneous rock are laid down in flat, level beds. Once this happens, however, these level beds may become tilted or offset by the processes of folding and faulting.

Look at Figure 6-41. Can you list the rock layers from oldest to youngest? If you said that **H** is the oldest layer and **A** is the youngest layer, you are correct. However, this is not the only information that can be determined from this diagram. Notice that the lower layers lie at an angle and are broken and offset, while the top layers remain level and unbroken. The boundary between layers **E** and **D** indicates that pages are missing from the earth's diary in this area. Such a boundary is called an **unconformity**. There is also a body marked "magma" that cuts across the lower layers. Do you notice anything else? How can these patterns be interpreted?

Layer **H** was deposited first, followed by **G**, **F**, and then **E**. Next, these layers were broken by a fault and tilted by mountain-building, and then invaded by a magma intrusion. A period of erosion then wore down the mountains to a flat, level plain. Finally, the process of deposition began again, laying down layers **D**, and then **C**, **B**, and finally **A**.

Figure 6-40 Geologists can study the exposed rocks of the Grand Canyon to learn about the geologic history of the region.

Unconformity A boundary between rock layers that indicates that layers are missing due to erosion.

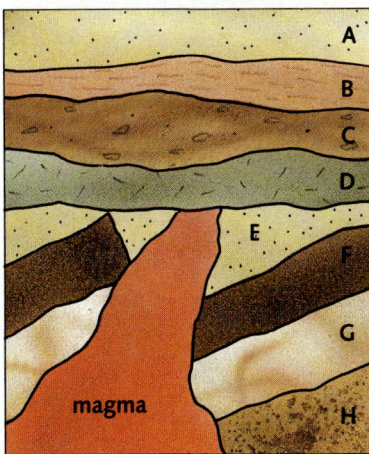

Figure 6-41 List the relative ages of these rocks from oldest to youngest.

Relative Age The age of a rock (older or younger) in comparison to other rocks.

Figure 6-42 This sequence of events leads to the kind of unconformity in which horizontal layers overlie tilted layers. This process requires millions of years.

Using the law of superposition, geologists can determine the **relative age** of rocks by studying the order in which they were laid down. It is important to remember that the relative age of a rock cannot be given in years; it can only state that one rock is older or younger than another rock. You might think that this information is not very useful; however, it is very important to a geologist who is trying to put together the geologic history of an area. By determining the relative age of rock layers and studying the way they are tilted or broken, a sequence of geologic events can be developed. See Figure 6-42.

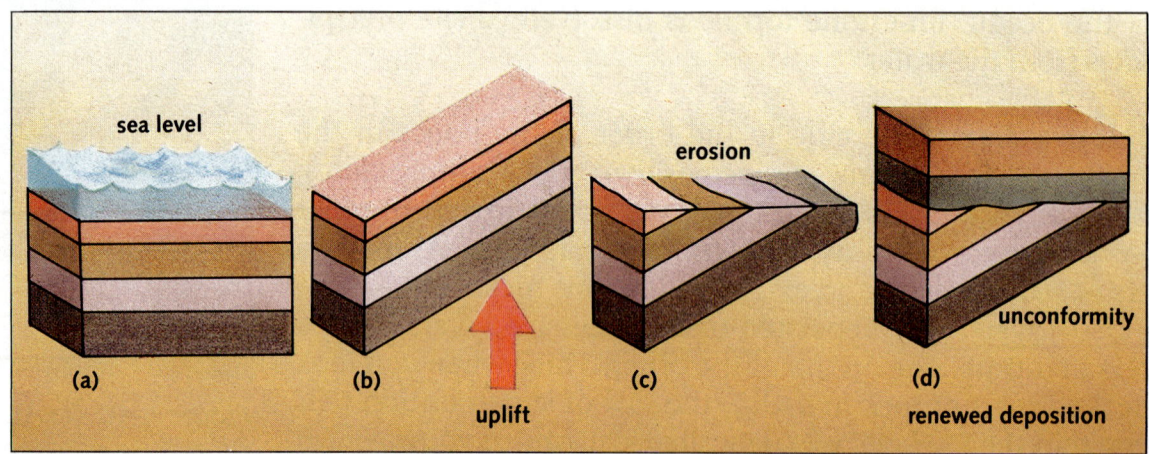

Half-life The time it takes for half of the atoms of a radioactive isotope to decay.

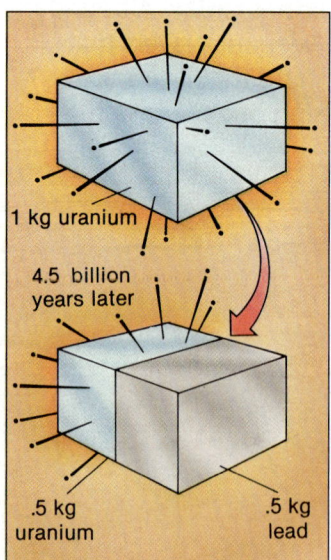

Figure 6-43 After 4.5 billion years, half of all the uranium atoms will have changed into lead.

Numerical Dating You may recall that some elements tend to be unstable and therefore undergo *radioactive decay*. By giving off atomic particles, these radioactive elements change into more stable elements. For example, uranium decays to eventually form lead—a very stable, nonradioactive element.

Some radioactive isotopes of elements are very unstable and decay rapidly; others decay more slowly. The time it takes for half of the atoms of a radioactive isotope to decay is called its **half-life**. Each successive half-life reduces the remaining number of atoms by one-half, turning them into atoms of a more stable element. See Figure 6-43. Some isotopes have a half-life shorter than one second; others have a half-life of billions of years.

Since radioactive isotopes decay at known rates, they act like an internal clock to mark the passage of time. Thus they can be used to find the approximate age of a material. *Numerical dating* (also know as absolute dating) is the process of estimating the age of a sample using the radioactive isotopes in the sample. Rocks, minerals, fossils, water, and ice are dated using this method. The isotopes most commonly used in numerical dating are listed in Table 6-2.

ISOTOPES COMMONLY USED IN NUMERICAL DATING		
Isotope	Half-life (yrs)	Material to which applied
Tritium	12	Ground water, sea water, ice
Carbon-14	5,730	Wood, bones, shells
Potassium-40	1,300,000,000	Rocks, minerals
Rubidium-87	48,000,000,000	Rocks, minerals
Thorium-232	14,000,000,000	Rocks, minerals
Uranium-235	704,000,000	Rocks, minerals
Uranium-238	4,500,000,000	Rocks, minerals

Table 6-2

One problem with dating rocks is that scientists do not know how much radioactive isotope was originally in a rock when it was formed. Therefore, the age of a rock cannot be found by determining the amount of isotope remaining. Scientists can, however, measure the amounts of both the parent isotope (such as uranium) and the decay product (lead). The ratio of decay product to parent isotope gives the approximate age of the sample. The older the mineral or rock, the more decay product there will be.

Numerical dating, however, cannot be used to date sedimentary or metamorphic rocks, which are made of pieces of other, much older rocks. In sedimentary rock, measuring radioactive elements and their decay products gives the age of these older rock pieces, not the age of the younger sedimentary rock itself. Furthermore, as a metamorphic rock forms, heat and pressure alters the rock's original minerals and "resets" the numerical clock, affecting the amounts of both the original isotopes and the decay products.

Sedimentary and metamorphic rocks must be dated by relative dating. A nearby igneous formation can be dated numerically, and then the age of the neighboring sedimentary or metamorphic formations can be estimated by their position relative to the igneous rock. See Figure 6-44.

Figure 6-44 This igneous rock intrusion may be helpful in determining the relative ages of the surrounding rocks.

Fossil Correlation Sometimes rocks are moved far away from where they were first formed—for example, when a continent breaks apart and then becomes separated by an ocean. When this happens, it becomes difficult to compare layers of rock between the two sections. However, by using the fossil remains of plants and animals, the age of rock layers can be determined.

On a geologic scale, plant and animal species exist for only a limited time. Because life forms change over time, fossils can be used as time indicators. Such fossils are called *index fossils*. Different index fossils appear in sedimentary rock of different ages. By numerically dating the fossils, or by dating similar fossils that occur elsewhere, the approximate age of sedimentary rock can be determined.

Fossil correlation is the use of index fossils to determine the relative age of widely separated rock layers. If similar fossils are found in rock layers separated by an ocean, the two rock layers would still be the same age. By comparing index fossils, and the rock layers in which they are found, geologists have been able to combine their observations to make a **geologic column**. See Figure 6-45. In this way, fossils and the rock layers they are found in can be used to sort out the pages of the earth's diary.

Geologic Column An arrangement showing rock layers in the order in which they were formed.

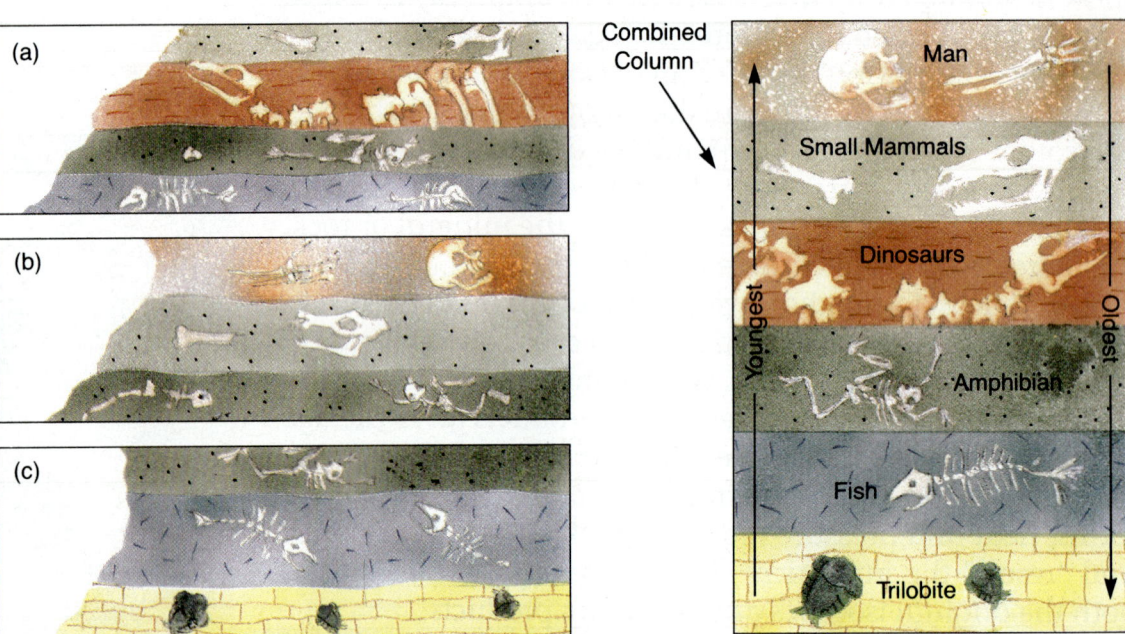

Figure 6-45 Layers having the same color and fossils (left) are the same geologic age no matter where they are found. A combined geologic column (right) can be constructed with each layer in order by its age.

Summary

Earth's history is written in layers of rock. Geologists piece this history together using the underlying principle of uniformity. Certain rocks contain clues that geologists can use to figure out what the environment was like when a rock was formed and how the conditions have changed. Using the law of superposition, geologists can determine the relative age of rocks. Radioactive isotopes can be used to find the approximate numerical age of geologic material. By using index fossils, the relative age of widely separated rock layers can be correlated.

Unit 7

In This Unit

7.1 Upward and Outward, p. S112

7.2 Stars: A Universe of Suns, p. S118

7.3 One Small Step, p. S126

Reading *Plus*

Now that you have been introduced to astronomy and human exploration of the universe, let's take a closer look at the members of our Solar System and beyond, and how humans further their investigation of space. Read pages S112 to S132 and write a short science-fiction story about one aspect of space or space exploration. Your story can be based on ideas taken from the following questions.

1. What are the differences and similarities among the various bodies of the Solar System?
2. How do astronomers get information from the stars?
3. In what way is our knowledge of the universe increasing?

7.1 UPWARD AND OUTWARD

The Earth, our home, is only one member among the multitudes of objects in space. Looking at the Earth from space, we see it as an entire planet. See Figure 7-1. From this vantage point, we have been able to get first-hand information about our own planet in space. But for most of history, people have looked up and out into the blackness of space and could only wonder at the marvels they saw.

Recently, however, we have been able to get a better look at some of our closest cosmic companions. The new knowledge we have gained, and are continuing to learn, is sometimes stranger than fiction. Let's look at some of this information our ancestors only dreamed about.

Figure 7-1 View of Earth from the first Apollo mission to the Moon in 1969.

The Moon, Our Nearest Neighbor

In some ways, the Moon resembles the Earth since both have a crust and mantle. The Moon may also have a hot core like the Earth, although it is probably much smaller. The astronauts who visited the Moon, however, were more impressed by the differences between the surfaces of the Moon and the Earth. First, there is no atmosphere on the Moon to shield it from the Sun's heat or help hold the heat during the night. During the day, the temperature on the Moon's surface rises to above 100°C; during the night it drops below –100°C. There is also no water, nor is there any evidence of living things on the Moon.

The surface features of the Moon, therefore, have not been changed by wind or water. Many of the features that can be seen on the Moon today have been there since its early history. The Moon is not free from all change, however. The lack of an atmosphere allows its surface to be hit constantly by small particles from space. This breaks up some of the rocks on the surface. Since the small pieces of rock are not washed or blown away, a fine dust covers almost all the Moon's surface. See Figure 7-2.

If it were not already the Earth's only natural satellite, the Moon could qualify as a small planet. In fact, some astronomers think that at one time the Moon actually was a small planet that was captured by the pull of Earth's gravity and pulled into orbit around the planet. Others say that the Earth and the Moon were formed at the same time from the same cloud of dust and gas. Still other astronomers claim that a third body hit the Earth and knocked material out into orbit around the planet. This material later came together to form the Moon.

Figure 7-2 Moon dust is stirred up by this astronaut on the Moon.

It is unlikely that the Moon was once part of the Earth, however, because there are too many differences in the chemical makeup of the two bodies. Furthermore, if the Moon was captured by the Earth, the principles of gravity should have given the Moon a different orbit than it has today. The theory that the Earth and Moon were both formed together is the most likely theory. However, the information so far gained from exploring the Moon has not settled the question about its origin.

The Sun: Source of Life

The Sun is the source of most of the energy used on the Earth. Not only does it heat our atmosphere, drive our climates, and warm our oceans, the Sun's energy is used by plants to make food—food that eventually ends up on our tables. Without the Sun, there would be no life on Earth as we know it. But what is this burning sphere in the sky? And why does it shine?

Composition of the Sun No one has ever been able to see into the interior of the Sun. No space probes from Earth have ever penetrated the Sun's surface. Yet by mathematical models, the inner structure of the Sun can be deduced.

The Sun is thought to be made up of several layers that blend into one another. Because all the layers are gases, there are no sharp dividing lines between them. Because the Sun is so hot, the gases in the Sun may well be a mixture of bare atomic nuclei and free electrons—a form of matter called *plasma*. On Earth, plasmas are found only where the temperatures are very high, as in the flame of a welding torch.

The Sun can be divided into six layers. See Figure 7-3. The changes that produce the Sun's huge supply of energy take place in its center, or *core*. The temperature in the core has never been measured directly, but scientists have

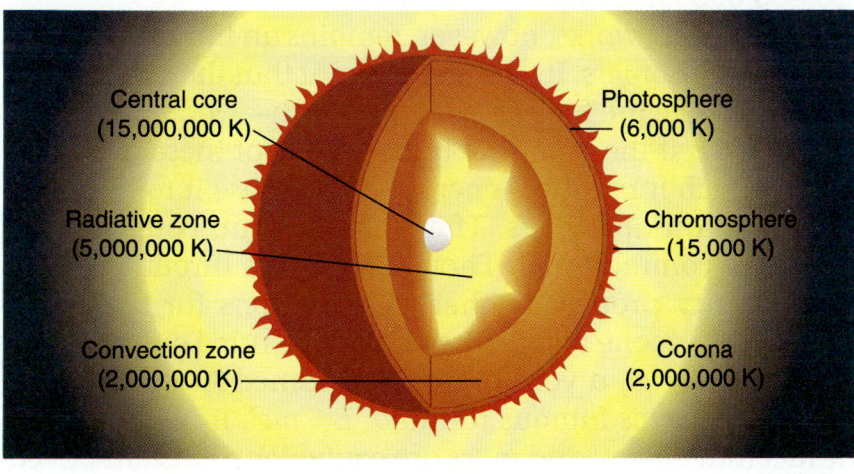

Figure 7-3 The three layers of the Sun's interior and its three atmospheric layers.

estimated the temperature to be at least 15,000,000 K. Energy coming from this central power plant is carried outward through the *radiative* and *convection* zones until it reaches the Sun's surface.

Energy from the Sun's interior reaches the surface where it causes a thin layer of gases to give off a brilliant light. This glowing layer, called the *photosphere*, produces the visible light that comes from the Sun. The photosphere, the lowest part of the Sun's atmosphere and the part of the Sun that we see, has a temperature of about 6000 K.

Just above the photosphere are two more layers that make up the Sun's atmosphere. The first layer, the *chromosphere*, gives off a faint red light that cannot be seen against the bright background of the photosphere. The chromosphere blends into a much less dense layer, called the *corona*, that surrounds the Sun like a halo. There is no outer boundary for the corona and some of the Sun's material escapes completely. Streams of particles from the Sun's corona make up the *solar wind*. This "wind" moves past the Earth at speeds of 300 to 700 km/s, much faster than a speeding bullet.

Fortunately for living organisms on Earth, the Earth's magnetic field pushes the solar wind to either side as it passes and causes the wind to flow around the planet like a river flowing around an island. Some of the charged particles composing the solar wind, however, pass through the magnetic field and become trapped. When these particles strike the gases in the Earth's atmosphere, light is given off. These greenish-white and blue lights are known as *auroras*, or northern lights and southern lights. See Figure 7-4.

Figure 7-4 *The northern lights—aurora borealis—as seen from Alaska.*

Energy Production by the Sun Each second, the Sun gives off an amount of energy equal to 200 billion hydrogen bombs. We are fortunate that this solar furnace is 150,000,000 km away!

How the Sun works, however, remains an intriguing question for astronomers. It was first thought that the Sun burned fuel, much like the fuel we know on Earth—wood, coal, and oil. But even if the Sun were burning one of these fuels, the rate at which it produces energy would cause it to burn out in a few thousand years. Therefore, the Sun could not be using any common fuel. Though scientists throughout the 19th century proposed other explanations for the Sun's energy output, one by one they had to be ruled out.

Then in 1905, a young physicist named Albert Einstein came up with his famous formula, $E = mc^2$. By establishing the fact that energy and mass are equivalent, Einstein paved the way for the discovery of a new source of energy—fusion. The process of fusion, in which hydrogen is converted to helium, can account for the Sun's tremendous output of energy. Even though it would take 4 million tons of hydrogen each second to produce the energy output of the Sun, the Sun contains so much hydrogen that it should continue to burn another 10 billion years before the hydrogen runs out.

This seems to be the answer to how the Sun and other stars produce energy. But this is not the only way that energy is produced on a large scale in the universe. In fact, objects have been discovered that seem to produce energy on such a scale that not even fusion can account for it.

The Solar System: Our Planetary Family

The Solar System is made up of the Sun, the planets, and billions of smaller bodies. Each of these bodies follows an orbit around the Sun. These orbits are not perfectly circular. Each orbit is in the shape of an ellipse. Many orbits, such as the Earth's, are only slightly elliptical.

The bodies of the Solar System are held in orbit by the force of gravity exerted by the huge mass of the Sun. In fact, the Sun makes up more than 99 percent of the Solar System's mass. Thus, the Sun is the controlling body of the Solar System. The nine planets are the best known and, in most cases, the largest bodies of the Solar System. See Figure 7-5.

Figure 7-5 The planets compared with each other and the Sun.

The Planets The planets can be separated into two major groups—the inner planets and the outer planets. The inner planets include Mercury, Venus, Earth, and Mars. These planets are the smaller ones and consist of solid rock with metal cores. The four large outer planets include Jupiter, Saturn, Uranus, and Neptune. These giant planets are made of ice and gases and have small solid cores. Pluto is usually the farthest planet from the Sun. It is very small and seems to be different from all the other planets. Table 7-1 compares some of the major characteristics of the planets.

CHARACTERISTICS OF THE PLANETS								
Planet	Distance from Sun (AU)	Revolutions (years)	Radius at Equator (km)	Rotation Period (days)	Mass Compared to Earth	Density (g/cm³)	Surface Gravity (Earth = 1)	Surface Temperature (°C)
Mercury	0.39	0.24	2,439	58.6	0.055	5.43	0.38	−193 to 342
Venus	0.72	0.62	6,052	−243.0	0.82	5.24	0.91	455
Earth	1.00	1.00	6,378	0.997	1.00	5.52	1.00	−88 to 58
Mars	1.52	1.88	3,397	1.026	0.11	3.9	0.38	−124 to −31
Jupiter	5.20	11.86	71,398	0.41	317.8	1.3	2.53	−160 to −149
Saturn	9.54	29.46	60,000	0.43	94.3	0.7	1.07	−176
Uranus	19.19	84.07	26,200	−0.65	14.6	1.3	0.92	−216
Neptune	30.06	164.82	25,225	0.72	17.2	1.5	1.18	−240
Pluto	39.53	248.6	1,100	6.38	0.0025	2.0	0.09	−223 to −208

Table 7-1

Scientists believe that the way the Solar System was formed explains the differences between the inner and outer planets. The inner planets are made up mostly of solid rock and metals. They were formed close to the Sun's heat and had their lighter materials driven off. The outer planets were formed farthest from the Sun's heat and held on to the lighter materials. The strong gravity of the massive outer planets enable them to hold on to the lighter materials. These materials, in the form of ices and gases, make up most of the mass of the outer planets.

Smaller Members of the Solar System

There are other bodies in the Solar System, some of which are closely associated with the planets. These bodies, though much smaller than the planets, far outnumber the nine planets. While you have already learned about meteors and comets, there are two other important members of the Solar System—satellites and asteroids.

Satellites The Earth has a very familiar *satellite*—the Moon. Other planets have satellites, or moons, of their own. Mars has two small satellites, Phobos and Deimos, while the gas giants have many satellites. Jupiter has 16, Saturn has 19, while Uranus and Neptune have 15 and 8 respectively. See Figure 7-6. Mercury and Venus, however, do not have any natural satellites.

Figure 7-6 The moons of Saturn in a mosaic made of several photographs.

Jupiter's four large satellites—Io, Europa, Ganymede, and Callisto—were discovered by Galileo in 1610, using his primitive telescope. More recently, active volcanoes have been observed on Io. One of the moons of Saturn, named Titan, is the only satellite in our Solar System known to have a substantial atmosphere. Nine of the moons of Uranus were discovered by *Voyager 2* in 1986, while all but two of Neptune's moons remained undiscovered until the visit by *Voyager 2* in 1989. Neptune's largest moon, Triton, rotates around its planet in the opposite direction from Neptune's rotation. This has lead astronomers to believe Triton was captured by Neptune's gravitational pull. Pluto has one moon, named Charon, that is about half the diameter of Pluto. Thus they make almost a double-planet system, each one orbiting the other in a period of about six days. See Figure 7-7.

Figure 7-7 An artist's rendering of Pluto and its moon Charon.

Asteroids Between the orbits of Mars and Jupiter is a wide region that does not contain a planet. In this zone, a huge number of rocks and boulders, called *asteroids,* are found. See Figure 7-8. The total number of asteroids is not known, but at least 4000 have been discovered. Ceres, the largest asteroid, makes up about 30 percent of the total mass of all the known asteroids. At one time, scientists thought that the asteroids might be the wreckage of a small planet that was torn apart by Jupiter's tremendous gravity. However,

Figure 7-8 Asteroids are fragments of rock that orbit the Sun.

Figure 7-9 *This meteorite was found in Antarctica and is estimated to be 1.3 billion years old.*

scientists now think that asteroids are the parts of a planet that never formed. Some asteroids follow very circular orbits around the Sun, while others have very elliptical orbits that might cross the paths of the inner planets including Earth. While most meteors seem to be associated with the dust trails of comets, most meteorites are fragments of asteroids that have wandered out of their usual place in the Solar System. See Figure 7-9.

Summary

The Solar System is dominated by the Sun and contains nine planets, which are of two major types. The inner planets are composed of solid rock and metals, while the outer planets are made of ices and gases. Most of the planets of our Solar System have satellites, or moons. The most numerous of the smaller bodies in the Solar System are asteroids, some of which may fall to Earth as meteorites.

7.2 STARS: A UNIVERSE OF SUNS

There are more stars in space than all the drops of water in the oceans of the world or all the grains of sand on the beaches of the Earth. Each star you see in the night sky is similar to the Sun. See Figure 7-10. Whether or not those stars have planets of their own is not yet known. However, some scientists feel that chances are high that other planetary systems will be found. What has been found, though, is quite intriguing in its own right.

Except for the Sun, the stars are so distant from the Earth that even the largest and most powerful telescopes show them only as pinpoints of light. Since stars are so far away, it might seem difficult to learn about them. But information about stars is carried in their light. By interpreting this information, scientists have learned what stars are made of, how large they are, and how they are moving in space.

How We Measure Stars

How far away is a star? When looking at the night sky, you can see that not all stars have the same brightness. Actually, some of the brightest objects seen are in fact planets, not stars. Venus, Mars, Jupiter, and Saturn can easily be seen without a telescope. Planets, as you know, can be sorted out from stars by observing their positions from night to night. Since stars are much farther away from the Earth than the

Figure 7-10 *The stars visible from Earth are just a small fraction of those that actually exist.*

planets, individual stars appear to keep the same position among other stars.

When we look at the sky, we see the stars in constellations as if they were located on a kind of ceiling over our planet. For example, the stars in the Big Dipper all seem to be at the same distance from the Earth. This is because of their great distance. But the stars are actually widely scattered and at different distances from us. See Figure 7-11.

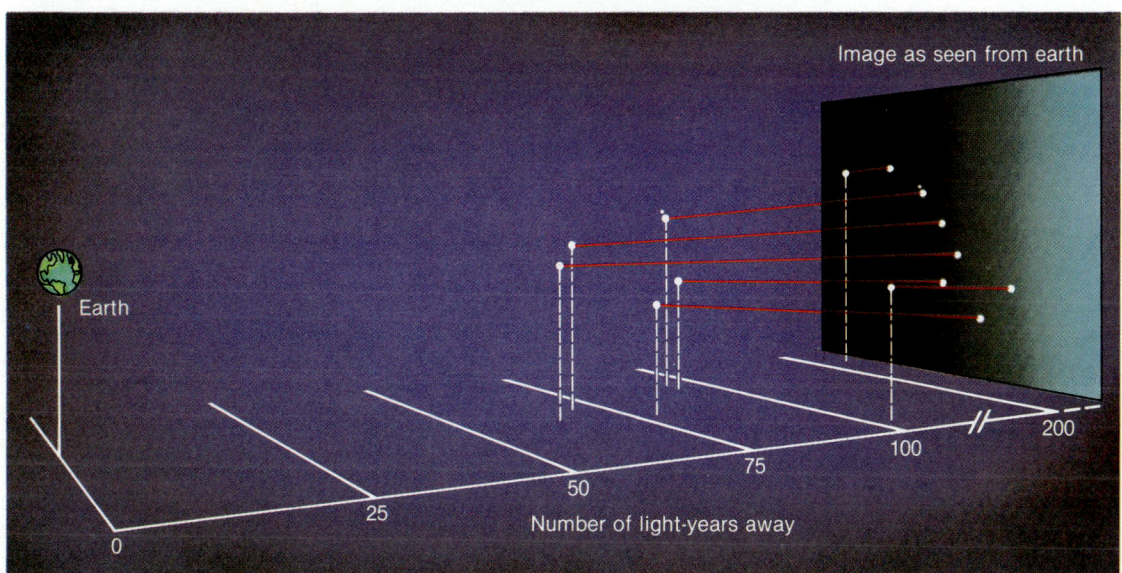

Figure 7-11 The stars that make up the Big Dipper are at different distances from the Earth.

Some stars *do* seem to move very slightly when observed over a six-month span. As the Earth moves in its orbit around the Sun, these stars seem to shift their positions in relation to more distant stars. This is similar to the way a nearby tree appears to move against the distant background when viewed from a moving car. Trees that are farther away appear to move less against a distant background. This apparent motion, called **parallax**, can be used to give an idea of how far away the trees are. By using parallax, the distances to nearby stars can also be measured. See Figure 7-12 on the next page.

Scientists use other methods to measure the distances to stars that are very far away. One method makes use of the brightness of the star. Have you ever judged the distance to a street light by how bright it looked? If you have, you were estimating distance using brightness. The actual amount of light a star sends out is called its **luminosity**. Stars have different luminosities; some are 10,000 times brighter than the Sun, while others are only a tiny fraction as bright. Once the luminosity of a star is known, its distance can be found by comparing its luminosity to its brightness as actually seen from the Earth. You might judge

Parallax The apparent shift of position of an object (such as a star) when seen from different locations or angles.

Luminosity A measure of the actual brightness of a star.

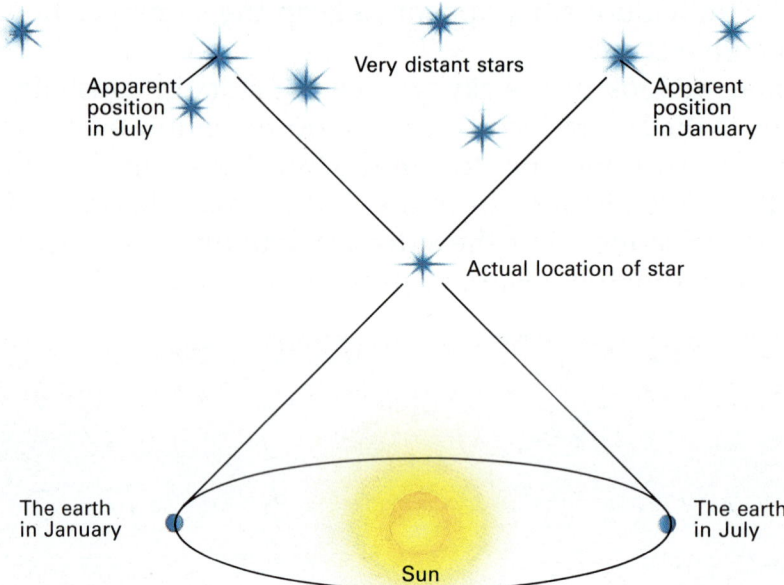

Figure 7-12 The movement of the Earth causes nearby stars to appear to move relative to those in the distant background.

the distance to a street light by making the same kind of comparison.

Astronomers have different ways of finding the luminosity of stars. One of the best ways makes use of a *variable star*, a kind of star whose brightness changes during periods of one day to many days. The length of time it takes for these stars to change their brightness is related to their luminosities. Once the luminosity of these stars is found, the distance to them (and to other stars close to them) can be found. This is done by comparing their perceived brightness with their luminosities. For example, if a star that has a high luminosity appears dim, it means that the star is probably very far away.

Types of Stars
The messages carried by the light from stars have shown scientists that there are many kinds of stars. Some stars are larger than the Earth's orbit around the Sun, while others are smaller than the Earth itself. There are stars made of gas thinner than air and others that are harder than diamond. One kind of star repeatedly blows up like a balloon, then shrinks and blows up again, in an endless cycle.

Red and Yellow, Black and White
One difference among stars is their color. You already know that the color of starlight is related to the temperature on the star's surface. The coolest red stars have surface temperatures of about 3000 °C, while the hottest blue stars have surface temperatures over 25,000 °C. See Table 7-2.

CLASSIFICATION OF STARS		
Color	Surface Temperature (°C)	Example
Violet	Above 28,000	10 Lacertae
Blue	10,000–28,000	Rigel, Algol
Blue	7500–10,000	Vega, Sirius
Blue–White	6000 to 7500	Canopus, Procyon
White–Yellow	5000 to 6000	Sol (Sun), Capella
Orange–Red	3500 to 5000	Arcturus, Aldebaran
Red	Less than 3500	Betelgeuse, Antares

Table 7-2

Astronomers have discovered that there is also a relationship between the star's surface temperature and its brightness. Generally, the higher the temperature of the star, the brighter it is. When stars are plotted on a chart according to their surface temperature and brightness, they fall into three main groups. See Figure 7-13. This chart is known as the Hertzsprung–Russell diagram, named after the Danish and American astronomers who first designed it. Most stars fall into a narrow band on the chart. The band begins with hot stars at the upper left and ends with cool stars at the lower right. Since most known stars fall into this group, they are called *main sequence* stars. A second group of stars appears at the upper right of the chart. These are bright stars, even though their red color indicates low surface temperatures. The stars in this group are called *red giants*. One red giant, Betelgeuse (BET ul jooz), is so large that if it were placed

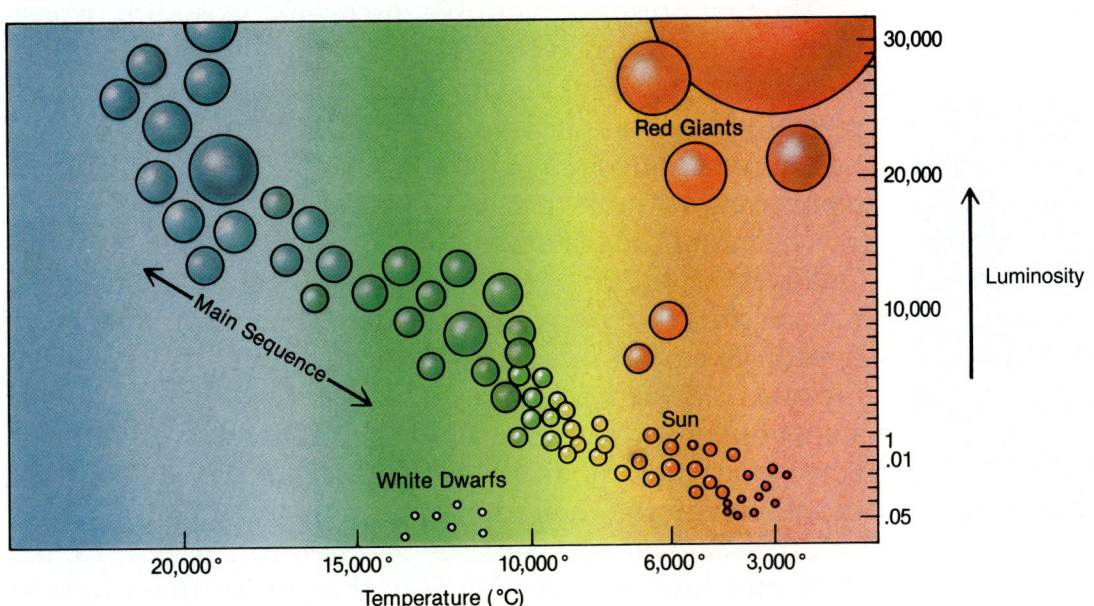

Figure 7-13 *Stars are grouped in this chart according to their surface temperature and luminosity.*

at the center of our Solar System, its surface would reach well past the orbit of Mars. A third group of stars appears at the bottom of the chart. These are called *white dwarfs*. They are very hot and produce white light, but they are very small and therefore not very bright. Some white dwarfs are only the size of the Earth.

Binary Stars Many of the stars you see in the night sky are not single stars, as is our own Sun. In fact, about 60 percent of all stars are actually double-star systems, or *binary stars*. In binary star systems, two stars revolve around each other. See Figure 7-14. Some binary systems are positioned so that the stars periodically eclipse one another from the point of view of an observer on Earth. Our closest neighboring star, Alpha Centauri, is actually a multiple-star system composed of the stars Alpha, Beta, and Proxima Centauri. Since it is located nearer the south celestial pole, it is not visible in the skies of North America. The nearest star visible from most of the United States is Sirius, which you already know is a binary star.

Variable Stars Variable stars, as you have seen, can be used to judge distance since their luminosities change. One type of variable star is the *pulsating variable*. Energy from below the surface of such a star heats the gases of the visible, shining surface. As the surface becomes hotter and brighter, it expands. Once the surface expands, however, the gases cool and the star again becomes dimmer. The cooler gases then contract, causing them to once again get hotter, and a new cycle begins.

The first pulsating variable discovered was Mira, whose luminosity changes every 331 days. The time from bright to dim and back to bright again is the star's period. Most known pulsating variables have a period of 100 days or less. Mira has an unusually long period. The North Star, in contrast, is a pulsating variable that has a period of only 4 days.

The magnitude of other stars varies for different reasons. One type of variable, called *eclipsing binaries*, are actually pairs of stars that move around each other. Their brightness appears to vary as one star periodically blocks the light from the other star, causing an eclipse. If, for example, a pair of stars happens to rotate around each other in line with the Earth, one will periodically block our view of the light from the other, varying their total brightness.

Exploding variables, such as the *nova* in Figure 7-15, are stars that have bursts of energy that make them appear

Figure 7-14 Revolution of a double star. These three photographs (covering a period of about 12 years) show the mutual revolution of the components of the nearby star, Kruger 60.

Figure 7-15 Novas occur when stars explode.

many thousands of times brighter for days or even years. The term *nova* comes from the Latin for "new." They were given this name because in ancient times they were thought to be new stars. A nova eventually returns to its original brightness, but a *supernova* is such a huge explosion that the star blows itself apart. See Figure 7-16.

Galaxies: Collections of Stars

Galaxies consist of billions of stars. There are many galaxies in the universe, including the Milky Way, of which our sun is a part. The Milky Way is a spiral galaxy of over 100 billion stars, stretching 100,000 light-years across. See Figure 7-17. Have you ever noticed, on a very clear night, a band of stars so dense that it looks like a starlit cloud? See Figure 7-18. This band is actually a portion of our galaxy, viewed from our position in one of its spiral arms.

Figure 7-16 This supernova, photographed in 1987, appeared in the sky of the Southern Hemisphere. It was the first supernova visible to the naked eye in nearly four centuries.

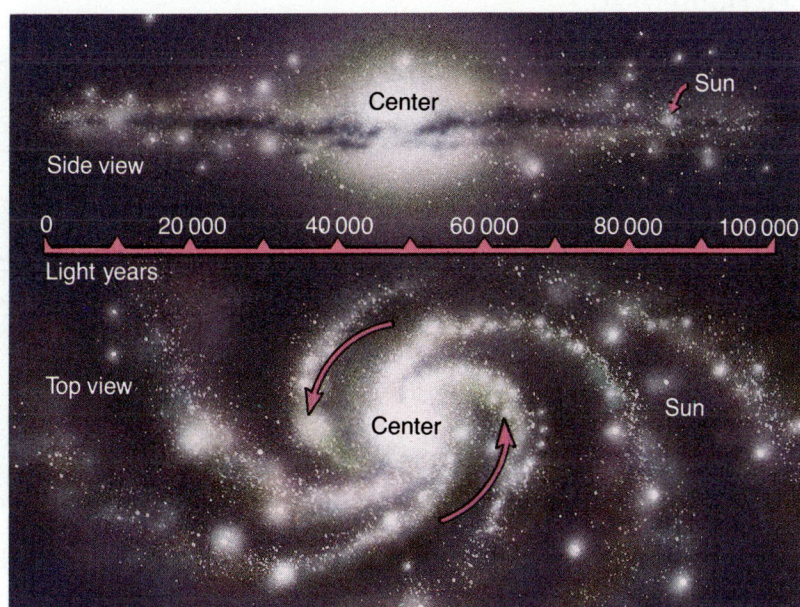

Figure 7-17 Earth is part of the Milky Way galaxy, shown here in relation to the Sun.

Figure 7-18 If you look up at the sky on a very dark, clear night, you can see the Milky Way.

Stars are held in a galaxy by gravity, which tends to pull the stars toward one another. The Milky Way is shaped like a disk with a bulge at its center. Our Solar System is located

Figure 7-19 *The spiral galaxy M31 in the constellation Andromeda looks very much like our own galaxy.*

closer to the edge than to the center of the galaxy. Thus, by looking in the direction of the galaxy's center, we see many more stars than when looking elsewhere. If you look at the constellation of Andromeda, it is possible to see a small, fuzzy patch of light among the stars. A telescope shows that this is a galaxy about 2.2 million light-years away. See Figure 7-19. Our own Milky Way would look similar if we could see it from far out in space.

Types of Galaxies Telescopes that allow us to look deep into space beyond our own galaxy show that the universe contains billions of other galaxies. Just as stars are seen to differ from each other, not all galaxies are alike. In addition to the spiral shape of the Milky Way and Andromeda galaxies, some galaxies are shaped like footballs, or like spheres. Other galaxies have no general shape. The major types of galaxies are spiral, barred spiral, elliptical, and irregular. See Figure 7-20.

Figure 7-20 *Types of galaxies: barred spiral (left), elliptical (middle), irregular (right).*

Except for the galaxies that share our "local group," all galaxies that can be observed have one thing in common: They are all moving away from each other at a very high speed. Scientists can tell that the galaxies are moving apart because of the information that can be read from light. The colors in the spectrum of the stars in these galaxies are shifted toward the redder colors. This is called the *red shift* of the starlight. The red shift in the light from the galaxies is caused when the light waves are spread out as the light source moves away. You may have heard the same effect in the pitch of sound waves that come from a moving source.

Strange Objects in Space

As if red and blue stars, novas and supernovas, spiral and elliptical galaxies were not enough, even stranger objects are among those to be found in the universe. Scientists have only begun to examine such objects, some of which are discussed below.

Pulsars At first, scientists thought certain stars were the source of alien radio messages. These stars, called *pulsars*, give off rapid pulses or bursts of radio waves, light, and X rays. As a pulsar spins on its axis, it sends out a stream of radiation in a pattern similar to that of the rotating light in a lighthouse. See Figure 7-21. When the Earth is in the path of one of these streams of radiation, radio telescopes can pick up the signals as rapid pulses. A pulsar at the center of the Crab Nebula rotates thirty times a second.

Figure 7-21 An artist's conception of a rotating pulsar.

Quasars One of the most puzzling of all objects detected by astronomers is the quasi-stellar radio source, or *quasar* for short. Quasars emit vast amounts of radio waves and light. They are several billion light-years away and may actually be the brightest objects in the universe, producing more energy than that produced by hundreds of galaxies combined. See Figure 7-22. Since they seem to be so far away, quasars may be among the earliest objects in the universe. Scientists still do not fully understand the nature of quasars.

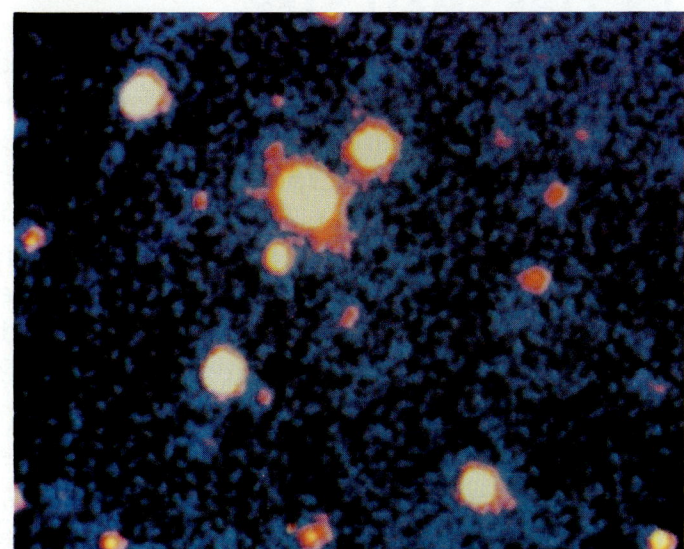

Figure 7-22 Quasar 3C 275.1, the first to be found at the center of a cluster of galaxies, appears as the brightest object near the center of this image. This quasar is 7 billion light-years away. The light we now observe left the quasar more than 2 billion years before our Solar System was formed.

Black Holes Astronomers have been searching for ways to explain the huge amounts of energy produced by quasars and other objects. Some have theorized that very large stars may collapse with such force that they pass beyond the stage of being a neutron star to become a "black hole." As you know, a black hole is an object so dense that its gravity will allow nothing—not even light—to escape from it. Because no energy of any kind can come from a black hole, there is no way to observe it directly. However, some astronomers hypothesize that they may be able to locate black holes by the effect they have on nearby objects. For example, a black hole close to a star would strip matter from its neighbor. The material lost from the star would then give off energy such as X rays as it disappeared into the black hole. This process could explain some of the powerful energy sources now detected in space. Yet, black holes remain unobserved theoretical objects. None have actually been discovered.

Summary

There are billions upon billions of stars in the universe. Astronomers use parallax and absolute magnitude to estimate the distance to stars. Astronomers estimate the surface temperature of a star by its color. Most stars are part of a binary system, and some stars are variables. Galaxies are huge collections of stars that come in a variety of shapes and sizes. Some of the more unusual objects in space are pulsars, quasars, and the elusive black holes.

7.3 ONE SMALL STEP

Stonehenge is the ruins of an ancient monument in southern England. See Figure 7-23. It consists of two rings of rectangular stone columns. Other stone rectangles lie on top of some of these columns. The tallest column is 7.3 m high. Each stone weighs 30 to 60 tons. Through the center of the ring is a broad avenue. At the end of the avenue is a stone 5.4 m high called the Head Stone. Why did people, beginning in 2,600 B.C., build this structure? Why did they carry these heavy stones from as far away as 380 km?

Figure 7-23 What was the purpose of Stonehenge?

How We See the Universe

For centuries, astronomers watched the evening skies. They noticed that the stars kept their same relative positions. However, five points of light, the visible planets, slowly moved with respect to the stars. Ancient people could only wonder at their nature. They thought of these planets as gods.

In the 1600s, the telescope was developed. The planets could then be seen as spheres that are illuminated by the Sun. Some surface features were visible through the new telescopes, but space probes gave us our first truly clear images of the planets. By 1989, eight of the nine known planets had been visited by probes sent from Earth.

The Telescope is Born The invention of the telescope may have occurred many times in different places. By the 1600s, it was being used as a field glass for viewing distant objects on Earth. In 1609, however, Galileo made his own telescope and pointed it toward the sky. See Figure 7-24. Although his instrument was primitive, Galileo made many discoveries, such as the craters and mountains on the Moon, the phases of Venus, and the four large satellites of Jupiter.

Isaac Newton constructed the first reflecting telescope. See Figure 7-25. He realized that concave mirrors have three advantages over lenses. First, they reflect light instead of refracting it and therefore produce no distortion. Second, a good mirror absorbs less light than even the most transparent glass, so the image will be brighter. Third, a mirror is not as heavy as a lens of the same size. Therefore, a mirror is easier to support, and much larger telescopes can be constructed. The largest optical telescopes used today are reflecting telescopes. Composed of 36 interlocking mirrors with a total diameter of 10 m, the new Keck telescope, located on top of Mt. Mauna Kea in Hawaii, is the largest reflector in the world—twice the size of the famous Hale telescope at Mt. Palomar. The images this giant telescope can produce rival those of the orbiting Hubble Space Telescope.

Figure 7-24 *Galileo's telescope used lenses to focus light from distant objects.*

Other Types of Telescopes Visible light is not the only type of radiation emitted by stars and galaxies. Radio waves, X rays, and ultraviolet waves are also produced by these objects. For example, stars, large planets, and the cores of many galaxies emit radio waves. Quasars emit many radio waves. Since radio waves are not absorbed by matter, as are visible light waves, radio waves from outer space may cross the entire universe with little loss through absorption.

Astronomers can capture radio waves with a special instrument called a radio telescope. Though some radio telescopes are like large television antennas, most are shaped like metallic dishes. Radio waves are reflected by metallic surfaces just as light is reflected by a mirror. At the focus of a radio telescope is a radio receiver. A radio astronomer uses a tuner to select the radio frequency he or she wishes to study.

A single radio telescope provides an image of the sky that is less precise than the image provided by an optical telescope. Radio telescopes can be linked together, however, so that the image received by one can be compared and mixed with the image received by another. Combining images from

Figure 7-25 *A replica of the reflecting telescope made by Isaac Newton.*

Figure 7-26 A radio image (left) of quasar 2300-189 at a distance of 1.2 billion light-years. This array of 27 radio-telescopes (right) provides a much clearer image of the universe than a single dish does.

several radio telescopes can produce a much clearer radio image of the sky. See Figure 7-26.

Other telescopes have been sent into orbit around the Earth. Ultraviolet and infrared telescopes, as well as X-ray and gamma-ray collectors now give astronomers a much broader look at the universe around us. See Figure 7-27.

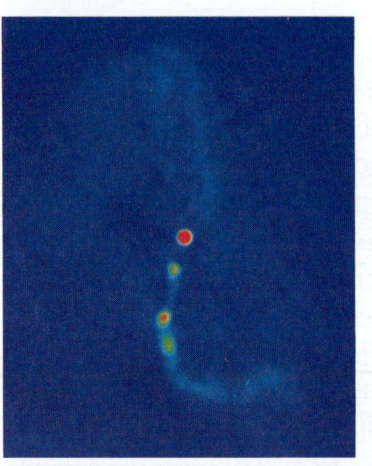

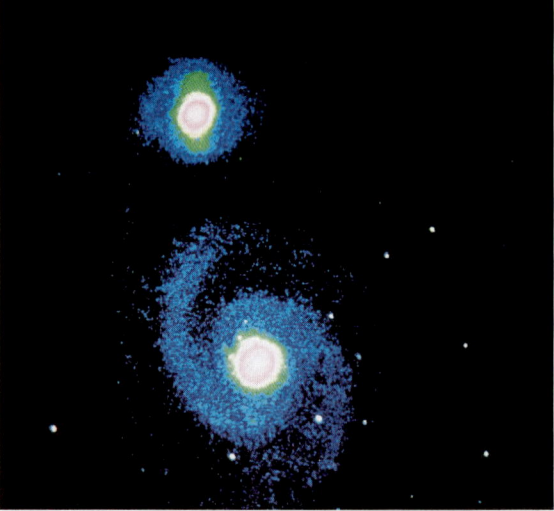

Figure 7-27 Optical images (left) and infrared images (right) of M51—the Whirlpool Galaxy.

The Exploration Begins

Suppose that observers from outer space have been watching the Earth since it was formed. For more than 4 billion years they would not have seen anything leave the Earth. Then suddenly, starting about 40 years ago, they would have noticed that the people of the Earth began to send off tiny spacecraft. At first, the spacecraft only flew around the Earth. Then they landed on the Moon. Later, the observers would have seen other spacecraft fly to the Earth's neighboring planets. The observers might conclude that the people of the Earth had begun a new age of exploration.

Rockets and Early Satellites Even though Jules Verne had written about a fictitious trip to the Moon in 1865, the first person to scientifically study the use of rockets for space travel was Konstantin Tsiolkovsky, a Russian inventor. In the 1920s, he designed a spaceship with a bullet-shaped passenger cabin in the nose and fuel tanks filled with liquid hydrogen and liquid oxygen in the tail. Although Tsiolkovsky never launched a rocket himself, his thinking was fundamental to the development of rocket flight.

At about the same time, an American physicist, Robert Goddard, was actually experimenting with rockets. In 1926 Goddard launched the world's first liquid-fueled rocket. See Figure 7-28. During World War II, the Germans constructed many rockets for delivering bombs. At the end of the war, scientists in the United States used captured German rockets to scientifically explore the upper atmosphere. An American two-stage rocket, with a captured German rocket as its first stage, reached a record altitude of 393 km in 1949.

Figure 7-28 Robert Goddard launched a gasoline-fueled rocket in 1926. The rocket was 1.2 m long, and it rose to a height of 56 m at an average speed of 103 km/h.

The development and use of rockets proceeded slowly until October 4, 1957, when the Soviet Union surprised the world by placing into orbit an artificial satellite called *Sputnik 1*. This 83.6-kg satellite carried instruments to measure the density and temperature of the upper atmosphere. See Figure 7-29. A month later, on November 3, 1957, the Soviet Union put another satellite into orbit. This second satellite carried a passenger—a dog named Laika. The purpose for orbiting the animal was to study the effects of space travel on a living organism.

Figure 7-29 Sputnik 1 *was the first artificial satellite to orbit Earth.*

The first American satellite, *Explorer 1*, was launched on January 31, 1958. The rocket was 2.05 m long and carried a scientific package for measuring temperature, cosmic rays, and meteors. *Explorer 1* discovered the Van Allen belts of radiation surrounding the Earth.

Human Exploration On April 12, 1961, Yuri Gagarin, a 27-year old cosmonaut, was launched into space. He orbited the Earth once, reaching an altitude of 327 km. American astronaut John Glenn was launched into space on February 20, 1962, and made three orbits around the Earth. The first woman to go into space was Valentina Tereshkova, who was launched in June of 1963. In 1965, Alexei Leonov became the first human to "walk" in space. Dressed in a protective spacesuit, he left his pressurized cabin and floated for 10 minutes in space. All these space firsts had been accomplished in less than 10 years after the launch of *Sputnik 1*; the next 10 years would be even more dramatic.

Figure 7-30 *Neil Armstrong and "Buzz" Aldrin spent 2½ hours walking on the Moon.*

Figure 7-31 *By landing on a runway, the Space Shuttle can be used again and again.*

Figure 7-32 *An artist's conception of the U.S. Space Plane.*

As a challenge to the Soviet Union's leadership in space, the United States determined to place an astronaut on the Moon before the end of the 1960s. In preparation for this journey to the Moon, many additional Earth-orbiting missions were launched. During the *Mercury* and *Gemini* missions, American astronauts learned to live and work in space. On July 16, 1969, *Apollo 11* was launched from Cape Canaveral, Florida toward the Moon. The *Apollo 11* lunar module landed on the surface of the Moon on July 20 at 8:17 P.M. See Figure 7-30. Neil Armstrong and Edwin "Buzz" Aldrin became the first humans to set foot on another world.

The Space Shuttle, a vehicle that can lift off like a rocket and return to Earth like an airplane, is now the United States' major vehicle for space travel. See Figure 7-31. It was designed to be used in the construction of a space station. Many missions involving the delivery of military, scientific, and telecommunications satellites as well as orbiting telescopes have been launched. A space plane that will be able to enter orbit after taking off horizontally from a runway is now under development by the United States. See Figure 7-32.

Interplanetary Probes You have already been introduced to several interplanetary probes—*Pioneer 10*, *Voyagers 1* and *2*, *Magellan*—that have sent back much more information about our neighbors in the Solar System than could ever have been gained by looking through telescopes. Another space probe currently in operation is the *Galileo*. Having passed Mars and the asteroid belt, it will reach Jupiter in 1995 where a smaller probe will be sent into the Jovian atmosphere. See Figure 7-33. This probe will send valuable information back to *Galileo* for only 75 seconds before it is crushed by Jupiter's atmospheric pressure.

Another probe, called the *Ulysses*, is on its way to an historic rendezvous with the Sun. It will be the first probe to ever journey outside the ecliptic plane. Rather than orbiting the Sun like the planets, it will travel at right angles to their paths and thus cross over the Sun's south pole. Once the

Ulysses reaches the Sun in 1994, it will complete a polar orbit, giving us our first view of the top and bottom of our own star, the Sun.

Theories of Cosmology

Since light travels at a certain speed, the farther we look out into space, the further back in time we can see. This ability may help us answer some very fundamental questions about our universe. **Cosmology** is the study of how the universe began, and how it may eventually end. There is no way to witness the actual events that occurred billions of years ago, but scientists have proposed several theories to account for the universe as we see it today.

The Big Bang The idea that the universe is expanding has led scientists to develop the *big-bang* theory. This theory states that, at one time, all the matter of the entire universe was concentrated in an infinitesimally small volume. Then, about 15 to 20 billion years ago, there was a tremendous explosion that caused the universe to begin expanding. About one billion years after the big bang, huge clouds of hydrogen and helium began to form. These clouds were the beginnings of galaxies. In time, stars started to develop in the galaxies. Within about 10 billion years after the big bang, all of the galaxies that can now be seen had formed. Today, we see the results of the big bang as a rapidly expanding universe with galaxies moving away from each other. See Figure 7-34.

As you know, scientists have found that all space is filled with weak radio waves. One explanation of this cosmic background radiation is that it is evidence left by the great explosion that was the big bang. But one thing the big-bang theory cannot explain, however, is the uniformity of this background radiation. This radiation is the same no matter

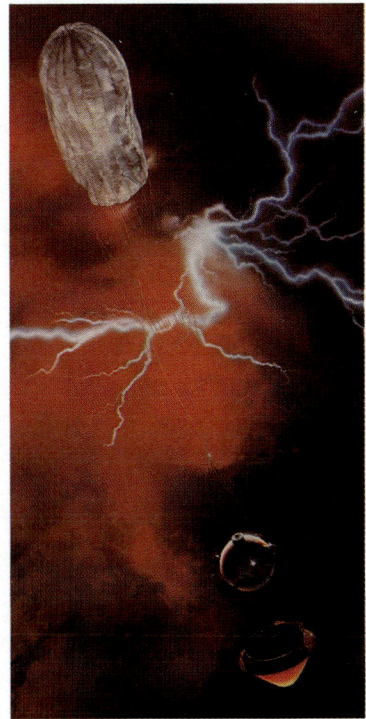

Figure 7-33 *Artist's impression of Galileo's entry probe descending through Jupiter's atmosphere.*

Cosmology The study of the structure, origin, and evolution of the universe.

Figure 7-34 *Before the big bang, all matter and energy in the universe was concentrated in a very tiny space. When the big bang occurred, matter was sent outwards in all directions.*

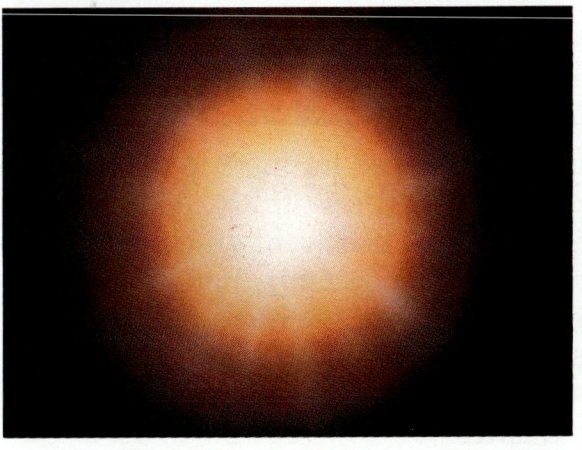

what direction we look. Because of the size of the universe, it is not likely that the radiation coming from one side has ever been in contact with the radiation from the opposite side. How then is this radiation the same temperature everywhere?

Pulsating Universe If there is enough matter in the universe, gravity should eventually stop the outward expansion of the galaxies caused by the big-bang explosion. A variation of the big-bang theory, called the *pulsating* or *oscillating* universe theory, states that the universe will eventually contract, bringing the galaxies back together into another hot mass, until another big bang sends them flying out again. This is repeated over and over, so the universe will continue to expand and shrink forever. See Figure 7-35. However, the estimated density of the universe is such that there is a balance between the two possibilities. It cannot, at this time, be determined whether expansion will stop, or if the universe will expand forever.

Figure 7-35 Most astronomers think that the universe will continue expanding (A–E) and eventually die. Others think there is enough matter in the universe to make it stop expanding and collapse back on itself, leading to another big bang, endlessly repeating this cycle.

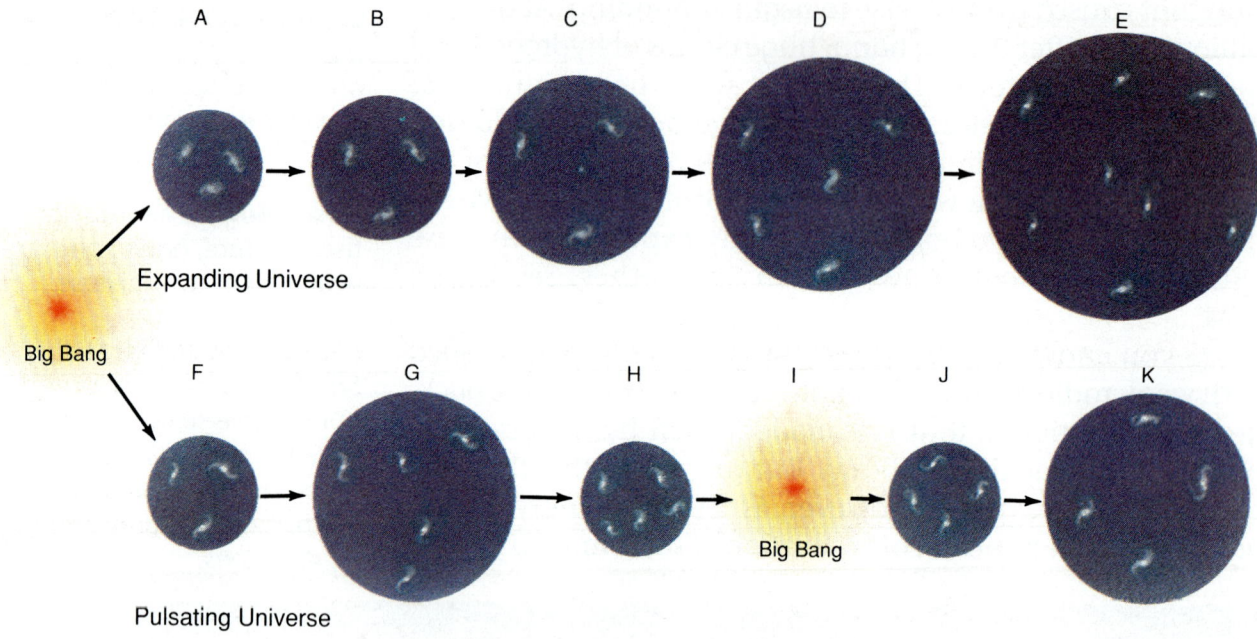

Summary

A variety of telescopes has enabled humans to explore the universe from their home on Earth. With the development of the rocket, space probes have been sent on close-up missions to the members of our Solar System. The universe contains billions of other galaxies that are apparently moving apart as the universe expands. The way the universe may have begun and how it might end is a question that scientists still are trying to answer.

Unit 8

Reading *Plus*

Now that you have learned how to grow plants, let's take a closer look at the parts of a plant, how plants grow, and how we use plants. Read pages S134 to S144 and prepare a presentation that explains how plants are adapted to their environments and why plants must be protected and studied. Your presentation should answer the following questions.

1. How do the leaves, stems, and roots work together in a healthy plant?
2. Why might plants be more affected by the environmental changes than are humans?
3. In what ways do animals and plants depend on each other?

In This Unit

8.1 Plant Structure, p. S134

8.2 Uses of Plants, p. S141

8-1 PLANT STRUCTURE

Have you ever wondered why a plant's body is so different from yours? These differences are related to the way plants and animals live. Being an animal, your way of life involves walking around, searching for and capturing other organisms for food, then eating that food and digesting it inside your body. In this way, you obtain the energy and nutrients you need for growth and the functioning of your cells. Having arms, legs, eyes, and a mouth, not to mention all your other body parts, makes you able to accomplish these necessary tasks. A plant's way of life, however, is very different from yours. Plants gather the *inorganic* nutrients carbon dioxide and water from their surroundings. Then, by capturing and using energy from the sun, plants convert those inorganic nutrients into *organic* substances such as sugar, which they use for food. Their bodies are adapted to the way they obtain energy and interact with the environment.

Figure 8-1 Land plants need vascular tissues in order to grow large.

Most of the plants you see and use every day are *vascular plants*. These plants have become adapted to living on land by developing a body that consists of leaves, stems, and roots. Inside the leaves, stems, and roots of a vascular plant are special tubes that transport water, minerals, and food throughout the plant. See Figure 8-1. These tubes form the vascular system that links the parts of the plant together into a functional organism. To understand how plants function, let's examine the basic parts of a plant's body.

Leaves

Leaves are the food-making organs of plants. Light, carbon dioxide, and water obtained by leaves are used during *photosynthesis*, the process by which plants use energy from the sun to convert carbon dioxide and water into sugar and oxygen. In order to carry out photosynthesis, leaves also exchange carbon dioxide and oxygen, and help water to move through the plant. As you know, leaf shapes and sizes vary greatly. These variations are often adaptations that allow leaves to carry out their activities.

Figure 8-2 The vase-shaped leaves of a pitcher plant (top) *and the jaw-like leaves of a Venus' flytrap* (bottom) *are adaptations for trapping insects.*

Some leaves have very special shapes, such as the jaw-like leaves of the Venus' flytrap or the vase-shaped leaves of the pitcher plant. See Figure 8-2. These carnivorous plants actually trap and digest insects to obtain some of their nutrients. Lucky for us, none of these plants are large enough to "eat" people!

Structure of Leaves The cells of a typical leaf are arranged into a flattened structure, called the *blade*, which has many raised *veins*. Leaves are usually connected to a stem by a stalk called a *petiole*. See Figure 8-3. Most leaves are thin and flat, which exposes as many cells as possible to sunlight. This shape also provides good contact with the air, from which leaves take in carbon dioxide. The veins of a leaf are made of vascular tissues that bring water into the leaf.

A leaf's veins also carry the sugar it has made to other parts of the plant. Each vein is a bundle of long, tubular cells that resemble small water pipes. The main vein into a leaf divides many times into smaller and smaller veins. As a result, water and food molecules do not have far to travel between a vein and any cell.

The cells of a typical leaf blade are organized into several specialized layers. See Figure 8-4. A layer of cells, called the *epidermis*, covers both the upper and lower surfaces of the leaf and prevents the loss of water from inner cells. The epidermis is tough, giving strength to a leaf, and its cells are clear, allowing light to pass through.

Figure 8-3 *A typical leaf has a blade with many veins and a petiole.*

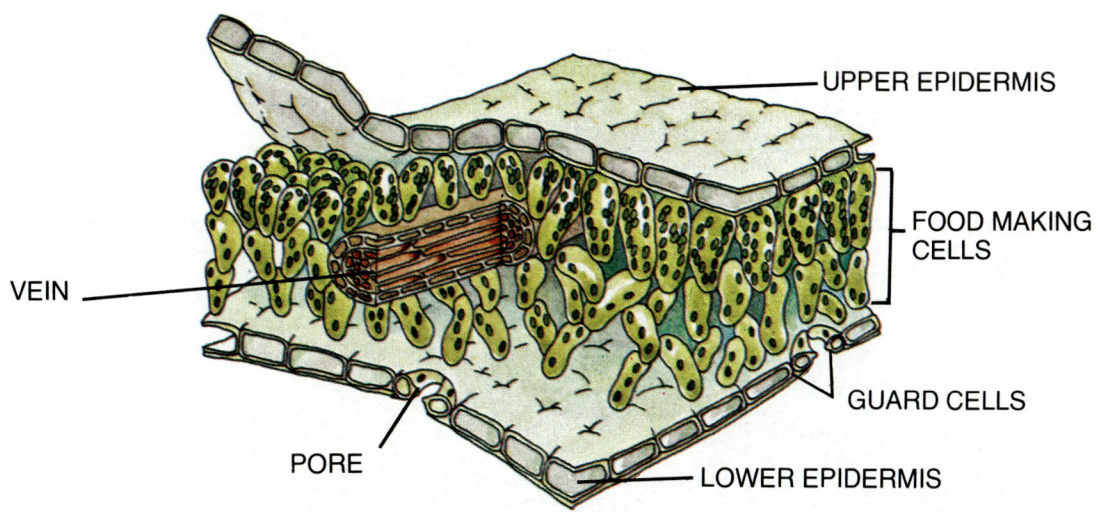

Figure 8-4 *The inner structure of a leaf.*

Chloroplasts Organelles of plant cells that contain chlorophyll and are the sites of photosynthesis.

Stomata Small pores in leaves that allow gases to enter and leave plants.

Inside the leaf are two kinds of food-making cells that contain special organelles called **chloroplasts**. The chloroplasts contain chlorophyll, which absorbs light for photosynthesis. The tightly packed upper layer of food-making cells performs most of the photosynthesis in the leaf. The food-making cells on the lower side of the leaf are loosely packed and have many air spaces between them. These air spaces are connected to the outside by small openings called **stomata**. Each stoma is surrounded by two *guard cells* that can change shape to open and close the stomata. The top surface of a leaf has very few stomata and is coated with a waxy material that prevents water from evaporating. But the lower surface of a leaf may have thousands of these

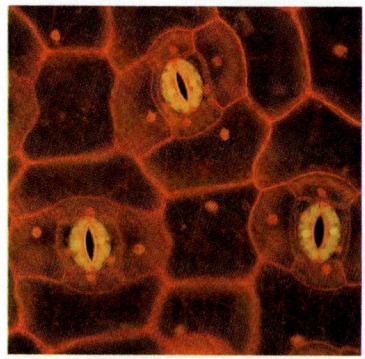

Figure 8-5 *Stomata occur primarily on the lower epidermis of leaves.*

openings that allow carbon dioxide, oxygen, and water vapor to go into and out of the leaf. See Figure 8-5.

Leaf Functions During a typical day, all leaf functions occur simultaneously and are linked to each other. The food-producing activities of a leaf begin when light first strikes it. Light causes the guard cells to open the leaf's stomata, allowing gases to move into and out of the leaf. Some of the sunlight is captured by chlorophyll and used to begin the process of making sugar. Once made, these sugars move into the leaf veins and are carried to other parts of the plant. Since plant cells also need energy for cell processes, plants use some of the sugar they make to supply this energy. Unused sugar is transported out of the leaf and is stored elsewhere as *starch*. Some of the oxygen made during photosynthesis is used by the plant to "burn" sugar during cellular respiration. Excess oxygen moves out of the leaf into the surrounding air. At the same time, more carbon dioxide is drawn back into the leaf through the stomata. This cycle of activity, shown in Figure 8-6, continues throughout the daylight hours. At night, the stomata close and photosynthesis stops. But the leaf can continue to make sugar by using energy stored during the daytime.

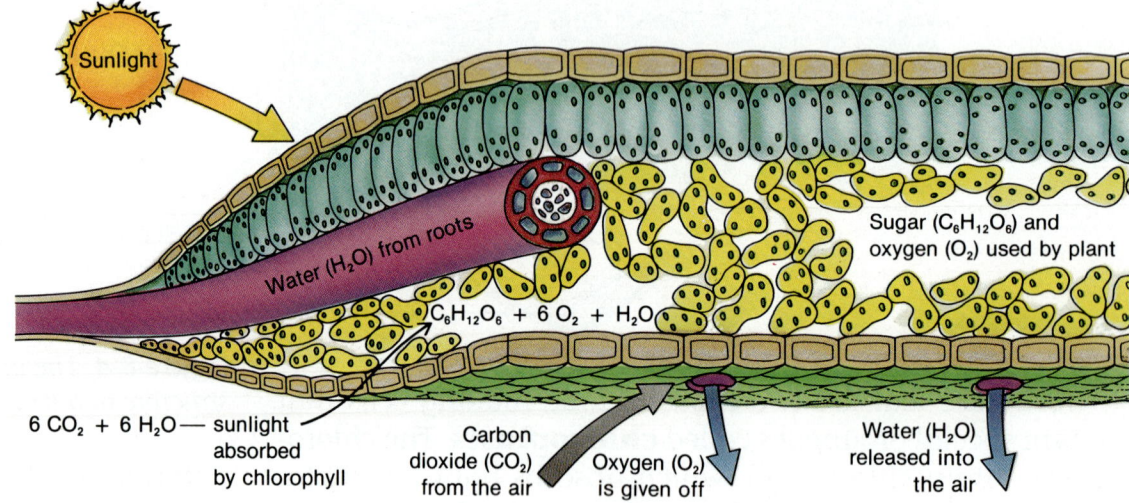

Figure 8-6 *The activities of a leaf include making sugar from water and carbon dioxide and exchanging gases.*

Water that is used during photosynthesis or lost through open pores is replaced by water from the veins. Leaf veins are connected to the plant's vascular system, which extends all the way down into its roots. When water is used or lost by the leaves, the roots of the plant take up more water to replace it. See Figure 8-7. As long as there is enough water available, it continues to move through a plant during the day. But if the soil around the roots becomes too dry, the leaves will not receive enough water. What do you think this does to the food-making activities of the leaf? Just as a factory must slow down or stop its production when there are not enough raw materials, photosynthesis also slows down or stops. To conserve water, the stomata close, which keeps water vapor from escaping. If it still does not get water, the leaf will eventually wilt and may even dry up.

Stems

What do a cactus pad, a potato, a stalk of asparagus, and a tree trunk have in common? They are all stems of plants. The stem is the midsection of a plant's body that in most plants grows above the ground. For many plants, stems have specialized functions. The potato plant, for example, stores food in specialized stems (the potatoes). In the cactus, the stem stores water. However, all plant stems have two basic functions—they support the leaves and transport food and water between the leaves and the roots.

Structure of Stems
You may have noticed that some plant stems are soft and flexible while others are stiff and hard. Plants that have stems that are soft and flexible, such as the stems of beans, tomatoes, grasses, and petunias, are called *herbaceous* plants. Plants whose stems are stiff and hard, such as those of trees and shrubs, are called *woody* plants. Although the stems of both types are able to support their leaves, there is a big difference in their sizes. Woody plants are able to grow much taller and larger because the woody stem provides more support.

Both herbaceous and woody stems contain two kinds of vascular tissues. One kind carries water and minerals upward from the roots to the leaves, while the other generally carries food downward from the leaves to the stems, roots, and any developing fruits and seeds. Both kinds of tissue are bundled together in long strands, called **vascular bundles**. These bundles run through the entire length of the stem. In some stems, they are scattered throughout a matrix of unspecialized cells, and in others, they are arranged in a ring. See Figure 8-8 on the next page.

Notice that in each bundle, the cells of the inner side are larger and have thicker walls than the cells on the outer side of the bundle. The larger, thick-walled cells are the water-carrying cells. When these cells are filled with water they help to stiffen the stems of an herbaceous plant. Wood is also made of these cells. The small, thin-walled cells are the food-carrying cells. The cells located between the two types of vascular cells are special dividing cells. These cells are able

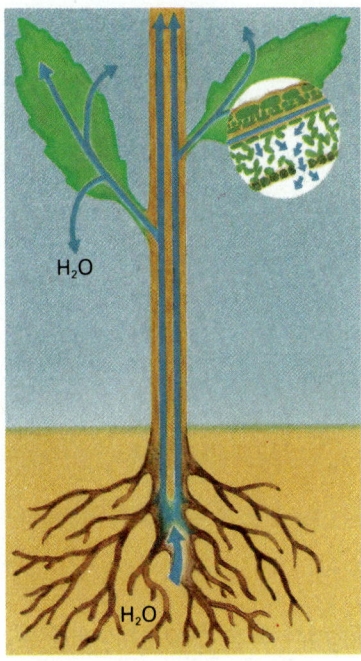

Figure 8-7 When water is lost from the leaves, water from the roots is pulled upward through the plant's vascular system.

Vascular Bundles Groups of water-carrying and food-carrying cells.

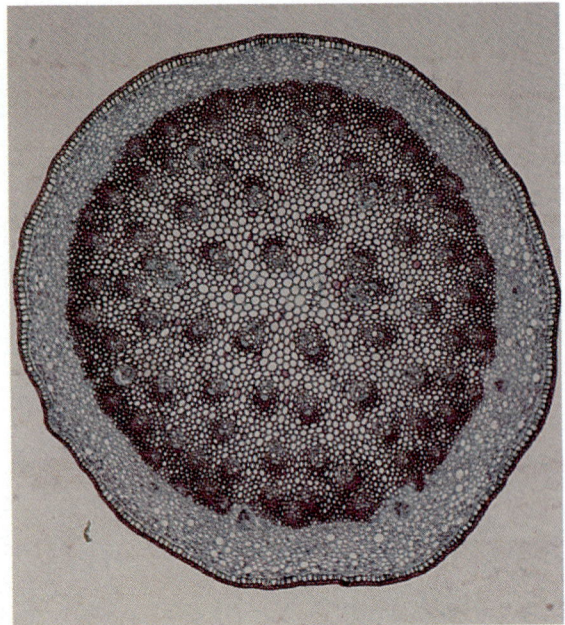
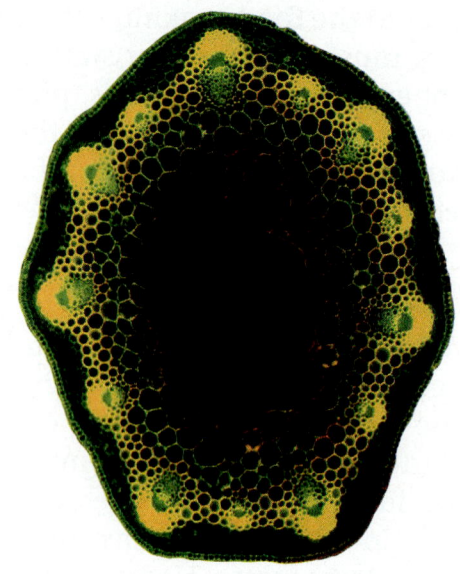

Figure 8-8 *Two ways that vascular bundles are arranged in stems.*

to produce new cells that mature into each of the other types of stem-tissue cells.

Stem Growth While you are growing, most of your body cells are able to divide to make more cells. However, in plant stems, only certain cells are able to divide to make more cells. Groups of these cells are found in two places in a stem—between the two types of vascular tissues in the vascular bundles and inside the *buds*.

The dividing cells inside the vascular bundles cause stems to grow in width. As layers of new cells are added, the stem becomes thicker. Herbaceous stems thicken only a limited amount and usually die at the end of a growing season. Woody stems, on the other hand, can become very thick since they do not die each year.

In woody stems, the vascular bundles are arranged in a solid ring that produces new vascular tissue and a new layer of wood each year. See Figure 8-9. The wood consists mostly of old water-carrying cells. As new water-carrying cells pro-

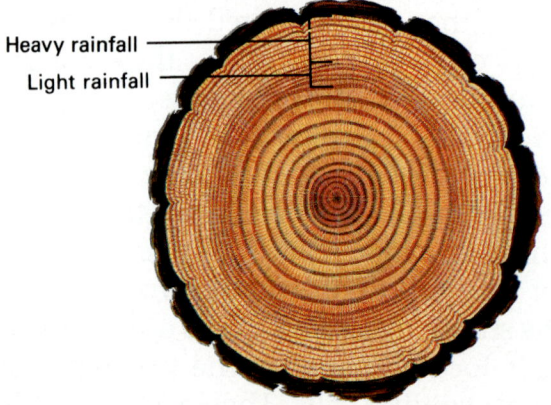

Figure 8-9 *You can determine the age of a tree by counting its rings.*

duced on the inside of the ring mature, the food-carrying and dividing cells of the ring are pushed outward. As a result, the outermost cells of a woody stem are crushed and die. These dead cells become part of the protective covering of a woody stem—its bark. Because the actively growing and transporting cells are located just under the bark, woody plants can be killed if their bark is severely damaged or removed.

Inside the buds of a stem are groups of dividing cells that cause stems to grow in length. See Figure 8-10. In woody plants, buds are usually formed at the end of the previous year's growing season. When conditions are favorable for stem growth, a plant produces chemicals that stimulate dividing cells to produce new cells rapidly. The stem becomes longer by adding new cells at its tip. Some of the new cells become new stem tissue, while others become the tissues that make up leaves. The new stem cells grow in length just after they form, but once they mature, their size does not change. Normally, only the bud at the end of a stem grows. If this bud dies or is broken off, buds on the side of the stem will begin to grow.

Figure 8-10 Stems grow only at their tips. The end of a stem where the dividing cells are located is called the apical meristem.

Roots

The roots are the part of a plant's body that, in most cases, grows underground. When you look at a plant, consider that this underground portion is about the same size as the above-ground portion. Just like stems and leaves, roots also have several functions. One function of the roots is to anchor the plant in the ground. Thus the plant is less likely to be blown away by wind or washed away by rain. Some roots, such as carrots, sweet potatoes, and turnips, also store food. However, the most important function of roots is to absorb water and minerals for the plant.

Types of Root Systems
If you have ever had to pull weeds in a garden or a lawn, you have probably noticed that some weeds come up easily, while others are held so firmly that

the top of the plant usually breaks off before the roots come out of the ground. One reason for this is that there are two basic types of root systems in plants.

Taproots are the large, central roots that grow almost straight down. See Figure 8-11. They can be as long as a plant is tall, anchoring it firmly into the ground. In some trees, the main taproot may reach down hundreds of feet into the ground. Taproots can be long and slender, thick like carrots, or bulb-shaped like radishes. However, the smooth, unbranched structure of carrots and radishes is not typical of most taproots. These shapes were developed through years of selective breeding. Most taproots branch, but the branches are much smaller than the main root.

Figure 8-11 *Taproot root systems grow deep into the soil, while fibrous root systems spread out over a large area under the surface of the soil.*

Fibrous root systems, on the other hand, consist of many thin roots that are all about the same size. Fibrous roots often form a dense network near the surface of the soil. This network holds soil particles together and helps to keep them from being carried away by water. Grasses, which play an important role in preventing soil erosion, have fibrous roots.

Root Growth Roots grow in length by making new cells at the ends of the roots, or *root tips*. See Figure 8-12. The dividing cells are located just inside the very end of the root tip, which is protected by the *root cap*. After they are formed, new root cells lengthen and become one of the specialized types of root tissue. Cells that are one centimeter or more from the end of a root tip cannot lengthen further. Thus for a root to continue to grow in length, new cells must be produced at the root tip.

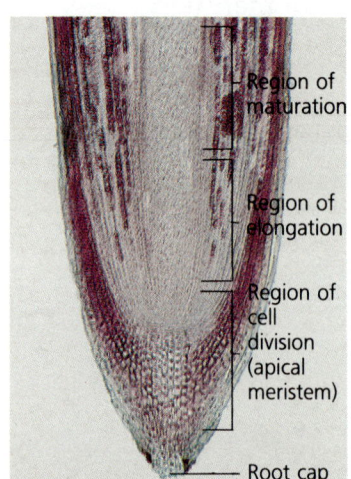

Figure 8-12 *The three parts of a root tip are the regions of division, elongation, and maturation (×150).*

Vascular tissues develop in the center of the root, while support and storage tissues develop around the vascular tissues to form the outer portion of the root. Some of the new cells that become part of the center of a root retain their ability to divide. These cells cause a root to thicken by producing more vascular and support cells. The outermost cells become the epidermis, through which water and minerals are absorbed. As the root thickens, the epidermal cells are crushed against surrounding soil particles forming an outer layer of dead cells. This part of the root can no longer absorb water.

Water and minerals are taken up by the actively growing tips of new root branches. These branch roots grow outward from the central cylinder. Most of the water and minerals taken in by the roots are absorbed through *root hairs*, which are extensions of the epidermal cells on root tips. The root system of a single plant may have billions of these tiny, threadlike structures, which are only a single cell wide. See Figure 8-13. As long as there is enough water in the soil, water is drawn into the root hairs by *osmosis*. This water continues to move through or between the root cells until it reaches the water-conducting cells. There, it can travel up to the leaves.

The addition and lengthening of new cells exerts enough pressure to drive a root tip through the soil and even into tiny cracks in rocks. If a root grows into a crack, the pressure from the thickening of a root enlarges the crack by forcing its sides apart. Roots are an important factor in weathering rocks and in forming soil.

Figure 8-13 *The fine, white fuzz on the root of this germinating corn seed consists of root hairs.*

Summary

The body of a vascular plant consists of roots, stems, and leaves. These structures all contain and are connected by two types of vascular tissue. One carries the water and minerals up to the leaves, while the other carries the food produced in the leaves down to the stems and roots or into the fruits and seeds. Leaves are adapted to produce food by photosynthesis. Plants use some of this food, while the rest is stored. Stems hold up the leaves and transport materials through continuous strands of vascular tissue called vascular bundles. Roots anchor a plant in the ground and absorb water and minerals through root hairs. Both stems and roots grow in length by producing new cells at their tips.

8-2 USES OF PLANTS

Plants are used in a multitude of ways, from providing food and covering for our bodies, to keeping us well. Plants are more important than you may think. Modern technology is largely based on the use of plant materials.

Plants for Food

Have you ever eaten a flower? If you have eaten broccoli or cauliflower, you have. In fact, many different plant parts make excellent food. See Figure 8-14. For example, an onion bulb is actually a bud, and celery is the petiole of a leaf. When you eat lettuce, spinach, or cabbage, you are eating whole, mature leaves. The tea in a tea bag and seasonings, such as parsley, chives, basil, rosemary, mint, and sage, are also leaves. Some herbal teas contain flower parts as well.

Our most important sources of food, however, are stems, roots, fruits, and seeds. Of course, you already know that potatoes and asparagus are stems. Potatoes, however, grow underground, while asparagus grows above ground, which is why asparagus is green and potatoes are not. Also, about half of the sugar you eat comes from the stems of a large grass called *sugar cane*. The rest of the sugar you eat comes from the roots of sugar beets. Carrots, turnips, radishes, and sweet potatoes are also plant roots. Tapioca, which we make into a pudding, comes from the roots of the cassava plant. In the tropics, the cassava root is the main food for millions of people.

Bananas, apples, oranges, and grapes are probably the fruits most familiar to you. But some of the foods you know as "vegetables" are really fruits, as well, because they develop from an ovary and contain seeds. Tomatoes, eggplant, squash, peppers, cucumbers, and green beans are examples of these fruits. However, it may surprise you to know that the cereal grains, such as wheat, rice, corn, oats, and barley, are also fruits.

The nutritious part of a cereal grain, such as rice, is the single, large seed contained in each fruit. See Figure 8-15. The cereal grains are the most important source of food for most of the people of the world. They were among the first plants to be cultivated and improved by humans.

Figure 8-14 *How many different plant parts are represented in this photograph?*

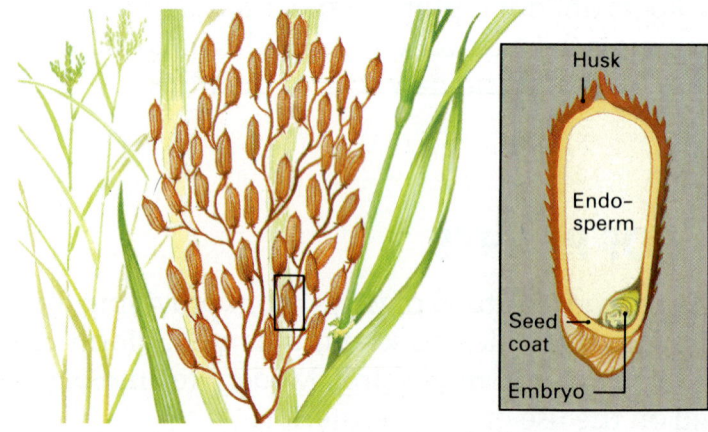

Figure 8-15 *Rice is the principal food in many parts of Asia. The rice grain is actually a fruit.*

Other nutritious seeds, such as beans, peas, and peanuts, are found in the fruits of a group of plants called *legumes*. These plants produce seeds that are rich in protein. In fact, the soybean is often used in place of meat. Peas and beans were also among the first types of plants to be cultivated, over 11,000 years ago.

Plant Fibers

You have probably heard the term "fiber" used in commercials for breakfast cereals and food supplements. But what is this fiber anyway? The fiber you eat is the material with which plants make their cell walls. This material, called **cellulose**, cannot be digested by your body. However, it helps to move food through your digestive system. Some plants are able to make very long strands of cellulose fibers. These fibers are used to make things such as cloth, rope, and paper. See Figure 8-16.

Cotton is the world's most important plant fiber. It has been grown by humans for over 5000 years. The thread used to make cotton cloth is spun from long fibers that grow from the seeds of the cotton plant. Cotton cloth is used to make clothing, bed sheets, towels, rugs, and diapers. Cotton cloth is very comfortable because it holds in body heat to keep you warm and absorbs perspiration, while allowing air flow to your skin, to keep you cool. Cotton fibers are also used to make bandages, stuffing for cushions, paper, and even explosives.

Linen is another important plant fiber that is used to make cloth. Linen fibers come from the stems of the flax plant. See Figure 8-17. Like cotton, linen is also very strong and absorbant. The ancient Egyptians used linen cloth to wrap mummies. Now it is used to make cool, summer clothing as well as table cloths, napkins, handkerchiefs, and paper. In the United States, paper money is 25 percent linen and 75 percent cotton fiber.

Hemp and sisal are two very long and coarse plant fibers that are used to make rope and twine. At one time hemp fibers were used to make paper. Another plant fiber used to make paper comes from the inner bark of a mulberry tree. See Figure 8-18. Most paper is made of ground-up wood.

Medicine from Plants

Before there were drug stores and hospitals, people obtained the medicines they needed from plants. Plants make a variety of chemicals, in addition to sugar and starch. Many of these chemicals have special properties that can help cure diseases or relieve certain kinds of discomfort. Although

Figure 8-16 Cotton cloth is made from fibers produced in the fruits of cotton plants.

Cellulose The material that makes up plant cell walls and plant fibers.

Figure 8-17 The stem of the flax plant is the source of a strong, durable fiber that is used to make linen cloth.

Figure 8-18 Kozo, a fiber that comes from mulberry trees, is the chief ingredient in a fine, very light-weight paper that is popular in Japan.

Figure 8-19 *The leaves of the foxglove contain digitalis, a powerful heart stimulant.*

many modern medicines are usually made synthetically, most of these medicines were first discovered in plants, and some are still extracted from those plants.

Aspirin, one medicine you have probably taken, contains *salicylic acid*, which originally came from the bark of willow trees. It is an effective pain reliever and also controls fever. *Ephedrine*, a drug you may have taken for a stuffy nose, comes from a rare type of gymnosperm called *ephedra*. Foxgloves and periwinkles, plants you may have in your garden or neighborhood, are another source of important medicines. See Figure 8-19. *Digitalis*, which is extracted from the leaves of the foxglove, is a stimulant that is used to treat heart disease. The drugs made from periwinkles help to relieve the symptoms of two types of cancer, Hodgkin's disease and leukemia. *Quinine*, the drug used to treat malaria, comes from the bark of the tropical chinchona tree and makes life in the tropics bearable for millions of people who suffer from this disease.

Taxol is a new drug that shows promise as a treatment for cancer. This drug comes from the bark of the Pacific Yew, a tree that grows in northern California. Unfortunately, the amount of extract required for treatment requires that many yew trees be cut down, and it takes many years for a yew tree to mature. Even though cancer is a dreaded human disease, collecting this useful medicine poses a threat to the existence of other species.

World-wide research continues to identify plants with medicinal value. Much of this research is concentrated on the tropical rain forests, where the richest variety of plant species on earth is found. Unfortunately, these forests are being cut down at an alarming rate. Untold numbers of plant species are becoming extinct as a result. The cures for many human diseases may have been found in some of these plant species. Perhaps the cure for AIDS is hidden within an undiscovered plant in the rain forest. But once these plants are gone, their secrets will be lost forever.

Summary

Plants are an important source of food for humans. Edible plant parts include roots, stems, leaves, fruits, and seeds. Plants also contain fibers that are used to produce cloth and other valuable products. Many plants produce chemicals that can be used as medicines. Many modern medicines were originally derived from plants. The destruction of the rain forests, however, may preclude the discovery of new medicines.

GLOSSARY

Abiotic nonliving, and having no living origin (8)

Acid a chemical that has a sour taste, turns litmus paper red, reacts with a base to form a salt, and has a pH value below 7 (53)

Acidic having a pH value of less than 7 (53)

Action the act of doing something, or the state of being in motion (252)

Adaptations features of organisms that arise over time and enable the organisms to survive in a given environment (74)

Aerated having oxygen added (49)

Alkaline having a pH value greater than 7 (53)

Appendage something that is attached to something else, such as a tail or a finger (84)

Arch a curved structure, often constructed of stone, that supports the weight of the material above an open space (285)

Asteroids rocks or boulders that revolve around the Sun, usually between the orbits of Mars and Jupiter (408)

Astronomy the science that studies the stars, planets, and other celestial objects (379)

Atmosphere the air around the earth (179)

Axis a straight line around which an object rotates (399)

Bacteria a large class of single-celled organisms, considered neither plant nor animal, belonging to Kingdom Monera. Their actions cause a number of diseases, as well as certain processes including decay, fermentation, and soil enrichment. (18)

Basalt a dark, fine-grained volcanic rock (339)

Base a chemical that has a bitter taste, turns litmus paper blue, reacts with an acid to form a salt, and has a pH value greater than 7 (53)

Beam a long, rigid piece of stone, wood, or metal used for horizontal support in a ceiling, roof, etc. (282)

Big bang a theorized explosion, believed to have occurred 15 to 20 billion years ago, resulting in the formation of the universe (436)

Biosphere an area or region where conditions are suitable for living things to survive (448)

Biotic living or having a living origin (8)

Black hole an object with a gravitational field so strong that neither matter nor light can escape once it has been drawn in (427)

Boiling the process in which liquid rapidly changes into a gas at its boiling point (the maximum temperature at which the liquid can exist as a liquid) (152)

Buoyant force the directional force exerted upward by a liquid on a submerged body (203)

Buttress a structure built against a wall to reinforce or support it (286)

Camouflage the method by which an organism disguises itself or blends with its surroundings to conceal itself from its enemies (76)

Cantilever a beam or structure that is fixed at one end only (284)

Carnivore a consumer of meat; a meat eater (18)

Cementation the process in which the spaces between loose particles are filled with a hardening or bonding agent (354)

Classify to sort into groups (94)

Comets small frozen masses of dust and gases that travel definite paths through the solar system. As they pass near the sun, they are vaporized, leaving trails resembling hair. (408)

Commensalism a relationship between organisms in which one organism benefits from the relationship, while the other neither benefits from nor is harmed by the relationship (12)

Compaction the process in which overlying particles compress the spaces between underlying particles (354)

Compressive force the force that causes material to become more compact and pressed together (273)

Concentrated solution a strong solution, having a relatively large amount of solute dissolved in the solvent (170)

Concentration the strength of a solution, determined by the amount of solute dissolved in the solvent (170)

Consumers organisms that depend upon other organisms as food sources (16)

S145

Counteract to undo the effect of something; to neutralize (53)

Crystallization a process in which the chemicals in a solution form solids. This is one form of lithification. (354)

D

Decomposers consumers that break down dead food sources into substances that enrich the soil (18)

Deflection the act of bending (276)

Desalination the removal of salt from a substance, especially sea water (156)

Dikes intrusions that form in vertical cracks in existing rocks (355)

Dilute solution a weak solution, having a relatively small amount of solute dissolved in the solvent (170)

Dissolving the spreading of the particles of a solute evenly throughout a solvent (140)

Distillation a two-step process involving (1) the heating of a solution to change one part of it into a gas or vapor and (2) cooling the vapor back into liquid form and collecting it (156)

Diversity differences or variety among living things (66)

Dome a hemispherical roof (285)

E

Elastic force the force that is exerted by a material when it is stretched (203)

Elastic rebound theory H. F. Reid's theory of the cause of earthquakes whereby energy is stored along faults and then released suddenly as the faults slip (327)

Elastic response the tendency of a material to return to its original shape or size after it has been stretched (275)

Electrical force the force that causes unlike charged materials to attract each other and like charged materials to repel each other (203)

Endangered species those species that are near extinction and would not survive in the wild without human protection (92)

Environment the physical surroundings of an organism, including all the conditions and circumstances that affect its development (5)

Epicenter the point on the surface of the earth directly above the focus or hypocenter of an earthquake (330)

Epoch a sub-division of a geologic period on the geologic time scale (366)

Era the largest division of the geologic time scale, denoting an interval of time when certain rocks were formed (366)

Evaporation the gradual process by which a liquid changes into a gas or vapor (152)

Extrusive rocks rocks produced by molten lava that flows onto the surface of the earth; also called volcanic rocks (349)

F

Fault the boundary between two rock sections that have been displaced relative to each other (324)

Focus the point where a rupture starts an earthquake; also called a hypocenter (328)

Food chain a chain-like linked diagram that shows the relationship between various organisms and their sources of food energy (21)

Food web a web-like linked diagram that shows as many food relationships as possible between various organisms and their sources of food energy (22)

Friction the rubbing of objects against one another. This produces heat and causes surfaces to wear away. (232)

Frictional force the force that resists the motion of objects rubbing against each other (203)

G

Geologic time scale the sequence of events of the earth's geologic story told in sedimentary rock (366)

Germinate to sprout; to begin to grow from a seed (450)

Gravitational force the mutual force of attraction exerted by particles of matter (203)

H

Habitat the place where an organism lives (10)

Hard water water that contains certain dissolved chemicals (146)

Herbivore a consumer of plants; a plant eater (18)

Humus a brown or black component of soil, resulting from the decay of material that was once living (454)

Hydroponics the process of growing plants without soil using solutions to provide the necessary nutrients for growth **(471)**

Hypocenter the point where a rupture starts an earthquake; also called a focus **(328)**

Hypothesis a possible explanation for why things happen the way they do. In an experiment, a hypothesis is tested. **(160)**

Igneous rock rock produced by the cooling and stratification of magma, either at or below the earth's surface **(344)**

Inertia the property of matter to remain at rest or resist change in its state of motion **(246)**

Inferences conclusions, drawn from observations, that attempt to explain or to make sense of the observations **(160)**

Insoluble unable to dissolve or pass into solution **(148)**

Intrusion the pushing up of magma into cracks in existing rock **(355)**

Intrusive rocks rocks produced by the cooling of magma that pushes up into the earth's crust; also called plutonic rocks **(350)**

Invertebrates animals without backbones **(100)**

Lava magma that emerges on the surface of the earth **(339)**

Lichens plant-like organisms that consist of a fungus and an alga existing together to the mutual benefit of each other **(13)**

Life cycle the series of developmental stages in an organism's life, from birth to reproduction **(116)**

Light-year the distance light travels through space in one year **(431)**

Lithification the process by which sediments become rock **(354)**

Lithosphere the outer shell of the earth containing the crust, the plates, and the continents **(321)**

Loam a mixture of sand, clay, and organic matter **(456)**

Local Group a family of galaxies, including the Milky Way, that are held together by mutual gravitational forces **(434)**

Magma melted rock deep within the earth **(339)**

Magnetic force the attractive or repelling forces between a magnet and iron, between two magnets, etc. **(203)**

Magnitude the strength, or measured size, of an earthquake **(329)**

Mass the measure of the amount of matter in an object **(207)**

Metamorphic rock rock that has undergone change as a result of intense heat and pressure **(344)**

Metamorphosis change; a transformation of the nature of a thing **(120)**

Meteorites meteors that reach the ground without burning up completely in the earth's atmosphere **(406)**

Meteors lumps of rock and/or metals that enter the earth's atmosphere and burn **(406)**

Milky Way galaxy a spiral-shaped system, of which our Solar System is a part, containing billions of stars, huge clouds of dust particles, and gases **(433)**

Mimicry the resemblance in behavior, look, sound, or smell of an organism to another organism or object in its surroundings. Mimicry helps to protect organisms from their natural enemies **(79)**

Mutualism a relationship between organisms in which all benefit from the relationship **(12)**

Natural selection the selecting or favoring of organisms that are best able to survive and reproduce in their environment. Charles Darwin used it to explain how the features of a species can change over many generations. **(86)**

Neutron star a collapsed star composed mostly of neutrons, compressed to the size of an asteroid; contains the mass of many Suns **(426)**

Newton the international metric unit (SI) used to measure force. Written as N, it is approximately equal to the weight of a 100-g mass. **(213)**

Niche an organism's role in its community, including its behavior and its place in the food chain **(10)**

S147

Noncontact forces forces exerted by agents that do not touch the receivers **(200)**

Nutrients chemicals needed for the functioning and growth of living things **(456)**

Omnivore an eater of both plants and animals **(18)**

Organism living thing with organs and/or parts that work together **(66)**

Osmosis the movement of water across a semipermeable membrane from areas where water particles are more concentrated to areas where they are less concentrated **(464)**

Paleontology the study of fossils, ancient life forms, and their evolution **(364)**

Parasitism the relationship between two organisms in which one organism, called a parasite, lives to the detriment or harm of the other, called the host **(13)**

Percolation rate the time it takes for a certain volume of water to pass through a sample of soil **(458)**

Period a sub-division of a geologic era on the geologic time scale **(366)**

pH scale a scale, ranging from 0 to 14, used to measure the alkalinity or acidity of a solution **(53)**

Photosynthesis the process by which green plants and plant-like organisms use energy from sunlight to convert water and carbon dioxide to sugars that the organisms can use for food **(16)**

Plane of the ecliptic the plane of Earth's orbit around the Sun. All of the other planets' orbits lie roughly in this plane. **(395)**

Plate tectonics the theory that large crustal plates of rock move on the earth's surface **(321)**

Plutonic rocks rocks produced by the cooling of bodies of magma in the earth; also called intrusive rocks **(350)**

Pollination the transfer of dust-like pollen from the stamen of one plant to the pistil of another plant of the same species in order to produce seeds **(12)**

Predator a consumer that hunts or captures a live food source **(18)**

Prey an organism that is hunted or captured and eaten by another consumer **(18)**

Producers Organisms that produce food for themselves and others **(16)**

Pumice a dark, porous, light-weight volcanic rock **(339)**

Quasars bright energy sources, about the size of our Solar System, that are billions of light years from Earth **(436)**

Reaction in physics, the action that opposes a given action **(252)**

Recycle to reuse the remains of things **(40)**

Red giant a star of great brightness, volume, and a relatively large diameter; emits red light **(426)**

Relative time scale an organization of events based only on their sequence. No dates are included. **(365)**

Retrograde motion as seen from Earth, the apparent backward motion of a planet before continuing its forward motion **(389)**

Revolution the motion of one body around another **(395)**

Richter Scale of Magnitude a scale of measurement in which the number expressing the magnitude of an earthquake is calculated by means of a formula **(329)**

Rotation the spinning motion of a planet or other object around its axis **(395)**

Saturation that point at which no more solute can dissolve in a solvent **(176)**

Scavenger a consumer of dead food sources **(18)**

Sedimentary rock rock produced by cemented mineral particles deposited by wind, water, ice, or chemical reactions **(344)**

Seismic waves intense vibrations in the earth during an earthquake; also called shock waves **(328)**

Seismograph the instrument used to record the vibrations of the earth during an earthquake **(329)**

Seismologists people who study earthquakes **(329)**

Shear force a force that tends to cause a material to tear or be cut apart **(273)**

Shock waves the release of stored energy in the form of intense vibrations; also called seismic waves **(328)**

Sills intrusions that form in horizontal cracks of existing rocks **(355)**

Soft water water that is mostly free of dissolved matter **(146)**

Solubility the amount of substance able to pass into solution under certain conditions; the measure of the amount of solute needed to saturate a solution **(177)**

Soluble able to dissolve or pass into solution **(148)**

Solute a substance that is dissolved in a solution **(141)**

Solution substances in which the solute particles are spread evenly throughout the mixture **(135)**

Solvent a substance that dissolves a solute to make a solution **(141)**

Stomata microscopic openings on the lower surface of leaves that regulate the passage of substances into and out of a plant **(468)**

Stratovolcano a cone-shaped volcano made up of layers of lava, ash, and cinders; also called a composite cone volcano **(336)**

Succession the process in which a new group of organisms, which are better suited to the changes that have taken place in a particular area, replaces an old group of organisms **(36)**

Supergiant a star of very great brightness and enormous size, at least 100 times larger and brighter than the sun **(426)**

Supernova an exploding star **(426)**

T

Tensile force the force that causes a material to stretch **(273)**

Texture a characteristic of soil based on particle size, roughness, and tactile properties **(455)**

Tolerance the ability to endure a set of conditions **(26)**

Transpiration the movement of water out of plants through the leaves **(466)**

Truss a framework of connected beams or bars built to add strength and support to a bridge, roof, or other structure **(282)**

Tyndall effect a test for determining true solutions. True solutions are transparent, but in a nonsolution, the particles are large enough to scatter or reflect light. **(135)**

V

Vertebrates animals with backbones **(100)**

Volcanic rock rock produced by molten lava that flows onto the surface of the earth; also called extrusive rock **(349)**

Volcano an opening in the earth's surface through which lava and other materials erupt **(332)**

W

Water cycle the process, involving evaporation and condensation, by which the earth's water is constantly purified **(158)**

Weight the measure of the gravitational force acting on an object **(205)**

Z

Zodiac in astronomy, twelve different constellations, named mostly for animals, that lie close to the ecliptic (NOTE: These constellations do not correspond to the "birth signs" you may have read about.) **(393)**

INDEX

Boldface numbers refer to an illustration on that page.

A

Abiotic factors, 8, 26–34, **34**, S2, S6
Acceleration
  description of, S63–S64
  force and, S70–S71
  due to gravity, S66
Acids
  in the atmosphere, 179
  characteristics of, S51–S52
  defined, S52
  formation of, S52
  strength of, S53
  uses of, S53–S54
Acid rain
  air pollution and, 53–54, S18
  in the United States, **53**
  effects of, 55, S19
  pH of, 53
Adaptation, S39–S40
Adaptations
  of animals, 76–79, 84, **76–79**
  defined, 74
  diversity and, 74
  of plants, 80, **80–81**
  of seeds, 82, **82–83**
African architecture, S87–S88
Agricola, 320, **320**
Agricultural engineering, 496–497
Aldrin, Edwin "Buzz," S130
Algae
  description of, S25–S26
  lichens and, 13, S10
Amphibians, S33
Ancient Greece
  architecture of, 296, **296**
  belief about mountains, 319
Ancient Rome
  architecture of, 297, **297**
  belief about mountains, 319
Andromeda, 434, **434**, S124
Angiosperms, S29
Animal Kingdom, 98–109
  classification of, 106, S30–S34
Ants, 15, **15**
Anther, 475, **475**, 476, **476**
Aphids, 15, **15**
Apollo 11 spacecraft, S130
Appendages, 84

Arch, 284–288, **285**
  making a model of an, 287
Arched bridge, 288, **288**
Architecture
  careers in, 312–313
  of dome of Florence, 314–315, **314**
  Gropius and, 301–302
  through history, 293–300
  in other cultures, S78–S88
  Lambert and, 306
  Wright and, 301–303
Aristarchus, 389
Aristotle
  gravitation and, S74–S75
  model of the universe, 388
  motion and, S65
Armstrong, Neil, S130
Arthropods, S32
Asteroids, 408, S117–S118
Astronomer, 381
Astronomical unit (AU), 393, 430
Astronomy
  careers in, 442–443
  defined, 379, 382
  history of, 382, 388–390
Atmosphere, 179
Auk, *See Great Auk*
Auroras, S114
Autumnal equinox, 401
Axis, 399
Aztec architecture, S84

B

Bacteria
  blue-green, S25
  as decomposers, 18, **19**
  description of, S25
  in human digestive tract, 15, **15**
  nitrogen-fixing, S7
  resistance to chemicals of, S41
Bark, S139
Barnacles, 15, **15**
Basalt, 339, **348**, 349, S94–S95
Bases
  described, 53, S54
  characteristics of, S54–S55
  uses of, S55–S56
Beam
  compared to arch, 286–287
  as a cantilever, 284, **284**
  defined, 282

I-, 283, **283**
  compared to truss, 283
Bear, 70, **70**
Big-bang theory, 436–437, S131–S132
Binary stars, S122
Biochemical similarities of organisms, S43
Biochemistry, 188
Biomes
  defined, S12
  types of, S12–S17
Biosphere
  in Arizona, **416**
  defined, 448, **448**, S2
  in space, 416
Biotic factors, 8–9, **8**, S2
Birds, S34
Black holes, 378–379, 427, 444, S125–S126
Blue-green bacteria, S25
Bridges
  cantilever, 284, **284**
  classification of, 288–289, **288–289**
  famous, 289, **290**
  Tacoma Narrows, 267, **267**, 291, **291**
Brown pelican, 92, **92**
Brunelleschi, Filippo, 314–315
Buoyant force, 203
Burdock seeds, 15, **15**
Byzantine architecture, S78–S79

C

Camouflage, 76–77, **76–77**
Cantilever, 284, **284**
Cantilever bridge, **288**
Carbon dioxide–oxygen cycle, S7
Careers
  in agricultural engineering, 496–497
  in architecture, 312–313
  in astronomy, 442–443
  in chemistry, 188–189
  in electrical engineering, 258–259
  in genetic engineering, 498–499
  in geology, 372–373
  in medicine, 126–127
  as seed analyst, 451
  in veterinary science, 60–61
Carnivore
  defined, 18, S6

  top, 21, S6
Cartier, Jacques, 43
Cathedral of Florence, 314–315, **314**
Cavendish, Lord, 207
Cellulose, S143
Cementation, 354
Cenozoic Era, 366, 368, S36
Chameleon, 76, **76**
Chemistry, 188–189
Chinese architecture, S81–S82
Chloroplasts, S135
Classification
  of animals, 98–100, 106, S30–S34
  biological system of, 102
  of fungi, S26–S27
  of invertebrates, 100–101, S30–S32
  of living things, 98, S24–S35
  of monerans, S24–S25
  of plants, 110–111, S27–S29
  of protists, S25–S26
  scientific names and, 112
  system of Linnaeus, 112
  of vertebrates, 100, S32–S34
Clay, 456
Cleavage, S103
Coal
  anthracite, S98
  bituminous, S97
  formation of, S96–S97
  lignite, S97
Coelenterates, S30
Colloids, S50–S51
Color
  of minerals, S101–S102
  of stars, S120–S121
Comets, 408–410, **408–410**
  Halley's observation of, 409–410
Commensalism, 12, **12**, 14, **15**
Community
  climax, S11
  defined, S5
  forest floor, 46
  habitat and niche in, 10
  of living things, 10–15
  movement of energy in, 16, **16**
  pond, 48, **48–49**
  roles in, 16–18
  rotting-log, 38–39
  interactions in, 12
  woodland, 44

S150

Compaction, 354
Compost pile, 19
Compressive force, 273–274
Concentration, of a solution, 170, 171–173
Conglomerate, 352, **352**, S95
Coniferous forest biome, S13
Coniferous tree, 44–45, **44–45**
Conservation
  defined, S18
  methods of, S20–S22
  of water, S22
  of wildlife, S22
Constellations, *see Stars*
Consumer
  in a community, 16
  primary, 21
  secondary, 21
Convergent boundary, 321, **321**
Copernicus
  Galileo and, 392
  model of planets, 393
Coquina, **352**
Cortés, Hernando, S84
Cosmology, S131
Cows, 14, **14**
Cross-breeding, 478
Cross-pollination, 477, **477**
  as plant breeding technique, 478
Crystallization, 354
Crystals
  defined, S91
  growing, 183–184, **183–184**
  shape of, S103
Cullinan Diamond, 374–375
Cycles
  carbon dioxide–oxygen, S7
  nitrogen, S7–S8
  rock, 357–359, S91–S93
  water, 158–159, S6–S7

Darwin, Charles, 88–89, **88**, S35–S36
  theory of natural selection, S36–S38
Deciduous forest biome, S14
Deciduous tree, 44–45, **44–45**
Decomposers
  bacteria and fungi as, 19, **19**
  as consumers, 18
  in a rotting-log community, 40
Deflection, 276
Desalination, 156, S59–S60

Desert biome, S14–S15
Design
  Bauhaus School of, 301, **301**
  in the environment, 304–308, **304, 305**
  of structures, 293–308
  through history, 293–300
  zoning laws and, 305
Diamonds, 374–375, **374–375**
Digitalis, S144
Dikes, 355
Dinosaurs, 363
Dissolving, 140
Distillation, 156–158
Distillery, 158, **158–159**
Divergent boundary, 321, **321**
Diversity
  defined, 66
  evolution and, S35
  extinction and, 90, S19–S20
  inherited characteristics and, 114–115
  reasons for, 74
  within a species, 114–115
Dodder plant, 14, **14**
Dome, 284–285, **285**
  cathedral of Florence, 314–315, **314**
Dragonfly, 71, **71**
Drying, 354

E

Early civilization, 295, **295**
Earth
  Aristarchus' theory of motion of, 389
  characteristics of, 399–401, S112, S115–S116
  crust of, S99–S101
  early atmosphere of, S43–S44
  as frame of reference, S64
  geologic history of, S105–S110
  history of life on, S35–S36
  journey from, 429–435
  orbit of, S115
  Ptolemy's theory of motion of, 389
Earthquakes, 326–331
  cause of (elastic rebound theory), 327–328, **327–328**
  seismograph measuring, 329–330
Echinoderms, S32
Ecliptic
  model of, 399–401, **399**

  plane of, 395
Ecosystem
  defined, S2
  parts of, S2–S3
Einstein, Albert, S114
Elastic force, 203, 275–277
Elastic rebound theory, 327–328, **327–328**
Electrical engineering, 258–259
Electrical force, 203, 205
Elements
  of the Greeks, S74
  native, S101
Embryo development, S42–S43
Endangered species, 92, S19
Endoskeleton, S32
Energy
  in black holes, S125–S126
  diagram, **16**
  in leaf functions, S136
  movement in community, 16
  pathways in ecosystems, S5–S6
  production by sun, S114–S115
  in quasars, S125
  sun as source of, S113, S134
  in variable stars, S122
Engineering
  agricultural, 496–497
  electrical, 258–259
  genetic, 498–499
Environment
  abiotic and biotic factors of, 9, **9**
  defined, 5
  design in the, 304–308, **304–307**
  interactions in, 7
  preventing problems in, 51, S20–S22
Environmentalists, S20
Ephedrine, S144
Epicenter, 330, **330**
Epidermis, S135
Epochs, 366, **368**
Eras, 366, **368**
Estuary, S15
Evolution, 86, S36–S40
Exoskeleton, S32
Experiments
  controlling variables, 148
  hypothesis in an, 160
  inference in an, 160
Extinction
  brown pelican and, 92, **92**
  caused by DDT, 92
  defined, S17
  of endangered species, 92
  game of, 90–91

  as loss of diversity, 90, S19–S20

F

Fault
  causing earthquakes, 326
  formation of, 324
  in elastic rebound theory, 327–328, **327–328**
Fertilizer
  N-P-K rating of, 469–470
  pH rating of, 470
Fiber, S143
Finch, 88–89, **88**
Fish, 70, **70**, S33
Flatworms, S30–S31
Flowers
  parts of, 473, 475, **475**
  in plant reproductive cycle, 476, **476**
  pollination of, 473–478
Focus, 328, **328**
Fold, 324
Food chain
  diagram of, 21
  food webs and, 22
  primary consumer in, 21
  secondary consumer in, 21
  top carnivore in, 21
Food pyramid
  defined, S5
  energy and, S5–S6
Food web, 22–23
Force
  agent, receiver, and effect of, 198, **198–199**
  balanced, 240, **240**, S66
  buoyant, 203
  centripetal, S67
  compressive, 273–274, **273**
  defined, S65
  describing, 197
  elastic, 203, 275–277
  electrical, 203
  estimating, 216–219
  frictional, 203, 225–235
  gravitational, 203, 205
  of inertia, 246, **246–247**
  magnetic, 203
  as mass times acceleration, S71
  measuring, 210–215, **210–213**
  motion and, 236–241, S65–S66
  in pairs, 251–254, **252–253**
  in sailing, 196
  shear, 273–274, **274**
  tensile, 273, **273**, 274
  unbalanced, 240

S151

Force meter, 210–213, **210–213**
Forest-floor community, 46
Fossil correlation, S109–S110
Fossils
  defined, 361, **361**
  dinosaur, 363
  evolution and, S35
  formation of, 361, **361**
  index, S110
  study of (paleontology), 364
  types of, 361
Fracture, S103–S104
Frame of reference, S64
Fresh water biome, S16–S17
Friction, S65
Frictional force
  case studies, 225–226, **225–226**
  cause of, 231
  characteristics of, 229–230
  harm of, 232
  introduced, 203
  measuring, 227–228
  effect on motion, 241
  reduction of, 231–233
Frog, 70, **70**, 116–117, **116**
Fruit, **476**
Fungi
  classification of, S26–S27
  club, S27
  in compost, 19
  kinds of, 19, **19**
  lichens and, 13, **13**
  sac, S27
  thread-like, S26

G

Gabbro, 348, 350, S94
Gagarin, Yuri, S129
Galapagos Islands, 88–89
Galaxy
  Andromeda, 434, **434**, S124
  defined, 433
  description of, S123
  formation of, 436–437
  Local Group, 434, **434**
  Milky Way, 433, **433**, S123–S124
  motion of, S124
  NGC 1275, 444, **444**
  quasars forming, 436
  types of, S124
Galileo
  Copernicus' model and, 393
  gravitation and, S74–S75
  Jupiter's moons and, S117
  telescopes and, S127
  thought experiments, 237
  trial of, 392

Galileo spacecraft, S130
Games
  Extinction, 90–91
  interaction string, 7–9
  Puffball, 242–245
Gateway of the Sun, S85
Genetic engineering, 498–499
Geocentric, 390, 440
Geologic column, S110
Geologic time scale
  dates on, 367
  described, 366
  diagram of, **368**
  subdivisions of, 366, S36
  table of, 364
Geology, 372–373
Germination
  defined, 450
  in plant reproductive cycle, **476**
Girder beam bridges, 289, **289**
Glenn, John, S129
Goddard, Robert, S129
Gold, 182, S101
Granite, 348, 350, S92, S94–S95
Grasshopper, 77, **77**
Grassland biome, S14
Gravitational force
  defined, 205
  galaxies and, S123
  introduced, 203
  Lord Cavendish's proof of, 207
  measured, 214, **214**
  on the moon, 206
  Newton's Law of, S75
  Newton's theory of, 206–207
  related to weight, 205, S76
  Solar System and, S115
  theories of, S74–S75
Great Auk, 43, **43**
Great Smoky Mountains National Park, 24
Great Wall of China, S81–S82
Gropius, Walter, 301–302
Gymnosperms, S29
Gypsum, 352, **352**, S96, S101
Gypsy moth caterpillar, 43

H

Habitat
  in community, 10
  defined, 10
Hagia Sophia, S78–S79
Half-life, S108
Halley, Edmund, 409
Halley's comet, 409–410, **409–410**
Hardness, of minerals, S102–S103

Heliocentric, 390, 440
Herbivore, 18, S6
Hertzsprung-Russell diagram, S121
Homologous structures, S42
Hornet fly, 79
Host, in parasitism, 13
Humus, 454
Hutton, James, 355–356, S105
Hydronium ion, S52
Hydroponics, 471–472
Hydroxide ion, S54
Hypocenter, 328
Hypothesis, 160

I

Ichneumon flies, 14, **14**
Igneous rock
  classification of, 348, S94–S95
  described, 344
  extrusive (volcanic), 349, **349**, S94
  formation of, 349–350, **350**, S94
  identification of, 345, **345**
  intrusive (plutonic), 350, **350**, S94
  simulated formation of, 346–347, **346–347**, S90–S91
Impatiens plant, 465, **465**
Incan architecture, S85
Indian architecture, S80
Indicators, pH, S57–S58
Inertia
  defined, 246, S69
  examples of, 246–247, **246–247**
  experiencing, **246–250**
  as momentum, S72
Inference, 160
Inks, 161
Insects
  diversity of, 122
  life cycles of, 118–120
  resistance to chemicals, S41
Insolubility, 148
Interactions
  examples of, 4–5, **4–5**
  in the environment, 7
  string game, 6–8, **6–8**
Intrusions, 355–356
  dikes, 355, **355**
  sills, 355, **355**
Invertebrates
  classification of, 102–103
  defined, 100, S30
  examples of, **100–101**
  subgroups of, 102–103, S30–S32
Islamic architecture, S79

J

Japanese architecture, S82–S83
Jewelry, 374–375
Jupiter, 412, **412**, 416, S116–S117
  red spot of, 413, **413**

K

Katydid, 76, **76**
Kepler, Johannes
  discovery of elliptical orbits, 402
Krakatoa, 336, **336**

L

Lambert, Phyllis, 306, **306**
Lamination, 280
Landfills, S21
Landscape, 488
Lava, 339, **339**, 349, **349**, S94
Laws
  of conservation of momentum, S73
  of Gravitation, Newton's, S75–S76
  of Motion, Newton's, S68–S72
  of superposition, S107
Leaves
  function of, S136–S137
  stomata of, 468, S135–S136
  structure of, 467–468, **468**, S134–S136
  water and, 468
Lemmings, 43, **43**
Leonov, Alexei, S129
Lichens
  in a forest-floor community, 46–47
  as fungus and algae, 13, **13**
  as mutualism, 12
  as pioneer species, S10
Life cycle
  of frogs, 116–117, **116**
  of insects, 118–120
  metamorphosis and, 120
Light
  levels for plant growth, 487
  speed of, 429
Light tolerance, 32, **32**
Light-year, 431
Limestone, 352, **352**, 356, **356**, S96
Limiting factors, S4
Linnaeus, Carolus, 112, **112**

Linnaeus' system of classification, 112
Lithification, 354
Lithosphere, 321
Litmus paper, 470, S57
Lizard, **66**, 76, **76**
Loam, 456
Luminosity, S119
Lungworm, 15, **15**
Luster, S102
Lyell, Charles, 355–356, **356**

Machu Picchu, S85
Magellan spacecraft, 415, 442–443, S130
Magma, 339, S91
Magnetic force, 203
Malthus, Thomas, S37–S38
Mammals, S34
Marble, 356, **356**, S98
Marine biome, S15–S16
Mars
    characteristics of, S115–S116
    colonizing, 416, **419**
    Ptolemy's theory about, 389–390
    retrograde motion of, 389, **389**, 396–397, **396**
Mass
    acceleration and, S71
    defined, 207
    inertia and, S69
    measuring, 223, **223**
    weight compared to, 208, S76
Mayan architecture, S85
Medical research, 126–127
Medieval Europe, 298, **298**
Meiji Shrine, S83
Mercury, **415**, S115–S116
Mesozoic Era, 366, **368**, S36
Metamorphic rock
    classification of, S98
    formation of, 355–356, **355**, S91, S97–S98
    Hutton and, 355–356
    identification of, 345
    introduced, 344, **344**
    Lyell and, 355, 356
Metamorphosis
    in rock, 355
    life cycles and, 120
Meteorites, 406–407, S118
Meteors
    defined, 406
    compared to meteorites, 407, S118
    as a shooting star, 407

Mexican milk snake, 79, **79**
Milky Way, 433, S123
Mimicry
    eyespots as, 79
    of the hawkmoth, 79, **79**
    of the hornet fly, 79
Minarets, S79
Minerals
    cleavage of, S103
    color of, S101–S102
    crystal shape of, S103
    defined, S91
    fracture of, S103–S104
    hardness of, S102–S103
    kinds of, S99–S101
    luster of, S102
    nonsilicates, S100–S101
    precipitation of, S96
    properties of, S101–S104
    silicates, S99–S100
    special properties of, S104
    streak of, S102
Mistletoe, 15, **15**
Mixtures, S46
Modern civilization, 300, **300**
Moisture tolerance, 29–30
Molecules
    biological, S44
    rate of dissolving and, S48
Mollusks, S31–S32
Momentum, S72–S74
Monerans, classification of, S24–S25
Mont Pelée, 337, **337**
Moon
    composition of, S112
    observations about, 383, **383**, 386, **386**, 405, **405**
    origin of, S112–S113
    surface of, 386, 406, **406**
Moons, *see Satellite*
Mosques, S79
Mosses, 71, **71**
Motion
    circular, S66–S67
    continual, 237
    defined, S62
    of earth, S64
    effect of force on, 240, S65
    effect of friction on, 241
    free fall, S66
    Galileo's experiments on, 237, **237**
    overcoming inertia, 246
    after initial force, 237
    laws of, S68–S72
    projectile, S67–S68
    thought experiments on, 237–241
    types of, S66–S68
    uniform circular, S67
Mound builders, S86

Mountains
    formation of, 324
    formed by erosion, 320, **320**
    plate tectonics and, 321
    rise and fall of, 319–321
Mount Everest, 318, **318**, 319
Mount Pinatubo, 335
Mount Saint Helens, 332–334, **332–334**
    location of, 338
    magma movement in, 339
    as a stratovolcano, 336
Mutation, S39
Mutualism, 12, **12**, 14, **15**

Native American architecture, S86–S87
Native elements, S101
Natural selection
    adaptation and, S39–S40
    Charles Darwin and, 86–89, S35–S39
    defined, 86, S38
    diversity and, 86
    finches and, 88–89, **88**
    peppered moths and, 87, **87**
Neptune, 411, 413, **413**, **414**, S116–S117
Neutralization, S55
Newton, Isaac
    friction and, S65
    gravity and, 206, **206**, S75–S76
    laws of force and motion, 207, S68–S72
    telescopes and, S127
    theory of force, 252
Newtons, 213
Newton's First Law of Motion, S69–S70
Newton's Law of Gravitation, S75
Newton's Second Law of Motion, S70–S71
Newton's Third Law of Motion, S71–S72
Niche
    in a community, 10, **10**
    defined, 10
    table, 11
Nitrogen, 469–470
Nitrogen cycle, S7–S8
Nomad, 294, **294**
Nonsilicate minerals, S100–S101
Nova, S123
N-P-K rating, 469–470
Numerical dating, S108–S109

Nutrients
    defined, 456
    in hydroponic solutions, 471
    in plants, 456

Obsidian, **348**
Ocean, 180
Oceanography, 62-63
Oil spill, 145, **145**
Omnivore, 18
Onion, 70, **70**
Oparin, Alexander, S44
Orchid, **14**
Organism
    defined, 66
    life cycles of, 116
Origin of life, S43–S44
Osmosis
    defined, 464, **464**
    in plants, 464–465, S141

Pagoda, S81
Paleontology, 364
Paleozoic Era, 366, **368**, S36
Parallax, S119
Parasite, 13, 18
Parasitism, 13, **14**, 15
Paricutin, 337
Pelicans, 71, **71**
    brown, 92, **92**
Peppered moth, 87, S39
Percolation rate, 458
Pesticides, 484–485
    alternatives to, 485
    resistance to, S41
Petiole, S135
pH
    defined, S56
    of fertilizer, 470
    of human body, S58–S59
    indicators of, S57–S58
    nature and, S58–S59
Phosphorous, 469–470
Photosynthesis
    algae and, 13, S25
    blue-green bacteria and, S25
    in chloroplasts, S135
    defined, 16, S134
    as an energy resource, 16
    in leaves, 466, S136
    by lichens, 13
    process of, 16
Physician, 126–127
Pioneer 10 spacecraft, S69, S71
Pistil, 473, 475, **475**
Planarian, S31

Planets
  elliptical orbit of, 402–403
  Kepler's theory of, 402
  model of motion of, 393–395
  orbits, 402
  in plane of the ecliptic, 395
  revolution of, 395
  rotation of, 395
  of Solar System, 411–416, **411–416,** 418–419, 421, S115–S116
Plant Kingdom
  classification of, 110, S27–S29
Plants
  adaptations in, 80, S134
  in a biosphere, 448–449, **448–449**
  classification of, S27–S29
  environmental needs of, 486
  flowers on, 473–478, **474**
  germination of seeds in, 450, **450**
  hydroponic growth of, 471–472
  light levels for growth, 487
  nutrients of, 456
  pesticide effect on, 484–485
  photosynthesis in, 466
  pollination of, 473–478, **477–478**
  reproduction of, 476, **476**
  root growth in, 481–482, **482,** S139–S141
  soil and, 456
  structure of, 467–468, S134–S141
  transpiration in, 466
  transplanting of, 461, **461**
  uses of, 487–489, **487–489,** S141–S144
  vegetative reproduction of, 479–480, **479**
  water and, 463–468
Plant reproductive cycle, 476, **476**
Plate tectonics
  plate boundaries and, 321, **321**
  rock recycling and, 358
  theories of, 321
  volcanoes and, 339
Pluto, 416, 421, S116–S117
Polar bear, 77, **77,** 86
Polaris, 391
Pollen, 473, 476
Pollination
  cross-, 477–478
  defined, 473
  in flowers, 473–478

as mutualism, 12
  in plant reproductive cycle, **476**
  self-, 477–478
Pollutants, S18
Pollution
  acid rain, 51–56
  consequences of, S19
  defined, S18
  effect on grebes, 145, **145**
  of the environment, S18–S19
Populations, 41–43, S3–S5
Porpoise, 86
Potassium, 469–470
Praying mantis, 77, **77**
Precambrian Era, 366, **368,** S36
Precipitation, S96
Pre-Columbian architecture, S84–S85
Predator, 18
Prey, 18
Primary consumer, 21
Principle of Uniformitarianism, S105
Producer, in a community, 16
Protists, classification of, S25
Protozoa, 14, **14**
Protozoan, description of, S25
Ptolemy, 389–390, **389**
Puffball, game of, 242–245
Pulsar, S125
Pumice, 339, **348**
Pyramid of the Sun, S84

Q

Quartz, S99–S100
Quartzite, **356**
Quasar, 436, 444, **444,** S125
Quinine, S144

R

Rabbits, 43
Radioactive decay, S108
Radio telescope, 444–445, **444,** S127–S128
Red shift, S124
Regeneration, S32
Relative age, S108
Relative time scale, 365, **365**
Renaissance, 299, **299,** 314–315
Reproductive group, S3
Reptiles, S33–S34
Resistance, S41
Resources
  abiotic, S6–S8
  natural, S17–S18
  nonrenewable, S18

renewable, S18
Retrograde motion, 389–390
Revolution, 395
Rhyolite, **348**
Richter Scale of Magnitude, 329–330
Rock cycle, 357–359, S91–S93
Rockets, S129
Rocks
  classification of, 343, S93–S98
  formation of S90–S91
  fossil correlation and, S109–S110
  groups of, 344
  identification key for, 345
  igneous, 344, **344,** 346–350, **348, 349,** S94–S95
  metamorphic, 344, **344,** 355–358, **356,** S97–S98
  numeric dating of, S108–S109
  plate tectonics and, 358
  recycled, 357
  relative age of, S107–S108
  rock cycle and, 357–359, S91–S93
  sedimentary, 344, **344,** 351–354, **352,** S95–S97
  telling time with, 365–368, **368**
  uses of, 341–342, **341–342**
Root
  cap, 482, S140
  growth, 481–482, **481–482,** S140–S141
  hairs, 463–464, **464,** S141
  systems, S139–S140
  tip, 482, S140
  in vegetative reproduction, 481
Rotation, 395
Roundworms, S31

S

Salt industry, 152–155
Salts
  characteristics of, S59
  defined, S59
  uses of, S59–S60
Sand, 456
Sandstone, **352,** 356, **356,** S95
Satellites, 413, S116–S117
Saturation, 176
Saturn, **413,** 416, S116–S117
Scavenger, 18
Secondary consumer, 21

Sedimentary rock
  classification of, S95–S97
  identification of, 345, **345**
  introduction to, 344
  formation of, 351–354, S95–S96
  lithification of, 354
  simulated formation of, 354, **354,** S91
Seed
  adaptations of, 82
  analyst, 451
  germination of, 450–452
  in plant reproductive cycle, 476, **476**
Segmented worms, S31
Seismic waves, 328
Seismograph, 329, **329**
Seismologist, 329
Selective breeding, S40
Self-pollination, 477–478
Sepal, 475, **475,** 476
Shale, **352,** 356, **356,** S95
Shear force, 273–274, **274**
Shock waves, 328
Silicates, S99–S100
Sill, 355
Silt, 456
Sirius, 431
Slate, 356, **356,** S98
Sliding boundary, 321, **321**
Snowshoe hare, 77, **77**
Soil
  contents of, 454, 456
  formation during succession, S10
  mixtures, 456, 458
  particles of, 455
  percolation rate of, 458
  plants and, 456
  sorting of, 456
  texture of, 455
  water and, 457
Solar System, 389, 395, S115–S118
Solar wind, S114
Solubility
  described, 148, S48
  graph of, 178
  table of, 177
Solutes
  effect on boiling point, 164–166
  crystals form in, 183, **184**
  concentrated, 170
  defined, 141, S46
  dilute, 170
  effect on freezing point, 169
  in hard water, 147
  osmosis and, 464
  salt as a, 152–155
  saturation with, 176
  solubility of, 177

Solution
  atmosphere as a, 179
  changes in, 152, 156
  concentrated, 170–171, S47
  defined, 135, S46
  desalination of, 156
  dilute, 170, S47
  in the environment, 179–188
  non-solution and, 134–139
  ocean as a, 180
  parts of, 140–141, **141**
  saturated, 176, S47
Solvents, 141, 144, S46
Space Plane, S130
Space Shuttle, S130
Species
  endangered, S19
  pioneer, S10
Speed
  defined, S62
  terminal, S66
Sponges, S30
Sputnik 1 spacecraft, S129
Stamen, 473, **475**
Stars
  absolute magnitude of, S119–S120
  binary, S122
  chart of, 387, **387**
  composition of, 423
  distance to, S118–S119
  formation of, 437
  forming black holes, 427
  history of, 425
  life of, 426–429, **428**
  nuclear reactions in, 423
  paralax and, S119
  Sun, 422–423, 426–427
  supernova, 426, **426**, 436
  temperature of, 424–425, S120–S121
  tracks of, 387
  types of, 426–427, S120–S123
  variable, S120, S122
Starling, 43
Stems
  growth of, S138–S139
  structure of, S137–S138
Stick insect, 76, **76**
Stigma, 475, **475**, 476
Stomata, 468, **468**, S135
Stonehenge, 341, **341**
Streak, of a mineral, S102
Strength
  of building materials, 278
  in engineering structures, 282–283
  measuring, 220–222
  in shapes, 279–281, **279–281**
  tensile, 278

Structure
  of an arch, 284–285, 286–287
  of a cantilever, 284
  of a dome, 284–285
Structures
  engineering, 282–291
  failure of, 267–268, **267–268**
  response to forces, 272–277
  strong shapes in, 279–281
  successful, 269, **269–270**
Stupa, S80
Style, 475
Succession
  defined, 36, S9
  environmental changes and, 36, **36**, 37
  at Mt. St. Helens, 36, **36**
  primary, S9–S11
  secondary, S11–S12
Suckerfish, 14, **14**
Sugar
  honey and, 166
  maple tree, 164
  as a solution, 164
Summer solstice, 401
Sun
  calendar wheel, 386, **386**
  characteristics of, 422–423, **422**
  composition of, S113–S114
  dial, 384
  energy production of, S114–S115
  formation of, 437
  life of, 426–427
Supernova, 426, 436, S123
Surtsey volcano, 337, **337**
Suspension, S49–S50
Suspension bridge, 288, **288**

Tacoma Narrows Bridge, 267, **267**, 291, **291**
Taj Mahal, S80–S81
Taxol, S144
Tectonic plates, S92
Telescope, S127–S128
Temperature
  of freezing water, 168
  range of tolerance for, 26–28
  saturation and, 177
  in solutions, 168–169, **168**
Tenochtitlán, S84
Tensile force, 273–274
Tensile strength, 278

Teotihuacán, S84
Tepee, S87
Tereshkova, Valentina, S129
Termite, 14, **14**
Terrestrial biomes, S12–S15
Texas coral snake, 79, **79**
Tihuanaco, S85
Time scale
  geologic, 366, **368**
  relative, 365, **365**
Tolerance
  light, 32–34
  moisture, 29–31
  temperature, 26–29
Top carnivore, 21
Torii, S83
Trajectory, S67
Transpiration, 466
Tropic of Cancer, 401
Tropic of Capricorn, 400
Tropical rain forest, S14
Truss, 282
  compared to beam, 283, **283**
Tsiolkovsky, Konstantin, S129
Tubular column, 281
Tundra, S12–S13
Turbellarians, S31
Tyndall Effect, 136–137, **136–137**, S51

Ulysses spacecraft, S130–S131
Unconformity, S107
Uniformitarianism, Principle of, S105
Uniformity, S105–S106
Universal indicator, S58
Uranus, 416, 421, S116–S117
Urban environment, 304–306, **304–305**

Variable stars
  absolute magnitude and, S120
  description of, S122
Variation, S38–S39
Vascular bundles, S137–S138
Vascular tissue, S28, S137
Vegetative reproduction, 479–480
Velocity, S63
Venus, 415, 416, 421, S115–S116
Vernal equinox, 400
Vertebrates
  classification of, 100
  defined, 100, S32
  examples of, **104–105**

  subgroups of, 104, S32–S34
Veterinary science, 60–61
Volcano
  famous eruptions of, 336–337, **336–337**
  kinds of, 336–337
  location of, 338
  Mount Pinatubo, 335
  Mount Saint Helens, 332–334, **332–334**
  plate tectonics causing, 339
Voyager spacecraft, 411, **412**, 414, S117

Wallace, Alfred Russel, S38
Water
  conservation of, S22
  cycle, 158–159, **158–159**, S6–S7
  freezing point of, 168
  hard, 146
  percolation rate, 458
  plant interaction with, 463–468
  pollution, S18–S19
  salt, 174
  soft, 146
  soil and, 457
Water cycle, 158–159, **158–159**, S6–S7
Wegener, Alfred, 321
Weight
  compared to mass, 205, S76
  defined, 205
  measuring, 223, **223**
Wigwam, S87
Wilson, J. Tuzo, 321, **321**
Winter solstice, 400
Wisakedjak, 113
Woodland community, 44
Woodpecker, 70, **70**
Wright, Frank Lloyd, 302–303

Zodiac, 393–394, **394**
Zoning laws, 305

PHOTO CREDITS

Abbreviated as follows: (t) top; (b) bottom; (l) left; (r) right; (c) center.

Table of Contents: iii(tr), Jeff Lepore/Photo Researchers, iii(b), HRW Photo by John Langford; iv(t), Robert Frerck/Woodfin Camp & Associates, iv(b), Nelson Max/Peter Arnold, Inc.; v(tr), Kip Peticolas/Photo Researchers, v(b), HRW Photo by John Langford; vi(t), L. Kemper/SuperStock International; vi(b), R. King/SuperStock International; vii(t), Robert Cassell/The Stock Market; vii(b), HRW Photo by John Langford; viii(t), FPG International; viii(b), Runk/Schoenberger/Grant Heilman Photography; ix, HRW Photo by John Langford; x(l), Telegraph Colour Library/FPG International; x(c), FPG International; x(r), HRW Photo by John Langford.

Unit 1: 2, John Livzey/Allstock; 9(tr), Stephen J. Krasemann/Valan Photos; 9(cr), Ontario Ministry of Natural Resources; 9(br), Philip Jon Bailey/SuperStock International; 9(bc), M. Keller//Superstock International; 10, Jim Simmen/Allstock; 11(l), Ontario Ministry of Natural Resources, Toronto; 11(c), Thomas Kitchin/Valan Photos; 11(r), Bill Ivy; 12(l), Bill Ivy; 12(r), Bill Ivy; 13(l), John Fowler/Valan Photos; 13(tl), Gilles Delisel/Valan Photos; 13(tr), Charlie Ott/Photo Researchers, 13(bl), Harold V. Green/Valan Photos; 13(br), Charlie Ott/Photo Researchers; 16tr, Ontario Ministry of Natural Resources, Toronto; 16(inset), J. A. Wilkinson/Valan Photos; 16(tc), R. C. Simpson/Valan Photos; 16(bc), Bill Ivy; 16(br), Ontario Ministry of Natural Resources, Toronto; 17(tl), Stephen J. Krasemann/Valan Photos; 17(tcl), Bill Ivy; 17(tcr) Kennon Cooke/Valan Photos; 17(tr), Pat Louis/Valan Photos; 17(c), Bill Ivy; 17(cr), R.K. LaVal/Animals, Animals; 17(bl), HRW Photo by Richard Haynes; 17(cbl) Val Whelan/Valan Photos; 17(cbr), Bill Ivy; 17(br), Bill Ivy; 18(t), Alan Carey/Photo Researchers; 18(tc), Peter Arnold, Inc.; 18(c), William & Marcia Levy/Photo Researchers; 18(bc), Stock Imagery/Canapress Photo Service; 18(b), Bill Ivy; 19(t), Runk & Schoenberger/Grant Heilman Photography; 19(b), C. C. Lockwood/Animals Animals; 20(tl), Ontario Ministry of Natural Resources, Toronto; 20(c), Harold V. Green/Valan Photos; 20(tr), Ontario Ministry of Natural Resources, Toronto; 20(bl), Joseph Collins/Photo Researchers; 20(br), Bill Ivy; 24(l), Pat Louis/Valan Photos; 24(tr), John R. MacGregor/Peter Arnold, Inc.; 24(tcr), Bill Ivy; 24(bcr), Ontario Ministry of Natural Resources, Toronto; 24(br), Gregory G. Dimijian, M.D./Photo Researchers; 25(tl), Egon Bock/Department of Regional Industrial Expansion; 25(c), John Fowler/Valan Photos; 25(cr), Gary Crandall/Envision; 25(bc), Sid Leszczynski/Animals, Animals; 25(bl), Ed Reschke/Peter Arnold, Inc; 25(br), Tom & Pat Leeson/Photo Researchers; 29(tr), Tom Murphy/SuperStock International; 29(cl), Harold V. Green/Valan Photos; 29(c), Stephen J. Krasemann/Valan Photos; 29(cr), Thomas Kitchin/Valan Photos; 29(bl), Stock Imagery/Canapress Photo Service; 29(bc), Alan D. Briere/SuperStock International; 29(br), Don McPhee/Valan Photos; 32(tl), Ontario Ministry of Agriculture and Food; 32(tr), Kenbrate/Photo Researchers; 32(cl), Emily Johnson/Envision; 32(cr), Emil Muench/Photo Researchers; 32(bl), Bill Ivy; 32(br), Bill Ivy; 34(tr), Richard Steedman/The Stock Market; 34(cr), A. Kaiser/SuperStock International; 34(bl), John Heseltine/Photo Researchers; 34(br), John M. Roberts/The Stock Market; 36(tr), Kirkendall/Spring; 36(bl), Curtis Willocks/Brooklyn Image Group; 36(bc), Robert Noonan/Photo Researchers; 36(br), Patti Murray/Earth Scenes; 37(tl), Jeff Henry/Peter Arnold, Inc.; 37(tr), Jeff Henry/Peter Arnold, Inc.; 37(bl), William E. Ferguson; 37(br), Ferrell Grehan/Photo Researchers; 40, Stephen J. Krasemann/Allstock; 43(t), B. Lyon/Valan Photos; 43(b), Victor Englebert/Canapress Photo Service; 44(l), Stephen J. Krasemann/Valan Photos; 44(c), Stephen J. Krasemann/Valan Photos; 44(r), Tom & Pat Leeson; 45(tl), Department of Regional Industrial Expansion; 45(tr), Ontario Science Centre; 45(rc), Bill Ivy; 45(c), Ontario Ministry of Natural Resources, Toronto; 45(bl), Michel Bourque/Valan Photos; 45(rbc), Ontario Science Centre; 45(bc), Ontario Ministry of Natural Resources, Toronto; 46(lc), Bill Ivy; 46(tr), Francois Morneau/Valan Photos; 46(bl), Michael P. Gadomski/Photo Researchers; 46(br), Bill Ivy; 47(tl), Bill Ivy; 47(tr), Bill Ivy; 47(br), Bill Ivy; 47(c), Bill Ivy; 47(cr), J. A. Wilkinson/Valan Photos; 51(t), Ontario Ministry of Natural Resources, Toronto; 51(c), Margaret McCarthy/Peter Arnold, Inc.; 51(b), Charlie Archambault/U.S. News & World Reports; 54(t), Ontario Ministry of Natural Resources, Toronto; 54(b), Canapress Photo Service; 55, Eric Beldowski/Canapress Photo Service; 56, AP/Wide World Photos, Inc.; 58(t), William & Marcia Levy/Photo Researchers; 58(bl), Alan D. Briece/SuperStock International; 58(cr), Emily Johnson/Envision; 58(br), Jim Simmen/Allstock; 59, John Livzey/Allstock; 60, Frank Siteman/Stock Boston; 61(t), David R. Frazier; 61(bl), Bohdan Hrynewych/Stock Boston; 61(br), Elena Rooraid/Photo-Edit; 62(t), Al Giddings/Images Unlimited; 62(b), Doug Menuez/Reportage; 63(t), Norbert Wu/Peter Arnold, Inc.; 63(b), Norbert Wu/Peter Arnold, Inc.

Unit 2: 64, D. Northcott/SuperStock International; 70(c), Bill Ivy; 70(cr), Michael Mitchell; 70bl, Breck P. Kent/Animals, Animals; 70(bcl), Michael Mitchell; 70(bcr), Gordon S. Smith/Canapress Photo Service; 70(br), Pat Morrow/Department of Regional Industrial Expansion; 71(bl), Michael Mitchell; 71(bcl), Steven Dalton/Canapress Photo Service; 71(bcr), Wayne Lankien/Valan Photos; 71(br), Canapress Photo Service; 72(tr), Bill Ivy; 72(c), Bill Ivy; 72(cr), Herman H. Giethoorn/Valan Photos; 72(bl), Kennon Cooke//Valan Photos; 72(bc), Stephen Dalton/AllStock; 72(br), Ted Grand/Department of Regional Industrial Expansion; 73(tl), Tom McHugh/AllStock; 73(tc), Milton Love/Peter Arnold, Inc.; 73(tr), Heather Angel Biofotos; 73(l), Bill Curtsinger/Photo Researchers; 73(cr), R. C. Simpson/Valan Photos; 73(br), James Watt/Animals Animals; 74(t), Peter Arnold, Inc.; 74(c), Luis Castaneda/The Image Bank; 74(b), Nuridsany et Perennou/Photo Researchers; 75(tl), Clayton Fogle/Allstock; 75(tr), Daniel Weiss/SuperStock International; 75(rc), Jeffrey L. Rotman/Peter Arnold, Inc.; 75(lc), S. Barrow/SuperStock International; 75(bl), SuperStock International; 75(br), Oxford Scientific Films/Canapress Photo Service; 76(tl), Anthony Bannister/Animals Animals; 76(tr), Bill Ivy; 76(bl), Bill Ivy; 76(br), Stouffer Production/Animals Animals; 77(tl), Mike Beedell/Department of Regional Industrial Expansion; 77(tc), K. G. Preston-Mafham/Animals Animals; 77(tr), Bill Ivy; 77(bl), Stephen J. Krasemann/Photo Researchers; 78(tl), Rob Talbot/TSW; 78(tr), Bill Ivy; 78(cr), Canapress Photo Service; 78(cl), Bill Ivy; 78(br), Hans Pfetschinger/Canapress Photo Service; 78(bl), Ontario Ministry of Natural Resources, Toronto; 79, Karl Maslowski/Canapress; 86(t), Ontario Ministry of Natural Resources, Toronto; 86(b), Ontario Ministry of Natural Resources, Toronto; 87(t), Breck P. Kent/Animals Animals; 87(b), Peter Parks/Animals Animals; 92, Jack Dermid; 93, Peter Menzel/Stock Boston; 97(tl), Ontario Ministry of Natural Resources, Toronto; 97(tc), Bill Ivy; 97(tr), M. I. Walker/Photo Researchers; 97(cl), Bill Ivy; 97(c), Bill Ivy; 97(cr), James Kent/Canapress Photo Service; 97(bl), Philip A. Harrington/Peter Arnold Inc.; 97(bc), Martin Kuhnigk/Valan Photos; 97(br), Manley Features/SuperStock International; 98(moray-eel), Norbert Wu/Peter Arnold, Inc.; 98(penguins), George Holton/Photo Researchers; 98(snail), Y. Momatiuk/Photo Researchers; 98(snake), Suzanne L. Collins & Joseph T. Collins/Photo Researchers; 98(mussel), Andrew J. Martinez/ Photo Researchers; 99(giraffe), Don Mason/The Stock Market; 99(frog), John R. MacGregor/Peter Arnold, Inc.; 99(dog), HRW Photo by Russell Dian; 99(grasshopper), HRW Photo by Russell Dian; 99(fish), S. Machado/SuperStock International; 99(chick), Schuster/SuperStock International; 99(lobster), Bob Abraham/The Stock Market; 99(salamander), D. Northcott/SuperStock International; 99(octopus), Stephen Frink/The Stock Market; 99(tiger), Reagan Bradshaw/The Image Bank; 99(crocodile), J. Desselliers/SuperStock International; 99(tarantula), Tom McHugh/Allstock; 100(millipede), Ricard Kolar/Animals Animals; 100(jellyfish), National Audubon Society/Photo Researchers; 100(starfish), Dale Calder/Royal Ontario Museum, Toronto; 100(clams), James Carmichael, Jr./Nature Photographer; 100(spider), Bill Ivy; 100(earthworm), Bill Ivy; 100(sand dollar), Biophoto Associate/Photo Researchers; 100(sea anemone), Stephen J. Krasemann/Valan Photos; 101(mollusk), Fred Braverman/Peter Arnold, Inc; 101(butterfly), Bill Ivy; 101(leech), J. Howard/Photo Researchers; 101(crayfish), John R. MacGregor/Peter Arnold, Inc.; 101(centipede), J. A. Wilkinson/Valan Photos; 101(spider), Bill Ivy; 101(clam), Zig Leszczynski/Animals Animals; 101(sea urchin & star fish), Doug Wechsler/Animals Animals; 101(sea creature), Sea Studios/Peter Arnold Inc.; 101(daddy long legs), Stephen Dalton/Animals Animals; 101(bettle), Pam Hickman/Valan Photos; 101(octopus), Paul J. Janosi/Valan Photos; 101(grasshopper), E. R. Degginger/Animals Animals; 101(crab), Dale Calder/Royal Ontario Museum, Toronto; 101(amoeba), E. R. Degginger/Animals Animals; 104(bat), Stephen Dalton/AllStock; 104(dragon fish), Bill Ivy; 104(loon), Stephen J. Krasemann/Peter Arnold, Inc.; 104(python), K. L. Switak/Photo Researchers; 104(tree frog), R. Andrew Odum/Peter Arnold, Inc.; 104(girl), R. Chen/SuperStock International; 104(frog), Bill Ivy; 105(seal), C. Allan Morgan/Peter Arnold, Inc.; 105(angel fish), C. Newbert/SuperStock International; 105(alligator), Don Brison/Envision; 105(owl), John Cancalosi/Peter Arnold, Inc.; 105(eft), Jeff Lepore/Photo Researchers; 105(duck), Department of Regional Industrial Expansion, Ottawa; 105(turtle), Ontario Ministry of Natural Resources, Toronto; 105(shark), Zig Leszczynski/Animals Animals; 107(mantee), Fred Bavendam/Peter Arnold, Inc.; 107(locust), Scott Camazine/Photo Researchers; 107(crab), Kevin Schafer/Martha Hill/Tom Stack & Associates; 107(snail), Y. Momautiuk/Photo Researchers; 107(horse), Gerard Lacz/Peter Arnold, Inc.; 108(millipede), G. I. Bernard/Animals Animals; 108(squirrel), Wayne Lankinen/Valan Photos; 108(mollusk), Zig Leszczynski/Animals Animals; 108(starfish), J. A. Wilkinson/Valan Photos; 108(toad), Bill Ivy; 108(salamander), Bill Ivy; 109(sea turtle), Luiz C. Marigo/Peter Arnold, Inc.; 109(caterpillar), Zig Leszczynski/Animals Animals; 109(amoeba), Biophoto Associates/Photo Researchers; 109(sea creature), Harold V. Green/Valan Photos; 109(starfish), Zig Leszczynski/Animals Animals; 109(porpoise), Kennon Cooke/Valan Photos; 109(snake), Bill Ivy; 109(duck), Ontario Ministry of Natural Resources, Toronto; 109(penquins), Luiz C. Marigo/Peter Arnold, Inc.; 109(ostrich), Stephen J. Krasemann/Valan Photos; 110(t), Ontario Ministry of Natural Resources, Toronto; 110(b), Bill Ivy; 111(all), Bill Ivy; 114(l), Canapress Photo Service; 114(r), UNICEF; 115(l), UNICEF; 115(r), Harvey Lloyd/The Stock Market; 124(tl), James Carmichael, Jr./Nature Photographer; 124(tcl), Gerard Lacz/Peter Arnold, Inc.; 124(tcr), Steinhart Aquarium/Tom McHugh/Photo Researchers; 124(tr), Suzanne L. Collins & Joseph T. Collins/Photo Researchers; 124(bl), E. S. Ross; 124(bcl), Tom McHugh/Photo Researchers; 124(bcr), Tim Gibson/Envision; 124(br), Alisa Schulman/Marine Mammal Images; 125(tl), Stephen J. Krasemann/Peter Arnold, Inc.; 125(tcl), A. Briere/SuperStock International; 125(tcr), John Mitchell/Photo Researchers; 125(tr), Gary Crandall/Envision; 125(bl), R. King/SuperStock International; 125(bcl), Ken Graham/Allstock; 125(bcr), D. Cavagnaro/Peter Arnold, Inc.; 125(br), Dr. E. R. Degginger, FPSA; 125(b), D. Northcott/SuperStock International; 126(t), R. Heinzen/SuperStock International; 126(c), Ken Straiton/The Stock Market; 126(b), Curtis Willocks/Brooklyn Image Group; 127(t), David R. Frazier; 127(b), Hal Stucker/Brooklyn Image Group; 128(t), Don & Pat Valenti/DRK Photo; 128(b), Grant Heilman/Grant Heilman Photography; 129(t), Photo by Peter B. Kaplan (c) 1974; all rights reserved; Peter B. Kaplan Photography, Inc.; 129(b), Stephen J. Krasemann/DRK Photo.

Unit 3: 130, Guy Motil/West Light; 145(l), Tom Myers/Hot Shot Stock Photos; 145(r), Robert

Landau/West Light; 174, A. B. Allen/Bruce Coleman, Inc.; 183(tl), Royal Ontario Museum, Toronto M24849; 183(c), Royal Ontario Museum, Toronto M3335; 183(cr), Royal Ontario Museum, Toronto M34014; 183(bl), Phillip A. Harrington/The Image Bank; 183(br), Royal Ontario Museum, Toronto M32303; 187, Guy Motil/West Light; 188(l), Hank Morgan/Photo Researchers; 188(r), Jay Freis/The Image Bank; 189, HRW Photo by Russell Dian; 190(t), Klaus D. Francke/Bilderberg/The Stock Market; 190(c), George H. Harrison/Grant Heilman Photography; 190(r), Anthony Bannister/Animals Animals; 191, Lefever/Grushow/Grant Heilman Photography.

Unit 4: 192, Tom Skrivan/The Stock Market; 194, Brent Bear/West Light; 196(all), C. McNeill/Miller Comstock; 198(t), Paul Kennedy/Leo de Wys, Inc.; 198(c), HRW photo by Bruce Buck; 198(cr), J. Imber/H. Armstrong Roberts, Inc.; 198(b), Focus on Sports; 199(tr), HRW photo by Bruce Buck; 199(b), Steve Smith/Wheeler Pictures; 207, Telegraph Colour Library/FPG International; 214, Michael Mitchell/Equipment Courtesy Sargent Welch Scientific Company; 215(t), HRW Photo by Peter Gonzales; 215(b), Michael Mitchell/Equipment Courtesy Sargent Welch Scientific Company; 216(l), Michael Mitchell/Equipment Courtesy Sargent Welch Scientific Company; 216(r), Michael Mitchell/Equipment Courtesy Sargent Welch Scientific Company; 218(l), Michael Mitchell/Equipment Courtesy Sargent Welch Scientific Company; 218(r), Michael Mitchell/Equipment Courtesy Sargent Welch Scientific Company; 219(t), Michael Newman/ PhotoEdit; 219(b), Michael Mitchell/Equipment Courtesy Sargent Welch Scientific Company; 220, Michael Mitchell/Equipment Courtesy Sargent Welch Scientific Company; 222(t), Michael Mitchell/Equipment Courtesy Sargent Welch Scientific Company; 222(b), Michael Mitchell/Equipment Courtesy Sargent Welch Scientific Company; 231(l), Michael Mitchell; 231(r), Michael Mitchell; 231(b), Pulp and Paper Research Institute of Canada.; 232(t), Michael Mitchell/Equipment Courtesy Pedlar Cycles; 232(b), Michael Mitchell/Equipment Courtesy Pedlar Cycles; 234(tr), Larry Sergeant/Comstock, Inc.; 234(tl), Nathan Bilow Photography; 234(cr), Thomas Zimmermann/FPG International; 234(bl), Marva Production/The Image Bank; 234(bc), Rothwell/FPG International; 234(br), Ira Block/The Image Bank; 235(t), Andre Sima /Miller Comstock; 235(b), NASA; 236, NASA; 248(t), HRW Photo by Peter Gonzales; 248(both), Michael Mitchell; 249(both), Michael Mitchell; 256, Brent Bear/West Light; 257, Tom Skrivan/The Stock Market; 258, Stephen Derr/The Image Bank; 259(t), The Hamilton Spectator; 259(c), Jeff Smith/The Image Bank; 259(br), Phil Degginger; 260(t), Peter Gridley/FPG International; 260(b), NASA; 261(t), Bruce Frisch/Photo Researchers; 261(b), NASA.

Unit 5: 262, SuperStock International; 267, UPI/Bettmann Newsphotos; 268(l), Canapress Photo Service; 268(cr), K. Leland; 268(br), Martin Roetker/Taurus Photos; 269(tr), Mitchel Osborne/The Image Bank; 269(cr), John Butterhill; 269(l), Ontario Ministry of Natural Resources; 269(br), Miller Comstock; 270(tl), Stephen Dalton/ Animals Animals; 270(tr), Masterfile; 270(cl), Miller Comstock; 270(cr), AP/Wide World Photos; 270(bl),Nick Pavloff/ The Image Bank; 270(br),Bard Martin/ The Image Bank; 282, K. Leland; 285(t), Miller Comstock; 285(bl), Dallas & Tom Heaton/Miller Comstock; 285(br), H. Armstrong Roberts; 290(tl), Shostal Associates/SuperStock International; 290(tr), George Hunter/Miller Comstock; 290(cl), R. Waldkirch/Miller Comstock; 290(cr), Jeff Gnass/The Stock Market; 290(br), M. Schneiders/H. Armstrong Roberts; 291, Forest and Whitmire Photography/Washington State Highway Department Photos; 292(tl), Masterfile; 292(tr), William Marvin; 292(bl), Higuchi/Miller Comstock; 292(cr), D. Person; 294(t), Dilip Mehta; 294(b), Canapress Photo Service; 295(t), Mike Yamashita/Woodfin Camp & Associate; 295(b), H. Higuchi/Miller Comstock; 298, Hans Wolfe/The Image Bank; 299(tl), D. Nuccio/SuperStock International; 299(tr), A. Friedlander/SuperStock International; 299(b), R. Manley /SuperStock International; 300(tl), Ronald R. Johnson/The Image Bank; 300(tr), Marvin E. Newman/The Image Bank; 300(b), R. Kord/H. Armstrong Roberts; 301, Sandak/A Division of G.K. Hall & Co.; 302(l), Todd White/The Hayward Gallery, London; 302(t), "Translucent Variation on Spheric Theme," 1951. Collection Solomon R. Guggenheim Museum, NY, Robert E. Mates; 303(tl), The Bridgeman Art Library; 303(tr), Canapress Photo Service; 303(b), Sandak/A Division of G.K. Hall & Co.; 304(t), R. Kord/H. Armstrong Roberts; 304(b), Ralph Krubner/H. Armstrong Roberts; 305, SuperStock International; 306(l), Canapress Photo Service; 306(r), Alan Becker/The Image Bank; 307, Geri Engberg/The Stock Market; 309, Comstock, Inc.; 310, Royal Ontario Museum, Toronto; 311, SuperStock International; 312(b), SuperStock International; 312(t), SuperStock International; 312(b), Henley & Savage/The Stock Market; 313(t), Burton McNeely/The Image Bank; 313(b), William E. Ferguson; 314(t), SuperStock International; 314(b), Filippo Brunelleschi/Scala/Firenze/Art Resource, NY; 315(t), SuperStock International; 315(b), Jan Lukas/Art Resource, NY.

Unit 6: 316, Tom Algire/H. Armstrong Roberts; 318, Sharon Chester/Comstock, Inc.; 320, John N.A. Lott/Biological Photo Service; 321, Jack Marshall and Co. Ltd/Canapress Photo Service; 323(tr), Bruce F. Molnia/Terraphotographics; 323(cr), Ernest S. Booth/Earth Images; 323(bl), H. Gritscher/Peter Arnold, Inc.; 323(br), Galen Rowell/Peter Arnold, Inc.; 324(t), John K. Nakata/Terraphotographics; 324(b), Michael Mitchell; 325, G. K. Gilbert/U.S. Geological Survey; 326, Michael Mitchell; 327(t), R. E. Wallace/U.S. Geological Survey; 327(b), CBC Picture Service/ Canapress Photo Service; 329, Tom McHugh/Photo Researchers; 330(t), Larry Smith/H. Armstrong Roberts; 330(b), Terry Domico/Earth Images; 332(all), Gary Rosenquist/Earth Images; 333(r), Gary Rosenquist/Earth Images; 334(t), Canapress Photo Service; 334(b), Jim Shelton/Canapress Photo Service; 335(l), George Wedding/Canapress Photo Service; 335(c), Canapress Photo Service; 335(r), Canapress Photo Service; 336(t), The Bettmann Archive; 336(b), Franz Lazi/FPG International; 337(t), Canapress Photo Service; 337(b), Canapress Photo Service; 338, Geological Survey of Canada; 339(t), Geological Survey of Canada; 339(b), Geological Survey of Canada; 340(l), Geological Survey of Canada; 340(r), Canapress Photo Service; 341, Bernard Silbertstein/Canapress Photo Service; 342(t), Alain Le Garsmeur/TSW; 342(bl), The Telegraph Colour Library/FPG International; 342(cr), E. Faure/SuperStock International; 342(br), Barbara J. Miller/Biological Photo Service; 344(tr), Breck Kent/Earth Science; 344(cr), Breck Kent; 344(l), William E. Ferguson; 345(all), Michael Mitchell; 348(all), Michael Mitchell; 349(bl), Geological Survey of Canada 85965; 349(inset), Geological Survey of Canada 202486-H; 349(tr), Joseph Holmes; 349(br), Mike Chuang/FPG International; 352(all), Michael Mitchell; 355(both), Geological Survey of Canada; 356(b), Dr. E. R. Degginger, FPSA; 361(tr), Royal Ontario Museum, Toronto; 361(bl), Ward's Natural Science Establishment; 361(bc), Ward's Natural Science Establishment; 361(br), Ward's Natural Science Establishment; 362(all), Bill Ivy; 367, Metropolitan Toronto Public Reference Library; 369, Grant Heilman; 370, G.R. Roberts; 371, Tom Algire/H. Armstrong Roberts; 372(t), Carl Bigras/Canapress Photo Service; 372(b), M. Thonig/H. Armstrong Roberts; 373, David Austen/AUSTRALIA/Stock Boston; 374, Murray Alcosser/The Image Bank; 375, SuperStock International.

Unit 7: 376, Henry Groskinsky/LIFE Magazine © Time Warner, Inc.; 378, Astrostock/Sanford; 380(t), David A. Hardy/Science Photo Library/Photo Researchers; 380(cl), NASA; 380(cr), Astronomical Society of the Pacific, The Planetary System; 380(c), Astronomical Society of the Pacific; 380(bl), Photri/The Stock Market; 380(br), Astronomical Society of the Pacific; 381(t), Dr. Seth Shostak/Science Photo Library/Photo Researchers; 381(c), Dr. Fred Espenak/Science Photo Library/Photo Researchers; 383, Lick Observatory; 386(t), Comstock; 386(br), Lick Observatory; 387(t), Lick Observatory; 387(b), Erich Lessing/Magnum Photos, Inc.; 388, Omikron Collection/Photo Researchers; 389(t), Dr. Jeremy Burgess/Science Photo Library/Photo Researchers; 389(b), George Lovi; 392, Bridgeman/Art Resource, NY; 395(l), Eunice Harris/Photo Researchers; 395(r), HRW Photo by Eric Beggs; 403, Ligature, Inc.; 406, Paolo Koch/Photo Researchers; 407, Astrostock/Sanford; 408(r), Astronomical Society of the Pacific; 408(c), Giraudon/Art Resource, NY; 409, Astronomical Society of the Pacific; 410(b), Otto Rogge/The Stock Market; 411(b), Astronomical Society of the Pacific; 412(t), Jet Propulsion Labs/NASA; 412(c), Astronomical Society of the Pacific; 413(l), Astronomical Society of the Pacific; 413(tr), Jet Propulsion Labs/NASA; 413(c), Astronomical Society of the Pacific; 413(rc), Jet Propulsion Labs/NASA; 413(br), Astronomical Society of the Pacific; 413(bl), Astronomical Society of the Pacific; 414(both), Astronomical Society of the Pacific; 415(l), NASA/Peter Arnold, Inc.; 415(r), Astronomical Society of the Pacific; 415(b), Astronomical Society of the Pacific; 416, Pay Pfortner/Peter Arnold, Inc.; 417, NASA; 419(both), Astronomical Society of the Pacific; 420, Art by Michael Whelan/National Geographic; 421(b), Astronomical Society of the Pacific; 422(c), NASA; 422(b), Hale Observatories/Science Source/Photo Researchers; 423(both), Astrostock/Sanford; 425, Astronomical Society of the Pacific; 426, Lick Observatory; 427(t), Frank P. Rossotto/The Stock Market; 427(b), Robin Scagell/Science Photo Library/Photo Researchers; 429, NASA; 432, Lick Observatory; 434(b), Science Source/Photo Researchers; 435(l), Jean Miele/The Stock Market; 436(b), Dana Berry/NASA; 440, Dennis Di Cicco/Peter Arnold, Inc.; 441(b), Henry Groskinsky/Life Magazine © Time Warner, Inc.; 442, HRW Photo by Michael Lyon; 443(tr), HRW Photo by John Langford; 443(bl), HRW Photo by John Langford; 444(tr), Bob Burch/Bruce Coleman, Inc.; 444(lc), Photri; 444(rc), Caltech Astronomy Library/Palomar Observatory.

Unit 8: 446, Photri, Inc.; 453(b), Kennon Cooke/Valan Photos; 454, Geological Survey of Canada 95660; 463, Clara Parsons/Valan Photos; 467(t), Barry L. Runk/Grant Heilman Photography; 467(b), Joyce Photographics/Valan Photos; 468(tl), Val and Alan Wilkinson; 468(r), Runk and Schoenberger/Grant Heilman Photography; 468(bl), Arthur Strange; 472, G. I. Bernard/Earth Scenes; 473(t), HRW Photo by Eric Beggs; 473(b), Lefever/Grushow/Grant Heilman Photography; 474(tl), Irwin Barrett/Valan photos; 474(tr), Professor R.C. Simpson/Valan Photos; 474(bl), Gerhard Karmann/Valan Photos; 474(br), Gerhard Karmann/Valan Photos; 478, University of Alberta Department of Plant Science; 481, Doug Wechsler/Earth Scenes; 486, E. Lewin/The Image Bank; 493(tl), Donald Specker/Earth Scenes; 493(tr), Donald Specker/Earth Scenes; 495, Photri, Inc.; 496(tl), Photri, Inc.; 496(tr), HRW Photo by Eric Beggs; 497, Grant Heilman; 498(t), R. L. Brinster and R. E. Hammer, School of Veterinary Medicine, University of Pennsylvania; 498(bl), Nelson Max/LLNL/Peter Arnold, Inc.; 498(br), Keith V. Wood, Ph.D.

SOURCEBOOK

Unit 1: S1, John Livzey/Allstock; S3, Art Wolfe/The Image Bank; S4, Steven Kaufman/Peter Arnold, Inc.; S5, Johnny Johnson/DRK Photos; S7, Brent Kent/Earth Scenes; S8, Stephen Krasemann/DRK Photos; S9(l), Larry Ulrich/DRK Photos; S9(r), Jim Brompton/Valan Photos; S10, Doug Allan/Earth Scenes; S11, Tom Bean; S13(l), Nancy Simmermann/TSW; S13(c), Leonard Rue/Animals Animals; S13(r), Breck Kent; S13(b), C. C. Lockwood; S14, Katherine Thomas/Taurus Photos; S15(t), Dudley Foster/Woods Hole Oceanographic Institute; S15(c), D. P. Wilson/Science Source/Photo Researchers; S15(b), D. P. Wilson/Science Source/Photo Researchers; S16, O S F/Animals Animals; S17(t), David Stoecklein/After-Image; S17(b), John James Audubon, National Audubon Society Photo Researchers; S18, George Hall/Woodfin Camp & Associates; S19(tr), Mark Sherman/Bruce Coleman, Inc.; S19(bc), Yoram Lehmann/Peter Arnold, Inc.; S20(tl), Australian News & Information Bureau; S20(tr), Tom McHugh/Photo Researchers; S20(b), HRW Photo/Yoav Levy/Phototake; S21(c), Bruce Kurosaki/After-Image; S21(br), Richard Choy/Peter Arnold, Inc.; S22(t), Jim Whitmer/Nawrocki Stock Photo; S22(b), Peregrine Fund.

Unit 2: S23, SuperStock International; S23(tl), S23(tr), David M. Phillips/Visual Unlimited; S23(bl), T. E. Adams; S23(br), T. E. Adams/Click/Chicago; S25(t), Manfred Kage/Peter Arnold, Inc.; S25(c), Eric Graves/Photo Researchers; S25(b), Arthur M. Siegelman; S26(l), Manfred Kage/Peter Arnold, Inc.;

S26(c), Walt Anderson/Tom Stack & Associates; S26(r), Walter Dawn; S27(cl), D. Smiley/Peter Arnold, Inc.; S27(cr), M. P. Kahl/Photo Researchers; S28(t), W. H. Hodge/Peter Arnold, Inc.; S28(b), Michael P. Gadomski/Bruce Coleman, Inc.; S28(bl), Rod Plank/Tom Stack & Associates; S28(bc), Stephen J. Krasemann; S28(br), E. S. Ross; S29(cl), Zig Leszczynski/Earth Scenes; S28(c), E. R. Degginger/Bruce Coleman, Inc.; S28(cb), E. R. Degginger; S30(t), Jeff Simon; S30(b), Kim Taylor/Bruce Coleman, Inc.; S31(t), Carl Roessler; S31(bl), Runk/Schoenberger/Grant Heilman Photography; S31(br), Oxford Scientific Films/Animals Animals; S32(l), Bob & Clara Calhoun/Bruce Coleman, Inc.; S32(c), Rod Borland/Bruce Coleman, Inc.; S32(r), Runk/Schoenberger/Grant Heilman Photography; S33(l), Runk/Schoenberger/Grant Heilman Photography; S33(c), Tom McHugh/Photo Researchers; S33(r), Dave Woodward/Tom Stack & Associates; S34(tl), Zig Leszczynski/Animals Animals; S34(tr), S34(bl), John Markham/Bruce Coleman, Inc.; S34(br), M. Timothy O'Keefe/Bruce Coleman, Inc.; S36, Breck Kent; S37, Summerhays/Photo Researchers; S38(cl), Andrew S. Dalsimer/Bruce Coleman, Inc.; S38(b), M. Austerman/Animals Animals; S39, Heather Angel Biofotos; S40(t), Ray Hillstrom/Root Resources; S40(b), HBJ Photo; S41(l), (c), Edith Hughes; S41(r), HBJ Photo; S44, Precision Graphics.

Unit 3: S45, Guy Motil/West Light; S46, HRW Photo by Dennis Fagan; S46(t), HRW Photo by Dennis Fagan; S49(b), Paul Silverman/Fundamental Photography; S50(tl), (tr), Yoav Levy/Phototake; S50(bl), HRW Photo by Richard Haynes; S51(t), HRW Photo by Richard Haynes; S51(b), HBJ Photo by Richard Haynes; S53, Gary Milburn/Tom Stack & Associates; S54(tl), Ted Horowitz/The Stock Market; S54(tc), Erik Anderson/Stock Boston; S54(tr), Brownie Harris/The Stock Market; S54(cl), E. R. Degginger; S56, Gabe Palmer/After Image; S57(tr), Grapes/Michaud/Photo Researchers; S57(b), HRW Photo by Dennis Fagan; S58(tl), HRW Photo by Dennis Fagan; S58(b), Bill Weedmark/Panographics; S59, E. R. Degginger; S60, H. Wendler/The Image Bank.

Unit 4: S61, Tom Skrivan/The Stock Market; S62, Bill Ross/Woodfin Camp & Associates; S63(cr), Joe McNally/Wheeler Pictures; S63(br), Daniel Zisinsky/Photo Researches; S64(t), Ken Biggs/The Stock Market; S64(b), The Stock Market; S65, HRW Photo by Eric Beggs; S66, Pat Rogers/The Image Bank; S67(t), Eunice Harris/Photo Researchers; S67(b), Park Street Photography; S69(t), (c), Loomis Dean, Life Magazine © 1954, Time, Inc.; S69(b), NASA; S70(t), General Motors Corp.; S71(t), NASA; S71(b), HBJ Photo by Rodney Jones; S72(both), HRW Photo by Elizabeth Hathon; S73(tr), Sportech Productions; S73(b), HBJ Photo; S75(t), The Bettmann Archive; S75(b), Richard Megna/Fundamental Photographs; S76, NASA.

Unit 5: S77, SuperStock International; S78, Roger Wood/TSW; S79(tr), Robert Frerck/The Stock Market; S79(bl), (br), SuperStock International; S80(tc), Paolo Koch/Photo Researchers; S80(bl), Dan DeWilde; S81(tc), Cameramann International, Ltd.; S81(cr), Richard Gorbun/Leo deWys, Inc.; S81(br), Steve Vidler; S82(t), Tony Stone Worldwide; S82(b), Susan McCartney/Photo Researchers; S83(tr), Cameramann International, Ltd.; S83(cr), SuperStock International; S83(br), Tony Stone Worldwide; S83(bc), SuperStock International; S84(tl), James Hackett/Leo deWys, Inc.; S84(bl), James Hackett/Leo deWys, Inc.; S84(br), SuperStock International; S85(tr), Van Phillips/Leo deWys, Inc.; S85(cr), Mark Sherman/Bruce Coleman, Inc.; S85(bl), Olaf Soot/TSW; S86(cl), L. K. Townsend; S86(c), Art Grossman; S86(b), David Muench Photography; S87, Marc & Evelyn Bernheim/Woodfin Camp & Associates; S88, Rhodesian National Tourist.

Unit 6: S89, Tom Algire/H. Armstrong Roberts; S90, J. H. Robinson; S91(l), E. R. Degginger; S91(r), Michael & Barbara Reed/Earth Scenes; S92, Bud Titlow/Naturegraphics/Tom Stack & Associates; S93(l), E. R. Degginger; S93(r), Breck P. Kent/Earth Scenes; S94(tl), Gianni Tortoli/Photo Researchers; S94(cl), Pat Lanza Field/Bruce Coleman, Inc.; S94(cr), E. R. Degginger; S94(bl), E. R. Degginger; S95(tr), Breck P. Kent/Earth Scenes; S95(c), E. R. Degginger; S95(cr), E. R. Degginger; S95(bl), E. R. Degginger; S95(br), Breck P. Kent/Earth Scenes; S96(t), E. R. Degginger; S96(c), E. R. Degginger; S96(cl), E. R. Degginger; S96(cr), Breck P. Kent/Earth Scenes; S96(bl), S. D. Halperin/Animals Animals; S97(tl), Horst Schafer/Peter Arnold, Inc.; S97(cl), Paolo Koch/Photo Researchers; S97(cr), Robert Whitney; S97(tr), Robert Whitney; S97(b), (all), E. R. Degginger; S98(tl), Art Kane/The Image Bank; S98(bl), E. R. Degginger/Earth Scenes; S98(cr), Breck Kent; S99, Mary Root/Root Resources; S100(cl), E. R. Degginger; S100(c), Mary Root/Root Resources; S100(cr), E. R. Degginger; S100(bl), Breck P. Kent/Earth Scenes; S101(tr), Paul Silverman/Fundamental Photographs; S101(c), J. Beckett/American Museum of Natural History; S102(t), Paul Silverman/Fundamental Photographs; S102(c), Paul Silverman/Fundamental Photographs; S102(bl), Fundamental Photographs; S103(t-b), E. R. Degginger; S103, Breck P. Kent/Earth Scenes; S103, E. R. Degginger; S103, E. R. Degginger; S103, E. R. Degginger; S103, E. R. Degginger; S104(tl), Breck P. Kent/Earth Scenes; S104(tc), Breck P. Kent/Earth Scenes; S104(cl), E. R. Degginger; S104(bl), E. R. Degginger; S105(t), HRW Photo by Eric Beggs; S105(b), Park Street Photography; S106, Craig Hamill/The Stock Market; S107(t), Vivian M. Peevers/Peter Arnold, Inc; S107(r), James Tallon; S109, Robert P. Carr/Bruce Coleman, Inc.

Unit 7: Henry Groskinsky/LIFE Magazine © Time Warner, Inc.; S112(t), NASA; S112(b), NASA; S114, Jack Finch/Science Photo Library/Photo Researchers; S117(t), Jet Propulsion Laboratory; S117(c), Science Source/Photo Researchers; S117(br), Julian Baum/Science Photo Library/Photo Researchers; S118(t), NASA; S118(b), Anglo-Australian Telescope Board, 1977; S122(t), Yerkes Observatory; S122(b), John Sanford/Science Photo Library/Photo Researchers; S123(t), National Optical Astronomy Observatories; S123(c), Dennis Diccicco/Peter Arnold, Inc.; S123(b), Bill Iburg/Science Photo Library/Photo Researchers; S124(t), NASA; S124(c), Hale Observatory; S124(c), National Optical Astronomy Observatories; S124(cr), Kitt Peak National Observatory; S124(tr), NASA; S125(t), Chris Butler; S125(c), National Optical Astronomy Observatories; S126, Lawrence Migdale; S127(t), Art Resource, NY; S127(b), The Granger Collection, New York; S128(tl), National Radio Astronomy Observatory/AU; S128(tr), Peter Menzel/Wheeler Pictures; S128(cl), (cr), NOAO; S129(cl), Culver Pictures; S129(b), Photri, Inc.; S130(t), NASA; S130(cl), (c), NASA; S131(t), NASA/ARC; S131(bl), (br), Precision Graphics.

Unit 8: S133, Photri, Inc.; S134(t), Biomedia Associates; S134(c), E. S. Ross; S134(b), Zig Leszczynski/Animals Animals; S136, P. Dayanandan/Photo Researchers; S138(tl), E. R. Degginger; S138(tr), P. Dayanandan/Photo Researchers; S140, Runk/Rannels/Grant Heilman Photography; S141, Runk/Schoenberger/Grant Heilman Photography; S142, HRW Photo by Russell Dian; S143(t), Billy E. Barnes/Stock Boston; S143(c), Michael Philips Manheim/Photo Researchers; S143(b), William Hopkins; S144, Zig Leszczynski/Animals Animals.

ART CREDITS

Victoria Bruck 6, 40, 50, 52, 142, 148, 168, 343, 354, 369, 381, 382, 402, 424, 435, 462, 485
Les Case 22, 50, 92, 138, 139, 152, 153, 268, 328, 379, 394, 397, 400, 401, 411, 423, 431, 432, 434, 439, 441
Brenda Clark 206
Alan Clarke 264, 265, 296, 297
Heather Collins 28, 90, 91, 147, 150, 157, 159, 161, 169, 171, 175, 176, 181, 185, 200, 201, 202, 219, 220, 228, 229, 230, 233, 234, 238, 239, 242, 243, 244, 245, 250, 251, 252, 253, 254, 319, 320, 353, 354, 357, 365
Holly Cooper 133, 152, 153, 167, 224, 255, 355, 359, 369, 384
Sam Daniel 346, 347, 350, 358, 364
Nancy Easun 38, 39, 94
Helen Fox 450, 461, 470, 485, 489
David Griffin 14, 15, 53, 79, 119, 122, 141, 151, 166, 167, 186, 275, 351, 375, 384, 390, 391, 399, 412-413, 421, 433, 456, 459, 471, 472, 477, 479, 484, 487, 488
Linda Hendry 21, 57
Andrew Hickson 452, 464
Francis Kagege 113
Michael Krone 84, 85, 93, 132, 134, 337, 351, 363, 365, 385, 391, 398, 404, 418, 433, 438, 492, 493
Vesna Krstanovich 26, 27, 146, 149, 181, 204, 205, 246, 247
Susan Krykorka 154, 155, 164, 165
Ligature, Inc. 387
Emmanuel Lopez 272, 278
Jock MacRae 271, 272, 274, 276, 277, 279, 280, 281, 283, 284, 286, 287, 288, 289, 448, 450, 469, 491
Laurie Marks 150, 151, 162, 163
Michael Martchenko 207, 208, 226, 344
Bernard Martin 463, 464, 465, 477, 483
Jon McKee 266, 293
Rebecca Merrilees 494
Julian Mulock 4, 5, 14, 15, 48, 49, 66, 67, 68, 69, 80, 81, 82, 83, 96, 115, 117, 118, 212, 213, 223, 254, 360, 368, 449, 457, 458, 460, 466, 469, 480, 482
Debi Perna 209
Larry Raymond 52
Ric Riordon 142, 158, 159, 201, 203, 210, 211
Joan Rivers S113
David Simpson 225, 241
TSI 385, 410, 428, 430
Gary Undercuffler 197, 203, 206, 227, 254
Angela Vaculik 453, 475, 476, 479
Henry Van Der Linde 120, 121
Peter Van Gulik 13, 33, 35, 136, 137, 138, 140, 141, 169, 170, 171, 172, 173, 179, 180, 184, 217, 233, 234, 237, 240, 321, 322, 330, 331, 341, 355
Edythe Wright 138, 139

ACKNOWLEDGMENTS

For permission to reprint copyrighted material, grateful acknowledgment is made to the following sources:

Dr. Ivan H. Crowell: "A Legend: Wisakedjak Names All Creatures" by Dr. Ivan H. Crowell.

Harcourt Brace Jovanovich, Inc.: From *HBJ Earth Science* by Cesare Emiliani, et.al. Copyright © 1989 by Harcourt Brace Jovanovich, Inc. From *HBJ Physical Science* by William G. Lamb, et al. Copyright © 1989 by Harcourt Brace Jovanovich, Inc.

Heritage Canada: From an article on citizens' group participation in conservation of old buildings and communities by Phyllis Lambert from *Canadian Heritage*, August 1980. Copyright © 1980 by Heritage Canada.

Holt, Rinehart and Winston, Inc.: From *Holt Earth Science* by William L. Ramsey, et al. Copyright © 1986, 1982, 1978 by Holt, Rinehart and Winston, Inc. From *Holt General Science* by William L. Ramsey, et al. Copyright © 1988, 1983, 1979 by Holt, Rinehart and Winston, Inc. From *Holt Life Science* by William L. Ramsey, et al. Copyright © 1986, 1982, 1978 by Holt, Rinehart and Winston, Inc. From *Holt Physical Science* by William L. Ramsey, et al. Copyright © 1988, 1986, 1982, 1978 by Holt, Rinehart and Winston, Inc.

Life Magazine Company: From "The Red Planet May be the Next Giant Step for Mankind" by Brad Darrach and Steve Petranek from *Life*, May 1991. Copyright © 1991 by Life Magazine Company.

Ray Lincoln Literary Agency, Elkins Park House, 107-B, Elkins Park, PA, as agent for Marian T. Place: From "Search and Rescue" from *Mount St. Helens: A Sleeping Volcano Awakes* by Marian T. Place. Copyright © 1981 by Marian T. Place.

Macmillan Publishing Company: "Bats" from *The Bat-Poet* by Randall Jarrell. Copyright © 1963, 1964 by Macmillan Publishing Company.

Scott Meredith Literary Agency on behalf of Carl Sagan: From *Cosmos* by Carl Sagan. Copyright © 1980 by Carl Sagan. All rights reserved.

National Geographic Society: "Mars" from *National Geographic Picture Atlas of Our Universe* by Roy A. Gallant. Copyright © 1980 by National Geographic Society.

OWNER'S MANUAL PHOTO CREDITS

T–13, David Young Wolff/PhotoEdit; T–14(t), HRW Photo; T–14(c), Art Montes de Oca/FPG International; T–14(b), HRW Photo by Tomás Pantin; T–15(tr), A. Mercieca/SuperStock International; T–15(b), Carl Rosenstein/Viesti Associates; T–16(t), David F. Malin/Anglo–Australian Telescope Board; T–16(b), HRW Photo by Ken Lax; T–17(tl), HRW Photo; T–17(tr), SuperStock International; T–17(b), HRW Photo by Tomás Pantin; T18(tl), HRW Photo by Richard Weiss; T18(tr), HRW Photo, T18(b), D. Northcott/Super Stock International T20 (tl), Super Stock International; T20(b), HRW Photo by Greg Schaler. T–21(t), HRW Photo by Tomás Pantin; T–21(b), David Young Wolff/PhotoEdit; T–22, David Young Wolff/PhotoEdit; T–23, Larry Hamill; T–24(t), HRW Photo; T–24(b), HRW Photo by Tomás Pantin; T–25(t), HRW Photo by Russell Dian; T–25(b), HBJ Photo by Sam Joosten; T–26, HRW Photo by Russell Dian; T–27(t), Diamond Promotion Service; T–27(bl), Myron J. Dorf/The Stock Market; T–27(br), David Young Wolff/PhotoEdit; T–28(t), Kenneth Hayden/Black Star; T–28(b), David Young Wolff/PhotoEdit; T–29(l), HRW Photo; T–29(tr), Telegraph Colour Library/FPG International; T–29(b), David Young Wolff/PhotoEdit; T–30(t), David Young Wolff/PhotoEdit; T–30(bl), HRW Photo by Martha Cooper; T–30(br), David Young Wolff/PhotoEdit; T–31(t), HRW Photo by Dennis Fagan; T–31(b), HRW Photo by Dennis Fagan; T–32(t), HRW Photo by Yoav Levy/Phototake; T–32(c), SuperStock International; T–32(b), Roger Ressmeyer/Starlight; T–33(tl), NASA; T–33(tr), HRW Photo by Russell Dian; T–33(b), Tom Till DRK Photo; T–34(t), David Young Wolff/PhotoEdit; T–34(c), Hans Pfetschinger/Peter Arnold, Inc.; T–34(b), HRW Photo by Tomás Pantin; T–36(t), HRW Photo; T–36(c), Michael D. Sullivan/TexaStock; T–37(t), Michael Powers/Adventure Photos; T–39(b), HRW Photo by Tomás Pantin; T–38(t), SuperStock International; T–38(c), National Optical Astronomy Observatories; T–38(b), HRW Photo by Tomás Pantin; T–39(tl), NASA/JPL Photo; T–39(tr), Roger Allyn Lee/SuperStock International; T–39(b), Peter Vandermark/Stock Boston; T–40(t), HRW Photo by Russell Dian; T–40(b), Tony Freeman/PhotoEdit; T–41(t), HRW Photo; T–41(b), David Young Wolff/PhotoEdit; T–42(t), HRW Photo, T–42(b), NASA; T–43(t), HRW Photo by Stephen McCarroll; T–43(c), HRW Photo by Russell Dian; T–43(b), David Young Wolff/PhotoEdit; T–44(t), HRW Photo by Dennis Fagan; T–44(b), HRW Photo by Richard Haynes; T–45, HRW Photo; T–46(t), HRW Photo by Russell Dian; T–46(b), HRW Photo by Tim Pannell; T–47, HRW Photo by Dennis Fagan; T–48(t), NASA; T–48(b), HRW Photo by Dennis Fagan; T–49(tl), HRW Photo; T–49(tr), HRW Photo by Dennis Fagan.

Unit 1: 1A, HRW Photo by Richard Haynes; 1B, Telegraph Colour Library/FPG International; 1C, Phil Dotson/Nature Source/Photo Researchers; 1D, Jeff LePore/Natures Source/Photo Researchers.

Unit 2: 63A, HRW Photo by Richard Haynes; 63B, Jonathan Blair/Woodfin Camp & Associates; 63C, Robert Frerck/Woodfin Camp & Associates; 63D, The Telegraph Colour Library/FPG International.

Unit 3: 129A, HRW Photo by Dennis Fagan; 129B, A. Mercieca/Superstock International; 129C, Hans Pfletschinger/Peter Arnold, Inc.; 129D, Art Montes de Oca/FPG International.

Unit 4: 191A, HRW Photo by Ken Karp; 191B, Westlight/Chuck O'Rear; 191C, HRW Photo by NASA; 191D, John Terrence Turner/Allstock.

Unit 5: 261A, Michael Lyon; 261B, Rita Bailey/Leo de Wys, Inc.; 261C, Michael Powers/Adventure Photos; 261D, The Telegraph Colour Library/FPG International.

Unit 6: 315A, Michael Lyon; 315B, The British Museum; 315C, HRW Photo by Russell Dian; 315D, Robert Cassell/The Stock Market.

Unit 7: 375A, HRW Photo by NASA; 375B, HRW Photo by Morgan Cain; 375C, NASA Collection/Superstock International; 375D, NASA Collection/Superstock International

Unit 8: 445A, HRW Photo by Rodney Jones; 445B, HRW Photo by Richard Haynes; 445C, HRW Photo by Carole Lyndre; 445D, HRW Photo by Yoav Levy.

Answers to Tables and Graphs

Unit 2, *Checking Your Understanding,* page 125, question 2.

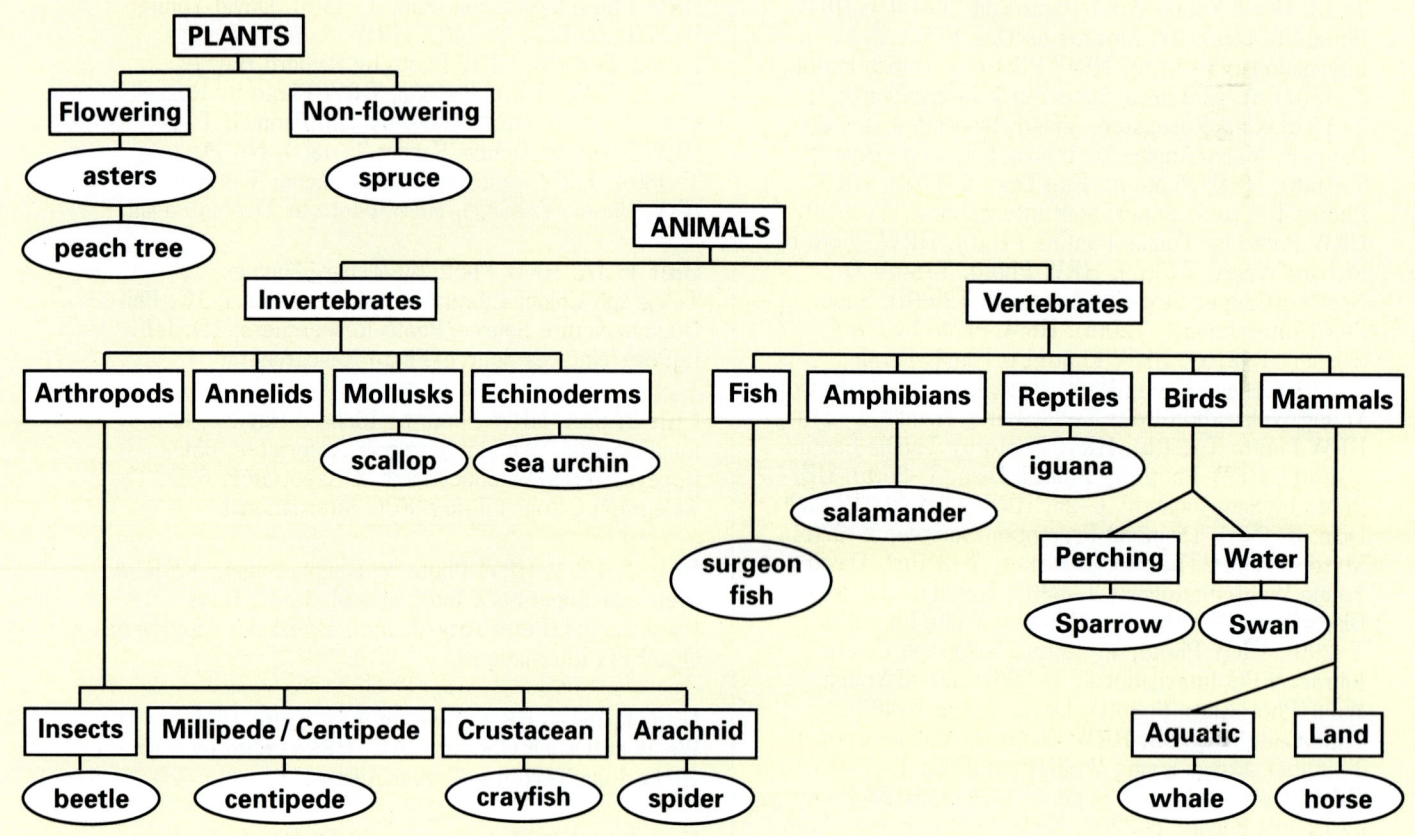

Unit 2, *Checking Your Understanding,* page 125, question 3.

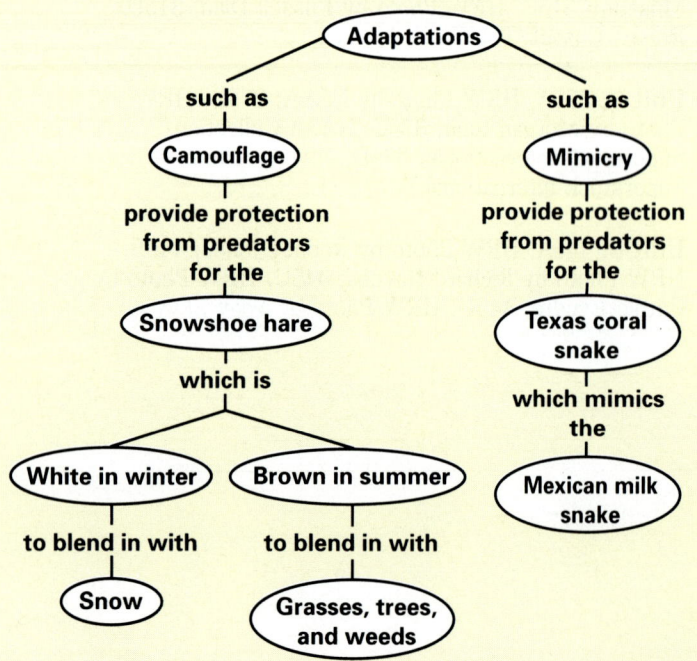

Unit 4, Checking Your Understanding, page 256, question 1.

Type of Force	Agent	Receiver	Effect
Frictional	the boat and the sand	the boat and the sand	the two objects act on each other, resisting movement
	the boat and the water	the boat and the water	same as above
Gravitational	the earth	the boat	weight of the boat due to earth's gravity pushes down on the water
	the earth	Kathleen	causes Kathleen to fall into the water
Buoyant	the water	the boat	keeps the boat from sinking due to the gravitational force
	the water	Kathleen	keeps Kathleen from sinking due to the force of gravity

Type of Force	Example	Effect
Balanced	the force of the wind on the sails = the frictional force between the water and the boat	the boat moves along at a steady speed
	the downward force of gravity on Kathleen and Kimiko = the upward force of the boat seats	Kathleen and Kimiko are at rest in the boat
Unbalanced	the force of Kathleen and Kimiko pulling the boat is greater than the frictional force between the boat and the sand	the boat slides into the water
	the force of the wind on the sails is greater than the friction between the boat and the water	the boat speeds up
	the force of friction between the boat and the water is greater than the force of the wind on the boat	the boat slows down

Unit 6, Measuring the Strength of Earthquakes, page 329.

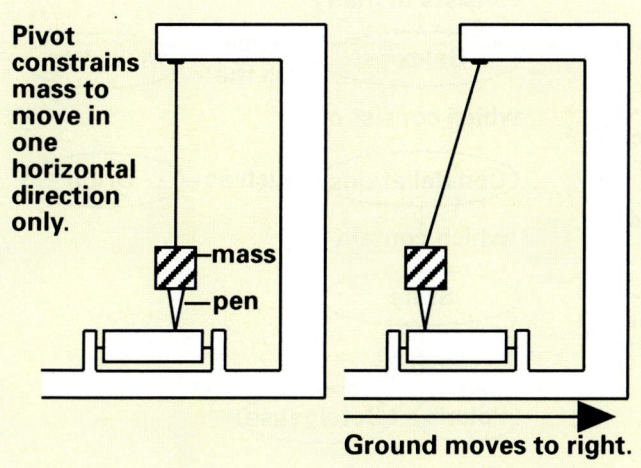

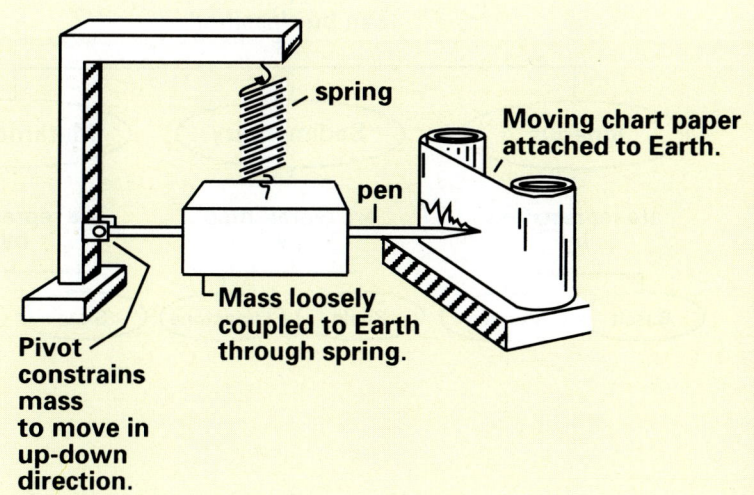

Unit 6, *The Formation of Igneous Rocks,* page 350.

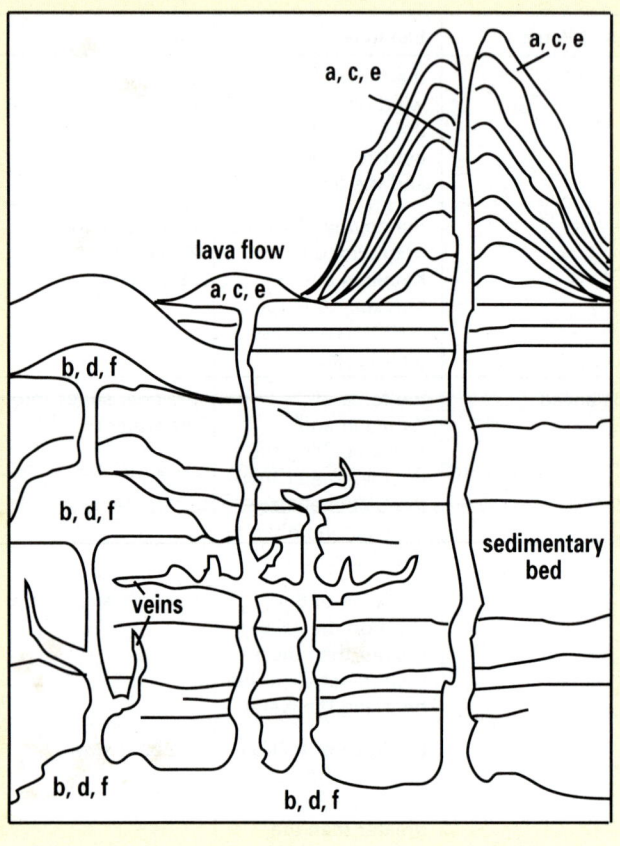

Unit 6, *Those Plates Again!,* page 358.

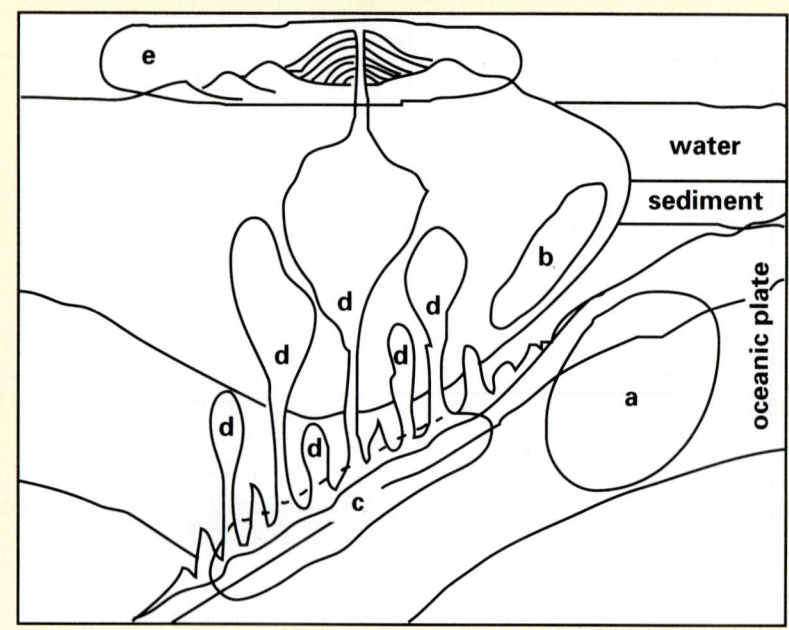

Unit 6, *Checking Your Understanding,* page 371, question 4.

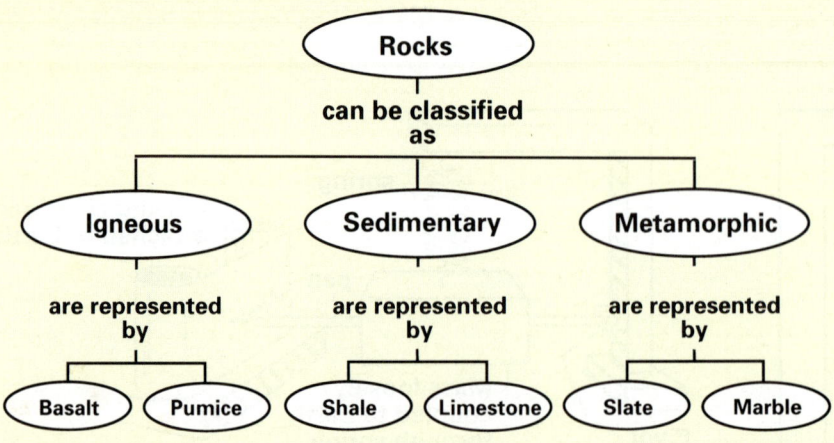

Unit 7, *Checking Your Understanding,* page 441, question 4.

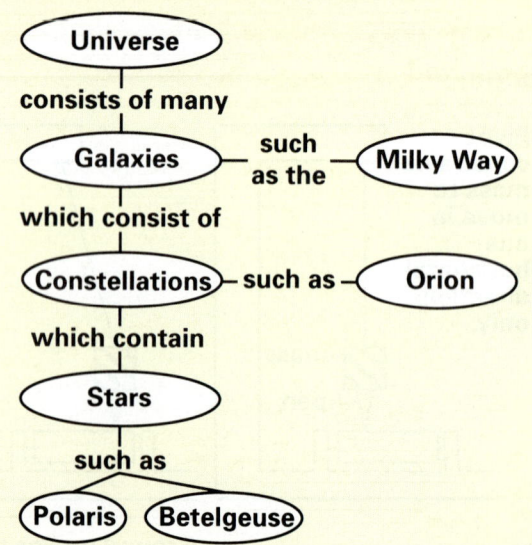